Lecture Notes in Mathematics

Edited by A. Dold and B. Eckmann

1299

S. Watanabe Yu. V. Prokhorov (Eds.)

Probability Theory and Mathematical Statistics

Proceedings of the Fifth Japan-USSR Symposium, held in Kyoto, Japan, July 8–14, 1986

Springer-Verlag

Berlin Heidelberg New York London Paris Tokyo

Editors

Shinzo Watanabe
Department of Mathematics, Kyoto University
Kitashirakawa, Sakyo-ku, Kyoto, Japan

Juriĭ Vasilievich Prokhorov
Steklov Mathematical Institute of the Academy of Sciences of the USSR
Vavilova 42, Moscow 117966, USSR

Mathematics Subject Classification (1980): 60XX, 62XX

ISBN 3-540-18814-2 Springer-Verlag Berlin Heidelberg New York
ISBN 0-387-18814-2 Springer-Verlag New York Berlin Heidelberg

Printed in Germany

Printing and binding: Druckhaus Beltz, Hemsbach/Bergstr.
2146/3140-543210

PREFACE

The Fifth Japan-USSR Symposium on Probability Theory was held at Kyoto University, July 8-14, 1986. Attendance from USSR numbered 25 and from Japan 190. This volume contains the papers presented at the symposium. Records of the meetings, lists of the Organizing Committee and a list of the local editorial committee of the Proceedings are attached at the end of the volume.

As well as the previous four, the fifth one was very fruitful not only in promoting probability and statistics in both countries but also in producing mutual understanding. We express our deep gratitude to all those who have contributed to the success of the symposium and who made efforts in preparing the proceedings. The support made by Mathematical Society of Japan, Kyoto University, and Japan World Exposition ('70) Commemorative Fund is greatly acknowledged.

Professor G. Maruyama deceased three days before the opening. He was chairman of the Japanese organizing committee and made immeasurable contribution to these symposia. In memory of him, Professors H. Tanaka and A.N. Shiryayev delivered two lectures that are contained at the beginning of this volume. We remind ourselves of another sad fact that Professor G.M. Mania passed away without seeing the fifth symposium. He had been chairman of the local organizing committee at Tbilisi for the fourth symposium. These bring great sorrow to all the participants.

S. Watanabe
Yu.V. Prokhorov

CONTENTS

PROFESSOR GISIRO MARUYAMA, IN MEMORIAM

Professor Gisiro Maruyama passed away on July 5, 1986, only three days before the opening of this symposium. His death is mourned over by all mathematicians, but especially at this symposium, which was organized by himself, his death is felt more deeply. The Japan-USSR Symposium on Probability Theory has been held five times in which Maruyama always played a leading part; we have been influenced by him through his work and through his person.

Since everyone in this audience knows Maruyama's work on probability theory, it is hardly necessary to review his work. But, by giving only a few comments on some of the highlights, I would like simply to remind you how important his contributions were and how wide his interests were in probability theory.

Professor Maruyama was born on April 4, 1916. He graduated from Tôhoku Imperial University in 1939, whose department of mathematics was very active in the field of Fourier analysis at that time. There he studied Fourier analysis, but he was also interested in physics and spent a year and a half in the department of physics in Hokkaido University. He began his mathematical work by writing a paper on Fourier series in 1939. But his interest was not limited to Fourier analysis; his main interest was probability theory, especially Wiener's work. Except for a few of his earlier works, most of his works are concerned with probability theory, but we still see that the method of Fourier analysis, which he acquired while he was in Tôhoku University, underlies his whole work as a fundamental tool. He was appointed as a research assistant in Kyushu University in 1941 and was promoted to a professor in 1949. Later he served as a professor in several universities, namely in Ochanomizu University, Kyushu University (for the second time), Tokyo University of Education, the University of Tokyo, the University of Electro-Communications, and finally in Tokyo Denki University where he remained until his death.

In an earlier period Maruyama was much interested in Wiener's generalized harmonic analysis and also in the works of E. Slutsky (1927, 1938), H. Wold (1937) and E. Hopf (1937: Ergodentheorie). Influenced by these works he studied stationary processes and wrote the paper [5] in 1949, for which he received the degree of Doctor of Science. The Japanese version of this paper appeared already in 1947. In this work he showed that the assumption of the Gaussian structure in addition to the stationarity leads to many fruitful results. Among them the following is known as Maruyama's theorem. Let

$$z(t) = x(t) + \sqrt{-1}\, y(t)$$

be a stationary Gaussian process and let F, G and H be the corresponding spectral distribution functions defined by

$$E\{x(t+\tau)x(t)\} = \int_{-\infty}^{\infty} \exp(\sqrt{-1}\,\lambda\tau)\, dF(\lambda)$$

and by similar formulas for $E\{y(t+\tau)y(t)\}$ and $E\{x(t+\tau)y(t)\}$. Then $z(t)$ *is ergodic if and only if* F, G *(and so* H*) are continuous.* He also gave a proof to the result of A. N. Kolmogorov: Sur l'interpolation et extrapolation des suites stationaires, C.R.(1939).

Ideas similar to those used in the paper of 1949 are seen in some of his subsequent papers. For example, about 20 years later he again considered stationary processes but he assumed that the finite dimensional distributions are infinitely divisible. He then obtained the following result ([23]): *An infinitely divisible stationary process* $x(t)$ *is mixing if and only if*

(i) $\lim_{t\to\infty} \rho(t) = 0,$

(ii) $\lim_{t\to\infty} Q_{0t}(|xy| > \delta) = 0 \quad (\forall\delta > 0), \quad \lim_{t\to\infty} \int_{0<x^2+y^2\le 1} xy\, Q_{0t}(dxdy) = 0,$

where $\rho(t)$ is the correlation function of the Gaussian component of $x(t)$ and $Q_{0t}(dxdy)$ is the 2-point Lévy measure of the Poisson component.

In 1942, K. Itô published his first paper on stochastic differential equations which was written in Japanese. Maruyama immediately recognized the importance of Itô's work. I have heard that Maruyama read Itô's Japanese paper in the military camp where he had been drafted for World War II. At that time Maruyama was the only person who read thoroughly Itô's paper of 1942. Soon he wrote two papers on stochastic differential equations [11],[14].

A simple form of Itô's stochastic differential equation is

$$y(t) = y_0 + \int_0^t a(\tau, y(\tau))d\tau + x(t),$$

where $x(t)$ is a Brownian motion. What Maruyama proved in [11] is that the transition function

$$F(s,x;t,y) = P\{y(t) \le y \,/\, y(s) = x\}, \quad s < t,$$

of the Markov process $y(t)$ described by the above stochastic differential equation is smooth and satisfies the parabolic equation

$$\{\frac{\partial}{\partial s} + a(s,x)\frac{\partial}{\partial x} + \frac{1}{2}\frac{\partial^2}{\partial x^2}\}F(s,x;t,y) = 0.$$

The proof is purely probabilistic and is based on the following formula, for which [11] may be more appreciated.

$$F(s, y_0; t, x)$$

$$= \int_{-\infty}^{x} E\{\exp[\int_s^t a(\tau, w(\tau))dw(\tau) - \frac{1}{2}\int_s^t a^2(\tau, w(\tau))d\tau]/w(s) = y_0, w(t) = y\}$$

$$\cdot \frac{1}{\sqrt{2\pi(t-s)}}\exp[-(y-y_0)^2/2(t-s)]dy.$$

Here $E\{-- /w(s) = y_0, w(t) = y\}$ is the expectation with respect to the conditional Wiener measure. This formula is now well-known as the Cameron-Martin formula or Girsanov's formula. Before Maruyama was engaged in this work, some mathematicians in analysis in Japan had been interested in a series of papers by Cameron and Martin. I believe, however, he was the first to apply the idea of Cameron-Martin for the study of Markov processes.

In 1938 stochastic difference equations were considered by S. Bernstein (Équations différentielles stochastiques) and then developed by I. I. Gihman. In the paper [14] Maruyama developed Bernstein's idea within the framework of Itô's stochastic differential equations with emphasis on limit theorems; in particular, he formulated an invariance principle in the framework of Markov processes and, as a special case, gave a proof to the Kolmogorov-Smirnov limit theorem concerning empirical distribution functions. Even though he did not formulate the problem as a weak convergence of probability measures on a suitable (Polish) space of trajectories, his work is highly appreciated as one of the earliest limit theorems for stochastic processes. A general and complete formulation of limit theorems of this type was given by Yu. V. Prohorov and A. V. Skorohod in 1956.

In 1956 Maruyama moved again to Kyushu University. I remember that he took a great interest in the work of Prohorov and Skorohod and suggested to his colleagues and students the study of this field. He wrote in Japanese a fine exposition [18] of this field in 1960. I must also say that he always had a great interest in the limit theorems in USSR, especially in the works of S. Bernstein, A. Ya. Khintchine, Yu. V. Linnik and S. H. Sirazdinov. He also recognized the importance of the theory of Markov processes which was rapidly progressing at that time due to the works of W. Feller, J. L. Doob, K. Itô, H. P. McKean and E. B. Dynkin, and encouraged us to work in this direction ([15], [16], [17]).

Since the middle of the 1960's Maruyama was interested again in stationary processes and flows and took a leadership in this field in Japan. In 1966 he showed that *any continuous flow can be realized as a measurable flow on a compact metric space* ([20]). Based on this fundamental result he discussed several topics among which are the following:

(i) Time change of flows.
(ii) Spectral properties of K-flows.
(iii) Natural extension of semigroups to flows.

Outside mathematics Maruyama was also active in his efforts to reflect aspirations of scientists in the governmental policy on science; he was a member of the Science Council of Japan for nine years from 1969 to 1978.

For the past several years he was engaged in the work on Wiener functionals and probability limit theorems. His ideas, some of which seem to have been in his mind ever since his work of 1949, are to consider a stationary Gaussian process as a basic field on which various limit theorems for subordinated stationary processes are discussed. Recently he concentrated on writing papers on this subject and, astonishingly, he actually wrote up three long papers [30], [31], [32]. The third paper [32] which became his last one was completed only last April.

It is obviously impossible to give a complete description of his work here. We have a plan to publish a volume of his selected papers in which all the (English) papers I mentioned here will be included.

Until quite recently everyone here expected to listen to him deliver his lecture "Gaussian limit theorems for Wiener functionals"([33]) in this hall, but now we can learn his results only through reading the three papers [30], [31], [32]. The author is no longer with us, but his works remain with us. To the end he really loved mathematics and dedicated his life to it.

H. Tanaka

List of papers of Gisiro Maruyama

[1] Determination of the jumps of a function by its Fourier series, Tôhoku Math. J. 46 (1939), 68-74.

[2] Summability of Fourier series, Tôhoku Math. J. 47 (1940), 255-260.

[3] Interpolation (I), Mem. Fac. Sci. Kyushu Imp. Univ. Ser.A 2 (1942), 205-215 (with T.Kawata).

[4] Interpolation (II), Mem. Fac. Sci. Kyushu Imp. Univ. Ser.A 3 (1943), 57-65 (with T.Kawata).

[5] The harmonic analysis of stationary stochastic processes, Mem. Fac. Sci. Kyushu Univ. Ser.A 4 (1949), 45-106.

[6] On an asymptotic property of a gap sequence, Kodai Math. Sem. Rep. 2 (1950), 31-32.

[7] Notes on Wiener integrals, Kodai Math. Sem. Rep. 2 (1950), 41-44.

[8] Note on the arc sine law in the theory of probability, Nat. Sci. Rep. Ochanomizu Univ. 2 (1951), 25-27.

[9] Some properties of infinitely divisible laws, Rep. Stat. Appl. Res. JUSE. Vol.1, No.3 (1951), 22-27 (with K. Kunisawa).

[10] Markov processes and stochastic equations, Nat. Sci. Rep. Ochanomizu Univ. 4 (1953), 40-43.

[11] On the transition probability functions of the Markov process, Nat. Sci. Rep. Ochanomizu Univ. 5 (1954), 10-20.

[12] On the Poisson distribution derived from independent random walks, Nat. Sci. Rep. Ochanomizu Univ. 6 (1955), 1-6.

[13] Fourier analytic treatment of some problems on the sums of ramdom variables, Nat. Sci. Rep. Ochanomizu Univ. 6 (1955), 7-24.

[14] Continuous Markov processes and stochastic equations, Rend. Circ. Mat. Palermo (2). 4 (1955), 48-90.

[15] Some properties of one-dimensional diffusion processes, Mem. Fac. Sci. Kyushu Univ. Ser.A 11 (1957), 117-141 (with H.Tanaka).

[16] On the strong Markov property, Mem. Fac. Sci. Kyushu Univ. Ser.A 13 (1959), 17-29.

[17] Ergodic property of N-dimensional recurrent Markov processes, Mem. Fac. Sci. Kyushu Univ. Ser.A 13 (1959), 157-172 (with H.Tanaka).

[18] Methods of functional analysis for the convergence of stochastic proceses, Seminar on Probability, Vol.4, 1960 (with H. Totoki, in Japanese).

[19] Infinitely divisible laws derived from independent random walks, Lecture Note at Columbia University, Mathematical Statistics, 1961.

[20] Trasnsformations of flows, J. Math. Soc. Japan. 18 (1966), 303 -330.

[21] A singular flow with countable Lebesgue spectrum, J. Math. Soc. Japan. 19 (1967), 359-365.

[22] Infinitely divisible processes, USSR-Japan Symposium on Probability. Habarovsk, August, 1969, Abstract of Cummunications, 240-246.

[23] Infinitely divisible processes, Teor. Verojatnost. i Primenen. 15 (1970), 3-23 (Theory of Prob. and its Appl. Vol.15,1-22).

[24] Some aspects of Ornstein's theory of isomorphism problems in ergodic theory, Publ. R. I. M. S. Kyoto Univ. 7(1972), 511-539.

[25] Discussion on Professor Ornstein's paper, Ann. of Prob. 1 (1973), 63-64.

[26] Applications of Ornstein's theory to stationary processes, Proc. Second. Japan-USSR Symp. on Prob. Theory (Kyoto 1972). Lecture Notes in Math. No.330, 304-309, Springer-Verlag, 1973.

[27] On regularity of linear stationary stochastic differential operators. International Conference on Probability Theory and Mathematical Statistics. Vilnius, 1973, Abstract of Communications , Vol.2, 57-58.

[28] Nonlinear functionals of Gaussian stationary processes and their applications, Proc. Third Japan-USSR Symp. on Prob. Theory (Tashkent,1975). Lecture Notes in Math. No.550, 375-378, Springer-Verlag, 1976.

[29] Applications of the multiplication of the Ito-Wiener expansions to limit theorems, Proc. Japan Acad. Ser.A 58 (1982), 388-390.

[30] Wiener functionals and probability limit theorems I: The central limit theorems, Osaka J. Math. 22 (1985), 697-732.

[31] Wiener functionals and probability theorems II: Term-wise multiplication and its application, Hokkaido Math. J. 15 (1986), 405-451.

[32] Wiener functionals and probability limit theorems III: Asymptotics of the solution of an ordinary SDE associated with a random field accompanied by stationarity and almost-periodicity, Yokohama Math. J. 34 (1986), 91-142.

[33] Gaussian limit theorems for Wiener functionals, in this volume.

SOME WORDS IN MEMORY OF PROFESSOR G. MARUYAMA

On the day of arrival of the Soviet delegation in Japan, July 5, 1986, to attend the 5-th Japan-Soviet Symposium on Probability Theory we learned with regret about the decease of Professor G. Maruyama.

Professor G. Maruyama was well-known to Soviet probabilists both as a distinguished scientist and a permanent co-chairman of our Symposia. All those who were immediately involved in the organization of the symposia are aware of his distinguished role in the realization of the concept of the regular joint Soviet-Japan meeting on probability theory and in the concrete organization of all the Soviet-Japan and Japan-Soviet Symposia.

In giving a tribute of gratitude and deep respect both for the personality and for the scientific achievement of Professor G. Maruyama one wants to share with the participants of the Symposium a notion of in what way the works of G.Maruyama exerted a large influence on the several lines of research in the theory of probability, especially, in our country.

In the famous paper of G. Maruyama, "The harmonic analysis of stationary stochastic processes", published in 1949 (the Japanese version was published in 1947) in addition to such major results as the spectral representation

$$\xi_t = \int_{-\infty}^{\infty} e^{it\lambda} X(d\lambda) \tag{1}$$

for stationary stochastic processes $\xi_t = \xi_0(T^t\omega)$, $-\infty < t < \infty$, were also included necessary and sufficient conditions for validity, in Gaussian case, of the properties

$$(M) = \mathit{Mixing}, \quad i.e. \quad P(A \cap T^{-s}B) - P(A)P(B) \longrightarrow 0 \quad s \to \infty,$$

$$(E) = \mathit{Ergodicity}, \quad i.e. \quad \frac{1}{t}\int_0^t [P(A \cap T^{-s}B) - P(A)P(B)]ds \longrightarrow 0, \quad t \to \infty,$$

$(MW) =$ *Mixing in the wide sense*, *i.e.*

$$\frac{1}{t}\int_0^t [P(A \cap T^{-s}B) - P(A)P(B)]^2 ds \longrightarrow 0, \quad t \to \infty,$$

If $E\xi_t = 0$ and $R_\xi(\tau) \equiv E\xi_{t+\tau}\xi_t = \int_{-\infty}^{\infty} e^{i\tau\lambda} F_\xi(d\lambda)$ then the results of Maruyama can be formulated as

$$(M) \Longleftrightarrow R_\xi(\tau) \longrightarrow 0, \quad \tau \to \infty;$$

$$(MW) \Longleftrightarrow (E) \Longleftrightarrow \frac{1}{t}\int_0^t |R_\xi(s)|^2 ds \longrightarrow 0, \quad t \to s$$

$$\Longleftrightarrow F_\xi(\lambda) \ \mathit{continuous\ in}\ \lambda$$

where $F_\xi(\lambda) = F_\xi(-\infty, \lambda]$.

In 1957, A. N. Kolmogorov in the context of nonlinear analysis of stochastic processes, posed the problem of extension of Maruyama's necessary and sufficient conditions for properties (M), (MW) and (E) to the case of non-Gaussian processes. As a first step in this direction it was natural to consider the case of stationary processes with finite moments of all order, $E|\xi_t|^k < \infty$, $k \geq 1$, for which one has uniqueness of the moment problem.

Starting from the spectral representation (1) for a process, one can obtain for the moments $m_\xi^{(k)}(t_1, \cdots, t_k) = E\xi_{t_1} \cdots \xi_{t_k}$ and corresponding semiinvariants $S_\xi^{(k)}(t_1, ..., t_k)$ the following spectral representations

$$m_\xi^{(k)}(t_1, \cdots, t_k) = \int e^{i(t,\lambda)} M_\xi^{(k)}(d\lambda),$$

$$S_\xi^{(k)}(t_1, \cdots, t_k) = \int e^{i(t,\lambda)} F_\xi^{(k)}(d\lambda),$$

where $M_\xi^{(k)}$ and $F_\xi^{(k)}$ are moments and semiinvariant measures of order k, concentrated on hyperspaces

$$L_k = \{(\lambda_1, \cdots, \lambda_k) : \quad \lambda_1 + \cdots + \lambda_k = 0\}.$$

The semiinvariants $S_\xi^{(k)}(t_1, \cdots, t_k)$ play the role of correlation functions of high orders and in some sense their structure is simpler than that of moments. Moreover, higher order moments can be expressed in symmetric form using semiinvariants. For example,

$$\begin{aligned} m_\xi^{(2)}(t_1, t_2) =& S_\xi^{(2)}(t_1, t_2) + S_\xi^{(1)}(t_1) S_\xi^{(1)}(t_2), \\ m_\xi^{(3)}(t_1, t_2, t_3) =& S_\xi^{(3)}(t_1, t_2, t_3) + S_\xi^{(2)}(t_1, t_2) S_\xi^{(1)}(t_3) \\ &+ S_\xi^{(2)}(t_1, t_3) S_\xi^{(1)}(t_2) + S_\xi^{(2)}(t_2, t_3) S_\xi^{(1)}(t_1). \end{aligned}$$

It turns out (V.P. Leonov, Doklady of the Acad. of Sci. USSR, 133,3 (1960), 523-536) that the results of Maruyama for the non-Gaussian case can be generalized as follows

$$(M) \Longleftrightarrow S_\xi^{(k+l)}(t_1, \cdots, t_k, t'_1+\tau, \cdots, t'_l+\tau) \longrightarrow 0, \quad \tau \to \infty, k, l \geq 1, \quad t_i, t'_j \in \mathbf{R};$$

$$(MW) \Longleftrightarrow \frac{1}{t} \int_0^t |S_\xi^{(k+l)}(t_1, \cdots, t_k; t'_1 + \tau, \cdots, t'_l + \tau)|^2 d\tau \longrightarrow 0, \quad t \to \infty,$$
$$k, l \geq 1, t_i, t'_j \in \mathbf{R};$$

$$(MW) \Longleftrightarrow var|F_\xi^{(k)}(L_k^l)| = 0, \quad 1 \leq l < k,$$
$$L_k^l = \{(\lambda_1, \cdots, \lambda_k) : \lambda_1 + \cdots + \lambda_k = 0, \quad \lambda_1 + \cdots + \lambda_l = \nu\}, \quad \nu \in \mathbf{R},$$

where $var|\cdot|$ denotes total variation. In terms of spectral characteristics, the results of G. Maruyama about conditions for ergodicity have been generalized as follows (A.N. Shiryayev, Theory of Probab., VIII, 4(1963),470-473 ;Ya.G. Sinai, Theory of Probab.,VIII,4(1963),463-470) :

$$(E) \quad \Longleftrightarrow \quad M_\xi^{(k+l)}(\Lambda_k \times \Lambda_l) = M_\xi^{(k)}(\Lambda_k) M_\xi^{(l)}(\Lambda_l), \quad k,l \leq 1$$

$$\Lambda_k \subseteq \{(\lambda_1, \cdots, \lambda_k) : \lambda_1 + \cdots + \lambda_k = 0\},$$

$$\Lambda_l \subseteq \{(\lambda'_1, \cdots, \lambda'_l) : \lambda'_1 + \cdots + \lambda'_l = 0\},$$

$$\Lambda_k \times \Lambda_l = \{(\lambda_1, \cdots, \lambda_k, \lambda'_1, \cdots, \lambda'_l) : \lambda + \cdots + \lambda_k = 0, \lambda'_1 + \cdots + \lambda'_l = 0\}.$$

The second paper of G. Maruyama upon which we wish to dwell is entitled "Continuous Markov processes and stochastic equations" (1955). In this paper Professor G. Maruyama studied the relation between construction of Markov diffusion process starting from parabolic equation for the transition function, and the construction of K. Itô based on regarding processes as solution of stochastic differential equations driven by Brownian motion (Wiener process). So for example in this paper it was stated that transition function $F(s,x;t,y) = P(Y_t \leq y | Y_s = x)$ of a Markov process which is a solution of the stochastic differential equation

$$dY_t = a(t, Y_t)dt + b(t, Y_t)dW_t$$

under mild smoothness conditions on the coefficients, is a solution of the parabolic equation

$$\frac{\partial F}{\partial s} + a(s,x)\frac{\partial F}{\partial x} + \frac{1}{2}b^2(s,x)\frac{\partial^2 F}{\partial x^2} = 0 .$$

In the same paper the problem of convergence of finite dimensional distributions of sequences of Markov processes to finite dimensional distributions of a Markov process which is a solution of a stochastic differential equation was studied. Moreover he studied the problem of validity of the invariance principle in the form

$$P(f(t) \leq Y_t^n \leq g(t), t \leq 1) \longrightarrow P(f(t) \leq Y_t \leq g(t), t \leq 1), \quad n \to \infty,$$

for sequences of Markov processes Y_t^n, $n \geq 1$, which converge to diffusion processes Y_t.

From this he obtained, in particular, a strong substantiation of the heuristic approach of J.L. Doob to the proof of the Kolmogorov-Smirnov limit theorem.

This paper of G.Maruyama played an essential role in the theory of optimal non-linear filtering for obtaining a stochastic differential equation for a posteriori characteristics. For illustration suppose that a stochastic process ξ_t, $t \geq 0$, is observed which has stochastic differential

$$d\xi_t = \theta_t dt + dW_t ,$$

where W_t is a Wiener process, θ_t is a Markov process with states 0 and 1 having only a single transition from 0 to 1, and independent of W_t, $t \geq 0$, with transition density $\lambda > 0$. Then for the estimate $\pi_t = E(\theta_t|\mathcal{F}_t^\xi)$, $\mathcal{F}_t^\xi = \sigma\{\omega : \xi_s , s \le t\}$ which is an optimal estimate of θ_t in the mean square sense, one can obtain the following stochastic differential equation

(2) $$d\pi_t = (1-\pi_t)(\lambda - \pi_t^2)dt + \pi_t(1-\pi_t)d\xi_t,$$

where the corresponding stochastic integral with respect to $d\xi_t$ can be understood as the sum of two integrals, with respect to $\theta_t dt$ and dW_t(A.N. Shiryayev, Theory of Probab., VIII, 1 (1963), 26-51). In the late of 1950's when the problem of understanding this equation arose, the notion of stochastic integral with respect to a semimartingale had not yet appeared. Hence it was natural to try to interpret this equation in terms of the Itô calculus with respect to a Wiener process. It was not difficult to see that the process π_t, with respect to the family $(\mathcal{F}_t^\xi)_{t\ge 0}$, is Markov with local characteristic

$$a(\pi) = \lambda(1-\pi), \qquad b^2(\pi) = \pi^2(1-\pi)^2 \quad .$$

Then from results of G. Maruyama it follows that, for some Wiener process $\bar{W}_t$, the process π_t admits the following stochastic differential

(3) $$d\pi_t = \lambda(1-\pi_t)dt + \pi_t(1-\pi_t)d\bar{W}_t \quad .$$

Direct comparison of the right sides of (2) and (3) shows that (at least in distribution)

$$d\bar{W}_t = d\xi_t - \pi_t dt.$$

In other word, the process

$$\bar{W}_t \equiv \xi_t - \int_0^t E(\theta_s|\mathcal{F}_s^\xi)ds$$

is a Wiener process and moreover, for each $t \ge 0$, $\mathcal{F}_t^{\bar{W}} = \mathcal{F}_t^\xi$. Using current terminology one can say that the innovation process $\bar{W}_t$, $t \ge 0$, is a Wiener process and the correct form for the stochastic equation is given by (2), where the stochastic integral is understood as the "classical" Itô integral with respect to a Wiener process (A. N. Shiryayev, Problemy peredachi Informatzii, II, 3 (1966), 3-22).

In finishing this tribute about papers of Professor G. Maruyama and their influence, one wishes to say some words about his personality. Many Soviet probabilists had repeated opportunities to meet with him. And they were unfailingly impressed with his modesty ,cordiality, affability and intelligence. All who knew him preserve this impression of Professor G. Maruyama - a distinguished scientist and remarkable person.

A.N. Shiryayev

SECOND AND THIRD ORDER ASYMPTOTIC COMPLETENESS OF THE CLASS OF ESTIMATORS

Masafumi Akahira, Fumiko Hirakawa, and Kei Takeuchi

1. Introduction

The concept of higher order asymptotic efficiency of estimator $\hat{\theta}_n^*$ of a parameter θ in an open subset Θ of R^p is usually defined by the property that for any other estimator $\hat{\theta}_n$ satisfying some condition we have

$$\lim_{n\to\infty} n^{(k-1)/2} [P_{\theta,n}\{\sqrt{n}(\hat{\theta}_n^* - \theta) \in C\} - P_{\theta,n}\{\sqrt{n}(\hat{\theta}_n - \theta) \in C\}] \geq 0$$

for all $\theta \in \Theta$ and any convex set C containing the origin. And usually the condition imposed is k-th order asymptotically median unbiased and the like, and $\hat{\theta}_n^*$ is usually "modified" maximum likelihood estimator (MLE) (see [4], [6], [7] and [8]).

However, the condition of the asymptotic median unbiasedness is rather arbitrary, and a more meaningful property will be that of higher order "asymptotic completeness" which is defined as follows: $\hat{\theta}_n^*$ is called k-th order asymptotically complete if for any estimator $\hat{\theta}_n$ within some class of estimators we can construct

$$\hat{\theta}_n^{**} = \hat{\theta}_n^* + h_n(\hat{\theta}_n^*),$$

$\{h_n\}$ being a sequence of functions depending on $\hat{\theta}_n$ so that we have

$$\lim_{n\to\infty} n^{(k-1)/2} [P_{\theta,n}\{\sqrt{n}(\hat{\theta}_n^{**} - \theta) \in C\} - P_{\theta,n}\{\sqrt{n}(\hat{\theta}_n - \theta) \in C\}] \geq 0$$

for all $\theta \in \Theta$ and any convex set C containing the origin. Actually, the definition was given in the monograph [4] by Akahira and Takeuchi. But there the discussion was limited to the class of asymptotically median unbiased estimators in its application. The concept is also discussed in [10], but again the class of estimators was limited to that of regular functions of sufficient statistics.

The purpose of this paper is to show that, for any class of estimators which admit Edgeworth expansions but are not necessarily asymptotically median unbiased, we get the second order asymptotic completeness of the MLE, and the third order asymptotic completeness of the MLE $\hat{\theta}_{ML}$ together with the second order derivative of the log-likelihood function evaluated at the MLE which is denoted by $Z_2(\hat{\theta}_{ML})$ later.

It follows from this higher order asymptotic completeness property that, in any "decision theoretic" set-up, where we consider a sequence of loss functions of the type $L_n(u) \sim L^*(\sqrt{n}u)$ for a sufficiently large n, for any Bayes decision rule δ_n with smooth prior we can construct a $\delta_n^* = \hat{\theta}_{ML} + h_n(\hat{\theta}_{ML}, Z_2(\hat{\theta}_{ML}))$, $\{h_n\}$ being a sequence of functions depending on $\hat{\theta}_n$, such that

$$\lim_{n \to \infty} n(E[L_n(\delta_n - \theta)] - E[L_n(\delta_n^* - \theta)]) \geq 0$$

for all $\theta \in \Theta$.

The concept of asymptotic completeness is related to that of asymptotic sufficency. A sequence $\{T_n\}$ of statistics is called to be higher order asymptotically sufficient if for any sequence $\{T_n'\}$ we can construct $T_n'' = g_n(T_n', U_n)$, $\{g_n\}$ being a sequence of functions depending on T_n, where U_n is a random variable independent of θ, such that the asymptotic distribution of T_n'' coincides with that of T_n' up to the order $n^{-(k-1)/2}$. (The concept of higher order asymptotic sufficiency may be given in other ways but basically it amounts to the above.)

Asymptotic sufficiency, however, has generally been derived from the statement that we have

$$T_n' - T_n'' = o_p(n^{-k/2}) ,$$

which is actually much stronger than the equivalence of asymptotic distributions up to the order $n^{-(k-1)/2}$, where $X_n = o_p(Y_n)$ means that X_n/Y_n converges to zero in probability as $n \to \infty$. For example in Takeuchi [9], Bickel, Götze and van Zwet [5], it was diccussed that for any Bayes decision procedure δ_n it is possible to get $\delta_n^* = g_n(\hat{\theta}_{ML})$ such that $\delta_n^* - \delta_n = o_p(n^{-3/2})$ only if the loss function is symmetric. However, when the loss is <u>not</u> symmetric, it was shown in [9] that even with Z_2 we can not construct $\delta_n^* = g_n(\hat{\theta}_{ML}, Z_2(\hat{\theta}_{ML}))$ such that $\delta_n^* - \delta_n = o_p(n^{-3/2})$. But, in order that δ_n^* and δ_n have asymptotically equivalent distribtions up to the same order, hence have asymptotically the same risk, it is not necessary that δ_n^* and δ_n are stochastically equivalent up to the same order. As a simple illustration of this fact, let us assume that

$$\delta_n = \hat{\theta}_n + \frac{1}{n} \varepsilon_n ,$$

where ε_n has the property that $E(\varepsilon_n \mid \hat{\theta}_n) = o(1)$. Then

$$\delta_n - \hat{\theta}_n = \frac{1}{n} O_p(\varepsilon_n) = O_p(\frac{1}{n}) ,$$

but

$$E[\exp(it\delta_n)] = E[\exp\{it(\hat{\theta}_n + \frac{1}{n}\varepsilon_n)\}]$$

$$= E[\exp(it\hat{\theta}_n)] + \frac{it}{n} E[\varepsilon_n \exp(it\hat{\theta}_n)] + o(\frac{1}{n})$$

$$= E[\exp(it\hat{\theta}_n)] + o(\frac{1}{n}) \quad ,$$

which means that the distributions of $\hat{\theta}_n$ and δ_n are asymptotically equivalent up to the order n^{-1}. Actually, from what is derived in the paper, it is shown that, under a sequence of decision problems with a proper set of regularity conditions, for any Bayes decision rule δ_n with some smooth prior there exists a $\delta_n^*(\hat{\theta}_{ML}, Z_2(\hat{\theta}_{ML}))$ such that

$$P_{\theta,n}\{\delta_n^* \in A\} = P_{\theta,n}\{\delta_n \in A\} + o(\frac{1}{n})$$

for all measurable set A in the decision space.

The main purpose of this paper is to establish that $\hat{\theta}_{ML}$ is generally second order asymptotically complete, and $\hat{\theta}_{ML}$ together with $Z_2(\hat{\theta}_{ML})$ isthird order asymptotically complete. Although this result is given in the framework of "point" estimation theory, it is also applicable to other problems of inference including testing hypothesis and interval estimation, which will be discussed in subsequent papers.

2. Preliminaries

Let χ be an abstract space, elements of which are denoted by x, and let β be a σ-field of subsets of χ. Let Θ be a parameter space, which is assumed to be an open subset of the Euclidean p-space R^p with the usual norm denoted by $\|\cdot\|$. We shall denote by $(\chi^{(n)}, \beta^{(n)})$ the n-fold direct product of (χ, β). We consider a sequence of classes of probability measures $\{P_{\theta,n} : \theta \in \Theta\}$ $(n=1, 2, \ldots)$, each defined on $(\chi^{(n)}, \beta^{(n)})$, such that for each n and each $\theta \in \Theta$ the following holds:

$$P_{\theta,n}(B^{(n)}) = P_{\theta,n+1}(B^{(n)} \times \chi)$$

for all $B^{(n)} \in \beta^{(n)}$.

An estimator of θ is defined to be a sequence $\{\hat{\theta}_n\}$ where $\hat{\theta}_n$ is a $\beta^{(n)}$-measurable function from $\chi^{(n)}$ into Θ $(n=1, 2, \ldots)$. For simplicity we denote an estimator as $\hat{\theta}_n$ instead of $\{\hat{\theta}_n\}$. For an increasing sequence of positive numbers $\{c_n\}$ (c_n tending to infinity) an estimator $\hat{\theta}_n$ is called to be c_n-consistent if for every η of Θ, there exists a sufficiently small positive number δ such that

$$\lim_{L\to\infty}\ \overline{\lim_{n\to\infty}}\ \sup_{\theta:\|\theta-\eta\|<\delta} P_{\theta,n}\{c_n\|\hat\theta_n-\theta\|\geq L\} = 0 \quad \text{(Akahira [1]).}$$

3. One parameter case

Suppose that Θ is an open subset of R^1. Let $X_1, X_2, \ldots, X_n, \ldots$ be a sequence of independently and identically distributed (i.i.d.) real random variables with a density function $f(x,\theta)$ with respect to a σ-finite measure μ, where $\theta\in\Theta$. In the subsequent discussion it is enough to consider only the case $c_n=\sqrt{n}$. For each $k=1, 2, \ldots$, $0<\alpha(\theta)<1$ and a continuously differentiable function $a(\theta)$ of θ, a c_n-consistent estimator $\hat\theta_n$ is called k-th order asymptotically (α,a)-biased (as. (α,a)-biased) estimator if, for any $\eta\in\Theta$, there exists a positive number δ such that

$$\lim_{n\to\infty}\ \sup_{\theta:|\theta-\eta|<\delta} n^{(k-1)/2}\left|P_{\theta,n}\{\sqrt{nI(\theta)}(\hat\theta_n-\theta)\leq a(\theta)\}-\alpha(\theta)\right|=0,$$

$$\lim_{n\to\infty}\ \sup_{\theta:|\theta-\eta|<\delta} n^{(k-1)/2}\left|P_{\theta,n}\{\sqrt{nI(\theta)}(\hat\theta_n-\theta)\geq a(\theta)\}-1+\alpha(\theta)\right|=0,$$

where $I(\theta)$ denotes the Fisher information of f where $\alpha(\theta)$ and $a(\theta)$ are determined from the outset. Now we assume that $\alpha(\theta)$ is a constant α. Alternatively, we may fix $a(\theta)$ to be constant letting $\alpha(\theta)$ depend on θ, and the subsequent discussions apply exactly in a similar way.

For each $k=1, 2, \ldots$ we denote by $B_k(\alpha,a)$ the class of k-th order as. (α,a)-biased estimators for which the distribution of $\sqrt{nI(\theta)}(\hat\theta_n-\theta)-a(\theta)$ admits the Edgeworth expansion up to the k-th order. If an estimator $\hat\theta_n$ belongs to the class $B_k(\alpha,a)$, then we call it a $B_k(\alpha,a)$-estimator.

For each $k=1, 2, \ldots$, a $B_k(\alpha,a)$-estimator $\hat\theta_n^*$ is called k-th order asymptotically efficient in the class $B_k(\alpha,a)$ if for any $B_k(\alpha,a)$-estimator $\hat\theta_n$

$$P_{\theta,n}\{-t_1<\sqrt{nI(\theta)}(\hat\theta_n^*-\theta)-a(\theta)<t_2\}\geq P_{\theta,n}\{-t_1<\sqrt{nI(\theta)}(\hat\theta_n-\theta)-a(\theta)<t_2\}+o(n^{-(k-1)/2})$$

for all $t_1, t_2>0$ and $\theta\in\Theta$.

We assume the following conditions.

(A.1) $\{x: f(x,\theta)>0\}$ does not depend on θ.

$(A.2)_k$ For almost all $x[\mu]$, $f(x,\theta)$ is k times continuously differentiable in θ. In the Taylor expansion

$$\log\frac{f(x,\theta+h)}{f(x,\theta)}=\sum_{i=1}^{k}\frac{h^i}{i!}\ell^{(i)}(\theta,x)+h^kR(x,h)$$

$R(x,h)$ is uniformly bounded by a function $G(x)$ which has moments up to the k-th order, where

$$\ell^{(i)}(\theta,x) = (\partial^i/\partial\theta^i)\,\ell(\theta,x) \qquad (i=1,\ \ldots,\ k)$$

with $\ell(\theta,x) = \log f(x,\theta)$.

(A.3) For each $\theta \in \Theta$

$$0 < I(\theta) = E_\theta[\{\ell^{(1)}(\theta,X)\}^2] = -E_\theta[\ell^{(2)}(\theta,X)] < \infty \quad ,$$

and $I(\theta)$ is differentiable in θ .

(A.4) There exist

$$J(\theta) = E_\theta[\ell^{(1)}(\theta,X)\,\ell^{(2)}(\theta,X)] \ ,$$

$$K(\theta) = E_\theta[\{\ell^{(1)}(\theta,X)\}^3] \ ,$$

and both $J(\theta)$ and $K(\theta)$ are differentiable in θ , and

$$E_\theta[\ell^{(3)}(\theta,X)] = -3J(\theta)-K(\theta).$$

In the following theorem we obtain the asymptotic distribution of a $B_2(\alpha,a)$-estimator up to the second, i.e., the order $n^{-1/2}$.

Theorem 3.1. Assume that the conditions (A.1), $(A.2)_3$, (A.3), and (A.4) hold and that the asymptotic cumulants of a biased best asymptotically normal (BAN) estimator $\hat\theta_n$ are given as follows : For $T_n = \sqrt{n}(\hat\theta_n - \theta)$

$$E_\theta(T_n) = C_0(\theta) + \frac{C_1(\theta)}{\sqrt{n}} + o\left(\frac{1}{\sqrt{n}}\right) \ ;$$

$$V_\theta(T_n) = \frac{1}{I(\theta)} + \frac{C_2(\theta)}{\sqrt{n}} + o\left(\frac{1}{\sqrt{n}}\right) \ ;$$

$$\kappa_{3,\theta}(T_n) = \frac{C_3(\theta)}{\sqrt{n}} + o\left(\frac{1}{\sqrt{n}}\right) \ ;$$

where $C_i(\theta)$ (i=0, 1, 2, 3) are differentiable functionsof θ . Then the second order asymptotic distribution of the second order as. (α,a)-biased BAN estimator $\hat\theta_n^*$, of the form $\hat\theta_n^* = \hat\theta_n - n^{-1/2}u_1(\hat\theta_n) - n^{-1}u_2(\hat\theta_n)$, is given by

(3.1) $P_{\theta,n}\{\sqrt{nI(\theta)}(\hat\theta_n^* - \theta) \le t + a(\theta)\}$

$$= \Phi(t+\nu_\alpha) + \frac{1}{\sqrt{n}}\phi(t+\nu_\alpha)\Big\{-\frac{1}{6}I^{3/2}(\theta)C_3(\theta)t^2 - \frac{1}{3}I^{3/2}(\theta)C_3(\theta)\nu_\alpha t + t\Big(C_0'(\theta) - \frac{I(\theta)C_2(\theta)}{2} - \frac{a'(\theta)}{\sqrt{I(\theta)}} + \frac{2J(\theta)+K(\theta)}{2I^{3/2}(\theta)}(a(\theta)-\nu_\alpha)\Big)\Big\} + o\left(\frac{1}{\sqrt{n}}\right) \ ,$$

where $\Phi(u) = \int_{-\infty}^{u}\phi(x)dx$ with $\phi(x) = (1/\sqrt{2\pi})e^{-x^2/2}$ and $\Phi(\nu_\alpha) = \alpha$.

Remark: In the above theorem, $u_1(\theta)$ and $u_2(\theta)$ are given as follows:

$$u_1(\theta) = C_0(\theta) - \frac{a(\theta)-\nu_\alpha}{I(\theta)} \quad ;$$

$$u_2(\theta) = C_1(\theta) - C_0(\theta)u_1'(\theta) + \frac{1}{2}\sqrt{I(\theta)}\{C_2(\theta) - \frac{2u_1'(\theta)}{I(\theta)}\}[a(\theta) - \sqrt{I(\theta)}\{C_0(\theta)-u_1(\theta)\}] + \frac{1}{6}I(\theta)C_3(\theta)[\{a(\theta) - \sqrt{I(\theta)}(C_0(\theta)-u_1(\theta))\}^2 - 1] \quad .$$

In the following theorem we shall obtain the bound for the asymptotic distributions of $B_2(\alpha, a)$-estimators up to the order $n^{-1/2}$.

Theorem 3.2. Assume that the conditions (A.1), $(A.2)_3$, (A.3), and (A.4) hold. Then the bound $F^*(t,\theta)$ for the asymptotic distributions of $\sqrt{nI(\theta)}(\hat\theta_n - \theta)$ for the all $B_2(\alpha, a)$-estimators $\hat\theta_n$ is given by

$$F^*(t,\theta) = \Phi(t+\nu_\alpha) + \frac{1}{\sqrt{n}}\phi(t+\nu_\alpha)[\frac{\{3J(\theta)+2K(\theta)\}t^2}{6I^{3/2}(\theta)} - \frac{t}{\sqrt{I(\theta)}}\{a'(\theta) - \frac{a(\theta)(2J(\theta)+K(\theta))}{2I(\theta)} - \frac{\nu_\alpha K(\theta)}{6I(\theta)}\}] + o(\frac{1}{\sqrt{n}})$$

in the sense that for any $B_2(\alpha, a)$-estimator $\hat\theta_n$

$$P_{\theta,n}\{\sqrt{nI(\theta)}(\hat\theta_n - \theta) \le t + a(\theta)\} \le F^*(t,\theta)$$

for all $t > 0$ and all $\theta \in \Theta$,

$$P_{\theta,n}\{\sqrt{nI(\theta)}(\hat\theta_n - \theta) \le t + a(\theta)\} \ge F^*(t,\theta)$$

for all $t < 0$ and all $\theta \in \Theta$.

From Theorems 3.1 and 3.2 we have the following :

Theorem 3.3. Under the conditions (A.1), $(A.2)_3$, (A.3), and (A.4) it holds that for the biased BAN estimator $\hat\theta_n^*$

$$c_3(\theta) = -\frac{3J(\theta)+2K(\theta)}{I^3(\theta)}, \quad C_0'(\theta) \le \frac{I(\theta)C_2(\theta)}{2} \quad \text{for all } \theta \in \Theta \quad ,$$

nd further if $C_0'(\theta) \equiv I(\theta)C_2(\theta)/2$, the second order as. (α, a)-biased BAN estimator $\hat\theta_n^*$ is second order asymptotically efficient in the class $B_2(\alpha, a)$.

Sketches of the proofs of Theorems 3.1, 3.2 and 3.3 are given below, since their detailed ones are done in Akahira [2].

Sketch of the proof of Theorem 3.1.

Let $\hat\theta_n^* = \hat\theta_n - \frac{u_1(\hat\theta_n)}{\sqrt{n}} - \frac{u_2(\hat\theta_n)}{n}$, where $u_1(\theta)$ is continuously differentiable in θ and $u_2(\theta)$ is a function of θ. Then we have the following asymptotic cumulants of $T_n^* = \sqrt{nI(\theta)}(\hat{\theta}_n^* - \theta)$:

(3.2) $E_\theta(T_n^*) = \sqrt{I}(C_0-u_1)+\sqrt{I/n}\,(C_1-C_0u_1'-u_2)+o\left(\frac{1}{\sqrt{n}}\right)$;

(3.3) $V_\theta(T_n^*) = 1+\frac{I}{\sqrt{n}}\left(C_2-\frac{2u_1'}{I}\right)+o\left(\frac{1}{\sqrt{n}}\right)$;

(3.4) $\kappa_{3,\theta}(T_n^*) = \frac{1}{\sqrt{n}} I^{3/2} C_3+o\left(\frac{1}{\sqrt{n}}\right)$,

where I, C_0, u_1, u_1', u_2, C_2 and C_3 denote $I(\theta)$, $C_0(\theta)$, $u_1(\theta)$, $u_1'(\theta)=du_1(\theta)/d\theta$, $u_2(\theta)$, $C_2(\theta)$ and $C_3(\theta)$.

From (3.2) to (3.4) it follows that the Edgeworth expansion of the distribution of $\sqrt{nI}(\hat{\theta}_n^*-\theta)$ is given by

$$(3.5)\quad P_{\theta,n}\{\sqrt{nI}(\hat{\theta}_n^*-\theta)\le t+\sqrt{I}(C_0-u_1)\}$$
$$=\Phi(t)-\frac{1}{\sqrt{n}}\phi(t)\left\{\sqrt{I}(C_1-C_0u_1'-u_2)+\frac{I}{2}\left(C_2-\frac{2}{I}u_1'\right)t+\frac{1}{6}I^{3/2}C_3(t^2-1)\right\}$$
$$+o\left(\frac{1}{\sqrt{n}}\right).$$

By the second order as. (α,a)-biased condition we have the following u_1 and u_2 :

$$(3.6)\quad u_1 = C_0-\frac{a-\nu_\alpha}{I} ;$$

$$(3.7)\quad u_2 = C_1-C_0u_1'+\frac{\sqrt{I}}{2}\left(C_2-\frac{2u_1'}{I}\right)\{a-\sqrt{I}(C_0-u_1)\}$$
$$+\frac{IC_3}{6}[\{a-\sqrt{I}(C_0-u_1)\}^2-1] ,$$

where $\Phi(\nu_\alpha)=\alpha$.

Since $I'(\theta)=2J(\theta)+K(\theta)$, it follows from (3.5), (3.6) and (3.7) that (3.1) holds. This completes the proof.

<u>Sketch of the proof of Theorem 3.2.</u> Let θ_0 be arbitrary but fixed in Θ. Consider the problem of testing the hypothesis $H: \theta=\theta_1$ against alternative $A: \theta=\theta_0$, where $\theta_1=\theta_0+O(1/\sqrt{n})$. In order to obtain the upper bound of

$$P_{\theta_0,n}\{\sqrt{nI(\theta_0)}(\hat{\theta}_n-\theta_0)\le t+a(\theta)\}$$

for each $t>0$ and all $B_2(\alpha)$-estimators $\hat{\theta}_n$, that is, under the second order as. (α,a)-biased condition

$$P_{\theta_1,n}\{\sqrt{nI(\theta_1)}(\hat{\theta}_n-\theta_1)\le a(\theta_1)\} = \alpha+o\left(\frac{1}{\sqrt{n}}\right) ,$$

we first take

$$\theta_1 = \theta_0 + \frac{t}{\sqrt{nI}}\left[1 - \frac{1}{\sqrt{nI}}\left\{a' - \frac{a(2J+K)}{2I}\right\}\right],$$

where a, a', I, J and K denote $a(\theta_0)$, $a'(\theta_0)$, $I(\theta_0)$, $J(\theta_0)$ and $K(\theta_0)$.

Setting

$$T_n = \sum_{i=1}^{n} \log \frac{f(x_i, \theta_0)}{f(x_i, \theta_1)},$$

by the Edgeworth expansion of the distribution of T_n we take the following a_n so that the test with a rejection region $\{T_n \geq a_n\}$ has the level $\alpha + o(1/\sqrt{n})$:

$$a_n = -\frac{t^2}{2} - t\nu_\alpha + \frac{t}{\sqrt{nI}}\left\{a' - \frac{a(2J+K)}{2I}\right\}(t+\nu_\alpha)$$
$$- \frac{t}{6\sqrt{n}I^{3/2}}\{(3J+2K)t^2 + 3(J+K)\nu_\alpha t + K(\nu_\alpha^2 - 1)\} + o\left(\frac{1}{\sqrt{n}}\right).$$

In a similar way to the above we have the asymptotic power of the test with the rejection region $\{T_n \geq a_n\}$ as follows:

(3.8) $P_{\theta_0, n}\{T_n \geq a_n\}$

$$= \Phi(t+\nu_\alpha) + \frac{1}{\sqrt{n}}\phi(t+\nu_\alpha)\left[\frac{(3J+2K)t^2}{6I^{3/2}} - \frac{t^2}{\sqrt{I}}\left\{a' - \frac{a(2J+K)}{2I} - \frac{\nu_\alpha K}{6I}\right\}\right]$$
$$+ o\left(\frac{1}{\sqrt{n}}\right)$$
$$= F^*(t, \theta_0) \quad \text{(say)}.$$

By the fundamental lemma of Neyman and Pearson it is seen that the asymptotic power series of the most powerful test of the level $\alpha + o(1/\sqrt{n})$ is given by (3.8). Since θ_0 is arbitrary, we have the desired result for all $t > 0$. In a similar way to the case $t > 0$, we also obtain the conclusion for all $t < 0$.

<u>Sketch of the proof of Theorem 3.3.</u> By Theorems 3.1 and 3.2 we have

(3.9) $F^*(t, \theta) - P_{\theta, n}\{\sqrt{nI(\theta)}(\hat{\theta}^*_n - \theta) \leq t + a(\theta)\}$

$$= \frac{1}{\sqrt{n}}\phi(t+\nu_\alpha)\left\{\frac{3J(\theta)+2K(\theta)+I^3C_3(\theta)}{6I^{3/2}(\theta)}(t^2+2\nu_\alpha t) - \left(C_0' - \frac{IC_2}{2}\right)t\right\}$$
$$+ o\left(\frac{1}{\sqrt{n}}\right).$$

If one assumes that $3J(\theta_0)+2K(\theta_0)+I^3(\theta_0)C_3(\theta_0) \neq 0$ for some $\theta_0 \in \Theta$, it leads to a contradiction to the bound $F^*(t, \theta)$. Hence it follows that

(3.10) $$C_3(\theta) = -\frac{3J(\theta)+2K(\theta)}{I^3(\theta)} \quad \text{for all } \theta \in \Theta .$$

Since $F^*(t,\theta)$ is the bound for the asymptotic distribtions of $\sqrt{nI(\theta)}(\hat{\theta}_n-\theta)$ for all $B_2(\alpha,a)$-estimators $\hat{\theta}_n$, it follows from (3.9) and (3.10) that

$$C_0'(\theta) \le I(\theta)C_2(\theta)/2 \quad \text{for all} \quad \theta \in \Theta .$$

If $C_0'(\theta) \equiv I(\theta)C_2(\theta)/2$, then it is seen that the asymptotic distribution of $\sqrt{nI(\theta)}(\hat{\theta}^*_n-\theta)$ attains the bound $F^*(t,\theta)$ for all t and all $\theta \in \Theta$. Hence the desired result follows.

Let $\hat{\theta}_{ML}$ be the maximum likelihood estimator (MLE) of θ. Then the second order asymptotic efficiency of the MLE is given as follows:

Theorem 3.4. Assume that the conditions (A.1), $(A.2)_3$, (A.3), and (A.4) hold. Then the estimator $\hat{\theta}^*_{ML}$ modified from the MLE to be in $B_2(\alpha,a)$ is second order asymptotically efficient in the class $B_2(\alpha,a)$.

The proof follows from Theorems 3.1 and 3.3 since $C_0(\theta)=C_2(\theta) \equiv 0$ and $C_3(\theta) \equiv -\{3J(\theta)+2K(\theta)\}/I^3(\theta)$ in the asymptotic cumulants of the MLE. (See Akahira and Takeuchi [4], page 90).

Theorem 3.5. Assume that the conditions (A.1), $(A.2)_3$, (A.3), and (A.4) hold. Suppose that $\hat{\theta}_n$ is any BAN estimator of which the Edgeworth expansion up to the order $n^{-1/2}$ is valid and the coefficients $C_i(\theta)$ $(i=0, 1, 2, 3)$ in its asymptotic cumulants as in Theorem 3.1 are continuously differentiable in θ. Then the MLE is second order asymptotically complete in the sense that there exists a modified MLE $\hat{\theta}^*_{ML}$, of the form

$$\hat{\theta}^*_{ML} = \hat{\theta}_{ML}+\frac{1}{\sqrt{n}}g_1(\hat{\theta}_{ML})+\frac{1}{n}g_2(\hat{\theta}_{ML}) ,$$

such that

$$P_{\theta,n}\{-t_1<\sqrt{nI(\theta)}(\hat{\theta}^*_{ML}-\theta)-a(\theta)<t_2\}$$
$$\ge P_{\theta,n}\{-t_1<\sqrt{nI(\theta)}(\hat{\theta}_n-\theta)-a(\theta)<t_2\}+o\left(\frac{1}{\sqrt{n}}\right)$$

for all $t_1, t_2>0$ and all $\theta \in \Theta$.

Sketch of the proof. For fixed α, define $a_n(\theta)$ by

$$P_{\theta,n}\{\sqrt{nI(\theta)}(\hat{\theta}_n-\theta) \le a_n(\theta)\} = \alpha+o\left(\frac{1}{\sqrt{n}}\right) ,$$

$$P_{\theta,n}\{\sqrt{nI(\theta)}(\hat{\theta}_n-\theta) \ge a_n(\theta)\} = 1-\alpha+o\left(\frac{1}{\sqrt{n}}\right) ,$$

locally uniformly in θ, then $a_n(\theta)$ can be expanded as $a_n(\theta)=$

$a(\theta)+n^{-1/2}b(\theta)+o(n^{-1/2})$, where $a(\theta)$ and $b(\theta)$ are continuously differentiable. Consider the class $B_2(\alpha, a_n)$ with $a_n(\theta)$ thus defined. Analogously to the proof of Theorem 3.4 we can show that within the class there exists a modified MLE $\hat{\theta}^*_{ML}$, of the form

$$\hat{\theta}^*_{ML} = \hat{\theta}_{ML}+\frac{1}{\sqrt{n}} g_1(\hat{\theta}_{ML})+\frac{1}{n} g_2(\hat{\theta}_{ML}),$$

which is second order asymptotically efficient in the class $B_2(\alpha, a)$. This implies the desired result.

For the third order asymptotic completeness we further assume the following:

(A.5) There exist $E_\theta[\ell^{(1)}(\theta,X)\ell^{(3)}(\theta,X)]$, $E_\theta[\{\ell^{(2)}(\theta,X)\}^2]$,

$E_\theta[\{\ell^{(1)}(\theta,X)\}^2\ell^{(2)}(\theta,X)]$ and $E_\theta[\{\ell^{(1)}(\theta,X)\}^4]$.

Theorem 3.6. Assume that the conditions (A.1), $(A.2)_4$, (A.3), (A.4), and (A.5) hold. Suppose that $\hat{\theta}_n$ is second order asymptotically efficient in $B_2(\alpha, a)$, of which the Edgeworth expansion up to the order n^{-1} is valid. Then the pair of statistics $(\hat{\theta}_{ML}, Z_2(\hat{\theta}_{ML}))$ is third order asymptotically complete in the sence that there exists a modified MLE $\hat{\theta}^*_{ML}$, of the form

$$\hat{\theta}^*_{ML} = \hat{\theta}_{ML}+\frac{1}{\sqrt{n}} g_1(\hat{\theta}_{ML})+\frac{1}{n} g_2(\hat{\theta}_{ML}, Z_2(\hat{\theta}_{ML}))+\frac{1}{n\sqrt{n}} g_3(\hat{\theta}_{ML}, Z_2(\hat{\theta}_{ML})),$$

such that

$$P_{\theta,n}\{-t_1<\sqrt{nI(\theta)}(\hat{\theta}^*_{ML}-\theta)-a(\theta)<t_2\}$$
$$\geq P_{\theta,n}\{-t_1<\sqrt{nI(\theta)}(\hat{\theta}_n-\theta)-a(\theta)<t_2\}+o\left(\frac{1}{n}\right)$$

for all $t_1, t_2>0$ and $\theta \in \Theta$, where $Z_2(\theta)=\sum_{i=1}^n \ell^{(2)}(\theta, X_i)/\sqrt{n} + \sqrt{n}I(\theta)$.

Sketch of the proof. The asymptotic cumulants κ_i (i=1, 2, 3, 4) of $T_n=\sqrt{n}(\hat{\theta}_n-\theta)$ have the following form:

$$\kappa_1 = E_\theta(T_n) = \mu_{10}(\theta)+\frac{\mu_{11}(\theta)}{\sqrt{n}}+\frac{\mu_{12}(\theta)}{n}+o\left(\frac{1}{n}\right),$$

$$\kappa_2 = V_\theta(T_n) = \frac{1}{I(\theta)}+\frac{\mu_{21}(\theta)}{\sqrt{n}}+\frac{\mu_{22}(\theta)}{n}+o\left(\frac{1}{n}\right),$$

$$\kappa_3 = \kappa_{3,\theta}(T_n) = E_\theta[\{T_n-E_\theta(T_n)\}^3]$$
$$= \frac{\mu_{31}(\theta)}{\sqrt{n}}+\frac{\mu_{32}(\theta)}{n}+o\left(\frac{1}{n}\right),$$

$$\kappa_4 = \kappa_{4,\theta}(T_n) = E_\theta[\{T_n - E_\theta(T_n)\}^4] - 3\{V_\theta(T_n)\}^2$$
$$= \frac{\mu_{42}(\theta)}{n} + o\left(\frac{1}{n}\right).$$

Then it is seen that the coefficients $\mu_{10}(\theta)$, $\mu_{11}(\theta)$, $\mu_{21}(\theta)$, $\mu_{31}(\theta)$ and $\mu_{42}(\theta)$ in the above are determined from the second order asymptotic efficiency of $\hat{\theta}_n$, but $\mu_{22}(\theta)$ and $\mu_{32}(\theta)$ are not so (e.g. see Akahira and Takeuchi, [4]). And it follows that, in the Edgeworth expansion of the distribtion $P_{\theta,n}\{\sqrt{nI(\theta)}(\hat{\theta}_n - \theta) - a(\theta) \le t\}$, the only undecided term has the form $3\mu_{32}(\theta) + \{\mu_{22}(\theta)/\sqrt{I(\theta)}\}w(t)$, where $w(t)$ is some linear function of t. It follows by the discretized likelihood method in Akahira and Takeuchi [3], [4] that for any $\hat{\theta}_n \in B_2(\alpha, a)$ with $\mu_{32}(\theta)$ and $\mu_{22}(\theta)$ in its asymptotic cumulants, any fixed $\theta_0 \in \Theta$ and each real t, there exists a second order as. (α, a)-biased estimator $\hat{\theta}^t$, with $\mu_{32}^t(\cdot)$ and $\mu_{22}^t(\cdot)$ in its asymptotic cumulants, such that

$$(3.11)\qquad 3\mu_{32}^t(\theta_0) + \frac{\mu_{22}^t(\theta_0)}{I(\theta_0)} w(t) \le 3\mu_{32}(\theta_0) + \frac{\mu_{22}(\theta_0)}{I(\theta_0)} w(t).$$

For any given $\hat{\theta}_n$, let $\mu_{22}^0(\theta)$ be the corresponding $\mu_{22}(\theta)$, then it is shown that there exists $t_0 = t_0(\theta_0)$ such that $\mu_{22}^0(\theta_0) = \mu_{22}^{t_0}(\theta_0)$. Since, by (3.8), for any $\mu_{32}^0(\cdot)$

$$3\mu_{32}^{t_0}(\theta_0) + \frac{\mu_{22}^{t_0}(\theta_0)}{\sqrt{I(\theta_0)}} w(t_0) \le 3\mu_{32}^0(\theta_0) + \frac{\mu_{22}^0(\theta_0)}{\sqrt{I(\theta_0)}} w(t_0),$$

it follows that

$$(3.12)\qquad \mu_{32}^{t_0}(\theta_0) \le \mu_{32}^0(\theta_0).$$

From (3.11) and (3.12) we have

$$3\mu_{32}^{t_0}(\theta_0) + \frac{\mu_{22}^0(\theta_0)}{\sqrt{I(\theta_0)}} w(t) \le 3\mu_{32}^0(\theta_0) + \frac{\mu_{22}^0(\theta_0)}{\sqrt{I(\theta_0)}} w(t)$$

for all real t.

Hence, for any second order asymptotically efficient estimator $\hat{\theta}_n$ in $B_2(\alpha, a)$

$$(3.13)\qquad P_{\theta_0,n}\{-t_1 < \sqrt{n}(\hat{\theta}^{t_0} - \theta_0) - a(\theta_0) < t_2\}$$
$$\ge P_{\theta_0,n}\{-t_1 < \sqrt{n}(\hat{\theta}_n - \theta_0) - a(\theta_0) < t_2\} + o\left(\frac{1}{n}\right)$$

for all $t_1, t_2 > 0$.

Now define

$$\hat{\theta}^{0}_{ML} = \hat{\theta}_{ML} + \frac{1}{\sqrt{n}} h_1(\hat{\theta}_{ML}) + \frac{1}{n} h_2(\hat{\theta}_{ML}) + \frac{1}{n\sqrt{n}} h_3(\hat{\theta}_{ML})$$

such that $E_\theta(\hat{\theta}^{0}_{ML})$ is asymptotically equal to $E_\theta(\hat{\theta}_n)$ up to the order n^{-1}, then $\hat{\theta}^{0}_{ML}$ and $\hat{\theta}_n$ have the same values for $\mu_{10}(\theta)$, $\mu_{11}(\theta)$, $\mu_{21}(\theta)$, $\mu_{31}(\theta)$ and $\mu_{42}(\theta)$, but the coefficient $\hat{\theta}^{*}_{32}(\theta)$ of the order n^{-1} in the third order cumulant for $\hat{\theta}^{0}_{ML}$ is identically equal to zero.

By the discretized likelihood method in Akahira and Takeuchi [3], [4] it follows that

$$\hat{\theta}^{t_0} = \hat{\theta}^{0}_{ML} + \frac{t_0}{2nI(\theta_0)} \{ Z_2(\theta_0) - \frac{J(\theta_0)}{I(\theta_0)} Z_1(\theta_0) \} + o_p\left(\frac{1}{n\sqrt{n}}\right) .$$

Now define

$$\hat{\theta}^{\hat{t}_0} = \hat{\theta}^{0}_{ML} + \frac{\hat{t}_0}{2nI(\hat{\theta}_{ML})} Z_2(\hat{\theta}_{ML}) ,$$

where $\hat{t}_0 = t_0(\hat{\theta}_{ML})$.

Then it can be shown that

$$\text{(3.14)} \qquad \hat{\theta}^{\hat{t}_0} = \hat{\theta}^{t_0} + \frac{1}{n\sqrt{n}} W + o_p\left(\frac{1}{n\sqrt{n}}\right) ,$$

where W is some stochastic quantity of magnitude of order 1 and $E_{\theta_0}(W) = o(1)$. Since $\hat{\theta}^{t_0}$ can be stochastically expanded as

$$\sqrt{n}(\hat{\theta}^{t_0} - \theta_0) = \frac{1}{I(\theta_0)} Z_1(\theta_0) + \frac{1}{\sqrt{n}} Q + \frac{1}{n} R + o_p\left(\frac{1}{n}\right) ,$$

it follows from (3.14) that

$$\sqrt{n}(\hat{\theta}^{\hat{t}_0} - \theta_0) = \frac{1}{I(\theta_0)} Z_1(\theta_0) + \frac{1}{\sqrt{n}} Q + \frac{1}{n}(R+W) + o_p\left(\frac{1}{n}\right) ,$$

where both Q and R are certain stochastic quantities of magnitude of order 1. Then it is shown in Akahira and Takeuchi [4] that the asymptotic distributions of the above two estimators $\hat{\theta}^{t_0}$ and $\hat{\theta}^{\hat{t}_0}$ coincide up to the order n^{-1} if $E_{\theta_0}(W) = 0$. Hence it follows from (3.13) that for any second order asymptotically efficient estimator $\hat{\theta}_n$ in $B_2(\alpha, a)$

$$P_{\theta_0,n}\{-t_1 < \sqrt{n}(\hat{\theta}^{\hat{t}_0} - \theta_0) - a(\theta_0) < t_2\}$$

$$\geq P_{\theta_0,n}\{-t_1 < \sqrt{n}(\hat{\theta}_n - \theta_0) - a(\theta_0) < t_2\} + o\left(\frac{1}{n}\right)$$

for all $t_1, t_2 > 0$.

Since $\hat{\theta}^{\hat{t}_0}$ is independent of θ_0 and θ_0 is arbitrary, we have the conclusion of the theorem.

4. Multiparameter case.

Assume that Θ is an open subset of R^p. Let $X_1, X_2, \ldots, X_n, \ldots$ be a sequence of i.i.d. real random variables with a density function $f(x, \theta)$ with respect to a σ-finite measure μ, where $\theta \in \Theta$. Denote by $\hat{\theta}_n = (\hat{\theta}_{n1}, \ldots, \hat{\theta}_{np})'$ an estimator of $\theta = (\theta_1, \ldots, \theta_p)' \in \Theta$. We assume that the distribution of $\sqrt{n}(\hat{\theta}_n - \theta)$ admits the Edgeworth expansion up to the order n^{-1}, and $\hat{\theta}_n$ has the stochastic expansion

$$\sqrt{n}(\hat{\theta}_{n\alpha} - \theta_\alpha) = U_\alpha + \frac{1}{2\sqrt{n}} Q_\alpha + \frac{1}{n} W_\alpha + o_p\left(\frac{1}{n}\right) \quad (\alpha = 1, \ldots, p),$$

with $U_\alpha = \sum_{\beta=1}^{p} I^{\alpha\beta}(\theta) \sum_{i=1}^{n} (\partial/\partial\theta_\beta) \log f(X_i, \theta)/\sqrt{n}$, $Q_\alpha = O_p(1)$ and $W_\alpha = O_p(1)$ $(\alpha = 1, \ldots, p)$, whose asymptotic cumulants are the following: for $Y_\alpha = \sqrt{n}(\hat{\theta}_{n\alpha} - \theta_\alpha)$ $(\alpha = 1, \ldots, p)$

$$E_\theta(Y_\alpha) = \mu_{0\alpha}(\theta) + \frac{\mu_{1\alpha}(\theta)}{\sqrt{n}} + \frac{\mu_{2\alpha}(\theta)}{n} + o\left(\frac{1}{n}\right);$$

$$\mathrm{Cov}(Y_\alpha, Y_\beta) = I^{\alpha\beta}(\theta) + \frac{C^{(1)}_{\alpha\beta}(\theta)}{\sqrt{n}} + \frac{C^{(2)}_{\alpha\beta}(\theta)}{n} + o\left(\frac{1}{n}\right);$$

$$\kappa_{3,\theta}(Y_\alpha, Y_\beta, Y_\gamma) = \frac{\kappa^{(1)}_{\alpha\beta\gamma}(\theta)}{\sqrt{n}} + \frac{\kappa^{(2)}_{\alpha\beta\gamma}(\theta)}{n} + o\left(\frac{1}{n}\right);$$

$$\kappa_{4,\theta}(Y_\alpha, Y_\beta, Y_\gamma, Y_\delta) = \frac{\kappa_{\alpha\beta\gamma\delta}(\theta)}{n} + o\left(\frac{1}{n}\right)$$

for $\alpha, \beta, \gamma, \delta = 1, \ldots, p$, where $I^{\alpha\beta}(\theta)$ is an (α, β)-element of the inverse matrix of the Fisher information matrix.

For each $k = 1, 2, \ldots,$ an estimator $\hat{\theta}^*_n$ is called k-th order asymptotically efficient in the class A_k of the estimators $\hat{\theta}_n$ with the same bias

$$E_\theta(\hat{\theta}_{n\alpha}) = \theta_\alpha + \sum_{i=0}^{k-1} \mu_{i\alpha}(\theta) n^{-(i+1)/2} + o(n^{-k/2}) \quad (\alpha = 1, \ldots, p)$$

up to the order $n^{-k/2}$ if for any estimator $\hat{\theta}_n \in A_k$

$$P_{\theta,n}\{\sqrt{n}(\hat{\theta}^*_n - \theta) - \mu_0(\theta) \in C\} \geq P_{\theta,n}\{\sqrt{n}(\hat{\theta}_n - \theta) - \mu_0(\theta) \in C\} + o(n^{-(k-1)/2})$$

for any convex set C of R^p containing the origin and all $\theta \in \Theta$.

We shall use the following notations: for $\alpha, \beta, \gamma, \delta = 1, \ldots, p$

$$J_{\alpha\beta\cdot\gamma} = E_{\theta}[\{\frac{\partial^2}{\partial\theta_\alpha\partial\theta_\beta}\log f(X,\theta)\}\{\frac{\partial}{\partial\theta_\gamma}\log f(X,\theta)\}] ;$$

$$K_{\alpha\beta\gamma} = E_{\theta}[\{\frac{\partial}{\partial\theta_\alpha}\log f(X,\theta)\}\{\frac{\partial}{\partial\theta_\beta}\log f(X,\theta)\}\{\frac{\partial}{\partial\theta_\gamma}\log f(X,\theta)\}] ;$$

$$J^{\alpha\beta\cdot\gamma} = \sum_{\alpha'=1}^{p}\sum_{\beta'=1}^{p}\sum_{\gamma'=1}^{p} I^{\alpha\alpha'} I^{\beta\beta'} I^{\gamma\gamma'} J_{\alpha'\beta'\cdot\gamma'} ;$$

$$K^{\alpha\beta\gamma} = \sum_{\alpha'=1}^{p}\sum_{\beta'=1}^{p}\sum_{\gamma'=1}^{p} I^{\alpha\alpha'} I^{\beta\beta'} I^{\gamma\gamma'} K_{\alpha'\beta'\gamma'} ;$$

$$M_{\alpha\beta\cdot\gamma\delta} = E_{\theta}[\{\frac{\partial^2}{\partial\theta_\alpha\partial\theta_\beta}\log f(X,\theta)\}\{\frac{\partial^2}{\partial\theta_\gamma\partial\theta_\delta}\log f(X,\theta)\}]-I_{\alpha\beta}I_{\gamma\delta} ;$$

$$M^{\alpha\beta\cdot\gamma\delta} = \sum_{\alpha'}\sum_{\beta'}\sum_{\gamma'}\sum_{\delta'} I^{\alpha\alpha'} I^{\beta\beta'} I^{\gamma\gamma'} I^{\delta\delta'} M_{\alpha'\beta'\cdot\gamma'\delta'} ;$$

$$J^{\alpha\beta}_{\cdot\gamma} = \sum_{\alpha'}\sum_{\beta'} I^{\alpha\alpha'} I^{\beta\beta'} J_{\alpha'\beta'\cdot\gamma} ; \qquad J^{\alpha\cdot\gamma}_{\beta} = \sum_{\alpha'}\sum_{\gamma'} I^{\alpha\alpha'} I^{\gamma\gamma'} J_{\alpha'\beta\cdot\gamma'} ;$$

$$K^{\alpha\beta}_{\gamma} = \sum_{\alpha'}\sum_{\beta'} I^{\alpha\alpha'} I^{\beta\beta'} K_{\alpha'\beta'\gamma} .$$

Theorem 4.1. Assume that $\hat{\theta}_n$ is any asymptotically efficient estimator in the class of the estimators with the same bias $\mu_{0\alpha}(\theta)$ $(\alpha=1, \ldots, p)$ up to the order $n^{-1/2}$ in the above, which admits the Edgeworth expansion of its distribution up to the order $n^{-1/2}$. Then the MLE $\hat{\theta}_{ML}$ is second order asymptotically complete in the sense that there exists a modified MLE $\hat{\theta}^*_{ML}$, with the same asymptotic bias as $\hat{\theta}_n$ up to the order n^{-1}, of the form

$$\hat{\theta}^*_{ML} = \hat{\theta}_{ML}+\frac{1}{\sqrt{n}}\underset{\sim}{g}_1(\hat{\theta}_{ML})+\frac{1}{n}\underset{\sim}{g}_2(\hat{\theta}_{ML}) ,$$

such that

$$P_{\theta,n}\{\sqrt{n}(\hat{\theta}^*_{ML}-\theta)-\mu_0(\theta)\in C\} \geq P_{\theta,n}\{\sqrt{n}(\hat{\theta}_n-\theta)-\mu_0(\theta)\in C\}+o(\frac{1}{\sqrt{n}})$$

for any convex set C of R^p containing the origin and all $\theta\in\Theta$, where $\mu_0(\theta)=(\mu_{01}(\theta), \ldots, \mu_{0p}(\theta))'$.

The proof is omitted since it is done in a similar way to that of Theorem 3.4.

Theorem 4.2. Assume that $\hat{\theta}_n$ is second order asymptotically efficient in the class of the estimators with the same bias up to the order n^{-1} in the above, which admits the Edgeworth expansion of its distribution up to the order n^{-1}. Then the pair of statistics $(\hat{\theta}_{ML}, \underset{\sim}{Z}_2(\hat{\theta}_{ML}))$ is third order asymptotically complete in the sense that there exists a modified MLE $\hat{\theta}^*_{ML}$, with the same asymptotic bias as $\hat{\theta}_n$ up to the order $n^{-3/2}$, of the form

$$\hat{\theta}^*_{ML} = \hat{\theta}_{ML} + \frac{1}{\sqrt{n}} g_1(\hat{\theta}_{ML}) + \frac{1}{n} g_2(\hat{\theta}_{ML}, Z_2(\hat{\theta}_{ML})) + \frac{1}{n\sqrt{n}} g_3(\hat{\theta}_{ML}, Z_2(\hat{\theta}_{ML}))$$

such that

$$P_{\theta,n}\{\sqrt{n}(\hat{\theta}^*_{ML} - \theta) - \mu_0(\theta) \in C\} \geq P_{\theta,n}\{\sqrt{n}(\hat{\theta}_n - \theta) - \mu_0(\theta) \in C\} + o(\frac{1}{n})$$

for any convex set C of R^p containing the origin and all $\theta \in \Theta$, where

$$Z_2(\theta) = \frac{1}{\sqrt{n}} \sum_{i=1}^{n} \frac{\partial^2}{\partial\theta\partial\theta'} \log f(X_i, \theta) + \sqrt{n} I(\theta)$$

with the Fisher information matrix $I(\theta)$.

Sketch of the proof. First it is noted that the coefficients $\mu_{0\alpha}(\theta)$, $\mu_{1\alpha}(\theta)$, $C^{(1)}_{\alpha\beta}(\theta)$, $\kappa^{(1)}_{\alpha\beta\gamma}(\theta)$ and $\kappa_{\alpha\beta\gamma\delta}(\theta)$ $(\alpha, \beta, \gamma, \delta = 1, \dots, p)$ in the above asymptotic cumulants of $\hat{\theta}_n$ are determined from the second order asymptotic efficiency of $\hat{\theta}_n$. In order to obtain $\hat{\theta}^*_{ML}$ with more concentration probability than $\hat{\theta}_n$ up to the order n^{-1} by the Edgeworth expansion of the distribution of $\hat{\theta}_n$, it is enough to find an estimator $\hat{\theta}_n = \hat{\theta}^*_n$ which minimizes $C^{(2)}_{\alpha\beta}(\theta) = \mathrm{Cov}_\theta(Q_\alpha, Q_\beta)$, given $\kappa^{(2)}_{\alpha\beta\gamma}(\theta)$.

Putting $V_{\alpha\beta} = \sum_{\alpha=1}^{p} I^{\alpha\beta}(\partial/\partial\theta_\gamma)U_\beta$ with $I^{\alpha\beta} = I^{\alpha\beta}(\theta)$ $(\alpha, \beta = 1, \dots, p)$, we have for $\alpha, \beta, \gamma = 1, \dots, p$

(4.1) $\kappa^{(2)}_{\alpha\beta\gamma}(\theta) = E_\theta(Q_\alpha V_{\beta\gamma}) + E_\theta(Q_\beta V_{\alpha\gamma}) + E_\theta(Q_\gamma V_{\alpha\beta})$.

We also have for $\alpha, \beta, \gamma, \delta = 1, \dots, p$,

(4.2) $E_\theta(U_\beta U_\gamma Q_\alpha) = 2(\mu_{1\alpha} I^{\beta\gamma} - K^{\alpha\beta\gamma} - J^{\beta\gamma\cdot\alpha}) + o(1)$;

(4.3)
$$E_\theta(U_\beta V_{\gamma\delta} Q_\alpha) = 2\{-\mu_{1\alpha}(K^{\gamma\delta\beta} + J^{\gamma\beta\cdot\delta}) + \sum_{\delta'=1}^{p}(K^{\beta\alpha}_{\delta'} + J^{\beta\cdot\alpha}_{\delta'}) \cdot (K^{\gamma\delta\cdot\delta'} + J^{\gamma\delta'\cdot\delta}) - \sum_{\delta'=1}^{p} J^{\beta\alpha}_{\cdot\delta'} J^{\gamma\delta\cdot\delta'} + M^{\beta\alpha\cdot\gamma\delta}\} + o(1) ,$$

where for each $\alpha = 1, \dots, p$, $\mu_{1\alpha}$ denotes $\mu_{1\alpha}(\theta)$.

From the second order asymptotic efficiency of $\hat{\theta}_n$ we obtain

(4.4) $E_\theta(U_\alpha Q_\alpha) = o(1) \quad (\alpha = 1, \dots, p)$.

Then it follows by the Lagrange method that, under the conditions (4.1) to (4.4) and $E_\theta(Q_\alpha) = 2\mu_{1\alpha} + o(1)$, $Q_\alpha = Q^*_\alpha$ minimizing $\mathrm{Cov}_\theta(Q_\alpha, Q_\beta)$ have the following: for $\alpha = 1, \dots, p$

$$(4.5)\quad Q_\alpha^* = 2\mu_{1\alpha} + \sum_{\beta=1}^{p}\sum_{\gamma=1}^{p} \eta_\alpha^{\beta\gamma}(\theta)\{V_{\beta\gamma} - \sum_{\beta=1}^{p}\sum_{\gamma=1}^{p}\mu_{\beta\gamma}^{\alpha}(\theta)U_\alpha\}$$
$$+ \sum_{\beta=1}^{p}\sum_{\gamma=1}^{p}\lambda_\alpha^{\beta\gamma}(\theta)(U_\beta U_\gamma - I^{\beta\gamma}) + \sum_{\beta=1}^{p}\sum_{\gamma=1}^{p}\sum_{\delta=1}^{p}\lambda_\alpha^{\beta\gamma\delta}(\theta)$$
$$\cdot(U_\beta V_{\gamma\delta} + K^{\alpha\beta\gamma} + J^{\alpha\gamma\cdot\beta}),$$

where $\eta_\alpha^{\beta\gamma}(\theta)$, $\mu_{\beta\gamma}^{\alpha}(\theta)$, $\lambda_\alpha^{\beta\gamma}(\theta)$ and $\lambda_\alpha^{\beta\gamma\delta}(\theta)$ $(\alpha, \beta, \gamma, \delta = 1, \ldots, p)$ are Lagrangian multipliers determined so that the conditions are satisfied. Hence it is seen the desired estimator $\hat\theta_n^*$ has Q_α^* in its stochastic expansion.
Substituting $\hat\theta_{ML}$ for θ in (4.5) and making it to have the same asymptotic bias as $\hat\theta_n$ up to the order $n^{-3/2}$, we obtain the desired result.

References

[1] M, Akahira: Asymptotic theory for estimation of location in non-regular cases, I: Order of convergence of consistent estimators, Rep. Stat. Appl. Res., JUSE 22 (1975), 8-26.

[2] M. Akahira: The structure of asymptotic deficiency of estimators, to appear in Queen's papers in Pure and Applied Mathematics.

[3] M. Akahira and K. Takeuchi: Discretized likelihood methods —— Asymptotic properties of discretized likelihood estimators (DLE's), Ann. Inst. Statist. Math., 31 (1979), 39-56.

[4] M. Akahira and K. Takeuchi: Asymptotic Efficiency of Statistical Estimators: Concepts and Higher Order Asymptotic Efficiency, Lecture Notes in Statistics 7, Springer-Verlag, New York-Heidelberg-Berlin (1981).

[5] P. J. Bickel, F. Götze and W. R. van Zwet : A simple analysis of third-order efficiency of estimates, Proceedings of the Berkeley Conference in Honor of Jerzy Neyman and Jack Kiefer, Vol. II (L. M. LeCam and R. A. Olshen eds.), Wadsworth (1985), 749-768.

[6] J. K. Ghosh, B. K. Sinha and H. S. Wieand: Second order efficiency of the mle with respect to any bounded bowl-shaped loss function, Ann. Statist., 8 (1980), 506-521.

[7] J. Pfanzagl and W. Wefelmeyer: A third order optimum property of the maximum likelihood estimator, J. Multivariate Anal., 8 (1978), 1-29.

[8] J. Pfanzagl and W. Wefelmeyer: Asymptotic Expansions for General Statistical Models, Lecture Notes in Statistics 31, Springer-Verlag, Berlin-Heidelberg-New York-Tokyo (1985).

[9] K. Takeuchi: Higher order asymptotic efficiency of estimators in in decision procedures, Proc. III Purdue Symp. on Decision Theory and Related Topics, Vol. II, (J. Berger and S. Gupta eds.), Academic Press, New York (1982), 351-361.

[10] K. Takeuchi and K. Morimune: Third-order efficiency of the extended maximum likelihood estimators in a simultaneous equation system, Econometrica 53, (1985), 177-200.

M. Akahira
Department of Mathematics
University of Electro-
Communications
Chofu, Tokyo 182
Japan

F. Hirakawa
Department of Mathematics
Science University of Tokyo
Yamazaki, Noda-shi
Chiba 278, Japan

K. Takeuchi
Faculty of Economics
University of Tokyo
Hongo, Bunkyo-ku
Tokyo 113, Japan

AN ACCURACY OF GAUSSIAN APPROXIMATION OF SUM DISTRIBUTION OF INDEPENDENT RANDOM VARIABLES IN BANACH SPACES

I. S. Borisov

1. Introduction and formulation of the main results. Let $\{X_{ni};\ i\le n\}$, $n=1,2,\dots$, be a triangular array of row-wise independent random variables (r.v.-s) taking values in a Banach space $(\mathfrak{X},\|\cdot\|)$. We suppose that the distribution of X_{ni} has a separable support and $EX_{ni}=0$, $\Sigma_{i\le n}\sigma^2_{ni}=1$ for each i and n, where $\sigma^2_{ni}=E\|X_{ni}\|^2$, $\lim_{n\to\infty}\max_{i\le n}\sigma_{ni}=0$. Suppose also that for all i and n there exists Gaussian r.v. W_{ni} with the same covariance as r.v. X_{ni}.

Denote $S_n=\sum_{i\le n}X_{ni}$, $W_n=\sum_{i\le n}W_{ni}$. The goal of the paper is the estimation of the distance

$$d_n(F) = \sup_{z\in R} |P(F(S_n)<z)-P(F(W_n)<z)|$$

where F is a regular functional on $\mathfrak{X}$.

We shall write $F\in\mathcal{G}(m,\beta,\alpha)$ where $m=0,1,\dots$, $\beta\in[0,1]$, $\alpha>0$, if

$$\|F^{(k)}(x)\|^* \le C_0(F)\exp\{\alpha\|x\|\}, \qquad k=0,1,\dots,m,$$
$$\|F^{(m)}(x)-F^{(m)}(y)\|^* \le C_0(F)\exp\{\alpha(\|x\|+\|y\|)\}\|x-y\|^\beta \tag{1}$$

where $F^{(k)}(x)[\cdot]$ is the k-th Frechét derivative of F (the k-linear continuous functional on $\mathfrak{X}^k$); $\|\cdot\|^*$ is the norm of a multilinear continuous functional.

Absolute positive constants in the paper are denoted by C, C_k. The dependence on some parameters is fixed by the corresponding arguments from $C(\cdot)$ or will be denoted by other letters.

Introduce some additional restrictions on r.v.-s $\{X_{ni}\}$, $\{W_{ni}\}$ and F. Let for each subset $N\subseteq\{1,\dots,n\}$

$$E\|\sum_{i\in N}X_{ni}\|^2 \le A\sum_{i\in N}\sigma^2_{ni},$$
$$E\|\sum_{i\in N}W_{ni}\|^2 \le A\sum_{i\in N}\sigma^2_{ni}, \tag{2}$$

where the constant A does not depend on n and N (but may depend on $\mathfrak{X}$ and some probability characteristics of $\{X_{ni}\}$ and $\{W_{ni}\}$).

Remark 1. If $\mathfrak{X}$ is the type 2 space, then (2) holds and $A=A(\mathfrak{X})$. If $\mathfrak{X}$ is not the type 2 space, then the first inequality in (2) does not contain the second one. But (2) may hold in Banach spaces which are no type 2 spaces. For example, let

$$X_{ni} = X_i(nE\|X_1\|^2)^{-1/2}, \tag{3}$$

where $\{X_i\}$ are i.i.d. r.v.-s in $\mathfrak{X}$. Suppose that r.v.-s $\{X_i\}$ satisfy the central limit theorem, that is, the distributions of S_n converge weakly to that of the corresponding Gaussian r.v. γ. From here it follows (see [1])

$$A_1 = \sup_n E\|S_n\|^2 < \infty \tag{4}$$

that is, if $m = \mathrm{Card}\, N$ then

$$E\|\sum_{i\in N} X_{ni}\|^2 = mn^{-1}\|\sum_{i=1}^{m} X_i(mE\|X_1\|^2)^{-1/2}\|^2 \le A_1 mn^{-1} = A_1\sum_{i\in N}\sigma_{ni}^2 ,$$

$$E\|\sum_{i\in N} W_{ni}\|^2 = mn^{-1}E\|\gamma\|^2 \le A_1\sum_{i\in N}\sigma_{ni}^2 .$$

Now put in (2) $A=A_1$.

It is known (see [7]) that one cannot obtain a good estimate for $d_n(F)$ from (1) only (even for sequence $\{X_{ni}\}$ defined in (3) for bounded r.v.-s $\{X_i\}$). Therefore, we add the following mixed condition on $\{X_{ni}\}$, $\{W_{ni}\}$ and F (cf. [3]): for any subset $N\subseteq\{1,\dots,n\}$

$$\sup_{z>0} z^{-M}P(\sum_{i\in N} D_{ni}(W_n) < z\sum_{i\in N}\sigma_{ni}^2) \le B ,$$
$$M = \max\{(5+\beta)\beta^{-1}, 25\} , \tag{5}$$

where $D_{ni}(y)=DF^{(1)}(y)[\overline{X}_{ni}]$, $\overline{X}_{ni}=X_{ni}I(\|X_{ni}\|\le 2\sigma_{ni})$, $I(A)$ is the indicator of an event A.

Theorem. Let $F\in\mathscr{E}(4,\beta,\alpha)$, $\beta>0$, and let the conditions (2), (5) be fulfilled. Then

$$d_n(F) \le C(F,A,B)\sum_{i\in n} E\min\{\|X_{ni}\|^3, C\alpha^{-1}\|X_{ni}\|^2\} . \tag{6}$$

Remark 2. The sum in (6) (introduced by W. Feller) is more convenient than the traditional expression

$$\varepsilon(n) = \sum_{i\le n} E\|X_{ni}\|^3 I(\|X_{ni}\|\le C\alpha^{-1}) + \sum_{i\le n} E\|X_{ni}\|^2 I(\|X_{ni}\|>C\alpha^{-1}).$$

The theorem improves the corresponding results of [3] and [10]. The method can be extended on the case of symmetric functionals (see [10],[11]).

Remark 3. Hereinafter the dependence of $C(\cdot)$ on F means the dependence only on $C_0(F)$ and α in (1).

Corollary 1. Let $\{X_{ni}\}$ be defined in (3) and $E\|X_1\|^3<\infty$. Then under the assumptions of the theorem

$$d_n(F) \le C(F,A_1,B)n^{-1/2}E\|X_1\|^3(E\|X_1\|^2)^{-3/2}, \tag{7}$$

where A_1 is defined in (4).

Denote by $B_{\mathfrak{X}}[0,1]$ the Banach space of all measurable bounded $\mathfrak{X}$-valued functions on $[0,1]$ with the norm

$$\|y\|_\infty = \sup_{0\le t\le 1} \|y(t)\| .$$

Consider in $B_{\mathfrak{X}}[0,1]$ the following r.v.-s.

$$Y_{ni} \equiv Y_{ni}(t) = X_{ni}I_{i/n}(t), \quad G_{ni} \equiv G_{ni}(t) = W_{ni}I_{i/n}(t),$$

where $\{X_{ni}\}$ and $\{W_{ni}\}$ satisfy (2), $I_u(t)=1$ if $u<t$ and $I_u(t)=0$ if $u\ge t$. In this case r.v.-s S_n and W_n are called random broken lines defined on sequences $\{X_{ni}\}$ and $\{W_{ni}\}$ correspondingly.

Corollary 2. Let $F:B_{\mathfrak{X}}[0,1]\to R$ and $F\in\mathscr{C}(4,\beta,\alpha)$. Let $\{X_{ni}\}$ and $\{W_{ni}\}$ satisfy (2) and moreover, for $\{Y_{ni}\}$, $\{G_{ni}\}$ and F the condition (5) is fulfilled. Then estimate (6) holds.

Corollary 3. Let $\mathfrak{X}=R$. If for some $\Delta_0, v\in(0,1)$

$$\sup_{0<\Delta\le\Delta_0} P(\Delta^{-1}\int_v^{v+\Delta} (F^{(1)}(W_n)[I_u])^2 du < z) \le Bz^M \tag{8}$$

then

$$d_n(F) \le C(F) \sum_{i \le n} E|X_{ni}|^s, \qquad 2<s\le 3. \tag{9}$$

The estimate (9) improves results of [2].

Remark 4. The estimate (9) is true if in (8) the process $W_n(t)$ will be replaced by standard Wiener process.

Corollary 4. Let $F(y)=\int_0^1 f(t,y(t))dt$ where $\frac{\partial^k}{\partial x^k} f(t,x)$ satisfy conditions (1) with respect to x uniformly on $t\in[0,1]$. If, moreover,

$$\inf_{0\le t\le 1} \frac{\partial}{\partial x} f(t,x) \ge C(f)|x|^r$$

for all $x\in R$ and some $r>0$, then (8) and (9) hold.

Denote by $F_n(z)$, $z\in[0,1]$, the empirical distribution function based on the sample from the uniform distribution on $[0,1]$. Consider two-parametric random field

$$S_n \equiv S_n(t,z) = [nt]n^{-1/2}(F_{[nt]}(z)-z), \qquad t,z\in[0,1].$$

Let $W_n \equiv W_n(t,z)$ be the Gaussian random field with the same covariance as $S_n(t,z)$. We will consider S_n and W_n as elements of Hilbert space $L_2([0,1]^2,\lambda)$ where λ is the Lebesgue measure on the unit square.

Corollary 5. Let $F:L_2([0,1]^2,\lambda)\to R$ be defined by the formula

$$F(y) = \int_0^1 \int_0^1 f(t,z,y(t,z))dtdz.$$

If $\frac{\partial^k}{\partial x^k} f(t,z,x)$, $k\le 4$, satisfy the conditions (1) uniformly on $t,z\in [0,1]$, and, moreover,

$$\inf_{t,z\in[0,1]} \frac{\partial}{\partial x} f(t,z,x) \ge C(f)|x|^r, \qquad r>0,$$

then $d_n(F)\le C(F)n^{-1/2}$.

2. Preliminary results. Denote by (X'_{ni}), $(X^{(1)}_{ni})$, $(X^{(2)}_{ni})$ three independent triangular arrays of row-wise independent r.v.-s defined

by the formulas

$$X'_{ni} = X_{ni} I(\|X_{ni}\| \le C^*/\alpha) \qquad (10)$$

where C^* will be chosen later;

$$P(X^{(1)}_{ni} \in A) = P(X'_{ni} \in A \mid \|X'_{ni}\| \le 2\sigma_{ni}),$$

$$P(X^{(2)}_{ni} \in A) = 2P(X'_{ni} \in A) - P(X^{(1)}_{ni} \in A).$$

Note that the distribution of $X^{(2)}_{ni}$ is defined correctly, since

$$P(\|X'_{ni}\| \le 2\sigma_{ni}) \ge 1 - \frac{E\|X'_{ni}\|^2}{4\sigma^2_{ni}} \ge 3/4 .$$

Let $\{\xi_i\}$ be the Bernoulli sequence of i.i.d. r.v.-s ($\xi_i = 0$ or 1 with probability 1/2). We suppose that for each n the sequences $\{X_{ni}\}$, $\{X'_{ni}\}$, $\{X^{(1)}_{ni}\}$, $\{X^{(2)}_{ni}\}$ and $\{\xi_i\}$ are independent. In this case

$$X'_{ni} \overset{d}{=} X^{(1)}_{ni}\xi_i + X^{(2)}_{ni}(1-\xi_i) . \qquad (11)$$

This representation will be used many times in the proof of the theorem (cf. [11],[12]).

Denote by $W^{(2)}_{ni}$ the Gaussian r.v. with the same covariance as $X^{(2)}_{ni}$. The existence of r.v. $W^{(2)}_{ni}$ follows from inequality $E(L(X^{(2)}_{ni}))^2 \le 2E(L(X_{ni}))^2$, $L \in \mathfrak{X}^*$ (see [9]).

Below the symbol E_ξ denotes the conditional expectation with respect to r.v. ξ. If we consider a sequence $\{\xi_i\}$, then E_ξ denotes the same with respect to $\{\xi_i\}$.

__Lemma 2.1__. For any subset $N \subseteq \{1,\dots,n\}$ and $\mu \in [0, \alpha(8C^*)^{-1}]$

$$E \exp\{\mu \|\sum_{i \in N} X'_{ni}\|\} \le 3 \exp\{\alpha\mu/C^* + \mu A^{1/2} + 8\mu^2\} \qquad (12)$$

and for any $\mu \ge 0$

$$E \exp\{\mu \|\sum_{i \in N} W_{ni}\|\} \le (1 + 2\mu(2\pi A)^{1/2}) \exp\{\mu^2 A^2/2 + \mu A^{1/2}\}. \qquad (13)$$

__Proof__. Using inequality (4) from [6] we have

$$E \exp(\mu|\|\sum_{i\in N} X'_{ni}\|-E\|\sum_{i\in N} X'_{ni}\||)$$

$$= 1 + \mu \int_0^\infty e^{\mu z} P(|\|\sum_{i\in N} X'_{ni}\|-E\|\sum_{i\in N} X'_{ni}\||\geq z)dz$$

$$\leq 1+ 2\mu \int_0^\infty \exp\left\{\mu z - \frac{z^2}{2(1+z(8\mu)^{-1})}\right\}dz \tag{14}$$

$$= 1 + 2\mu(\int_0^{8\mu} + \int_{8\mu}^\infty)$$

$$\leq 1 + 2(e^{8\mu^2}-1) + 2\mu \int_{8\mu}^\infty \exp\left\{\mu z - \frac{z^2}{2(1+z(8\mu)^{-1})}\right\}dz.$$

Since for $z\in[8\mu,\infty)$

$$\frac{z^2}{2(1+z(8\mu)^{-1})} \geq 2\mu z$$

then we can obtain finally from (14)

$$E \exp(\mu|\|\sum_{i\in N} X'_{ni}\|-E\|\sum_{i\in N} X'_{ni}\||) \leq 3e^{8\mu^2}. \tag{15}$$

It follows from (2)

$$\begin{aligned} E\|\sum_{i\in N} X'_{ni}\| &\leq E\|\sum_{i\in N} X_{ni}\| + \sum_{i\in N} E\|X_{ni}\| I(\|X_{ni}\|>C^*/\alpha) \\ &\leq A^{1/2} + \alpha/C^*. \end{aligned} \tag{16}$$

Now inequality (12) follows from (15), (16) and

$$E\exp(\mu\|\sum_{i\in N} X'_{ni}\|) \leq E\exp(\mu|\|\sum_{i\in N} X'_{ni}\|-E\|\sum_{i\in N} X'_{ni}\||)\exp(\mu E\|\sum_{i\in N} X'_{ni}\|).$$

The proof of (13) is analogous to the considered one by the use of the inequality (see [6])

$$P(|\|\gamma\|-E\|\gamma\||>z) \leq 2\exp\left\{-\frac{z^2}{2E\|\gamma\|^2}\right\} \tag{17}$$

where γ is an arbitrary centered Gaussian r.v. in $\mathfrak{X}$. Lemma is proved.

<u>Lemma 2.2</u>. Let Ψ be arbitrary convex even function on $\mathfrak{X}$. Then for any $N\subseteq\{1,\dots,n\}$

$$E\Psi(\sum_{i\in N}(X_{ni}^{(1)}-EX_{ni}^{(1)})\xi_i) \le E\Psi(2\sum_{i\in N}X_{ni}'),$$

$$E\Psi(\sum_{i\in N}(X_{ni}^{(2)}-EX_{ni}^{(2)})(1-\xi_i)) \le E\Psi(2\sum_{i\in N}X_{ni}'),$$

$$E\Psi(\sum_{i\in N}(\xi_i EX_{ni}^{(1)}+(1-\xi_i)EX_{ni}^{(2)})) \le E\Psi(\sum_{i\in N}X_{ni}').$$

Proof. Denote by $\{y_{ni}^{(k)}\}$, $k=1,2$, the independent copies of $\{X_{ni}^{(k)}\}$, $k=1,2$. Then it follows from Jensen's inequality that

$$E_\xi\Psi(\sum_{i\in N}(X_{ni}^{(1)}-EX_{ni}^{(1)})\xi_i)$$

$$\le E_\xi\Psi(\sum_{i\in N}(X_{ni}^{(1)}-EX_{ni}^{(1)})\xi_i+\sum_{i\in N}(X_{ni}^{(2)}-EX_{ni}^{(2)})(1-\xi_i))$$

$$\le E_\xi\Psi(\sum_{i\in N}(X_{ni}^{(1)}\xi_i+X_{ni}^{(2)}(1-\xi_i))-\sum_{i\in N}(Y_{ni}^{(1)}\xi_i+Y_{ni}^{(2)}(1-\xi_i)))$$

$$\le E_\xi\Psi(2\sum_{i\in N}(X_{ni}^{(1)}\xi_i+X_{ni}^{(2)}(1-\xi_i))).$$

Using representation (11) it is easy to obtain from here the first statement of the lemma.

The proof of the second inequality is analogous to that above. The third one follows from the inequality

$$\Psi(\sum_{i\in N}(\xi_i EX_{ni}^{(1)}+(1-\xi_i)EX_{ni}^{(2)})) \le E_\xi\Psi(\sum_{i\in N}(X_{ni}^{(1)}\xi_i+X_{ni}^{(2)}(1-\xi_i)))$$

which is obtained very easily from Jensen's inequality and from (11). Lemma is proved.

As examples one can use Lemma 2.2 for the functions $\Psi(X)=\|X\|^C$, $C\ge1$, and $\Psi(X)=\exp\{C\|X\|\}$.

Lemma 2.3. For any $r\ge1$,

$$E\|\sum_{i\in N}(X_{ni}^{(1)}-EX_{ni}^{(1)})\xi_i\|^r \le C(A,r,\alpha)(\sum_{i\in N}\sigma_{ni}^2)^{r/2}, \qquad (18)$$

$$E\|\sum_{i\in N}((EX_{ni}^{(1)}-EX_{ni}^{(2)})\xi_i+EX_{ni}^{(2)})\|^r \le C(A,r,\alpha)(\sum_{i\in N}\sigma_{ni}^2)^{r/2}. \qquad (19)$$

Proof. It is proved in [5] that for independent r.v.-s $Y_1,\ldots,$

$Y_m \in \mathfrak{X}$ and $r \geq 1$

$$E\|\sum_{i\leq m} Y_i\|^r \leq C(r)\max\{(E\|\sum_{i\leq m} Y_i\|)^r,\ (\sum_{i\leq m} E\|Y_i\|^2)^{r/2},\ \sum_{i\leq m} E\|Y_i\|^r\}.$$

Put $Y_i=(X_{ni}^{(1)}-EX_{ni}^{(1)})\xi_i$. Then from Lemma 2.2 and (2) we can obtain

$$E\|\sum_{i\in N} Y_i\| \leq 2E\|\sum_{i\in N} X'_{ni}\| \leq 2E\|\sum_{i\in N} X_{ni}\|+2\sum_{i\in N} E\|X_{ni}\|I(\|X_{ni}\|>C^*/\alpha)$$

$$\leq C(A,\alpha)(\sum_{i\in N}\sigma_{ni}^2)^{1/2}.$$

Moreover, $E\|Y_i\|^p \leq C(p)\sigma_{ni}^p$ for any $p\geq 0$. Therefore,

$$\sum_{i\in N} E\|Y_i\|^r \leq C(r)\sum_{i\in N}\sigma_{ni}^r \leq C(r)(\sum_{i\in N}\sigma_{ni}^2)^{r/2}$$

for any $r\geq 2$. In the case $1\leq r<2$ to prove (18) we must use Hölder's inequality which allows us to reduce the estimation of $E\|\Sigma Y_i\|^r$ to the case considered above.

Inequality (19) is proved in a similar way. In this case we only note that

$$\|EX_{ni}^{(2)}\|^r \leq 2^r\|\int_{\mathfrak{X}} zP(X'_{ni}\in dz)\|^r + 2^{r-1}(E\|X_{ni}^{(1)}\|)^r$$

$$\leq 2^r(E\|X_{ni}\|I(\|X_{ni}\|>C^*/\alpha))^r + 2^{2r-1}\sigma_{ni}^r \leq C(r,\alpha)\sigma_{ni}^r.$$

Lemma is proved.

<u>Lemma 2.4</u>. For any centered Gaussian r.v. $\gamma\in\mathfrak{X}$ and for any $r\geq 1$

$$E\|\gamma\|^r \leq 2^{r-1}(r2^{r/2}\Gamma(r/2)+1)(E\|\gamma\|^2)^{r/2}.$$

The proof follows directly from (17).

<u>Lemma 2.5</u>. For any $N\subseteq\{1,\dots,n\}$

$$E\|\sum_{i\in N} W_{ni}^{(2)}\| \leq 2\sqrt{2}\, E\|\sum_{i\in N} W_{ni}\|. \tag{20}$$

The proof is based on the following representation of an arbitrary centered Gaussian r.v. γ (see [8]):

$$\gamma = \int_{\mathfrak{X}} x\, W_{P_\gamma}(dx) \tag{21}$$

where $P_\gamma(\cdot)$ is the distribution of γ, W_{P_γ} is the so-called white noise, that is, independently scattered Gaussian random measure with the covariance

$$EW_{P_\gamma}(A)W_{P_\gamma}(B) = P_\gamma(A\cap B) \ .$$

Using the definition of the stochastic integral (21) (see [8]) and the comparison theorem (see [1], p.108) we can obtain (20).

Let $\{Y_i\}$ be a sequence of independent bounded centered r.v.-s in $\mathfrak{X}$, and let N be a finite subset of natural numbers. Let $N=\cup_{j\le m+1}N_j$ where $\{N_j\}$ are pairwise disjoint subsets. Denote $S(N_k)=\sum_{i\in N_k} Y_i$ and

$$\gamma_m(L,N) = \sum_{j\le m+1}\ \sum_{i\in N_j} E|L(Y_i)|^3(EL(S(N_j))^2)^{-1}, \qquad L\in\mathfrak{X}^*. \tag{22}$$

Some of the statements below improve the corresponding results in [3].

<u>Lemma 2.6</u>. Let L and L_m be continuous linear and m-linear functionals on $\mathfrak{X}$ and $\mathfrak{X}^m$ correspondingly. If $\gamma_m(L,N)|t|\le 1/4$, then

$$\begin{aligned}&|EL_m(S(N),\dots,S(N))\exp\{itL(S(N))\}|\\ &\le m^m\|L_m\|^*(\sum_{j\le m+1} E\|S(N_j)\|^m)\sum_{j\le m+1}\exp\{-\frac{t^2}{3}DL(S(N_j))\}.\end{aligned} \tag{23}$$

<u>Proof</u>. Denote by J the left-hand side of (23). Then we have

$$J = \Big|\sum_{i_1,\dots,i_m=1}^{m+1} EL_m(S(N_{i_1}),\dots,S(N_{i_m}))\exp\{it\sum_{j\le m+1}L(S(N_j))\}\Big|.$$

In the sum $\sum_{j\le m+1}L(S(N_j))$ there is a summand $L(S(N_{j*}))$ where $j^*=j^*(i_1,\dots,i_m)$, which does not depend on $S(N_{i_1}),\dots,S(N_{i_m})$. Then

$$J \le \sum_{i_1,\dots,i_m=1}^{m+1} E|L_m(S(N_{i_1}),\dots,S(N_{i_m}))||E\exp\{itDL(S(N_{j*}))\}|. \tag{24}$$

It is well known (see [4]) that if

$$|t|\sum_{k\in N_{j*}} E|L(Y_k)|^3(EL(S(N_{j*}))^2)^{-1} \le 1/4$$

then

$$|E\exp(itL(S(N_{j*})))| \le \exp(-\frac{t^2}{3}DL(S(N_{j*}))). \qquad (25)$$

Since for each $j \le m+1$

$$\sum_{k \in N_j} E|L(Y_k)|^3 (EL(S(N_{j*}))^2)^{-1} \le \gamma_m(L,N)$$

then it is easy to obtain (23) from (24) and (25).

Corollary. Inequality (23) holds if

$$\|L\|^* |t| \max_{k \in N} \text{ess.sup}|Y_k| \le 1/4. \qquad (26)$$

The statement follows from (22) and from the simple inequality

$$\sum_{i \in N} u_i (\sum_{i \in N} v_i)^{-1} \le \max_{i \in N} u_i/v_i,$$

where $\{u_i\}$, $\{v_i\}$ are arbitrary positive numbers.

Remark. If $\{Y_i\}$ are centered Gaussian r.v.-s in $\mathfrak{X}$, then (23) holds for all $t \in R$.

Denote for any subset $N \subseteq \{1,\dots,n\}$

$$S_{nk}(N) = \sum_{i<k, i \notin N} X'_{ni} + \sum_{i>k, i \notin N} W_{ni} + \sum_{i<k, i \in N} (X^{(2)}_{ni} - EX^{(2)}_{ni})(1-\xi_i)$$

$$+ \sum_{i<k, i \in N} ((EX^{(1)}_{ni} - EX^{(2)}_{ni})\xi_i + EX^{(2)}_{ni}),$$

$$\delta_{nk}(N) = \sum_{i<k, i \in N} (X^{(1)}_{ni} - EX^{(1)}_{ni})\xi_i + \sum_{i>k, i \in N} W_{ni}.$$

Lemma 2.7. Let $F \in \mathcal{E}(m,\beta,\alpha)$. Then for any $r \le m$

$$E|F^{(r)}(S_{nk}(N))[\delta_{nk}(N)^r]| \le C(F,A,r)(\sum_{i \in N} \sigma^2_{ni})^{r/2}.$$

The proof follows from 2.1-2.4 and Hölder's inequality. Note that for arbitrary $u>0$ we can choose C^* in (10) such that

$$\sup_n E \exp\{u\|\sum_{i \le n} X'_{ni}\|\} < \infty.$$

Introduce the following notations

$$S_{nk} = \sum_{i<k} X'_{ni} + \sum_{i>k} W_{ni},$$

$$\Delta(N) = \Big(\sum_{i\in N} \sigma^2_{ni}\Big)^{1/2},$$

$$\sigma_n(N) = \max_{i\in N} \sigma_{ni}.$$

Lemma 2.8. Let $H\in\mathscr{C}(m,\beta,\alpha)$, $F\in\mathscr{C}(r,\beta,\alpha)$, $m\geq 1$, $r\geq 2$, and let $\{N_j;\ j\leq 4r+m+1\}$ be pairwise disjoint subsets such that $N=\bigcup N_j\subset\{1,\ldots,n\}$. If $\Delta(N)^2|t|\leq 1$ and $C_0(F)\sigma_n(N)^{1-\nu}|t|\leq 1/16$ for some $\nu\in(0,1)$, then for any $p>1+c$

$$|EH(S_{nk})e^{itF(S_{nk})}| \leq C(F,H,A,\nu,m,r)\Big\{\Delta(N)^{m+\beta}+|t|\Delta(N)^{r+\beta}+|t|^5\Delta(N)^{10} \tag{27}$$
$$+ \max_{j\leq 4r+m+1} E^{1/p}\exp\{-\frac{t^2p}{3}E_{S_{nk}(N),\xi}(F^{(1)}(S_{nk}(N))[\delta_{nk}(N_j)])^2\}\Big\}.$$

Proof. Using the Taylor formula we obtain

$$|EH(S_{nk})e^{itF(S_{nk})}| \leq |E\sum_{\ell=0}^{m}\frac{1}{\ell!}H^{(\ell)}(S_{nk}(N))[\delta_{nk}(N)^{\ell}]$$
$$\times\exp\{it(F(S_{nk}(N))+\sum_{s=1}^{r}\frac{1}{s!}F^{(s)}(S_{nk}(N))[\delta_{nk}(N)^{s}])\}|$$
$$+C(H)\Delta(N)^{m+\beta}+C(H,F)|t|\Delta(N)^{r+\beta}$$

$$\leq \sum_{\ell=0}^{m} E\Big|E_{S_{nk},\xi}\frac{1}{\ell!}H^{(\ell)}(\cdot)[\cdot]\exp\{itF^{(1)}(S_{nk}(N))[\delta_{nk}(N)]\} \tag{28}$$
$$\times\{1+\sum_{d=1}^{4}\frac{1}{d!}(it\sum_{s=2}^{r}\frac{1}{s!}F^{(s)}(S_{nk}(N))[\delta_{nk}(N)^{s}])^{d}\}\Big|$$
$$+C(H,F)(\Delta(N)^{m+\beta}+|t|\Delta(N)^{r+\beta}+|t|^5\Delta(N)^{10}).$$

Now consider the conditional expectation $E_{S_{nk},\xi}$ on the sets $\Omega_1=\{\|S_{nk}(N)\|\leq\frac{\nu}{\alpha}|\log\sigma_n(N)|\}$ and $\Omega_2=\{\|S_{nk}\|>\frac{\nu}{\alpha}|\log\sigma_n(N)|\}$. This expectation is estimated on Ω_1 by Lemmas 2.6 and 2.7 because estimated in the sum of the right-hand side of (28) are, in fact, expressions of the type

$$|E_{S_{nk}(N),\xi}L_{\ell}(\delta_{nk}(N)^{\ell})\exp\{itL(\delta_{nk}(N))\}|$$

where $L_{\ell}(x^{\ell})=L_{\ell}(x,\ldots,x)$ and $\ell\leq 4r+m$,

$$\|L\|^* \le C_0(F)\ \exp\ \{\alpha\|S_{nk}(N)\|\}\ ,$$

$$\|L_\ell\|^* \le C(F,H)\ \exp\ \{(4r+m)\alpha\|S_{nk}\|\}\ .$$

Moreover, the following inequality holds on Ω_1

$$\|L\|^* \max_{i\in N}\ \text{ess.sup}\ |(X_{ni}^{(1)}-EX_{ni}^{(1)})\xi_i|$$

$$\le 4\sigma_n(N)\|F^{(1)}(S_{nk}(N))\|^* \le 4C_0(F)\sigma_n(N)^{1-\nu}.$$

Hence, under the conditions of the lemma

$$|E_{S_{nk}(N),\xi}L_\ell(\delta_{nk}(N)^\ell)\exp\{itL(\delta_{nk}(N))\}|$$

$$\le C(F,H,m,r)\exp\{(4r+m)\alpha\|S_{nk}(N)\|\}\{\ \sum_{j=1}^{4r+m+1} E_\xi\|\delta_{nk}(N_j)\|^\ell$$

$$\times \sum_{j=1}^{4r+m+1} \exp\{-\frac{t^2}{3}E_{S_{nk}(N),\xi}(F^{(1)}(S_{nk}(N))[\delta_{nk}(N_j)])^2\} \tag{29}$$

$$+I(\|S_{nk}(N)\|>\frac{\nu}{\alpha}|\log\sigma_n(N)|)\}\ .$$

We need the estimate of probability $P(\Omega_2)$. Using Hölder's and Jensen's inequalities and Lemmas 2.1, 2.2 we obtain

$$P(\Omega_2) \le E^{1/2}\exp\left\{\frac{\alpha}{8C^*}\|S_{nn+1}\|\right\}E^{1/2}\exp\left\{\frac{\alpha}{8C^*}\|W_n\|\right\}\exp\left\{-\frac{\nu}{16C^*}|\log\sigma_n(N)|\right\}$$

$$\le C(A,\alpha)\sigma_n(N)^{\nu(16C^*)^{-1}}\ . \tag{30}$$

Taking the expectation from both sides of (29) and using Hölder's inequality, Lemmas 2.1-2.6, (30) and choosing sufficiently small $C^*=C^*(\nu,m,r,p)$ we can obtain from (28) the statement of the lemma.

<u>Lemma 2.9</u>. Let $|t|\Delta(N)\ge1$, $t^2\Delta(N)^3\le1$ and $|t\Delta(N)|^6\varepsilon(n)\le1$. Then

$$E\exp\{-t^2E_{S_{nk}(N),\xi}(F^{(1)}(S_{nk}(N))[\delta_{nk}(N_j)])^2\}$$

$$\le C(F,A)\{E^{3/4}\exp\{-t^2E_{W_n,\xi}(F^{(1)}(W_n)[\delta_{nk}(N_j)])^2\}$$

$$+(t^2\Delta(N))^6+(|t\Delta(N)|^6\varepsilon(n))^{7/2}\}\ .$$

Proof. Denote

$$\Phi_{kj}(x) = E_\xi(F^{(1)}(x)[\delta_{nk}(N_j)])^2,$$

$$e_{nk}(N) = \sum_{i<k,\,i\in N}((EX_{ni}^{(1)}-EX_{ni}^{(2)})\xi_i+EX_{ni}^{(2)}),$$

$$S_{nkr}(N) = \sum_{i<r,\,i\in N}(X_{ni}^{(2)}-EX_{ni}^{(2)})(1-\xi_i)+\sum_{i<r,\,i\notin N}X_{ni}^{'} \tag{31}$$

$$+\sum_{r<i\le k,\,i\in N}W_{ni}^{(2)}(1-\xi_i)+\sum_{i>r,\,i\notin N}W_{ni}, \qquad r\le k.$$

Hereinafter it is supposed that the sequences $\{X_{ni}^{'}\}$, $\{X_{ni}^{(2)}\}$, $\{W_{ni}\}$, $\{W_{ni}^{(2)}\}$ and $\{\xi_i\}$ are independent.

We shall use Lindeberg's operator method of proving the central limit theorem (cf. [3]). We need the following Taylor's formula:

$$\exp\{-t^2\Phi_{kj}(x+\delta)\} = \exp\{-t^2\Phi_{kj}(x)\}(1-t^2\Phi_{kj}^{(1)}(x)[\delta]+\frac{t^4}{2}(\Phi_{kj}^{(1)}(x)[\delta])^2$$

$$-\frac{t^2}{2}\Phi_{kj}^{(2)}(x)[\delta^2])+\frac{1}{2}\int_0^1(1-\theta)^2\Psi(x+\theta\delta,t)\exp\{-t^2\Phi_{kj}(x+\theta\delta)\}d\theta, \tag{32}$$

where

$$\Psi(x,t) = t^6(\Phi_{kj}^{(1)}(x)[\delta])^3-3t^4\Phi_{kj}^{(1)}(x)[\delta]\Phi_{kj}^{(2)}(x)[\delta^2]+t^2\Phi_{kj}^{(3)}(x)[\delta^3].$$

Note that if $F\in\mathscr{C}(4,\beta,\alpha)$, then $\Phi_{kj}\in\mathscr{C}(3,\beta,\alpha)$. Further we have

$$|E\exp\{-t^2\Phi_{kj}(S_{nk}(N))\}-E\exp\{-t^2\Phi_{kj}(W_n)\}|$$

$$\le \sum_{r\le k}E|E_\xi\exp\{-t^2\Phi_{kj}(S_{nkr}(N)+e_{nk}(N)+\tilde{X}_{nr})\}$$

$$-E_\xi\exp\{-t^2\Phi_{kj}(S_{nkr}(N)+e_{nk}(N)+\tilde{W}_{nr})\}| \tag{33}$$

$$+|E\exp\{-t^2\Phi_{kj}(S_{nk0}(N)+e_{nk}(N))\}-E\exp\{-t^2\Phi_{kj}(W_n)\}|,$$

where $\tilde{X}_{nr}=X_{nr}^{'}$, $\tilde{W}_{nr}=W_{nr}$ if $r\notin N$ and $\tilde{X}_{nr}=(X_{nr}^{(2)}-EX_{nr}^{(2)})(1-\xi_r)$, $\tilde{W}_{nr}=W_{nr}^{(2)}(1-\xi_r)$ if $r\in N$. Now we apply the formula (32) to each term of difference in the first sum of the right-hand side of (33), for $x=S_{nkr}(N)+e_{nk}(N)$ and $\delta=\tilde{X}_{nr}$ or $\delta=\tilde{W}_{nr}$ correspondingly. Note that for fixed $\{\xi_i\}$ r.v. $S_{nkr}(N)$ consists of independent r.v.-s and it follows from the conditions of the theorem

$$E_\xi\Phi_{kj}^{(1)}(x)[X_{nr}^{'}] = E_\xi\Phi_{kj}^{(1)}(x)[X_{nr}I(\|X_{nr}\|>C^*/\alpha)],$$

$$|E_{\xi,x}L_2(x)[X'^2_{nr}]-E_{\xi,x}L_2(x)[W^2_{nr}]|$$

$$= |E_{\xi,x}L_2(x)[X'^2_{nr}]-E_{\xi,x}L_2(x)[X^2_{nr}]| \qquad (34)$$

$$\le 2\|L_2(x)\|^* E\|X_{nr}\|^2 I(\|X_{nr}\|>C^*/\alpha),$$

$$E_\xi L_2(x)[(X^{(2)}_{nr}-EX^{(2)}_{nr})^2] = E_\xi L_2(x)[W^{(2)2}_{nr}],$$

where $L_2(x)$ is arbitrary bilinear continuous functional; for example, $L_2(x)[z,y]=\Phi^{(2)}_{kj}(x)[z,y]$ or $L_2(x)[z,y]=\Phi^{(1)}_{kj}(x)[z]\Phi^{(1)}_{kj}(x)[y]$.

Hence by (32) and (34) we can obtain

$$E|E_\xi \exp\{-t^2\Phi_{kj}(x+\tilde{X}_{nr})\}-E_\xi \exp\{-t^2\Phi_{kj}(x+\tilde{W}_{nr})\}|$$

$$\le C(F)|t\Delta(N)|^4 E\|X_{nr}\|^2 I(\|X_{nr}\|>C^*/\alpha)E^{(q-1)/q}\exp\{\frac{4q\alpha}{q-1}(\|S_{nkr}(N)\|$$

$$+\|e_{nk}(N)\|)\}E^{1/q}\{-qt^2\Phi_{kj}(x)\}+C(F)|t\Delta(N)|^6$$

$$\times E^{(q-1)/q}\exp\{\frac{6\alpha q}{q-1}(\|S_{nkr}(N)\|+\|e_{nk}(N)\|+2C^*/\alpha)\}$$

$$\times \sup_{\theta\in[0,1]} E\|\tilde{X}_{nr}\|^3 E^{1/q}_{\tilde{X}_{nr}}\exp\{-qt^2\Phi_{kj}(S_{nkr}(N)+e_{nk}(N)+\theta\tilde{X}_{nr})\} \qquad (35)$$

$$+C(F)E^{(q-1)/q}\|\tilde{W}_{nr}\|^{3q/(q-1)}\exp\{\frac{6\alpha q}{q-1}\|\tilde{W}_{nr}\|\}$$

$$\times E^{(q-1)/q}\exp\{\frac{6\alpha q}{q-1}(\|S_{nkr}(N)\|+\|e_{nk}(N)\|)\}$$

$$\times \sup_{\theta\in[0,1]} E^{1/q}\exp\{-qt^2\Phi_{kj}(S_{nkr}(N)+e_{nk}(N)+\theta\tilde{W}_{nr})\}.$$

Using Lemmas 2.1-2.5 and (2) we have

$$E^{(q-1)/q}\|\tilde{W}_{nr}\|^{3q/(q-1)}\exp\{\frac{6\alpha q}{q-1}\|\tilde{W}_{nr}\|\}$$

$$\le E^{(q-1)/2q}\|\tilde{W}_{nr}\|^{6q/(q-1)}E^{(q-1)/2q}\exp\{\frac{12\alpha q}{q-1}\|\tilde{W}_{nr}\|\}$$

$$\le C(q)\sigma^3_{nr} \le C_1(q)(E\|X'_{nr}\|^3+E\|X_{nr}\|^2 I(\|X_{nr}\|>C^*/\alpha)).$$

Finally, from (35) there follows the estimate of the first sum in the right-hand side of (35)

$$\sum_{r\le k} \le C(F,A,q)|t\Delta(N)|^6$$

$$\times \{\varepsilon(n) \sup_{\theta\in[0,1],r\le k} E^{1/q}\exp\{-qt^2\Phi_{kj}(S_{nkr}(N)+e_{nk}(N)+\theta\tilde{W}_{nr})\} \qquad (36)$$

$$+\sum_{r\le k}\sup_{\theta\in[0,1]} E\|\tilde{X}_{nr}\|^3 E^{1/q}_{\tilde{X}_{nr}}\exp\{-qt^2\Phi_{kj}(S_{nkr}(N)+e_{nk}(N)+\theta\tilde{X}_{nr})\}.$$

Here it is supposed that C^* in (10) is sufficiently small. In this case there exists $\sup_{r,k,n}\exp\{C\|S_{nkr}(N)\|\}$ for sufficiently large C.

To estimate the second summand we shall use the following simple inequality

$$|e^{-x}-e^{-y}| \le e^{-y}\sum_{j=1}^{m}\frac{1}{j!}|x-y|^j + \frac{|x-y|^{m+1}}{(m+1)!},$$

where m is an arbitrary natural number, $x,y\ge 0$. Then, by Lemmas 2.1-2.5, condition (2) and Hölder's inequality we obtain

$$|E\exp\{-t^2\Phi_{kj}(S_{nk0}(N)+e_{nk}(N))\}-E\exp\{-t^2\Phi_{kj}(W_n)\}|$$

$$\le C(F,m)E^{1/q}\exp\{-qt^2\Phi_{kj}(W_n)\}E^{(q-1)/q}\{\sum_{\ell=1}^{m}E_\xi\|\delta_{nk}(N_j)\|^{2\ell}|t|^{2\ell}$$

$$\times(\|e_{nk}(N)\|^{\ell}+E_\xi\|\sum_{i\in N}W^{(2)}_{ni}(1-\xi_i)\|^{\ell}+E\|\sum_{i\in N}W_{ni}\|^{\ell})\}^{q/(q-1)}$$

$$\times\exp\{\frac{2\ell q}{q-1}(\|S_{nk0}(N)\|+\|e_{nk}(N)\|+\|W_n\|)\} \tag{37}$$

$$+C(F,m)|t|^{2(m+1)}E^{1/2}\exp\{4(m+1)(\|S_{nk0}(N)\|+\|e_{nk}(N)\|+\|W_n\|)\}$$

$$\times E^{1/2}(\|e_{nk}(N)\|^{m+1}+\|\sum_{i\in N}W^{(2)}_{ni}(1-\xi_i)\|^{m+1}+\|\sum_{i\in N}W_{ni}\|^{m+1})E_\xi\|\delta_{nk}(N_j)\|^{4(m+1)}$$

$$\le C(F,A,m,q)\{E^{1/q}\exp\{-qt^2\Phi_{kj}(W_n)\}+|t\Delta(N)|^{2(m+1)}\Delta(N)^{m+1}\}.$$

Hence by (33)-(37) it follows

$$E\exp\{-t^2\Phi_{kj}(S_{nk}(N))\} \le C(F,A,m,q)\{E^{1/q}\exp\{-t^2\Phi_{kj}(W_n)\}$$

$$+|t\Delta(N)|^6\Big\{\varepsilon(n)\sup_{\theta\in[0,1],r\le k}E^{1/q}\exp\{-qt^2\Phi_{kj}(S_{nkr}(N)+e_{nk}(N)+\theta\tilde{W}_{nr})\}$$

$$+\sum_{r\le k}\sup_{\theta\in[0,1]}E\|\tilde{X}_{nr}\|^3E^{1/q}_{\tilde{X}_{nr}}\exp\{-qt^2\Phi_{kj}(S_{nkr}(N)+e_{nk}(N)+\theta\tilde{X}_{nr})\} \tag{38}$$

$$+|t\Delta(N)|^{2(m+1)}\Delta(N)^{m+1}\}\Big\}.$$

Inequality (38) is recurrent because r.v.-s $S_{nkr}(N)$ and $S_{nkr}(N)+\theta\tilde{W}_{nr}$ for fixed $\{\xi_i\}$ and θ consist of independent r.v.-s. Hence for each expectation of the right-hand side of (38) containing r.v. $S_{nkr}(N)$ the estimate (37) holds (uniformly on $\theta\in[0,1]$). Using this recurrent inequality thrice we obtain under the conditions of the lemma

$$E\exp\{-t^2\Phi_{kj}(S_{nk}(N))\} \le C(F,A,m,q)\{E^{1/q^4}\exp\{-t^2\Phi_{kj}(W_n)\}$$
$$+(t^2\Delta(N)^3)^{(m+1)/q^3}+(|t\Delta(N)|^6\varepsilon(n))^{1+1/q+1/q^2+1/q^3}\}.$$

Putting $q=(4/3)^{1/4}$, $m=7$ we obtain the statement of the lemma.

Lemma 2.10. If $C_0(F)|t|\sigma_n(N)^{1-\nu}\le 1/8$, then

$$E\exp\{-t^2\Phi_{kj}(W_n)\} \le E\exp\{-\frac{3t^2}{8}\Phi^*_{kj}(W_n)\}+C(A,\alpha,\nu)\sigma_n(N)^4,$$

where $\Phi^*_{kj}(X)=E(F^{(1)}(X)[\sum_{i\in N_j,i<k}(X^{(1)}_{ni}-EX^{(1)}_{ni})+\sum_{i\in N_j,i>k}W_{ni}])^2$.

Proof. Denote $W'_n=W_nI(\|W_n\|\le\frac{\nu}{\alpha}|\log\sigma_n(N)|)$. Then we have

$$E\exp\{-t^2\Phi_{kj}(W_n)\} \le E\exp\{-t^2\Phi_{kj}(W'_n)\}+P(\|W_n\|>\frac{\nu}{\alpha}|\log\sigma_n(N)|)$$
$$\le E\exp\{-t^2E_{W_n}(F^{(1)}(W_n)[\sum_{i\in N_j,i>k}W_{ni}])^2\}$$
$$\times\prod_{i\in N_j,i<k}\{1-\frac{1}{2}(1-\exp\{-t^2E_{W_n}(F^{(1)}(W'_n)[X^{(1)}_{ni}-EX^{(1)}_{ni}])^2\})\}$$
$$+P(\|W_n\|>\frac{\nu}{\alpha}|\log\sigma_n(N)|).$$

By the simple inequalities

$$1-e^{-x}\ge x-x^2/2\ ,$$
$$\prod_{i\in N}\{1-y_i\}\le\exp\{-\sum_{i\in N}y_i\}$$

we obtain

$$E\exp\{-t^2\Phi_{kj}(W_n)\} \le E\exp\{-t^2E_{W_n}(F^{(1)}(W_n)[\sum_{i\in N_j,i>k}W_{ni}])^2\}$$
$$\times\exp\{-\frac{t^2}{2}\sum_{i\in N_j,i<k}E_{W_n}(F^{(1)}(W'_n)[X^{(1)}_{ni}-EX^{(1)}_{ni}])^2 \tag{39}$$
$$+\frac{t^4}{4}\sum_{i\in N_j,i<k}(E_{W_n}(F^{(1)}(W'_n)[X^{(1)}_{ni}-EX^{(1)}_{ni}])^2)^2\}+P(\|W_n\|>\frac{\nu}{\alpha}|\log\sigma_n(N)|)$$
$$\le E\exp\{-t^2E_{W_n}(\cdots)^2\}\exp\{-\frac{t^2}{2}(1-C_0(F)^2t^2\sigma_n(N)^{2(1-\nu)})$$

$$\times \sum_{i\in N_j, i<k} E_{W_n}(F^{(1)}(W_n)[X_{ni}^{(1)}-EX_{ni}^{(1)}])^2) + 2P(\|W_n\| > \frac{\nu}{\alpha}|\log\sigma_n(N)|).$$

Note that the last term of the right-hand sides of (39) is estimated by inequality (17) and condition (2). Lemma is proved.

Corollary. Under the conditions of Lemmas 2.9 and 2.10

$$\max_{k,j} E \exp\{-t^2\Phi_{kj}(S_{nk}(N))\}$$
$$\le C(F,A,B)(|t\Delta(N)|^{-3M/2} + (t^2\Delta(N)^3)^6 + (\varepsilon(n)(t\Delta(N))^6)^{7/2}), \tag{40}$$

if only $\min_{k,j}\Delta(N_j\setminus\{k\}) \ge C\Delta(N)$.

Proof. By the equality of the covariances of r.v.-s X_{ni} and W_{ni}

$$\Phi^*_{kj}(W_n) = E_{W_n}(F^{(1)}(W_n)[\sum_{i\in N_j, i<k}(X_{ni}^{(1)}-EX_{ni}^{(1)})])^2$$
$$+ \sum_{i\in N_j, i>k} E_{W_n}(F^{(1)}(W_n)[(X_{ni}^{(1)}\xi_i + X_{ni}^{(2)}(1-\xi_i)])^2 \equiv J.$$

Using the simple inequality $E(\xi+C)^2 \ge E(\xi-E\xi)^2$ and Jensen's inequality we obtain

$$J \ge \sum_{i\in N_j, i<k} E_{W_n}(F^{(1)}(W_n)[X_{ni}^{(1)}-EX_{ni}^{(1)}])^2$$
$$+\frac{1}{2}\sum_{i\in N_j, i>k} E_{W_n}(F^{(1)}(W_n)[X_{ni}^{(1)}-EX_{ni}^{(1)}])^2 \ge \frac{1}{2}\sum_{i\in N_j\setminus\{k\}} D_{ni}(W_n). \tag{41}$$

Inequality (40) follows from Lemmas 2.8 and 2.10, (41) and from the estimate

$$Ee^{-t^2\xi} = t^2\int_0^\infty P(\xi<x)e^{-t^2x}\,dx$$
$$\le e^{-t^2} + Bt^2\int_0^1 x^M e^{-t^2x}\,dx \le C(B)|t|^{-2M}. \tag{42}$$

Lemma 2.11. Let $G(z)$ be an arbitrary bounded smooth function on R which has twelve bounded derivatives. Then

$$EG(\Delta(N)^{-2}\Phi_{kj}(S_{nk}(N))) \tag{43}$$

$$\leq C(F,A) \sup_{z\in R, s\leq 12} \left|\frac{d^s}{dz^s}G(z)\right| (EG(\Delta(N)^{-2}\Phi_{kj}(W_n))^{3/4}+\Delta(N)^6+\varepsilon(n)^{7/2})$$

Proof. The method of proving Lemma 2.9 is not changed if we consider $G(t^2z)$ instead of $\exp\{-t^2z\}$. To obtain (43) we must put $t=\Delta(N)^{-1}$.

Lemma 2.12. Let H and F satisfy the conditions of Lemma 2.8. Then for any $t\in R$

$$|EH(W_n)\exp\{itF(W_n)\}| \leq C(F,H,A,B)|t|^{-\mu}, \qquad \mu>0. \tag{44}$$

The proof is analogous to that of Lemma 2.8 where we must use the fact that W_{ni} is infinitely divisible.

Lemma 2.13. For any $z>0$ and $\gamma\in(0,1)$

$$P(\Delta(N)^{-2}\Phi_{kj}(S_{nk}(N))<z)$$

$$\leq C(F,A,B,\gamma)z^{-12}\{z^{3M/4}+(\Delta(N)^2\sigma_n(N)^{\gamma-2})^{-3M/4}+\Delta(N)^6+\varepsilon(n)^{7/2}\},$$

where M and $\Phi_{kj}(X)$ are defined in (5) and (31) correspondingly, $\min_{k,j}\Delta(N_j\setminus\{k\})\geq C\Delta(N)$.

Proof. Let a non-negative function $G(z)$ satisfy the conditions of Lemma 2.11 and

$$\begin{aligned} G(z) &\geq C_0 \quad \text{if} \quad |z|\leq 1, \\ G(z) &= 0 \quad \text{if} \quad |z|\geq 2. \end{aligned} \tag{45}$$

Then it follows from (43) and (45)

$$P(\Delta(N)^{-2}\Phi_{kj}(S_{nk}(N))<z) \leq C_0^{-1}G(z^{-1}\Delta(N)^{-2}\Phi_{kj}(S_{nk}(N)))$$

$$\leq C(F,G,A)z^{-12}\{P^{3/4}(\Delta(N)^{-2}\Phi_{kj}(W_n)<2z)+\Delta(N)^6+\varepsilon(n)^{7/2}\}.$$

Denote $W_n'=W_nI(\|W_n\|\leq\frac{\gamma}{2\alpha}|\log\sigma_n(N)|)$. We have

$$P(\Delta(N)^{-2}\Phi_{kj}(W_n)<2z) \leq P(\Phi_{kj}(W_n)-\tfrac{1}{2}E_{W_n}\Phi_{kj}(W_n)<0)$$

$$+P(\tfrac{1}{4}\sum_{i\in N_j\setminus\{k\}} D_{ni}(W_n)<2z\Delta(N)^2) \tag{46}$$

$$\leq P(\Phi_{kj}(W_n')-\tfrac{1}{2}E_{W_n}\Phi_{kj}(W_n')<0)+C(F,B)(z^M+\sigma_n(N)^8) .$$

To estimate the first summand in the right-hand side of (46) we shall use the Bernstein's inequality and (42):

$$P(\Phi_{kj}(W_n')-\tfrac{1}{2}E_{W_n}\Phi_{kj}(W_n')<0)$$

$$= EP_{W_n}(-\Phi_{kj}(W_n')+E_{W_n}\Phi_{kj}(W_n')>\tfrac{1}{2}E_{W_n}\Phi_{kj}(W_n'))$$

$$\leq E\exp\left\{-\frac{C(E_{W_n}\Phi_{kj}(W_n'))^2}{\sum_{i\in N_j}(E_{W_n}(F^{(1)}(W_n')[X_{ni}^{(1)}-EX_{ni}^{(1)}])^2)^2+\sigma^2(N)\|F^{(1)}(W_n')\|^{*2}E_{W_n}\Phi_{kj}(W_n')}\right\}$$

$$\leq E\exp\{-\sigma_n(N)^{\gamma-2}C(F)E_{W_n}\Phi_{kj}(W_n')\}$$

$$\leq C(F,B)\{(\Delta(N)^2\sigma_n(N)^{\gamma-2})^{-M}+\sigma_n(N)^8\}.$$

Lemma is proved.

Denote

$$\gamma_{nm}(x,N,\xi) = \sum_{j=1}^{m}\frac{\sum_{i\in N_j}E_\xi|F^{(1)}(x)[X_{ni}^{(1)}-EX_{ni}^{(1)}]\xi_i|^3}{\sum_{i\in N_j}E_\xi(F^{(1)}(x)[X_{ni}^{(1)}-EX_{ni}^{(1)}]\xi_i)^2},$$

where $N=\bigcup_{j\leq m}N_j$, $N_k\cap N_j=\emptyset$.

<u>Lemma 2.14</u>. For any fixed $\{\xi_i\}$ satisfying the condition $\sum_{i\in N_j}\xi_i>0$ the functional $\gamma_{nm}(x,N,\xi)$ has two continuous Frechét derivatives and the second one satisfies the Lipschitz condition. Moreover,

$$\gamma_{nm}(x,N,\xi)\leq C(F)e^{\alpha\|x\|}\sigma_n(N)$$

$$\|\gamma_{nm}^{(1)'}(x,N,\xi)\|^*\leq C(F)e^{2\alpha\|x\|}\sigma_n(N)\{1+\sum_{j=1}^{m}\Delta(N_j)^2(\Phi_{n+1,j}(x))^{-1}\},$$

$$\|\gamma_{nm}^{(2)}(x,N,\xi)\|^*\leq C(F)e^{4\alpha\|x\|}\sigma_n(N)\{1+\sum_{k=1}^{2}\sum_{j=1}^{m}\Delta(N_j)^{2k}(\Phi_{n+1,j}(x))^{-k}\},$$

$$\|\gamma_{nm}^{(2)}(x+h,N,\xi)-\gamma_{nm}^{(2)}(x,N,\xi)\|^*$$

$$\le C(F)\|h\|e^{6\alpha\|x\|}\sigma_n(N)\{1+\sum_{k=1}^{3}\sum_{j=1}^{m}\Delta(N_j)^{2k}(\Phi_{n+1,j}(x))^{-k}\} .$$

3. Proof of the theorem. Introduce the following notations

$$\sigma'_{nk} = (E\|X'_{nk}\|^2)^{1/2}, \qquad N^* = \{j(n,k);\ k=1,\dots,m(n)\},$$

where the natural numbers $\{j(n,k)\}$ and $m(n)$ are defined by the conditions $\sigma'_{nj(n,1)} \le \sigma'_{nj(n,2)} \le \dots \le \sigma'_{nj(n,n)}$ and

$$m(n) = \max\{k:\ \sum_{i\le k}\sigma'^2_{j(n,i)} \le 1/2\} .$$

Denote by N° a subset of N^* such that

$$|\Delta(N^\circ)^2 - \varepsilon(n)^{1.88}| \le \sigma_n^2(N^*) ,$$

where $\varepsilon(n)$ is defined in Remark 2 of the first part of the paper; $\Delta(N)$, $\sigma_n(N)$ are defined in Lemma 2.8. Denote also

$$S_n^\circ = \sum_{i\le n,\, i\notin N^\circ} X'_{ni} + \sum_{i\in N^\circ}(X_{ni}^{(2)} - EX_{ni}^{(2)})(1-\xi_i) + e_{n,n+1}(N^\circ),$$

$$\tilde{\gamma}_n(X) = \varepsilon(n) + \gamma_{n20}(X, N^\circ, \xi) .$$

Further we shall suppose that $N^\circ = \bigcup_{j\le 20} N_j^\circ$ where $\{N_j^\circ\}$ are chosen so that $N_j^\circ \cap N_k^\circ = \phi$, $j \ne k$,

$$|\Delta(N_j^\circ) - \varepsilon(n)^{1.88}(20)^{-1}| \le \sigma_n^2(N^*) .$$

Moreover, we choose subsets N, $\{N_j\}$ in Lemmas 2.8-2.13 to be such that $N \subseteq N^* \setminus N^\circ$ and

$$|\Delta(N_j)^2 - \Delta(N)^2\ (\mathrm{card}(\{N_j\}))^{-1}| \le \sigma_n^2(N^*) .$$

Denote by τ the r.v. with the following density of distribution

$$K(dx) = Cx^{-10}(\sin(x/40))^{10}dx .$$

Note that $E\tau^8 < \infty$ and

$$\hat{K}(t) = \int_R e^{itx}K(dx) = \begin{cases} \ge C_0 & \text{if } |t| \le 1/8, \\ 0 & \text{if } |t| \ge 1/4. \end{cases}$$

The method of proving the theorem was inspired by the remarkable paper of F. Götze [3]. We have (pair $(\tilde{\gamma}_n(S_n^\circ), F(S_{nn+1}))$ and τ are independent)

$$d_n(F) \le \sup_{x\in R} |P(F(S_{nn+1})+\tau\tilde{\gamma}_n(S_n^\circ)<x)-P(F(W_n)<x)|$$
$$+ \sup_{x\in R} |P(F(S_{nn+1})+\tau\tilde{\gamma}_n(S_n^\circ)<x)-P(F(S_{nn+1})<x)| + \sum_{i\le n} P(\|x_{ni}\|>C^*/\alpha). \tag{47}$$

Denote by I_1 and I_2 correspondingly the first and the second summands of the right-hand side of (47), and by $f_{n1}(t)$, $f_{n2}(t)$, $f_{n3}(t)$, respectively, the characteristic functions of the r.v.-s $F(S_{nn+1})$, $F(S_{nn+1})+\tau\tilde{\gamma}_n(S_n^\circ)$, $F(W_n)$.

By Barry's formula (see [4]) and Lemma 2.12 we obtain

$$I_1 \le \int_{|t|\le b\varepsilon(n)^{-1}} |t|^{-1}|f_{n2}(t)-f_{n3}(t)|dt + C(F,A,B,)\varepsilon(n). \tag{48}$$

Introduce the following notations

$$I_1^{(1)} = \int_{|t|\le b\varepsilon(n)^{\nu-1}} |t|^{-1}|f_{n1}(t)-f_{n3}(t)|dt, \qquad b=(16C_0(F))^{-1},$$

$$I_1^{(2)} = \int_{|t|\le b\varepsilon(n)^{\nu-1}} |t|^{-1}|f_{n2}(t)-f_{n1}(t)|dt,$$

$$I_1^{(3)} = \int_{b\varepsilon(n)^{\nu-1}\le|t|\le b\varepsilon(n)^{-1}} |t|^{-1}|f_{n2}(t)|dt,$$

$$I_1^{(4)} = \int_{b\varepsilon(n)^{\nu-1}\le|t|\le b\varepsilon(n)^{-1}} |t|^{-1}|f_{n3}(t)|dt,$$

where ν will be chosen later. It follows from (47) that

$$I_1 \le \sum_{k=1}^{4} I_1^{(k)} + C(F,A,B)\varepsilon(n). \tag{49}$$

The estimate $I_1^{(4)}\le C(F,A,B)\varepsilon(n)$ follows immediately from Lemma 2.12.

Lemma 3.1.

$$I_1^{(1)} \le C(F,A,B)\varepsilon(n) .$$

Proof. Using the Lindeberg operator method (see the proof of Lemma 2.9) we obtain

$$|f_{n1}(t)-f_{n3}(t)| \le \sum_{k\le n} (|E\varphi_t^{(1)}(S_{nk})[X_{nk}']|$$
$$+\frac{1}{2}|E\varphi_t^{(2)}(S_{nk})[X_{nk}'^2]-E\varphi_t^{(2)}(S_{nk})[W_{nk}^2]| \tag{50}$$

$$+\frac{1}{2}\int_0^1(1-\theta)^2\{|E\varphi_t^{(3)}(S_{nk}+\theta X'_{nk})[X'^3_{nk}]|+|E\varphi_t^{(3)}(S_{nk}+\theta W_{nk})[W^3_{nk}]|\}d\theta\},$$

where $\varphi_t(X)=\exp\{itF(X)\}$. By (40) and Lemmas 2.8, 2.9 where we put $\Delta(N)\sim|t|^{\nu_1-1}$ it is easy to obtain

$$\begin{aligned}|E\varphi_t^{(1)}(S_{nk})[X'_{nk}]| &= |E\varphi_t^{(1)}(S_{nk})[X_{nk}-X'_{nk}]|\\ &\le E|E_{X_{nk}}F^{(1)}(S_{nk})[X_{nk}-X'_{nk}]\exp\{itF(S_{nk})\}||t| \qquad (51)\\ &\le C(F,A,B)|t|g(t)E\|X_{nk}\|I(\|X_{nk}\|>C^*/\alpha)\end{aligned}$$

if only $|t|\le b\varepsilon(n)^{\nu-1}$, where

$$g(t)=|t|^{-3\nu_1 M/2p}+|t|^{\nu_1-(1-\nu_1)(3+\beta)}+(|t|^{21\nu_1}\varepsilon(n)^{7/2})^{1/p}, \qquad p\in(1,2).$$

As in the proof of Lemma 2.9 we have

$$|E\varphi_t^{(2)}(S_{nk})[X'^2_{nk}]-E\varphi_t^{(2)}(S_{nk})[W^2_{nk}]|$$

$$\le 2|E\varphi_t^{(2)}(S_{nk})[X_{nk}-X'_{nk},X_{nk}]\le 2|t|E|E_{X_{nk}}F^{(2)}(S_{nk})[\cdots]\exp\{itF(S_{nk})\}|$$

$$+2t^2E|E_{X_{nk}}F^{(1)}(S_{nk})[X_{nk}-X'_{nk}]F^{(1)}(S_{nk})[X_{nk}]\exp\{itF(S_{nk})\}$$

and by (40) it follows that

$$\begin{aligned}&E|E_{X_{nk}}F^{(2)}(S_{nk})[X_{nk}-X'_{nk},X_{nk}]\exp\{itF(S_{nk})\}|\\ &\le C(F,A,B)(g(t)+|t|^{-(1-\nu_1)(2+\beta)})E\|X_{nk}\|^2I(\|X_{nk}\|>C^*/\alpha),\end{aligned} \qquad (52)$$

$$\begin{aligned}&E|E_{X_{nk}}F^{(1)}(S_{nk})[X_{nk}-X'_{nk}]F^{(1)}(S_{nk})[X_{nk}]\exp\{itF(S_{nk})\}|\\ &\qquad\le C(F,A,B)g(t)E\|X_{nk}\|^2I(\|X_{nk}\|>C^*/\alpha).\end{aligned} \qquad (53)$$

The estimation of $\varphi_t^{(3)}(\cdot)[\cdot]$ in (50) is analogous to that considered above

$$|E\varphi_t^{(3)}(S_{nk}+\theta X'_{nk})[X'^3_{nk}]| \le C(F,A,B)|t|(g(t)+|t|^{-(1-\nu_1)(1+\beta)})E\|X'_{nk}\|^3,$$

$$|E\varphi_t^{(3)}(S_{nk}+\theta W_{nk})[W^3_{nk}]| \le C(F,A,B)|t|(g(t)+|t|^{-(1-\nu_1)(1+\beta)})E\|W_{nk}\|^3.$$

Note that by Lemma 2.4 and (2) the following estimation holds

$$E\|W_{nk}\|^3 \le C(A)\sigma_{nk}^3 \le C_1(A)(E\|X'_{nk}\|^3 + E\|X_{nk}\|^2 I(\|X_{nk}\|>C^*/\alpha)).$$

Finally if $|t|\le b\varepsilon(n)^{\nu-1}$ and

$$p = \frac{16}{15}, \qquad \nu_1 = \min\left\{\frac{2\beta}{3(4+\beta)}, \frac{1}{15}\right\},$$
$$M > \frac{2p}{3\nu_1} = \max\left\{\frac{16(4+\beta)}{15\beta}, \frac{32}{3}\right\} \tag{54}$$

then

$$|f_{n1}(t) - f_{n3}(t)| \le C(F,A,B)|t|^{-\mu}\varepsilon(n) . \tag{55}$$

Lemma is proved.

<u>Lemma 3.2</u>. For any $N\subseteq N^*$

$$\sigma_n(N) \le 4\varepsilon(n) .$$

<u>Proof</u>. We can assume that $\varepsilon(n)\le 1/4$. Then $\sum_{i\le n}\sigma'^2_{ni}\ge 3/4$. By the definition of N^* it follows that $\sum_{i\in N^*}\sigma'^2_{ni}\le 1/2$. Hence $\sum_{i\notin N^*}\sigma'^2_{ni}\ge 1/4$. Further we have the following simple inequalities

$$\varepsilon(n) \ge 3\sum_{i\le n} E\|X'_{ni}\|^3 (4\sum_{i\le n}\sigma'^2_{ni})^{-1} \ge 3\sum_{i\notin N^*} E\|X'_{ni}\|^3 (12\sum_{i\notin N^*}\sigma'^2_{ni})^{-1}$$
$$\ge \frac{1}{4}\min_{i\notin N^*}(\sigma'_{ni})^{-2} E\|X'_{ni}\|^3 \ge \frac{1}{4}\min_{i\notin N^*}\sigma'_{ni} \ge \sigma_n(N^*)/4.$$

Lemma is proved.

<u>Lemma 3.3</u>.

$$I_1^{(2)} \le C(F,A,B)\varepsilon(n) .$$

<u>Proof</u>. We have

$$I_1^{(2)} = \int_{|t|\le b\varepsilon(n)^{\nu-1}} |t|^{-1} |E\exp\{itF(S_{nn+1})\}\{\hat{K}(t\tilde{\gamma}_n(\mathring{S}_n))-1\}|dt. \tag{56}$$

Using Taylor formula for $\hat{K}$

$$\hat{K}(t\tilde{\gamma}_n(\mathring{S}_n)) = 1 + t\int_0^1 d\theta \int_R \exp\{i\theta ut\tilde{\gamma}_n(\mathring{S}_n)\}u\tilde{\gamma}_n(\mathring{S}_n)K(du)$$

we obtain from (56) and from the Fubini theorem

$$I_1^{(2)} \le \int_{|t|\le b\varepsilon(n)^{\nu-1}} dt \int_R |u|K(du)$$

$$\times\int_0^1 d\theta|E_{\delta_{nn+1}(N^\circ)}\exp\{it(F(S_n^\circ+\delta_{nn+1}(N^\circ))+\theta u\tilde{\gamma}_n(S_n^\circ))\}$$

$$\times\tilde{\gamma}_n(S_n^\circ)I(E_{S_n^\circ,\xi}(F^{(1)}(S_n^\circ)[\delta_{nn+1}(N^\circ)])^2>|t|^{-\nu_2}\Delta(N^\circ)^2) \qquad (57)$$

$$+C(F,A)E^{1/26}\tilde{\gamma}_n(S_n^\circ)^{26}$$

$$\times\int_{|t|\le b\varepsilon(n)^{\nu-1}} P^{25/26}(E_{S_n^\circ,\xi}(F^{(1)}(S_n^\circ)[\delta_{nn+1}(N^\circ)])^2 \le |t|^{-\nu_2}\Delta(N^\circ)^2).$$

From Lemma 2.13 there follows

$$\|\tilde{\gamma}_n^{(k)}(x)\| \le C(F)|t|^{k\nu_2}\sigma_n(N^\circ)\exp\{2k\|x\|\}$$

if only $E_\xi(F^{(1)}(x)[\delta_{nn+1}(N^\circ)])^2>|t|^{-\nu_2}\Delta(N^\circ)^2$ where $k=1,2,3$ (for $k=3$ the value $\tilde{\gamma}_n^{(k)}$ denotes the constant in the Lipschitz condition). Hence for any fixed t, θ, u, y the functions $\sigma_n(N^\circ)^{-1}\tilde{\gamma}_n(x)$ and $\psi_{\theta,u,y}(x)=F(x+y)+\theta u\tilde{\gamma}_n(x)$ belong to $\mathcal{E}(2^n,1,3\alpha)$ with the constants of the type $C(F)\theta|u||t|^{3\nu_2}\exp\{\alpha\|y\|\}$. Therefore, for the case $|t|\le b\varepsilon(n)^{\nu-1}$ from Lemma 2.9 we obtain ($\tilde{S}_n^\circ=S_n^\circ-\delta_{nn+1}(N)$, $N\subseteq N^*\setminus N^\circ$)

$$J(t)\equiv|E_{\delta_{nn+1}(N^\circ)}\tilde{\gamma}_n(S_n^\circ)\exp\{it\psi_{\theta,u,\delta_{nn+1}(N^\circ)}(S_n^\circ)\}$$

$$\times I(E_{S_n^\circ,\xi}(F^{(1)}(S_n^\circ)[\delta_{nn+1}(N^\circ)])^2\le|t|^{-\nu_2}\Delta(N^\circ)^2)|$$

$$\le C(F,A)\varepsilon(n)(1+|u|^5)|t|^{3\nu_2}\exp\{\alpha\|\delta_{nn+1}(N^\circ)\|\}\{\max_{j\le 12} E^{1/p}_{\delta_{nn+1}(\cdot)}\exp(-\frac{t^2}{3}$$

$$\times E_{\tilde{S}_n^\circ,\xi}(\psi^{(1)}_{\theta,u,\delta_{nn+1}(\cdot)}(\tilde{S}_n^\circ)[\delta_{nn+1}(N_j)])^2)I(\cdots)+|t|\Delta(N)^3+|t|^5\Delta(N)^{10}\}$$

$$(58)$$

$$\le C(F,A)\varepsilon(n)(1+|u|^5)|t|^{3\nu_2}\exp\{\alpha\|\delta_{nn+1}(\cdot)\|\}\{\max_{j\le 12} E^{1/p}_{\delta_{nn+1}(\cdot)}\exp(-\frac{t^2}{3}$$

$$\times E_{\tilde{S}_n^\circ,\xi}(F^{(1)}(\tilde{S}_n^\circ)[\delta_{nn+1}(N_j)+\delta_{nn+1}(N^\circ)])^2)+(|t|^{1+\nu_2}\Delta(N)\varepsilon(n))^{2/p}$$

$$+|t|\Delta(N)^3+|t|^5\Delta(N)^{10}\}.$$

Since

$$E_{\tilde{S}_n^\circ,\xi,\delta_{nn+1}(N^\circ)}(F^{(1)}(\tilde{S}_n^\circ)[\delta_{nn+1}(N_j)+\delta_{nn+1}(N^\circ)])^2$$

$$\geq E_{\tilde{S}_n^\circ,\xi}(F^{(1)}(\tilde{S}_n^\circ)[\delta_{nn+1}(N_j)])^2$$

then, putting $\Delta(N)\sim|t|^{\nu_3-1}$, we finally obtain from (58)

$$J(t) \leq C(F,A,B)\varepsilon(n)(1+|u|^5)|t|^{3\nu_2}\exp\{\alpha\|\delta_{nn+1}(N^\circ)\|\}$$

$$\times\{|t|^{-3\nu_3 M/2p}+(|t|^{\nu_2+\nu_3}\varepsilon(n))^{2/p}+|t|^{3\nu_3-2}\},$$

where M satisfies condition (54) and, moreover,

$$5\nu_2 + 2\nu_3 \leq (1+\nu)(1-\nu)^{-1},$$

$$\nu_2 + \nu_3 < 1/3, \tag{59}$$

$$3\nu_3 M/2p - 3\nu_2 > 1.$$

Hence it follows that the first summand in the right-hand side of (58) is estimated by the value $C(A,B,F)\varepsilon(n)$.. For the estimation of the second summand we use Lemmas 2.12 and 3.2

$$P(E_{S_n^\circ,\xi}(F^{(1)}(S_n^\circ)[\delta_{nn+1}(N^\circ)])^2\leq|t|^{-\nu_2}\Delta(N^\circ)^2) \tag{60}$$

$$\leq C(F,A,B,\gamma)\{|t|^{-3\nu_2 M/4}+\varepsilon(n)^{3(2\nu_3-\gamma)M/4}+\varepsilon(n)^{7/2}+|t|^{-9(1-\nu_3)/2}\}|t|^{12\nu_2}.$$

Now put $\nu_2=1/6$, $\nu_3=5\nu=1/10$, $\gamma=1/40$, $p=16/15$, $M\geq 25$. Then (59) will hold and, moreover,

$$3\nu_2 M/4 - 12\nu_2 \geq 25/24,$$

$$3(2\nu_3-\gamma)M/4 - 12\nu_2 \geq 5/4. \tag{61}$$

Hence the statement of the lemma follows from (56)-(61). Note that to estimate the second summand in the right-hand side of (58) we assume in (10) $C^*\leq 1/52$ (for the existence of the corresponding exponential moment of r.v. $\|S_{nk}\|$).

Lemma 3.4.

$$I_1^{(3)} \leq C(F,A,B)\varepsilon(n).$$

Proof. From the definition of function $\hat{K}(t)$ it follows

$$I_1^{(3)} = \int_{b\varepsilon(n)^{\nu-1}\le|t|\le b\varepsilon(n)^{-1}} |t|^{-1} |E\exp\{itF(S_{nn+1})\}\hat{K}(t\tilde{\gamma}_n(S_n^\circ))|dt$$

$$\le E\int_{\substack{b\varepsilon(n)^{-0.98}\le|t|\le b\varepsilon(n)^{-1}\\ |t|\tilde{\gamma}_n(S_n^\circ)\le 1/4}} |t|^{-1}\int_R K(du)|E_{S_n^\circ}\exp\{itF(S_{nn+1})\}|dt.$$

Due to Lemmas 2.6-2.10 we have for $|t|\tilde{\gamma}_n(S_n^\circ)\le 1/4$

$$|E_{S_n^\circ}\exp\{itF(S_{nn+1})\}| \le C(F,A,B)\exp\{5\alpha\|S_n^\circ\|\}\{|t|\Delta(N^\circ)^4$$

$$+(\Delta(N^\circ)\varepsilon(n)^{-0.98})^{-3M/2p}+|t|^5\Delta(N^\circ)^{10}+(t^2\Delta(N^\circ)^3)^6+(\varepsilon(n)t^6\Delta(N^\circ)^6)^{7/2p}\}.$$

Setting $\Delta(N^\circ)=\varepsilon(n)^{0.94}$, $M\ge 25$, $p=16/15$ we easily obtain the statement of the lemma.

Further let us estimate I_2. Denote $r(u)=\max\{1,|u|\}$. Repeating the proof of the corresponding statement in [3] we obtain

$$I_2 \le \sup_{z\in R} P(|F(S_{nn+1})-z| \le r(\tau)\tilde{\gamma}_n(S_n^\circ))$$

$$= \sup_z \int K(du)P(|F(S_{nn+1})-z| \le r(u)\tilde{\gamma}_n(S_n^\circ))$$

$$\le C \sup_z \int_R K(du)K\left(\frac{F(S_{nn+1})-z}{r(u)\tilde{\gamma}_n(S_n^\circ)}\right)$$

$$= C_1 \sup_z \int_R K(du)E\int_R \exp\left\{-\frac{itF(S_{nn+1})-z}{r(u)\tilde{\gamma}_n(S_n^\circ)}\right\}\hat{K}(t)dt \tag{62}$$

$$= C_1\sup_z\int_R K(du)E\int_R \exp\{-iy(F(S_{nn+1})-z)\}\hat{K}(r(u)y\tilde{\gamma}_n(S_n^\circ))r(u)\tilde{\gamma}_n(S_n^\circ)dy$$

$$\le C_1 \sup_z |E\iiint_{r(u)|y|\tilde{\gamma}_n(S_n^\circ)\le 1/4} dy\ r(u)K(du)K(dv)\gamma_n(S_n^\circ)$$

$$\times\exp\{-iyF(S_{nn+1})+r(u)vy\tilde{\gamma}_n(S_n^\circ)\}|$$

$$\le C_1\left\{\iiint_{|y|\le b\varepsilon(n)^{\nu-1}} dyK(du)K(dv)|E(\cdots)|+\left|E\iiint_{\substack{|y|\tilde{\gamma}_n(S_n^\circ)\le 1/4\\ |y|>b\varepsilon(n)^{\nu-1}}}\cdots\right|\right\}.$$

The estimate of the type $C(F,A,B)\varepsilon(n)$ for each of the two integrals in the right-hand side of (62) is contained in fact in Lemmas 3.3 and 3.4.

Theorem is proved.

References

[1] Araujo A., Gine E. The central limit theorem for real and Banach valued random variables. Wiley, New York, 1980.

[2] Borisov I. S. On the rate of convergence of distributions of functionals of integral type. Theory Probab. Appl., 1976, v.21, No2, p.283-299 (English translation).

[3] Götze F. On the rate of convergence in central limit theorem in Banach spaces. Preprints in Statistics, No68, University of Cologne: 1981.

[4] Petrov V. V. Sums of independent random variables. Springer, New York, 1975.

[5] Pinelis I. F. On distribution of sums of independent random variables with values in Banach space. Theory Probab. Appl., 1978, v.23, No3, p.630-637 (In Russian).

[6] Pinelis I. F., Sakhanenko A. I. Remarks on inequalities of large deviations. Theory of Probab. and its Appl., 1985, v.30, No1, p.127-131 (In Russian).

[7] Rhee W. S., Talagrand M. Bad rates of convergence for central limit theorem in Hilbert space. Ann. Probab. 1984, v.12, No3, p.843-850.

[8] Rosinski J., Suchanecki Z. On the space of vector-valued functions integrable with respect to the white noise. Colloquium Mathematicum, 1980, v.43, No1, p.183-201.

[9] Tortra A. Lois indéfinment divisibles ($\mu \in I$) dans un group topologique abélian metrisable X. Cas des espaces vectoriels. C. R. Acad. Sci. Paris, 1965, v.261, No23, p.4973-4975.

[10] Ulyanov V. V. Asymptotic expansions for the distributions of sums of independent random variables in Hilbert space. Theory Probab. Appl., 1986, v.31, No1, p.31-46 (In Russian).

[11] Vinogradova T. R. On the accuracy of normal approximation on sets defined by a smooth function. I, II. Theory Probab. Appl., 1985, v.30, No2, p.219-229, No3, p.554-557 (In Russian).

[12] Yurinskii V. V. On the accuracy of normal approximation in a Hilbert space. Theory Probab. Appl., 1982, v.27, No2, p.280-289 (English translation).

Institute of Mathematics,
Novosibirsk 630090 USSR

ON THE WEAK CONVERGENCE TO BROWNIAN LOCAL TIME

A. N. Borodin

In this paper we consider the sequences of stochastic processes which converge weakly to Brownian local time. These processes are generated by a recurrent random walk with finite variance.

The local time of Brownian motion $w(s)$ $(Ew^2(s)=Ds,\ 0<D<\infty)$ is defined to be the density of the measure $\mu_t(E)=\int_0^t 1\!\!1_E(w(s))ds$, where $1\!\!1_E$ is the indicator function of a Borel subset E of R^1. The continuous with respect to (t,x) modification of this density will be denoted by $\hat{\ell}(t,x)$, $(t,x)\in[0,\infty)\times R^1$.

Let ν_k, $k=0,1,\dots$ be a recurrent random walk, i.e. $\nu_0=0$, $\nu_k=\sum_{\ell=1}^{k}\xi_\ell$ where $\{\xi_\ell\}_{\ell=1}^{\infty}$ be independent identically distributed random variables, $E\xi_1=0$, $E\xi_1^2=D<\infty$. We shall use the following definition of the weak convergence of processes. The processes $\xi_n(s)$, $s\in S$ converge weakly as $n\to\infty$ to the process $\xi_\infty(s)$, if it is possible on some probability space to construct such processes $\xi'_n(s)$, $n=1,2,\dots,\infty$, that for each n the finite dimensional distributions of the process $\xi'_n(s)$ coincide with those of the process $\xi_n(s)$ and

$$\sup_{s\in S}|\xi'_n(s)-\xi'_\infty(s)|\longrightarrow 0$$

in probability. For the convergence of the processes considered in the paper this definition is equivalent to the classical one. For each particular case it is possible to determine such complete separable functional space with uniform metric that the weak convergence of processes imply the convergence of the measures generated by these processes in the functional space and conversely. The proof of the converse statement is based on the lemma from [1], p.14. It is well known that the process $w_n(t)=\frac{1}{\sqrt{n}}\nu_{[nt]}$, $t\in[0,T]$, converges weakly to Brownian motion $w(t)$. The classical definition of the weak convergence of the processes $w_n(t)$ is equivalent to that given above for the functional space $RD[0,T]$, where $RD[0,T]$ is the space of right continuous and having left-hand side limits functions defined on $[0,T]$ which may have a discontinuities only in the rational points of the interval $[0,T]$. For a wide class of functionals the weak convergence of $w_n(t)$ to $w(t)$ imply the convergence of a distribution of a functional of random walk ν_k to a distribution of the corresponding functional of Brownian motion. However there exist such functionals of random walk for which the convergence of distributions can not be obtained from the weak convergence of $w_n(t)$. The convergence of a distributions of such functionals is investigated in the monograph of A. V. Skorokhod and N. P. Slobodenyuk [2]. We do not dwell on the history of this question since one can find it in [2].

A. V. Skorokhod and N. P. Slobodenyuk considered the asymptotic behaviour as $n\to\infty$ of the functionals

$$\eta_n = \sum_{k=1}^{n-\ell} f_n(\nu_k,\dots,\nu_{k+\ell}) ,$$

where $\{f_n\}$ is a some sequence of functions and ℓ is a fixed number. The asymptotic behaviour of η_n as it was shown in [2] coincide with that of $\bar\eta_n = \sum_{k=1}^{n} g_n(\nu_k)$, where $g_n(y)=Ef_n(y,y+\nu_1,\dots,y+\nu_\ell)$. The first our aim is to demonstrate how the asymptotic behaviour of the functional $\bar\eta_n$ can be derived with the help of the weak convergence of some processes, generated by a random walk ν_k, to Brownian local time. This method essentially differs from that used in [2]. We consider the limit behaviour of the process

$$\bar\eta_n(t) = \sum_{k=1}^{[nt]} g_n(\nu_k) , \qquad t \in [0,T] ,$$

where ν_k is the random walk with integer values. Let $\varphi(n,r)=\sum_{k=1}^{n} 1\!\!1_{\{r\}}(\nu_k)$ be the number of times the random walk ν_k hits the point r up to time n. Denote $\hat t_n(t,x)=n^{-1/2}\varphi([nt],[x\sqrt{n}])$. In [3] it was proved that the process $\hat t_n(t,x)$, $(t,x)\in[0,T]\times R^1$, converges weakly to the process $\hat t(t,x)$. Let

$$G_n(x) = \sqrt{n}\ \mathrm{sign}\ x \sum_{\frac{\sqrt{n}}{2}(x-|x|)<\ell\le\frac{\sqrt{n}}{2}(x+|x|)} g_n(\ell) .$$

Then

$$\bar\eta_n(t) = \sum_{r=-\infty}^{\infty} \varphi([nt],r)g_n(r) = \int_{-\infty}^{\infty} \hat t_n(t,x)dG_n(x) .$$

Suppose that $G_n(x)\longrightarrow G(x)$ and for each $A>0$

$$\sup_n \sqrt{n} \sum_{|\ell|\le A\sqrt{n}} |g_n(\ell)| < \infty .$$

Then using the weak convergence of $\hat{\xi}_n(t,x)$ to $\hat{\xi}(t,x)$ it is not hard to prove the weak convergence of the process $\bar{\eta}_n(t)$, $t\in[0,T]$, to the process

$$\eta(t) = \int_{-\infty}^{\infty} \hat{\xi}(t,x)dG(x) .$$

The solution of the problem of asymptotic behaviour of the processes $\bar{\eta}_n(t)$ for the random walks with continuous values can be based on the weak convergence for each $\delta>0$ of the process

$$\hat{\xi}_{n,\delta}(t,x) = \frac{1}{\sqrt{n}\delta} \sum_{k=1}^{[nt]} \mathbb{1}_{(-\delta,0]}(\nu_k - x\sqrt{n}) , \qquad (t,x)\in[0,T]\times R^1 .$$

This process also converges weakly to the process $\hat{\xi}(t,x)$. The limit behaviour of biadditive functions

$$\mu_n(t,s) = \sum_{k=1}^{[nt]} \sum_{\ell=1}^{[ns]} g_n(\nu_k, \nu_\ell) , \qquad (t,s) \in [0,T]^2 ,$$

and in general polyadditive functionals can be investigated in just the same way. Biadditive functionals were considered by G. N. Sytaya [4] with the help of another method.

Let us consider one more problem, which is solved by means of application of the weak convergence of the process $\hat{\xi}_n(t,x)$. The problem is to describe the asymptotic behaviour as $n\to\infty$ of the sums $S_n = \sum_{k=1}^{n} X_{\nu_k}$, where $(X_\ell)_{\ell=-\infty}^{\infty}$ is the sequence of independent random variables, which is independent of the random walk ν_k with integer values. For different random walks this problem was investigated by the author [5],[6],[7] and independently by H. Kesten and F. Spitzer [8]. We consider the case when $E\nu_1=0$, $E\nu_1^2=D<\infty$. Assume that X_ℓ are identically distributed random variables, $EX_\ell=0$, $EX_\ell^2=1$. Let $Z_n(x) = n^{-1/4}\sum_{\ell=0}^{[nt]} X_\ell$ if $x\ge 0$ and $Z_n(x) = -n^{-1/4} \sum_{\ell=[x\sqrt{n}]+1}^{-1} X_\ell$ if $x<0$. The process

$$S_n(t) = n^{-3/4} S_{[nt]}$$

can be represented in the form

$$S_n(t) = \sum_{r=-\infty}^{\infty} \frac{1}{\sqrt{n}} \varphi([nt],r) \frac{X_r}{\sqrt[4]{n}} = \int_{-\infty}^{\infty} \hat{\xi}_n(t,x)dZ_n(x)$$

Let $[x_1,x_2]$ be arbitrary interval from R^1. By the invariance principle the process $Z_n(x)$, $x\in[x_1,x_2]$, converges weakly as $n\to\infty$ to the process $Z(x)$, where $Z(x)$ and $Z(-x)$, $x>0$, are independent Brownian motion processes. Using the weak convergences of $\hat{t}_n(t,x)$ to $\hat{t}(t,x)$, $Z_n(x)$ to $Z(x)$ and the independence of the processes $\hat{t}(t,x)$ and $Z(x)$ it is not hard to prove the weak convergence of the process $S_n(t)$, $t\in[0,T]$, to the process

$$S(t) = \int_{-\infty}^{\infty} \hat{t}(t,x)dZ(x) .$$

So the problem under consideration essentially be reduced to the problem of convergence of $\hat{t}_n(t,x)$. H. Kesten and F. Spitzer [8] used another approach which is not so universal as this one.

There are some other applications of the weak convergence of the process $\hat{t}_n(t,x)$. It thus becomes of interest to investigate the problem of convergence of some processes generated by a recurrent random walk to Brownian local time. We shall distinguish the cases of discrete (D) and continuous (C) values of the random walk ν_k. Let $\varphi(t)=E\exp(it\xi_1)$. One of the following conditions is assumed:

(D) $|\varphi(t)|=1$ if and only if t is a multiple of 2π,

(C) $\varphi(t)$, $t\in R^1$, is square integrable function.

Let $f(y,z)$ be arbitrary function. In case (D) suppose that $f(y,z)$ is defined in $\mathbb{Z}\times\mathbb{Z}$ and the function $h(y)=Ef(y,y+\xi_1)$ satisfies $\sum_{\ell=-\infty}^{\infty} |h(\ell)|<\infty$. Denote $h=\sum_{\ell=-\infty}^{\infty} h(\ell)$. In case (C) suppose that $f(y,z)$ is a Borel function, $(y,z)\in R^2$ and $h(y)$ satisfies $\int_{-\infty}^{\infty}|h(y)|dy<\infty$. Denote $h=\int_{-\infty}^{\infty} h(y)dy$. Let ν_k^n, $k=0,1,2,\ldots$, be some sequence of random walks such that for each fixed n random walk ν_k^n has the same distributions as the random walk ν_k.

Consider the process

$$q_n(t,x) = \frac{1}{\sqrt{n}} \sum_{k=1}^{[nt]} f(\nu_{k-1}^n - x(n), \nu_k^n - x(n)) , \qquad (t,x)\in[0,T]\times R^1 ,$$

where $x(n)=[x\sqrt{n}]$ in case (D) and $x(n)=x\sqrt{n}$ in case (C). The results considered below can also be carried over to the case when in the definition of the process $q_n(t,x)$ the variables $f(\nu_{k-1}^n-x(n), \nu_k^n-x(n),\ldots,\nu_{k+\ell}^n-x(n))$ are taken instead of $f(\nu_{k-1}^n-x(n),\nu_k^n-x(n))$.

By choosing particular functions f we obtain various characteristics of the behaviour of a random walk ν_k^n near the level $x(n)$. In case (D), for example, if $f(y,z)=\mathbb{1}_{\{0\}}(z)$ then the variable $\sqrt{n}q_n(t,x)$ is the number of times the random walk ν_k^n hits the point

$[x\sqrt{n}]$ after $[nt]$ steps. In both cases if $f(y,z)=1\!\!1_{[-\alpha,\beta]}(y)$, $\alpha>0$, $\beta>0$, then the variable $\sqrt{n}q_n(t,x)$ is the number of visits of ν_k^n at the interval $[x(n)-\alpha, x(n)+\beta]$ before the time $[nt]$: if $f(y,z)=1\!\!1_{(-\infty,0)}(yz)$ then $\sqrt{n}q_n(t,x)$ is the number of times ν_k^n crosses the level $x(n)$ in $[nt]$ steps; and if $f(y,z)=|z-y|1\!\!1_{(-\infty,0)}(yz)$ then $\sqrt{n}q_n(t,x)$ is the total length of those steps of ν_k^n which cross the level $x(n)$ before the time $[nt]$.

The main result of this paper is the weak convergence of the process $q_n(t,x)$ to the process $h\hat{\ell}(t,x)$, $(t,x)\in[0,T]\times R^1$. For the random walk ν_k we assume only that it has a finite variance. Under the more restrictive moment conditions in [9] there is given an uniform estimate of the rate of convergence of $q_n(t,x)-h\hat{\ell}(t,x)$ to zero and the convergence of the finite dimensional distributions of the process

$$Q_n(t,x) = \sqrt[4]{n}(q_n(t,x)-h\hat{\ell}(t,x(n)/\sqrt{n})) , \quad (t,x)\in[0,\infty)\times R^1 ,$$

is proved.

For some particular functions f the rates of convergence of the processes $q_n(1,x)$ to the process $h\hat{\ell}(1,x)$, $x\in R^1$, were obtained by P. Révész [10], [11], E. Csáki and P. Révész [12]. For different particular functions f and for more general random walks the convergence of the processes $q_n(1,x)$, $x\in R^1$, were investigated by Yu. A. Davydov [13], E. Perkins [14], M. V. Petrova [15], the author [16] and M. Czörgo and P. Révész [17]. The convergence of the distributions of the functionals η_n for a wide class of random walks was proved by I. A. Ibragimov [18].

Denote $w_n(t)=\frac{1}{\sqrt{n}}\nu^n_{[nt]}$, $t\geq0$.

<u>Theorem 1</u>. Let condition (D) holds. Suppose that

$$\sum_{v=-\infty}^{\infty} E|f(v,v+\xi_1)| < \infty , \tag{1}$$

$$\sum_{v=-\infty}^{\infty} Ef^2(v,v+\xi_1) < \infty . \tag{2}$$

Then by Brownian motion $w(s)$ one can construct such random walks ν_k^n for each n having the same distributions as the random walk ν_k that the processes $w_n(t)$, $q_n(t,x)$ satisfy for any $T>0$ and $\varepsilon>0$ the relations

$$\lim_{n\to\infty} P(\sup_{t\in[0,T]} |w_n(t)-w(t)|>\varepsilon) = 0 , \tag{3}$$

$$\lim_{n\to\infty} P\Big(\sup_{(t,x)\in[0,T]\times R^1} |q_n(t,x)-h\hat{\ell}(t,x)|>\varepsilon\Big) = 0 . \tag{4}$$

Relation (3) is a well-known weak invariance principle for random walks. The main result is relation (4). It is important that the both relations hold simultaneously since it allows to establish the convergence of a distributions for a wide class of functionals of random walk.

Consider the example. Let $\psi_n(r)$ be the number of times the random walk ν_k hits the point r up to the first exit time from the interval $(-a\sqrt{n},b\sqrt{n})$, $a>0$, $b>0$, and let m be the first exit time of Brownian motion $w(t)$ from the interval $(-a,b)$.

Proposition 1. The process $n^{-1/2}\psi_n([x\sqrt{n}])$, $x\in[-a,b]$, converges weakly as $n\to\infty$ to the process $\hat{\ell}(m,x)$.

Indeed from (3) it is easy to obtain that

$$m_n \longrightarrow m$$

in probability, where m_n is the first exit time of the process $w_n(t)$ from the interval $(-a,b)$. Since

$$n^{-1/2}\psi_n([x\sqrt{n}]) = \hat{\ell}_n(m_n,x)$$

then using (4) for $f(y,z)=\mathbb{1}_{\{0\}}(z)$ it is not hard to prove the desired result.

Theorem 1 can be formulated otherwise. It is connected with the way of construction of the random walk ν_k^n (the Skorokhod embedding scheme) and the scaled property of Brownian motion process: for any fixed $c>0$ the process $\frac{1}{\sqrt{c}}w(ct)$ is also Brownian motion and the process $\frac{1}{\sqrt{c}}\hat{\ell}(tc,x\sqrt{c})$ is its local time. Consider the first representative ν_k^1, $k=0,1,2,\ldots$, of the sequence of random walks ν_k^n, $k=0,1,2,\ldots$. Let

$$v_n(t) = \nu^1_{[nt]}, \qquad t \geq 0 ,$$

$$r_n(t,x) = \sum_{k=1}^{[nt]} f(\nu^1_{k-1}-x(n),\ \nu^1_k-x(n)) , \qquad (t,x)\in[0,\infty)\times R^1.$$

In view of the scaled property of Brownian motion process and the way of construction of the random walks ν_k^n the finite dimensional distributions of the process $(v_n(s)-w(ns),\ r_n(t,x)-h\hat{\ell}(nt,x\sqrt{n}))$ coincide with those of the process $\sqrt{n}(w_n(s)-w(s),\ q_n(t,x)-h\hat{\ell}(t,x))$. This allows to formulate the following variant of Theorem 1.

Theorem 1'. Let conditions (D), (1) and (2) hold. Then by Brownian motion $w(s)$ one can construct such random walk ν_k^1 having the same distributions as the random walk ν_k that the processes $v_n(t)$, $r_n(t,x)$ satisfy for any $T>0$ and $\varepsilon>0$ the relations

$$\lim_{n\to\infty} P\Big(\sup_{t\in[0,T]} |v_n(t)-w(nt)|>\sqrt{n}\varepsilon\Big) = 0 , \tag{5}$$

$$\lim_{n\to\infty} P\Big(\sup_{(t,x)\in[0,T]\times R^1} |r_n(t,x)-h\hat{\ell}(nt,x\sqrt{n})|>\sqrt{n}\varepsilon\Big) = 0 . \tag{6}$$

To obtain the analogy of Theorem 1 for continuous case one should introduce some additional assumptions on function $f(y,z)$. We suppose that for all $(y,z)\in R^2$, all $\Delta\le\Delta_0$ and some points α_i, β_i, $i=1,2,\ldots,r$, $r<\infty$,

$$\begin{aligned}&\sup_{0\le v\le\Delta} |f(y+v,z+v)-f(y,z)|\\ &\le C(y,z)\Delta + \sum_{i=1}^{r} D_i(y,z)(\mathbb{1}_{[\alpha_i-\Delta,\alpha_i]}(y)+\mathbb{1}_{[\beta_i-\Delta,\beta_i]}(z)) ,\end{aligned} \tag{7}$$

where $C(y,z)$ and $D_i(y,z)$ are some non-negative Borel functions. Under this condition function f may have a discontinuities which are parallel to the coordinate axis.

Theorem 2. Let condition (C) hold. Suppose that

$$\int_{-\infty}^{\infty} E|f(v,v+\xi_1)|dv < \infty , \qquad \int_{-\infty}^{\infty} Ef^2(v,v+\xi_1) < \infty , \tag{8}$$

$$P\Big(\sup_{v\in R^1}|f(v,v+\xi_1)|>L\Big) \longrightarrow 0 , \quad L\longrightarrow\infty , \tag{9}$$

$$\int_{-\infty}^{\infty} \sqrt{|v|}\ EC(v,v+\xi_1)dv < \infty , \qquad \int_{-\infty}^{\infty} EC^2(v,v+\xi_1)dv < \infty , \tag{10}$$

and that for $i=1,\ldots,r$ and some constant $C>0$

$$EC^2(v,v+\xi_1)>C, \quad ED_i^2(v-\xi_1,v)<C, \quad ED_i^2(v,v+\xi_1)<C . \tag{11}$$

Then by Brownian motion $w(s)$ one can construct such random walks

ν_k^n for each n having the same distributions as the random walk ν_k that the processes $w_n(t)$, $q_n(t,x)$ satisfy for any $T>0$ and $\varepsilon>0$ relations (3) and (4).

Remark 1. Note that (9) is the necessary condition for (4).

As in the case of integer valued random walk it is possible to give another variant of Theorem 2.

Theorem 2'. Let conditions (C) and (8)-(11) hold. Then by Brownian motion $w(s)$ one can construct such random walk ν_k^1 having the same distributions as the random walk ν_k that the processes $v_n(t)$, $r_n(t,x)$ satisfy for any $T>0$ and $\varepsilon>0$ relations (5) and (6).

The proof of Theorems 1 and 2 is based on the methods developed in [9]. It is carried out in two steps. At first we prove the convergence of $q_n(t,x)$ to $h\hat{\ell}(t,x)$ in probability and then establish the weak compactness of the processes $q_n(t,x)$.

References

[1] Skorokhod, A. V. Studies in the theory of random processes. Kiev: Kiev University Press 1961.

[2] Skorokhod, A. V., Slobodenyuk, N. P. Limit theorems for random walks. Kiev: Naukova Dumka 1970.

[3] Borodin, A. N. An asymptotic behaviour of local times of a recurrent random walk with finite variance. Theory Probab. Appl. vol.26, 769-783 (1981).

[4] Sytaya, G. N. Limit theorems for some functionals of random walks. Theory Probab. Appl. vol.12, 483-492 (1967).

[5] Borodin, A. N. A limit theorem for sums of independent random variables defined on a recurrent random walk. Dokl. USSR Academy of Sciences vol.246, N4, 786-788 (1979).

[6] Borodin, A. N. Limit theorems for sums of independent random variables defined on a nonrecurrent random walk. Zapiski Nauchnych Seminarov Leningradskogo Otdeleniya Matematicheskogo Instituta im. V. A. Steklova AN SSSR vol.85, 17-29 (1979).

[7] Borodin, A. N. Limit theorems for sums of independent random variables defined on a recurrent random walk. Theory Probab. Appl. vol.28, 98-114 (1983).

[8] Kesten, H., Spitzer, F. A limit theorem related to a new class of self similar processes. Z. Wahrscheinlichkeitstheor. verw. Geb. B.50, 5-25 (1979).
[9] Borodin, A. N. On the character of convergence to Brownian local time. II. Probab. Th. Rel. Fields vol.72, 251-277 (1986).
[10] Révész, P. Local time and invariance. Lecture Notes in Math. vol.861, 128-145 (1981).
[11] Révész, P. A strong invariance principle of the local time of R.V.'s with continuous distribution. Carleton Math. Lect. Note vol.37, (1982).
[12] Csaki, E., Révész, P. Strong invariance for local times. Z. Wahrscheinlichkeitstheor. verw. Geb. B.62, 263-278 (1983).
[13] Davydov, Yu. A. Sur une classe des fonctionnelles des processus stables et des marches aléatoires. Ann. Inst. Henri Poincaré, sect.B vol.X, 1-29 (1974).
[14] Perkins, E. Weak invariance principles for local time. Z. Wahrscheinlichkeitstheor. verw. Geb. B.60, 437-451 (1982).
[15] Petrova, M. V. On weak convergence of functionals of random walks. Dokl. USSR Academy of Sciences vol.278, 806-809 (1984).
[16] Borodin, A. N. An asymptotic behaviour of local times of a recurrent random walk with infinite variance. Theory Probab. Appl. vol.29, 312-326 (1984).
[17] Czörgo, M., Révész, P. On strong invariance for local time of partial sums. Stochastic Proc. Appl. vol.20, 59-84 (1985).
[18] Ibragimov I. A. Théorèmes limites pour les marches aléatoires. Lecture Notes in Math. vol.117, 199-297 (1985).

Leningrad Branch Steklov Institute of Mathematics Academy of Sciences of the USSR, Fontanka 27, 191011, Leningrad, USSR

ON OPTIMAL STOPPING WITH INCOMPLETE DATA

V. M. Dochviri

Let $(\Omega,\mathcal{F},P)$ be a complete probability space and assume that on this space independent Wiener processes $W=(W_t,\mathcal{F}_t^W)$ and $\widetilde{W}=(\widetilde{W}_t,\mathcal{F}_t^{\widetilde{W}})$, $0\le t\le T$ are given. For every $\varepsilon\ge 0$ consider a partially observable random process $(\theta,\xi^\varepsilon)=(\theta_t,\xi_t^\varepsilon)$, $0\le t\le T$, satisfying the following system of stochastic differential equations

$$\begin{aligned} d\theta_t &= dW_t\,, \qquad && \theta_0=0, \\ d\xi_t^\varepsilon &= W_t dt + \varepsilon d\widetilde{W}_t\,, \qquad && \xi_0^\varepsilon = 0\,. \end{aligned} \tag{1}$$

Let a reward function of the following form

$$g(t,x) = \frac{u+x}{b+t}\,, \tag{2}$$

be also given, where $-\infty<u<\infty$, $b>0$ are constants. Introduce the costs S^0 and S^ε in the so called "0-problem" and "ε-problem", respectively [3]

$$S^0 = \sup_{\tau\in\mathfrak{M}^W} Eg(\tau,W_\tau)\,, \qquad S^\varepsilon = \sup_{\tau\in\mathfrak{M}^{\xi^\varepsilon}} Eg(\tau,W_\tau)\,. \tag{3}$$

For the random process $X=(X_t,\mathcal{F}_t^X)$, $0\le t\le T$ here and in the forthcoming $\mathfrak{M}^X$ will denote a class of all stopping times (Markov moments) with respect to the family of σ-algebras of $F^X=(\mathcal{F}_t^X)$ where $\mathcal{F}_t^X=\sigma(\omega\colon X_s,\ 0\le s\le t)$. The cost S^0 corresponds to the case of complete observation of the process W and when the cost S^ε is considered it is assumed that the process W is a partially observable one or, to be more precise, we only observe the process ξ^ε containing partial (incomplete) information about the process W. When $\varepsilon=0$ we have the case of complete observation of the process W and, hence, the cost S^ε coincides with the cost S^0 here. One would naturally think that if the "obstacle" is small, then S^ε will approach S^0, i.e. when $\varepsilon\to 0$, then, seemingly, $S^\varepsilon\to S^0$. However this fact does not always hold. In particular, let the process (θ,ξ^ε) be defined by system (1) and $g(t,x)=1$ when $x=x_0\ne 0$ and $g(t,x)=0$ when $x\ne x_0$. Then we can prove that $S^0=1$ and $S^\varepsilon\to 0$ as $\varepsilon\to 0$.

In paper [4] explicit forms of the optimal stopping time and the cost S^0 are obtained for system (1) and the reward function (2). In [3] for the same case explicit expressions of these values are obtained in the "ε-problem" and the convergence of the order $\varepsilon^{1/2}$ of the cost S^ε to the cost S^0 is proved for $\varepsilon \to 0$. In the present work the method proposed in [2] is used, the convergence of costs is proved to be of the order ε and some generalizations of this result are given.

2. Let $m_t^\varepsilon = E(W_t | \mathcal{F}_t^{\xi^\varepsilon})$, $\gamma_t^\varepsilon = E(W_t - m_t^\varepsilon)^2$, $0 \le t \le T$. It can be easily shown that

$$m_t^\varepsilon = \int_0^t \frac{\gamma_s^\varepsilon}{\varepsilon} \, d\overline{W}_s^\varepsilon , \qquad \gamma_t^\varepsilon = \varepsilon \operatorname{th} \frac{t}{\varepsilon} , \tag{4}$$

where $\overline{W}^\varepsilon = (\overline{W}_t^\varepsilon, \mathcal{F}_t^{\overline{W}^\varepsilon})$, $0 \le t \le T$ is the so called innovation Wiener process for which σ-algebras of $\mathcal{F}_t^{\overline{W}^\varepsilon} = \sigma\{\omega : \overline{W}_s^\varepsilon, \ 0 \le s \le t\}$ coincide with $\mathcal{F}_t^{\xi^\varepsilon}$, $0 \le t \le T$ [1]. Now let $\eta = \eta(\omega)$ be a standard normal random variable independent of W and $\widetilde{W}$.

Define the class of the functions $\delta = (\delta_1(t), \delta_2(t))$ where $\delta_1(t)$ is an arbitrary non-negative continuous increasing function and $\delta_2(t)$ is a non-negative continuous function, $0 \le t \le T$. For any function $\delta(t)$ in this class we can define a random process W^δ by the equality

$$W_t^\delta = W_t + \eta \sqrt{\delta_2(t)} . \tag{5}$$

<u>Lemma</u>. Let system (1) and reward function (2) be given. Assume that for some $\varepsilon_0 > 0$ the following conditions

$$|\delta_1(t) - t| \le \varepsilon_0 , \qquad \delta_2(t) \le \varepsilon_0 , \tag{6}$$

$$E \sup_{t \le T} |g(\delta_1(t), W_t^\delta)| < \infty \tag{7}$$

hold.

Then for any stopping time $\tau \in \mathfrak{M}^{\xi^\varepsilon}$ for $\varepsilon \le \varepsilon_0$ the following equality

$$Eg(\tau, W_\tau) = Eg(\tau, m_\tau^\varepsilon + \eta \sqrt{\gamma_\tau^\varepsilon}) \tag{8}$$

holds.

Proof. For the stopping time $\tau\in\mathfrak{M}^{\xi^\varepsilon}$ we can define τ_n in the following manner

$$\tau_n = \begin{cases} \frac{k}{2^n}, & \frac{k-1}{2^n} \le \tau \le \frac{k}{2^n}, \\ T, & \text{otherwise}, \end{cases}$$

where $k/2^n<T$, $k=1,2,\ldots$. By virtue of the usual mathematical expectations and the independence of the random variable η of σ-algebra $\mathcal{F}_t^{\xi^\varepsilon}$ it can be easily obtained that

$$Eg(\tau_n, W_{\tau_n}) = Eg(\tau_n, m_{\tau_n}^\varepsilon + \eta\sqrt{\gamma_{\tau_n}^\varepsilon}). \tag{9}$$

By virtue of Lemma 3 from [2] we have

$$E\sup_{t\le T} |g(t,W_t)| < \infty,$$

$$E\sup_{t\le T} |g(t,m_t^\varepsilon+\eta\sqrt{\gamma_t^\varepsilon})| < \infty.$$

Therefore in equality (9) we can pass to the limit when $n\to\infty$ and thus obtain the assertion of the lemma.

Now we can define the function $\delta(t)=(\delta_1(t),\delta_2(t))$ by the equality

$$\delta_1(t) = \begin{cases} h_\varepsilon^{-1}(t), & t \le T_\varepsilon, \\ T, & T_\varepsilon \le t \le T, \end{cases} \tag{10}$$

$$\delta_2(t) = \begin{cases} \gamma^\varepsilon_{h_\varepsilon^{-1}(t)}, & t \le T_\varepsilon, \\ \gamma_T^\varepsilon, & T_\varepsilon \le t \le T, \end{cases} \tag{11}$$

where

$$h_\varepsilon(t) = \int_0^t \left(\frac{\gamma_s^\varepsilon}{\varepsilon}\right)^2 ds$$

and T_ε is defined by the equality $T_\varepsilon=h_\varepsilon(T)$.

Theorem. Let system (1) be given, reward function (2) and S^0 and S^ε are defined by (3). Then beginning with $\varepsilon\le\varepsilon_0$ the following estimate

$$0 \le S^0-S^\varepsilon \le \varepsilon \{ \sup_{t\le T} E|g_t'(t,W_t)|$$

$$+ \sup_{\delta} E \sup_{t\le T} |g_t'(\delta_1(t), W_t^{\delta})| + \frac{1}{b} E \sup_{t\le T} |W_t|\}$$

holds.

Proof. The inequality $S^0 \ge S^{\varepsilon}$ can be easily obtained by virtue of $\mathfrak{M}^{\xi^{\varepsilon}} \subseteq \mathfrak{M}^{(W,\tilde{W})}$. Further, note that by Lemma 2 from [2] we have

$$\sup_{t\le T_{\varepsilon}} |h_{\varepsilon}^{-1}(t)-t| \le \varepsilon_0 .$$

Then the second part of the theorem results from (7) and the theorem from [2]. Note that functions (10) and (11) satisfy conditions (6).

3. Now suppose that a partially observable random process $(\theta,\xi^{\varepsilon_1,\varepsilon_2})=(\theta_t,\xi_t^{\varepsilon_1,\varepsilon_2})$, $0\le t\le T$ is given having the following form:

$$d\theta_t = [a_0(t)u_t + a_1(t)\theta_t + a_2(t)\xi_t^{\varepsilon_1,\varepsilon_2}]dt + b_1(t)dW_t + b_2(t)d\tilde{W}_t ,$$

$$d\xi_t^{\varepsilon_1,\varepsilon_2} = [A_0(t) + A_1(t)\theta_t + A_2(t)\xi_t^{\varepsilon_1,\varepsilon_2}]dt + \varepsilon_1 dW_t + \varepsilon_2 d\tilde{W}_t ,$$

where $a_i(t)$, $A_i(t)$, $i=0,1,2$, $b_i(t)$, $i=1,2$ are deterministic continuous functions, $\varepsilon_i\ge 0$, $i=1,2$ are constants, $u=u_t=u(t,\xi_t^{\varepsilon_1,\varepsilon_2})$ is a control and W and $\tilde{W}$, as in the above, are independent Wiener processes [1]. Consider the square loss functional of the form

$$V(u,\theta) = E\,\{\theta_T^2 h(T) + \int_0^T [\theta_t^2 H(t) + u_t^2 R(t)]dt\} ,$$

where $h(t)$, $H(t)$, $R(t)$, $0\le t\le T$ are deterministic continuous bounded functions [1]. Denote

$$V^0 = \inf_{u_t\in U} V(u_t,\theta_t), \qquad V^{\varepsilon_1,\varepsilon_2} = \inf_{u_t\in U^{\varepsilon_1,\varepsilon_2}} V(u_t,\theta_t)$$

where U and $U^{\varepsilon_1,\varepsilon_2}$ are classes of admissible controls for which $E\{u^4(t,\theta_t)\}<\infty$ and $E\{u^4(t,\xi_t^{\varepsilon_1,\varepsilon_2})\}>\infty$ respectively. Under the assumption that $u_t\equiv 1$ we also denote $S^0=\sup Eg(\tau,\theta_\tau)$, $\tau\in\mathfrak{M}^{\theta}$ and $S^{\varepsilon_1,\varepsilon_2}=\sup Eg(\tau,\theta_\tau)$, $\tau\in\mathfrak{M}^{\xi^{\varepsilon_1,\varepsilon_2}}$ where $g(t,x)$, $-\infty<x<\infty$, $0\le t\le T$ is a continuous function. Using the results of [2] we can show that $V^{\varepsilon_1,\varepsilon_2}\to V^0$ and $S^{\varepsilon_1,\varepsilon_2}\to S^0$ as $\varepsilon_1,\varepsilon_2\to 0$. In this case the convergence in both cases is of the order $\varepsilon_1+\varepsilon_2$. Besides, it should be noted that similar results also hold in the discrete case [5] and when the continuous scheme is approximated by discrete schemes [6].

References

[1] Liptzer R. Sh., Shiryayev A. N. Statistics of random processes. M., "Nauka", 1974, 696p.

[2] Dochviri V. M., Shashiashvili M. A. On the convergence of costs in the problem of optimal stopping of stochastic processes in the scheme of Kalman-Bucy. Commun. Statist. Sequential Analysis, 1 (3), 1982, p.163-176.

[3] Fahrmann H. H. On optimal stopping of a Wiener process with incomplete data. Teor. Veroyatn. i Primen., 1978, v.23, No1, p.143-148.

[4] Shepp L. A. Explicit solutions to some problems of optimal stopping. Ann. Math. Statist., 40,3 (1969), p.993-1010.

[5] Dochviri V. M., Shashiashvili M. A. On the convergence of cost functions in a discrete problem of optimal stopping with incomplete data. Bulletin of the Acad. of Sci. of the Georgian SSR, 109, No1, 1983.

[6] Dochviri V. M. On the convergence of costs in the case of approximation of a continuous scheme of Kalman-Bucy by discrete schemes. Proceedings of Tbilisi University, 1983, v.15, p.65-76.

Tbilisi State University

BELLMAN EQUATION WITH UNBOUNDED COEFFICIENTS AND ITS APPLICATIONS

Masatoshi Fujisaki

0. Introduction. Consider the following Bellman equation with degenerate diffusion coefficients:

$$(0.1)\begin{cases} \partial_s v + 1/2 \sum_{i,j=1}^{\nu} a_{ij}(s)\partial_i\partial_j v + \inf_{\alpha\in R^d}\{\sum_{i=1}^{d} b(\alpha,s,x)\partial_i v + f(\alpha,s,x)\} = 0, \\ 0<s<T,\ x\in R^d, \\ v(T,x) = g(x),\ x\in R^d \end{cases}$$

where $1\leq\nu<d$, $a=(a_{ij}), 1\leq i,j\leq\nu$, is a positive definite matrix, which is written as $a=\sigma\sigma^*$ (σ^* denotes the transposed matrix of σ) and σ is a $\nu\times\nu$ matrix. Furthermore, the coefficients b and f are of the forms: $b(\alpha,t,x) =\tilde{b}(t)\alpha+\tilde{b}(t,x)$ and $f(\alpha,t,x)=|\alpha|^2+\tilde{f}(t,x)$. Under adequate regularity conditions on the coefficients, we shall prove the existense and uniqueness of "generalized solution" v of Eq.(0.1), that is, v is continuous on $[0,T]\times R^d$ and has generalized derivatives (in distribution sense) $\partial_s v(=\partial v/\partial s)$, $\partial_i v(=\partial v/\partial x_i, 1\leq i\leq d)$ and $\partial_i\partial_j v(=\partial^2 v/\partial x_i\partial x_j, 1\leq i,j\leq\nu)$, which are locally integrable, and it satisfies Eq.(0.1) for almost all $(s,x)\in(0,T)\times R^d$.

We shall also consider its applications to differential equations and moreover, to the theory of nonlinear filtering. In fact, we can show that there exists a unique solution of the filtering equation and that this solution is the density with respect to the Lebesgue measure on R^d of the filter. W.H.Fleming and S.K.Mitter ([2]) already worked these problems in the case when $\nu=d$ in Eq.(0.1)(a is not singular) and $\tilde{f}$ is bounded from below. In [2], they proved that there exists a unique solution of Eq.(0.1) belonging to $C^{1,2}((0,T)\times R^d)$ and, moreover, they obtained some estimates about this unique solution.

In [3], §5, the author also studied Eq.(0.1) whose coefficients have more restrictive forms than those given in this paper.

1. Formulations and notations. Let T be a finite positive number which is fixed throughout this paper. Let $\{A_n\}, n=1,2,\ldots$, be an increasing sequence of nonempty bounded subsets of R^d such that $R^d=\cup_{n=1}^{\infty}A_n$ (we often write A instead of R^d). Put $Q_T=(0,T)\times R^d$ and also $\bar{Q}_T=[0,T]\times R^d$. For each $(s,x)\in Q_T$, consider the following stochastic control problem for a

system described by stochastic differential equations of the type:

$$(1.1)\quad \begin{cases} dX_t = b(\alpha_t, s+t, X_t)dt + \sigma(s+t)dB_t, & 0<t\leq T-s, \\ X_0 = x, \end{cases}$$

where $(B_t), 0\leq t\leq T$, is a d-dimensional Brownian motion process. Assume that the coefficients σ and b satisfy the following conditions:

(A.1) σ is an $R^d\times R^d$- valued (d×d-matrix) continuous function of t. b is an R^d-valued function defined on $R^d\times\bar{Q}_T$, which is represented as follows:

$$(1.2)\quad b(\alpha,t,x) = \bar{b}(t)\alpha + \tilde{b}(t,x),$$

where $\bar{b}(t)$ is an $R^d\times R^d$-valued bounded function of t, and $\tilde{b}(t,x)$ is an R^d-valued continuous function on $\bar{Q}_T$, satisfying the following conditions: for some nonnegative constant k, for all $(t,x)\varepsilon\bar{Q}_T$, $(t,x')\varepsilon\bar{Q}_T$,

$$(1.3)\quad |\tilde{b}(t,x) - \tilde{b}(t,x')| \leq k|x - x'|, \quad \text{and}$$

$$(1.4)\quad |\tilde{b}(t,x)| \leq k(1+|x|).$$

Assume further that $\tilde{b}(t,.)\varepsilon C^2(R^d)$ and its derivatives $\partial_i\tilde{b}(1\leq i\leq d)$, $\partial_i\partial_j\tilde{b}(1\leq i,j\leq d)$ are all bounded functions uniformly with respect to (t,x). Let $1\leq\nu<d$, and finally suppose that σ is a d×d-matrix such that $\sigma(t)=\begin{pmatrix}\bar{\sigma}(t) & 0\\ 0 & 0\end{pmatrix}$, where $\bar{\sigma}$ is a $\nu\times\nu$-matrix such that $\bar{a}=\bar{\sigma}\bar{\sigma}^*$ is not singular, i.e. there is a constant $\eta>0$ such that $(\bar{a}(t)\xi,\xi)\geq\eta|\xi|^2$, for all $\xi\varepsilon R^\nu$. □

Now we introduce the concept of strategy following [3], §5.

Definition 1.1 Let $n\geq 1$. We write $\alpha\varepsilon\mathfrak{A}_n$ if the process $\alpha=\alpha_t(\omega), 0\leq t\leq T$, is defined on a probability space $(\Omega,\mathcal{F},P;\mathcal{F}_t)$ satisfying the usual conditions, is progressively measurable with respect to $\{\mathcal{F}_t\}$ and takes values from A_n for $t\varepsilon[0,T]$. Let $\mathfrak{A}=\cup_{n\geq 1}\mathfrak{A}_n$ and call an element of $\mathfrak{A}$ a strategy . □

By means of the assumption (A.1), for each strategy $\alpha\varepsilon\mathfrak{A}$ and $(s,x)\varepsilon\bar{Q}_T$, there exists a unique solution of Eq.(1.1) and we denote it by $(X_t^{\alpha,s,x})$. Next, define the cost by the following way: for each $\alpha\varepsilon\mathfrak{A}$ and $(s,x)\varepsilon\bar{Q}_T$, put

$$(1.5)\quad v^\alpha(s,x) = E\left[\int_0^{T-s} f(\alpha_t, s+t, X_t^{\alpha,s,x})dt + g(X_{T-s}^{\alpha,s,x})\right],$$

where $(X_t^{\alpha,s,x})$ is the solution of Eq.(1.1) associated with (α,s,x). Suppose that f and g satisfy the following conditions;

(A.2) f is a real valued function defined on $R^d\times\bar{Q}_T$, which is written as follows:

(1.6) $f(\alpha,t,x) = |\alpha|^2 + \tilde{f}(t,x)$,

where $\tilde{f}$ is continuous with respect to (t,x) and, for all $(t,x)\in\bar{Q}_T$,

(1.7) $-k'(1+|x|) \leq \tilde{f}(t,x) \leq k(1+|x|)^m$.

Here k' is a constant depending on k, which will be given later, and m is a constant (≥ 1). g is a nonnegative function on R^d such that

(1.8) $0 \leq g(x) \leq k(1+|x|)^m$, for all $x\in R^d$.

Assume further that for each t, $\tilde{f}(t,.)\in C^2(R^d)$ and $g\in C^2(R^d)$, and, moreover, all their derivatives, $\partial\tilde{f}/\partial x_i$, $\partial^2\tilde{f}/\partial x_i\partial x_j$, $\partial g/\partial x_i$ and $\partial^2 g/\partial x_i\partial x_j$ $(1\leq i,j\leq d)$, are bounded uniformly with respect to (t,x). □

For each $(s,x)\in\bar{Q}_T$ and $n\geq 1$, define the <u>costs</u> v_n and v by the formulas:

(1.9) $$v_n(s,x) = \inf_{\alpha\in\mathfrak{U}_n} v^\alpha(s,x),$$

and

(1.10) $$v(s,x) = \inf_{\alpha\in\mathfrak{U}} v^\alpha(s,x).$$

Then, from (A.1) and (A.2), it is easy to see that $\lim_{n\to\infty} v_n(s,x)=v(s,x)$ and also that, for fixed n, each v_n has the following properties (for the proof, see e.g.[3]).

<u>Proposition</u> 1.1 Let n be fixed. Then (a) there exists a function u_n defined on $\bar{Q}_T$ such that $u_n\geq 0$, $\in C^{1,2}(\bar{Q}_T)$ and, for all (s,x),

(1.11) $|v_n(s,x)| \leq u_n(s,x)$,

(b) $v_n\in W^{1,2,\nu}_{\lambda,loc}(Q_T)\cap C(\bar{Q}_T)$,

(c) v_n satisfies the following Bellman equation for almost all $(s,x)\in Q_T$:

(1.12) $$\partial_s v_n + 1/2 \sum_{i,j=1}^{\nu} a_{ij}(s)\partial_i\partial_j v_n + \inf_{\alpha\in A_n}\{\sum_{i=1}^{d} b_i(\alpha,s,x)\partial_i v_n + f(\alpha,s,x)\}$$
$$= 0,$$

(d) $v_n(T,x) = g(x)$, $x\in R^d$,

where $a=\sigma\sigma^*$. Here we say that, for each $1\leq\gamma\leq d$ and $\lambda\geq 1$, a function $u(s,x)$ over $\bar{Q}_T$ belongs to $W^{1,2,\gamma}_{\lambda,loc}(Q_T)$ if u has a first order generalized derivative with respect to s, first order generalized derivatives with respect to x_i, $1\leq i\leq d$, and second order generalized derivatives with respect to $x_i x_j$, $1\leq i,j\leq\gamma$, on Q_T and, moreover, for any bounded open set $Q\subset Q_T$ the foergoing derivatives belong to $L_\lambda(Q)$ (we write $W^{1,2}_{\lambda,loc}(Q_T)$ when $\gamma=d$).□

It is shown that the function v of (1.10) is also well-defined by the following result.

Proposition 1.2 There exist functions $\underline{u}$ and $\bar{u}$ of (s,x) which are both elements of $C^{1,2}(\bar{Q}_T)$ such that for all $(s,x)\in\bar{Q}_T$,

(1.13) $\underline{u}(s,x) \leq v(s,x) \leq \bar{u}(s,x)$. □

Proof It is easy to see that there exists a fuction $\bar{u}$ such that $v\leq\bar{u}$, because, for example, $v(s,x)\leq v_1(s,x)$ by the definition of v, and $|v_1(s,x)|\leq u_1(s,x)$ by Proposition 1.1 (a). In order to prove the inverse relation, define a real valued function $\underline{u}$ on $\bar{Q}_T$ by the formula:

(1.14) $\underline{u}(t,x) = -Ne^{M(T-t)}(1+|x|^2)^{1/2}$,

where M and N are nonnegative constants. Then, it follows from (A.1) and (A.2) (especially (1.7)) that we can choose a triplet (M,N,k'), depending on k, such that (1.13) holds (for the detail, see [4]). □

Remark 1.1 (a) Because of the defintion (1.10) of v, it is easily seen that if $\tilde{f}$ is dominated from below by a constant then the assertion of the proposition is obvious. If $\tilde{f}$ is not so, then we cannot take an arbitrary k' in (1.7) but it depends on k.
(b) It is not difficult to extend the above result to the case where $\tilde{f}$ satisfies the following condition:

(1.15) $-k'(1+|x|)^2 \leq \tilde{f}(t,x) \leq k(1+|x|)^m$,

where k' also depends on k (see also [4]). □

2. Bellman equation. In this section we shall prove that the cost v of (1.10) is "generalized solution" of Eq.(0.1). The method we adopt is the same as [1] or [3]. Let $\varepsilon>0$, and define d×d-matrix σ^ε by the formula:

$$(2.1)\quad \sigma^\varepsilon(t) = \begin{pmatrix} \bar{\sigma}(t) & 0 \\ 0 & \begin{matrix}\varepsilon & & 0\\ & \ddots & \\ 0 & & \varepsilon\end{matrix}\end{pmatrix} \begin{matrix}\}\nu \\ \}d-\nu\end{matrix}$$

(column blocks: $\underbrace{\quad}_{\nu}$, $\underbrace{\quad}_{d-\nu}$)

Remark that $\sigma^\varepsilon(\sigma^\varepsilon)^*$ is nonsingular d×d-matrix because of the assumption (A.1). For each $\varepsilon>0$ and $n\geq1$, consider the following nondegenerate Bellman equation:

$$(2.2)\quad \begin{cases} 0 = \partial_s v + 1/2 \sum_{i,j=1}^{\nu} a_{i,j}(s)\partial_i\partial_j v + \varepsilon^2/2 \sum_{i=\nu+1}^{d} \partial_i^{\,2} v + \inf_{\alpha\in A_n}\{(b^\alpha(s,x),\nabla v) \\ \qquad + f^\alpha(s,x)\} = 0, \quad \text{on } Q_T, \\ v(T,x) = g(x), \quad x\in R^d. \end{cases}$$

It is well known in [3] that in this case the cost v_n^ε of (1.9), associated with $(\sigma^\varepsilon,n,s,x)$, has the following properties.

<u>Proposition</u> 2.1 (a) For all $\lambda\geq 1$, $v_n^{\varepsilon}\in W^{1,2}_{\lambda,loc}(Q_T)\cap C(\bar{Q}_T)$.

(b) There exists a constant M depending only on (k,d,T) such that

(2.3) $\sum_{i=1}^{d} |\partial_i v_n^{\varepsilon}(s,x)|\leq M$, for all ε, n and $(s,x)\in Q_T$.

(c) v_n^{ε} satisfies Eq.(2.2) for almost all $(s,x)\in Q_T$.

<u>Proof</u> Since the assertions (a) and (c) can be proved in a way similar to [3] (Theorem 2.1), it is sufficient to show (b) only. For each $\varepsilon>0$, for any $(\alpha,s,x)\in\mathfrak{A}\times\bar{Q}_T$, put

(2.4) $$v^{\alpha,\varepsilon}(s,x)=E[\int_0^{T-s} f^{\alpha_t}(s+t,X_t^{\alpha,s,x,\varepsilon})dt + g(X_{T-s}^{\alpha,s,x,\varepsilon})],$$

where $(X_t^{\alpha,s,x,\varepsilon})$ is a solution of Eq.(1.1) in which σ is replaced by σ^{ε}. It is shown that for each t, $v^{\alpha,\varepsilon}(t,.)\in C^2(R^d)$ by virtue of the assumptions (A.1) and (A.2), and,moreover,

(2.5) $$\partial_i v^{\alpha,\varepsilon}(s,x) = E[\int_0^{T-s} \sum_{j=1}^{d} \partial_j f^{\alpha_t}(s+t,X_t^{\alpha,s,x,\varepsilon})\partial_i X_{t,j}^{\alpha,s,x,\varepsilon}\, dt + \sum_{j=1}^{d} \partial_j g(X_{T-s}^{\alpha,s,x,\varepsilon})\partial_i X_{T-s}^{\alpha,s,x,\varepsilon}].$$

But, by means of (A.1) and (A.2), it is easily shown that (2.5) is uniformly bounded with respect to (α,s,x,ε), i.e. there exists a constant M such that for all $(\alpha,s,x,\varepsilon)\in\mathfrak{A}\times Q_T\times R_+$,

(2.6) $$\sum_{i=1}^{d} |\partial_i v^{\alpha,\varepsilon}(s,x)| \leq M.$$

Since $v_n^{\varepsilon}=\inf_{\alpha\in\mathfrak{A}_n} v^{\alpha,\varepsilon}$ and, for each s, v_n^{ε} is absolutely continuous with respect to x, generalized derivatives $\partial_i v_n^{\varepsilon}$, $1\leq i\leq d$, exist and now the inequality (2.3) is obvious because of (2.6). □

It is also proved (cf.(1.13)) that there exist two functions $\underline{u}$ and $\bar{u}$ such that they belong to $C^{1,2}(Q_T)$ and for all $(s,x)\in\bar{Q}_T$, $\varepsilon>0$ and $n\geq 1$,

(2.7) $\underline{u}(s,x) \leq v_n^{\varepsilon}(s,x) \leq \bar{u}(s,x)$.

Define v^{ε} by the following formula:

(2.8) $$v^{\varepsilon}(s,x) = \inf_{\alpha\in\mathfrak{A}} v^{\alpha,\varepsilon}(s,x)$$

Then we can show the following easily.

<u>Lemma</u> 2.2 (a) For all $\lambda\geq 1$, $v^{\varepsilon}\in W^{1,2}_{\lambda,loc}(Q_T)\cap C(\bar{Q}_T)$, (b) $\lim_{n\to\infty} v_n^{\varepsilon}(s,x)=v^{\varepsilon}(s,x)$ uniformly on each compact subset of Q_T. (c) For each i $(1\leq i\leq d)$,

$\partial_i v_n^\varepsilon \to \partial_i v^\varepsilon$ as $n\to\infty$ uniformly on each compact subset of Q_T. (d) $\partial_s v_n^\varepsilon \to \partial_s v^\varepsilon$, $\partial_i\partial_j v_n^\varepsilon \to \partial_i\partial_j v^\varepsilon (1\leq i,j\leq d)$ as $n\to\infty$ weakly in $L_\lambda(Q)$, $Q\subset Q_T$ is a bounded open set . □

The proof of this lemma is based on the general theory of differential equations and in addition on the inequalities (2.3) and (2.7). Roughly speaking, it is shown that, for each $\varepsilon>0$ and for any $\lambda(1<\lambda<\infty)$, bounded $Q\subset Q_T$, the $W^{1,2}_{\lambda,Q}$-norm of v_n^ε is uniformly bounded with respect to n and, therefore, the Hölder norm of $\partial_i v_n^\varepsilon$ is also uniformly bounded with respect to n (for the detail, see e.g. [1], pp.207).

Since v_n^ε satisfies Eq.(2.2), it holds that

$$(2.9)\quad \partial_s v_n^\varepsilon + 1/2 \sum_{i,j=1}^{\nu} a_{ij}(s)\partial_i\partial_j v_n^\varepsilon + \varepsilon^2/2 \sum_{i=\nu+1}^{d} \partial_i{}^2 v_n^\varepsilon + H_n(s,\nabla v_n^\varepsilon)$$
$$+ \sum_{i=1}^{d} \tilde{b}_i(s,x)\partial_i v_n^\varepsilon + \tilde{f}(s,x) = 0 \quad \text{a.e.}(Q_T),$$

where for any $(t,p,\alpha)\in[0,T]\times R^{2d}$, $H(t,p,\alpha)$, $H_n(t,p)$ and $H(t,p)$ are defined by the formulas:

$$(2.10)\quad H(t,p,\alpha) = \sum_{i,j=1}^{d} \bar{b}_{ij}(t)\alpha_j p_i + |\alpha|^2,$$

$$(2.11)\quad H_n(t,p) = \inf_{\alpha\in A_n} H(t,p,\alpha),$$

and

$$(2.12)\quad H(t,p) = \inf_{\alpha\in A} H(t,p,\alpha).$$

It is clear that $H(t,p)$ is finite and $\lim_{n\to\infty} H_n(t,p)=H(t,p)$ because $H(t,p,\alpha)\geq -|\bar{b}^*(t)p|^2/4$ for all (t,p,α), and, furthermore, equality holds if and only if $\alpha= -\bar{b}^*p/2$ ($\bar{b}^*$ means the transposed matrix of $\bar{b}$). Now letting $n\to\infty$ in Eq.(2.9), we have the following.

<u>Theorem</u> 2.3 v^ε satisfies the following equation for almost all $(s,x)\in\bar{Q}_T$.

$$(2.13)\quad \partial_s v^\varepsilon + 1/2 \sum_{i,j=1}^{\nu} a_{ij}(s)\partial_i\partial_j v^\varepsilon + \varepsilon^2/2 \sum_{i=\nu+1}^{d} \partial_i{}^2 v^\varepsilon + H(s,\nabla v^\varepsilon)$$
$$+ \sum_{i=1}^{d} \tilde{b}_i(s,x)\partial_i v^\varepsilon + \tilde{f}(s,x) = 0.$$

<u>Proof</u> Because of Lemma 2.2, it is sufficient to show that $H_n(s,\nabla v_n^\varepsilon)\to H(s,\nabla v^\varepsilon)$ as $n\to\infty$ a.e.(Q_T). Remark first that, for sufficiently large n,

$$(2.14)\quad H_n(t,\nabla v_n^\varepsilon) = -|\bar{b}^*(t)\nabla v_n^\varepsilon|^2/4,$$

for ∇v_n^ε is uniformly bounded by means of (2.3). Indeed, if p is bounded then $H_n(t,p)=H(t,p)$ for sufficiently large n. See (2.10)∿(2.12) and the note following them. Then the assertion follows immediately from Lemma 2.2 (c). □

Remark 2.1 (a) If $H(s,p)$ is Hölder continuous on Q_T, then it is well known (see e.g.[1], pp.208) that $v^\varepsilon \varepsilon C^{1,2}(Q_T)$ (more precisely, $v^\varepsilon \varepsilon C^{1,2}(Q_T) \cap C(\bar{Q}_T)$. Therefore, in this case v^ε satisfies Eq.(2.13) for all $(s,x)\varepsilon Q_T$ (cf.[2]).
(b) It is shown that Eq.(2.13) has a unique solution, that is, for all $\lambda>1$, if $u\varepsilon W^{1,2}_{\lambda,loc}(Q_T)\cap C(\bar{Q}_T)$ is a solution of Eq.(2.13) such that $|u(t,x)| \leq k(1+|x|^2)$ for some constant k, then $u=v^\varepsilon$ on $\bar{Q}_T$ ([5], pp.228∿238). □

By the well known method (see [1]), it is shown that v^ε has the following properties.

Lemma 2.4 (a) $\partial_i v^\varepsilon (1\leq i\leq d)$ is bounded uniformly with respect to (ε,s,x).
(b) For any bounded open set Q ($\subset Q_T$) and for any $\lambda\varepsilon(1,\infty)$, there is a constant $M_{Q,\lambda}$ such that

$$(2.15)\quad \int_Q [|\partial_s v^\varepsilon|^\lambda + \sum_{i,j=1}^{\nu} |\partial_i\partial_j v^\varepsilon|^\lambda]dsdx \leq M_{Q,\lambda}. \quad \square$$

In fact, the assertion (a) is due to (2.3) and also Lemma 2.2 (c). Moreover, the assertion (b) can be obtained by the same way as [1]. The following result is also verified by the way similar to [3], where v is given by (1.10).

Lemma 2.5 Let $\varepsilon\to 0$. Then (a) v^ε converges to v uniformly on each compact subset of Q_T, (b) $\partial_s v^\varepsilon$ and $\partial_i\partial_j v^\varepsilon, 1\leq i,j\leq\nu$, converge weakly in $L_\lambda(Q)$ to $\partial_s v$ and $\partial_i\partial_j v$, $1\leq i,j\leq\nu$, respectively, where λ is any number >1, and Q is an arbitrary bounded set of Q_T, (c) $\partial_i v^\varepsilon(1\leq i\leq d)$ converges to $\partial_i v(1\leq i\leq d)$ for a.e.(s,x). □

Proof The assertion (a) is easily verified, and (b) follows from Lemma 2.4 (see also [1]). In order to show (c), it is sufficient to remark that, for each s, v is absolutely continuous with respect to x, and that there is a nonnegative function $u\varepsilon C^{1,2}(Q_T)$ such that the function $v(s,x)-u(s,x)$ is concave with respect to x (for the detail, see e.g.[3]) . □

Finally we obtain the following main result.

Theorem 2.6 The cost v, given by (1.10), is a generalized solution of Eq.(0.1), that is,

(a) for any $\lambda(1<\lambda<\infty)$, $v\varepsilon W^{1,2,\nu}_{\lambda,loc}(Q_T)\cap C(\bar{Q}_T)$,

and

(b) v satisfies Eq.(0.1) for almost all $(s,x)\varepsilon Q_T$. □

3. Uniqueness. We shall discuss in this section the uniqueness of the solution of Bellman's equation (0.1). Since we know already that the function v of (1.10) is a generalized solution of this equation, we shall show that any generalized solution is equal to v. For technical reasons, we assume hereafter that the coefficient a is constant, i.e. a is independent of t in Eq.(0.1).

Let $u \in W^{1,2,\nu}_{\lambda,loc}(Q_T) \cap C(\bar{Q}_T)$ for any $\lambda > 1$, and, moreover, assume the following conditions:

(A.3) (a) for some positive constants k and m,

(3.1) $|u(t,x)| \leq k(1+|x|)^m$, for all $(t,x) \in \bar{Q}_T$.

(b) $\partial_i u (1 \leq i \leq d)$ is uniformly bounded, i.e.

(3.2) $\sum_{i=1}^{d} |\partial_i u(t,x)| \leq k$, for $\forall (t,x) \in Q_T$.

(c) For all $\delta > 0$, $l \in R^d$ such that $|l| = 1$, and $(t,x) \in \bar{Q}_T$,

(3.3) $D^{2,l}_{\delta} u(t,x) \equiv \{u(t,x+\delta l) + u(t,x-\delta l) - 2u(t,x)\}/\delta^2 \leq k(1+|x|)^m$. □

Note first that the function v also satisfies this assumption (A.3). Indeed, the inequalities (3.1) and (3.2) are deduced from (1.13) and (2.3) respectively. As to (3.3), see [3] (Remark 4.1). For all $u \in W^{1,2,\nu}_{\lambda,loc}(Q_T)$ and $(\alpha,t,x) \in A \times Q_T$, define $F^{\alpha}[u](t,x)$ and $F[u](t,x)$ by the following:

(3.4) $F^{\alpha}[u](t,x) = \partial_t u + 1/2 \sum_{i,j=1}^{\nu} a_{ij} \partial_i \partial_j u + \sum_{i=1}^{d} b_i^{\alpha}(t,x) \partial_i u + f^{\alpha}(t,x)$,

and

(3.5) $F[u](t,x) = \inf_{\alpha \in A} F^{\alpha}[u](t,x)$.

Then we obtain the following result.

Theorem 3.1 Let $u \in W^{1,2,\nu}_{\lambda,loc}(Q_T) \cap C(\bar{Q}_T)$ for any $\lambda > 1$. Also, let (A.3) be satisfied . If $F[u](t,x) = 0$ a.e.(Q_T), and $u(T,x) = g(x)$ for all $x \in R^d$, then $u = v$ in $\bar{Q}_T$. □

The method of proof is the same as [3]. That is, we can prove the following two lemmas by the way similar to [3], §3~§4.

Lemma 3.2 Let $u \in W^{1,2,\nu}_{\lambda,loc}(Q_T) \cap C(\bar{Q}_T)$ for any $\lambda > 1$. Also, let (a) and (b) of (A.3) be satisfied. If $F[u](t,x) \geq 0$ a.e.(Q_T), and $u(T,x) \leq g(x)$ for all $x \in R^d$, then $u \leq v$ in $\bar{Q}_T$. □

In fact, although we didn't assume that the second derivatives $\partial_i \partial_j u$, $1 \leq i,j \leq \nu$, are locally bounded, it is easily seen that the proof of [3] (Theorem 3.1) is still applicable to this lemma, because the coefficient a is constant in this case.

Next, let us demonstrate the inverse relation.

Lemma 3.3 Let $u \in W^{1,2,\nu}_{\lambda,loc}(Q_T) \cap C(\bar{Q}_T)$ for any $\lambda > 1$, and also let all of the conditions of (A.3) be satisfied. If $F[u] \leq 0$ a.e.(Q_T), and $u(T,x) \geq g(x)$ for $x \in R^d$, then $u \geq v$ in $\bar{Q}_T$.

Proof Recall first the fact (see (2.10)∿(2.14)) that

$$(3.6)\ F[u](t,x) = \partial_t u + 1/2 \sum_{i,j=1}^{\nu} a_{ij}\partial_i\partial_j u + \sum_{i=1}^{d} \tilde{b}_i(t,x)\partial_i u + \tilde{f}(t,x)$$
$$- |\bar{b}^*(t)\nabla u|^2/4.$$

Define a function $\alpha(t,x)$ over $\bar{Q}_T$ by the formula:

$$(3.7)\ \alpha_i(t,x) = -\sum_{j=1}^{d} \bar{b}^*_{ij}(t)\partial_j u/2, \quad 1 \leq i \leq d.$$

Then it follows from (A.1) and (3.2) that, for some nonnegative constant $N=N(k)$,

$$(3.8)\ |\alpha(t,x)| \leq N \quad \text{for } (t,x) \in \bar{Q}_T.$$

Consider the following stochastic differential equation: for $s \leq t \leq T$, $x \in R^d$,

$$(3.9)\ d\xi_t = \{\bar{b}(t)\alpha(t,\xi_t) + \tilde{b}(t,\xi_t)\}dt + \sigma^\varepsilon d\beta_t, \quad \xi(s)=x,$$

where σ^ε is given in (2.1) and (β_t) is d-dimensional Brownian motion process. In Eq.(3.9) above, we assume that $\bar{b}$ and α are bounded functions (see (A.1) and (3.8)), and $\tilde{b}$ satisfies a linear growth condition. Since σ^ε is not degenerate, it follows from the theorem of Girsanov that for each (s,x) there exists a solution of Eq.(3.9). Then, the rest of proof is quite the same as [3] (Theorem 4.1). □

Now the assertion of Theorem 3.1 follows immediately from the above two lemmas.

4. Applications. We can apply the preceding results to differential equations and also the nonlinear filtering. We begin with discussing linear differential equations of parabolic type. Consider the following Cauchy problem on $\bar{Q}_T$:

$$(4.1)\ \partial u/\partial s = \mathcal{L}u, \quad u(T,x)=g(x),$$

where $\mathcal{L}$ is second order differential operator given by

$$(4.2)\ \mathcal{L} = 1/2 \sum_{i,j=1}^{d} a_{ij}(s,x)\partial_i\partial_j + \sum_{i=1}^{d} b_i(s,x)\partial_i + c(s,x).$$

Then, Eq.(4.1) can be formally transferred to the Bellman equation by simple transformations as follows. Suppose that there exists a solution u of Eq.(4.1) which is positive in $\bar{Q}_T$, and put

(4.3) $v(s,x) = -\log u(T-s,x).$

It follows from (4.1) that v satisfies the following equation:

(4.4) $$0 = \partial_s v + 1/2 \sum_{i,j} a_{ij}(s,x)\partial_i\partial_j v - 1/2 \sum_{i,j} a_{ij}(s,x)\partial_i v \partial_j v + \sum_i b_i(s,x)\partial_i v - c(s,x).$$

But, it is easy to see that the above equation is equivalent to the following:

(4.5) $$0 = \partial_s v + 1/2 \sum_{i,j}^{d} a_{ij}(s,x)\partial_i\partial_j v + \inf_{\alpha\in R^d}\{2^{1/2}(\sigma(s,x)\alpha,\nabla v) + |\alpha|^2\} + \sum_i^d b_i(s,x)\partial_i v - c(s,x), \quad (a=\sigma\sigma^*),$$

which is Bellman's equation of the type (0.1).

Using these facts we are able to solve the nonlinear filtering problem whose system is given by the stochastic differential equation:

(4.6) $$\begin{cases} dX_t = \rho(t,X_t)dt + \sigma(t,X_t)dw_t, \\ dY_t = g(t,Y_t)h(t,X_t,Y_t)dt + g'(t,Y_t)dt + g(t,Y_t)dw'_t, \end{cases}$$

with initial conditions $X_0=\xi$ and $Y_0=0$. Here, (w_t) and (w'_t), $0\leq t\leq T$, are independent d_1-dimensional and d_2-dimensional Brownian motions respectively and $(X_t,Y_t)\in R^{d_1}\times R^{d_2}$. Our fundamental situation is that it is impossible to observe the process (X_t) directly but the process (Y_t) is observable. Let $(\Omega,\mathcal{F},P,X_t,Y_t)$ be a solution of Eq.(4.6) and denote by $\{\mathcal{F}_t\}$ the σ-field generated by the sets $\{Y_s, s\leq t\}$. Then the conditional distribution of X_t with respect to $\mathcal{F}_t$ is called (nonlinear) "filter". For any bounded continuous function f on R^d , put

(4.7) $\pi_t(f) = E[f(X_t)|\mathcal{F}_t],$

where E means the mathematical expectation with respect to P. Then it is known that the (modified) formal density q of the measure $\pi_t(.)$ satisfies the "pathwise filtering equation", which is equal to Eq.(4.1), in which the coefficients $a=\sigma\sigma^*$, b and c are functions of (σ,ρ,g,g',h) (for the detail, see e.g.[2] or [4]).

W.H.Fleming and S.K.Mitter proved in [2] that if a is not singular and c is bounded from above then the Bellman equation (4.5) has a unique solution belonging to $C^{1,2}(Q_T)$ and, moreover, they obtained some estimates about this unique solution. As we have seen in the preceding sections, we can extend their results to the case where a is degenerate and c is not necessarily bounded. By combining the theory of filtering with the preceding results, we can summarize as follows (for details, see [4]).

Proposition 4.1 Under the assumptions (A.1)~(A.3), there exists a unique generalized solution of the pathwise filtering equation (4.1) and, moreover, this unique solution is constructed from the filter. □

Remark 4.1 It is remarkable that we may take an unbounded function as g.h in Eq.(4.6), while there are a lot of results about the filtering equation in which g.h is assumed to be bounded (cf. [2], [4]). □

Remark 4.2 Theorem 2.6 implies that there exists a generalized solution of the equation:

$$(4.8)\quad \partial_s v + 1/2 \sum_{i,j=1}^{\nu} a_{ij}(s)\partial_i\partial_j v - \sum_{i,j=1}^{d} \bar{b}_{ij}(s)\partial_i v\partial_j v + \sum_{i=1}^{d} b_i(s,x)\partial_i v$$
$$+ f(s,x) = 0,$$

where $(\bar{b}_{ij})$, $1\leq i,j\leq d$, is a symmetric nonnegative matrix which is bounded in $[0,T]$, and a, b and f satisfy the same conditions as a, $\tilde{b}$ of (A.1) and $\tilde{f}$ of (A.2), respectively. Note that Eq.(4.8) is an extension of (0.7) in [3] (see Example 5.2). □

References

[1] W.H.Fleming, R.W.Rishel: Deterministic and Stochastic Optimal Control, Appl. Math. No.1, Springer, 1975.

[2] W.H.Fleming, S.K.Mitter: Optimal control and nonlinear filtering for nondegenerate diffusion processes, Stochastics, 8(1982), 63-77.

[3] M.Fujisaki: Control of diffusion processes in R^d and Bellman equation with degeneration, Osaka J. Math. 23(1986), 117-149.

[4] M.Fujisaki: Nonlinear filtering and Bellman equation, Shōdai Ronshū, 38(1986), 102-120 (in Japanese).

[5] N.V.Krylov: Controlled Diffusion Processes, Appl. Math. No.14, Springer, 1980.

Kobe University of Commerce
Tarumi, Kobe, 655, Japan

A NOTE ON CAPACITIES IN INFINITE DIMENSIONS

Masatoshi Fukushima

§1 Capacities on the Wiener space and properties of Brownian paths.

Given a metric space X, a σ-finite measure m on X and a strongly continuous contraction semigroup $\{T_t, t > 0\}$ of symmetric Markovian operators on $L^2(X;m)$, the Dirichlet space $(\mathcal{F},\mathcal{E})$ is defined by $\mathcal{F} = \mathcal{D}[\sqrt{-L}]$, $\mathcal{E}(u,v) = (\sqrt{-L}\,u, \sqrt{-L}v)$, where L is the infinitesimal generator of T_t and $(\ ,\ )$ denotes the L^2-inner product. The associated capacity on X is then introduced by

(1) $C(A) = \inf\{\ \mathcal{E}_1(u,u)\ ;\ u \in \mathcal{F}\ ,\ u \geq 1\ \ m\text{-a.e. on } A\}$

for open set A and extended to any subset on X as an outer capacity. Here $\mathcal{E}_1(u,u) = \mathcal{E}(u,u) + (u,u)$.

The semigroup $\{T_t, t > 0\}$ also operates on $L^p(X,m)$, $p > 1$, and its Gamma transformation

$$V_r = \frac{1}{\Gamma(r/2)} \int_0^\infty t^{\frac{r}{2}-1} e^{-t}\, T_t\, dt \quad , \qquad r > 0,$$

decides a Banach space $(\mathcal{F}_{r,p}, \|\ \|_{r,p})$ by $\mathcal{F}_{r,p} = V_r(L^p)$, $\|u\|_{r,p} = \|f\|_p$, $u = V_r f$, $f \in L^p$. In particular $\mathcal{F}_{1,2} = \mathcal{F}$ and $\|u\|_{1,2}^2 = \mathcal{E}_1(u,u)$. The (r,p)-capccity $C_{r,p}$ can then be introduced in the analogous way to (1). $C_{r,p}$ dominates m and increases as r increases ([2]). A set $A \subset X$ is said to be slim (with respect to the semigroup T_t) if $C_{r,p}(A) = 0$ for any $r > 0$, $p > 1$.

When X is the Euclidean n-space, m is the Lebesgue measure and $\{T_t, t > 0\}$ is the Brownian transition semigroup, $(\mathcal{F}_{r,p}, \|\ \|_{r,p})$ is just the Sobolev space of order (r,p) and X admits no non-empty slim set. The situation becomes quite different in the infinite dimensional case that X is $W = \{ w\ ;\ [0,\infty) \to R^d\}$ thc Wiener space, m is P the Wiener measure and $\{T_t, t > 0\}$ is the Ornstein-Uhlenbeck semigroup with infinitesimal generator $L = -\sum_{n=0}^{\infty} \frac{n}{2} P_n$, P_n being the projection of L^2 on the Wiener chaos of order n.

W then admits many slim sets including each one point set ([9]). The notion of slim sets was first introduced by Malliavin in this case [7]. Many important functionals on W such as solutions of stochastic differential equations are not continuous with respect to the usual metric on W, but they are quasi-continuous with respect to $C_{r,p}$ for any $r > 0$, $p > 1$. This was one of the reasons for introducing slim sets.

It is also interesting to see, in the latter case just mentioned, whether well known basic properties of Brownian motions $b(t,w)$, $t > 0$, $w \in W$, holding for P-a.e $w \in W$ could be refined to hold except on a finer set of $C_{r,p}$-capacity zero or a slim set. Concerning the nowhere differentiability, Lévy's Hölder continuity and the law of the iterated logarithm of the one dimensional Brownian motion, the author [1] established this refinement to zero sets of $C = C_{1,2}$ capacity and Takeda [9] further refined to slim sets.

Among capacities $C_{r,p}$, $r > 0$, $p > 1$, on W, $C = C_{1,2}$ is distinguished because C-zero sets admit a probabilistic interpretation that they are exactly polar sets of the Ornstein-Uhlenbeck (diffusion) process on W ([1],[9]). Hence C-quasi-everywhere properties of the Brownian motion can be investigated as almost sure properties of the Ornstein-Uhlenbeck process, or equivalently, of a scale changed Brownian sheet without refering to the function space $(\mathcal{F},\mathcal{E})$. Refined statements in this sense on the Hausdorff dimension and the local time were thus proven by Komatsu-Takashima [3],[4] and Shigakawa [8].

Reccurence and double points are more interesting properties since their dependence on the dimension d shift by 2 if we pass from P-a.e. statements to C-q.e. ones. Indeed C-q.e. d-dimensional Brownian path is transient if $d \geq 5$ ([1]), while Kono [5] proved that, if $d \leq 4$,

$$C(\underline{\lim}_{t\to\infty} |b(t) - a| < \varepsilon, \quad a \in R^d, \quad \varepsilon > 0) > 0.$$

C-q.e. d-dimensional Brownian path $b(t)$ has no double point if $d \geq 6$, while Lyons [6] proved that, if $d = 4$ or 5, $b(t)$ has double points with positive C-capacity and, if $d = 3$, it has quadruple point with positive C-capacity.

P-a.e. properties of Brownian paths $b(t)$ depend only on the finite dimensional distributions of $b(t)$ and accordingly does not depend on the specific choice of the metric space W on which the Wiener measure P sits. Professor K. Ito raised to the author a question whether this is also true for $C_{r,p}$-q.e. statements or not. In the next section, we formulate the question in a general setting and give an affirmative answer only for $C = C_{1,2}$. Note that the function space $(F_{r,p}, \| \|_{r,p})$ depends only on the measurable structure of the underlying space (X,m), but the capacity $C_{r,p}$ apparently depends upon the topology of X in view of the definition (1).

§2 An invariance property of capacities

Let X be a Polish space, m be an everywhere dense probability measure on X and $\{T_t, t > 0\}$ be a strongly continuous contraction semigroup of symmetric Markovian operators on $L^2(X;m)$. Suppose further that we are given a Polish space Y such that

(2) $\quad Y \hookrightarrow X, \quad m(X - Y) = 0,$

where $\hookrightarrow$ means that "continuously embedded onto a dense subset". Y is then a Borel subset of X and $L^2(Y;m)$ can be identified with $L^2(X;m)$. Hence (1) provides us with set functions C_X and C_Y on X and Y respectively via Dirichlet space (F,E) of $\{T_t, t > 0\}$. We want to know whether

(3) $\quad C_X(A) = C_Y(A \cap Y) \quad \forall A \subset X.$

This particularly means $C_X(X - Y) = 0$.

Denote by $C_b(X)$ the family of all bounded continuous functions on X. We say a Markov process $\underline{\underline{M}}_Y = (Y_t, P_x)_{x \in Y}$ on Y to be a Hunt process if $Y_t \in Y$ is right continuous in $t \geq 0$, has left limit and quasi-left continuous.

Theorem We assume the following two conditions :

(A.1) there is a countable set $S \subset F \cap C_b(X)$ which is dense in F and separates points of X.

(A.2) there exists a Hunt process $\underline{\underline{M}}_Y$ on Y whose transition function p_t satisfies $p_t f = T_t f$ m-a.e. for any $t > 0$ and $f \in C_b(Y)$.

Then (3) is valid and moreover

(4) $C_X(N) = 0$ iff $P_m(\sigma_N < \infty) = 0$

for Borel $N \subset X$. Here σ_N denotes the hitting time of N of the process $\underline{\underline{M}}_Y$.

<u>Proof</u>. We use a method of compactification. Let Z be the Gelfand space of the closed subalgebra of $C_b(X)$ generated by S. Z is then a compactification of X : Z is a compact metric space and $X \subsetneq Z$. Extende m from X to Z by setting $m(Z - X) = 0$. (F, E) can be then viewed as a regular Dirichlet space on $L^2(Z;m)$ ($= L^2(X;m)$) and (1) provides us with a set function C_Z on Z.

On the other hand, $\underline{\underline{M}}_Y$ in condition (A.2) can be viewed as a Hunt process on Y ($\subset Z$) with respect to the relative topology of Z. Extend this to a Hunt process $\underline{\underline{M}}_Z$ on Z by letting each point of $Z - Y$ be a trap. Since $m(Z - Y) = 0$, the resulting Hunt process $\underline{\underline{M}}_Z$ is associated with the regular Dirichlet space (F, E).

We can then prove exactly in the same manner as §6 of Takeda [9] that

(5) $C_Z(A) = C_X(A \cap X), \quad \forall A \subset Z$

and

(6) $C_Z(N) = 0$ iff $P_m(\sigma_N < \infty) = 0$

for Borel $N \subset Z$. A key observation in proving these is that, for any finely open set $A \subset Z$ with respect to the process $\underline{\underline{M}}_Z$, $C_Z(A)$ can be evaluated by the same formula as (1).

Since Z is also a compactification of Y, we have (5) with X being replaced by Y completing the proof of Theorem.

<u>Example</u> Let H be a separable Hilbert space, B_1 and B_2 be separable Banach spaces such that $H \subsetneq B_1 \subsetneq B_2$ with B_1 and B_2 being equipped with measurable norms with respect to H. Let be a Gaussian measure on B_2 such that

$$\int_{B_2} e^{i\langle \ell, x\rangle} \mu(dx) = \exp\left(-\tfrac{1}{2}\|\ell\|_H\right), \quad \ell \in B_2^*$$

then $\mu(B_2 - B_1) = 0$. Condition (2) is thus satisfied for $X = B_2$,

$Y = B_1$, $m = \mu$.

There exists a diffusion process $\underline{\underline{M}} = (X_t, P_x)_{x \in B_1}$ on B_1 with transition function

(7) $p_t f(x) = \int_{B_1} f(e^{-t}x + \sqrt{1 - e^{-2t}}y)\, \mu(dy),, \quad x \in B_1, \quad f \in C_b(B_1).$

$\underline{\underline{M}}$ is called the standard Ornstein-Uhlenbeck process on B_1 and decides a strongly continuous contraction semigroup $\{T_t, t > 0\}$ of symmetric Markovian operators on $L^2(B_2, \mu)$.

The tame functionals

$$u(x) = f(<\ell_1, x>, \cdots, <\ell_n, x>), \quad f \in C_0(R^n), \ \ell_1, \cdots, \ell_n \in B_2^* ,$$

are dense in the associated Dirichlet space (F, E) and Theorem applies in getting

$$C_2(A) = C_1(A \cap B_1), \ \forall A \subset B_2,$$

$$C_2(N) = 0 \ \text{ iff } \ P_\mu(\sigma_N < \infty) = 0 \ \text{ for Borel } N \subset B_2,$$

where C_1 (resp. C_2) is the set function on B_1 (resp. B_2) defined by (1) via the Dirichlet space (F, E).

References

[1] M. Fukushima, Basic properties of Brownian motion and a capacity on the Wiener space, J. Math. Soc. Japan, 36(1984), 161-176.

[2] M. Fukushima and H. Kaneko, On (r,p)-capacities for general Markovian semigroups,in "Infinite dimensional analysis and stochastic processes" ed. by S. Albeverio, Pitman, 1985.

[3] T. Komatsu and K. Takashima, The Hausdorff dimension of quasi-all Brownian paths, Osaka J. Math., 21(1984), 613-619.

[4] T. Komatsu and K. Takashima, On the existence of intersectional local time except on zero capacity set, Osaka J. Math.,21(1984), 913-929.

[5] N. Kôno, 4-dimensional Brownian motion is recurrent with positive capccity, Proc.Japan Acad., 60, Ser.A(1984), 57-59.

[6] T.J. Lyons, The critical dimension at which quasi-every Brownain path is self-avoiding, Adv. in Appl. Probab,suppl.,87-99,1986.

[7] P. Malliavin, Implicit functions in finite corank on the Wiener space, in "Stochastic Analysis" ed. by K. Ito, Kinokuniya, 1984.

[8] I. Shigekawa, On the existence of the local time of the 1-dimen sional Brownian motion in quasi-everywhere, Osaka J. Math., 21 (1984), 621-627.

[9] M. Takeda, (r,p)-capccities on the Wiener space and properties of Brownian motion, Z. Wahrscheinlichkeitsteorie verw.Gebiete, 68(1984), 149-162.

Department of Mathematics
College of General Education
Osaka University
Toyonaka, Osaka, Japan

ON DIFFUSIVE MOTION OF CLOSED CURVES

Tadahisa Funaki

1. Introduction. Let M be a complete and compact C^∞-Riemannian manifold with metric $g = \{g_{\alpha\beta}\}$. In this paper a diffusive motion of closed curves on M will be discussed by investigating a quasilinear stochastic partial differential equation (SPDE). The closed curves $f \in C(S,M)$ will be parametrized by $\sigma \in S = R/Z$. Consider the following SPDE for $\{f_t\,;\, t \geqq 0\}$:

$$(1) \qquad df_t(\sigma) = \Delta f_t(\sigma)dt + V_0(f_t(\sigma))dt + \circ dW_t(\sigma, f_t(\sigma)),\ t > 0,\ \sigma \in S.$$

Here $V_0 \in C^\infty(TM) = \{C^\infty\text{-vector fields on } M\}$ and $W_t(\sigma,x)$ is a $C^\infty(S\times TM)$-valued Wiener process, namely, $W_t = \sum_{i=1}^\infty w_t^i V_i$ converging in a proper sense with independent real-valued Brownian motions $\{w_t^i\}$ and $V_i \in C^\infty(S\times TM) = \{C^\infty\text{-sections of the product bundle } S\times TM \to S\times M\}$ (see Baxendale [3],[4]). We write $\circ dW_t$ to denote the Stratonovich stochastic differential and $\circ dW_t(\sigma,f_t(\sigma)) = \sum_{i=1}^\infty V_i(\sigma,f_t(\sigma))\circ dw_t^i$. The nonlinear Laplacian Δ acting on a smooth mapping f of S into M determines $\Delta f(\sigma) = \{(\Delta f(\sigma))^\alpha\}_\alpha \in T_{f(\sigma)}M$, $\sigma \in S$, by the formula

$$(\Delta f(\sigma))^\alpha = \frac{d^2 f^\alpha}{d\sigma^2}(\sigma) + \Gamma^\alpha_{\beta\gamma}(f(\sigma))\frac{df^\beta}{d\sigma}(\sigma)\frac{df^\gamma}{d\sigma}(\sigma)\ ,\quad f = \{f^\alpha\}_\alpha\ ,$$

in terms of local coordinates (see Hamilton [9]), where $\Gamma^\alpha_{\beta\gamma}$ stand for the Christoffel symbols on M. We use Einstein's convention. Funaki [8] investigated an equation of the form (1) in a Euclidean space R^d with random noise $W_t(\sigma,x)$ being non-smooth in σ.

In physical literatures equations similar to (1) appear especially concerning the kinetic theory of phase transitions (see Hohenberg and Halperin [11] and Kawasaki and Ohta [17]) and the theory of stochastic quantization (see Parisi and Wu [23] and Namiki, Ohba and Okano [22]). We remark that $\Delta f(\sigma)$ appears as a functional derivative $-\delta E(f)/\delta f(\sigma)$ of an energy $E(f) = \int_S |\frac{df}{d\sigma}(\sigma)|^2 d\sigma$ associated with every physical configuration $f \in C^1(S,M)$. Faris [7] and Wick [24] discussed a stochastic Heisenberg model which may be regarded as a discrete version of our model

From the mathematical point of view the equation (1) gives an example of quasilinear SPDE of the new type. A rather general theory of such equations was developed by Krylov and Rozovskii [19]; however, their method does not work in our situation.

2. Main results. The compact manifold M can be embedded into a suitable Euclidean space R^N. We extend the metric g of M smoothly to R^N ([6],[9]). We also extend the vector fields V_0 and W_t, i.e. $\{V_i\}_{i=1}^{\infty}$, properly, at least smoothly, to R^N respectively $S\times R^N$ and then consider the SPDE (1) globally in the Euclidean coordinate system of R^N.

Let $W^{s,p}(R)$, $s \geq 0$, $p \geq 1$, be the usual Sobolev space, more precisely, the space of all $f \in S'(R)$ = {Schwartz distributions on R} satisfying $((1+\xi^2)^{s/2}\hat{f})^{\vee} \in L^p(R)$, where $\hat{f}$ and $\check{h}$ denote the Fourier transform and its inverse, respectively. The Sobolev space $W^{s,p}(S)$ over S is also defined as usual in the following manner: Consider an open covering $S = U_1 \cup U_2$ with isometries ψ_i from U_i onto open intervals I_i of R, $i=1,2$. Take a partition of unity $\{\phi_i \in C^{\infty}(S); i=1,2\}$ corresponding to the covering $\{U_i\}$ and extend functions $f_i := (f\cdot\phi_i)\circ\psi_i^{-1}$ on I_i to R by putting $f_i = 0$ outside I_i. We call $f \in W^{s,p}(S)$ if and only if $f_i \in W^{s,p}(R)$ for $i=1,2$. Let $W^{s,p}(S,R^N)$ be a product space $\{W^{s,p}(S)\}^N$. The norm $\|\cdot\|_{s,p}$ of the space $W^{s,p}(S)$ or $W^{s,p}(S,R^N)$ can be defined in a natural way. The Sobolev's imbedding theorem tells that $W^{s,p}(S)$ is continuously imbedded in the space $C^n(S)$ if $s > \frac{1}{p} + n$, $n = 0,1,2,\cdots$. Let $W^{s,p}(S,M)$, $s > \frac{1}{p}$, $p > 1$, be the space of all $f \in W^{s,p}(S,R^N)$ satisfying $f(\sigma) \in M$ for every $\sigma \in S$.

Definition By a solution of the SPDE (1) we mean a stochastic process $f_t = \{f_t(\sigma;\omega)\}$, $t \in [0,T]$, $T > 0$, defined on a suitable probability space (Ω,F,P) equipped with reference family $\{F_t\}$ such that

(i) f_t is an $\{F_t\}$-adapted $W^{s,p}(S,M)$-valued continuous process with some $s > \frac{1}{p} + 1$ and $p > 1$,

(ii) there exists an $\{F_t\}$-Wiener process $W_t(\sigma,x) = \sum_{i=1}^{\infty} w_t^i V_i(\sigma,x)$, $(\sigma,x) \in S\times R^N$, defined on (Ω,F,P) and

(iii) with probability one, f_t and W_t satisfy

$$f_t(\sigma) = f_0(\sigma) + \int_0^t \Delta f_s(\sigma)ds + \int_0^t V_0(f_s(\sigma))ds + \int_0^t \circ dW_s(\sigma,f_s(\sigma)),$$
$$t \in [0,T], \quad \sigma \in S,$$

in the sense of generalized functions on S.

Here, by $\{F_t\}$-Wiener process, we mean as usual that W_t is $\{F_t\}$-adapted and its increment $W_t - W_s$ is independent of F_s for every

$0 \leq s < t$.

It should be pointed out that $f_t \in W^{s,p}(S,M)$, $s > \frac{1}{p} + 1$, implies $f_t \in C^1(S,M)$ and therefore the nonlinear term of $\Delta f_t(\sigma)$, which consists of products of first order derivatives of f_t, is well-defined in a classical sense. In the following the initial data f_0 of the SPDE (1) is always taken from the space $C^\infty(S,M)$, on which we consider the usual topology of uniform C^n-convergence for every $n \geq 0$. We now formulate our main theorems. These results are probabilistic extensions of those due to Eells and Sampson [6].

Theorem 1 (Existence of global solutions) Suppose that M has non-positive sectional curvature K_M. Then there exists a solution f_t, $t \in [0,\infty)$, of the SPDE (1).

Theorem 2 (Regularization of solutions) Every solution f_t, $t \in [0,T]$, of the SPDE (1) belongs to the space $C([0,T],C^\infty(S,M))$ with probability one.

Theorem 3 (Pathwise uniqueness of solutions) Let f_t and f_t', $t \in [0,T]$, be two solutions of the SPDE (1) defined on a same probability space $(\Omega,F,P;\{F_t\})$ with a same Wiener process W_t. If $f_0 = f_0'$, then we have $f_t = f_t'$, $t \in [0,T]$, with probability one.

3. Proof of Theorem 1. Here only an outline of the proof of Theorem 1 will be given. We omit the proofs of Theorems 2 and 3 since they can be carried out applying Itô's stochastic calculus in a similar line to the non-random case [6],[9]. We begin this section by referring to the following result.

Proposition 1 ([6]) If $K_M \leq 0$, then there exists a unique solution $f_t \in C([0,\infty),C^\infty(S,M))$ of the nonlinear heat equation:

$$\frac{\partial f_t}{\partial t}(\sigma) = \Delta f_t(\sigma)\ , \ t > 0,\ \sigma \in S\ ;\quad f_0 = f \in C^\infty(S,M). \tag{2}$$

We shall sometimes denote the solution of (2) by $f_t(\sigma;f)$ in order to specify its initial data. Consider the following SDE on M with parameter $\sigma \in S$:

$$\begin{cases} d\xi_t = V_0(\xi_t)dt + \circ dW_t(\sigma,\xi_t)\ , & t > s, \\ \xi_s = x \in M\ , & s \geq 0. \end{cases} \tag{3}$$

We also denote the solution ξ_t of (3) by $\xi_{s,t}(\sigma,x)$ to make its dependence on (s,σ,x) clear. Noting that $\overline{\xi}_{s,t}(\sigma,x) = (\sigma,\xi_{s,t}(\sigma,x)) \in S\times M$ solves a "nice" SDE on a product manifold $S\times M$, we see that $\{\overline{\xi}_{s,t}\}$ determines a stochastic flow of diffeomorphisms on $S\times M$. Therefore $\xi_{s,t}(\cdot,\cdot)$ is a $C^\infty(S\times M,M)$-valued continuous process with two time parameters (s,t), $0 \leqq s \leqq t < \infty$. Here we consider the usual topology on the space $C^\infty(S\times M,M)$.

Roughly speaking, the solution $f_t(\sigma)$ of (1) can be constructed as an infinitesimal composition of two solutions of (2) and (3) like Trotter's product formula. More precisely, for every partition $\Pi = \{0 = t_0 < t_1 < \cdots < t_n = T\}$, $T > 0$, of the time interval $[0,T]$, we define a $C^\infty(S,M)$-valued continuous process f_t^Π inductively by

$$(4) \qquad f_t^\Pi(\sigma) = \xi_{t_k,t}(\sigma, f_{t-t_k}(\sigma; f_{t_k}^\Pi)) \ , \ \sigma \in S, \ t \in [t_k,t_{k+1}], \quad k = 0,1,2,\cdots,n-1 \ ; \ f_0^\Pi = f_0.$$

The proof of Theorem 1 consists of the following three steps: (a) We show the tightness of the family of distributions $\{P^\Pi\}$ of the processes $\{f_t^\Pi\}$ on the space $C([0,T],W^{s,p}(S,M))$, $p > 1$, $1 + \frac{1}{p} < s < 2$. (b) It is shown that every limit P of $\{P^\Pi\}$ as $|\Pi| := \max_k |t_{k+1}-t_k| \to 0$ solves the martingale problem corresponding to the SPDE (1). (c) We prove the equivalence between the martingale problem and the SPDE (1). However, since the steps (b) and (c) are standard (see Iwata [16] for the step (c)), we give only a sketch of the step (a) in the rest of this section.

Two kinds of energies of a mapping $f = \{f^\alpha(\cdot)\}_{\alpha=1}^N \in C^\infty(S,M)$ are introduced as follows:

$$e(f)(\sigma) = \frac{1}{2} g_{\alpha\beta}(f(\sigma)) \frac{df^\alpha}{d\sigma}(\sigma) \frac{df^\beta}{d\sigma}(\sigma),$$

and

$$\kappa(f)(\sigma) = \frac{1}{2} g_{\alpha\beta}(f(\sigma)) (\Delta f)^\alpha(\sigma) (\Delta f)^\beta(\sigma) \quad , \quad \sigma \in S.$$

The first task is to derive the following energy estimates.

<u>Lemma 1</u> For every positive integer p, there exists a positive constant K such that

$$(5) \qquad E[1 + \|e(f_{t_{k+1}}^\Pi)\|_p^p \mid F_{t_k}] \leqq e^{K(t_{k+1}-t_k)} \{1 + \|e(f_{t_k}^\Pi)\|_p^p\} ,$$

$$(6) \qquad E[\ \|\kappa(f_{t_{k+1}}^\Pi)\|_p^p \mid F_{t_k}] \leqq$$

$$\leq K(t_{k+1}-t_k)\{1+\|e(f^{\Pi}_{t_k})\|^{2p}_{2p}\} + e^{K(t_{k+1}-t_k)}\|\kappa(f^{\Pi}_{t_k})\|^p_p ,$$

where $\|\cdot\|_p$ is the norm of the space $L^p(S)$. We also have

(7) $$\sup\{E[e^p(f^{\Pi}_t)(\sigma)] + E[\kappa^p(f^{\Pi}_t)(\sigma)] ;\ \Pi, \sigma \in S, t \in [0,T]\} < \infty.$$

This lemma is shown by combining the estimates on the solutions of (2) and (3), which can be derived by the method of [6] and by using Itô's formula, respectively. We should note the independence of $\xi_{t_k,t}$ and $f_{t-t_k}(\sigma;f^{\Pi}_{t_k})$. From the estimates (5) and (6), we can take C_1 and $C_2 > 0$ in such a manner that

$$X_k = e^{-Kt_k}\|\kappa(f^{\Pi}_{t_k})\|^p_p + C_1 e^{-C_2 t_k}\{1+\|e(f^{\Pi}_{t_k})\|^{2p}_{2p}\} , \quad k = 0,1,2,\cdots,n,$$

is an $\{F_{t_k}\}$-supermartingale. Therefore Doob's inequality yields two bounds on random variables $\sup_{0\leq k\leq n}\|e(f^{\Pi}_{t_k})\|_p$ and $\sup_{0\leq k\leq n}\|\kappa(f^{\Pi}_{t_k})\|_p$, from which we obtain the following:

Lemma 2 For every $p \geq 1$, we have

(8) $$\lim_{\lambda\to\infty} \sup_{\Pi} P(\sup_{0\leq t\leq T}\|e(f^{\Pi}_t)\|_p > \lambda) = 0,$$

and

(9) $$\lim_{\lambda\to\infty} \sup_{\Pi} P(\sup_{0\leq t\leq T}\|\kappa(f^{\Pi}_t)\|_p > \lambda) = 0.$$

These two estimates can be combined into

Lemma 3 $\lim_{\lambda\to\infty} \sup_{\Pi} P(\sup_{0\leq t\leq T}\|f^{\Pi}_t\|_{2,p} > \lambda) = 0 , \quad p \geq 1.$

The estimate (7) with the help of Itô's formula proves:

Lemma 4 For every $p \geq 1$, there exists a positive constant C satisfying

$$\sup_{\Pi,\sigma\in S} E[|f^{\Pi}_t(\sigma) - f^{\Pi}_s(\sigma)|^p] \leq C|t-s|^{p/2} , \quad 0 \leq s \leq t \leq T.$$

The following lemma is just a conclusion of functional analytic arguments.

Lemma 5 Assume that f^k_t and $f_t \in C([0,T],W^{1,2}(S,R^N))$, $k=1,2,\cdots,$

satisfy the following conditions:

(i) f_t^k converges to f_t as $k \to \infty$ in the space $C([0,T],W^{1,2}(S,R^N))$ having the topology of uniform convergence,

(ii) $\sup_k \sup_{0\leq t\leq T} \|f_t^k\|_{2,p} < \infty$, with some $p \geq 2$.

Then, for every $s < 2$, f_t^k and f_t belong to the space $C([0,T], W^{s,p}(S,R^N))$ and f_t^k converges to f_t as $k \to \infty$ in this space.

We can now complete the step (a) of the proof of Theorem 1 by showing the following proposition.

Proposition 2 The family of distributions $\{P^\Pi\}_\Pi$ of $\{f_t^\Pi\}_\Pi$ on the space $C([0,T],W^{s,p}(S,R^N))$, $0 < s < 2$, $p \geq 1$, is weakly relatively compact.

Proof. First we prove the tightness of $\{P^\Pi\}$ on the space $C([0,T], W^{1,2}(S,R^N))$. To this end, since $W^{1,2}(S,R^N)$ is a Hilbert space, it suffices to check the conditions of Holley and Stroock [12]. However these can be checked from Lemmas 3 and 4 by noting that an imbedding of $W^{2,2}(S,R^N)$ into $W^{1,2}(S,R^N)$ is a compact operator (Rellich's lemma, see Adams [1]). Now we put $X^\Pi = \sup_{0\leq t\leq T} \|f_t^\Pi\|_{2,p}$. Then Lemma 3 shows that the family of distributions of $\{X^\Pi\}_\Pi$ on R_+ is tight. Therefore we see that the family of joint distributions of $\{(X^\Pi,f_t^\Pi)\}_\Pi$ on the space $R_+\times C([0,T],W^{1,2}(S,R^N))$ is also tight. Now Lemma 5 concludes the proof with the help of Skorohod's theorem which represents a weak convergence by an almost-everywhere convergence on a proper probability space (see, e.g., Ikeda and Watanabe [14,p9]). □

Remark We can remove the assumption that M is compact. See Hartman [10].

4. Examples. (i) Let $M = G$ be a connected Lie group with a right-invariant Riemannian metric. Such metric exists certainly, e.g., for compact G. For $V_0, V_1,\cdots,V_r \in \mathcal{G}$ = {left-invariant vector fields on G}, we consider an SPDE:

$$(1)' \qquad df_t(\sigma) = \Delta f_t(\sigma)dt + \sum_{i=0}^{r} V_i(f_t(\sigma))\circ dw_t^i ,$$

where $w_t^0 = t$ and $\{w_t^i\}_{i=1}^r$ are independent real-valued Brownian motions. Let $g_t = g_t(g)$ be a solution of an SDE on G:

$$(3)' \qquad \begin{cases} dg_t(g) = \sum_{i=0}^{r} V_i(g_t(g))\circ dw_t^i , \quad t > 0, \\ g_0(g) = g \in G. \end{cases}$$

The process g_t is usually called a left-invariant Brownian motion on G and has the property: $g_t(g) = g \cdot g_t(e)$, $t \geq 0$, a.s., where e is a unit element of G (see Ikeda and Watanabe [15]). We can prove the following theorem.

Theorem 4 Assume that there exists a (local) solution of the nonlinear heat equation on G:

$$(2)' \qquad \frac{\partial \bar{f}_t}{\partial t}(\sigma) = \Delta \bar{f}_t(\sigma) \ , \ t \in [0,T], \ T > 0 \ ; \ \bar{f}_0 = f_0 \in C^\infty(S,G).$$

Then the unique solution of the SPDE (1)' with initial data f_0 is given by

$$f_t(\sigma) = g_t(\bar{f}_t(\sigma)) \ , \quad t \in [0,T].$$

This composition theorem should be compared with (4) and also with results of Kunita [20] and Bismut [5].

Now assume that G is compact and let μ be a Haar probability measure on G, which is an invariant measure of g_t. Fet's theorem [18] asserts that there exists a non-trivial closed geodesic $\bar{f}(\cdot)$ on G parametrized as $\Delta \bar{f} \equiv 0$. It is then an easy consequence of Theorem 4 that an image measure $\nu(\bar{f})$ on the space $C^\infty(S,G)$ of μ under the mapping $\psi: g \in G \mapsto \bar{f}(\cdot)g \in C^\infty(S,G)$ is invariant for the SPDE (1)'. Especially when the fundamental group $\pi_1(G)$ of G is non-trivial, the invariant measure of (1)' is not unique.

We assume $K_G \leqq 0$ in addition. For example, Euclidean (flat) torus satisfies this assumption and see also Azencott and Wilson [2]. Under all these assumptions, (a) there exists a global solution $\bar{f}_t$ of (2)' and $\bar{f}_t$ converges to a closed geodesic $\bar{f}$ as $t \to \infty$ in the space $C^\infty(S,G)$ (Hartman [10]) and (b) for each $a \in \pi_1(G,e)$, there exists a unique closed geodesic $\bar{f}$ which represents a (Lawson and Yau [21, Fact 3]). We also assume that the family $\{V_i\}_{i=0}^r$ satisfies a condition of Hörmander's type (condition (E) of Ichihara and Kunita [13]). Then the process g_t is ergodic in a sense that, for every $g \in G$, the distribution of $g_t(g)$ on G converges weakly to μ as $t \to \infty$ [13, Section 6]. We can prove the following fact immediately: The distribution on the space $C^\infty(S,G)$ of the solution f_t of (1)' with initial data $f_0 \in C^\infty(S,G)$ converges weakly to $\nu(\tilde{f}_0)$ as $t \to \infty$, where $\tilde{f}_0$ is a closed geodesic which is parametrized as $\Delta \tilde{f}_0 \equiv 0$ and belongs to the same homotopy class as f_0.

(ii) The problem of singular perturbations for the SPDE like (1) was investigated by Funaki [8, Sections 3 and 4]. Here we give an example of unstable solutions for this problem. On a 2-dimensional sphere

$S^2 = \{x=(x_1,x_2,x_3) \in R^3; |x|=1\}$, consider an SPDE with a vector field $V(x) = (x_2,-x_1,0) \in T_xS^2$, $x \in S^2$:

$$(1)'' \qquad df_t(\sigma) = \frac{1}{\varepsilon}\Delta f_t(\sigma)dt + V(f_t(\sigma))\circ dw_t, \quad t > 0, \sigma \in S, \varepsilon > 0,$$

where w_t is a real-valued Brownian motion. We find an unstable solution:

$$f_t(\sigma) = (\sin\sigma\cdot\cos w_t, -\sin\sigma\cdot\sin w_t, \cos\sigma), \quad \sigma \in S \ (\cong [0,2\pi)),$$

and a stable solution

$$f_t(\sigma) = (c\cos w_t, -c\sin w_t, \pm\sqrt{1-c^2}), \quad \sigma \in S,$$

for every $c \in [0,1]$. These are solutions of (1)'' for every $\varepsilon > 0$.

References

[1] R.A. Adams, Sobolev spaces, Academic Press, 1975.

[2] R. Azencott and E.N. Wilson, Homogeneous manifolds with negative curvature, Part II, Mem. Amer. Math. Soc., vol. 8, 178 (1976), 1-102.

[3] P.H. Baxendale, Brownian motions in the diffeomorphism group I, Compositio Math., 53 (1984), 19-50.

[4] P.H. Baxendale, Asymptotic behaviour of stochastic flows of diffeomorphisms, Proceedings of 15th SPA conference, Lecture Notes in Math., 1203, 1-19, Springer, 1986.

[5] J.M. Bismut, Mécanique aléatoire, Lecture Notes in Math.,866, Springer, 1981.

[6] J. Eells and J.H. Sampson, Harmonic mappings of Riemannian manifolds, Amer. J. Math., 86 (1964), 109-160.

[7] W.G. Faris, The stochastic Heisenberg model, J. Funct. Anal., 32 (1979), 342-352.

[8] T. Funaki, Random motion of strings and related stochastic evolution equations, Nagoya Math. J., 89 (1983), 129-193.

[9] R.S. Hamilton, Harmonic maps of manifolds with boundary, Lecture Notes in Math., 471, Springer, 1975.

[10] P. Hartman, On homotopic harmonic maps, Canadian J. Math., 19 (1967), 673-687.

[11] P.C. Hohenberg and B.I. Halperin, Theory of dynamic critical phenomena, Rev. Mod. Phys., 49 (1977), 435-479.

[12] R.A. Holley and D.W. Stroock, Generalized Ornstein-Uhlenbeck processes and infinite particle branching Brownian motions, Publ. RIMS Kyoto Univ., 14 (1978), 741-788.

[13] K. Ichihara and H. Kunita, A classification of the second order degenerate elliptic operators and its probabilistic characterization, Z. Wahr. verw. Geb., 30 (1974), 235-254.

[14] N. Ikeda and S. Watanabe, Stochastic differential equations and diffusion processes, North-Holland/Kodansha, 1981.

[15] N. Ikeda and S. Watanabe, Stochastic flows of diffeomorphisms, Stochastic Analysis and Applications, Advances in Prob. and related Topics 7, 179-198, Marcel Dekker, 1984.

[16] K. Iwata, An infinite dimensional stochastic differential equation with state space C(R), Prob. Th. Rel. Fields, 74 (1987), 141-159.

[17] K. Kawasaki and T. Ohta, Kinetic drumhead model of interface I, Prog. Theoret. Phys., 67 (1982), 147-163.

[18] W.Klingenberg, Lectures on closed geodesics, Springer, 1978.

[19] N.V. Krylov and B.L. Rozovskii, Stochastic evolution equations (English translation), J. Soviet Math., 16 (1981), 1233-1277.

[20] H. Kunita, On the decomposition of solution of stochastic differential equations, Stochastic integrals, Lecture Notes in Math.,851, 213-255, Springer, 1981.

[21] H.B. Lawson and S.T. Yau, Compact manifolds of nonpositive curvature, J. Differential Geom., 7 (1972), 211-228.

[22] M. Namiki, I. Ohba and K. Okano, Stochastic quantization of constrained systems —— general theory and nonlinear sigma model, Prog. Theoret. Phys., 72 (1984), 350-365.

[23] G. Parisi and Y. Wu, Perturbation theory without gauge fixing, Scientia Sinica, 24 (1981), 483-496.

[24] W.D. Wick, Monotonicity of the free energy in the stochastic Heisenberg model, Commun. Math. Phys., 83 (1982), 107-122.

Department of Mathematics
Faculty of Science
Nagoya University
Nagoya, 464, Japan

NON-LINEAR FILTERING OF STOCHASTIC PROCESSES AND OPTIMAL SIGNAL TRANSMISSION THROUGH A FEEDBACK CHANNEL

O. A. Glonti

Methods and results of the theory of non-linear filtering of stochastic processes enables to construct optimal schemes for the transmission of Gaussian signals through a feedback channel (both with noise and noiseless) in a class of codings satisfying certain power conditions. A rather extensive study of this problem is given in [1]-[2].

The aim of the present work is to investigate the methodical possibilities of the filtering theory of stochastic processes in the problem of optimal transmission of random variables without any Gaussian restrictions in the case of noiseless feedback. The solution of the problem follows the traditional path due to the restrictions of the class of admissible codings which are defined by power type condition (3) proposed here and not the usual power condition (5). In the class of such codings the minimal error of signal reproduction and the optimal coding functionals have the same form as in the case of Gaussian signals. The absence of Gaussian properties, however, causes additional difficulties (but these difficulties belong to the filtering theory) depending on the establishment of the optimal decoding.

1. Let $(\Omega, \mathcal{F}, (\mathcal{F}_t)_{t\in[0,T]}, P)$ be a complete probability space. Let a useful signal θ be some random variables with a distribution density $p(x)$. Assume that the transmission of θ is carried out according to the following scheme

$$d\xi_t = [A_0(t,\xi) + A_1(t,\xi)\theta]dt + dw_t, \quad \xi_0 = 0 \tag{1}$$

where $W=(w_t, \mathcal{F}_t)$ is a Wiener process independent of θ, $A_0(t,\xi)$ and $A_1(t,\xi)$ are functionals non-anticipating with respect to ξ which define the coding.

The problem is to construct codings A_0^*, A_1^* and a decoding $\hat\theta_t=\hat\theta_t(\xi)$ optimal in the sense of the square criterion

$$\Delta(t) = \inf_{(A_0, A_1, \hat{\theta})} E[\theta - \hat{\theta}_t(\xi)]^2 \tag{2}$$

during the transmission according to the noiseless feedback channel (1) in the class of codings A_0 and A_1, for which equation (1) has a unique strong solution and which satisfies the power type condition

$$(P\text{-a.s.}) \quad E\{[A_0(t,\xi) + A_1(t,\xi)\theta]^2 \mid \mathcal{F}_t^\xi]\} \le p \tag{3}$$

(p is some constant)

and the decoding $\hat{\theta}$ which satisfies the condition

$$E\hat{\theta}_t^2 < \infty . \tag{4}$$

Note that condition (3) is more strict than the usual power condition

$$E[A_0(t,\xi) + A_1(t,\xi)\theta]^2 \le p . \tag{5}$$

We rewrite condition (3) in the following way

$$\begin{aligned} p &\ge E\{[A_0(t,\xi) + A_1(t,\xi)\theta]^2 \mid \mathcal{F}_t^\xi\} \\ &= A_1^2(t,\xi)\gamma_t + [A_0(t,\xi) + A_1(t,\xi)m_t]^2 , \end{aligned} \tag{6}$$

where $m_t = E[\theta | \mathcal{F}_t^\xi]$, $\gamma_t = E[(\theta - m_t)^2 | \mathcal{F}_t^\xi]$, $\mathcal{F}_t^\xi = \sigma\{\xi_s, s \le t\}$.

From (6) (P a.s.) we have

$$A_1^2(t,\xi)\gamma_t \le p . \tag{7}$$

It can be easily seen from (2) that the error of signal reproduction $\Delta(t)$ has the form

$$\Delta(t) = \inf_{(A_0, A_1)} E\gamma_t . \tag{8}$$

We obtain for γ_t the following representation using the results of the filtration theory of stochastic processes (see [1]).

$$\begin{aligned} \gamma_t = \gamma_0 &+ \int_0^t A_1(s,\xi)\, [E(\theta^3 | \mathcal{F}_s^\xi) - 3m_s\gamma_s - m_s^3]\, d\bar{w}_s \\ &- \int_0^t A_1^2(s,\xi)\gamma_s^2\, ds , \end{aligned} \tag{9}$$

where $\bar{W} = (\bar{w}_t, \mathcal{F}_t^\xi)$ is a Wiener (innovation) process.

From (9) we obtain

$$E\gamma_t = E\gamma_0 - \int_0^t E[A_1^2(s,\xi)\gamma_s^2]\,ds$$
$$= E\gamma_0 - \int_0^t E\gamma_s \left(\frac{E[A_1^2(s,\xi)\gamma_s^2}{E\gamma_s} \right) ds \tag{10}$$

(here $E\gamma_t > 0$ for each $t \in [0,T]$), and

$$E\gamma_t = E\gamma_0 \exp\left(- \int_0^t \frac{E[A_1^2(s,\xi)\gamma_s^2]}{E\gamma_s}\,ds\right) \tag{11}$$

Using inequality (7) from (11) we obtain

$$E\gamma_t \geq E\gamma_0 \exp\left(- \int_0^t \frac{E[p\gamma_s]}{E\gamma_s}\,ds\right) = E\gamma_0 e^{-pt} . \tag{12}$$

The equality in (12) is obtained on $A_1^* = (\frac{p}{\gamma_t^*})^{\frac{1}{2}}$ and $A_0^* = -A_1^* m_t^*$ which is evident from (16).

Hence the following theorem is proved.

<u>Theorem</u>. Let the useful signal θ, a random variable with the distribution density $p(x)$ and with $E\theta^4 < \infty$ be transmitted according to scheme (1) with the coding functionals

$$A_1^*(t,\xi^*) = (\frac{p}{\gamma_t^*})^{\frac{1}{2}} , \tag{13}$$

$$A_0^*(t,\xi^*) = -A_1^*(t,\xi) m_t^* \tag{14}$$

and the decoding functional $\hat{\theta}_t^* = m_t^*$ where $m_t^* = E[\theta | \mathcal{F}_t^{\xi^*}]$ and $\gamma_t^* = E[(\theta - m_t^*)^2 | \mathcal{F}_t^{\xi^*}]$ i.e. according to the scheme

$$d\xi_t^* = \left(\frac{p}{\gamma_t^*}\right)^{\frac{1}{2}} (\theta - m_t^*)dt + dw_t . \tag{15}$$

If (15) has a unique strong solution, then in the transmission according to scheme (1) the coding functionals A_1^*, A_0^* of form (13), (14) and the decoding functional m_t^* are optimal in the sense of the square criterion (2) in a class of codings and decodings satisfying conditions (3), (4) and the minimal error of reproduction has the form

$$\Delta(t) = \gamma_0 e^{-pt} \tag{16}$$

where

$$\gamma_0 = E(\theta - E\theta)^2 .$$

Remark 1. The question of evaluating of m_t^* and γ_t^* will be investigated later in II. The establishment of the uniqueness of the strong solution of (15) depends on the specific forms of m_t^* and γ_t^*.

Remark 2. From (15) it follows that

$$d\xi_t^* = d\overline{w}_t^* ,$$

where $\overline{W}^* = (\overline{w}_t^*, \mathcal{F}_t^{\xi^*})$ is a Wiener (innovation) process, i.e. the optimal transmission constructed in the theorem is such that the transmission of the innovation process $\overline{W}^*$ takes place.

Remark 3. The optimal transmission of θ in the sense of square criterion (2) defined by (15), is optimal in the sense of the maximum of mutual information as well in the class of transmission (1) satisfying power condition (5) which follows from the results of chapter 16 in [1].

Remark 4. The error of reproduction (16) also can be obtained on the codings

$$\hat{A}_1(t) = \left[p \frac{E\gamma_t}{E\gamma_t^2}\right]^{\frac{1}{2}} ,$$

$$\hat{A}_0(t,\xi) = -\left[p \frac{E\gamma_t}{E\gamma_t^2}\right]^{\frac{1}{2}} m_t + \left(p - p \frac{E\gamma_t}{E\gamma_t^2} \gamma_t\right)^{\frac{1}{2}} , \tag{17}$$

For such codings not the condition of power type (3) is fulfilled, but power condition (5), namely, on $(\hat{A}_0, \hat{A}_1)$ the equality on (5) takes place. In the case of the Gaussian random variable θ, since γ_t is non-random, scheme (1) with of form (17) becomes optimal with

$$A_1^*(t) = \left(\frac{p}{\gamma_t^*}\right)^{\frac{1}{2}} , \qquad A_0^* = -A_1^* m_t^*$$

but in a class of admissible codings satisfying condition (5). In a non-Gaussian case it is not clear whether the constructed codings (17) are optimal in the sense of criterion (2), since we do not know whether the error of reproduction has the form of (16) in the class of admissible codings satisfying (5).

II. Now we investigate how to find $m_t^*=E[\theta|\mathcal{F}_t^{\xi^*}]$ and $\gamma_t^*=E[(\theta-m_t^*)^2|\mathcal{F}_t^{\xi^*}]$.

Let $p(x,t)$ be a density of the conditional distribution $P[\theta\le x|\mathcal{F}_t^{\xi}]$ and $\rho(x,t)$ be the corresponding unnormalized density, i.e.

$$p(x,t) = \frac{\rho(x,t)}{\int\rho(x,t)dx}, \quad x \in R^1 .$$

Then

$$m_t^* = \frac{\int x\rho(x,t)dx}{\int\rho(x,t)dx},$$
$$\gamma_t^* = \frac{\int x^2\rho(x,t)dx}{\int\rho(x,t)dx} - \left(\frac{\int x\rho(x,t)dx}{\int\rho(x,t)dx}\right)^2 . \tag{18}$$

Hence, the problem is to find $\rho(x,t)$. The equation for the unnormalized density $\rho(x,t)$ in our case, i.e. the case of the partially observable process

$$d\theta_t \equiv 0,$$
$$d\xi_t = [A_0(t,\xi) + A_1(t,\xi)\theta]dt + dw_t$$

apparently, has the following form (see [3], [4])

$$d\rho(x,t) = [A_0(t,\xi) + A_1(t,\xi)x]\ \rho(x,t)\ d\xi_t . \tag{19}$$

Introduce the notation

$$u_t(x) = \int_0^t [A_0(s,\xi) + A_1(s,\xi)x]\ d\xi_s .$$

For each $x\in R^1$ the equation

$$d\rho(x,t) = \rho(x,t)\ du_t(x) \tag{20}$$

is Dolean-Dad's equation, a unique solution of which is the following

$$\rho(x,t) = \rho(x,0)\ \exp\ (u_t(x)-\tfrac{1}{2}\langle u(x)\rangle_t)$$

or

$$\rho(x,t)$$
$$= p(x)\exp\left(\int_0^t [A_0(s,\xi)+A_1(s,\xi)x]d\xi_s - \frac{1}{2}\int_0^t [A_0(s,\xi)+A_1(s,\xi)x]^2 ds\right) \tag{21}$$

In (18) substituting $\rho(x,t)$ from (21) we obtain formulas to establish the filter m_t^* and the error of tracing back γ_t^*.

Note that

$$\rho^*(x,t) = p(x)\exp\left(\int_0^t \sqrt{p/\gamma_s^*}(x-m_s^*)d\xi_s^* - \frac{1}{2}\int_0^t \frac{p}{\gamma_s^*}(x-m_s^*)^2 ds\right) . \tag{22}$$

III. The schemes of the transmission defined by the theorem has the form

$$d\xi_t^* = \left(\frac{p}{\gamma_t^*}\right)^{\frac{1}{2}} (\theta - m_t^*)dt + dw_t . \tag{15}$$

If (15) has a unique strong solution, then the transmission realized by (15) is optimal both in the sense of the square criterion in a class of transmission for which the condition of power type (3) is fulfilled and in the sense of the maximum of mutual information in a wider class of transmission, for which the usual power condition (5) is fulfilled. In the Gaussian case when θ is a Gaussian random variable $N(m,\gamma)$, equation (15) has a unique strong solution. Then the class of transmissions for which the condition of power type (3) and power condition (5) are fulfilled coincide and the scheme defined by (15) is optimal, indeed. For a non-Gaussian signal θ with the density $p(x)$ the question of existence and uniqueness of the strong solution of (15) is not an easy problem. What can we do if we fail to determine whether (15) has a unique strong solution? Naturally, ε-optimal schemes are to be determined. We propose such a scheme below.

Let

$$t_n(t) = \frac{[nt]}{n} = \begin{cases} 0, & 0 \le t < \frac{1}{n}, \\ \frac{1}{n}, & \frac{1}{n} \le t < \frac{2}{n}, \\ \frac{2}{n}, & \frac{2}{n} \le t < \frac{3}{n}, \\ \dots\dots\dots\dots & . \end{cases}$$

For $t \in [0,\frac{1}{n})$ put

$$A_1^{(n)}(t,\xi^{(n)}) = \sqrt{\frac{p}{\gamma_0}} , \qquad A_0^{(n)}(t,\xi^{(n)}) = -\sqrt{\frac{p}{\gamma_0}}\, m_0 ,$$

where $m_0=E\theta$, $\gamma_0=E(\theta-E\theta)^2$.

Then the transmission scheme will have the following form

$$\xi_t^{(n)} = \int_0^t [A_0^{(n)}(s,\xi^{(n)}) + A_1^{(n)}(s,\xi^{(n)})\theta]ds + w_t$$

or

$$\xi_t^{(n)} = \sqrt{\frac{p}{\gamma_0}}(\theta-m_0)t + w_t .$$

Let

$$\xi_{\frac{1}{n}}^{(n)} = \lim_{t\uparrow\frac{1}{n}} \xi_t^{(n)}, \qquad m_{\frac{1}{n}}^{(n)} = \lim_{t\uparrow\frac{1}{n}} m_t^{(n)}, \qquad \gamma_{\frac{1}{n}}^{(n)} = \lim_{t\uparrow\frac{1}{n}} \gamma_t^{(n)},$$

where $m_t^{(n)}=E[\theta|\mathcal{F}_t^{\xi^{(n)}}]$ and $\gamma_t^{(n)}=E[(\theta-m_t^{(n)})^2|\mathcal{F}_t^{\xi^{(n)}}]$ is determined by (18), (21).

Similarly, for $t\in[\frac{1}{n},\frac{2}{n})$: $A_1^{(n)}(t,\xi^{(n)})=\sqrt{\frac{p}{\gamma_t^{(n)}}}$, $A_0^{(n)}(t,\xi^{(n)})=$ $-\sqrt{\frac{p}{\gamma_{\frac{1}{n}}^{(n)}}}\, m_{\frac{1}{n}}^{(n)}$,

$$\xi_t^{(n)} = \xi_{\frac{1}{n}}^{(n)} + \int_{\frac{1}{n}}^t [A_0^{(n)}(s,\xi^{(n)}) + A_1^{(n)}(s,\xi^{(n)})\theta]ds + w_t - w_{\frac{1}{n}}$$

or

$$\xi_t^{(n)} = \sqrt{\frac{p}{\gamma_0}}(\theta-m_0)\frac{1}{n} + \sqrt{\frac{p}{\gamma_{\frac{1}{n}}^{(n)}}}(\theta-m_{\frac{1}{n}}^{(n)})(t-\frac{1}{n}) + w_t ,$$

$$\xi_{\frac{2}{n}}^{(n)} = \lim_{t\uparrow\frac{2}{n}} \xi_t^{(n)}, \quad m_{\frac{2}{n}}^{(n)} = \lim_{t\uparrow\frac{2}{n}} m_t^{(n)}, \quad \gamma_{\frac{2}{n}}^{(n)} = \lim_{t\uparrow\frac{2}{n}} \gamma_t^{(n)} .$$

It can be easily seen that

$$\begin{aligned} &E[A_0^{(n)}(t,\xi) + A_1^{(n)}(t,\xi^{(n)})\theta]^2 \\ &= E\{E[[A_0^{(n)}(t,\xi^{(n)}) + A_1^{(n)}(t,\xi^{(n)})\theta]^2|\mathcal{F}_{t_n(t)}^{\xi^{(n)}}]\} \\ &= E\{[A_0^{(n)}(t,\xi^{(n)})+A_1^{(n)}(t,\xi^{(n)})m_{t_n(t)}^{(n)}]^2+A_1^{(n)}(t,\xi^{(n)})^2\gamma_{t_n(t)}^{(n)}\} \\ &= p . \end{aligned}$$

i.e. the constructed coding functionals belong to the class which is defined by power condition (5). Besides, they have the following property

$$E[[A_0^{(n)}(t,\xi^{(n)}) + A_1^{(n)}(t,\xi^{(n)})\theta]^2|\mathcal{F}_t^{\xi}]$$

$$= [A_0^{(n)}(t,\xi^{(n)}) + A_1^{(n)}(t,\xi^{(n)})m_t^{(n)}]^2 + A_1^{(n)}(t,\xi^{(n)})^2\gamma_t^{(n)}$$

$$= \left[\sqrt{\frac{p}{\gamma_{t_n(t)}^{(n)}}}(m_t^{(n)}-m_{t_n(t)}^{(n)})\right]^2 + p\,\frac{\gamma_t^{(n)}}{\gamma_{t_n(t)}^{(n)}} \xrightarrow[n\to\infty]{} p \qquad \text{(P-a.s.)}$$

by virtue of the continuity of m_t and γ_t and the form of their representation.

Since

$$E\gamma_t^{(n)} = \gamma_0\exp\left(-\int_0^t \frac{E[A_1^{(n)}(s,\xi^{(n)})^2\gamma_s^{(n)^2}]}{E\gamma_s^{(n)}}\,ds\right)$$

$$= \gamma_0\exp\left(-\int_0^t \frac{pE\left(\dfrac{\gamma_s^{(n)^2}}{\gamma_{t_n(s)}^{(n)}}\right)}{E\gamma_s^{(n)}}\,ds\right) \xrightarrow[n\to\infty]{} \gamma_0e^{-pt}$$

(again by virtue of the continuity of γ_t (see (9))) the ε-optimality of the constructed transmission scheme virtually holds.

References

[1] Liptzer R. Sh., Shiryayev A. N. Statistics of random processes. M.: "Nauka", 1974, 696p. or Springer-Verlag, New York, 1977, I,II.

[2] Glonti O. A. Investigations in the theory of conditionally Gaussian processes. "Metsniereba", Tb.: 1985, 196p.

[3] Krylov N. V., Rozovsky B. L. On the conditional distribution of diffusion processes. Bul. of the Acad. of Sci. of the USSR, Ser. Math. 42, No2, 1978, 356-378.

[4] Purtukhia O. G. On conditional distributions of diffusion processes with unbounded coefficients. 1st World Congress of Bernoulli Society. Abstracts of Communications. 1986.

A. M. Razmadze Mathematical Institute
of the Academy of Sciences
of the Georgian SSR, Tbilisi

ON BESSEL POTENTIALS IN LINEAR SPACES

Z. G. Gorgadze

Fractional powers of a Laplace operator and its versions, being in a sense a formal machinery, find wide applications when solving analytic problems in Euclidean spaces. They take an essential part in a number of probabilistic problems as well, in Markov processes in particular. Potentials arise when converting them. The Bessel potentials through their nice analytic properties turn out to be more convenient and so, preferable.

A detailed expound of the Bessel potential theory in finite dimensional spaces is contained by [2] and [1]. For the analytic purposes they are introduced and studied in [7].

In this paper a possible approach to this circle of problems in Banach spaces is suggested. We do not resort to the concept of the abstract Wiener space, usually associated with the infinite dimensional potential theory (c.f. [6]), giving preference to the direct usage of Gaussian measures and related bounded operators. In these and other topics of the theory of measure in Banach spaces we are guided by the book [9]. Gaussian measures play a crucial part: one can define the Bessel kernel as a mixture of those. We would not like to ignore the evident opportunity to define a potential (convolution "smoothing") not only for a point function but also for a set function. Thus, some problems of differentiation of such functions in Banach spaces are discussed first of all. In these topics we refer to the work [3] on the whole, though modifying the approach developed there. In general measure-theoretical facts we follow [5], IV, and also [4] for vector measures.

1. Let X and Y be the real Banach spaces, $\mathcal{L}(X,Y)$ denotes the space of all bounded linear mappings from X into Y. Let further $\mathcal{A}$ be a σ-algebra of subsets of X. We call a measure a (strongly) countably additive set function $\mu:\mathcal{A}\to Y$. Note that for a measure μ the set $\{\mu(A):A\in\mathcal{A}\}$ is bounded in Y.

Let us assume that the σ-algebra $\mathcal{A}$ has the following property: if $A\in\mathcal{A}$ then $A+x\in\mathcal{A}$ for all $x\in X$. Accept the notation: $\mu_x(A)=\mu(A+x)$, $\nu_t(A)=t^{-1}(\mu_{tx}(A)-\mu(A))$, $t\in\mathbb{R}-\{0\}=\mathbb{R}_0$ and $\mathbb{R}$ stands for reals.

A measure $\mu:\mathcal{A}\to Y$ is said to be differentiable on a set $A\in\mathcal{A}$ in the direction $x\in X$ if the limit $\lim_{t\to 0}\nu_t(A;x)=\nu(A;x)$ exists in the weak sense. If so, the scalar set function $\langle\nu(\cdot;x),y^*\rangle$ is countably additive for every $y^*\in Y^*$ as a limit on each $A\in\mathcal{A}$ of such functions and since the countably additivity in weak and strong senses coincide, we obtain that $\nu(\cdot;x)$ is countably additive on $\mathcal{A}$. Here and below X^* denotes the dual space.

We shall be interested in the tight measures defined on the Borel σ-algebra $\mathcal{B}=\mathcal{B}(X)$ of the space X, which are differentiable in all directions $x\in RX$, where $R\in\mathcal{L}(X^*,X)$. In this case a measure μ shall be called differentiable, assuming additionally that the correspondence $x^*\longmapsto\nu(A;Rx^*)$ is linear and continuous on X^*. Namely, $\nu(A;Rx^*)=\mu'(A)x^*$, where $\mu':\mathcal{A}\to\mathcal{L}(X^*,Y)$. The set function μ' is said to be the derivative of the measure μ. In the other words, the derivative μ' of the measure μ is countably additive in the weak and strong operator topologies of $\mathcal{L}(X^*,Y)$ and if in addition one uses the Dieudonne-Grothendieck theorem, the finite additivity of μ' in the uniform operator topology of $\mathcal{L}(X^*,Y)$ and the boundedness of the set of its values could be obtained.

As a consequence we get that the derivative $\mu':\mathcal{A}\to X$ of a scalar measure $\mu:\mathcal{A}\to\mathbb{R}$ (in case of differentiability, of course) is countably additive and the second derivative $\mu'':\mathcal{A}\to\mathcal{L}(X^*,X)$ in addition to the properties listed above is countably additive in weak* topology of the Banach space $\mathcal{L}(X^*,X)$ if X^* has the metric approximation property.

In the present work we are going to deal with the twice differentiable scalar measures, whose second derivatives take their values from the space $N(X^*,X)$ of nuclear operators. One can see that in this situation μ'' is countably additive set function. So, for such a measure the scalar set function $\mathrm{tr}(T\mu''T^*)$ is countably additive for arbitrary $T\in\mathcal{L}(X,H)$, H a Hilbert space. It seems to be natural to call the correspondence $\mu\longmapsto\mathrm{tr}(T\mu''T^*)$ a Laplace operator.

As usually we denote by $ca(X;\mathcal{A})$ the Banach space of measures $\mu:\mathcal{A}\to\mathbb{R}$, endowed with the variational norm $\|\mu\|=\mathrm{var}(\mu;X)$.

If a measure μ is differentiable in the direction $x_0\in X$ on each $A\in\mathcal{A}$, then it follows from the definition that the net $\{\langle\nu_t(A;x_0),y^*\rangle:t\in\mathbb{R}_0\}$ is bounded around the origin for every $y^*\in Y^*$ and each $A\in\mathcal{A}$. The Nikodym boundedness theorem gives, that the net $\{\langle\nu_t(\cdot;x_0),y^*\rangle:t\in\mathbb{R}_0\}$ is bounded around the origin for every $y^*\in Y^*$ and consequently, $\{\nu_t(\cdot;x_0):t\in\mathbb{R}_0\}$ is bounded in Y. Since $\langle\nu_t(A;x_0),y^*\rangle$ converges as $t\to 0$ for every $A\in\mathcal{A}$, it follows that the latter converges to $\langle \nu(\cdot;x_0),y^*\rangle$ weakly in the space $ca(X;\mathcal{A})$. For a Borel measure μ this means that for arbitrary bounded continuous function $f:X\to\mathbb{R}$ and for every $y^*\in Y^*$ one has the convergence of the integrals:

$$\lim_{t\to 0}\int_X f(x)\langle \nu_t(dx;x_0),y^*\rangle = \int_X f(x)\langle \nu(dx;x_0),y^*\rangle.$$

The latter, in its turn, implies that if a Borel measure μ is differentiable on RX^* and the derivative is countably additive, then for arbitrary $y^*\in Y^*$ and all $x^*\in X^*$ we have:

$$\lim_{t\to 0}\int_X f(x)\langle \nu_t(dx;Rx^*),y^*\rangle = \langle U_f x^*,y^*\rangle, \tag{1}$$

where

$$U_f = \int_X f(x)\mu'(dx)$$

and $U_f\in\mathcal{L}(X^*,Y)$.

2. Denote by $\mathcal{L}_s^+(X^*,X)$ the class of symmetric positive definite operators $R\in\mathcal{L}(X^*,X)$ i.e. the mappings R with the following properties: $\langle Rx_1^*,x_2^*\rangle=\langle Rx_2^*,x_1^*\rangle$, $\langle Rx^*,x^*\rangle\geq 0$ for all x^*,x_1^*,x_2^* in X^*. Denote moreover $\mathcal{R}(X)$ the class of covariance operators of tight Gaussian measures in X. For a Hilbert space H the equality $\mathcal{N}(H)\cap\mathcal{L}_s^+(H)=\mathcal{R}(H)$ is well known. The strong integrability property of the Gaussian measure γ with zero mean and covariance operator $R\in\mathcal{R}(X)$ implies that the expressions

$$m(A) = \int_A x\gamma(dx)\ ,$$

$$M(A) = \int_A x\otimes x\gamma(dx)\ , \qquad A\in\mathcal{B}$$

are well-defined. Here $x_1\otimes x_2\in\mathcal{L}(X^*,X)$ is defined by the equality $(x_1\otimes x_2)x^*=\langle x_1,x^*\rangle x_2$, $x^*\in X^*$, $x_1,x_2\in X$, and obviously $x\otimes x$ is in $\mathcal{L}_s^+(X^*,X)$. It is also clear that $M(A)\in\mathcal{L}_s^+(X^*,X)$, $A\in\mathcal{B}$.

Easy to get the estimations:

$$|\langle m(A),x^*\rangle| \leq \gamma^{1/2}(A)\langle Rx^*,x^*\rangle^{1/2}\ , \tag{2}$$

$$|\langle M(A)x_1^*,x_2^*\rangle| \leq \langle Rx_1^*,x_1^*\rangle^{1/2}\langle Rx_2^*,x_2^*\rangle^{1/2}\ . \tag{3}$$

Particularly,

$$\langle M(A)x^*, x^*\rangle \le \langle Rx^*, x^*\rangle , \qquad (4)$$

and consequently $R-M(A)\in\mathcal{L}_s^+(X^*,X)$ for every $A\in\mathcal{B}$. Note that for $R\in\mathcal{R}(X)$ (4) implies $M(A)\in\mathcal{R}(X)$ for every $A\in\mathcal{B}$. We recall that for $R\in\mathcal{L}_s^+(X^*,X)$ there exists a Hilbert space H and an operator $U\in\mathcal{L}(X^*,H)$ such that $R=U^*U$ and UX^* is dense in H. The inequalities (2) and (3) immediately imply that $m(A)\in U^*H$ and $M(A)x^*\in U^*H$ for all $x^*\in X^*$, $A\in\mathcal{B}$. The countable additivity of the set function m is evident. Moreover, as we have that $\mathcal{R}(X)$ is contained by the intersection $N(X^*,X)\cap\mathcal{L}_s^+(X^*,X)$ (see [9], p.144), the set function $M:\mathcal{B}\to\mathcal{N}(X^*,X)$ is countably additive as well.

Now, one can see that the Gaussian measure γ with the covariance operator $R\in\mathcal{R}(X)$ is differentiable on RX^* and for the first two derivatives one has:

$$\gamma'(A) = -m(A) ,$$

$$\gamma''(A) = M(A) - \gamma(A)R , \qquad A \in \mathcal{B} .$$

The first follows from the fact ([8]), that the measure γ_{tx} is absolutely continuous with respect to γ and the corresponding Radon-Nikodym derivative is of the form:

$$\frac{d\gamma_{tx}}{d\gamma}(\cdot) = \exp\{t\langle\cdot,x^*\rangle - \frac{t^2}{2}\langle Rx^*,x^*\rangle\} ,$$

where $x=Rx^*$.

The second could be proved in the same way.

3. Let us denote $\overline{Y}$ the complexification of a Banach space Y. Given a measure $\mu:\mathcal{B}\to Y$. The Fourier transform of μ is called the functional $\hat{\mu}:X^*\to\overline{Y}$ defined as follows:

$$\hat{\mu}(x^*) = \int_X \exp\{-i\langle x,x^*\rangle\}\mu(dx), \qquad x^*\in X^*.$$

If we take into consideration that Y^* (or its subspace) can be regarded as the space of (real) linear forms on $\overline{Y}$ and the identity

$$\langle\bar{y},\bar{y}^*\rangle = \langle\bar{y},y^*\rangle - i\langle i\bar{y},y^*\rangle , \qquad \bar{y}\in\overline{Y},$$

defines an isomorphism of $\overline{Y}$ and $\overline{Y}^*$, we get:

$$\langle\hat{\mu}(x^*),y^*\rangle = \langle\mu,y^*\rangle^{\wedge}(x^*), \qquad x^*\in X^*. \qquad (5)$$

If a measure $\mu:\mathcal{B}\to Y$ is differentiable in the direction $Rx_0^*\in X$ on every $A\in\mathcal{B}$, taking $f(x)=\exp(-i\langle x,x^*\rangle)$ in (1) and using (5), we obtain:

$$\langle\mu'(\cdot)x_0^*,y^*\rangle\hat{}(x^*) = i\langle Rx_0^*,x^*\rangle\langle\hat{\mu}(x^*),\bar{y}^*\rangle, \qquad x^*\in X^*,$$

which for the first two derivatives in the direction Rx_0^* of a measure $\mu:\mathcal{B}\to\mathbb{R}$ (assuming differentiability) gives:

$$\langle\mu'(\cdot),x_0^*\rangle\hat{}(x^*) = i\langle Rx_0^*,x^*\rangle\hat{\mu}(x^*) ,$$

$$\langle\mu''(\cdot)x_0^*,x_1^*\rangle\hat{}(x^*) = -\langle Rx_0^*,x^*\rangle\langle Rx_1^*,x^*\rangle\hat{\mu}(x^*)$$

$$= -\langle Rx_0^*,(x^*\otimes x^*)Rx_1^*\rangle\hat{\mu}(x^*), \qquad x^*,x_1^*\in X^*.$$

In other words, for a differentiable in the directions from RX^* measure μ, assuming the countable additivity of the derivatives we obtain

$$\hat{\mu}'(x^*) = iRx^*\hat{\mu}(x^*) , \tag{6}$$

$$\hat{\mu}''(x^*) = -(Rx^*\otimes Rx^*)\hat{\mu}(x^*) . \tag{7}$$

Let now $T\in\mathcal{L}(X,H)$ and let H be a Hilbert space. If a measure $\mu:\mathcal{B}\to\mathbb{R}$ is twice differentiable in the directions from RX^* and the second derivative μ'' is countably additive, then as we have mentioned above, the transformation (Laplace operator)

$$\Delta_T\mu = \operatorname{tr}(T\mu''T^*)$$

is well defined. From (5) and (7) we obtain:

$$(\Delta_T\mu)\hat{}(x^*) = -\langle RT^*TRx^*,x^*\rangle\hat{\mu}(x^*), \qquad x^*\in X^*. \tag{8}$$

Note that if R is represented in the form $R=U^*U$, $U\in\mathcal{L}(X^*,H)$, then for each $x^*\in X^*$ we shall have

$$\begin{aligned}\langle RT^*TRx^*,x^*\rangle &= \|TRx^*\|_H^2 = \|TU^*Ux^*\|_H^2 \\ &\le \|TU^*\|^2\|Ux^*\|_H^2 = \|TU^*\|\langle Rx^*,x^*\rangle.\end{aligned} \tag{9}$$

This, particularly, implies that if $R\in\mathcal{R}(X)$ then for arbitrary $T\in\mathcal{L}(X,H)$ we have $RT^*TR\in\mathcal{R}(X)$.

4. Take again $R\in\mathcal{R}(X)$ and fix the operator $T\in\mathcal{L}(X,H)$. Here H is the Hilbert space through which R is factorized. Denote

$K=RT^*TR$; $K\in\mathcal{R}(X)$ as we have noticed. In what follows γ_K is the Gaussian measure with zero mean and covariance operator K. g_p stands for the function

$$g_p(t) = 1/\Gamma(p)\exp(-t)t^{p-1}, \qquad p>0, \quad t>0.$$

We intend to construct a measure on $\mathcal{B}$ which has the Fourier transform $(1+\frac{1}{2}\langle Kx^*,x^*\rangle)^{-p}$, $p>0$, $x^*\in X^*$. For this we put

$$G_{p,K}(A) = \int_0^\infty \gamma_{tK}(A)g_p(t)dt , \qquad A\in\mathcal{B}.$$

It is clear that $G_{p,K}$ is a probability measure ("elliptically contoured") and for its Fourier transform we have:

$$\begin{aligned}\hat{G}_{p,K}(x^*) &= \int_0^\infty \left\{ \int_X \exp(-i\langle x,x^*\rangle)\gamma_{tK}(dx)\right\} g_p(t)dt \\ &= \int_0^\infty \exp\ (-\tfrac{t}{2}\langle Kx^*,x^*\rangle)g_p(t)dt = (1+\tfrac{1}{2}\langle Kx^*,x^*\rangle)^{-p},\end{aligned} \tag{10}$$

where the expression of the Fourier transform of the Gaussian measure

$$\hat{\gamma}_K(x^*) = \exp(-\tfrac{1}{2}\langle Kx^*,x^*\rangle) , \qquad x^*\in X^*,$$

is used. Put for convenience $G_{0,K}=\delta_0$ (the unit measure concentrated at the origin). It is easy to see that the family $(G_{p,K}: p\geq 0)$ has a semigroup property $G_{p+q,K}=G_{p,K}*G_{q,K}$, where $*$ stands for the convolution.

Let us now define a Bessel potential of a measure $\nu\in ca(X;\mathcal{B})$ as the convolution

$$\mathfrak{I}_{p,K}(\nu) = G_{p,K}*\nu .$$

The weak convergence criteria in ca together with the definition gives that $G_{p,K}$ converges weakly to δ_0 as $p\to 0$. Moreover,

$$\sum_{1\leq j\leq n} |(G_{p,K}*\nu)(A_j)| \leq \int_X \sum_{1\leq j\leq n} |\nu(A_j-x)|G_{p,K}(dx) \leq var(\nu;X) ,$$

i.e. $var(G_{p,K}*\nu;X)\leq var(\nu;X)$. When $p\to 0$, for a bounded continuous $f:X\to\mathbb{R}$ the integral

$$\int_X f(x)(G_{p,K}*\nu)(dx) = \int_X \left\{ \int_X f(x+y)\nu_x(dy)\right\} G_{p,K}(dx) = \int_X h(x)G_{p,K}(dx)$$

converges to

$$h(0) = \int_X f(y)\nu(dy) .$$

This is the weak convergence of $G_{p,K}*\nu$ to ν. Of course, continuity arguments of the convolution could give us the other way to see this.

If we use the differentiability property of Gaussian measure together with the strong integrability, we can show that for $\nu \in ca(x;\mathcal{B})$ the measure $G_{p,K}*\nu$ is twice differentiable, the set function $(G_{p,K}*\nu)'':\mathcal{B}\to\mathcal{N}(X^*,X)$ is countably additive and from (7) and (8) one has:

$$(\Delta_T(G_{p,K}*\nu))^{\wedge}(x^*) = - \frac{\langle Kx^*,x^*\rangle}{(1+\frac{1}{2}\langle Kx^*,x^*\rangle)^p}\hat{\nu}(x^*), \qquad x^* \in X^*.$$

By passing to Fourier transforms it is easy to see that the Bessel potential $\mu = G_{1,K}*\nu$ of a measure $\nu \in ca(X:\mathcal{B})$ gives a solution of the equation

$$\mu - \frac{1}{2}\,\Delta_T\mu = \nu .$$

Finally, we define the fractional power of the operator $I-\frac{1}{2}\Delta_T$ by the equality

$$(I - \frac{1}{2}\Delta_T)^p\,\mathcal{T}_{p,K}(\nu) = \nu , \qquad \nu \in ca(X;\mathcal{B}),$$

i.e. as an "inverse" of the transformation $\mathcal{T}_{p,K}$.

References

[1] Adams R., Aronszajn N., Smith K. T. Theory of Bessel potentials, II, Ann. Inst. Fourier, 1967, 17, 1-135.

[2] Aronszajn N., Smith K. T. Theory of Bessel potentials, I.-ibid., 1961, 11, 385-475.

[3] Авербух В. И., Смолянов О. Г., Фомин С. В. Обобщенные функции и дифференциальные уравнеия в линеиных пространствах. I. Дифференцируемые меры, Труды Московского математического общества, 1971, т.24, 133-174. II. Дифференциальные операторы и их преобразования Фурье, ibid., 1972, т.27, 247-262.

[4] Diestel J., Uhl J. J. Vector measures, AMS, Providence, R.I., 1977.

[5] Dunford N., Schwartz J. T. Linear operators, General theory, Interscience Publishers, New-York, London, 1958.

[6] Kuo H.-H. Gaussian measures in Banach spaces, Springer-Verlag, Berlin, Heidelberg, New-York, 1975.

[7] Stein E. M. Singular integrals and differentiability properties of functions, Princeton univ. Press, Princeton, N.J., 1970.

[8] Тариеладзе В. И. Эквивалентность гауссовских мер в банаховых пространствах, Сообщ. АН ГССР., 1974, 73:3, 529-532.

[9] Вахания Н. Н., Тариеладзе В. И., Чобанян С. А. Вероятностные распределения в банаховых пространствах, М., "Наука" 1985.

Tbilisi State University,
Tbilisi, USSR

A TIME CHANGE RELATING CONTINUOUS SEMI-MARKOV AND MARKOV PROCESSES

B. P. Harlamov

How can one transform a random process with Cantor type paths into a random process with the same sequence of states and without any intervals of constancy? We mean the Cantor type function is a continuous and non-constant function on real line such that the set of points on the real axis without all intervals of constancy of this function is a set of zero Lebesgue measure. This paper answers this question and other ones about continuous semi-Markov processes. The problem which stays open since Lévy is to represent a semi-Markov process as a Markov one with time change. A rather simple decision can be found in the case of stepped semi-Markov processes [7,8]. Investigation of this question for continuous semi-Markov processes has begun in the paper [3] where there was an analytical approach with a lot of assumptions. In this paper the approach to the problem is quite different. One can call it probabilistic. The aim of this paper is to prove the theorem about the Markov representation with more simple assumptions.

Firstly the paper contains a list of preliminaries such as definitions and properties of some spaces, functions, measures and their relations (1), outleading sequences of various ordering types (2), correct exit sets and related equivalent random times (3), sequences of states and conditional distributions of time pass processes (4). Some results about Lévy decomposition of the time pass process and its interpretation are proposed and the class of semi-Markov processes without fixed intervals of constancy is defined (5). Some information about continuity of families of measures (6), and about the space of individual time change functions (7) is communicated.

The main result is formulated in §2, Theorem 1. It treats the Markov representation of the semi-Markov process. Conditions of the theorem look more simple than the similar ones in the paper [3]. But this theorem is not stronger than the previous result since there is a condition about absence of fixed intervals of constancy. I don't know whether one can prove the Markov representation without this condition. I don't know either whether the condition about fixed intervals of constancy can be formulated without terms of conditional distributions of time pass processes, for example, in terms of unconditional probabilities.

From the practical point of view the class of continuous semi-Markov processes is important because it allows to stand and decide statistical problems in terms of transition functions. In the stationary case one can test semi-Markov hypothesis and estimate parameters of the process from a single path. Besides the continuous semi-Markov process is a natural model for some processes of the real world. One of the processes is a chromatographical process which combines diffusion and absorption properties. Just here such an exotic feature of the process as intervals of constancy of its paths has a real and important sense.

§1 Notations and Preliminaries

1. Semi-Markov Families of Probability Measures. Let X be a complete separable locally compact metric space, $\mathcal{D}$ the set of all functions $\xi: R_+ \to X$ which are right continuous on R_+ and have limit from the left in each point $t>0$. A random process with paths in $\mathcal{D}$ will be regarded as a measurable family $(P_x)_{x\in X}$ of probability measures on the σ-algebra $\mathcal{F}=\sigma(\pi_t, t\in R_+)$ where $\pi_t\xi=\xi(t)$ and $P_x(\pi_0\xi=x)=1$. We shall denote $\mu(f,A)=\int_A f d\mu$ and $\mu(f)=\mu(f,\Omega)$, where μ is a (sub)probability measure on a measurable space $(\Omega,\mathcal{A})$, $A\in\mathcal{A}$ and f is a measurable real function on the space. Put $(\forall t\in R_+)\mathcal{F}_t=\sigma(\pi_s, s\le t)$. Let $\mathcal{T}$ be the class of all mappings $\tau:\mathcal{D}\to\overline{R}_+=R_+\cup\{\infty\}$ and set $\mathcal{T}_+=\{\tau\in\mathcal{T}: (\forall t\in R_+)\{\tau<t\}\in\mathcal{F}_t\}$, $\mathcal{T}_0=\{\tau\in\mathcal{T}: (\forall t\in R_+)\{\tau\le t\}\in\mathcal{F}_t\}$. Then $(\forall\tau\in\mathcal{T}_0)\mathcal{F}_\tau =\{B\in\mathcal{F}: (\forall t\in R_+)B\cap\{\tau\le t\}\in\mathcal{F}_t\}$ is the σ-algebra of all events anterior to the stopping time τ. Put $(\forall\tau\in\mathcal{T}_+)$ $\mathcal{F}_\tau=\sigma(\tau,\alpha_\tau)$ where $(\forall t\in\overline{R}_+)\alpha_t:\mathcal{D}\to\mathcal{D}$ is a stopping operation: $(\forall s\in R_+)(\alpha_t\xi)(s)=\xi(s\wedge t)$ and $(\forall\tau\in\mathcal{T})\alpha_\tau\xi=\alpha_{\tau(\xi)}\xi$. We have $\mathcal{T}_0\subset\mathcal{T}_+$ and $(\forall\tau\in\mathcal{T}_0)\mathcal{F}_\tau=\sigma(\alpha_\tau)=\alpha_\tau^{-1}\mathcal{F}$ [2, p.22]. Let $\mathfrak{A}(\widetilde{\mathfrak{A}})$ be the class of all open (closed) subsets of the set X. We have $(\forall\Delta\in\mathfrak{A})\tau_\Delta\in\mathcal{T}_0$ and $(\forall\Delta\in\widetilde{\mathfrak{A}})\tau_\Delta\in\mathcal{T}_+$ [1, p.194]. $(\tau_\Delta(\xi)=\inf(t\ge 0;\ \xi(t)\notin\Delta))$ Let $(\forall t\in R_+)\theta_t:\mathcal{D}\to\mathcal{D}$ be a shift operation: $(\forall s\in R_+)(\theta_t\xi)(s)=\xi(t+s)$, and $(\forall\tau\in\mathcal{T})(\forall\xi\in\{\tau<\infty\})\theta_t\xi=\theta_{\tau(\xi)}\xi$. For all $\tau_1,\tau_2\in\mathcal{T}$ let $\tau_1\dot{+}\tau_2=\tau_1+\tau_2\circ\theta_{\tau_1}$ when $\tau_1<\infty$, and $\tau_1\dot{+}\tau_2=\infty$ when $\tau_1=\infty$. We denote $\tau_{(\Delta_1,\ldots,\Delta_n)}= \tau_{\Delta_1}\dot{+}\ldots\dot{+}\tau_{\Delta_n}$ where $n\in N=\{1,2,\ldots\}$, $\Delta_i\subset X$. If every Δ_i is open then

$\tau_{(\Delta_1,\ldots,\Delta_n)} \in \mathcal{F}_0$ and has the following property:

$$(\forall t\in R_+)\ \{t<\tau_{(\Delta_1,\ldots,\Delta_n)}\} = \{\tau_{(\Delta_1,\ldots,\Delta_n)}\circ\alpha_t=\infty\} .$$

A stopping time $\tau\in\mathcal{F}_+$ is called a regenerative time of the process (P_x) $(\tau\in RT)$ if $(\forall x\in X)$ P_x-a.s. on the set $\{\tau<\infty\}$

$$(\forall B\in\mathcal{F})\ P_x(\theta_\tau^{-1}B|\mathcal{F}_\tau) = P_{\pi_\tau}(B) ,$$

where $\pi_\tau\xi=\pi_{\tau(\xi)}\xi$. The family (P_x) is called a semi-Markov (SM) process if $(\forall\Delta\in\mathfrak{A})\tau_\Delta\in RT$. We have also $(\forall n\in N)(\forall\Delta_i\in\mathfrak{A})\tau_{(\Delta_1,\ldots,\Delta_n)}\in RT$. The family (P_x) is called a Markov process if $(\forall t\in R_+)\tau_t\in RT$ where $\tau_t\equiv t$, and a strong Markov process if $(\forall t\in\mathcal{F}_0)\tau\in RT$. The SM class contains the class of strong Markov processes and also the class of all semi-Markov step processes of Lévy and Smith, but the SM class is essentially wider than the union of these subclasses

2. Outleading Sequences. For a given sequence $\delta=(\Delta_n)_1^\infty(\Delta_n\in\mathfrak{A}\cup\tilde{\mathfrak{A}})$ the sequence $(\tau_\delta^n)_{n=1}^\infty$ $(\tau_\delta^n=\tau_{(\Delta_1,\ldots,\Delta_n)})$ is non-decreasing. The sequence δ is said to be outleading $(\delta\in OS)$ if $(\forall\xi\in\mathcal{D})\lim_{n\to\infty}\tau_\delta^n\xi=\infty$. It is known that for every open covering $\mathfrak{A}_1\subset\mathfrak{A}$ of the set X there exists an outleading sequence whose terms belong to this covering. Denote by $OS(\mathfrak{A}_1)$ the class of these sequences. For every outleading sequence δ the mapping $L_\delta:\mathcal{D}\to\mathcal{D}$ is well defined, where $L_\delta\xi(t)=\pi_{\tau_\delta^n}\xi$ for $\tau_\delta^n\xi\le t<\tau_\delta^{n+1}(\xi)$. Define rank $\delta=\sup\{\mathrm{diam}\Delta:\Delta\in\delta\}$. If rank δ is equal to $r>0$ then $(\forall t\in R_+)\rho(\xi(t),L_\delta\xi(t))\le r$ (ρ is the metric in X). Thus a sequence $(\delta_n)_1^\infty(\delta_n\in OS$, rank $\delta_n\le r_n$, $r_n\to 0)$ corresponds to a converging sequence uniformly on R_+ of step functions $L_{\delta_n}\xi\to\xi$.

Let $(\mathfrak{A}_n)_1^\infty$ be a sequence of countable open coverings of the set X with decreasing ranks: $(\forall\Delta\in\mathfrak{A}_n)\mathrm{diam}\Delta\le r_n(r_n\to 0)$ and let $\delta_n\in OS(\mathfrak{A}_n)$. Define the composition of k outleading sequences $\delta_1\times\ldots\times\delta_k$, where $\delta_i=(\Delta_{i1},\Delta_{i2},\ldots)$, as a sequence of an ordering type ω^k of the form $(\Delta_{1i_1}\cap\ldots\cap\Delta_{ki_k})$ $(i_1,\ldots,i_k)\in N^k$ (k-dimensional index) [5]. The rank of this sequence is not more than the least rank of the components. Convenience of using of the sequence $(\delta_1\times\ldots\times\delta_k)_{k=1}^\infty$ follows the property: $(\forall\xi\in\mathcal{D})$ the sequence of stopping times $(\tau_{i_1,\ldots,i_k}(\xi))$ defined by the k-th composition is a subset of stopping times

$(\tau_{i_1,\ldots,i_{k+1}}(\xi))$ defined by the (k+1)-th composition. The following property is the basis of the definition of indexes: $(\forall\Delta\in\mathfrak{A})$ $(\forall\delta\in OS)\tau_{\Delta_1\cap\Delta}\dot{+}\ldots\dot{+}\tau_{\Delta_k\cap\Delta}\to\tau_\Delta$ as $k\to\infty$, where $\delta=(\Delta_1,\Delta_2,\ldots)$, and $(\exists n\in N)\tau_\Delta(\xi)=(\tau_{\Delta_1\cap\Delta}\dot{+}\ldots\dot{+}\tau_{\Delta_k\cap\Delta})(\xi)$ as $\tau_\Delta(\xi)<\infty$. We define the following rule of indexation: 1) $(\forall m\in N)\tau(\underbrace{0,\ldots,0}_{m})=0$, 2) $(\forall m,k,i_k\in N)$ $\tau(i_1,\ldots,i_k,\underbrace{0,\ldots,0}_{m})=\tau(i_1,\ldots,i_k)$, 3) $(\forall k,i_k\in N)\tau(i_1,\ldots,i_k)=\tau(i_1,\ldots,$ $i_{k-1},i_k-1)\dot{+}\tau(\Delta_{1i_1+1}\cap\cdots\cap\Delta_{k-1,i_{k-1}+1}\cap\Delta_{ki_k})$. Under this rule the order relation on the time scale between times corresponds to the lexicographical ordering of their indexes in the alphabet $N_0=N\cup\{0\}$ and $\tau(i_1)\to\infty(i_1\to\infty,i_1\in N)$, $\tau(i_1,\ldots,i_k,i_{k+1})\to\tau(i_1,\ldots,i_{k-1},i_k+1)(i_{k+1}\to\infty)$. This is a rule for finite dimensional indexes. We shall consider infinite dimensional indexes, too. Times with such indexes naturally appear when localizing of an arbitrary "point of a sequence of states". A point $t\in R_+$ and a function $\xi\in\mathfrak{D}$ correspond to the infinite sequence $i=(i_n)_1^\infty$ $(i_n\in N_0)$ where $(\forall n\in N_0)$ $\tau_{i_1,\ldots,i_n}(\xi)\le t<$ $\tau_{i_1,\ldots,i_{n-1},i_n+1}(\xi)$. Thus the mapping $N_0^\infty\to X$ is defined. It is so called sequence of states which is considered below. Obviously the index $(i_1,\ldots,i_k)$ may be identified with the index $(i_1,\ldots,i_k,0,0,$ $\ldots)\in N_0^\infty$. Universality of the outleading sequences follows that different indexes can correspond to the same time.

3. Correct Exit. For arbitrary $\Delta\in\mathfrak{A}\cup\overline{\mathfrak{A}}$ we have $\tau_{\Delta^{-r}}\uparrow\tau\le\tau_\Delta$ where $\Delta^{-r}=\{x\in\Delta:\ \rho(x,\partial\Delta)>r\}$, $\partial\Delta=[\Delta]\setminus\mathrm{int}\Delta$, $[\Delta]$ is the closure of Δ, $\mathrm{int}\Delta$ is the set of all interior points of Δ. For $\Delta\in\mathfrak{A}$ we have also $\tau_{\Delta^{+r}}\downarrow$ $\tau_{[\Delta]}\ge\tau_\Delta$ where $\Delta^{+r}=\{x\in X:\ \rho(x,\Delta)<r\}$. This holds for $\Delta\in\overline{\mathfrak{A}}$ too, but in this case $[\Delta]=\Delta$. We are interested in the property $\tau(\xi)=\tau_\Delta(\xi)=$ $\tau_{[\Delta]}(\xi)$. Let $\Pi(\Delta)$ be the set of all functions $\xi\in\mathfrak{D}$ such that: 1) $\xi(0)\notin\partial\Delta$, 2) $\beta_{\tau_{\Delta^{-r}}}\xi\to\beta_{\tau_\Delta}\xi$, 3) $\tau_{\Delta^{+r}}(\xi)\to\tau_\Delta(\xi)$ $(r\to 0)$ where

$$\beta_\tau\xi=\begin{cases}\bar{\infty}, & \text{if } \tau(\xi)=\infty \quad (\bar{\infty}\in R_+\times X)\\ (\tau(\xi),\ \pi_\tau\xi), & \text{otherwise.}\end{cases}$$

We say that $\Pi(\Delta)$ is a correct exist set from Δ. For every probability measure P on $\mathcal{F}$ an arbitrary set $\Delta\subset X$ can be "slightly" deformed: $\Delta\to\Delta'$ to have $P(\Pi(\Delta'))=1$. For example, $(\forall x\in X)$ there are only denumerably many $r>0$ such that $P(\Pi(S(x,r)))<1$, $S(x,r)$ is an open ball with the radius r and the centre x. The definition of the correct exit set can be extended to finite and infinite sequences of open sets [6] including outleading sequences of the ordering type ω^k. In the paper [6] the method of equivalent sets based on the property of first exiting was used. Because of this property in all those cases when one needn't use all the class of open sets but its sufficiently rich part, first exit times from open sets and their iterations can be changed by first exit times from the closures of these open sets and their iterations and inversely. Further in this paper we don't stop for basing of such mutual changes. It can be easy done as in the mentioned paper.

4. Sequence of States. Let Φ be the class of all continuous non-decreasing functions $\varphi:R_+\to R_+$ for which $\varphi(0)=0$, $\varphi(t)\to\infty$ $(t\to\infty)$. The functions $\xi_1,\xi_2\in\mathcal{D}$ are said to have identical sequences of states if $\xi_1\circ\varphi_1=\xi_2\circ\varphi_2$ [4]. Under this relation we mean by a sequence of states $L\xi$ of a function ξ the corresponding equivalence class. This definition is a natural one, but it is inconvenient for investigating conditional distributions. For this reason we shall consider a rather coarser equivalence class partition of $\mathcal{D}$. Let

$$\gamma_\tau\xi = \begin{cases} \tilde{\infty}, & \text{if } \tau(\xi) = \infty \ (\tilde{\infty}\in X) \\ \pi_\tau\xi, & \text{otherwise.} \end{cases}$$

Let $T(\mathcal{T}_1)$ be a class closed with respect to the operation $\dot{+}$, generated by the class $\mathcal{T}_1\subset\mathcal{T}$. We say the functions $\xi_1,\xi_2\in\mathcal{D}$ have identical sequences of essential states if $(\forall\tau\in T(\tau_\Delta,\Delta\in\tilde{\mathfrak{A}}))\gamma_\tau\xi_1=\gamma_\tau\xi_2$. The corresponding equivalence class $\mathcal{L}\xi$ of a function ξ is called its sequence of essential states. Let $\tilde{\mathcal{F}}=\sigma(\gamma_\tau,\tau\in T(\tau_\Delta,\Delta\in\tilde{\mathfrak{A}}))$. Then $\tilde{\mathcal{F}}$ is known to be a countably generated σ-algebra [6]. We shall treat the conditional distribution $P_x(\cdot|\tilde{\mathcal{F}})$ which depends on a parameter $x\in X$. From countable generating of $\tilde{\mathcal{F}}$ and $\mathcal{B}(X)$-measurability with respect to the Borel σ-algebra $\mathcal{B}(X)$ of the family of measures $(P_x)_{x\in X}$ the existence of a $\mathcal{B}(X)\otimes\tilde{\mathcal{F}}$-measurable version of the conditional distribution which depends on the parameter (i.e. $(\forall B\in\mathcal{F})$ $P_\cdot(B|\tilde{\mathcal{F}})$ is a $\mathcal{B}(X)\otimes\tilde{\mathcal{F}}$-measurable function) follows. Since π_0 is an $\tilde{\mathcal{F}}$-measurable function the functions $P_{\pi_0}(B|\tilde{\mathcal{F}})$ $(B\in\mathcal{F})$ and $P_{\pi_0}(f|\tilde{\mathcal{F}})$ (f is an $\mathcal{F}$-measurable function) are $\tilde{\mathcal{F}}$-measurable.

5. Lévy Decomposition. Denote $P_{\pi_0}(e^{-\lambda\tau}|\tilde{\mathcal{F}})=\exp(-b(\lambda,\tau))$ $(\lambda\geq 0)$. It is known [6] that $(\forall x\in X)$ P_x-a.s. there exists a sufficiently rich subclass of the class $T(\tau_\Delta,\Delta\in\mathfrak{A})$ such that for each τ_1 from this

subclass on the set $\{\tau_1<\infty\}$ $b(\lambda,\tau_1\dot{+}\tau)=b(\lambda,\tau_1)+b(\lambda,\tau)\circ\theta_{\tau_1}$ where $\tau\in T$ $(\tau_\Delta,\Delta\in\mathfrak{A}\cup\tilde{\mathfrak{A}})$. Besides one can easily prove that in this case $b(\lambda,\tau_1)$ is an $\mathcal{F}_{\tau_1}$-measurable function. It is known [5] that $(\forall x\in X)$ P_x-a.s. there exist $\mathcal{F}$-measurable additive functionals $a(\tau)$, $n(du,\tau)$, and, may be, a denumerable family of indexes $\mathcal{I}$ such that on the set $\{\tau<\infty\}$

$$b(\lambda,\tau) = \lambda a(\tau)+\int_{0+}^{\infty}(1-e^{-\lambda u})n(du,\tau)-\sum_{(i\in\mathcal{I},\tau_i<\tau)}\log P_x(e^{-\lambda\tau_0\circ\theta_{\tau_i}}|\tilde{\mathcal{F}})$$

where $\tau_0(\xi)=\inf\{t\geq 0,\xi(t)\neq\xi(0)\}$ the first exit time from the initial state, $\tau_i=\lim_{n\to\infty}\tau_{i_n}$, $i_n\in N_0^m$, $i_n\leq i$, $i_n\uparrow i$. We call this formula the Lévy decomposition of the process and $n(du,\tau)$ its Lévy measure. From this decomposition the representation of the time τ follows:

$$\tau = a(\tau) + \int_{0+}^{\infty}u\,\mathcal{P}(du,\tau) + \sum_{\tau_i<\tau}\tau_0\circ\theta_{\tau_i}$$

where $\mathcal{P}(du,\tau)$ is an $\tilde{\mathcal{F}}$-measurable conditional Poisson random measure on $R_+\times[0,\tau)$. The atom (t_i,s_i) of the measure $\mathcal{P}$ is interpreted as the length (t_i) and the initial point (s_i) of an interval of constancy of a path with a given sequence of states. This is so called a conditional Poisson interval of constancy. The measure $n(du,\tau)$ is the intensity measure of the random measure $\mathcal{P}(du,\tau)$. It is its expectation with respect to the measure $P_x(\cdot|\tilde{\mathcal{F}})$. The times τ_i in the Lévy decomposition are initial points of intervals of constancy of another type. We call them conditionally fixed intervals of constancy with random lengths $\tau_0\circ\theta_{\tau_i}$ which are positive with a positive probability $P_x(\tau_0\circ\theta_{\tau_i}>0|\tilde{\mathcal{F}})>0$. In this paper we study semi-Markov processes without (conditionally) fixed intervals of constancy:

$$(\forall x\in X)P_x\text{-a.s. }(\forall i\in N_0^\infty)P_x(\tau_0\circ\theta_{\tau_i}>0|\tilde{\mathcal{F}}) = 0.$$

Besides for simplicity of investigation we only consider the case when paths don't finish with an infinite interval of constancy:

$$(\forall x\in X)P_x\text{-a.s. }(\exists(i_n),i_n\in N_0^\infty)(\forall n\in N)\tau_{i_n}<\infty,\quad \tau_{i_n}\to\infty\ (n\to\infty).$$

This property is not reflected in the Lévy decomposition which is obtained on a finite interval. Besides the event "an infinite interval of constancy is absent" is not $\tilde{\mathcal{F}}$-measurable.

6. λ-Continuity. It requires some properties of continuity of

the family (P_x) for proving its Markov property. The family (P_x) is called λ-continuous if $(\forall n\in N)(\forall\lambda_i>0)(\forall\varphi_k\in B(X))$ the function

$$P_x\left(\int_{R_+^n}\left(\prod_{k=1}^{n} e^{-\lambda_k t_k}\varphi_k(\pi_{s_k})dt_k\right)\right)$$

is continuous as a function of $x\in X$ [3] where $s_k=t_1+\ldots+t_k$, $B(X)$ is a class of all continuous and bounded functions on X. In particular as $n=1$ it is a λ-potential operator

$$R_\lambda(\varphi|x) = P_x\left(\int_{R_+} e^{-\lambda t}\varphi(\pi_t)dt\right)$$

and its continuity at all λ and φ means continuity of the operator $T_t(\varphi|x)=P_x(\varphi(\pi_t))$ at almost all t with respect to the Lebesgue measure (the Feller property). The λ-continuity of the family (P_x) itself follows from its weak continuity with respect to the Skorohod metric on $\mathcal{D}$ [1] and weak continuity of the family (P_x) follows from weak continuity of transition functions $(f_{\tau_\Delta})_{\Delta\in\mathfrak{A}_1}$ on X:

$$f_{\tau_\Delta}(\lambda,S|x) = P_x(e^{-\lambda\tau_\Delta},\ \{\pi_{\tau_\Delta}\in S,\ \tau_\Delta<\infty\}) \qquad (S\in\mathcal{B}(X))$$

of the semi-Markov process for a sufficiently rich countable subclass $\mathfrak{A}_1$ of the class $\mathfrak{A}$. Because of λ-continuity the class of regenerative times can be extended. For example, all first exit times from closed sets would be regenerative times. A λ-continuous SM family is a Markov family if $(\forall x\in X)P_x(\tau_0>0)=0$ and P_x-a.s. the process has no intervals of constancy (the proof is evident).

7. Time Change. The space Φ does not contain all functions which transforming the time scale for Markov representation of SM processes. For this aim we shall consider the space Ψ of all non-decreasing left continuous functions $\varphi:R_+\to R_+$ where $\varphi(0)=0$, $\varphi(t)\to\infty$ $(t\to\infty)$ [4]. Let $\varphi^*(t)=\inf\varphi^{-1}[t,\infty)$ $(t\in R_+,\varphi\in\Psi)$. The mapping $(\cdot)^*$ is one-to-one on Ψ and $(\varphi^*)^*=\varphi$. Let $\tilde{\theta}_t:\Psi\to\Psi$ be a shift operator on Ψ, $(\tilde{\theta}_t\varphi)(s)=\varphi(t+s)-\varphi(t)$. We have $(\forall\varphi\in\Phi)L\xi=L(\xi\circ\varphi)$. A function $\varphi\in\Psi$ having discontinuities rejects some states from a sequence of states $L\xi$ when mapping $\xi\mapsto\xi\circ\varphi$ if discontinuities of φ are not compensated by corresponding intervals of constancy of the function ξ. We call the pair $(\varphi,\xi)\in\Psi\times\mathcal{D}$ is coordinated if $(\forall t\in R_+)\alpha_t\xi\circ\varphi^*\in\mathcal{D}$ and $L(\alpha_t\xi\circ\varphi^*)=L\alpha_t\xi$. Let $\mathcal{K}$ be a set of all coordinated pairs (φ,ξ). The criterion of the coordination is that: $(\varphi,\xi)\in\mathcal{K}\Leftrightarrow(\forall\Delta\in IC(\varphi))$ $[\Delta]\in IC(\xi)$, where $IC(f)$ is the set of all intervals of constancy of

f, i.e. all connected sets where f is constant. Let $(\varphi,\xi)\in\mathcal{K}$ and put $u(\varphi,\xi)=\xi\circ\varphi^*$ the transformation of time change. Each probability measure Q on the σ-algebra $\mathfrak{B}=\mathcal{B}(\Psi)\otimes\mathcal{F}$ of subsets of the space $\Psi\times\mathcal{F}$ concentrated on $\mathcal{K}$ induces two probability measures on $\mathcal{F}$: $P=Q\circ q^{-1}$ the measure of the initial process where $q(\varphi,\xi)=\xi$, and $\overline{P}=Q\circ u^{-1}$ the measure of the transformed process. Thus Q defines the transformation $P\mapsto\overline{P}$. One can easily prove the property: $(\varphi,\xi)\in\mathcal{K}\Rightarrow(\varphi^*,\xi\circ\varphi^*)\in\mathcal{K}$, $\xi\circ\varphi^*\circ\varphi=\xi$. It defines a dual measure $\overline{Q}$: $\overline{Q}=Q\circ g^{-1}$ where $g(\varphi,\xi)=(\varphi^*,\xi\circ\varphi^*)$. Evidently $g=g^{-1}$.

In the paper [4] the family of measures (Q_x) which preserves the SM property is constructed with the aid of additive functionals. In the present paper a straight method of construction of (Q_x) for one of the general forms of SM families is proposed. The proper random time change transforms a semi-Markov process into a Markov one and inversely.

§2 The Main Result.

Theorem 1. Let $(P_x)\in SM$, (P_x) be a λ-continuous family and $(\forall x\in X)P_x$-a.s. paths of the process have no infinite and also no fixed intervals of constancy. Then there exists a Markov family $(\overline{P}_x)$ connected with the family (P_x) by time change. For each $\lambda_0>0$ the family $(\overline{P}_x)$ can be constructed as so $R_{\lambda_0}=\overline{R}_{\lambda_0}$, i.e. $(\forall x\in X)(\forall\varphi\in\mathbb{B}(X))$

$$P_x\left(\int_0^\infty e^{-\lambda_0 t}\varphi(\pi_t)dt\right)=\overline{P}_x\left(\int_0^\infty e^{-\lambda_0 t}\varphi(\pi_t)dt\right).$$

The semi-Markov transition function $\overline{f}_\tau$ of the process $(\overline{P}_x)$ is of the form: $(\forall x\in X)(\forall\lambda>0)(\forall S\in\mathcal{B}(X))(\forall\tau\in T(\tau_\Delta,\Delta\in\mathfrak{A}))$ $\overline{f}_\tau(\lambda,S|x)\equiv\overline{P}_x(e^{-\lambda\tau},\{\pi_\tau\in S,\tau<\infty\})=P_x((P_{\pi_0}(e^{-\lambda_0\tau}|\mathcal{F}))^{\lambda/\lambda_0},\{\pi_\tau\in S,\tau<\infty\})$.

Proof. Firstly we shall construct a time change which transforms the family (P_x) into the semi-Markov family $(\overline{P}_x)$ with the transition function $\overline{f}_\tau$. It will define the inverse time change transformation $(\overline{P}_x)\mapsto(P_x)$. Secondly we shall prove that the family $(\overline{P}_x)$ is a Markov one and has the mentioned properties. At the same time we shall prove that paths of this Markov process have no intervals of constancy.

1. For given $\lambda_0>0$ $(\forall x\in X)(\exists(\delta_n)_1^\infty$, $\delta_n\in OS$, $\operatorname{rank}\delta_n\to0)(\forall i\in\bigcup_{k=1}^\infty N_0^k\equiv\mathcal{I}_0)$

the function $b(\lambda_0,\tau_i)=-\log P_{\pi_0}(e^{-\lambda_0\tau_i}|\mathfrak{F})$ is defined for P_x-almost all $\xi\in\mathfrak{D}$. Because of countability of $\mathcal{I}_0$ the sequence $(b(\lambda_0,\tau_i))_{i\in\mathcal{I}_0}$ is defined P_x-a.s., too. This sequence does not decrease as a function of $i\in\mathcal{I}_0$. Consequently, for P_x-almost each $\xi\in\mathfrak{D}$ the function $y_\xi:R_+\to R_+$ is defined where $y_\xi(t)=\sup_{\tau_i(\xi)\le t}\frac{1}{\lambda_0}b(\lambda_0,\tau_i)(\xi)$.

It is continuous because of absence of fixed intervals of constancy. Besides $y_\xi(0)=0$ and $y_\xi(t)\to\infty$ $(t\to\infty)$ because of absence of infinite intervals of constancy. Hence $y_\xi\in\Phi$. Evidently each closed from the left interval of constancy of the function ξ is an interval of constancy of the function y_ξ. Let the function ξ is not constant on the interval (t_1,t_2). Then $(\exists i\in\mathcal{I}_0)(\exists\Delta\in\tilde{\mathfrak{A}})\,t_1<\tau_1(\xi)<(\tau_1\dot{+}\tau_\Delta)(\xi)<t_2$. Hence $\pi_{\tau_i}\xi_1\in\Delta$ and $\tau_\Delta\theta_{\tau_i}\xi_1>0$ for all ξ_1 with the same sequence of states as ξ has. Therefore $P_x(e^{-\lambda\tau_\Delta}\circ\theta_{\tau_i}|\mathfrak{F})(\xi)=P_{\pi_0}(e^{-\lambda\tau_\Delta}|\mathfrak{F})(\theta_{\tau_i}\xi)<1$ and $b(\lambda_1\tau_i\dot{+}\tau_\Delta)(\xi)-b(\lambda,\tau_i)(\xi)=b(\lambda,\tau_\Delta)(\theta_{\tau_i}\xi)>0$. Hence y_ξ is not constant on (t_1,t_2) and $(t_1,t_2)\in IC(y_\xi)\Rightarrow(t_1,t_2)\in IC(\xi)$. Further, right continuity of ξ follows $(t_1,t_2)\in IC(y_\xi)\Rightarrow[t_1,t_2)\in IC(\xi)$. On the other hand the right end of an interval of constancy $P_x(\cdot|\mathfrak{F})$-a.s. is not a point of discontinuity. It follows from the facts: 1) the set of all points of discontinuity is countable, 2) the k-th jump time with a jump size $\ge\varepsilon$ is an $\mathfrak{F}$-measurable function which is equal to some $\tau_i(i\in\mathcal{I}_0)$, 3) $(\forall i\in\mathcal{I}_0)P_x$ (an interval of constancy of ξ has $\tau_i(\xi)$ as its right end $|\mathfrak{F})=0$; it is a consequence of absence of fixed intervals of constancy. Hence $P_x(\cdot|\mathfrak{F})$.-a.s. $(t_1,t_2)\in IC(y_\xi)\Rightarrow[t_1,t_2]\in IC(\xi)$ and by the criterion of coordination $(y_\xi,\xi)\in\mathcal{K}$. From here, $\xi\circ y_\xi^*\in\mathfrak{D}$ and has the same sequence of states (in strong sense) as ξ has. The transformation $\xi\mapsto(y_\xi,\xi)$ defines a measure Q_x on $\mathfrak{B}$ as $Q_x(A\times B)=P_x(y_\xi\in A,\xi\in B)$ (degenerated measure) and a transformed process $\bar{P}_x=P_x\circ F^{-1}=Q_x\circ u^{-1}$ where $F(\xi)=\xi\circ y_\xi^*$, F is defined on $\mathfrak{D}$ P_x-almost everywhere. Since the sets of intervals of constancy of ξ and y_ξ are identical the path $\xi\circ y_\xi^*$ has no intervals of constancy. The dual measure $\bar{Q}_x=Q_x\circ g^{-1}$ is not yet degenerated.

2. We are going to prove that $(\bar{P}_x)$ is a Markov family. Define the random time $\sigma_t(\xi)=\inf\{s\ge0,y_\xi(s)\ge t\}=y_\xi^*(t)$. Show that it is a stopping time $(\sigma_t\in\mathfrak{T}_+)$ and such that $\sigma_t\in RT$. Because of λ-continuity of the family (P_x) it is sufficiently to show that σ_t can be approximated from above by regenerative times. It follows from that σ_t is not the left end of an interval of constancy of the

functions y_ξ and ξ. In this case $\sigma_t=\lim \sigma_{t,\delta}$ (rank $\delta\to 0$) where $\sigma_{t,\delta}=\tau_\delta^k$ as $\tau_\delta^{k-1}\le\sigma_t<\tau_\delta^k$, $\delta\in OS$. Show that $(\forall t\ge 0)P_x($t is a point of continuity of the function $y_\xi^*|\tilde{\mathcal{F}})=1$. Note that $(\forall\varepsilon>0)$ all points of discontinuity of the function y_ξ^* with jump sizes $\ge\varepsilon$ form a Poisson process with the intensity measure $N([\varepsilon,\infty),\cdot)$ which is defined by the condition:

$$(\forall\tau\in T(\tau_\Delta,\ \Delta\in\mathfrak{U}))\quad N([\varepsilon,\infty),\ \frac{1}{\lambda_0}\,b(\lambda_0,\tau)) = n([\varepsilon,\infty),\tau)\ ,$$

where $n(\cdot,\tau)$ is the measure of Lévy of the SM process. Boundedness of the intensity of this point process follows from the relations:

$$\frac{n([\varepsilon,\infty),\tau)}{\frac{1}{\lambda_0}b(\lambda_0,\tau)} = n([\varepsilon,\infty)\tau)(a(\tau)+\int_{0+}^{\infty}(1-e^{-\lambda_0 u})\frac{1}{\lambda_0}n(du,\tau))^{-1} \le \frac{\lambda_0}{1-e^{-\lambda_0\varepsilon}}\ .$$

Consequently $P_x($t is not a point of discontinuity of the function y_ξ^* with a jump size $\ge\varepsilon|\tilde{\mathcal{F}})=1$, and since (t is a point of discontinuity of $y_\xi^*)=\bigcup_{n=1}^{\infty}$(t is a point of discontinuity of y_ξ^* with a jump size $\ge\frac{1}{n}$) it is proved that P_x-a.s. y_ξ^* is continuous at the point t.

Show that the Markov property of the process $(\bar{P}_x)$ at the point t follows from the Markov property of the process (P_x) at the point σ_t. The first means that $(\forall x\in X)\bar{P}_x(\theta_t^{-1}B,\alpha_t^{-1}A)=\bar{P}_x(\bar{P}_{\pi_t}(B),\alpha_t^{-1}A)$ for all $A,B\in\mathcal{F}$. From the theorem about the extension of measures it follows that this property is sufficient to prove for A and B of the form $\bigcap_{k=1}^{m}\beta_{\tau_k}^{-1}(T_k\times S_k)$ where $\tau_k\in T(\tau_\Delta,\Delta\in\mathfrak{U})$, $T_k\in\mathcal{B}(R_+)$, $S_k\in\mathcal{B}(X)$. We have $F^{-1}\alpha_t^{-1}\beta_\tau^{-1}(T\times S)=\{\tau\circ\alpha_t\circ F\in T,\pi_\tau\circ\alpha_t\circ F\in S\}$. Further

$$(\tau\circ\alpha_t\circ F)(\xi) = \tau\alpha_t(\xi\circ y_\xi^*) = \tau(\alpha_{y_\xi^*(t)}\xi\circ y_\xi^*)$$

$$= y_\xi(\tau\alpha_{\sigma_t}(\xi)) = \begin{cases}\infty\ , & \tau>\sigma_t\\ y_\xi(\tau(\xi)), & \tau\le\sigma_t\ ,\end{cases}$$

$$(\pi_\tau\circ\alpha_t\circ F)(\xi) = \pi_\tau(\alpha_{\sigma_t}(\xi))\ .$$

Hence $F^{-1}\alpha_t^{-1}\beta_\tau^{-1}(T\times S)\in\mathcal{F}_{\sigma_t}$. We have also $F^{-1}\theta_t^{-1}\beta_\tau^{-1}(T\times S)=\{\tau\circ\theta_t\circ F\in T,$ $\pi_\tau\circ\theta_t\circ F\in S\}$, where

$$(\tau\circ\theta_t\circ F)(\xi) = \tau\theta_t(\xi\circ y_\xi^*) = \tau(\theta_{y_\xi^*(t)}\xi\circ\tilde{\theta}_t y_\xi^*)$$

$$= y_\xi((\sigma_t \dot{+} \tau)(\xi)) - y_\xi(\sigma_t(\xi)) = y_\xi((\sigma_{t,\delta} \dot{+} \tau)(\xi)) - y_\xi(\sigma_t(\xi))$$

$$= y_\xi \sigma_{t,\delta}(\xi) + y_\xi \tau \theta_{\sigma_{t,\delta}}(\xi) - y_\xi(\sigma_t(\xi))$$

$$= \lim y_\xi \tau \theta_{\sigma_{t,\delta}}(\xi) \quad (\text{rank } \delta \to 0).$$

Here we use the property $(\forall \Delta \in \mathfrak{A}) \sigma_t \le \sigma_{t,\delta} \le \sigma_t \dot{+} \tau_\Delta \Rightarrow \sigma_t \dot{+} \tau_\Delta = \sigma_{t,\delta} \dot{+} \tau_\Delta$. Further $y_\xi \tau \theta_{\sigma_{t,\delta}}(\xi) = (\frac{1}{\lambda_0} b(\lambda_0, \tau) \circ \theta_{\sigma_{t,\delta}})(\xi)$ and $(\frac{1}{\lambda_0} b(\lambda_0,\tau) \circ \theta_{\sigma_{t,\delta}} \in T) \in \theta^{-1}_{\sigma_{t,\delta}} \mathcal{F} \subset \theta^{-1}_{\sigma_t} \mathcal{F}$. Besides $(\pi_\tau \circ \theta_t \circ F)(\xi) = \pi_\tau(\theta_{\sigma_t} \xi \circ \vartheta_t y^*_\xi) = \pi_\tau \theta_{\sigma_t} \xi$. Hence $F^{-1} \alpha_t^{-1} \bigcap_{k=1}^{m} \beta^{-1}_{\tau_k}(T_k \times S_k) \in \mathcal{F}_{\sigma_t}$ and $F^{-1} \theta_t^{-1} \bigcap_{k=1}^{m} \beta^{-1}_{\tau_k}(T_k \times S_k) = \theta^{-1}_{\sigma_t} \bigcap_{k=1}^{m} (\frac{1}{\lambda_0} b(\lambda,\tau) \in T_k, \pi_\tau \in S_k)$. From here $\overline{P}_x(\theta_t^{-1} B, \alpha_t^{-1} A) = P_x(\theta^{-1}_{\sigma_t} \overline{B}, \overline{A})$ for some $\overline{B} \in \mathcal{F}$, $\overline{A} \in \mathcal{F}_{\sigma_t}$. On the other hand $\overline{P}_x(\overline{P}_{\pi_t}(B), \alpha_t^{-1} A) = P_x(P_{\pi_t \circ F}(F^{-1}B), \overline{A})$. We have $\pi_t F(\xi) = \pi_t(\xi \circ y^*_\xi) = \pi_{y^*_\xi(t)} \xi = \pi_{\sigma_t}(\xi)$ and besides $\tau(F(\xi)) = \tau(\xi \circ y^*_\xi) = y_\xi(\tau(\xi))$ and $\pi_\tau F(\xi) = \pi_\tau \xi$. Then $F^{-1} B = F^{-1} \bigcap_{k=1}^{m} \beta^{-1}_{\tau_k}(T_k \times S_k) = \overline{B}$. Hence $\overline{P}_x(\overline{P}_{\pi_t}(B), \alpha_t^{-1} A) = P_x(P_{\pi_{\sigma_t}}(\overline{B}), \overline{A})$ and the Markov property for $(\overline{P}_x)$ at the point t follows from the equality $P_x(\theta^{-1}_{\sigma_t} \overline{B}, \overline{A}) = P_x(P_{\pi_{\sigma_t}}(\overline{B}), \overline{A})$.

It remains to prove that for given $\lambda_0 > 0$ $\overline{R}_{\lambda_0} = R_{\lambda_0}$. Let $\varphi \in \mathbb{B}(X)$. We have

$$P_x\left(\int_{0+}^{\infty} e^{-\lambda_0 t} \varphi(\pi_t) dt\right)$$

$$= \lim_{\text{rank } \delta \to 0} P_x\left(\int_{0+}^{\infty} e^{-\lambda_0 t} \varphi(\pi_t \circ L_\delta) dt\right)$$

$$= \lim P_x\left(\sum_{k=1}^{\infty} \int_{\tau_\delta^{k-1}}^{\tau_\delta^k} e^{-\lambda_0 t} dt \cdot \varphi(\pi_{\tau_\delta^{k-1}})\right)$$

$$= \lim P_x\left(\sum_{k=1}^{\infty} \frac{1}{\lambda_0} (e^{-\lambda_0 \tau_\delta^{k-1}} - e^{-\lambda_0 \tau_\delta^k}) \varphi(\pi_{\tau_\delta^{k-1}})\right)$$

$$= \lim P_x\left(\sum_{k=1}^{\infty} \frac{1}{\lambda_0} (e^{-b(\lambda_0, \tau_\delta^{k-1})} - e^{-b(\lambda_0, \tau_\delta^k)}) \varphi(\pi_{\tau_\delta^{k-1}})\right);$$

and

$$\overline{P}_x\left(\int_{0+}^{\infty} e^{-\lambda_0 t} \varphi(\pi_t) dt\right)$$

$$= \lim \overline{P}_x\left(\int_{0+}^{\infty} e^{-\lambda_0 t} \varphi(\pi_t \circ L_\delta) dt \right)$$

$$= \lim \overline{P}_x\left(\sum_{k=1}^{\infty} \frac{1}{\lambda_0} \left(e^{-\lambda_0 \tau_\delta^{k-1}} - e^{-\lambda_0 \tau_\delta^{k}}\right) \varphi(\pi_{\tau_\delta^{k-1}}) \right);$$

but $(\tau \circ F)(\xi) = \frac{1}{\lambda_0} b(\lambda_0, \tau)(\xi)$ and $e^{-\lambda_0 \tau \circ F} = e^{-b(\lambda_0,\tau)}$. And the equality follows from here. Furthermore

$$\overline{f}_\tau(\lambda, \varphi | x) \equiv P_x(e^{-\lambda \tau \circ F} \varphi(\pi_\tau), \ \tau < \infty)$$

$$= P_x(e^{-\lambda \frac{1}{\lambda_0} b(\lambda_0, \tau)} \varphi(\pi_\tau), \ \tau < \infty)$$

$$= P_x((e^{-b(\lambda_0, \tau)})^{\lambda/\lambda_0} \varphi(\pi_\tau), \ \tau < \infty)$$

$$= P_x((P_{\pi_0}(e^{-\lambda_0 \tau} | \mathcal{F}))^{\lambda/\lambda_0} \varphi(\pi_\tau), \ \tau < \infty).$$

The theorem is proved.

Now it is clear how one can answer the question which has put in the beginning of the paper for the continuous SM process with the Lévy extension

$$P_{\pi_0}(e^{-\lambda \tau} | \mathcal{F}) = \exp\left(- \int_{0+}^{\infty} (1 - e^{-\lambda u}) \, n(du, \tau) \right)$$

which has paths of Cantor type. It is necessary to subject it to time change of the special form which transforms the above conditional generating function of time pass into the corresponding function of the form

$$\overline{P}_{\pi_0}(e^{-\lambda \tau} | \mathcal{F}) = (P_{\pi_0}(e^{-\lambda_0 \tau} | \mathcal{F}))^{\lambda/\lambda_0}$$

leaving distributions of sequences of states without changes. This transformed process has also some other properties which are told about in the conditions of Theorem 1. Note that λ-continuity of the initial family of measures was only necessary for proving of Markovness of the transformed SM family.

References

[1] Gihman, I. I., Skorohod, A. V., Theory of Stochastic Processes, Vol.2. Moscow, Nauka, 1976. (Russian)

[2] Shirjaev, A. N., Statistical Sequential Analysis. Moscow, Nauka, 1976. (Russian)

[3] Harlamov, B. P., Representation of a semi-Markov process as a time changed Markov process, Teor. Verojatnost. i Primen., 28, N4, 653-667 (1983). (Russian)

[4] Harlamov, B. P., Additive functionals and time change which preserves the semi-Markov property of a process, Zap. Nauchn. Sem Leningrad. Otdel. Mat. Inst. Steklov (LOMI), 97, 203-216 (1980). (Russian)

[5] Harlamov, B. P., Conditional distribution of time move given a sequence of states in semi-Markov processes, Zap. Nauchn. Sem. LOMI, 142, 167-173 (1985) (Russian)

[6] Harlamov, B. P., Continuous semi-Markov processes and an extremal property of Markov processes, In: Proceedings of the Seventh Conference on Probability Theory, Brasov, Romania (1982), Ed. Ac. 1984, 423-430.

[7] Serfozo, R. F., Random time transformations of semi-Markov processes, Ann. Math. Statist., v.42, N1, 176-188 (1971).

[8] Yackel, J., A random time change relating semi-Markov and Markov processes, Ann. Math. Statist., v.39, N2, 358-364 (1968).

Leningrad Branch of
Steklov institute of Mathematics,
Leningrad, USSR

BOUNDED SOLUTIONS AND PERIODIC SOLUTIONS OF A LINEAR STOCHASTIC EVOLUTION EQUATION

Akira Ichikawa

1. The stochastic evolution equation. Recently Morozan [8] has studied periodic solutions of a linear stochastic differential equation. In this paper we consider bounded solutions and periodic solutions of a linear stochastic evolution equation. Control problems for periodic systems are discussed in [2], [3], [4]. Let Y be a Hilbert space and let $A(t)$, $-\infty < t < \infty$, be a linear operator which generates an evolution operator $U(t,s)$, $-\infty < s \leq t < \infty$, in Y. A sufficient condition for this is the following [9]:

(i) there exist $M > 0$, $\theta \in (\pi/2, \pi)$ such that

$$S_\theta = \{\lambda \in C : |\arg \lambda| < \theta\} \subset \rho(A(t)) \quad \text{(resolvent set)}$$

$$|(\lambda - A(t)^{-1})| \leq \frac{M}{|\lambda|}, \quad \lambda \in S_\theta$$

(ii) there exist a Hilbert space $Y_0 \subset Y$ dense in Y such that $D(A(t)) = Y_0$ and an $\alpha \in (0,1]$ s.t.

$$A(\cdot) \in C^\alpha([0,T] ; L(Y_0, Y)).$$

We assume that $A(t)$ has a Yosida type approximation $A_m(t)$ and that $U(t,s)$ is approximated by $U_m(t,s)$ [2], [3], [4].

Let f, $g_i \in L_\infty(-\infty,\infty ; Y)$ and let $G_i \in L_\infty(-\infty,\infty ; L(Y))$, $i=1,2,\cdots,n$, be strongly measurable. Let $(\Omega, F, F_t, -\infty < t < \infty, P)$ be a probability space and w_i, $i=1,2,\cdots,n$, are independent Wiener processes with respect to F_t.

Consider

$$dy = [A(t)y + f(t)]dt + [g_i(t) + G_i(t)y]dw_i(t), \tag{1.1}$$

where the repeated subscript i implies the summation from $i=1$ to n. A mild solution on $(-\infty,\infty)$ of (1.1) is a process adapted to F_t which satisfies

$$y(t) = U(t,s)y(s) + \int_s^t U(t,r)f(r)dr + \int_s^t U(t,r)[g_i(r) + G_i(r)y(r)]dw_i(r) \tag{1.2}$$

for any $-\infty < s \leq t < \infty$. A mild solution $y(t)$ is said to be

(i) bounded if $y \in C_B(-\infty,\infty ; L_2(\Omega, F, P ; Y))$,

(ii) weakly T-periodic if its expectation and variance are T-periodic functions,

(iii) T-periodic if it has a T-periodic distribution.

Let $y(t)$ be a solution of (1.1) and set $m(t) = Ey(t)$ and $Q(t) = \mathrm{Cov}[y(t)]$.

Then we formally have

$$\dot{m} = A(t)m + f(t), \tag{1.3}$$

$$\dot{Q} = A(t)Q + QA^*(t) + G_i(t)QG_i^*(t) + [g_i(t) + G_i(t)m]\circ[g_i(t) + G_i(t)m], \tag{1.4}$$

where $(h\circ k)y = h\langle k,y\rangle$, $h, k, y \in Y$. For (1.3) and (1.4) we also take mild solutions i.e. solutions of

$$m(t) = U(t,s)m(s) + \int_s^t U(t,r)f(r)dr, \quad t \geq s , \tag{1.5}$$

$$Q(t) = \int_s^t U(t,r)\{G_i(r)Q(r)G_i^*(r) + [g_i(r) + G_i(r)m(r)]\circ[g_i(r) + G_i(r)m(r)]\}U^*(t,r)dr + U(t,s)Q(s)U^*(t,s). \tag{1.6}$$

2. The special case : $G_i = 0$.

First we consider the special case $G_i = 0$. We assume that $U(t,s)$ is exponentially stable i.e. there exist $M \geq 1$ and $a > 0$ such that

$$|U(t,s)| \leq Me^{-a(t-s)} \quad \text{for any} \quad s \leq t. \tag{2.1}$$

Proposition 2.1.

There exists a unique bounded mild solution of (1.1) given by

$$y(t) = \int_{-\infty}^t U(t,r)f(r)dr + \int_{-\infty}^t U(t,r)g_i(r)dw_i(r). \tag{2.2}$$

The mean and the covariance of $y(t)$ are given by

$$m(t) = \int_{-\infty}^t U(t,r)f(r)dr \tag{2.3}$$

$$Q(t) = \int_{-\infty}^t U(t,r)g_i(r)\circ g_i(r)U^*(t,r)dr \tag{2.4}$$

which are unique bounded mild solutions of (1.3), (1.4).

Proof. Boundedness of $y(t)$ follows easily from (2.1). To show that $y(t)$ is a mild solution we use the semigroup property

$$U(t,r)U(r,s) = U(t,s) \quad \text{for any} \quad s \leq r \leq t. \tag{2.5}$$

In fact

$$y(t) = \int_{-\infty}^s U(t,r)f(r)dr + \int_s^t U(t,r)f(r)dr + \int_{-\infty}^s U(t,r)g_i(r)dw_i(r) + \int_s^t U(t,r)g_i(r)dw_i(r)$$

$$= U(t,s)y(s) + \int_s^t U(t,r)f(r)dr + \int_s^t U(t,r)g_i(r)dw_i(r) \quad \text{for any} \quad s \leq t.$$

To show the uniqueness let y_1 be another bounded solution. Then $\tilde{y} = y - y_1$ satisfies

$$\tilde{y}(t) = U(t,s)\tilde{y}(s) \quad \text{for any} \quad s \leq t.$$

Letting $s \to -\infty$, we obtain $\tilde{y}(t) = 0$.

Next we assume that $A(t)$, $f(t)$ and $g_i(t)$ are T-periodic i.e., $A(t+T) = A(t)$, $f(t+T) = f(t)$ etc. The system (1.1) is said to be T-periodic if all functions and operators are T-periodic.

Proposition 2.2.

Suppose that (1.1) is T-periodic and that (2.1) holds. Then the mild solutions $y(t)$, $m(t)$ and $Q(t)$ given by (2.2) - (2.4) are T-periodic (i.e. unique T-periodic solutions).

Proof.

Note that if $A(t)$ is T-periodic $U(t+T,s+T) = U(t,s)$ for any $s \leq t$.

Now

$$\begin{aligned} y(t+T) &= \int_{-\infty}^{t+T} U(t+T,r)f(r)dr + \int_{-\infty}^{t+T} U(t+T,r)g_i(r)dw_i(r) \\ &= \int_{-\infty}^{t} U(t+T,s+T)f(s+T)ds + \int_{-\infty}^{t} U(t+T,s+T)g_i(s+T)dw_i(s+T) \\ &= \int_{-\infty}^{t} U(t,s)f(s)ds + \int_{-\infty}^{t} U(t,s)g_i(s)dw_i(s+T). \end{aligned}$$

Hence $y(t+T)$ has the same distribution with $y(t)$.
Similarly we obtain

$$m(t+T) = m(t), \quad Q(t+T) = Q(t) \quad \text{for any} \quad t.$$

3. Stability of the homogeneous equation.

To obtain a mild solution for (1.1) with $G_i \neq 0$ we need some preliminary results. We consider the homogeneous equation

$$\begin{aligned} dy &= A(t)y\,dt + G_i(t)y\,dw_i(t), \\ y(t_0) &= y_0. \end{aligned} \tag{3.1}$$

$y(t)$ is said to be a mild solution of (3.1) if it is adapted to F_t and

$$y(t) = U(t,t_0)y_0 + \int_{t_0}^{t} U(t,r)G_i(r)y(r)dw_i(r). \tag{3.2}$$

Since (3.2) is linear, there exists a unique solution in $C([t_0,t_1]\,;\,L_2(\Omega,F,P\,;\,Y))$ for any $t_0 < t_1$ which is given by

$$y(t) = \Phi(t,t_0)y_0, \quad y_0 \in L_2(\Omega,F_{t_0},P). \tag{3.3}$$

$\Phi(t,t_0)$ is called the stochastic fundamental solution of (3.1) [1]. We say that (3.1) is exponentially stable in mean square if there exist $N \geq 1$ and $b>0$ such that

$$E|\Phi(t,t_0)y_0|^2 \leq Ne^{-b(t-t_0)}E|y_0|^2 \quad \text{for any } y_0 \text{ and } t_0 \leq t. \tag{3.4}$$

For (3.1) $R(t) = E[y(t)\circ y(t)]$ formally satisfies

$$\dot{R} = A(t)R + RA^*(t) + G_i(t)RG_i^*(t) \tag{3.5}$$
$$R(t_0) = R_0 ,$$

where $R_0 = E[y_0 \circ y_0]$. For each $R_0 \geq 0$ there exists a unique mild solution of (3.5) which is given by

$$R(t) = \Psi(t,t_0)[R_0]. \tag{3.6}$$

$\Psi(t,t_0)$ is called the fundamental solution of (3.5).

Proposition 3.1.

The following statements are equivalent.

(i) $E|\Phi(t,t_0)y_0|^2 \leq Ne^{-b(t-t_0)}E|y_0|^2$.

(3.7)

(ii) $|\Psi(t,t_0)[R_0]| \leq \bar{N}e^{-b(t-t_0)}|R_0|$.

Concerning the exponential stability of (3.1) we have

Proposition 3.2.

The following statements are equivalent :

(i) There exists $K>0$ such that $\int_{t_0}^{\infty} E|\Phi(t,t_0)y_0|^2 dt \leq K|y_0|^2$ for any y_0 and t_0.

(ii) There exist $N \geq 1$ and $b>0$ such that $E|\Phi(t,t_0)y_0|^2 \leq Ne^{-b(t-t_0)}|y_0|^2$ for any y_0 and $t_0 \leq t$.

(iii) There exists a bounded nonnegative mild solution to

$$\dot{P} + A^*(t)P + PA(t) + M(t) + G_i^*(t)PG_i(t) = 0 \tag{3.8}$$

for any bounded $M(t) \geq 0$.

Proof.

(i)$\Longrightarrow$(ii) : This is shown in [7]. (ii)$\Longrightarrow$(iii) : As in [5] let P_n be the unique solution of the backward equation

$$\dot{P} + A^*(t)P + PA(t) + M(t) + G_i(t)PG_i^*(t) = 0, \quad P(n) = 0. \tag{3.9}$$

Then by comparison theorem

$$P_n(t) \leq P_{n+1}(t) \quad \text{for any} \quad t \leq n.$$

Moreover, by Ito's formula [3], [4], [6] we have

$$\langle P_n(t_0)y_0, y_0\rangle = \int_{t_0}^{n} E\langle My(t), y(t)\rangle dt \leq \int_{t_0}^{\infty} E\langle M(t)y(t), y(t)\rangle dt \leq K_1 |y_0|^2,$$

where $y(t)$ is a mild solution of (3.1). Since $P_n(t)$ is monotone increasing and bounded from above, $P_n(t) \nearrow P(t)$. Note that $P_n(t)$ satisfies

$$P_n(t) = U^*(s,t)P_n(s)U(s,t) + \int_t^s U^*(r,t)[M(r) + G_i^*(r)P_n(r)G_i(r)]U(r,t)dr$$

for any $t \leq s \leq n$.

Let $s \leq t$ be fixed but arbitrary. Then passing to the limit $n \to \infty$ we obtain

$$P(t) = U^*(s,t)P(s)U(s,t) + \int_t^s U^*(s,t)[M(r) + G_i^*(r)P(r)G_i(r)]U(r,t)dr$$

(iii) $\Longrightarrow$ (i) : Set $M(t) = I$ in (3.8) and apply Ito's formula.

Corollary 3.1.

Suppose that $|U(t,s)| \leq Me^{-a(t-s)}$, $|G(t)| \leq c$ for some $M \geq 1$, $a > 0$, $c > 0$ and $c^2M^2 < 2a$.

Then there exists a unique bounded solution to (3.8).

4. Existence of bounded and periodic solutions.

Now we consider

$$dy = [A(t)y + f(t)]dt + [g_i(t) + G_i(t)y]dw_i(t) \qquad (1.1)$$

$$\dot{Q} = A(t)Q + QA^*(t) + G_i(t)QG_i^*(t) + [g_i(t) + G_i(t)m]\circ[g_i(t) + G_i(t)m]. \quad (1.3)$$

We assume that

$$E|\Phi(t,t_0)|^2 \leq Ne^{-b(t-t_0)}, \quad -\infty < t_0 \leq t < \infty, \text{ for some } N \geq 1,\ b > 0. \qquad (4.1)$$

or equivalently

$$|\Psi(t,t_0)| \leq \bar{N}e^{-b(t-t_0)}, \quad -\infty < t_0 \leq t < \infty, \text{ for some } \bar{N} \geq 1,\ b > 0.$$

Then $E[\Phi(t,t_0)y_0] \leq Me^{-a(t-t_0)}|y_0|$ for some $M \geq 1$ and $a > 0$. Moreover, (1.3) has always bounded solution (2.3).

Now we give our main results.

Theorem 4.1. Assume that (4.1) holds. Then (1.3) has a unique bounded mild solution given by

$$Q(t) = \int_{-\infty}^{t} \Psi(t,r)\{[g_i(r) + G_i(r)m(r)]\circ[g_i(r) + G_i(r)m(r)]\}dr. \tag{4.2}$$

If (1.1) (and hence (1.3)) is T-periodic, then $Q(t)$ is also T-periodic.

Proof. Since (1.3) is linear, (4.1) and a contraction mapping theorem gives existence and uniqueness. To show that $Q(t)$ is the solution we use the fact that $R(t)$ given by (3.6) satisfies

$$R(t) = U(t,t_0)R_0U^*(t,t_0) + \int_{t_0}^{t} U(t,r)G_i(r)R(r)G_i^*(r)U^*(t,r)dr$$

and proceed as the proof of Proposition 2.1. If (1.1) is T-periodic, then $\Psi(t+T,s+T) = \Psi(t,s)$ and the T-periodicity of $Q(t)$ follows as Proposition 2.2.

Corollary 4.1. Suppose (1.1) is T-periodic and that it has a mild solution $y(t)$. Then $y(t)$ is weakly T-periodic.

Theorem 4.2. Assume that (4.1) holds. Then (1.1) has a unique bounded mild solution. If (1.1) is T-periodic, then the solution is also T-periodic.

Proof. Existence and uniqueness follow from a contraction mapping theorem. Suppose (1.1) is T-periodic. Then $\tilde{y}(t) = y(t+T)$ satisfies (1.2) with $w_i(t)$ replaced by $\tilde{w}_i(t) = w_i(t+T)$. Hence $y(t)$ and $\tilde{y}(t)$ have the same law.

Theorem 4.3. If $U(t,s)$ is a two-parameter group or if $g_i = 0$ or $G_i = 0$ for each i, then the mild solution of (1.1) is given by

$$y(t) = \int_{-\infty}^{t} \Phi(t,r)[f(r) - G_i(r)g_i(r)]dr + \int_{-\infty}^{t} \Phi(t,r)g_i(r)dw_i(r). \tag{4.3}$$

Proof. We note $\Phi(t,r) = \Phi(t,0)\Phi^{-1}(r,0)$ in the former case [10]. We proceed as Theorem 8.5.2.[1, p.141].

Corollary 4.2. Suppose (1.1) is time-invariant i.e., $A(t) = A$, $f(t) = f$, $g_i(t) = g_i$, $G_i(t) = G_i$. Suppose also A is stable.

(i) If $G_i = 0$, then there exist stationary solutions for (1.3) and (1.4) :

$$m = -A^{-1}f$$
$$AQ + QA^* + g_i\circ g_i = 0$$

(ii) If (3.1) is exponentially stable in mean square, then there exists a unique invariant measure of the Markov process associated with (1.1) with mean

$$m = -A^{-1}f$$

and the covariance Q given by

$$AQ + QA^* + G_iQG_i^* + (g_i + G_im)\circ(g_i + G_im) = 0.$$

Remark 4.1. (3.1) is exponentially stable in mean square if and only if there exists $P \geq 0$ such that

$$A^*P + PA + M + G_i^*PG_i = 0 \quad \text{for each } M \geq 0.$$

References

[1] L. Arnold: Stochastic Differential Equations; Theory and Applications, John Wiley & Sons, New York (1974).

[2] G. Da Prato: Synthesis of optimal control for an infinite dimensional periodic problem, report, Scuola Normale Superiore, Pisa (1985). (SIAM J. Control Optim., to appear)

[3] G. Da Prato and A. Ichikawa: Filtering and control of linear periodic systems, report, Scuola Normale Superiore, Pisa (1985). (submitted for publication).

[4] ________________________: Optimal control of linear systems with almost periodic inputs, SIAM J. Control Optim., to appear.

[5] ________________________: Bounded solutions on the real line to non-autonomous Riccati equations, Atti della Accademia Nationale dei Lincei, to appear.

[6] A. Ichikawa: Semilinear stochastic evolution equations; Boundedness, stability and invariant measure, Stochastics, 12 (1984), 1-39.

[7] __________: Equivalence of L_p stability and exponential stability for a class of nonlinear semigroups, Nonlinear Analysis, TMA, 8 (1984), 805-815.

[8] T. Morozan: Periodic solutions of affine stochastic differential equations, Stoch. Anal. Appl. 4 (1986), 87-110.

[9] H. Tanabe: Equations of evolution, Pitman, London (1979).

[10] J. Zabczyk: A private communication.

Faculty of Engineering
Shizuoka University
Hamamatsu
432 Japan

RENORMALIZATION GROUP METHOD ON A HIERARCHICAL LATTICE OF DYSON-WILSON TYPE

Keiichi R. Ito

1. Introduction. Renormalization group is a pragmatical method to carry out integrations over many degrees of freedom which one encounters in many fields of mathematical physics. One of such examples is the two-dimensional O(N)-invariant statistical mechanical model described by the Gibbs measure

$$\mu_\Lambda(d\vec{\phi}) = Z_\Lambda^{-1} \exp\Big[-\frac{1}{2} \sum_{\substack{(x,y)\in\Lambda \\ |x-y|=1}} (\vec{\phi}(x)-\vec{\phi}(y))^2\Big] \prod_{x\in\Lambda} e^{-V(\vec{\phi}^2(x))} d\vec{\phi}(x), \tag{1}$$

where $\Lambda\subset Z^2$ is a set of lattice points contained in a finite rectangular region, $\vec{\phi}(x)=(\phi_1(x),\ldots,\phi_N(x))\in R^N$ is a spin (random) variable at the lattice point $x\in\Lambda$, Z_Λ is the normalization constant (partition function) and $V(\vec{\phi}^2(x)) = \lambda(\vec{\phi}^2(x)-\kappa)^2$ ($\lambda>0$, $\kappa\geq 0$) is the (bare or original) potential which defines the single spin distribution function $g_0(\vec{\phi}^2)=\exp[-V(\vec{\phi}^2)]$.

Our interests are the properties of the thermodynamic limits of the Gibbs measures $\{\mu_\Lambda(d\vec{\phi});\ \Lambda\subset Z^2\}$, for which it is convenient to know the long-range behaviors of the correlation functions $\langle\prod \phi_{k_i}(x_i)\rangle = \lim_{\Lambda\uparrow Z^2} \int\prod \phi_{k_i}(x_i)\ \mu_\Lambda(d\vec{\phi})$, where suitable boundary conditions at $\partial\Lambda$ should be understood when the thermodynamic limits are taken.

Renormalization group is a method to carry out the integration by decomposing all configurations $\{\vec{\phi}(x)\in R^N;\ x\in\Lambda\}\subset R^{N|\Lambda|}$ into many tiny subsets over which the integrations are easy. The conventional way of doing this in physics is called the block spin transformation [13,15], and consists of the following steps which transform the Gibbs distribution functions recursively [4,8,9] :

$$\mu^{(n)}(\vec{\phi}^{(n)}) = R(\mu^{(n-1)})(\vec{\phi}^{(n)}) = (Z_n)^{-1}\int \prod_{x\in\Lambda_n} \delta(\vec{\phi}^{(n)}(x)-(C\vec{\phi}^{(n-1)})(x))\ \mu^{(n-1)}(d\vec{\phi}^{(n-1)}) \tag{2}$$

where $\mu^{(n-1)}(d\vec{\phi}^{(n-1)}) = \mu^{(n-1)}(\vec{\phi}^{(n-1)}) \prod_{x\in\Lambda_{n-1}} d\vec{\phi}^{(n-1)}(x)$,

$$\mu^{(0)}(\vec{\phi}) = (Z_\Lambda)^{-1}\exp\Big[-\frac{1}{2}\langle\vec{\phi},(-\Delta_\Lambda)\vec{\phi}\rangle - \sum_{x\in\Lambda} V(\vec{\phi}^2(x))\Big] \tag{3}$$

the starting Gibbs distribution function (Δ_Λ is the lattice laplacian satisfying the suitable boundary condition at $\partial\Lambda$) and $\Lambda_n = L^{-n}\Lambda\cap Z^2$ with $\Lambda=\Lambda_0=[-\frac{1}{2}L^K, \frac{1}{2}L^K]^2\cap Z^2$ (K is an arbitrarily large integer). Moreover Z_n is the normalization constant and

$$(4) \qquad (C\vec{\phi}^{(n-1)})(x) = L^{-2} \sum_{y \in \Box_{Lx}} \vec{\phi}^{(n-1)}(y)$$

is the averaging operator which makes the averaged variable (block spin) $(C\vec{\phi}^{(n-1)})(x)$ from $\{\vec{\phi}^{(n-1)}(y);\ y \in \Box_{Lx}\}$, where L is an integer larger than 2 and $\Box_{Lx}$ is the square of center Lx ($x \in \Lambda_n$) and size L×L. So Eq.(4) consists of averaging and scaling. Therefore it is understood that the formulas (2) are a series of steps in which fluctuations around $\vec{\phi}^{(n)}(x)=(C\vec{\phi}^{(n-1)})(x)$ are integrated out.

Thus the natural problem in this approach is to know and control the trajectories $\{\mu^{(n)}\}$ by which the system is determined.

2. Main Results and Basic Ideas. A very hard barrier in this approach is that the distribution functions $\{\mu^{(n)}\}$ are very complicated functions of the block spins and so are the effective Hamiltonians $H^{(n)} = -\ln \mu^{(n)}$. Usually $H^{(n)}$ contains not only single spin potentials $V^{(n)}(\vec{\phi}^{(n)}(x)^2)$ but also non-local many-body interaction terms, which makes this approach formidable.

Therefore in this paper, we consider a hierarchical model due to Dyson, Wilson and Ma [4,13,15] in which some non-local terms are omitted in a physically reasonable way. Hierarchical models have been intensively studied recently by Gawedzki and Kupiainen [6,7,8] . See also [1,2,10,11] .

To make the hierarchical model in the present system, we replace $\frac{1}{2}\langle\vec{\phi},(-\Delta_\Lambda)\vec{\phi}\rangle = \frac{1}{2}\sum_{|x-y|=1}(\vec{\phi}(x)-\vec{\phi}(y))^2$ in Eq.(1) by:

$$(5) \qquad \frac{1}{2}\langle\vec{\phi},(-\Delta_{\text{hierarchical}})\vec{\phi}\rangle = \frac{1}{2L^4}\sum_{x\in\Lambda_1}\Big(\sum_{y\in\Box_{Lx}} A(y-Lx)\vec{\phi}(y)\Big)^2$$

$$+ \frac{1}{2L^4}\sum_{x\in\Lambda_2}\Big(\sum_{y\in\Box_{Lx}} A(y-Lx)(C\vec{\phi})(y)\Big)^2 + \cdots$$

$$+ \frac{1}{2L^4}\Big(\sum_{y\in\Box_0} A(y)(C^{K-1}\vec{\phi})(y)\Big)^2 + \frac{1}{2}(C^K\vec{\phi})^2(0) ,$$

where $A(y)=\pm1$ for $y\in\Box_0$ ($A(y)=0$ for $y\notin\Box_0$) and $\sum_y A(y) = 0$. So $A(y)=1$ for $L^2/2$ points and $A(y)=-1$ for $L^2/2$ points. Though each $\Box_{Lx}\subset\Lambda_n$ contains L^2 points, we assume (just for simplicity) that there exist only two independent variables $(C^{n+1}\vec{\phi})(x)$ and $\vec{z}^{(n)}(x)$ in $\Box_{Lx}$ so that $(C^n\vec{\phi})(y)=(C^{n+1}\vec{\phi})(x)+A(y-Lx)\vec{z}^{(n)}(x)$. $A(y-Lx)z^{(n)}(x)$ plays the role of the fluctuation of $(C^n\vec{\phi})(y)$ around $(C^{n+1}\vec{\phi})(x)$. Then the recursion formulas (2) are now reduced to

$$(6) \qquad g_n(\vec{\phi}^2)=(\mathcal{N})^{-1}\int\big[g_{n-1}((\vec{\phi}-\vec{z})^2)g_{n-1}((\vec{\phi}+\vec{z})^2)\big]^{L^2/2}\, e^{-\vec{z}^2/2}\,d\vec{z} ,$$

where $g_0(\vec{\phi}^2)=(\mathcal{N})^{-1}\exp[-\lambda(\vec{\phi}^2-\kappa)^2]$ and the normalization constants $\mathcal{N}$ are chosen so that $g_n(0)=1$. $V^{(n)}(\vec{\phi}^2)=-\ln g_n(\vec{\phi}^2)$ is called the effective interaction at the distance scale L^n.

<u>Theorem 1.</u> If $N \geq 3$, then $\lim \partial V^{(n)}(x)/\partial x = \infty$ for all $x>0$, for any $g_0(\vec{\phi}^2)$ which decreases sufficiently rapidly for large $\vec{\phi}^2$. Namely in this case, $g_n(\vec{\phi}^2)$ converge to the function $g_c(\vec{\phi}^2)$ which is 1 for $\vec{\phi}^2=0$ and 0 otherwise.

<u>Theorem 2.</u> If $N \geq 3$, the thermodynamic limit of the Gibbs states of the present hierarchical model is unique and there exists no phase transition.

<u>Conjecture.</u> If $N \geq 3$, then for the recursion formulas (2) of the real systems (1), the effective Gibbs distribution functions $\{\mu_n\}$ converge to $\Pi g_c(\vec{\phi}(x)^2)$. In other words, the original Gibbs measure is driven to the product measure by the block spin transformations (2), for any $V(\vec{\phi}^2(x))$.

The main idea behind Theorem 1 (and the Conjecture) is a central limit-like theorem. To see this, take a square $\Box_0$ (the square of size $L\times L$ with center being at the origin) and consider the first step of (2) neglecting other spin variables outside $\Box_0$. So

$$\exp[-V^{(1)}(\vec{\phi}^2)] \cong \exp[-E(\vec{\phi}^2)]\,\Pr(\vec{\phi}^2) \tag{7}$$

where $\Pr(\vec{\phi}^2)$ is the probability density for $L^{-2}\sum_y \vec{\phi}(y)=\vec{\phi}^{(1)}(0)=\vec{\phi}$, and $E(\vec{\phi}^2)$ is the most probable value of $\frac{1}{2}\langle\phi,(-\Delta)\phi\rangle$ when $(C\vec{\phi})(0)$ is fixed at $\vec{\phi}$. The distribution of each spin $\vec{\phi}(y)$ is given by $g_0(\vec{\phi}^2)=\exp[-\lambda(\vec{\phi}^2-\kappa)^2] \cong \delta(\vec{\phi}^2-\kappa)$. Then

$$\Pr(\vec{\phi}^2)=\text{const}\int d\vec{\zeta}\ \exp[iL^2(\vec{\zeta},\vec{\phi})]E_0(\exp[i\sqrt{\kappa}(\vec{\zeta},\vec{s})])^{L^2}$$

where E_0 is the expectation with respect to the N-1 dimensional unit sphere $S^{N-1}=\{\vec{s}\in R^N\ ;\ \|\vec{s}\|=1\}$. Thus [3,14]

$$\Pr(\vec{\phi}^2)=\text{const}\int_0^\infty \zeta^{N-1}d\zeta\, J_{\frac{N}{2}-1}(L^2\zeta\varphi)(L^2\zeta\varphi)^{1-\frac{N}{2}}[J_{\frac{N}{2}-1}(\sqrt{\kappa}\zeta)(\sqrt{\kappa}\zeta)^{1-\frac{N}{2}}]^{L^2} \tag{8}$$

$$\cong \begin{cases} \text{const}\exp[\text{const}\ L^2(N-1)\ln(1-\frac{\varphi^2}{\kappa})], & \text{for } \varphi^2/\kappa=O(1) \\ \text{const}\exp[-\text{const}\ \frac{NL^2}{\kappa}\varphi^2], & \text{for } \varphi \leq O(L^{-\varepsilon}). \end{cases} \tag{9a,b}$$

where $\varphi=\|\vec{\phi}\|$ and J_k denotes the Bessel function of order k. Eq.(9b) is nothing but the standard central limit theorem [3] applied to L^2 N-1 dimensional spheres. But what is important now is Eq.(9a) which is expected to dominate $E(\varphi^2) \simeq -\frac{2}{L^2}(\sum_x \vec{\phi}(x))^2 = -2L^2\varphi^2$. The energy term $E(\varphi^2)$ enhances the probability that φ takes larger values and on the other hand $\Pr(\varphi^2)$ works in the converse direction. Eq.(9a) implies that if N is large, then $\Pr(\varphi^2)$ wins for large φ driving $g_0(\varphi^2) \simeq \delta(\varphi^2-\kappa)$ to a function $g_1(\varphi^2)$ which takes its maximum around at $\varphi^2=\kappa-\text{const}(N-1)$.

What is expected is that this is kept through the recursion formulas (2), and this conjecture is justified in the hierarchical models. (See also [9,10,11].)

3. Sketch of the Proof. Without loss of generality, we set $L^2=2$ in (6) and assume $\operatorname{supp} g_0 \subset [0,\kappa]$, $\kappa \gg 1$. Thus $\operatorname{supp} g_n \subset [0,\kappa]$ for all n. To see the O(N)-invarinace explicitly, set

$$\vec{\phi}=(\varphi, \vec{0}) \in R_+ \times R^{N-1}, \qquad \vec{z}=(s, \vec{u}) \in R \times R^{N-1}$$

and rewrite (6) as

$$(10) \qquad g_n(x)=\frac{1}{\mathcal{N}}\int_0^\kappa dp \int_0^\kappa dq \int_{-\infty}^{\infty} ds \int_0^\infty u^{N-2}\, du\; \delta(p-(\varphi+s)^2-u^2)\,\delta(q-(\varphi-s)^2-u^2)$$

$$\times\, g_{n-1}(p) g_{n-1}(q) \exp[-\tfrac{1}{2}(s^2+u^2)]$$

$$=\frac{e^{x/2}}{\mathcal{N}}\frac{1}{\sqrt{x}}\int_0^\kappa dp \int_0^\kappa dq\; \theta(\mu(p,q;x))\,\mu^k(p,q;x)\, g_{n-1}(p)e^{-p/4} g_{n-1}(q)e^{-q/4}$$

where $x=\varphi^2 \in [0,\kappa]$, $k=(N-3)/2$,

$$(11) \qquad \mu(p,q;x)=\frac{p+q}{2}-x-\frac{(p-q)^2}{16x}$$

and θ is the Heaviside step function : $\theta(x)=1$ for $x \geq 0$ and $\theta(x)=0$ for $x<0$. Set

$$(12) \qquad D=\{(p,q) \in [0,\kappa]^2 ;\ \mu(p,q;x) \geq 0,\ 0 \leq q \leq p \leq \kappa\}.$$

Proposition 1. (A priori bound) (1) If $N \geq 3$, then for any g_{n-1}, $g_n(x)e^{-x/2}$ is monotone decreasing in $x \geq \frac{\kappa}{4}$.

(2) Assume $N \geq 4$. Then for any g_{n-1}, $g_n(x)e^{-x/2}$ obeys the following bound for $\frac{\kappa}{4} \leq x \leq \kappa$:

$$(13) \qquad -[\ln g_n(x)e^{-x/2}]' \geq \frac{1}{2x}+\frac{k}{\kappa-x}.$$

Proof. (1) Since $(p-q)^2 \leq \kappa^2$, $\mu(p,q;x)$ is monotone decreasing in $x \geq \frac{\kappa}{4}$.

Then so are $\mu(p,q;x)^k$ and $\theta(\mu(p,q;x))$, and thus the conclusion follows immediately from the expression (10).

(2) An easy calculation shows that

$$(14)\qquad -[\ln g_n(x)e^{-x/2}]' = \frac{1}{2x} + k\,\frac{\iint_D dpdq\ \mu^{k-1}(p,q;x)\left(1-\frac{(p-q)^2}{16x^2}\right)f_{n-1}(p)f_{n-1}(q)}{\iint_D dpdq\ \mu^k(p,q;x)\ f_{n-1}(p)f_{n-1}(q)}$$

$$(15)\qquad \geq \frac{1}{2x} + k \inf_{(p,q)\in D} \frac{1-\frac{(p-q)^2}{16x^2}}{\mu(p,q;x)}$$

$$= \frac{1}{2x} + \frac{k}{\kappa - x},$$

where $f_{n-1}(p)=g_{n-1}(p)e^{-p/4}$, $k=(N-3)/2$ and the infimum is taken at $p=q=\kappa$ since x is larger than or equal to $\kappa/4$. Q.E.D.

The meaning of this proposition is somewhat geometrical (overlap of two balls of diameters $2\sqrt{\kappa}$ separated by distance 2φ) and obviously reflects Eq.(9). It is easily checked that $g_n(x)e^{-x/2}$ is not necessarily decreasing for $x < \kappa/4$. In repeating the recursion relations (6), one difficulty is that D contains a set such that $q < \kappa/4$, even $x > \kappa/4$. (See fig.1 in the next page .) In fact $f_{n-1}(q)$ may be out of control for $q < \kappa/4$, especially after many non-linear recursions.

Assume that

$$-[\ln g_{n-1}(x)e^{-x/2}]' \geq \begin{cases} \alpha_{n-1}(x) + \frac{1}{2}, & \text{for } \kappa_{n-1} \leq x, \\ \frac{1}{2x} + \frac{k}{\kappa - x}, & \text{for } \frac{\kappa}{4} \leq x \leq \kappa_{n-1}, \end{cases}$$

where $\alpha_{n-1}(x)$ is a positive monotone increasing function in $x \geq \kappa_{n-1}$, and satisfies $\alpha_{n-1}(\kappa_{n-1}) \geq C_1 k$, $C_1=O(1)$. For simplicity in this note, we restrict ourselves to a large field region in which $(\kappa/4 <)\ \kappa_{n-1} \leq x$ and the analysis is rather easy. In this case, set $D= D_1 \cup D_2$ where

$$(16a)\qquad D_1= \{(p,q)\in D;\ x\leq q\leq p\}$$

$$(16b)\qquad D_2= \{(p,q)\in D;\ x \leq p \leq \kappa,\ (2\sqrt{x}-\sqrt{p})^2 \leq q \leq x\}.$$

See Figure 1. Then

(17) $$-[\ln g_n(x)e^{-x/2}]' = \frac{1}{2x} + k\,\frac{B_1+B_2}{A_1+A_2} \geq \frac{1}{2x} + k\,\min\{\frac{B_i}{A_i}\}_{i=1,2}$$

where $A_i = \int_{D_i} dpdq\, \mu(p,q;x)^k f_{n-1}(p) f_{n-1}(q)$ and B_i's are similarly defined. B_1/A_1 is easy to estimate because $-[\ln f_{n-1}(p)]' \geq \alpha_{n-1}(x) + \frac{1}{4}$ for $p \geq x$ and $\mu(x+\zeta, x+\xi; x) = \frac{1}{2}(\zeta+\xi) - (16x)^{-1}(\zeta-\xi)^2 \simeq \frac{1}{2}(\zeta+\xi)$. Thus an easy calculation yields :

(18a) $$k\,\frac{B_1}{A_1} \geq \frac{2k}{k+1}(1-\varepsilon)(\alpha_{n-1}(x) + \frac{1}{4})$$

where $\varepsilon = O(\alpha_{n-1}^{-1}(x))$. On the other hand, as for B_2/A_2 we use the following trick which is an extension of the inequality used in (17) :

$$k\,\frac{B_2}{A_2} = k\,\frac{\int_{(2\sqrt{x}-\sqrt{\kappa})^2}^{x} dq\, f_{n-1}(q)\,\hat{B}(q)}{\int_{(2\sqrt{x}-\sqrt{\kappa})^2}^{x} dq\, f_{n-1}(q)\,\hat{A}(q)} \geq k \inf_{\substack{(2\sqrt{x}-\sqrt{\kappa})^2 \leq q, \\ q \leq x}} \frac{\hat{B}(q)}{\hat{A}(q)} .$$

where

$$\hat{A}(q) = \int d\zeta\, \mu((2\sqrt{x}-\sqrt{q})^2+\zeta, q; x)^k f_{n-1}((2\sqrt{x}-\sqrt{q})^2+\zeta),$$

$$\hat{B}(q) = \int d\zeta\, \mu((2\sqrt{x}-\sqrt{q})^2+\zeta, q; x)^{k-1}\left(1 - \frac{[(2\sqrt{x}-\sqrt{q})^2+\zeta-q]^2}{16x^2}\right) f_{n-1}((2\sqrt{x}-\sqrt{q})^2+\zeta).$$

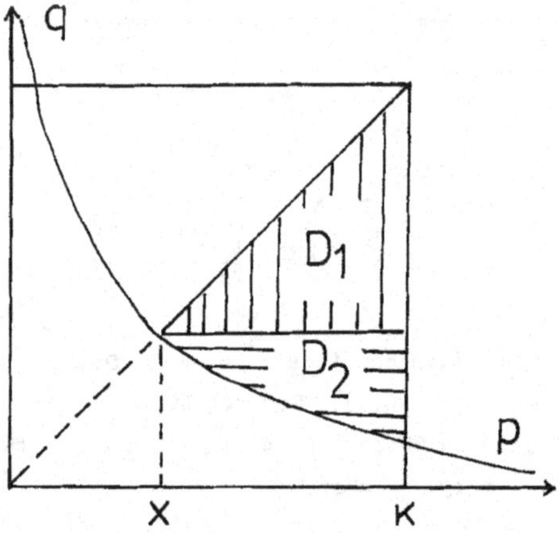

Figure 1. The domain D and its decomposition into D_1 and D_2. The parabola in the figure is $q^{1/2} = (2\sqrt{x} - \sqrt{p})^2$ along which $\mu(p,q;x) = 0$. By the symmetry $p \leftrightarrow q$, we can assume $q \leq p$.

Here $\mu((2\sqrt{x}-\sqrt{q})^2+\zeta,q;x)=(\zeta/16x)(16x-8\sqrt{xp}-\zeta)$ and $0\leq\zeta\leq\kappa-(2\sqrt{x}-\sqrt{q})^2$. Note that $\mu((2\sqrt{x}-\sqrt{q})^2+\zeta,q;x)\simeq\zeta/2$ if x is close to κ. Thus it is not difficult to find :

(18b) $$k\frac{B_2}{A_2}\geq 2(1-\varepsilon)(\alpha_{n-1}(x)+\frac{1}{4})$$

where $\varepsilon=O(\alpha_{n-1}^{-1}(x))$. Thus we have

$$-[\ln g_n(x)]'\geq\frac{2k}{k+1}(1-\varepsilon)(\alpha_{n-1}(x)+\frac{1}{4})-\frac{1}{2}.$$

Proposition 2. Assume $k>1$ (then $N\geq 6$) and $-[\ln g_{n-1}(x)]'\geq\alpha_{n-1}(x)$ where $\alpha_{n-1}(x)$ is a positive monotone increasing function in $x\in[\kappa_{n-1},\kappa]$ such that $\alpha_{n-1}(\kappa_{n-1})\geq C_1k$ ($C_1=O(1)$). Then for $x\in[\kappa_{n-1},\kappa]$

(19) $$-[\ln g_{n-1+\ell}(x)]'\geq[\frac{2k}{k+1}(1-\varepsilon)]^{\ell}(\alpha_{n-1}(x)-\beta)+\beta$$

where $\beta=\frac{1+\varepsilon k}{2[k(1-2\varepsilon)-1]}$ and $\varepsilon=O(\alpha_{n-1}^{-1}(x))$. The ε converges to zero as n (or ℓ) $\to\infty$.

This proposition implies that if N is larger than 5, then $-[\ln g_{n-1+\ell}(x)]'=\partial V^{(n-1+\ell)}(x)/\partial x$ tends to ∞ as $\ell\to\infty$ for large x. We can improve this proposition in the form of the following theorem by some additional tricks and efforts [12] :

Theorem 3. (1) Assume $N\geq 4$ and $x\geq 1$. Then there exist strictly positive constants C_1, C_2 and ω such that

(20a,b) $$-[\ln g_n(x)]'\geq\begin{cases}\frac{1}{2x}+C_1k\exp[C_2(x-\kappa_n)]-\frac{1}{2}, & \text{for } \kappa_n\leq x\leq\kappa,\\[2ex] \frac{1}{2x}+\frac{k}{\kappa_{n-1}-x}-\frac{1}{2}, & \text{for } \frac{1}{4}\kappa_n\leq x\leq\kappa_n,\end{cases}$$

where $\kappa_n=\kappa-n\omega$ and $n=1,2,\ldots,[(\kappa-1)/\omega]$. Moreover $C_2\omega\lesssim 2$.
(2) Let $n_0=[(\kappa-1)/\omega]$ and assume that $x<1$ and $n>n_0$. Then there exist strictly positive constants C_1', C_2' and δ ($\delta<1$) such that

(21) $$-[\ln g_n(x)]'\geq C_1'k\exp\{C_2'[(n-n_0)(-\ln\delta)+\ln x]\}-\frac{1}{2}$$

for $1>x\geq(\delta)^{n-n_0}$.

This theorem can be extended to contain the case of N=3. Thus it has been now established that there exist no non-trivial fixed points in the recursion formulas (6) if N is larger than or equal to 3. This theorem 3 leads us to theorems 1 and 2.

Acknowledgements. The main ideas of this work were inspired through stimulating discussions with T.Miwa, S.Kotani, Y.Higuchi and S.Kusuoka. The author also would like to thank T.Hattori, K-I.Kondo, T.Hara and H.Tasaki for their interests and encouragements. Part of this work was completed while he was a participant of the Research Program BiBoS in which he benefitted by discussions with K.Gawedzki and A.Kupiainen. He would like to thank L.Streit, S.Albeverio and Ph.Blanchard for giving him this opportunity.

References

[1] P.Bleher and Ya.Sinai: Critical indices for Dyson's asymptotically hierarchical models, Commun.Math.Phys. 45 (1973) 249-293.

[2] P.Bleher and P.Major : Renormalizaiton of Dyson's hierarchical vector valued ϕ^4 model at low temperature, Commun.Math.Phys. 95 (1984) 487-532.

[3] F.Dunlop and C.Newman : Multicomponent field theories and classical rotators, Commun.Math.Phys. 44 (1975) 223-235.

[4] F.J.Dyson : An Ising ferromagnet with distributions of long range order, Commun.Math.Phys. 21 (1971) 269-283.

[5] K.Gawedzki and A.Kupiainen: A rigorous block spin approach to massless lattice theories, Commun.Math.Phys. 77 (1980) 31-64.

[6] K.Gawedzki and A.Kupiainen : Non-Gaussian fixed points of the block spin transformation, hierarchical model approximation, Commun.Math.Phys. 89 (1983) 191-220.

[7] K.Gawedzki and A.Kupiainen : Continuum limit of the hierarchical non-linear O(N) sigma-model, Commun.Math. Phys. 106 (1986).

[8] K.Gawedzki : The renormalization, from magic to mathematics , talk delivered at the International Congress of Mathematics (1986, Berkley).

[9] Y.Higuchi and R.Lang : On the convergence of the Kadanoff transformation toward trivial fixed points, Z.Wahrscheinlichkeitstheorie verw.Gebiete, 58 (1981) 109-123.

[10] K.R.Ito: Analytic study of the Migdal-Kadanoff recursion formlula, Commun.Math. Phys., 95 (1984) 247-255.

[11] K.R.Ito; Permanent quark confinement in four-dimensional hierarchical lattice gauge theories of Migdal-Kadanoff type, Phys.Rev.Letters, 55 (1985) 558-562.

[12] K.R.Ito : paper in preparation.

[13] S.K.Ma : The modern theory of critical phenomena, Ch.8 , (Benjamin, Massachusetts 1976).

[14] V.Petrov : On the probabilities of large deviations for sums of independent random variables, Theory of Prob. and its Appl. 10 (1965) 287-298.

15] K.Wilson and J.Kogut : The renromalization group and ε-expansion, Phys.Rep. 12C (1974) 75-199.

Department of Mathematics
College of Liberal Arts
Kyoto University
Sakyo-ku, Kyoto 606
Japan

CONTIGUITY OF DISTRIBUTIONS OF MULTIVARIATE POINT PROCESSES

Yu. M. Kabanov

At the Third Soviet-Japan symposium on probability and statistics (Tashkent, 1975) R. Sh. Liptser, A. N. Shiryayev and the author of this notes presented a report on the criterion of the absolute continuity of distributions of the counting processes on the half line. Since then a noticeable progress has been achieved in understanding the relations between characteristics and distributions of the processes. For the counting and the multivariate point processes (m.p.p.) the defining characteristics are the compensators (also called dual predictable projections). In many questions concerning distributions (on absolute continuity, distance, contiguity, etc) we can assert that "properly" formulated statement about the compensators implies the analogous statement about the distributions. The notions of the Hellinger distance and the Hellinger integrals made the theory quite simple and transparent. It is curious to note that just this instrument has been used by S.Kakutani in his pioneering work on absolute continuity of the infinite Bernoulli sequences. Here we try to demonstrate its power for the problem of contiguity of two sequences of distributions of m.p.p. Although the main results of the paper are not new, we suggest that the formulations and the scheme of the proof give deeper understanding of the criteria of contiguity.

The paper is organized as follows. Section 1 contains results on contiguity of σ-finite measures and random measures. In §2 we give an information about the local density (density process) for m.p.p., in §3 we discuss properties of the Hellinger integrals. Section 4 contains the criteria of contiguity. At last, in §5 we give a brief review of the literature.

§1. The contiguity of σ-finite measures and random measures.

1. Let $(\Omega^n,\mathcal{F}^n)_{n\in\mathbb{N}}$ be a sequence of measurable spaces with a pair of σ-finite measures $(p^n,\tilde{p}^n)$ on each.

Definition. The sequence $(\tilde{p}^n)$ is called contiguous with respect to (p^n) (in symbols: $(\tilde{p}^n)\triangleleft(p^n)$) if $\lim_n \tilde{p}^n(\Gamma^n)=0$ for any $\Gamma^n\in\mathcal{F}^n$ such that $\lim_n p^n(\Gamma^n)=0$.

Definition. The sequences $(\tilde{p}^n)$ and (p^n) are called completely asymptotically separable (in symbols: $(\tilde{p}^n)\Delta(p^n)$) if there exists a sequence of sets $\Gamma^n\in\mathcal{F}^n$ such that

$$\underline{\lim}\,(\tilde{p}^n(\Gamma^n)+p^n(\bar{\Gamma}^n)) = 0 ,$$

i.e. for a some subsequence (n')

$$\lim \tilde{p}^{n'}(\Gamma^{n'}) = 0, \qquad \lim p^{n'}(\bar{\Gamma}^{n'}) = 0.$$

If $(\Omega^n,\mathcal{F}^n,p^n,\tilde{p}^n)=(\Omega,\mathcal{F},p,\tilde{p})$ and we have a unique measurable space with a pair of measures on it, then the contiguity implies the absolute continuity, and both definitions are equivalent when $\tilde{p}$ is finite. The notions of singularity and complete asymptotic separability are equivalent without additional assumptions.

2. The Hellinger distance and the Hellinger integrals. Let p and $\tilde{p}$ be σ-finite measures defined on the same space, $q:=(p+\tilde{p})/2$, $\mathfrak{z}:=dp/dq$, $\tilde{\mathfrak{z}}:=d\tilde{p}/dq$, $Z:=\tilde{\mathfrak{z}}/\mathfrak{z}$.

Let $\varphi_\alpha(u,v):=\alpha u+(1-\alpha)v-u^\alpha v^{1-\alpha}$, $u,v\geq 0$, $\alpha\in]0,1[$.

Put

$$d_H^2(\alpha,p,\tilde{p}) := E_q\varphi_\alpha(\mathfrak{z},\tilde{\mathfrak{z}})$$

where E_q denotes the integral with respect to q.

It is easily seen that the above definition does not change if we take as q any σ-finite measure dominating p and $\tilde{p}$ (so, we need only the relations $p\ll q$, $\tilde{p}\ll q$).

The function $d_H(p,\tilde{p}):=d_H(1/2,p,\tilde{p})$ is the distance on the set of σ-finite measures called the Hellinger distance. To prove the triangular inequality for p_1, p_2, p_3, it suffices to take as q any measure dominating all above measures and use the triangular inequality in $L^2(q)$ for the functions dp_i/dq, $i=1,2,3$.

It is useful to note that for any $\alpha\in]0,1[$

$$\alpha(1-\alpha)d_H^2(p,\tilde{p}) \leq d_H^2(\alpha,p,\tilde{p}) \leq d_H^2(p,\tilde{p}) . \qquad (1)$$

These relations follow from corresponding inequalities between φ_α and $\varphi:=\varphi_{1/2}$, which we prove in the next

Lemma 1. For any $\alpha\in]0,1[$

$$\alpha(1-\alpha)\varphi(u,v) \le \varphi_\alpha(u,v) \le 8\varphi(u,v) = 4(\sqrt{u}-\sqrt{v})^2. \tag{2}$$

Proof. Due to identity $\varphi_\alpha(u,v)=\varphi_{1-\alpha}(v,u)$ we need to consider only the case $\alpha\in]0,1/2]$. For $u=0$ inequalities (2) are trivial. Thus, it rests to prove that for any $y\ge 0$

$$\alpha(1-\alpha)f(y) \le f_\alpha(y) \le 8f(y) = 4(1-\sqrt{y})^2, \tag{3}$$

where $f_\alpha(y):=\varphi_\alpha(1,y)=\alpha+(1-\alpha)y-y^{1-\alpha}$, $f(y):=f_{1/2}(y)$.

It is clear that

$$\frac{\partial f_\alpha(y)}{\partial\alpha} = 1-y+y^{1-\alpha}\ln y \le 1-y+y\ln y$$

implies

$$f_\alpha(y) \le \alpha(1-y+y\ln y) .$$

It is easy to check that $1-y+y\ln y\le 4(1-\sqrt{y})^2$ for $y\in]0,4]$. But for $y\ge 4$ and $\alpha\in]0,1/2]$ we have evidently

$$f_\alpha(y) \le y-1 \le 4(1-\sqrt{y})^2 .$$

So, the right inequality in (3) holds. To prove the left inequality in (3) we introduce the function $F(y):=c_\alpha f_\alpha(y)-f(y)$ where $1/c_\alpha:=\alpha(1-\alpha)$. Note that $F(0)\ge 1/2$, $F(1)=F'(1)=0$, $F''(y)\ge 0$ for $y\ge 1/2$ and F changes sign only at one point. This implies that $F\ge 0$.

Remark. For any $y\in[4,e^c]$, $c\ge 2\ln 2$, we have

$$1-y+y\ln y \le (c-1)y+1 \le cy \le 4c(1-\sqrt{y})^2$$

and the right inequality in (3) can be strengthened. Namely, for any $K\ge 0$ we have

$$f_\alpha(y) \le 8\alpha(1\vee\ln K)f(y)I_{[0,K]}(y)+8f(y)I_{]K,\infty[}(y) .$$

Thus, for $K\ge e$,

$$\varphi_\alpha(u,v) \le 8\alpha\ln K\varphi(u,v)I_{[0,K]}(v/u)+8\varphi(u,v)I_{]K,\infty[}(u/v) . \tag{4}$$

When P and $\tilde{P}$ are probability measures the following characteristics called the Hellinger integrals of order $\alpha\in]0,1[$ is frequently used.

By definition,

$$H(\alpha,P,\tilde{P}) := E_Q \mathfrak{z}^{\alpha}\tilde{\mathfrak{z}}^{1-\alpha} .$$

It follows from the Iensen inequality that $H(\alpha,P,\tilde{P})\le 1$. Evidently,

$$d_H^2(\alpha,P,\tilde{P}) = 1-H(\alpha,P,\tilde{P}) .$$

The variation distance $Var(p-\tilde{p}):=E_q|\mathfrak{z}-\tilde{\mathfrak{z}}|$ and the Hellinger distance are related by the following inequalities:

$$2d_H^2(p,\tilde{p}) \le Var(p-\tilde{p}) \le 2\sqrt{2}\sqrt{p(\Omega)+\tilde{p}(\Omega)}\,d_H(p,\tilde{p}). \tag{5}$$

The proof is obvious because

$$(\sqrt{\mathfrak{z}}-\sqrt{\tilde{\mathfrak{z}}})^2 \le |\mathfrak{z}-\tilde{\mathfrak{z}}| = |\sqrt{\mathfrak{z}}-\sqrt{\tilde{\mathfrak{z}}}||\sqrt{\mathfrak{z}}+\sqrt{\tilde{\mathfrak{z}}}|.$$

In particular, for the probability measures P and $\tilde{P}$

$$Var(P-\tilde{P}) \le 4d_H(P,\tilde{P})$$

<u>3. Relation between contiguity and the Hellinger integrals</u>. Let us consider the following conditions:

(a) $(\tilde{p}^n)\triangleleft(p^n)$;

(b) $\lim_{K\uparrow\infty} \overline{\lim_n}\, \tilde{p}^n(z^n\ge K)=0$;

(c) $\lim_{\alpha\downarrow 0} \overline{\lim_n}\, d_H(\alpha,p^n,\tilde{p}^n)=0$;

(d) $\overline{\lim_n}\, d_H(p^n,\tilde{p}^n)<\infty$.

<u>Lemma 2</u>. The following statements are true:

$$(a)\Leftarrow(b)\Leftarrow(c)\Rightarrow(d).$$

If (d) holds, then (a),(b),(c) are equivalent.

<u>Proof</u>. (c)⇒(b). From the definition it follows that for any $\alpha\in]0,1/2]$ there exists $K=K(\alpha)\ge 1$ such that $f_\alpha(y)\ge f(y)$ for all $y\ge K$. Thus, for all u,v we have

$$\varphi_\alpha(u,v)I_{\{v\ge Ku\}} \ge \varphi(u,v)I_{\{v\ge Ku\}}$$

and

$$d_H^2(\alpha,p^n,\tilde{p}^n) \geq E_{q^n}\varphi_\alpha(\mathfrak{z},\tilde{\mathfrak{z}})I_{\{\tilde{\mathfrak{z}}\geq K\mathfrak{z}\}} \geq E_{q^n}\varphi(\mathfrak{z},\tilde{\mathfrak{z}})I_{\{\tilde{\mathfrak{z}}\geq K\mathfrak{z}\}}$$

$$= (1/2)E_{\tilde{p}^n}(\sqrt{1/Z}-1)^2 I_{\{Z\geq K\}} \geq (1/2)(\sqrt{1/K}-1)^2\tilde{p}^n(Z\geq K)$$

(for the sake of simplicity we retain the superscript n only in the notations for measures). As a result, for any $\alpha\in]0,1/2]$ we have

$$\overline{\lim_{K\uparrow\infty}}\ \overline{\lim_{n}}\ \tilde{p}^n(Z\geq K) \leq 2\ \overline{\lim_{n}}\ d_H^2(\alpha,p^n,\tilde{p}^n) ,$$

and (c)⇒(b).

(b)⇒(a). Assume that $\lim p^n(\Gamma^n)=0$. The Lebesgue decomposition

$$\tilde{p}^n = Z\ p^n + I_{\{Z=\infty\}}\tilde{p}^n$$

of the measure $\tilde{p}^n$ with respect to p^n implies that

$$\tilde{p}^n(\Gamma^n) = E_{p^n}ZI_{\{Z\leq K\}\cap\Gamma^n}+\tilde{p}^n(\Gamma^n,Z\geq K) \leq Kp^n(\Gamma^n)+\tilde{p}^n(Z\geq K).$$

So, we have

$$\overline{\lim_{n}}\ \tilde{p}^n(\Gamma^n) \leq \overline{\lim_{n}}\ \tilde{p}^n(Z\geq K)$$

and the implication (b)⇒(a) becomes evident.

(c)⇒(d). This implication is true because the left inequality in (1) implies

$$\overline{\lim_{n}}\ d_H(p^n,\tilde{p}^n) \leq c_\alpha\ \overline{\lim_{n}}\ d_H(\alpha,p^n,\tilde{p}^n) .$$

Assume that the condition (d) is satisfied.

(b)⇒(c). Making use of (4) for $K\geq e$ we have that

$$d_H^2(\alpha,p^n,\tilde{p}^n) \leq 8\alpha\ \ln K\ E_{q^n}\varphi(\mathfrak{z},\tilde{\mathfrak{z}})I_{\{\tilde{\mathfrak{z}}\leq K\mathfrak{z}\}} + 8\ E_{q^n}\ \varphi(\mathfrak{z},\tilde{\mathfrak{z}})I_{\{\tilde{\mathfrak{z}}\geq K\mathfrak{z}\}}$$

$$\leq 8\ \alpha\ \ln K\ d_H^2(p^n,\tilde{p}^n) + 4\ E_{\tilde{p}^n}(\sqrt{1/Z}-1)^2 I_{\{Z\geq K\}}$$

$$\leq 8\ \alpha\ \ln K\ d_H^2(p^n,\tilde{p}^n) + 4\tilde{p}^n(Z\geq K).$$

Now the desired statement follows if we take sequentially $\overline{\lim_{n}}$, $\overline{\lim_{\alpha}}$ and $\overline{\lim_{K}}$.

(a)⇒(b). Note that (a) is equivalent to the following

condition:

(a') $\overline{\lim\limits_{k}}\ \overline{\lim\limits_{n}}\ \tilde{P}^n(\Gamma_k^n)=0$ for any double sequence of sets (Γ_k^n), $\Gamma_k^n \in \mathcal{F}^n$, such that

$$\overline{\lim_{k}}\ \overline{\lim_{n}}\ P^n(\Gamma_k^n) = 0 .$$

Evidently,

$$d_H^2(P^n,\tilde{P}^n) \geq E_{q^n}(\sqrt{\mathfrak{z}}-\sqrt{\tilde{\mathfrak{z}}})^2 I_{\{\mathfrak{z}>0\}} \geq E_{P^n}(\sqrt{Z}-1)^2 I_{\{Z\geq K\}} \geq (\sqrt{K}-1)^2 P^n(Z\geq K).$$

So, (d) implies that

$$\overline{\lim_{K}}\ \overline{\lim_{n}}\ P^n\ (Z\geq K) = 0 .$$

Making use of (a') we come to (b).

Note that for probability measures $d_H^2(\alpha,P,\tilde{P})\leq 2$ and (d) holds automatically.

<u>4.Contiguity for random measures</u>. The notion of contiguity and the statements of Lemma 2 can be extended to sequences of random measures.

Assume that we are given probability spaces $(\Omega^n,\mathcal{F}^n,P^n)$ with random measures $\eta^n=\eta^n(\omega,dx)$ and $\tilde{\eta}^n=\tilde{\eta}^n(\omega,dx)$ defined on measurable spaces $(E^n,\mathcal{E}^n)$. We say that the sequence $(\tilde{\eta}^n)$ is contiguous to (η^n) modulo (P^n) (in symbols: $(\tilde{\eta}^n)\lhd(\eta^n)$ (mod (P^n))) if for any sequence of sets $\Gamma^n\in\mathcal{E}^n$ such that $P^n\text{-}\lim \eta^n(\Gamma^n)=0$ the equality $P^n\text{-}\lim \tilde{\eta}^n(\Gamma^n)=0$ also holds. (By definition, $P^n\text{-}\lim \xi^n=a$, $a\in R$ if $\lim\limits_{n} P^n(|\xi^n-a|\geq\varepsilon)=0$ for any $\varepsilon>0$.)

Put $\beta^n=(\eta^n+\tilde{\eta}^n)/2$, $y^n=d\eta^n/d\beta^n$, $\tilde{y}^n=d\tilde{\eta}^n/d\beta^n$, $Y^n=\tilde{y}^n/y^n$.

Let us consider the conditions:

(a) $(\tilde{\eta}^n)\lhd(\eta^n)$ (mod (P^n));

(b) $P^n - \overline{\lim\limits_{K\uparrow\infty}}\ \overline{\lim\limits_{n}}\ \tilde{\eta}^n\ (Y^n\geq K) = 0$;

(c) $P^n - \overline{\lim\limits_{\alpha\downarrow 0}}\ \overline{\lim\limits_{n}}\ d_H\ (\alpha,\eta^n,\tilde{\eta}^n) = 0$;

(d) $\underline{\lim}\ P^n\ (d_H(\eta^n,\tilde{\eta}^n)<\infty) = 1$.

<u>Lemma 3</u>. The following statements are true:

$$(a) \Leftarrow (b) \Leftarrow (c) \Rightarrow (d)$$

If (d) is satisfied, then (a), (b), (c) are equivalent.

Due to properties of the limit in probability, the proof of Lemma 3 is essentially the same as of Lemma 2.

5. For sequences of probability measures we have the following simple criterion of the complete asymptotical separability:

$$(\tilde{P}^n)\Delta(P^n) \quad \text{iff} \quad \underline{\lim}_n H(P^n,\tilde{P}^n) = 0.$$

We leave the proof to the reader.

§2. The local density for distributions of multivariate point process.

1. Let $(\Omega,\mathcal{F})$ be a measurable space with an increasing right-continuous family of σ-algebras $F=(\mathcal{F}_t)_{t\geq 0}$ and a multivariate point process $\Pi=(T_n,X_n)_{n\in\mathbb{N}}$. Here (T_n) is a sequence of stopping times with respect to F such that $0<T_n\leq T_{n+1}\leq\infty$ and $\{T_n<\infty\}=\{T_n<T_{n+1}\}$. It is assumed that X_n is a $\mathcal{F}_{T_n}$-measurable random variables with values in the extension of a Borel space $(E,\mathcal{E})$ by an extra point Δ and $\{X_n=\Delta\}=\{T_n=\infty\}$. Put $T_\infty=\lim T_n$.

The jump measure μ of the multivariate point process is defined by the formula

$$\mu(dt,dx) = \sum_{n\in\mathbb{N}} \varepsilon_{(T_n,X_n)}(dt,dx)$$

where ε_a is the Dirac measure at a. Let P be a probability measure on $(\Omega,\mathcal{F})$. Then there exists predictable random measure $\nu=\nu(dt,dx)$ called the compensator of μ such that for any nonnegative function $X=X(\omega,t,x)$ measurable with respect to $\tilde{\mathcal{P}}=\mathcal{P}\otimes\mathcal{E}$ ($\mathcal{P}$ is the predictable σ-algebra)

$$EX*\mu_\infty = EX*\nu_\infty$$

where

$$X*\mu_t = \int_{[0,t]}\int_E X(t,x)\mu(dt,dx) .$$

From now on we suppose that $\mathcal{F}_t=\mathcal{F}_0\vee\mathcal{F}^\mu_t$, where $\mathcal{F}^\mu_t=\sigma\{\mu([0,s]\times\Gamma),\ s\leq t,\ \Gamma\in\mathcal{E}\}$. Under this assumption the probability measure is uniquely defined by the compensator: if ν is the compensator of μ with respect to P and $\tilde{P}$ then $P=\tilde{P}$.

Let introduce the notion of the extended compensators of μ.

The using of it makes formulas essentially simpler.

Let o be another extra point, $(E^0,\mathcal{E}^0)=(E\cup o,\ \mathcal{E}\vee\{o\})$, $a_t=\nu(\{t\},E)$. Let us consider the random measure ν^0 on $(E^0,\mathcal{E}^0)$ defined by

$$\nu^0(dt,dx) = I_E\nu(dt,dx)+I_{\{a_t>0\}}(1-a_t)\varepsilon_{(t,o)}(dt,dx).$$

It is clear that ν^0 is the compensator of the random measure

$$\mu^0(dt,dx) = I_E\mu(dt,dx)+I_{\{a_t>0\}}(1-\mu(\{t\},E))\varepsilon_{(t,o)}(dt,dx).$$

We call ν^0 the extended compensator of μ.

2. Let ν and $\tilde{\nu}$ be the compensators of μ with respect to probability measures P and $\tilde{P}$. For the sake of simplicity we assume that the space $(\Omega,\mathcal{F},Q)$ (where $Q=(P+\tilde{P})/2$) is complete and $\mathcal{F}_0$ contains all Q-null sets from $\mathcal{F}$. Put $P_t=P|\mathcal{F}_t$, $\tilde{P}_t=\tilde{P}|\mathcal{F}_t$.

Now we formulate well-known results on local densities and absolute continuity in terms of the extended compensators.

Proposition 1. Let $\tilde{P}\ll P$. Then $\tilde{P}_0\ll P_0$ and there exists $\mathcal{P}\otimes\mathcal{E}^0$-measurable function $Y^0=Y^0(\omega,t,x)$ such that

$$\tilde{\nu}^0 = Y^0\nu^0 \qquad \tilde{P}\text{ - a.s.}$$

The local density of $\tilde{P}$ with respect to P, i.e. the right-continuous process $Z=(Z_t)$ with $Z_t=d\tilde{P}_t/dP_t$ is given by

$$Z_t = \begin{cases} Z_0 \exp\{\ln Y^0*\mu_t+(1-Y^0)*\nu_t^{0c}\}, & t<T_\infty \\ \varliminf_{s\uparrow T_\infty} Z_s, & t\geq T_\infty. \end{cases} \tag{6}$$

We have that $Z_-|Y^0-1|*\nu_t^0<\infty$ on $\{t<T_\infty\}$ P-a.s. and

$$Z_t = Z_0+Z_-(Y^0-1)*(\mu^0-\nu^0)_t \qquad \text{on } \{t<T_\infty\}\text{ P-a.s.} \tag{7}$$

Here $\nu^{0c}=I_{\{a=0\}}\nu^0$. Recall that $T_\infty=\inf\{t:\nu([0,t],E)=\infty\}$ P-a.s.

Define the Hellinger process $B=B(P,\tilde{P})$ as

$$B_t = \tfrac{1}{2}(\sqrt{d\nu^0/d\eta^0} - \sqrt{d\tilde{\nu}^0/d\eta^0})^2*\eta^0 \tag{8}$$

where $\eta^0=(\nu^0+\tilde{\nu}^0)/2$ (or any other random measure "dominating" ν^0 and $\tilde{\nu}^0$).

The Hellinger process $B(\alpha)=B(\alpha,P,\tilde{P})$ of order $\alpha\in]0,1[$ is defined by similar way:

$$B_t(\alpha) = \varphi_\alpha(d\nu^0/d\eta^0,\ d\tilde{\nu}^0/d\eta^0)*\eta^0_t. \tag{9}$$

If $\nu^0_t(\omega)$ denotes the restriction of $\nu^0(\omega,\cdot)$ on $([0,t]\times E^0,\ \mathcal{B}_{[0,t]}\otimes\mathcal{E}^0)$ then $B_t(\alpha)=d^2_H(\nu^0_t,\tilde{\nu}^0_t)$.

Theorem 1. $\tilde{P}\ll P$ iff the following conditions are satisfied

1. $\tilde{P}_0\ll P_0$;
2. $\tilde{\nu}^0(\omega,\cdot)\ll\nu^0(\omega,\cdot)$ $\tilde{P}$-a.s.
3. $B_\infty<\infty$ $\tilde{P}$-a.s.

The original proof of this result used the fact that $\tilde{P}\ll P$ iff $\tilde{P}(Z_\infty<\infty)=1$ (following from the Lebesgue decomposition). The main idea of the proof was to check the equality $\{Z_\infty<\infty\}=\{B_\infty<\infty\}$ $\tilde{P}$-a.s. using results on the sets of convergence of submartingales.

Now there exists other proofs of the above criterion. The most simple one is based on the study of the limit behaviour of the Hellinger integrals $H(\alpha,P,\tilde{P})$ as $\alpha\downarrow 0$ (it is easy to understand that $\tilde{P}\ll P$ iff $\underline{\lim}_{\alpha\downarrow 0} H(\alpha,P,\tilde{P})=1$).

The proof of the criterion of contiguity given below is an extension of the last idea. It is based on the criteria of contiguity of Lemmas 2 and 3.

§3. The Hellinger process and the Hellinger integrals.

1. Continue to study the situation of the last section when we have two probability measures corresponding to the compensators ν and $\tilde{\nu}$.

We construct the process $X(\alpha)=\mathfrak{z}^\alpha\tilde{\mathfrak{z}}^{1-\alpha}$ from the right-continuous Q-martingales $\mathfrak{z}_t=dP_t/dQ_t$ and $\tilde{\mathfrak{z}}_t=d\tilde{P}_t/dQ_t$. The Iensen inequality implies that $X(\alpha)$ is a Q-supermartingale. Since $0\le\mathfrak{z},\tilde{\mathfrak{z}}\le 2$, $\mathfrak{z}+\tilde{\mathfrak{z}}=2$, we have $0\le X(\alpha)\le 2$.

The Doob-Meyer decomposition of $X(\alpha)$ is

$$X(\alpha) = X_0(\alpha) - A_t(\alpha) + M_t(\alpha) \tag{10}$$

where $A(\alpha)$ is an increasing right-continuous predictable process, $M(\alpha)$ is a Q-martingale, $A_0(\alpha)=M_0(\alpha)=0$. We have

$$E_Q A_\infty(\alpha) \le E_Q X(\alpha) \le 2 , \tag{11}$$

$$E_Q \sup_t |M_t(\alpha)| \le E_Q \sup_t X_t(\alpha) + E_Q A_\infty(\alpha) \le 4. \tag{12}$$

Proposition 2. The process $A(\alpha)$ in (10) has the following form:

$$A_t(\alpha) = X_-(\alpha)\circ B(\alpha)_t = \int_{[0,t]} X_{s-}(\alpha)dB_s(\alpha) . \tag{13}$$

Proof. Let us consider the process

$$B'(\alpha) = \varphi_\alpha(y_1^0,\tilde{y}_1^0) * \kappa_t^0$$

where $\kappa^0=(\mathfrak{z}_-\nu^0+\tilde{\mathfrak{z}}_-\tilde{\nu}^0)/2$ is the Q-compensator of μ, $y_1^0=d\nu^0/d\kappa^0$ and $\tilde{y}_1^0=d\tilde{\nu}^0/d\kappa^0$ are $\mathcal{P}\otimes\mathcal{E}^0$-measurable functions (densities of the absolutely continuous components of ν^0 and $\tilde{\nu}^0$ with respect to κ^0). It is easily seen that $B'(\alpha)$ and $B(\alpha)$ coincide on $\{X_-(\alpha)>0\}$.

Let $\sigma=\inf\{t:X_t(\alpha)=0\}$, $\sigma_n=\inf\{t:\mathfrak{z}_t\vee\tilde{\mathfrak{z}}_t\le 1/n\}$. As zero is the absorbing state for nonnegative martingales, $\sigma=\lim\sigma_n$ and $\{X_-(\alpha)>0\}=\cup[\![0,\sigma_n]\!]$. On $[\![0,\sigma_n]\!]$ the functions y_1^0 and $\tilde{y}_1^0$ do not exceed $2/n$. Put $\tau_n=T_n\wedge n\wedge\sigma_n$. The local densities have bounded variation on $[\![0,\sigma_n]\!]$ and can be expressed as follows:

$$\mathfrak{z}_t = \mathfrak{z}_0 + \mathfrak{z}_-(y_1^0-1)*(\mu^0-\kappa^0)_t ,$$

$$\tilde{\mathfrak{z}}_t = \mathfrak{z}_0 + \tilde{\mathfrak{z}}_-(\tilde{y}_1^0-1)*(\mu^0-\kappa^0)_t .$$

Thus,

$$\Delta\mathfrak{z}_t/\mathfrak{z}_{t-} = \int_{E^0}(y_1^0(t,x)-1)\mu^0(\{t\},dx) ,$$

$$\Delta\tilde{\mathfrak{z}}_t/\tilde{\mathfrak{z}}_{t-} = \int_{E^0}(\tilde{y}_1^0(t,x)-1)\mu^0(\{t\},dx) .$$

By virtue of the formula of the change of variables we have

$$X_t^{\tau_n}(\alpha) = X_0(\alpha) + \alpha\mathfrak{z}_-^{\alpha-1}\tilde{\mathfrak{z}}_-^{1-\alpha}\circ\mathfrak{z}_{t\wedge\tau_n} + (1-\alpha)\mathfrak{z}_-^{\alpha}\tilde{\mathfrak{z}}_-^{-\alpha}\circ\tilde{\mathfrak{z}}_{t\wedge\tau_n}$$

$$+ \sum_{s\le t\wedge\tau_n} X_{s-}(\alpha)[(1+\Delta\mathfrak{z}_s/\mathfrak{z}_{s-})^{\alpha}(1+\Delta\tilde{\mathfrak{z}}_s/\tilde{\mathfrak{z}}_{s-})^{1-\alpha}-\alpha\Delta\mathfrak{z}_s/\mathfrak{z}_{s-}-(1-\alpha)\Delta\tilde{\mathfrak{z}}_s/\tilde{\mathfrak{z}}_{s-}]$$

$$= X_0(\alpha) + X_-(\alpha)[\alpha y_1^0 + (1-\alpha)\tilde{y}_1^0 - 1]*(\mu^0-\kappa^0)_{t\wedge\tau_n} - X_-(\alpha)\varphi_\alpha(y_1^0,\tilde{y}_1^0)*\mu^0_{t\wedge\tau_n}$$

$$= X_0(\alpha) - X_-(\alpha)\varphi_\alpha(y_1^0,\tilde{y}_1^0)*\kappa^0_{t\wedge\tau_n} + X_-(\alpha)[y_1^{0\alpha}\tilde{y}_1^{0\,1-\alpha} - 1]*(\mu^0-\kappa^0)_{t\wedge\tau_n} .$$

The right-hand side of the last equality is the Doob-Meyer decomposition of $X^{\tau_n}(\alpha)$. Put $m(\alpha)=X(\alpha)-X_0(\alpha)+X_-(\alpha)\circ B'(\alpha)$. We have proved that $m^{\tau_n}(\alpha)$ is a Q-martingale. At the moment T_∞ the processes $X(\alpha)$ and $B'(\alpha)$ are continuous. At σ we have two possibilities: $X_{\sigma-}(\alpha)=X_\sigma(\alpha)=0$, or $\sigma=\sigma_n$ for some n. Thus, $m(\alpha)=\lim_n m^{\tau_n}(\alpha)$ and $E_Q \sup_t|m_t(\alpha)|=\lim_n E_Q \sup_t|m_t^{\tau_n}(\alpha)|\le 4$ (see (12)). It follows that $m(\alpha)$ is a Q-martingale. So the defining equality for $m(\alpha)$ gives the Doob-Meyer decomposition of the Q-supermartingale $X(\alpha)$. Thus, the desired statement is true because $X_-(\alpha)\circ B'(\alpha)=X_-(\alpha)\circ B(\alpha)$.

The general properties of the Doob-Meyer decomposition for bounded semimartingales imply that

$$E_Q[A(\alpha)]^2 \le 4 . \tag{14}$$

Recall that the Doléans exponential $\mathcal{E}(-B(\alpha))$ has the following form:

$$\mathcal{E}_t(-B(\alpha)) = \exp\{-B_t(\alpha)\} \prod_{s\le t}(1-\Delta B_s(\alpha))e^{\Delta B_s(\alpha)} .$$

<u>Proposition 3</u>. The following equality holds:

$$X(\alpha) = \mathcal{E}(-B(\alpha))S(\alpha) \tag{15}$$

where $S(\alpha)$ is a nonnegative Q-supermartingale with the following properties:

(a) $S^\tau(\alpha)$ is a local Q-martingale for any stopping time τ such that $\mathcal{E}_\tau(-B(\alpha))>0$;

(b) if $\mathcal{E}_\tau(-B(\alpha))\ge c>0$ (c=const) then $S^\tau(\alpha)$ is a bounded Q-martingale.

<u>Proof</u>. Put $\sigma=\inf\{t:X_t(\alpha)=0\}$, $\theta=\theta_1\wedge\theta_2$, where $\theta_1=\inf\{t:\Delta B_t(\alpha)=1\}$, $\theta_2=\inf\{t:B_t(\alpha)=\infty\}$.

We show at first that $\theta\ge\sigma$. The process $A(\alpha)=X_-(\alpha)\circ B(\alpha)$ is finite. Thus, $\theta_2\ge\sigma$. Put $f=y_1^{0\alpha}\tilde{y}_1^{0\,1-\alpha}$, $\rho=\inf\{t:\Delta B_t'(\alpha)=1\}$. On the set $\{\rho<\infty\}$ we have

$$\int_{E^0} f(\rho,x)\kappa^0(\{\rho\},dx)$$

$$= 1-\int_{E^0}\varphi_\alpha(y_1^0(\rho,x),\tilde{y}_1^0(\rho,x))\kappa^0(\{\rho\},dx) = 1-\Delta B'_\rho(\alpha) = 0 .$$

So,

$$EI_{[\![\rho]\!]} f*\mu^0_\infty = EI_{[\![\rho]\!]} f*\kappa^0_\infty = 0 ,$$

and we have that

$$\int_{E^0} \ln[y_1^{0\alpha}(\rho,x)\tilde{y}_1^{0\,1-\alpha}(\rho,x)]\mu^0(\{\rho\},dx) = -\infty \quad \text{on} \quad \{\rho<\infty\}.$$

This means that $\rho\geq\sigma$. Thus, $\theta_1\geq\sigma$.

Put $S_t(\alpha)=X_t(\alpha)\mathcal{E}_t^{-1}(-B(\alpha))$ if $\mathcal{E}_t(-B(\alpha))>0$ and $S_t(\alpha)=0$ otherwise. We just proved that this definition leads to (15).

Assume that $\mathcal{E}_\tau(-B(\alpha))>0$. Then on $[0,\tau]$ the process $\Psi=\mathcal{E}^{-1}(-B(\alpha))$ satisfies the linear equation

$$\Psi = 1 + \Psi_-(1-\Delta B(\alpha))^{-1} \circ B(\alpha) .$$

Taking account this formula and the additive decomposition for $X(\alpha)$, we can calculate the product of $X(\alpha)$ and Ψ, and obtain that

$$S_{t\wedge\tau}(\alpha) = X_0(\alpha) + \Psi_-(\alpha) \circ M(\alpha)_{t\wedge\tau}.$$

It follows from above that $S(\alpha)$ is a Q-martingale and (a) is true.

The assertion (b) follows from (a) evidently.

Let (θ_n) be a sequence of stopping times announcing the predictable stopping time θ. We have proved that $S^{\theta_n}(\alpha)$ is a nonnegative local Q-martingale and, consequently, by Fatou's lemma, a Q-supermartingale. But Fatou's lemma implies also that the process $\bar{S}(\alpha)=S(\alpha)I_{[\![0,\theta[\![}+S_{\theta-}I_{[\![\theta,\infty[\![}$ is a Q-supermartingale as well. The equality $S(\alpha)=\bar{S}(\alpha)I_{[\![0,\theta[\![}$ shows that $S(\alpha)$ is also a Q-supermartingale.

<u>Corollary</u>. Assume that ν and $\tilde{\nu}$ are nonrandom, $T\in\mathbb{R}_+\cup\{+\infty\}$. Then

$$H_T(\alpha) = E_Q X(\alpha) = H_0(\alpha)\mathcal{E}_T(-B(\alpha)) \tag{16}$$

where $H_T(\alpha)=H(\alpha,P_T,\tilde{P}_T)$.

In general case we cannot obtain the explicit formula. Nevertheless, applying the Hölder inequality we easily obtain from the multiplicative decomposition the following statement.

Proposition 4. Let T be a stopping time. Then for $0<\alpha<\beta<1$ we have

$$H_T(\beta) \le H_0^{1/p}(\alpha) \; (E\mathcal{E}_T^{q/p}(-B(\alpha)))^{1/q} \tag{17}$$

where $p=(1-\alpha)/(1-\beta)$, $q=(1-\alpha)/(\beta-\alpha)$, and

$$H_T(\alpha) \le H_0^{1/r}(\beta) \; (\tilde{E}\mathcal{E}_T^{s/r}(-B(\beta)))^{1/s} \tag{18}$$

where $r=\beta/\alpha$, $s=\beta/(\beta-\alpha)$.

The above result can be used to obtain sufficient conditions for complete asymptotic separability.

§4. The criteria of contiguity.

1. Assume that for each $n\in\mathbb{N}$ we have a space $(\Omega^n,\mathcal{F}^n,P^n)$ with a multivariate point process with the jump measure μ^n. The compensator of μ^n with respect to probability measures P^n and $\tilde{P}^n$ are ν^n and $\tilde{\nu}^n$. We will use notations $H_T^n(\alpha) = H(\alpha,P_T^n,\tilde{P}_T^n)$, $B^n(\alpha)= B(\alpha,P^n,\tilde{P}^n)$ etc.

At first, we consider the case of nonrandom compensators.

Proposition 5. Assume that ν^n and $\tilde{\nu}^n$ are deterministic for all n. Then $(\tilde{P}^n)\triangleleft(P^n)$ iff the following conditions are satisfied:

I. $(\tilde{P}_0^n)\triangleleft(P_0^n)$;

II. $(\tilde{\nu}^{on})\triangleleft(\nu^{on})$;

III. $\overline{\lim_n} B_\infty^n<\infty$.

Proof. Note that the equality

$$\lim_{\alpha\downarrow 0} \lim_n H^n(\alpha) = 1$$

(the necessary and sufficient condition for the contiguity of probability measures $(\tilde{P}^n)\triangleleft(P^n)$, see Lemma 2(c)) holds iff

$$\underline{\lim_{\alpha\downarrow 0}}\ \underline{\lim_{n}}\ H_0^n(\alpha) = 1\ , \qquad \underline{\lim_{\alpha\downarrow 0}}\ \underline{\lim_{n}}\ \mathcal{E}_\infty(-B^n(\alpha)) = 1\ .$$

The first relation is equivalent to the condition I and the second one holds iff

$$\overline{\lim_{\alpha\downarrow 0}}\ \overline{\lim_{n}}\ B_\infty^n(\alpha) = 0\ . \tag{19}$$

Since $B_\infty^n(\alpha)=d_H^2(\alpha,P^n,\tilde{P}^n)$, it follows from Lemma 2 that (19) can be expressed as two conditions II and III.

Proposition 6. Assume that ν^n and $\tilde{\nu}^n$ are deterministic for all n. Then $(\tilde{P}^n)\Delta(P^n)$ iff the following conditions hold:

(a) $(\tilde{P}_0^n)\Delta(P_0^n)$;

(b) there exist a subsequence (n_k) and points $t_k\in\mathbb{R}$ such that

$$d_H(\nu^{on_k}(\{t_k\},\cdot),\tilde{\nu}^{on_k}(\{t_k\},\cdot)) = 1 \qquad (\text{i.e. } \Delta B_{t_k}^{n_k}=1);$$

(c) $\overline{\lim_{n}}\ B_\infty^n = \infty$.

The proof follows easily from the relation

$$(\tilde{P}^n)\Delta(P^n) \leftrightarrow \underline{\lim_{n}}\ H^n = 0.$$

2. The general case.

Theorem 2. $(\tilde{P}^n)\triangleleft(P^n)$ iff the following conditions are fulfilled:

I. $(\tilde{P}_0^n)\triangleleft(P_0^n)$;

II. $(\tilde{\nu}^{on})\triangleleft(\nu^{on})$ (mod $\tilde{P}^n$);

III. $\underline{\lim_{n}}\ \tilde{P}(B_\infty^n<\infty) = 1$.

Proof. Necessity. The necessity of I is evident. The multiplicative decomposition of the process $X(1/2)=\mathfrak{z}^{1/2}\tilde{\mathfrak{z}}^{1/2}$ implies that

$$\{\mathfrak{z}_\infty^{1/2}\tilde{\mathfrak{z}}_\infty^{1/2}=0\} \supseteq \{B_\infty=\infty\} \qquad Q\text{-a.s.}$$

As $Z_\infty=\tilde{\mathfrak{z}}_\infty\mathfrak{z}_\infty^{-1}$ and $\mathfrak{z}_\infty>0$ $\tilde{P}$-a.s., it follows that

$$\tilde{P}^n(Z_\infty<\infty) \le \tilde{P}^n(B_\infty<\infty).$$

But if $(\tilde{P}^n)\triangleleft(P^n)$ then $\lim\limits_{n} \tilde{P}^n(Z_\infty<\infty)=1$ (note that the condition (d) of Lemma 2 holds automatically) and condition III holds.

Put $\xi(\alpha)=\mathfrak{z}_-^{\alpha}\tilde{\mathfrak{z}}_-^{-\alpha}\circ B(\alpha)$. Note that the additive decomposition (10) and Proposition 2 imply formula

$$H^n(\alpha) = H_0^n(\alpha) - \tilde{E}^n\xi_\infty(\alpha) . \tag{20}$$

Using the criterion of contiguity we obtain that necessary

$$\overline{\lim_{\alpha\downarrow 0}}\ \overline{\lim_{n}}\ \tilde{E}^n\xi_\infty(\alpha) = 0 . \tag{21}$$

But for $\delta>0$, $\varepsilon\in]0,1[$ we have that

$$\tilde{E}^n\xi_\infty(\alpha) \ge \delta\varepsilon^{\alpha}(2-\varepsilon)^{-\alpha}\tilde{P}^n(B_\infty(\alpha)\ge\delta, \inf_t \mathfrak{z}_t\ge\varepsilon)$$

$$\ge \delta\varepsilon^{\alpha}(2-\varepsilon)^{-\alpha}[\tilde{P}^n(B_\infty(\alpha)\ge\delta)-P^n(\inf_t \mathfrak{z}_t<\varepsilon)] .$$

Thus,

$$\overline{\lim_{\alpha\downarrow 0}}\ \overline{\lim_{n}}\ \tilde{E}^n\xi_\infty(\alpha) \ge \delta \lim_{\alpha\downarrow 0}\ \overline{\lim_{n}}\ [\tilde{P}^n(B_\infty(\alpha)\ge\delta)-\tilde{P}^n(\inf_t \mathfrak{z}_t<\varepsilon)] .$$

But

$$\overline{\lim_{\varepsilon\downarrow 0}}\ \overline{\lim_{n}}\ \tilde{P}^n(\inf_t \mathfrak{z}_t<\varepsilon) = \overline{\lim_{\varepsilon\downarrow 0}}\ \overline{\lim_{n}}\ \tilde{P}^n(\sup_t Z_t>1/\varepsilon)$$

$$= \overline{\lim_{\varepsilon\downarrow 0}}\ \overline{\lim_{n}}\ \tilde{P}^n(Z_\infty>1/\varepsilon) = 0$$

and we have from above that

$$\overline{\lim_{\alpha\downarrow 0}}\ \overline{\lim_{n}}\ \tilde{E}^n\xi_\infty(\alpha) \ge \delta\ \overline{\lim_{\alpha\downarrow 0}}\ \overline{\lim_{n}}\ \tilde{P}^n(B_\infty(\alpha)\ge\delta) .$$

So the condition (21) is fulfilled only if

$$\overline{\lim_{\alpha\downarrow 0}}\ \overline{\lim_{n}}\ P(B_\infty(\alpha)\ge\delta) = 0, \qquad \forall\delta\ge 0, \tag{22}$$

and it rests to note that the condition III provides the equivalence of (22) and I (Lemma 2).

Sufficiency. As I holds iff $\underline{\lim}_{\alpha\downarrow 0}\ \underline{\lim}_{n}\ H_0^n(\alpha)=1$, the formula (20) shows that we need to check only the equality (21).

Let $\alpha\in]0,1/4[$, $\eta\in]0,1[$, $\Gamma=\{\tilde{\mathfrak{z}}\le\eta\}$. By virtue of the right inequality in (2) the difference $8B - B(\alpha)$ is an increasing process.

On Γ we have:

$$\mathfrak{z}^{\alpha}\tilde{\mathfrak{z}}^{-\alpha} \le 2\tilde{\mathfrak{z}}^{-\alpha-1/4}\eta^{1/4} \le 2\eta^{1/4}\tilde{\mathfrak{z}}^{-1/2} \le 2\eta^{1/4}\mathfrak{z}^{1/2}\tilde{\mathfrak{z}}^{-1/2} .$$

Thus,

$$\tilde{E}^n I_{\Gamma}\xi(\alpha) \le 16\eta^{1/4}\tilde{E}^n\xi_{\infty}(1/2) \le 16\eta^{1/4} . \tag{23}$$

By virtue of (14),

$$E_{Q^n}[\mathfrak{z}^{\alpha}\tilde{\mathfrak{z}}^{1-\alpha}\circ B(\alpha)_{\infty}]^2 \le 4$$

and as the result we have that

$$\tilde{E}^n(I_{\overline{\Gamma}}\circ\xi(\alpha)_{\infty})^2 \le 2E_{Q^n}[I_{\overline{\Gamma}}\mathfrak{z}^{\alpha}\tilde{\mathfrak{z}}^{-\alpha}(\tilde{\mathfrak{z}}/\eta)\circ B(\alpha)_{\infty}]^2 \le 8\eta^{-2}$$

Thus,

$$\begin{aligned}\tilde{E}^n I_{\overline{\Gamma}}\circ\xi(\alpha)_{\infty} &= \tilde{E}^n I_{\overline{\Gamma}\cap(B_{\infty}(\alpha)<\delta)}\mathfrak{z}^{\alpha}\tilde{\mathfrak{z}}^{-\alpha}\circ B(\alpha)_{\infty} + \tilde{E}^n I_{\overline{\Gamma}\cap(B_{\infty}(\alpha)\ge\delta)}\mathfrak{z}^{\alpha}\tilde{\mathfrak{z}}^{-\alpha}\circ B(\alpha)_{\infty} \\ &\le 2\delta_{\eta}^{-\alpha} + (\tilde{E}^n(I_{\overline{\Gamma}}\circ\xi(\alpha)_{\infty})^2)^{1/2}(\tilde{P}^n(B_{\infty}(\alpha)\ge\delta))^{1/2} \\ &\le 2\delta_{\eta}^{-\alpha} + 2\sqrt{2}\eta^{-1}(\tilde{P}^n(B_{\infty}(\alpha)\ge\delta))^{1/2}\end{aligned} \tag{24}$$

((22) is equivalent to the condition I in presence of III).

But (22)-(24) imply that

$$\varlimsup_{\alpha\downarrow 0}\ \varlimsup_{n}\ \tilde{E}^n\xi_{\infty}(\alpha) \le 16\eta^{1/4} + 2\delta .$$

Because the left-hand side of this inequality can be done arbitrarily close to zero, (21) holds. Theorem 2 is proved.

For the complete asymptotic separability, in general, it is impossible to prove a criterion similar to that of Proposition 4. Sufficient conditions are given by

<u>Proposition 7</u>. Assume that at least one of the following conditions are fulfilled:

(a) $(P_0^n)\Delta(P_0^n)$;

(b) $\varlimsup_{n} P^n(\mathcal{E}_{\infty}(-B)\le\delta)=1,\ \forall\delta>0.$

Then $(P^n)\Delta(P^n)$.

To prove it we need to note that (b) holds iff

$$\underline{\lim}_{n} E^n \mathcal{E}_\infty(-B) = 0 ,$$

and the inequality (18) shows that $\lim H^n(3/4)=0$ if any of the above conditions is fulfilled.

§5. Historical remarks. We mention here only a few "basic" works.

1. Absolute continuity. The famous paper due to S. Kakutani was the first publication on the problem of absolute continuity of measures, corresponding to random processes on the infinite time interval. Absolute continuity of point processes have been studied with the help of the martingale theory, e.g., in [2],[3],[4].

2. The notion of contiguity, playing an important role in asymptotical statistics, have been introduced by L. LeCam, [5]. Among works devoting to conditions of contiguity for random processes we point out [6],[7],[8],[9]. The Hellinger integrals actually appeared in [1], in modern setting they have been used intensively by F. Liese, see, [10],[11]. The additive and multiplicative decomposition of $X(\alpha)$ have been obtained in [12] in the setting of the general theory of the processes with applications to the variance distance between distributions.

3. The basic facts concerning multivariate point processes can be found in [3], see also [13]. As a reference book on the general theory of processes and martingales we recommend [14].

References

[1] Kakutani S. On equivalence of infinite product measures. Ann. Math., 49, 211-224 (1948).

[2] Kabanov Yu. M., Liptser R. Sh., Shiryayev A. N. Criteria of absolute continuity of measures corresponding to multivariate point processes (3rd USSR-Japan Symposium). Lect. Notes Math., 550, 232-252 (1976).

[3] Jacod J. Multivariate point processes: predictable projection, Radon-Nikodym derivatives, representation of martingales. Z. W-theorie verw. Geb., 31, 235-253 (1975).

[4] Kabanov Yu. M., Liptser R. Sh., Shiryayev A. N. Absolute

continuity and singularity of locally absolute continuous probability distributions, I, II. Math.. Sbornik, 107, 364-415 (1978), 108, 30-61 (1979) (in Russian). English translation: Math. USSR Sbornik, 35, 631-680 (1979), 36, 31-58 (1980).

[5] Le Cam L. Locally asymptotically normal families of distributions. Univ. of Calif. Publ. Statist., 3, 37-98 (1960).

[6] Eagleson G. K., Mémin J. Sur la contiguité de deux suites de mesures; généralisation d'un thèoréme de Kabanov-Liptser-Shiryayev. Lect. Notes Math., 920, 319-337 (1982).

[7] Liptser R. Sh., Shiryayev A. N. On the problem of "predictable" criteria of contiguity (4th USSR-Japan Symposium). Lect. Notes Math., 1021, 386-418 (1983).

[8] Mémin J. Distance en variation et condition de contiguité pour les lois de processus ponctuels. Preprint.

[9] Mémin J., Shiryayev A. N. Distance de Hellinger-Kakutani des lois correspondant à deux processus à accroisements independants. Z. W-theorie verw. Geb., 70, 67-89 (1985).

[10] Liese F. Hellinger integrals of Gaussian processes with independent increments. Stochastics, 6, 81-96 (1982).

[11] Liese F. An estimation of the Hellinger integrals of point processes. Preprint (1983).

[12] Kabanov Yu. M., Liptser R. Sh., Shiryayev A. N. On the variation distance for probability measures defined on a filtered space. Prob. Theory Rel. Fields, 71, 19-35 (1986).

[13] Brémaud P., Jacod J. Processus ponctuels et martingales: résultats récents sur la modélisation et le filtrage. Adv. Appl. Probab., 9, 362-416 (1977).

[14] Dellacherie C., Meyer P.-A. Probabilités et potentiels. Paris: Hermann, vol.1 (1975), vol.2 (1980).

Central Economical and Mathematical
Institute of the Academy of Science,
Moscow, USSR

ON BENFORD'S LAW: THE FIRST DIGIT PROBLEM

Shigeru Kanemitsu, Kenji Nagasaka, Gérard Rauzy, and Jau-Shyong Shiue

1. Introduction. It has long been known that many statistical tables with numerical data expressed in decimal form exhibit the peculiar phenomenon that the first significant digits k $(k=1,2,\ldots,9)$ of data, without regard to the position of their decimal points, do not occur with equal frequency $1/9$, as should be expected for a truly random table, but rather with frequency approximately equal to $\log \frac{k+1}{k}$. Here and throughout the paper, the base of the logarithm is taken to be 10, unless otherwise specified. Indeed, in many observed tables the first digit 1 occurs about 30% of all the data, 2 somewhat less and so on, with 9 occuring as a first digit less than 5% (see Table 1 of Ralph A. Raimi [22], which obeys this law rather badly, though). This particular logarithmic law, expressed in the suggestive language of probability theory as

$$\text{Prob}\{x \in D_k\} = \log \frac{k+1}{k}, \tag{1.1}$$

with D_k denoting the set of all real numbers whose standard decimal expansion begins with k, is called Benford's law, under suitable understanding of the left-hand side member. This law of Benford is so common and yet so remarkable that it is not surprising that many authors from varied fields of research tried to offer (both empirical and scientifical, not necessarily mathematical) explanations of it. All these explanations up to 1976, except for 4 papers to be alluded to later, are thoroughly reviewed in the aforementioned paper of Raimi to which we refer the reader for an exposition of Benford's law. For more recent references, see Nagasaka [19], [20].

Our standpoint in this paper is based, on the one hand, on B. J. Flehinger's philosophy [10] that since the smallest population containing the set of significant figures of all possible physical constants must be the set $\mathbf{N}$ of all positive integers, one should look for the explanation of Benford's law in the properties of $\mathbf{N}$ as represented in the radix number system; and, on the other, on Johann Cigler's heuristic observation that if for a sequence $\{a_n\}$ of positive reals, $\{\log a_n\}$ is uniformly distributed mod 1 (abbreviated as u.d. mod 1 or u.d. mod $\mathbf{Z}$), then $\{a_n\}$ may be said to satisfy Benford's law in the strictest sense, the sense of the natural density of D_k in $\{a_n\}$. Thus we adopt $\mathbf{N}$ as a model of the population (hence $D_k \cap \mathbf{N}$ is meant for D_k) and determine whether Benford's law holds or not for the integer sequences sampled according to a certain sampling procedure, mainly by checking whether $\{\log a_n\}$ is u.d. mod 1 or not.

We shall make an overview of the papers on Benford's law which are listed in [22] but not reviewed in detail or which have been published subsequently.

The geometrically sampled integer sequence $\{2^n\}$ has long been known to obey Benford's law (Arnol'd and Avez [2], or Rauzy [23]) as an application of an ergodic dynamical system. More generally, any geometrical sequence $\{ar^n\}$ is seen to satisfy Benford's law except for some special cases.

Linear recurrence sampling procedures (for more details, see Section 2) have been considered by several authors with a special reference to Fibonacci numbers $\{F_n\}$. R. L. Duncan [8] proved that $\{\log F_n\}$ is u.d. mod 1, of which another proof was given by Lauwerens Kuipers [15]. Brown and Duncan [4] and Kuipers and Shiue [13], extending the above results, proved that the sequence $\{u_n\}$ obeys Benford's law if $\{u_n\}$ satisfies a linear recurrence with some constraints. Nagasaka [19] succeeded in extending these results and removing unnecessary restriction on the linear recurrence formula, by introducing more essential conditions on the roots of maximum modulus of the corresponding characteristic equation. Nagasaka and Shiue [21] further developed the idea of [16] to use one of van der Corput's difference theorems [6] in proving the u.d. mod 1 property of $\{\log u_n\}$ (or rather of $\{\log |u_n|\}$) for a linear recurrence integer sequence of arbitrary order, and succeeded in giving not only another proof of, but also an extension of, Nagasaka's results. However, what have been settled in [19], [21] are relatively easier special cases in which either (a) the modulus of one of the roots dominate others or (b) there are two roots of maximum modulus (with both real or both purely imaginary).

Our main aim is to prove Theorem 2.1 which asserts that, under suitable conditions, the linear recurrence sampling procedures obey Benford's law if the characteristic equations have two distinct roots of maximum modulus. In particular, our theorem with $\theta \in Q$ completely contains Case (b) and gives a better description of it, while our theorem in the case of the second order linear recurrence and with $\theta \notin Q$ has been proved by Peter Schatte [27], independently of the authors. This will be done in Section 2. And in Section 3 we shall consider n-th convergents of simple continued fraction-expansions and, as an application of Theorem 4.1 in [19], prove that the denominators of certain quadratic irrationals obey Benford's law.

The reader may have observed that all the integer sequences considered up here are rapidly increasing. Another sequence of this type obeying Benford's law is $\{n!\}$ studied by Goto and Kano [11].

If, on the other hand, the sequence in question is slowly increasing, we cannot interpret the left-hand side of (1.1) neither as the natural density nor even as the Cesàro mean of any order of $\alpha_k(n)$, the characteristic function of D_k in $\mathbf{N}$. In this connection, Flehinger [10] considered successive cumulative averages (Hölder sums) of $\alpha_k(n)$ calling their limit the Banach limit, which is proved by Adler and Konheim [1] to be a finitely additive measure on $\mathbf{N}$ assigning measure 0 to every finite subset of $\mathbf{N}$. In particular, she has proved that $\{n\}$ obeys Benford's law in the sense of the Banach limit. This has been generalized by Nagasaka [19] to the case of polynomial sequences.

Schatte named the Banach limit, more symbolically, the H_∞-summation method, and considered the Mantissa distribution, a general notion including Benford's law. He remarked [26] that the above result of Nagasaka's is a consequence of one of his theorems, and also obtained a quantitative result as to the speed of convergence.

Flehinger's method, being a regular method with the property that it is stronger than all iterations of Cesàro methods, is weaker than the logarithmic summability method which yields the Dirichlet density. In our case it gives, as proved by Duncan [9],

$$\lim_{n\to\infty} (\log_e n)^{-1} \sum_{j=1}^{n} \alpha_k(j)/j = \log \frac{k+1}{k} .$$

It has been shown by R. E. Whitney [29] that the same holds for the characteristic function of D_k in the sequence of primes. Since the sequence of primes does not obey Benford's law in the strictest sense, this motivates us to consider various summation methods to interpret the left-hand side of (1.1) properly. We shall confine ourselves to the density found and proved to lie, in a way, between the natural and Dirichlet densities by Murata [18] and prove in Section 4 that this intermediate density also supplies the same value.

2. Benford's law for linear recurrence sequences. By $L(d,\underline{a},\underline{c},)$ we mean the linear recurrence sequence $\{u_n\}_{n=1,2,\ldots}$ satisfying the recurrence of order d, as well as the sampling procedure itself:

$$(2.1) \qquad u_{n+d} = \underline{u}_{n+d-1} \cdot {}^t\underline{a} = \sum_{j=0}^{d-1} a_j u_{n+j}, \quad n \in \mathbf{N},$$

with the initial condition

$$(2.2) \qquad \underline{u}_d := (u_d, u_{d-1}, \ldots, u_1) = \underline{c} := (c_d, c_{d-1}, \ldots, c_1),$$

where $\underline{a} = (a_{d-1}, a_{d-2}, \ldots, a_0)$ and $\underline{c}$ are d-dimensional integral vectors. As a result u_n may also take negative integral values, and we agree to make a convention that, in such a case, by $\{u_n\}$ we understand the sequence $\{|u_n|\}$.

The characteristic equation of (2.1) is given by

$$(2.3) \qquad \lambda^d = \sum_{j=0}^{d-1} a_j \lambda^j .$$

We suppose that (2.3) has p distinct roots $\alpha_1, \ldots, \alpha_p$ with multiplicity $m_1, \ldots, m_p$, respectively. Then u_n can be expressed in the form

$$(2.4) \qquad u_n = \sum_{j=1}^{p} b_j(n) \alpha_j^{n-1} ,$$

where $b_j(n)$ is a polynomial of degree $\leq m_j - 1$, depending only on $\underline{a}$, $\underline{c}$, and $\alpha_1, \ldots, \alpha_p$. We may suppose that all α_j's are non-zero, and so, arranging them in decreasing order of magnitude of their moduli

$$|\alpha_1| \geq |\alpha_2| \geq \cdots \geq |\alpha_p| \ (>0) ,$$

we note that the largest one must be >1.

Let us consider two cases where either (i) $|\alpha_1|$ dominates others (this contains the case where α_1 is a unique root of multiplicity d), in which case α_1 is real, or (ii) $|\alpha_1| = |\alpha_2|$ are strictly greater than others, in which case α_1 and α_2 are either both real and $\alpha_1 = -\alpha_2$ or imaginary conjugate pairs: $\overline{\alpha}_1 = \alpha_2$

We may suppose that $b_1(n) \neq 0$ for sufficiently large n, or that $v_n := b_1(n)\alpha_1^{n-1} + b_2(n)\alpha_2^{n-1} \neq 0$ for infinitely many n, as the case may be. Case (i), and Case (ii) with $\alpha_2 = -\alpha_1$ have been fully settled, while Case (ii) with $\alpha_2 = \bar{\alpha}_1$ has been treated only partially, in [21] (where α_1, α_2 are supposed purely imaginary). We shall deal with Case (ii). Note that in Case (ii) with $\alpha_2 = \bar{\alpha}_1$, we must have $b_2(n-1) = \overline{b_1(n-1)}$ since u_n is real and $u_{n_k} = r^{n_k}(v_{n_k} r^{-n_k} + o(1))$ for an infinite subsequence of $\mathbf{N}$ (where $r = |\alpha_1|$). Writing

$$(2.5) \qquad \alpha_1 = re^{2\pi i\theta} \qquad (r > 1) ,$$

we distinguish two cases according as (I) $\theta = \frac{k}{\ell} \in \mathbf{Q}$ or (II) $\theta \notin \mathbf{Q}$.

In Case (I) we partition v_n into ℓ subsequences $\{v_{n\ell+j}\}$, $0 \leq j \leq \ell-1$. Then

$$v_{n\ell+j} = r^{j-1}(r^{\ell})^n(b_1(n\ell+j)e^{2\pi i(j-1)k/\ell}+b_2(n\ell+j)e^{-2\pi i(j-1)k/\ell}) .$$

We further suppose

$$(2.6) \qquad b_1(n\ell+j)e^{2\pi i(j-1)k/\ell}+b_2(n\ell+j)e^{-2\pi i(j-1)k/\ell} \neq 0$$

for all $0 \leq j \leq \ell-1$. Then, as $n \to \infty$,

$$u_{n\ell+j} = r^{j-1}(r^{\ell})^n(v_{n\ell+j}r^{1-j-\ell n} + o(1)) ,$$

whence it follows that $\log u_{n\ell+j}$ is u.d. mod 1 for all j, $0 \leq j \leq \ell-1$ if $\log r$ is irrational (cf. the proof of Theorem 4.3 in [19]). Hence, $\log u_n$ is u.d. mod 1, so that $\{u_n\}$ obeys Benford's law unless r is of the form 10^m for any $0 \leq m \in \mathbf{Z}$ (cf. the proof of Theorem 3.3 in [21]).

In Case (II), u_n is expressed as

$$u_n = b_1(n)\alpha_1^{n-1} + \overline{b_1(n)}\,\bar{\alpha}_1^{\,n-1} + \sum_{j=3}^{p} b_j(n)\alpha_j^{n-1} ,$$

and so we deduce, on writing $b_1(n) = an^{s-1}e^{i\phi}$ + terms of lower degree, with $a > 0$, that

(2.7) $$u_n = v'_n + o(n^{s-1}r^n) ,$$

where

$$v'_n = an^{s-1}r^n\cos(2\pi n\theta + \phi) .$$

Since $\theta \notin \mathbb{Q}$, there exists a constant $C > 0$ such that for every $\varepsilon > 0$, the set $J(\varepsilon)$ of those n's satisfying

$$|\cos(2\pi n\theta + \phi)| \leq \varepsilon$$

has an upper density bounded by $C\varepsilon$. With this ε , the o-term in (2.7) can be estimated as

$$\leq \varepsilon^2 n^{s-1}r^n$$

if only n is sufficiently large, and so, for $n \notin J(\varepsilon)$ sufficiently large,

$$|u_n| = |v'_n|(1 + \eta_n) , \quad |\eta_n| < \varepsilon$$

(actually, $\eta_n \to 0$). By passage to the limit, we see that the distribution of the first digits of $|u_n|$ is the same as that of $|v'_n|$.

Note that θ being irrational, there is at most one value of n for which v'_n is zero, and therefore, $\log |v'_n|$ is always well-defined for n large enough.

Now, suppose the numbers $(1, \log r, \theta)$ are $\mathbb{Q}$-linearly independent (which, in particular, implies that $\log r$ is irrational). Then the sequence $((s-1) \log n + n \log r, n\theta)$ is u.d. mod $\mathbb{Z}^2$ in $\mathbb{R}^2$ ([17], Example 6.1), which implies ([17], Theorem 6.1) that for any bounded Riemann integrable function $F: \mathbb{R}^2 \to \mathbb{C}$, periodic mod $\mathbb{Z}^2$, one has

$$\frac{1}{N} \sum_{n<N} F((s-1)\log n + n \log r, n\theta) \to \int_0^1\int_0^1 F(x,y)\, dxdy.$$

Putting, in particular,

$$F_h(x,y) := e^{2\pi ih(\log a + x + \log |\cos(2\pi y + \phi)|)} ,$$

for $h \in \mathbb{Z} \setminus \{0\}$, we deduce that

$$\frac{1}{N}\sum_{n<N} e^{2\pi ih \log|v_n'|} \to \int_0^1\int_0^1 F_h(x,y)\,dxdy.$$

The last integral is seen to be zero by an application of Fubini's theorem. Hence, Weyl's criterion shows that $\log|v_n'|$ is u.d. mod 1, and thence that $\log|u_n|$ is u.d. mod 1. Thus we have proved

Theorem 2.1. Suppose the characteristic equation (2.3) has two distinct roots α_1, α_2 of maximum modulus:

$$|\alpha_1| = |\alpha_2| > |\alpha_3| \geq \cdots \geq |\alpha_p| ,$$

and write $\alpha_1 = re^{2\pi i\theta}$. Then, if $\theta = \frac{k}{\ell} \in \mathbb{Q}$ and (2.6) is satisfied, $\{u_n\}$ obeys Benford's law; if $\theta \notin \mathbb{Q}$ and $(1, \log r, \theta)$ are linearly independent over $\mathbb{Q}$, then $\{u_n\}$ obeys Benford's law.

Remark 2.1. In Case (I), for $k=\ell=1$, Condition (2.6) reduces to $b_1(n) \neq \pm b_2(n)$, and for $k=1$, $\ell=2$, it reduces to $\mathrm{Re}(b_1(n)i^n) \neq 0$, i.e. $\mathrm{Re}\, b_1(n) \neq 0$, $\mathrm{Im}\, b_1(n) \neq 0$. Hence our theorem covers Theorem 4.3 in [21] which in turn contains Theorem 3.3 as well as part of Theorem 3.1 in [21] as special cases (since (2.6) should be satisfied in the case $d=2$).

Remark 2.2. In Case (II) with $d=2$, our theorem coincides with Corollary 2 of Schatte [27]. For examples of r, θ satisfying the condition of Theorem 2.1, see, e.g., Example 1 of Schatte [27].

3. Benford's law for continued-fraction expansions. At the recent symposium held in Kyoto on "Transcendental number theory etc.", Professor H. Jager posed the following two questions after our talk on Benford's law: 'Do denominators of n-th convergents of continued-fraction expansions of almost all real numbers obey Benford's law?' and 'How about quadratic irrationals?' We can give a complete answer to the latter question, but the former one seems rather difficult.

Before stating the theorem let us recall some notation of simple continued-fraction expansions. The expansion algorithm is as follows. For a real number ω, we denote its fractional part by $\{\omega\}$. Define the first coefficient $a_1(\omega)$ of the simple continued-fraction expansion of ω to be the integral part of $1/\{\omega\}$, i.e. $a_1(\omega) = [1/\{\omega\}]$. Then define $a_2(\omega) = [1/\omega_1]$, where $\omega_1 := 1/\{\omega\} - a_1(\omega)$. Further define $a_3(\omega)$ from $\omega_2 := 1/\omega_1 - a_2(\omega)$ in the same manner. Continuing this process, say n times, we arrive at

$$(3.1) \qquad \omega = [\omega] + \cfrac{1}{a_1(\omega) + \cfrac{1}{a_2(\omega) + \cfrac{}{\ddots + \cfrac{1}{a_n(\omega) + \omega_n}}}} = \langle[\omega];a_1,\ldots,a_n,\omega_n\rangle \quad ,$$

say, where $p_n(\omega)/q_n(\omega) = \langle[\omega];\ a_1,\ldots,a_n,0\rangle$ is called the n-th convergent. In this notation our theorem reads:

Theorem 3.1. Let θ be a quadratic irrational whose simple continued fraction-expansion is ultimately periodic. Then $\{q_n(\theta)\}$ obeys Benford's law.

Proof. First recall that, by Lagrange's theorem, the sequence $\{a_n(\theta)\}$ is ultimately periodic, and that Benford's law is an asymptotic property. This justifies our assumption that $\{a_n(\theta)\}$ is purely periodic, and, for the sake of simplicity, we suppose that the period is equal to 1, putting $a_n(\theta) = s$ for all n with an $s \in \mathbf{N}$. The discussion of the general case will appear elsewhere.

The recurrence formula for $q_n(\theta)$ is

$$(3.1) \qquad q_{n+2}(\theta) = sq_{n+1}(\theta) + q_n(\theta) \ , \qquad n \in \mathbf{N} \ ,$$

and so the corresponding characteristic equation takes the form

$$(3.2) \qquad \lambda^2 = s\lambda + 1 \ .$$

Since (3.2) has two distinct real roots, it follows from Theorem 4.1 in [19] that $\{q_n(\theta)\}$ obeys Benford's law except when the root of larger modulus is of the form 10^m for some $0 \leq m \in \mathbf{Z}$. But $(s + \sqrt{s^2+4})/2$ cannot be 10^m for any $0 \leq m \in \mathbf{Z}$, which completes the proof of Theorem 3.1

4. Benford's law for prime numbers. It is an interesting problem to examine what summation methods are applicable to interpret the left-hand side of (1.1). Of course, the weaker the summation method is, in the stricter sense Benford's law holds. We shall restrict ourselves to considering the (0,1)-density $d_{0,1}$ in the sense of Murata [18], which

is weaker than the Dirichlet density $d_{1,0}$. Speaking of (α,β)-density, $\alpha-1$, β signifies the exponent of x and $\log x$ of the weight function, respectively. Namely, by $d_{0,1}(D_k)$ we understand

$$(4.1) \qquad \lim_{x\to\infty}\Big[\sum_{p\in D_k,\ p\leq x} \frac{\log_e p}{p}\Big] \Big/ \sum_{p\leq x} \frac{\log_e p}{p}$$

whenever the limit exists, where p runs through the set of primes with prescribed conditions. The Mertens asymptotic formula

$$\sum_{p\leq x} \frac{\log_e p}{p} = \log_e x + O(1)$$

is well-known; this can be sharpened, by the prime number theorem, into

$$\sum_{p\leq x} \frac{\log_e p}{p} = \log_e x + E + o(1) ,$$

where E is a constant. Using this, we can easily show that

$$(4.2) \qquad \sum_{p\in D_k,\ p\leq (k+1)10^m} \frac{\log_e p}{p} = (m+1)\log(1+\tfrac{1}{k}) + o(m) .$$

As in [29], we compute the lower and upper densities separately, with the aid of (4.2):

$$\underline{d_{0,1}(D_k)} \geq \lim_{n\to\infty}\Big[\sum_{p\in D_k,\ p\leq k10^n} \frac{\log_e p}{p}\Big]\Big/\log_e k10^n \geq \lim_{n\to\infty}\Big[\sum_{p\in D_k, p\leq (k+1)10^{n-1}} \frac{\log_e p}{p}\Big]$$

$$\cdot(\log_e k10^n)^{-1} = \lim_{n\to\infty}[n \log_e(1+\tfrac{1}{k}) + o(n)]\Big/[n\log_e 10 + \log_e k]$$

$$\overline{d_{0,1}(D_k)} \leq \lim_{n\to\infty}\Big[\sum_{p\in D_k, p\leq (k+1)10^n} \frac{\log_e p}{p}\Big]\Big/\log_e k10^n = \log(1+\tfrac{1}{k}) .$$

Thus we have

Theorem 4.1. The sequence $\{p\}$ of primes obeys Benford's law in the sense of the $(o,1)$-density, i.e.

$$\operatorname{prob}\{p\in D_k\} = d_{0,1}(D_k) = \log\frac{k+1}{k} ,$$

References

[1][§] R. L. Adler and A. G. Konheim, Solution of Problem 4999, Amer. Math. Monthly 70 (1963), 218-219.

[2][§] V. I. Arnol'd and A. Avez, Problèmes ergodiques de la mécanique classique, Gauthier-Villars, Paris 1967.

[3] P. Billingsley, Ergodic theory and information, John Wiley and Sons, Inc., New York-London-Sydney 1965.

[4] W. G. Brady, More on Benford's law, Fibonacci Quart. 16 (1978), 51-52.

[5] J. L. Brown and R. L. Duncan, Modulo one uniform recurrence of the second order, Proc. Amer. Math. Soc. 50 (1975), 101-106.

[6] J. G. van der Corput, Diophantische Ungleichungen, Acta Math. 56 (1931), 373-456.

[7] P. Diaconis, The distribution of leading digits and uniform distribution mod 1, Ann. Prob. 5 (1977), 72-81.

[8] R. L. Duncan, An application of uniform distributions to the Fibonacci numbers, Fibonacci Quart. 5 (1967), 137-140.

[9] R. L. Duncan, Note on the initial digit problem, ibid. 7 (1969), 474-475.

[10] B. J. Flehinger, On the probability that a random integer has initial digit A, Amer. Math. Monthly 73 (1966), 1056-1061.

[11] K. Goto and T. Kano, Uniform distribution of some special sequences, Proc. Japan Acad. Ser. A Math. Sci. 61 (1985), 83-86.

[12] G. H. Hardy, Divergent series, Oxford UP, London 1949.

[13] A. Ja. Hincin, Continued fractions, The University of Chicago Press, Chicago, Ill.-London 1964.

[14] P. Kiss and R. Tichy, Distribution of the ratios of the terms of a second order linear recurrence, to appear in Indag. Math.

[15] L. Kuipers, Remark on a paper by R. L. Duncan concerning uniform distribution mod 1 of the sequence of logarithms of Fibonacci numbers, Fibonacci Quart. 7 (1969), 465-466, 473.

[16] L. Kuipers and J.-S. Shiue, Remark on a paper by Duncan and Brown on the sequence of logarithms of certain recursive sequences, ibid. 11 (1973), 292-294.

[17] L. Kuipers and H. Niederreiter, Uniform distribution of sequences, John Wiley and Sons, Inc., New York-London-Sydney-Toronto 1974.

[18] L. Murata, On certain densities of sets of primes, Proc. Japan Acad. Ser. A Math. Sci. 56 (1980), 351-353; On some fundamental relatives among certain asymptotic densities, Math. Rep. Toyama Univ. 4 (1981), 47-61.

[19] K. Nagasaka, On Benford's law, Ann. Inst. Stat. Math. 36 (1984), 337-352.

[20] K. Nagasaka, Statistical properties of arithmetical sequences, Doctoral thesis, Tokyo Inst. Technology 1987.

[21] K. Nagasaka and J.-S. Shiue, Benford's law for linear recurrence sequences, submitted to Tsukuba J. Math.

[22] R. A. Raimi, The first digit problem, Amer. Math. Monthly 83 (1976), 521-538.

[23]§ G. Rauzy, Propriétés statistiques de suites arithmétiques, Presses Univ. France, Paris 1976.

[24]§ P. Schatte, Zur Verteilung der Mantissa der Gleichkommadarstellung einer Zerfallsgrösse, Z. Angew. Math. Mech. 53 (1973), 553-565.

[25] P. Schatte, On H_∞'-summability and the uniform distribution of sequences, Math. Nachr. 113 (1983), 237-243.

[26] P. Schatte, Estimates for the H_∞-uniform distribution, preprint.

[27] P. Schatte, On the uniform distribution of certain sequences and Benford's law, to appear in Math. Nachr.

[28] L. C. Washington, Benford's law for Fibonacci and Lucas numbers, Fibonacci Quart. 19 (1981), 175-177.

[29] R.E. Whitney, Initial digits for the sequence of primes, Amer. Math. Monthly 79 (1972), 150-152.

The references with § are those published before 1976 but not listed in [22].

·Department of Mathematics
Faculty of Science
Kyushu University
Fukuoka 812
Japan

Department of Mathematics
Faculty of Education
Shinshu University
Nishi-Nagano, Nagano 380
Japan

Faculté des Sciences de Luminy
Mathématique-Informatique
Case 901
70 Route Léon-Lachamp
13288 Marseille Cédex 9
France

Department of Mathematical Sciences
University of Nevada, Las Vegas
4505, Maryland Parkway
Las Vegas, Nevada 89154
U. S. A.

ONE-DIMENSIONAL DIFFUSIONS AND RANDOM WALKS IN RANDOM ENVIRONMENTS

K.Kawazu, Y.Tamura and H.Tanaka

Introduction

Let $\Xi = \{\xi(x), x \in Z\}$ be a sequence of i.i.d. random variables with values in $(0, 1)$. An integer-valued process $\{X_n, n = 0,1,...\}$ is called a random walk in the random environment Ξ if the transition probability in the frozen environment $\xi = (\xi(x), x \in Z) \in (0, 1)^Z$ is given by

$$P_\xi\{X_{n+1} = x + 1 | X_n = x\} = \xi(x) ,$$

$$P_\xi\{X_{n+1} = x - 1 | X_n = x\} = 1 - \xi(x) , \quad x \in Z .$$

The full distribution of $\{X_n\}$ is $\mathcal{P} = \int Q(d\xi)P_\xi$, where Q denotes the distribution of the random environment Ξ . Sinai [6] proved that if

$$E^Q\left\{\log \frac{1 - \xi(x)}{\xi(x)}\right\} = 0 ,$$

$$0 < E^Q\left\{\left|\log \frac{1 - \xi(x)}{\xi(x)}\right|^2\right\} = \sigma^2 < \infty ,$$

then $\sigma^2(\log n)^{-2}X_n$ is convergent in distribution as $t \to \infty$.

Recently Brox [1] obtained a similar result for the one-dimensional diffusion process $X(t, W)$ described by the stochastic differential equation

$$(1) \qquad dX(t) = dB(t) - \tfrac{1}{2}W'(X(t))dt , \quad X(0) = 0 ,$$

where $\{W(x), x \in \mathbb{R}\}$ is a Brownian environment independent of the Brownian motion $B(t)$. Schumacher [5] also stated similar results for a considerably wider class of asymptotically self-similar random environments including symmetric stable ones; he also mentioned that Sinai's result can be derived from the diffusion case by the aid of optional samplings.

The problems we discuss in this paper are the following:

(A) The limiting behavior of the diffusion process $X(t, W)$ for an asymptotically self-similar random environment.

(B) Derivation of a result of Sinai type for a random walk from the diffusion setup.

In each of the problems (A) and (B) our discussions are divided into two cases: special Case (I) and general Case (II). As in [1][4][6] the notion of suitably defined valleys of environments plays a fundamental role in the proof. The main difference between the case of previous works and our general Case (II) is that the bottom of a valley in our case is not necessarily a single point. In Case (II) our definition of a valley is somewhat complicated and the result is less explicit. Therefore we discuss separately special Case (I) in which the bottom of a valley is always a single point and the result obtained is quite analogous to those of Brox and Sinai. This case is almost the same as Schumacher's case; however, since [5] does not contain details of proof it will be worth while discussing this case. In this paper the results are stated in both cases but the proof is outlined only in Case (I). The outline of the proof in Case (II) is similar to that in Case (I). Details of the proof in Case(II) will appear elsewhere.

I. Main results

Denote by $\mathbb{W}$ the space of real-valued right continuous functions on $\mathbb{R}$ having left limits and vanishing at 0. We consider the Skorohod topology on $\mathbb{W}$. The meaning of a solution of (1) for $W \in \mathbb{W}$ is not clear since W is not differentiable in general. Instead of considering the exact meaning of a solution of (1) we deal only with diffusion processes with generator $\frac{1}{2}e^{W(x)}\frac{d}{dx}(e^{-W(x)}\frac{d}{dx})$. We construct such a diffusion as follows. Denote by Ω the space of continuous paths $\omega:[0,\infty)\to\mathbb{R}$ satisfying $\omega(0)=0$ and by P the Wiener measure on Ω. The value of $\omega(t)$ of ω at time t is often denoted by $B(t)$. Thus $\{B(t), P\}$ is a Brownian motion starting from 0. Let

$$L(t,x) = \lim_{\varepsilon\downarrow 0}\frac{1}{\varepsilon}\int_0^t \mathbb{1}_{[x,x+\varepsilon)}(B(s))ds \qquad \text{(local time)},$$

$$S(x) = \int_0^x e^{W(y)}dy ,$$

$$A(t) = \int_0^t e^{-2W(S^{-1}(B(s)))}ds = \int_{\mathbb{R}} e^{-2W(S^{-1}(x))}L(t,x)dx , \qquad t \geq 0 ,$$

where S^{-1} and A^{-1} are the inverse functions. Then the process $X(t, W) = S^{-1}(B(A^{-1}(t)))$ defined on the probability space (Ω, P) is a diffusion with generator $\frac{1}{2}e^{W(x)}\frac{d}{dx}(e^{-W(x)}\frac{d}{dx})$ starting from 0. If we set $(W^x)(\cdot) = W(\cdot + x)$, then $X^x(t, W) = x + X(t, W^x)$ is a diffusion with the same generator starting from x. The Brownian motion used here is not the same as the one in (1) but we use the same notation $B(t)$.

We consider a probability measure on $\mathbb{W}$ but it is always assumed that the Brownian motion $B(t)$ and the (random) environment W are independent. So the full probability distribution is the product of P and a probability measure considered on $\mathbb{W}$. Probability measures on $\mathbb{W}$ we consider are often supported by smaller spaces. We consider two examples of such smaller spaces, say $\mathbb{W}_I$ and $\mathbb{W}_{II}$. Our discussions are divided to Case (I) and Case (II) according as we deal with $\mathbb{W}_I$ or $\mathbb{W}_{II}$. Case (I) is somewhat special but it is still general enough to contain the case of symmetric stable environments.

Remark and notation:—— The local time defined in Itô-McKean's book[3] is $L(t,x)/2$. We use the notation:

$a \vee b = \max\{a, b\}$, $\sup_G W = \sup\{W(x): x \in G\}$;

$a \wedge b$ and $\inf_G W$ are defined similarly;

$W(x-)$ = the left limit of W at x, $W(x+) = W(x)$;

$W^*(x) = W(x) \vee W(x-)$, $W_*(x) = W(x) \wedge W(x-)$;

the map $\tau_\lambda^\alpha : \mathbb{W} \to \mathbb{W}$ is defined by $(\tau_\lambda^\alpha W)(x) = \lambda^{-1} W(\lambda^\alpha x)$, $x \in \mathbb{R}$, where $\alpha > 0$ and $\lambda > 0$;

$\tau_\lambda^\alpha \nu$ = the image measure of a probability measure ν on $\mathbb{W}$ under the map τ_λ^α;

ν is self-similar with exponent $\alpha > 0 \Leftrightarrow \tau_\lambda^\alpha \nu = \nu$, $\lambda > 0$.

We now proceed to our main results.

Case (I). The subspace $\mathbb{W}_I$ is defined as follows. Let $W \in \mathbb{W}$. By definition, W is said to be <u>oscillating</u> at x if

$$\sup_{I(+\varepsilon)} W > W(x), \quad \sup_{I(-\varepsilon)} W > W(x-), \quad \inf_{I(+\varepsilon)} W < W(x), \quad \inf_{I(-\varepsilon)} W < W(x-).$$

for any $\varepsilon > 0$, where $I(+\varepsilon) = (x, x+\varepsilon)$ and $I(-\varepsilon) = (x-\varepsilon, x)$. W is said to have a <u>local maximum</u> (resp. <u>local minimimum</u>) at x if $\sup_{I(\varepsilon)} W = W^*(x)$ (resp. $\inf_{I(\varepsilon)} W = W_*(x)$) for some $\varepsilon > 0$ where $I(\varepsilon) = (x-\varepsilon, x+\varepsilon)$. Let $\mathbb{W}_I$ be the set of $W \in \mathbb{W}$ with the following (i)-(iv).

(i) $\sup\{W(x): x \geq 0\} = \sup\{W(x): x \leq 0\} = \infty$.

(ii) If W is discontinuous at x, then W is oscillating at x.

(iii) For any open set G in $\mathbb{R}$ each of the sets

$$\left\{x \in G: W(x) = \sup_G W\right\}, \quad \left\{x \in G: W(x) = \inf_G W\right\}$$

contains at most one point.

(iv) W does not take a local maximum at $x = 0$.

Notice that (ii) implies the following:

(v) If W has a local maximum or local minimun at x, then W is continuous at x.

Also notice that $\mathbb{W}_I$ is a Borel subset of $\mathbb{W}$.

Let $W \in \mathbb{W}_I$. Then $V = (a,b,c)$ is called a <u>valley</u> of W if

(i) $a < b < c$,

(ii) W is continuous at a, b and c ,

(iii) $W(a) > W(x) > W(b)$ for every $x \in (a, b)$,
$W(c) > W(x) > W(b)$ for every $x \in (b, c)$.

For a valley $V = (a, b, c)$, $D = (W(a)-W(b)) \wedge (W(c)-W(b))$ is called the <u>depth</u> of V. A valley $V = (a, b, c)$ is said to be <u>proper</u> if

$$H_{a,b} < W(c) - W(b) , \quad H_{c,b} < W(a) - W(b) ,$$

where $H_{x,y} = \sup\{W(y') - W(x')\}$ wherein the supremum is taken over all pairs of x' , y' such that $x \le x' \le y' \le y$ or $y \le y' \le x' \le x$ according as $x < y$ or $x > y$. For a proper valley $V = (a, b, c)$, $A = H_{a,b} \vee H_{c,b}$ is called the <u>inner directed ascent</u> of V . It is clear that $A < D$. A valley $V = (a, b, c)$ is said to contain 0 if $a < 0 < c$.

<u>Proposition I-1</u>. For any $W \in \mathbb{W}_I$ and $r > 0$ there exists a proper valley $V = (a, b, c)$ of W containing 0 and satisfying $A < r \le D$. The bottom of such a proper valley is uniquely determined by W and r, and for each fixed r the map : $W \in \mathbb{W}_I \longrightarrow$ the unique bottom $(\in \mathbb{R})$ is Borel measurable.

<u>Proposition I-2</u>. If ν is a self-similar probability measure on $\mathbb{W}$ with $\nu(\mathbb{W}_I) = 1$, then for each $r > 0$ and W in some subset of $\mathbb{W}_I$ with full ν-measure there exists a proper valley $V = (a, b, c)$ of W containing 0 and satisfying $A < r < D$.

We are now in a position to state our results for the problem (A) in Case (I).

<u>Theorem I-A-1</u>. Let $W \in \mathbb{W}_I$ and let $V = (a, b, c)$ be a proper valley of W containing 0 with the depth D and the inner directed ascent A . Let $W_\lambda \in \mathbb{W}$, $\lambda > 0$, and assume that $W_\lambda \longrightarrow W$ (Skorohod topology) as $\lambda \to \infty$. Then for any $\varepsilon > 0$ and for any closed interval $I = [r_1, r_2]$ such that $A < r_1 < r_2 < D$ we have

$$\limsup_{\lambda\to\infty} \sup_{r\in I} P\{|X(e^{\lambda r}, \lambda W_\lambda) - b| > \varepsilon\} = 0 .$$

<u>Theorem I-A-2</u>. Let μ be a probability measure on $\mathbb{W}$ and let ν be a probability measure on $\mathbb{W}$ with $\nu(\mathbb{W}_I) = 1$. If $\tau_\lambda^\alpha \mu$ converges weakly to ν as $\lambda \to \infty$ for some α , then there exists Borel map

$$b_\lambda : \mathbb{W} \to \mathbb{R} \quad (\lambda > 0)$$

such that

(i) for any $\varepsilon > 0$

$$P\{|\lambda^{-\alpha}X(e^{\lambda}, w) - b_{\lambda}(W)| > \varepsilon\} \to 0$$

in probability with respect to μ as $\lambda \to \infty$,

(ii) the random variable $b_{\lambda}(W)$ defined on the probability space $(\mathbb{W}, \mu)$ converges in law to the random variable $b(W)$ defined on $(\mathbb{W}_I, \nu)$ as $\lambda \to \infty$.

To proceed to the random walk problem (B), let $\mathbb{W}_{\lambda}^{\alpha}$ be the space of step functions W in $\mathbb{W}$ which are flat on each interval $(n\lambda^{-\alpha}, (n+1)\lambda^{-\alpha})$, $n = 0, \pm 1, \cdots$. When $\lambda = 1$ we may suppress α in $\mathbb{W}_{\lambda}^{\alpha}$. In what follows $\alpha > 0$ is fixed. Given $W_{\lambda} \in \mathbb{W}_{\lambda}^{\alpha}$, we consider a random walk $Y_{\lambda}(n, \lambda W_{\lambda})$, $n = 0, 1, \cdots$, which is imbedded in the processes $X(t, \lambda W_{\lambda})$ as follows. Let

$\Gamma_{\lambda,0} = 0$,

$\Gamma_{\lambda,n}$ = the minimum of $t \geq 0$ such that

$$|X(t + \sum_{k=0}^{n-1} \Gamma_{\lambda,k}, \lambda W_{\lambda}) - X(\sum_{k=0}^{n-1} \Gamma_{\lambda,k}, \lambda W_{\lambda})| = \lambda^{-\alpha}, \quad n \geq 1,$$

$$Y_{\lambda}(n, \lambda W_{\lambda}) = X(\sum_{k=0}^{n} \Gamma_{\lambda,k}, \lambda W_{\lambda}), \quad n \geq 0$$

$$Y(n, W_1) = Y_1(n, W_1), \quad n \geq 0 .$$

Roughly the following Theorem I-B-1 follows from Theorem I-A-1 through optimal samplings but rigorously we still need some arguments even though they are quite similar to those for Theorem I-A-1.

<u>Theorem I-B-1.</u> Let the assumptions for W and V be the same as in Theorem I-A-1. Let $W_{\lambda} \in \mathbb{W}_{\lambda}^{\alpha}$, $\lambda > 0$, and assume that $W_{\lambda} \to W$ as $\lambda \to \infty$. Then for any $\varepsilon > 0$ and for any closed interval $I = [r_1, r_2]$ with $A < r_1 < r_2 < D$ we have

$$\lim_{\lambda\to\infty} \sup_{r\in I} P\{|Y_{\lambda}([e^{\lambda r}], \lambda W_{\lambda}) - b| > \varepsilon\} = 0 .$$

<u>Theorem I-B-2.</u> Let μ be a probability measure on $\mathbb{W}_1$ and let μ_n^{α} , $n \geq 2$, be the image measure on $\mathbb{W}_{\log n}^{\alpha}$ of μ under the map $\tau_{\log n}^{\alpha}$. Let ν be a probability measure on $\mathbb{W}$ with $\nu(\mathbb{W}_I) = 1$. If μ_n^{α} converges weakly to ν as $n \to \infty$ for some $\alpha > 0$, then there exists a sequence of Borel maps

$$b_n : \mathbb{W} \longrightarrow \mathbb{R}$$

such that

(i) for any $\varepsilon > 0$

$$P\{|\frac{Y(n,W)}{(\log n)^{\alpha}} - b_n(W)| > \varepsilon\} \to 0$$

in probability with respect to μ as $n \to \infty$,

(ii) the random variable $b_n(W)$ on the probability space $(\mathbb{W}_1, \mu)$ converges in law to the random variable $b(W)$ defined on $(\mathbb{W}_I, \nu)$ as $n \to \infty$.

Given $W \in \mathbb{W}_1$, we see easily that $Y(n, W)$ is a random walk on Z with

$$P\{Y(n+1, W) = x+1 \mid Y(n, W) = x\} = \xi(x),$$
$$P\{Y(n+1, W) = x-1 \mid Y(n, W) = x\} = 1 - \xi(x),$$

where

$$\xi(x) = \frac{1}{1 + e^{W(x)-W(x-1)}}, \quad x \in Z.$$

Therefore Theorem I-B-2 can be rephrased as in the following theorem which is an extension of Sinai's theorem.

Theorem I-B-3. Let $\{\xi(x), x \in Z\}$ be a family of random variables with values in $(0, 1)$. Let $\{Y(n, \xi), n = 0, 1, \cdots\}$ be a random walk on Z in the environment $\{\xi(x), x \in Z\}$ (in the sense explained in the beginning of Introduction). Define $W_\xi \in \mathbb{W}_1$ by

$$(\#) \qquad \begin{cases} W_\xi(x) \text{ is flat on each interval } (x, x+1), & x \in Z, \\ W_\xi(x) - W_\xi(x-) = \log\dfrac{1-\xi(x)}{\xi(x)}, & x \in Z. \end{cases}$$

Let μ be the probability distribution of W_ξ and let μ_n^α be the image measure of μ under the map $\tau_{\log n}^\alpha$. If μ_n^α converges weakly to a probability measure ν on $\mathbb{W}$ with $\nu(\mathbb{W}_I) = 1$ as $n \to \infty$ for some $\alpha > 0$, then there exists a sequence of Borel functions $b_n(\xi)$ on the space $(0, 1)^Z$ of realizations of environments such that

(i) for any $\varepsilon > 0$

$$P\{\left|\frac{Y(n,\xi)}{(\log n)^\alpha} - b_n(\xi)\right| > \varepsilon\} \to 0$$

in probability (w.r.t. ξ) as $n \to \infty$,

(ii) the random variable $b_n(\xi)$ converges in law to the random variable $b(W)$ defined on $(\mathbb{W}_I, \nu)$ as $n \to \infty$.

Remark. Brox [1] proved Theorem I-A-1 in the case $W_\lambda = W \in \mathbb{W}_I \cap \{$continuous environments$\}$ for every $\lambda > 0$ and then derived Theorem I-A-2 in the case where $\alpha = 2$ and μ is the Wiener measure.

Example. If ν is a probability measure on $\mathbb{W}$ with respect to which $\{W(x), x \geq 0\}$ and $\{W(-x), x \geq 0\}$ are independent symmetric stable processes with the same exponent, then ν is self-similar and $\nu(\mathbb{W}_I) = 1$. Thus the results of Brox and Sinai can be extended to the case of a symmetric stable environment.

Case (II). This case is considerably general since we deal with $\mathbb{W}_{II}$ which is defined as the set of $W \in \mathbb{W}$ satisfying only the condition

$$\overline{\lim_{x\to\infty}}\left\{W(x) - \inf_{[0,x]} W\right\} = \overline{\lim_{x\to-\infty}}\left\{W(x) - \inf_{[x,0]} W\right\} = \infty .$$

Obviously $\mathbb{W}_{II}$ is a Borel subset of $\mathbb{W}$.

Before stating our main theorems in Case (II) we introduce the space $\mathbb{K}$ of nonempty compact subsets of $\mathbb{R}$. If we set

$$\rho(K_1, K_2) = \inf\{\varepsilon > 0 : K_1 \subset U_\varepsilon(K_2),\ K_2 \subset U_\varepsilon(K_1)\}$$

for $K_1, K_2 \in \mathbb{K}$ where $U_\varepsilon(K)$ denotes the ε-neighborhood of K , then $\mathbb{K}$ becomes a (locally compact) metric space with metric ρ and hence we can consider the Borel structure on $\mathbb{K}$.

<u>Theorem II-A.</u> Let μ and ν be probability measures on $\mathbb{W}$ and assume that $\nu(\mathbb{W}_{II}) = 1$. If $\tau_\lambda^\alpha \mu$ converges to ν as $\lambda \to \infty$ for some $\alpha > 0$, then there eixst Borel maps

$$\mathbb{b}_\lambda : \mathbb{W} \longrightarrow \mathbb{K} , \quad (\lambda > 0)$$
$$\mathbb{b} : \mathbb{W}_{II} \rightarrow \mathbb{K} ,$$

such that the following (i) and (ii) holds.

(i) For any $\varepsilon > 0$

$$\lim_{\lambda\to\infty} P\{\lambda^{-\alpha} X(e^\lambda, W) \in U_\varepsilon(\mathbb{b}_\lambda)\} = 1 .$$

(ii) The probability distribution of the random set $\mathbb{b}_\lambda(W)$ on $(\mathbb{W}, \mu)$ converges to that of $\mathbb{b}(W)$ defined on $(\mathbb{W}_{II}, \nu)$ as $\lambda \to \infty$.

<u>Remark.</u> This theorem corresponds to Theorem I-A-2. We can obtain a theorem which corresponds to Theorem I-A-1 and which is used even for proving Theorem II-A. However, to state it we still need a suitably defined valley. We do not give the definition of a valley in Case (II) but we remark that the set $\mathbb{b}(W)$ in the above theorem is the bottom of a suitable valley. In Case (II) the bottom of a valley is not necessarily a single point but a compact set. A similar remark is also kept in mind for the following theorem which corresponds to Theorem I-B-3

<u>Theorem II-B.</u> Let $\{\xi(x), x \in Z\}$ be a family of random variables with values in $(0, 1)$. Let $\{Y(n, \xi), n = 0,1,\cdots\}$ be a random walk on Z in the environment $\{\xi(x), x \in Z\}$. Define $W_\xi \in \mathbb{W}_1$ by (#) of Theorem I-B-3. Let μ be the probability distribution of W_ξ and let μ_n^α be the image measure of μ under the map $\tau_{\log n}^\alpha$. If μ_n^α converges weakly to a probability measure ν on $\mathbb{W}$ with $\nu(\mathbb{W}_{II}) = 1$ as $n \to \infty$, for some $\alpha > 0$, then there exist $\mathbb{K}$-valued Borel functions $\mathbb{b}_n(\xi)$, $n \geq 3$, defined on the space $(0, 1)^Z$ of environments such that

(i) for any $\varepsilon > 0$

$$P\left\{\frac{Y(n,\xi)}{(\log n)^{\alpha}} \in U_{\varepsilon}(\mathbb{b}(\xi))\right\} \to 0$$

in probability (w.r.t. ξ) as $n \to \infty$;

(ii) the probability distribution of the random set $\mathbb{b}_n(\xi)$ converges to that of $\mathbb{b}(W)$ defined on $(\mathbb{W}_{II}, \nu)$ as $n \to \infty$, where $\mathbb{b}$ is the same as the one in Theorem II-A.

II. Discussions in Case (I)

Fundamentally we follow the methods of Brox [1] and Sinai [6] but there are many nontrivial points to be modified carefully.

1. Existence of proper valleys. Given any $W \in \mathbb{W}_I$ and $r > 0$ we prove that there exists a proper valley of W containing 0 with $A < r \leq D$ (Proposition I-1). Using the properties stated in the definition of $\mathbb{W}_I$ it is not hard to prove that there exists $c > 0$ such that W is continuous at c, $r \leq W(c) = \sup_{[0,c]} W$. Similarly there exists $a < 0$ such that W is continuous at a, $r \leq W(a) = \sup_{[a,0]} W$. Then there exists a unique $b \in (a, c)$ such that $W(b) = \inf_{(a,c)} W$ and W is continuous at b. Obviously $V=(a,b,c)$ is a valley of W containing 0 with depth $\geq r$. Now we prove the existence of a proper valley following [6] and [4]. If $V = (a,b,c)$ is a valley of W and if there exist a' and b' such that $b < a' < b' < c$ and $0 < W(a') - W(b') = H_{c,b}$, then $V' = (a,b,a')$ and $V = (a',b',c)$ are again valleys of W with depth $\geq H_{c,b}$. We say that they are obtained by a <u>right refinement</u> of V. A left refinement is defined similarly. Starting from a valley V containing 0 with depth $\geq r$ we can obtain proper valleys with $A < r \leq D$ after a finite number of refinements. Among such proper valleys there is one containing 0 and this proves Proposition I-1.

Another consideration on proper valleys due to Brox[1] is useful for proving Proposition I-2. Given $W \in \mathbb{W}_I$ and $r > 0$ we consider the set

$$\mathbb{B}_r = \{b \in \mathbb{R} \,|\, {}^{\exists}V = (a,b,c)\text{:valley with depth} \geq r\} .$$

Then it is easy to see that for any $b_1, b_2 \in \mathbb{B}_r$ with $b_1 < b_2$

$$\sup\{W(x) - W(b_i) : b_1 \leq x \leq b_2\} \geq r , \quad i = 1, 2 ,$$

from which it is also not hard to prove that $\mathbb{B}_r$ is a locally finite set. Next, to each $b \in \mathbb{B}_r$ we attach a proper valley $V^b = (a,b,c)$ as follows. Define a and c by

$$a = \begin{cases} \text{the unique point at which } \sup_{[b',b]} W \text{ is attained,} & \text{if } \mathbb{B}_r \cap (-\infty,b) \neq \phi, \\ -\infty , & \text{if } \mathbb{B}_r \cap (-\infty,b) = \phi , \end{cases}$$

$$c = \begin{cases} \text{the unique point at which } \sup_{[b,b'']} W \text{ is attained}, & \text{if } \mathbb{B}_r \cap (b,\infty) \neq \emptyset, \\ \infty, & \text{if } \mathbb{B}_r \cap (b,\infty) = \emptyset, \end{cases}$$

where b' (resp. b'') is the point of $\mathbb{B}_r \cap (-\infty, b)$ (resp. $\mathbb{B}_r \cap (b,\infty)$) which is closest to b. Then V^b becomes a proper valley*) of W. Also we have $A < r \leq D$ for each V^b. Among the proper valleys V^b, $b \in \mathbb{B}_r$, there exists one which contains 0. We denote it by $V_r = V_r(W) = (a_r, b_r, c_r)$. Then we can see that the map

(1.1) $$r \in (0, \infty) \longrightarrow (a_r, b_r, c_r) \in \mathbb{R}^3$$

is left continuous. Noting that $\mathbb{B}_r$ is a locally finite set which is decreasing in r, we see that the function (1.1) is a step function which has only finitely many jumps in each bounded interval away from 0. Also we have the scaling property:

(1.2a) $$V_r(\tau_\lambda^\alpha W) = \lambda^{-\alpha} V_{\lambda r}(W),$$

(1.2b) $$\{V_r(W), r > 0\} \overset{d}{=} \{\lambda^{-\alpha} V_{\lambda r}(W), r > 0\},$$

where $\overset{d}{=}$ means that the left and the right hand sides have the same distribution (as processes with time parameter r). With these preparations we can now prove Proposition I-2. It suffices to prove, for ν given as in the proposition, that

(1.3) $$\nu\{A_r < r < D_r\} = 1, \quad \forall r > 0,$$

where D_r (resp. A_r) is the depth (resp. the inner directed ascent) of $V_r = V_r(W)$. To prove (1.3) it is enough to show that

$$p(r) \equiv \nu\{r \text{ is a jump point of the map (1.1)}\}$$

is zero, because $A_r < r < D_r$ holds if r is not a jump point of the map (1.1). Since the set of jump points of (1.1) is locally finite (ν-a.s.),

$$J \equiv \inf\{r > 1 : (1.1) \text{ has a jump at } r\}$$

is strictly larger than 1 (ν-a.s.) and hence

$$p(r) \leq \nu\{J \leq r\} \downarrow 0 \quad \text{as} \quad r \downarrow 1.$$

But the scaling property (1.2b) implies that $p(r)$ does not depend on $r > 0$ and hence $p(r) \equiv 0$.

2. Exit times from proper valleys. In this section, following the method of Brox [1], we give the rate of growth, as $\lambda \to \infty$, of an exit

*) When we say that $V = (a, b, \infty)$ is a proper valley with $A < r \leq D$ we mean that there exists a sequence $\{c_n\}$ with $\lim c_n = \infty$ such that $V_n = (a, b, c_n)$ is a proper valley with $A_n < r \leq D_n$. A similar interpretation is done for $V = (-\infty, b, c)$ and $(-\infty, b, \infty)$.

time from a proper valley of W for the diffusion process with environment λW_λ under the assumption that

(2.1) $W_\lambda \in \mathbb{W},\ \lambda > 0,\ W \in \mathbb{W}_I$ and $W_\lambda \longrightarrow W$ (Skorohod topology) as $\lambda \to \infty$.

Proposition I-3. Under the above assumption let $V = (a,b,c)$ be a proper valley of W with depth D. Then for any $\delta > 0$ and a closed interval $I = [u, v] \subset (a, c)$

$$\lim_{\lambda\to\infty} \inf_{x\in I} P\left\{e^{\lambda(D - \delta)} < T_\lambda^x < e^{\lambda(D + \delta)}\right\} = 1 ,$$

where T_λ^x is the exit time from (a, c) for the diffusion process $X_\lambda^x(t) = x + X(t, \lambda W_\lambda^x)$.

In addition to the notation we already introduced (in the above and in the beginning of Part I), we use the following:

$$\tau(x_1, x_2) = \inf\{t \geq 0 : B(t) \notin (x_1, x_2)\} ,$$

$$L(x_1, x_2, x) = L(\tau(x_1, x_2), x) , \quad x_1 < x_2 , \ x \in \mathbb{R} ,$$

$$S_\lambda(x) = \int_0^x e^{\lambda W_\lambda(y)} dy , \quad \hat{S}_\lambda(y) = \int_x^y e^{\lambda W_\lambda(z)} dz \ (x \text{ is fixed}).$$

We state several lemmas without proof. The first three of them are found in [1] almost in the same form and the rest can be prove by modifying carefully the proof of the corresponding assertions in [1] (1° and 2° are used for this).

1° $\{L(\lambda x_1, \lambda x_2, \lambda x),\ x \in \mathbb{R}\} \stackrel{d}{=} \{\lambda L(x_1, x_2, x),\ x \in \mathbb{R}\}, \quad \lambda > 0$.

2° $T_\lambda^x \stackrel{d}{=} \int_a^c L(\hat{S}_\lambda(a), \hat{S}_\lambda(c), \hat{S}_\lambda(y)) e^{-\lambda W_\lambda(y)} dy , \quad a < x < c$.

3° For any fixed $\alpha > 0$, $\lambda > 0$ and $W \in \mathbb{W}$

$$\{X(t, \lambda\tau_\lambda^\alpha W),\ t \geq 0\} \stackrel{d}{=} \{\lambda^{-\alpha} X(\lambda^{2\alpha} t, W),\ t \geq 0\}.$$

4° $\lim_{\lambda\to\infty} P\left\{e^{\lambda(D - \delta)} < T_\lambda^b < e^{\lambda(D + \delta)}\right\} = 1$ for $\forall \delta > 0$,

5° $\lim_{\lambda\to\infty} \inf_{a<x<c} P\left\{\tilde{T}_\lambda^x < (c - a)^2 e^{\lambda(A + \delta)}\right\} = 1$ for $\forall \delta > 0$,

where $\tilde{T}_\lambda^x$ is the exit time from $(a,b) \cup (b,c)$ for the process $X_\lambda^x(t)$.

6° $\lim_{\lambda\to\infty} \inf_{x\in I} P\{\tilde{T}_\lambda^x = T_\lambda^x(b)\} = 1$,

where $I = [u,v] \subset (a,c)$, $\tilde{T}_\lambda^x$ is the same as above and $T_\lambda^x(b)$ is the passage time through b for the process $X_\lambda^x(t)$.

Once these lemmas are taken for granted, the proof of the proposition is given as follows. If we set

$$\hat{T}_\lambda^x = \inf\{t \geq 0 : X_\lambda^x(T_\lambda^x(b) + t) \notin (a, c)\},$$

then for $0 < \delta' < D - A$ we have

$$\begin{aligned}
&\inf_{x\in I} P\{e^{\lambda(D-\delta)} < T_\lambda^x < e^{\lambda(D+\delta)}\} \\
\geq &\inf_{x\in I} P\{e^{\lambda(D-\delta)} < T_\lambda^x(b) + \hat{T}_\lambda^x < e^{\lambda(D+\delta)},\ \tilde{T}_\lambda^x = T_\lambda^x(b),\ \tilde{T}_\lambda^x < (c-a)^2 e^{\lambda(A+\delta')}\} \\
\geq &\inf_{x\in I} P\{e^{\lambda(D-\delta)} < \hat{T}_\lambda^x < e^{\lambda(D+\delta)} - (c-a)^2 e^{\lambda(A+\delta')},\ \tilde{T}_\lambda^x = T_\lambda^x(b),\ \tilde{T}_\lambda^x < (c-a)^2 e^{\lambda(A+\delta')}\} \\
= &\inf_{x\in I} E\{f_\lambda(X_\lambda^x(T_\lambda^x(b)));\ \tilde{T}_\lambda^x = T_\lambda^x(b),\ \tilde{T}_\lambda^x < (c-a)^2 e^{\lambda(A+\delta')}\} \\
&\quad (\text{where } f_\lambda(y) = P\{e^{\lambda(D-\delta)} < T_\lambda^y < e^{\lambda(D+\delta)} - (c-a)^2 e^{\lambda(A+\delta')}\}) \\
= &f_\lambda(b)\cdot\inf_{x\in I} P\{\tilde{T}_\lambda^x = T_\lambda^x(b),\ \tilde{T}_\lambda^x < (c-a)^2 e^{\lambda(A+\delta')}\} \to 1 \quad \text{as} \quad \lambda\to\infty
\end{aligned}$$

by virtue of 4°, 5° and 6°.

3. Proof of Theorem I-A-1. We employ a coupling technique due to Brox [1]. As a matter of convenience we change the notation slightly. Let $V = (a,b,c)$ be a given proper valley of W containing 0 and set

$$\Omega = C([0,\infty)\to\mathbb{R})\ ,\ \hat{\Omega} = C([0,\infty)\to[a,c])\ .$$

$\omega(t)$ and $\hat{\omega}(t)$ denote the values of $\omega(\in\Omega)$ and $\hat{\omega}(\in\hat{\Omega})$, respectively, at time t. Let P_λ^x, $x\in\mathbb{R}$, be the probability measure on Ω induced by the diffusion process $X_\lambda^x(t)$ and let $\hat{P}_\lambda^y$, $a\le y\le c$, be the probability measure on $\hat{\Omega}$ induced by the diffusion process on $[a, c]$ starting from y, with (local) generator $\frac{1}{2}e^{\lambda W_\lambda(x)}\frac{d}{dx}(e^{-\lambda W_\lambda(x)}\frac{d}{dx})$ and with reflecting barriers at a and c. The latter diffusion has a unique invariant probability measure m_λ, for which the following holds:

(3.1) $\quad m_\lambda(G)\to 1$ as $\lambda\to\infty$ for any open set G containing b.

We set
$$\hat{P}_\lambda = \int_a^b \hat{P}_\lambda^y m_\lambda(dy)\ ,\quad \mathbb{P}_\lambda^{x,y} = P_\lambda^x\otimes\hat{P}_\lambda^y\ ,\quad \mathbb{P}_\lambda^x = P_\lambda^x\otimes\hat{P}_\lambda .$$

For a subset K of $\mathbb{R}$ let $T(K)$ (resp. $\hat{T}(K)$) be the first time at which $\omega(t)$ (resp. $\hat{\omega}(t)$) hits K and set $T = T((a,c)^c)$, $\hat{T} = \hat{T}((a,c)^c)$. We also set

$$\begin{aligned}
R &= \inf\{t\ge 0 : \omega(t) = \hat{\omega}(t)\}\ ,\\
T_R &= \inf\{t\ge R : \omega(t)\notin (a,c)\}\ ,\\
\hat{T}_R &= \inf\{t\ge R : \hat{\omega}(t)\notin (a,c)\ .
\end{aligned}$$

Then, R, T_R and $\hat{T}_R$ are random variables defined on the product space $\Omega\times\hat{\Omega}$. T and $\hat{T}$ are random variables on Ω and $\hat{\Omega}$, respectively, but they are also considered as random variables on $\Omega\times\hat{\Omega}$.

We first prove that

(3.2) $$\lim_{\lambda\to\infty}\mathbb{P}_\lambda^0\{R < e^{\lambda(A+\delta)}\} = 1 \quad \text{for } \forall\delta > 0\ .$$

We prove this in the case $b \neq 0$. Without loss of generality we may assume that $a < 0 < b < c$. Take a constant $\delta > 0$ such that $A + \delta < D$ and choose $c_1 \in (b, c)$ so that W is continuous at c_1 and

$$W(c_1) = \sup_{[b,c_1]} W, \quad W(c_1) - W(b) < A + \frac{\delta}{2}.$$

We then define $W'_\lambda \in \mathbb{W}$ by $W'_\lambda(x) = W_\lambda(x)$ for $x \le c_1$ and $= x - c_1 + W_\lambda(c_1)$ for $x > c_1$, and also define $W' \in \mathbb{W}$ similarly. Then $W' \in \mathbb{W}_I$ and $W'_\lambda \to W'$(Skorohod top.) as $\lambda \to \infty$. Next we define c' $(> c_1)$ by $W'(c') - W(b) = A + \frac{\delta}{2}$, that is, $c' = A + \frac{\delta}{2} + c_1 - W(c_1) + W(b)$. Then $V' = (a,b,c')$ is a proper valley of W' with depth $A + \frac{\delta}{2}$. We put

$$T_0 = T(\{c_1\}), \quad T_1 = T((a,c_1)^c), \quad T' = T((a,c')^c).$$

Then it can be proved that

$$\text{(3.3)} \qquad P^0_\lambda\{T_0 < \infty\} \ge P^0_\lambda\{T_0 = T_1\} \to 1 \quad \text{as} \quad \lambda \to \infty,$$

and hence

$$\begin{aligned}
&\mathbb{P}^0_\lambda\{R \le T_0\} \\
&\ge \mathbb{P}^0_\lambda\{\hat{\omega}(0) \in [0,c],\ \hat{\omega}(T_0) \in [a,c_1]\} \\
&\ge \mathbb{P}^0_\lambda\{\hat{\omega}(0) \in [0,c]\} + \mathbb{P}^0_\lambda\{\hat{\omega}(T_0) \in [a,c_1]\} - 1 \\
&= m_\lambda([0,c]) + m_\lambda([a,c_1])P^0_\lambda\{T_0 < \infty\} - 1 \\
&\to 1 \quad \text{as} \quad \lambda \to \infty.
\end{aligned}$$

Therefore, denoting by P'_λ the probability measure on Ω induced by the diffusion process $X(t, \lambda W'_\lambda)$ we have

$$\begin{aligned}
&\mathbb{P}^0\{R < e^{\lambda(A+\delta)}\} \\
&\ge P^0_\lambda\{T_0 < e^{\lambda(A+\delta)}\} - o(1) \quad \text{(by the above)} \\
&\ge P^0_\lambda\{T_1 < e^{\lambda(A+\delta)}\} - o(1) \quad \text{(by (3.3))} \\
&= P'_\lambda\{T_1 < e^{\lambda(A+\delta)}\} - o(1) \\
&\ge P'_\lambda\{T' < e^{\lambda(A+\delta)}\} - o(1) \\
&\to 1 \quad \text{as} \quad \lambda \to \infty
\end{aligned}$$

by Proposition I-3 since T' is the exit time from a proper valley with depth $A + \frac{\delta}{2}$. The proof of (3.2) in the case $b=0$ is omitted (see[1]).

We now give a proof of Theorem I-A-1. By Proposition I-3 and (3.1) we have

$$\begin{aligned}
&\hat{P}_\lambda\{e^{\lambda(D-\delta)} < \hat{T} < e^{\lambda(D+\delta)}\} \\
&= \int_a^c P^x_\lambda\{e^{\lambda(D-\delta)} < T < e^{\lambda(D+\delta)}\} m_\lambda(dx) \to 1, \quad \lambda \to \infty,
\end{aligned}$$

and this combined with (3.2) implies

$$\begin{aligned} p_\lambda &= \mathbb{P}^0_\lambda\{R < e^{\lambda r_1} < e^{\lambda r_2} \leq T_R\} \\ &\geq \mathbb{P}^0_\lambda\{R < e^{\lambda r_1} < e^{\lambda r_2} \leq \hat{T}\} \quad (\text{use } \hat{T} \leq \hat{T}_R) \\ &\to 1 \quad \text{as} \quad \lambda \to \infty \end{aligned}$$

for any r_1 and r_2 with $A < r_1 < r_2 < D$. Therefore for any $r \in [r_1, r_2]$ and for any neighborhood U of b

$$\begin{aligned} P\{X(e^{\lambda r}, \lambda W_\lambda) \in U\} &= P^0_\lambda\{\omega(e^{\lambda r}) \in U\} \\ &\geq \mathbb{P}^0_\lambda\{R < e^{\lambda r_1}, \omega(e^{\lambda r}) \in U, e^{\lambda r_2} \leq T_R\} \\ &= \mathbb{P}^0_\lambda\{R < e^{\lambda r_1}, \hat{\omega}(e^{\lambda r}) \in U, e^{\lambda r_2} \leq \hat{T}_R\} \\ &\geq p_\lambda + m_\lambda(U) - 1 \to 1 \quad \text{as} \quad \lambda \to \infty. \end{aligned}$$

4. Proof of Theorem I-A-2. Take a metric $m(\cdot,\cdot)$ in the space of 1-dim. probability distributions, compatible with the weak convergence, and write $M\{f(W)\} = m(p_f, p_b)$ where p_f is the distribution of $f(W)$ defined on $(\mathbb{W}, \mu)$ and p_b is that of $b(W)$ defined on $(\mathbb{W}, \nu)$. Denote by $\Delta(\lambda)$ the infimum of

$$J\{f(W)\} = \int \mu(dW) \int_0^1 P\{|\lambda^{-\alpha}X(e^\lambda, W) - b_\lambda(W)| > \varepsilon\}d\varepsilon + M\{f(W)\}$$

as $f(W)$ runs over all Borel functions on $\mathbb{W}$. Then for the proof of Theorem I-A-2 it is enough to show that

$$(4.1) \qquad \Delta(\lambda) \to 0 \quad \text{as} \quad \lambda \to \infty,$$

because, if (4.1) holds, the theorem is proved by choosing $b_\lambda(W)$ so that $J\{b_\lambda(W)\} < \Delta(\lambda) + \lambda^{-1}$ holds. To prove (4.1) take an arbitrary positive sequence $\{\lambda_n\}$ satisfying $\lim \lambda_n = \infty$. Since $\tau^\alpha_{\lambda_n} \to \nu$ weakly as $n \to \infty$, by Shorohod's theorem on realization of almost sure convergence we can find $\mathbb{W}$-valued random variables $\widetilde{W}$ and $\widetilde{W}^\alpha_{\lambda_n}$, $n \geq 1$, defined on a suitable probability space $(\widetilde{\Omega}, \widetilde{P})$ and with the following properties:

(i) The distributions of $\widetilde{W}$ and $\widetilde{W}^\alpha_{\lambda_n}$ are ν and $\tau^\alpha_{\lambda_n}\mu$, respectively.

(ii) $\widetilde{W}^\alpha_{\lambda_n} \to \widetilde{W}$ (Skorohod top.) as $n \to \infty$, $\widetilde{P}$-a.s.

Since ν is self-similar, Proposition I-2 implies that there exists a proper valley of $\widetilde{W}$ containing 0 and with $A < 1 < D$ for almost all $\widetilde{W}$ with respect to $\widetilde{P}$, and hence by Theorem I-A-1

$$(4.2) \qquad P\{|X(e^{\lambda_n r_n}, \lambda_n \widetilde{W}^\alpha_{\lambda_n}) - b(\widetilde{W})| > \varepsilon\} \to 0, \ \widetilde{P}\text{-a.s.}$$

for any sequence $\{r_n\}$ with $r_n \to 1$ and $\varepsilon > 0$. On the other hand it is not hard to prove that there exists a sequence of Borel maps $\widetilde{b}_n : \mathbb{W} \to \mathbb{R}$ such that

(4.3) $$\widetilde{b}_n(\widetilde{W}^{\alpha}_{\lambda_n}) \to b(\widetilde{W}) , \quad n \to \infty , \quad \widetilde{P}\text{-a.s.}$$

Then (4.2) implies

(4.4) $$P\{|X(e^{\lambda_n r_n}, \lambda_n \widetilde{W}^{\alpha}_{\lambda_n}) - \widetilde{b}_n(\widetilde{W}^{\alpha}_{\lambda_n})| > \varepsilon\} \to 0 , \quad \widetilde{P}\text{-a.s.}$$

Since $(\widetilde{W}^{\alpha}_{\lambda_n}, \widetilde{P}) \overset{d}{=} (W, \tau^{\alpha}_{\lambda_n}\mu) \overset{d}{=} (\tau^{\alpha}_{\lambda_n} W, \mu)$,

(4.4) implies

$$P\{|X(e^{\lambda_n r_n}, \lambda_n \tau^{\alpha}_{\lambda_n} \cdot W) - \widetilde{b}_n(\tau^{\alpha}_{\lambda_n} W)| > \varepsilon\} \to 0$$

in probability with respect to μ. Now we use the scaling property 3° of §2; the result is

$$P\{|\lambda_n^{-\alpha} X(\lambda_n^{2\alpha} e^{\lambda_n r_n}, W) - \widetilde{b}_n(\tau^{\alpha}_{\lambda_n} W)| > \varepsilon\} \to 0$$

in probability with respect to μ. Setting $r_n = 1 - 2\alpha\lambda_n^{-1}\log \lambda_n$, we have

$$P\{|\lambda_n^{-\alpha} X(e^{\lambda_n}, W) - \widetilde{b}_n(\tau^{\alpha}_{\lambda_n} W)| > \varepsilon\} \to 0$$

in prob. w.r.t. μ, which combined with (4.3) proves

$$\Delta(\lambda_n) \le J\{\widetilde{b}_n(\tau^{\alpha}_{\lambda_n} W)\} \to 0 \quad \text{as} \quad n \to \infty .$$

This proves (4.1).

5. <u>Sketch of the proof of Theorem I-B-1.</u> Instead of taking $X(\cdot, \lambda W_\lambda)$ in the definitions of $\Gamma_{\lambda,n}$ and $Y_\lambda(n, \lambda W_\lambda)$ we now take $X^x(\cdot, \lambda W_\lambda)$ to define $\Gamma^x_{\lambda,n}$ and $Y^x_\lambda(n)$. We then write N^x_λ for the exit time from (a_λ, c_λ) for $Y^x_\lambda(\cdot)$ where $a_\lambda = \min\{k\lambda^{-\alpha}: k\lambda^{-\alpha} \ge a\}$ and $c_\lambda = \max\{k\lambda^{-\alpha}: k\lambda^{-\alpha} \le c\}$. Since $\Gamma^x_{\lambda,n}$, $n \ge 1$, are i.i.d. for each fixed x and λ, Proposition I-3 implies (after an application of the law of large numbers) that for any $\delta > 0$ and for any closed interval $I \subset (a, c)$

$$P\left\{e^{\lambda(D-\delta)} < N^x_\lambda < e^{\lambda(D+\delta)}\right\} \to 1 \text{ uniformly in } x \in I \cap \lambda^{-\alpha}Z \text{ as } \lambda \to \infty .$$

Next consider a Markov chain $\hat{Y}_\lambda(n)$ on $[a_\lambda, c_\lambda] \cap \lambda^{-\alpha}Z$ with the same transition mechanism as that of $Y^x_\lambda(\cdot)$ inside (a_λ, c_λ) and with reflecting barriers at a_λ and c_λ. Let $P_\lambda(x, dy)$ be the transition function of $\hat{Y}_\lambda(\cdot)$. Then there exist a unique pair of probability measures m'_λ and m''_λ on $[a_\lambda, c_\lambda] \cap \lambda^{-\alpha}Z$ such that $\int P_\lambda(x,\cdot)m'_\lambda(dx) = m''_\lambda$, $\int P_\lambda(x,\cdot)m''_\lambda(dx) = m'_\lambda$ and $m''_\lambda(\{0\}) = 0$. We assume that the initial distribution of $\hat{Y}_\lambda(\cdot)$ is m'_λ and that $Y^x_\lambda(\cdot)$ and $\hat{Y}_\lambda(\cdot)$ are independent. Then as in (3.2), we can prove that $\text{Prob.}\left\{R' < e^{\lambda(A+\delta)}\right\} \to 1$ as $\lambda \to \infty$ where R' is the collision time between $Y^x_\lambda(\cdot)$ and $\hat{Y}_\lambda(\cdot)$. Once this is proved, Theorem I-B-1 can be proved by showing, as in §3, that for any $U = (b - \varepsilon, b + \varepsilon)$

$$P\{Y_\lambda([e^{\lambda r}], \lambda W_\lambda) \in U\} \sim p'_\lambda + m_\lambda(U) - 1 \sim 1, \quad \lambda \to \infty$$

where the definition of p'_λ is given similarly as for p_λ of §3 and $m_\lambda = m'_\lambda$ or m''_λ according as $[e^{\lambda r}]$ is even or odd.

References

[1] Th. Brox: A one-dimensional diffusion process in a Wiener medium, Ann Probab. 14 (1986), 1206-1218.

[2] A. O. Golosov: The limit distribution for random walks in random environments, Soviet Math. Dokl. 28 (1983), 18-22.

[3] K. Itô and H. P. McKean: Diffusion Processes and Their Sample Paths, Springer-Verlag (1965).

[4] H. Kesten: The limit distribution of Sinai's random walk in random environment, to appear in Physica.

[5] S. Schumacher: Diffusions with random coefficients, Contemporary Math. (Particle Systems, Random Media and Large Deviations, ed. by R. Durrett) 41 (1985), 351-356.

[6] Y. G. Sinai: The limiting behavior of a one-dimensional random walk in a random medium, Theory of Probab. and its Appl. 27 (1982) 256-268.

[7] H. Tanaka: Limit distributions for one dimensional diffusion processes in self-similar random environments, to appear in the IMA March (1986) Workshop Proceedings Volume, Springer-Verlag.

[8] H. Tanaka: Limit distribution for 1-dimensional diffusion in a reflected Brownian medium, to appear in Séminaire de Probabilités.

K. Kawazu:
Department of Mathematics
Faculty of Education
Yamaguchi University
Yamaguchi, 753 Japan

Y. Tamura and
H. Tanaka:
Department of Mathematics
Faculty of Science and Technology
Keio University
Yokohama, 223 Japan

THE DOMAIN OF ATTRACTION OF A NON-GAUSSIAN SELF-SIMILAR PROCESS WITH FINITE VARIANCE

Norio Kôno

§1.Introduction. Let $\{ Z(t); 0 \leqq t < +\infty \}$ be a real valued self-similar process with parameter $H>0$, i.e, for any $a>0$, $\{ Z(at) \}$ and $\{ a^H Z(t) \}$ have the same finite dimensional distributions. (We denote this by $Z(at) =$ (f.d.)$a^H Z(t)$). Furthermore, if $Z(t) =$(f.d.) $Z(t+b) - Z(b)$ holds for any $b>0$, $\{Z(t)\}$ is called a self-similar stationary increment process with parameter H (denoted by s.s.s.i.p.). We will say that a process $\{X(t); 0 \leqq t < +\infty\}$ belongs to the domain of attraction of a s.s.s.i.p. $Z(t)$ when there exists an increasing sequence a_n such that any finite dimensional distributions of $X(nt)/a_n$ converges in law to those of $Z(t)$.

In this paper we concern ourselves how rich the domain of attraction of a s.s.s.i.p. is, and we will give some results which give an example belonging to the domain of attraction of a non-Gaussian s.s.s.i.p.with finite variance. Trivially a s.s.s.i.p. $\{ Z(t) \}$ itself belongs to the domain of attraction of itself. Therefore we want to find other examples. Let $\{ X_n \}^{\infty}_{n=1}$ be a sequence of random variables and S_n be their partial sum ($S_0 = 0$). Set $X(t) = S_{[t]} + (t-[t]) X_{[t]+1}$, where $[t]$ is the integral part of t. If the process $\{ X(t) \}$ belongs to the domain of attraction of a s.s.s.i.p. $\{ Z(t) \}$, we say that $\{ X_n \}$ belongs to this domain of attraction. For example, by Donsker's theorem, any i.i.d. sequence with mean 0 and variance 1 belongs to the domain of attraction of a Brownian motion. On the other hand, there are many examples which belong to the domain of attraction of non-Gaussian s.s.s.i.p. with finite variance. Most of them are given by non-linear functional of a stationary Gaussian processes ([1],[2],[3],[5]). We will give an example of a sequence of random variables, not necessarily a functional of Gaussian random variables, which belongs to a very simple non-Gaussian s.s.s.i.p. with finite variance.

§2. Theorems. Let $\{B(t); 0 \leqq t < +\infty\}$ be a one dimensional Brownian motion and Y be a real valued non-constant random variable which is independent of $\{B(t)\}$. Then, obviously $X(t) = Y\,B(t)$ is a non-Gaussian s.s.s.i.p with parameter 1/2. In this case, any i.i.d. random sequence does not belong to the domain of attraction of this process. So, if a random sequence $\{X_n\}$ belongs to the domain of attraction of this process, $\{X_n\}$ must satisfy a kind of strong dependence.

Now we consider a random sequence $\{X_n\}$ satisfying the following conditions:

(i) $|X_n| \leqq M$ (uniformly bounded).

(ii) $E[X_n] = 0, \qquad E[X^2_n] = 1, \qquad E[X_n X_m] = 0 \ (n \neq m),$

(iii) there exists a nonnegative random variable Y (not necessarily constant) such that X^2_n converges to Y^2 weakly in L^2.

Remark. If $\{X_n\}$ is uniformly bounded in L^2 with the strong topology, then there exists a subsequence which converges in L^2 with the weak topology.

Theorem 1. Let $\{X_n\}$ be a random sequence satisfying the above conditions.

Then, there exists a subsequence $\{X_{n_k}\}_{k=1}^{\infty}$ which belongs to the domain of attraction of a s.s.s.i.p. $YB(t)$, where $\{B(t)\}$ is a Brownian motion independent of Y.

Set $\qquad c_k = |\alpha|\,\mathrm{sgn}\,k\,|k|^{-\alpha-1}, \quad k = \pm 1, \pm 2, \cdots \quad 0 < |\alpha| < 1/2, \quad c_0 = 0,$

and

$$Z_j = \sum_{k=-\infty}^{\infty} c_{j-k} X_{n_k},$$

where $\{n_k\}$ is the sequence obtained by rearranging $n_k, k = 1,2,\cdots$ in Theorem 1 such a way as $k = 1,-1,2,-2,\cdots$. Then, we have

Theorem 2. $\{Z_j\}$ belongs to the domain of attraction of a s.s.s.i.p. $Y\,B_H(t)$ where $\{B_H(t)\}$ is a fractional Brownian motion with parameter $H = 1/2 - \alpha$ independent of Y, where the fractional Brownian motion means a path continuous centered Gaussian process with $B_H(0) = 0$, $E[(B_H(t) - B_H(s))^2] = C_H|t-s|^H$, and

$$C_H = \int_{-\infty}^{\infty} \{|1-x|^{(1-H)/2} - |x|^{(1-H)/2}\}^2 \, dx .$$

Theorem 3. The convergence in Theorems 1 and 2 can be strengthened to the convergence in distribution on C[0,T] for any $T > 0$.

Our theorems are the extensions to functional limit theorems of the theorems due to Morgenthaler [4]. He also proved the existence of a sequence of real functions satisfying Assumptions (i)-(iii). ([4], p.300)

§3. **Proofs**. We need two lemmas which show how to choose a subsequence in Theorem 1.

Lemma 1. ([6] Theorem 5.1.2) Let $\{\xi_n\}$ be a sequence of real ramdom variables for which $E[|\xi_n|^2] \leqq K$, $n = 1,2,\cdots$. Then there exists a random variable η and a subsequence $n_1 < n_2 < \cdots$ such that

$$\lim_{N\to\infty} \sum_{k=1}^{N} \xi_{n_k} / N = \eta \quad \text{a.s.}$$

Lemma 2. Let $\{X_n\}$ be a real ramdom sequence satisfying the following conditions:

(i) $\forall n,\ |X_n| \leqq 1$.

(ii) $\forall n \neq \forall m,\ E[X_n X_m] = 0$.

Then, for a random variable with finite variance Y, there exists a subsequence $1 = n_1 < n_2 < \cdots$ such that for $k = 1,2,\cdots$

$$(*) \quad \sup_{0<t<\infty} \max |E[exp(-tY^2) X_{n_{n_1}} \cdots X_{n_{n_j}} X_{n_k})] \leqq 3^{-k+1}$$

where max is taken over all $1 \leqq j \leqq k-1$ and $1 \leqq n_1 < \cdots < n_j \leqq k-1$

Proof. Suppose that $1 = n_1 < n_2 < \cdots < n_{k-1}$ are already chosen in such a way that (*) is fulfilled. Set

$$f_{m, n_1, \cdots, n_j}(t) = E[exp(-tY^2) X_{n_{n_1}} \cdots X_{n_{n_j}} X_m],$$

where $1 \leqq n_1 < n_2 < \cdots < n_j \leqq k-1$. Then, by condition (ii), we have

$$\lim_{m\to\infty} f_{m,n_1,\cdots,n_j}(t) = 0 \quad \text{and also we have}$$

$$\lim_{t\to\infty} f_{m,n_1,\cdots,n_j}(t) = 0 \quad .$$

Since by the condition (i) we have

$$\begin{aligned} |f_{m,n_1,\cdots,n_j}(t) - f_{m,n_1,\cdots,n_j}(s)| &\leqq E[\,|\exp(-tY^2) - \exp(-sY^2)|\,] \\ &\leqq |t-s|E[Y^2] \end{aligned}$$

and

$$|f_{m,n_1,\cdots,n_j}(t)| \leqq 1,$$

$\{f_{m,n_1,\cdots,n_j}(t)\}_{m=n_{k-1}+1}^{\infty}$ is uniformly bounded equi-continuous which implies by Ascoli-Alzera's theorem that there exists a subsequence which converges uniformly to 0. So we can choose such n_k that (*) is fulfilled.

Now we start proving Theorem 1. First we choose subsequence satisfying the conclusions of Lemmas 1 and 2. Set for $0 = t_1 < t_2 < \cdots < t_p, u_1, \cdots, u_p \in R$,

$$\Phi_n = \Phi_n(u_1,\cdots,u_p) = E[exp(i/\sqrt{n} \sum_{r=1}^{p} u_r(X(nt_r) - X(nt_{r-1})))] \qquad \text{and}$$

$$\Phi = \Phi(u_1,\cdots,u_p) = E[\prod_{r=1}^{p} exp(-1/2\ (t_r - t_{r-1})u_r^2 Y^2)] \ .$$

To prove Theorem 1 it is sufficient to show that Φ_n converges to Φ for each $t_1, \cdots, t_p$ and $u_1, \cdots, u_p$. Since $|X_n| \leqq 1$, setting $X'(t) = S_{[t]}$ and denoting Φ'_n by replacing X' instead of X in Φ_n, it is sufficient to prove that Φ'_n converges to Φ. Using the elementary formula $e^{ix} = (1+ix)e^{-x^2/2+o(x^2)}$ as $|x|\to 0$, we have

$$exp(i/\sqrt{n} \sum_{r=1}^{p} u_r(X'(nt_r) - (X'(nt_{r-1}))$$

$$= exp(i/\sqrt{n} \sum_{r=1}^{p} u_r \sum_{[nt_{r-1}]<j\leq[nt_r]} X_{n_j})$$

$$= \prod_{r=1}^{p} \prod_{[nt_{r-1}]<j\le[nt_r]} (1 + iu_r X_{n_j} / \sqrt{n})\ exp(-u_r^2 X_{n_j}^2 / (2n) + o(u_r^2 / n))$$

$$= exp(o(\sum_{r=i}^{p} u_r^2 (t_r - t_{r-1})) \prod_{r=1}^{p} \prod_{[nt_{r-1}]<j\le[nt_r]} (1 + iu_r X_{n_j} / \sqrt{n}) exp(-u_r^2 X_{n_j}^2 / (2n))$$

Since $$| \prod_{r=1}^{p} \prod_{[nt_{r-1}]<j\le[nt_r]} (1 + iu_r X_{n_j} / \sqrt{n}) |$$

$$\leqq \prod_{r=1}^{p} \prod_{[nt_{r-1}]<j\le[nt_r]} (1 + u_r^2\, Xn_j^2 / n)^{1/2}$$

$$\leqq\ exp(1/2 \sum_{r=1}^{p} \sum_{[nt_{r-1}]<j\le[nt_r]} u_r^2 / n)\ \leqq\ exp(1/2\, t_p \sum_{r=1}^{p} u_r^2)$$

is bounded, it follows that for almost all ω

$$\lim_{n\to\infty} \{ \prod_{r=1}^{p} \prod_{[nt_{r-1}]<j\le[nt_r]} (1 + iu_r X_{n_j} / \sqrt{n}) exp(-u_r^2 X_{n_j}^2 / (2n)) -$$

$$\prod_{r=1}^{p} \{ \prod_{[nt_{r-1}]<j\le[nt_r]} (1 + iu_r X_{n_j} / \sqrt{n}) \} exp(-1/2\, u_r^2\, Y^2 (t_r - t_{r-1})) \} = 0 .$$

Taking account of Lemma 2 and the relation

$$\prod_{[nt_{r-1}]<j\le[nt_r]} (1 + iu_r X_{n_j} / \sqrt{n}) = 1 + \sum_{[nt_{r-1}]<j\le[nt_r]} iu_r R_j X_{n_j} / \sqrt{n} ,$$

where R_j does not involve the terms X_{n_k}, $n_k \geqq n_j$, we have

$$| E [\prod_{r=1}^{p} \{ \prod_{[nt_{r-1}]<j\le[nt_r]} (1 + iu_r X_{n_j} / \sqrt{n}) \} exp(-1/2\, u_r^2\, Y^2 (t_r - t_{r-1})] - \Phi |$$

$$\leqq 2^p \sum_{r=1}^{p} |u_r| \sum_{q=1}^{[nt_r]} (2/3)^q / \sqrt{n} \ \to\ 0 \quad as \quad n \to \infty \ .$$

This yierds $\lim_{n\to\infty} \phi'_n = \phi$. This completes the proof of Theorem 1.

To prove the uniform tightness of $\{X(nt)/\sqrt{n}\}$, which is a part of Theorem 3, first we prove the following lemma.

Lemma 3. For any subsequence $\{X_{n_j}\}$ which fulfuills the results of Lemmas 1 and 2, setting $S_k = X_{n_1} + \cdots + X_{n_k}$, we have

$$P(|S_k/\sqrt{k}| \geqq x) \leqq 2C\, e^{-x^2/4} \qquad \text{for all} \quad x \geqq 0 ,$$

where $\quad C = 1 + \sum_{j=1}^{\infty} (2/3)^{j-1}$.

Proof. By almost the same way as above we have for any $0 < a \leqq \sqrt{k}/2$

$$\prod_{j=1}^{k} exp(aX_{n_j}/\sqrt{k}) \leqq \prod_{j=1}^{k} (1 + aX_{n_j}/\sqrt{k})\, exp(a^2 \sum_{j=1}^{k} X_{n_j}^2/(2k) + o(a^2/2))$$

$$\leqq e^{a^2} \prod_{j=1}^{k} (1 + aX_{n_j}/\sqrt{k})$$

$$= e^{a^2} (1 + \sum_{j=1}^{k} a R_j X_{n_j}/\sqrt{k}) .$$

It follows by Lemma 2 that

$$E[\prod_{j=1}^{k} exp(aX_{n_j}/\sqrt{k})] \leqq e^{a^2} (1 + \sum_{j=1}^{k} (2/3)^{j-1})$$

$$\leqq C e^{a^2} .$$

By setting $a = x/2$ if $0 \leqq x \leqq \sqrt{k}$, and using Chebyshev' inequality we have

$$P(S_k/\sqrt{k} \geqq x) \leqq C\, e^{-x^2/4}.$$

Since $|S_k| \leqq k$, if $x > \sqrt{k}$ there is nothing to prove. This completes the proof.

Now we go back to the proof of the uniform tightness. To do this it is sufficient to show that

$$E[(X(nt)/\sqrt{n} - X(ns)/\sqrt{n})^4] \leqq \text{const.}\,|t-s|^2 .$$

but this estimation follows from Lemma 3.

Proof of Theorem 2.

Roughly speaking, we have

$$n^{-a} \sum_{1 \le j \le nt} c_{k\ j} \propto |k/n - t|^{-a} - |k/n|^{-a}.$$

Therefore we can expect that $n^{-H} \Sigma_{1 \leqq j \leqq nt} Z_j$ converges in law on the scace C[0,1] to $Y \int_{-\infty}^{\infty} (|t-s|^{-a} - |s|^{-a})\, dB(s) = YB_H(t)$. To justify this procedure we need several steps. It is not difficult to prove each step, so we omit it. We remark that the proof is slightly different for the cases $1/2 > a > 0$ and $0 > a > -1/2$. In case of positive a, $\Sigma_{-\infty}^{\infty} c_k = 0$ is crucial, because the kernel $|t-s|^{-a} - |s|^{-a}$ is not bounded with respect to s. We will give the steps only for the case of negative a.

Set

$$Z_j = \sum_{-\infty}^{\infty} c_{j-k} X'_k \quad \textit{and} \quad S(t) = \sum_{1 \le j \le t} Z_j ,$$

where $X'_k = X_{n_k}$.

Step 1. Set

$$T_n = \sum_{r=1}^{p} u_r S(nt_r) = \sum_{r=1}^{p} u_r \sum_{-\infty}^{\infty} \Big(\sum_{1 \le k \le nt_r} c_{k-j} \Big) X'_j \qquad \text{and}$$

$$T^*_n = \sum_{r=1}^{p} u_r \sum_{|j| \le Kn} \Big(\sum_{1 \le k \le nt_r} c_{k-j} \Big) X'_j .$$

Then for any $\varepsilon > 0$ there exist $K > 0$ and n_0 such that for any $n \geqq n_0$

$$E[|T_n - T^*_n|^2] \leqq \varepsilon n^{2H} .$$

Step 2. Set

$$A_{n,j} = \sum_{r=1}^{p} u_r(\sum_{1 \le k \le nt_r} c_{k-j}) \quad \text{and}$$

$$B_{n,j} = \sum_{r=1}^{p} u_r(|nt_r - j + 1|^{-a} - |1-j|^{-a}) .$$

Then
$$\lim_{n\to\infty} n^{-2H} \sum_{-\infty<j<\infty} |B_{n,j}||A_{n,j} - B_{n,j}| = 0 \qquad \text{and}$$

$$\lim_{n\to\infty} n^{-2H} \sum_{-\infty<j<\infty} (A_{n,j} - B_{n,j})^2 = 0 .$$

So we have

$$\lim_{n\to\infty} n^{-2H} \sum_{-\infty<j<\infty} A^2_{n,j} = \lim_{n\to\infty} n^{-2H} \sum_{-\infty<j<\infty} B^2_{n,j}$$

$$= \int_{-\infty}^{\infty} (\sum_{r=1}^{p} u_r f_{t_r}(s))^2 ds ,$$

where $f_t(s) = |t-s|^{-a} - |s|^{-a}$.

Step 3. There exists a step function $f^{\varepsilon}_r(s)$ such that

$$\sup_{-\infty<s<\infty} \sum_{r=1}^{p} |u_r||f_{t_r}(s) - f^{\varepsilon}_r(s)| \le \varepsilon/K \qquad \text{and}$$

$$| n^{-2H}B^2_{n,j} - n^{-1}(\sum_{r=1}^{p} u_r f^{\varepsilon}_r(j/n))^2| \le \varepsilon n^{-1}K^{-1} .$$

Step 4.

$$e^{iT^*_n n^{-H}} = \prod_{|j|\le Kn} (1+n^{-H}iA_{n,j}X'_j)e^{-\frac{1}{2}n^{-2H}A^2_{n,j}X'^2_j} + o(n^{-2H}A^2_{n,j}) .$$

Stpe 5.

$$\lim_{n\to\infty} |e^{iT^*_n n^{-H}} - \prod_{|j|\le Kn} (1+n^{-H}iA_{n,j}X'_j)e^{-\frac{1}{2}n^{-2H}B^2_{n,j}X'^2_j}| = 0 \qquad \text{a.s.}$$

Step 6. $\forall n \geqq n_0$

$$|exp(-\frac{1}{2}n^{-2H}\sum_{|j|\le Kn} B_{n,j}^2 X_j'^2) - exp(-\frac{1}{2n}(\sum_{|j|\le Kn}(\sum_{r=1}^{p} u_r f_r^\varepsilon(j/n))^2 X_j'^2))|$$

$$\leqq \text{const.}\,\varepsilon .$$

Step 7. (Use Lemma 1)

$$\lim_{n\to\infty} exp(-\frac{1}{2n}(\sum_{|j|\le Kn}(\sum_{r=1}^{p} u_r f_r^\varepsilon(j/n))^2 X_j'^2)$$

$$= exp(-\frac{1}{2}\int_{-K}^{K}(\sum_{r=1}^{p} u_r f_r^\varepsilon(s))^2 ds\, Y^2) \qquad \text{a.s.}$$

Step 8. (Use Lemma 2)

$$\lim_{n\to\infty} E[\prod_{|j|\le Kn}(1+n^{-H} iA_{n,j}X_j')exp(-\frac{1}{2}\int_{-K}^{K}(\sum_{r=1}^{p} u_r f_r^\varepsilon(j/n))^2 ds\, Y^2)]$$

$$= E[exp(-\frac{1}{2}\int_{-K}^{K}(\sum_{r=1}^{p} u_r f_r^\varepsilon(s))^2 ds\, Y^2)]$$

Step 9. Combining from Step 1 to Step 8, we have

$$\lim_{n\to\infty} E[e^{in^{-H}T_n}] = E[exp(-\frac{1}{2}\int_{-\infty}^{\infty}(\sum_{r=1}^{p} u_r f_{t_r}(s))^2 ds\, Y^2]$$

$$= E[exp(i\sum_{r=1}^{p} u_r \int_{-\infty}^{\infty} f_{t_r}(s)\, dB(s)\, Y)]$$

$$= E[exp(i\sum_{r=1}^{p} u_r B_H(t_r)\, Y)] \quad .$$

Convergence in C[0,1] can be shown by a similar way to Theorem 1.

References

[1] R.L.Dobrushin and P.Major; Non-central limit theorems for non-linear functionals of Gaussian fields. Z.Wahr.G. 50(1979), 27-52.

[2] L.Giraitis and D.Surgailis; CLT and other limit theorems for functionals of Gaussian processes. Z.Wahr.G. 7(1985), 191-212.

[3] P.Major; Limit theorems for non-linear functionals of Gaussian sequences Z.Wahr.G. (19891), 1-14.

[4] G.W.Morgenthaler; A central limit theorem for uniformly bounded orthoganal systems. Tran.A.M.S. vol. 79(1955), 281-311.

[5] M.S.Taqqu; Weak convergence to fractional Brownian motion and to the Rosenblatt process. Z.Wahr.G. 31(1975), 287-603.

[6] P.Révész; The laws of large numbers. Probability and Mathematical Statistics. Academic press, 1968.

Institute of Mathematics,
Yoshida College
Kyoto University
Kyoto 606, Japan

ABSOLUTELY CONTINUOUS SPECTRUM OF ONE-DIMENSIONAL RANDOM SCHRÖDINGER OPERATORS AND HAMILTONIAN SYSTEMS

S. Kotani

This is a continuation of S. Kotani [2]. In §4 of [2] we considered a Hamiltonian system associated with given random Schrödinger operator, which was first introduced by J. Moser [3] and applied to the determination of potentials with finite bands spectrum. In this note, we discuss the same problem more completely, and especially as one of its applications we give a sufficient condition for the smoothness of potentials by the vanishing set of the Lyapunov exponent: $\{\xi \in \mathbb{R}: \gamma(\xi) = 0\}$.

§1. Green functions and spectral measures.

Let $\Omega = L^2_{real}(\mathbb{R}, dx/1+|x|^3)$. Then it is known by S. Kotani [1] that for $q \in \Omega$, $L(q) = -\frac{d^2}{dx^2} + q$ defines a self-adjoint operator in $L^2(\mathbb{R}, dx)$, namely the boundaries $\pm\infty$ are of limit point type in Weyl's sense. For $q \in \Omega$, $\lambda \in \mathbb{C}$, let φ_λ, ψ_λ be unique solutions of

$$\begin{cases} L(q)\varphi_\lambda = \lambda\varphi_\lambda, \ \varphi_\lambda(0) = 1, \ \varphi'_\lambda(0) = 0 \\ L(q)\psi_\lambda = \lambda\psi_\lambda, \ \psi_\lambda(0) = 0, \ \psi'_\lambda(0) = 1. \end{cases}$$

Then it is known that there exist unique $h_\pm(\lambda, q) \in \mathbb{C}_+$ for any $\lambda \in \mathbb{C}_+$ such that

$$f_\lambda^\pm(x, q) = \varphi_\lambda(x, q) \pm h_\pm(\lambda, q)\,\psi_\lambda(x, q) \in L^2(R_\pm, dx).$$

Then the Green function $g_\lambda(x, y, q)$ (or in short $g_\lambda(x, y)$) of $L(q) - \lambda$ can be written by $f_\lambda^\pm$:

$$g_\lambda(x, y) = g_\lambda(y, x) = -(h_+(\lambda, q) + h_-(\lambda, q))^{-1} f_\lambda^+(x, q)\, f_\lambda^-(y, q)$$

for $x \geqq y$. Now for $\lambda \in \mathbb{C}_+$, set

$$H(\lambda) = \begin{bmatrix} \dfrac{-1}{h_+(\lambda) + h_-(\lambda)}, & \dfrac{-h_+(\lambda)}{h_+(\lambda) + h_-(\lambda)} + \dfrac{1}{2} \\ \dfrac{-h_+(\lambda)}{h_+(\lambda) + h_-(\lambda)} + \dfrac{1}{2}, & \dfrac{h_+(\lambda)h_-(\lambda)}{h_+(\lambda) + h_-(\lambda)} \end{bmatrix}.$$

Denoting

$$\mathcal{H} = \{h;\ \text{holomorphic on } \mathbb{C}_+ \text{ with positive imaginary}\},$$

we see for any $z \in \mathbb{C}^2$, $(H(0)z, \bar{z}) \in \mathcal{H}$. Here $(z, w) = z_1 w_1 + z_2 w_2$ for $z = z_1, z_2$, $w = w_1, w_2 \in \mathbb{C}^2$. Hence the Herglotz representation theorem says that there exists a self-adjoint matrix A, non-negative definite matrix B and a non-negative definite matrix-valued measure $\Sigma(d\xi)$ on $\mathbb{R}$ such that

$$H(\lambda) = A + B\lambda + \int \left(\frac{1}{\xi - \lambda} - \frac{\xi}{1 + \xi^2}\right) \Sigma(d\xi). \tag{1.1}$$

(Actually in this case $B = 0$.) On the other hand,

$$g_\lambda(x, y) = (H(\lambda)\Phi_\lambda(x), \Phi_\lambda(y)) + \frac{1}{2}\varphi_\lambda(x)\psi_\lambda(y) - \frac{1}{2}\psi_\lambda(x)\varphi_\lambda(y) \tag{1.2}$$

for $x \geq y$ with $\Phi_\lambda(x) = {}^t(\varphi_\lambda(x), \psi_\lambda(x))$.

Lemma 1.2. For $q \in \Omega$

$$\int \frac{1}{1 + |\xi|^\alpha} (\Sigma(d\xi)\Phi_\xi(x), \Phi_\xi(x)) < \infty \text{ for } \alpha > \frac{1}{2} \text{ and } x \in \mathbb{R}. \tag{1.3}$$

Moreover for $x, y \in \mathbb{R}$, $\lambda \in \mathbb{C}_+$,

$$g_\lambda(x, y) = \int \frac{1}{\xi - \lambda} (\Sigma(d\xi)\Phi_\xi(x), \Phi_\xi(y)). \tag{1.4}$$

Proof. First note the following resolvent relation:

$$(L(q) - \lambda)^{-1} = (L(0) - \lambda)^{-1} - (L(0) - \lambda)^{-1} q (L(0) - \lambda)^{-1} + (L(0) - \lambda)^{-1} q (L(q) - \lambda)^{-1} q (L(0) - \lambda)^{-1}.$$

From this we easily get

$$|g_\lambda(x, y)| \leqq C_1 |\lambda|^{-\frac{1}{2}} (1 + (\mathrm{Im}\,\lambda)^{-1}) \tag{1.5}$$

for $|\lambda| \geqq 1$. C_1 depends on $q \in \Omega$, $x, y \in \mathbb{R}$.

For fixed $x \in \mathbb{R}$, from (1.1) and (1.2) we see $g_\lambda(x, x) \in \mathcal{H}$ as a function of λ and

$$\mathrm{Im}\, g_\lambda(x, x) = \beta\, \mathrm{Im}\,\lambda + \int \frac{\mathrm{Im}\,\lambda}{|\xi - \lambda|^2} (\Sigma(d\xi)\Phi_\xi(x), \Phi_\xi(x))$$

with a constant $\beta \geq 0$ depending on x, q.

However applying (1.5) for $\lambda = i\eta$ ($\eta > 0$), we easily conclude $\beta = 0$ and the property

(1.3). Hence, again using (1.5) we obtain (1.4) at least for $x = y$. Now for $x \neq y$, we see from (1.1) and (1.2) that the difference

(1.6) $$f(\lambda, x, y) = g_\lambda(x, y) - \int \frac{1}{\xi - \lambda} (\Sigma(d\xi)\Phi_\xi(x), \Phi_\xi(y))$$

is holomorphic on $\mathbb{C}$ with respect to λ. Now for $\varphi \in C_0^\infty(\mathbb{R})$ set

$$\hat{\varphi}(\xi) = \int \Phi_\xi(x)\, \varphi(x)dx.$$

Then (1.6) implies

$$\int f(\lambda, x, y)\, \varphi(x)\, \overline{\varphi(y)}\, dx\, dy = ((L(q) - \lambda)^{-1} \varphi, \varphi) - \int \frac{1}{\xi - \lambda} (\Sigma(d\xi)\hat{\varphi}(\xi), \hat{\varphi}(\xi)).$$

Since the LHS is a difference of two $\mathcal{H}$-functions, the entire function $(f_\lambda\varphi, \varphi)$ should be linear. However, considering (1.5), we easily see $(f_\lambda\varphi, \varphi) \equiv 0$, which completes the proof.

§2. Estimate of Lyapunov exponent.

Let P be a probability measure on Ω satisfying

(2.1) shift-invariance: $P(T_x A) = P(A)\ x \in \mathbb{R},\ A \in \beta(\Omega)$.

(2.2) integrability: $$\int_\Omega \left\{\int_0^1 q(x)^2 dx\right\} p(dq) < +\infty.$$

(2.3) ergodicity: $P(T_x A\ \theta A) = 0$ for any $x \in \mathbb{R} \Rightarrow P(A) = 0$ or 1.

Here $T_x : \Omega \to \Omega$ defined by $T_x q(\cdot) = q(\cdot + x)$. For any P satisfying (2.1) ~ (2.3), one can define the Lyapunov exponent $\gamma(\lambda)$ for $\lambda \in \mathbb{C}$ in the following manner. Let $U_\lambda(x, q)$ be the fundamental matrix of

$$U'(x) = \begin{bmatrix} 0 & 1 \\ q-\lambda & 0 \end{bmatrix} U(x),\ U(0) = I.$$

Then

(2.4) $$\gamma(\lambda) = \lim_{x\to\infty} \frac{1}{x} \log \|U_\lambda(x, q)\| \geq 0.$$

The limit exists a.s. $q \in \Omega$ with respect to P and is independent of $q \in \Omega$ because of (2.3).

Lemma 2.1. For $\lambda \in \mathbb{C}$

(2.5) $$|\gamma(\lambda) - \operatorname{Im} \sqrt{\lambda}| \leq m_1 \sqrt{|\lambda|}^{-1}$$

where $m_1 = \int_\Omega \{\int_0^1 |q(x)| \, dx\} \, p(dq)$.

Proof. We follow the proof of Lemma 3.3 of [ ]. Let

$$A = \begin{bmatrix} 1 & 0 \\ 0 & \sqrt{\lambda} \end{bmatrix}, \; B = \begin{bmatrix} 0 & 1 \\ -1 & 0 \end{bmatrix}, \; U(x) = c^{-\sqrt{\lambda}xB} A^{-1} U_\lambda(x, q) A.$$

Then U satisfies

$$U'(x) = \sqrt{\lambda}^{-1} q(x) C \, U(x), \; U(0) = I,$$

where $C = \begin{bmatrix} 0 & 0 \\ 1 & 0 \end{bmatrix}$. We can easily check

$$\|U(x)\|, \; \|U(x)^{-1}\| \leqq \exp\{|\lambda|^{-\frac{1}{2}} \int_0^x |q(y)| dy\}.$$

Since $\|e^{\sqrt{\lambda}xB}\| = e^{x \mathrm{Im}\sqrt{\lambda}}$, we have

$$\gamma(\lambda) = \lim_{x \to \infty} \frac{1}{x} \log \|U_\lambda(x \; q)\| \leqq \mathrm{Im}\sqrt{\lambda} + \sqrt{|\lambda|}^{-1} m_1.$$

Conversely $e^{\sqrt{\lambda}xB} = A^{-1} U_\lambda(x, q) A \; U(x)^{-1}$, hence

$$\mathrm{Im} \sqrt{\lambda} \leqq \cdot \gamma(\lambda) + \sqrt{|\lambda|}^{-1} m_1,$$

which shows (2.5).

Corollary. $\gamma(-\xi) \geqq \sqrt{\xi} - m_1 \sqrt{\xi}^{-1} > 0$ if $\xi > m_1$.

This Corollary says that for a given random system $\{L(q), P\}$, there exists no absolutely continuous spectrum in $(-\infty, -m_1)$ a.s.

§3. Hamiltonian system associated with P.

We introduce one more condition on P.

(3.1) $$L(q) \geqq 0 \text{ for all } q \in \mathrm{supp}\, P.$$

Fix P satisfying (2.1) ~ (2.3) and (3.1). Define $\gamma(\lambda)$ by (2.4). γ depends on P and we denote γ_P if necessary. Let

$$N = N_P = \{\xi \in \mathbb{R}; \gamma(\xi) = 0\} \subset [0, \infty) = \mathbb{R}_+$$

$$u(\lambda) = \exp J(\lambda),\ J(\lambda) = \frac{1}{2}\int_N \left(\frac{1}{\xi - \lambda} - \frac{\xi}{1+\xi^2}\right)d\xi.$$

Then we see

(3.2) $$0 < \operatorname{Im} J(\lambda) < \frac{\pi}{2} \text{ on } \mathbb{C}_+$$

(3.3) $$\operatorname{Im} J(\xi + i0) = \begin{cases} 0 & \text{a.e. on } N^c \\ \frac{\pi}{2} & \text{a.e. on } N. \end{cases}$$

On the other hand, from Theorem 4.5 of [1] we have

(3.4) $$\operatorname{Im}(H(\xi + i0, q)z, \bar{z}) = \frac{\pi}{2} \text{ a.e. on } N$$

for any $q \in \operatorname{supp} p$ and $z \in \mathbb{C}^2$. Moreover for a. e. $q \in \Omega$ with respect to P

(3.5) $$\operatorname{Im}(H(\xi + i0, q)z, \bar{z}) = 0 \text{ or } \pi \cdot \text{ on } N^c$$

for any $z \in \mathbb{C}^2$ by virtue of Theorem 4.4 of [1]. In [2], we asserted that (3.5) is valid for any $q \in \operatorname{supp} P$, however this is wrong generally.

In view of (3.2) ~ (3.4), we can conclude

$$(u(\cdot)\, H(\cdot, q)z, \bar{z}) \in \mathcal{H}$$

for any $z \in \mathbb{C}^2$, $q \in \operatorname{supp} P$. Therefore there exists a self-adjoint matrix A_1, non-negative definite matrix B_1 and a non-negative definite matrix-valued measure $\Xi(d\xi, q)$ on $\mathbb{R}_+$ such that

(3.6) $$u(\lambda)H(\lambda, q) = A_1 + B_1\lambda + \int \left(\frac{1}{\xi - \lambda} - \frac{\xi}{1+\xi^2}\right) \Xi(d\xi, q).$$

Now from (3.1) it follows $N \subset [0, \infty) = R_+$, therefore we easily see that $u(\lambda)$ is founded on $\mathbb{C}_+$. This combined with the same argument done in the proof of Lemma 1.1 shows

(3.7) $$\int \frac{1}{1 + |\xi|^\alpha} (\Xi(d\xi, q)\, \Phi_\xi(x), \Phi_\xi(x)) < \infty$$

(3.8) $$u(\lambda)\, g_\lambda(x, x, q) = \int \frac{1}{\xi - \lambda} (\Xi(d\xi, q)\Phi_\xi(x), \Phi_\xi(x))$$

for any $\alpha > \frac{1}{2}$, $q \in \operatorname{supp} p$ and $x \in \mathbb{R}$. Now we need

<u>Lemma</u> <u>3.1</u>. Assume $0 \leq \delta(\xi) \leq 1$ $(0 \leq \tilde{\delta}(\xi) \leq 1)$ is related to a non-negative measure σ $(\tilde{\sigma})$ by

(3.9) $$\exp\left\{-\int_{\mathbb{R}_+}\frac{\delta(\xi)}{\xi-\lambda}\,d\xi\right\} = -\lambda\int_{\mathbb{R}_+}\frac{\sigma(d\xi)}{\xi-\lambda},$$

(3.10) $$\exp\left\{-\int_{\mathbb{R}_+}\frac{\tilde{\delta}(\xi)}{\xi-\lambda}\,d\xi\right\} = 1-\int_{\mathbb{R}_+}\frac{\tilde{\sigma}(d\xi)}{\xi-\lambda}$$

respectively. Then there exist universal polynomials $p_n(x_1, \ldots, x_n)$ with positive coefficients such that

(3.11) $$\begin{cases}\displaystyle\int_{\mathbb{R}_+}\xi^n\sigma(d\xi) \leqq a_{n-1} + p_{n-1}(a_0, a_1, \ldots, a_{n-2}) \quad (n = 1, 2, \ldots) \\ \displaystyle\int_{\mathbb{R}_+}\sigma(d\xi) = 1\end{cases}$$

(3.12) $$\int_{\mathbb{R}_+}\xi^n\tilde{\sigma}(d\xi) \leqq \tilde{a}_n + p_{n-1}(\tilde{a}_0, \tilde{a}_1, \ldots, \tilde{a}_{n-1}) \quad (n = 0, 1, \ldots).$$

where $a_n = \int_{\mathbb{R}_+}\xi^n\delta(\xi)d\xi$, $\tilde{a}_n = \int_{\mathbb{R}_+}\xi^n\,\tilde{\delta}(\xi)d\xi$.

Proof. If δ has a compact support, then (3.11) can be proved by comparing the coefficients of Taylor expansions of the RHS and LHS of (3.9). This case (3.11) becomes identity. For general δ, we have only to approximate it by δ_n with compact supports from below. The proof of (3.12) also is similar.

From now on assume

(3.13) $$\int_{\mathbb{R}_+\setminus N} d\xi < +\infty.$$

Then we redefine $u(\lambda)$:

(3.14) $$u(\lambda) = \frac{2}{\sqrt{-\lambda}}\exp\left\{-\frac{1}{2}\int_{\mathbb{R}_+\setminus N}\frac{d\xi}{\xi-\lambda}\right\}.$$

The new $u(\lambda)$ differs from the original one only by a positive multiple constant. Since

$$g_\lambda(x, x, q) \sim \frac{1}{2\sqrt{-\lambda}} \text{ as } \lambda \to -\infty$$

we easily see

(3.15) $$u(\lambda)\, g_\lambda(x, x, q) = -\frac{1}{\lambda}\exp\left\{-\int_{\mathbb{R}_+\setminus N}\frac{\delta(\xi)}{\xi-\lambda}\,d\xi\right\}$$

with $\delta(\xi) = 1 - \frac{1}{\pi}\arg g_{\xi+i0}(x, x, q)$. Since $g_\lambda(x, x, q) \in \mathcal{H}$ as a function of $\lambda \in \mathbb{C}_+$, δ

satisfies $0 \leqq \delta(\xi) \leqq 1$ a.e. $\xi \in \mathbb{R}$. Hence, from (3.13) we have

$$\int_{\mathbb{R}_+\backslash N} \delta(\xi)d\xi \leqq \int_{\mathbb{R}_+\backslash N} d\xi < +\infty.$$

Applying (3.11) of Lemma 3.1, we see

$$\int_{\mathbb{R}_+} (\Xi(d\xi, q)\, \Phi_\xi(x), \Phi_\xi(x)) = 1 \text{ identically,} \tag{3.16}$$

$$\int_{\mathbb{R}_+} \xi\, (\Xi(d\xi, q)\, \Phi_\xi(x), \Phi_\xi(x)) \leqq \int_{\mathbb{R}_+\backslash N} d\xi < \infty. \tag{3.17}$$

Now we introduce $\tilde{g}_\lambda(x, x, q)$ by

$$\tilde{g}_\lambda(x, x, q) = \lim_{y \downarrow x} \frac{\partial^2}{\partial x \partial y}\, g_\lambda(x, y, q).$$

Since

$$\tilde{g}_\lambda\, (x, x, q) = \frac{h_+(\lambda, Tx\, q)h_-(\lambda, Tx\, q)}{h_+(\lambda, Tx\, q) + h_-(\lambda, Tx\, q)} \in \mathcal{H}.$$

we can argue $u(\lambda)\tilde{g}_\lambda$ similarly and we have

$$u(\lambda)\tilde{g}_\lambda(x, x, q) = -\exp\left\{-\int_{\mathbb{R}_+\backslash N} \frac{\tilde{\delta}(\xi)}{\xi - \lambda}\, d\xi\right\}. \tag{3.18}$$

with $\tilde{\delta}(\xi) = 1 - \frac{1}{\pi} \arg \tilde{g}_{\xi+io}\, (x, x, q)$, which also satisfies $0 \leqq \tilde{\delta}(\xi) \leqq 1$ a.e. on $\mathbb{R}$. Hence applying (3.11) again, we have

$$\int_{\mathbb{R}_+} (\Xi(d\xi, q)\, \Phi'_\xi(x), \Phi'_\xi(x)) \leqq \int_{\mathbb{R}_+\backslash N} d\xi < +\infty. \tag{3.19}$$

Making use of the above discussion, we can show a fundamental relation between $q(x)$ and $\Xi(d\xi, q)$. (3.16) and (3.17) enables us to obtain

$$\begin{aligned} g_\lambda(x, x, q) &= \frac{\sqrt{-\lambda}}{2} \exp\left\{\frac{1}{2}\int_{\mathbb{R}_+\backslash N} \frac{d\xi}{\xi - \lambda}\right\} \int_{\mathbb{R}_+} \frac{1}{\xi - \lambda}\, (\Xi(d\xi, q)\, \Phi_\xi(x), \Phi_\xi(x)) \\ &= \frac{1}{2\sqrt{-\lambda}} + \frac{1}{4}(-\lambda)^{-\frac{3}{2}}\left\{\int_{\mathbb{R}_+\backslash N} d\xi - 2\int_{\mathbb{R}_+} \xi(\Xi(d\xi, q)\Phi_\xi(x), \Phi_\xi(x)\right\} \\ &\quad + O((-\lambda)^{-\frac{3}{2}}) \text{ as } \lambda \to -\infty. \end{aligned} \tag{3.20}$$

On the other hand, the resolvent relation used in the proof of Lemma 1.2, gives

$$g_\lambda(x, x, q) = \frac{1}{2\sqrt{-\lambda}} + \frac{1}{4\lambda}\int e^{-2\sqrt{-\lambda}|x-y|}q(y)dy + O(\lambda^{-2}),$$

(see (1.12) Of [1]). This together with (3.20) shows

$$(3.21)\quad \int e^{-2\sqrt{-\lambda}|x-y|}q(y)dy = \frac{1}{\sqrt{-\lambda}}\left\{2\int_{\mathbb{R}_+}\xi(\Xi(d\xi, q)\Phi_\xi(x), \Phi_\xi(x)) - \int_{\mathbb{R}_+\setminus N} d\xi\right\} + o(|\lambda|^{-\frac{1}{2}})$$

However

$$\text{the LHS of (3.21)} = \int_0^\infty e^{-2\sqrt{-\lambda}y}\{q(x - y) + q(x + y)\}dy,$$

and if $x \in \mathbb{R}$ is a point such that

$$\int_0^y \{q(x - y) + q(x + y)\}dy = 2q(x)y + o(y) \text{ as } y\downarrow 0,$$

then (3.21) yields

$$(3.22)\quad q(x) = 2\int_{\mathbb{R}_+}\xi(\Xi(d\xi, q)\Phi_\xi(x), \Phi_\xi(x)) - \int_{\mathbb{R}_+\setminus N} d\xi.$$

If we argue the same thing also for $\tilde{g}_\lambda$, we obtain

$$(3.23)\quad q(x) = \int_{\mathbb{R}_+\setminus N} d\xi - 2\int_{\mathbb{R}_+}(\Xi(d\xi, q)\,\Phi'_\xi(x), \Phi'_\xi(x)).$$

Consequently, we see an identity:

$$(3.24)\quad \int_{\mathbb{R}_+}\xi(\Xi(d\xi, q)\,\Phi_\xi(x), \Phi_\xi(x)) + \int_{\mathbb{R}_+}(\Xi(d\xi, q)\,\Phi'_\xi(x), \Phi'_\xi(x)) = \int_{\mathbb{R}_+\setminus N} d\xi.$$

Now introduce

$$\nu(d\xi, q) = \Xi(d\xi, q).$$

Then diagonalizing $(\Xi(d\xi, q)\,\Phi_\xi(x), \Phi_\xi(x))$ with respect to ν, we have

$$(\Xi(d\xi, q)\,\Phi_\xi(x), \Phi_\xi(x)) = \{f_1(x, \xi)^2 + f_2(x, \xi)^2\}\,\nu(d\xi, q),$$

where $f_j(x, \xi)$ $(j = 1, 2)$ satisfy

$$(3.25)\quad -f''_j(x, \xi) + q(x)\,f_j(x, \xi) = \xi f_j(x, \xi) \text{ on } \mathbb{R}$$

for a.e. $\xi \in \mathbb{R}$ with respect to ν. Define a non-negative self-adjoint operator A on $X = L^2(\mathbb{R}_+, \nu) \times L^2(\mathbb{R}_+, \nu)$ by

$$(3.26)\quad Af(\xi) = \xi f(\xi),\ f = (f_1, f_2) \in X.$$

Then (3.16), (3.17) and (3.19) show that for each $x \in \mathbb{R}$, $f(x) = (f_1(x, \cdot), f_2(x, \cdot)) \in X$ and

$$f(x) \in \mathcal{D}(\sqrt{A}), \; f'(x) \in X.$$

Summing up these facts, we have.

<u>Theorem 3.1</u>. Suppose P satisfies (2.1) ~ (2.3) and (3.1). Assume the Lyapunov exponent satisfies (3.13). Then for any $q \in \text{supp } P$ there exists a non-negative measure $\nu(d\xi, q)$ on $\mathbb{R}_+$ which is singular at least on N such that

$$u(\lambda)g_\lambda(x, x, q) = ((A - \lambda)^{-1}f(x), f(x)) \tag{3.27}$$

$$u(\lambda)\tilde{g}_\lambda(x, x, q) = ((A - \lambda)^{-1} f'(x), f'(x)) \tag{3.28}$$

for $\lambda \in \mathbb{C}_+$, and $f(x)$ satisfies

$$(f(x), f(x)) \equiv 1, \tag{3.29}$$

$$(\sqrt{A}\, f(x), \sqrt{A}\, f(x)) + (f'(x), f'(x)) = \int_{\mathbb{R}_+ \setminus N} d\xi, \tag{3.30}$$

$$q(x) = 2(Af(x), f(x)) - \int_{\mathbb{R}_+ \setminus N} d\xi. \tag{3.31}$$

In this case q(x) satisfies

$$|q(x)| \leqq \int_{\mathbb{R}_+ \setminus N} d\xi \text{ a.e. } x \in \mathbb{R}. \tag{3.32}$$

Moreover, for a.e. $q \in \Omega$ with respect to P, $\nu(d\xi, q)$ is singular on $\mathbb{R}_+$.

<u>Remark</u>. The above theorem was pointed out by J. Moser [3] in case N consists of finitely many disjoint closed intervals and L(q) has only absolutely continuous spectrum.

(3.25) says roughly that $-f''(x) + q(x)f(x) = Af(x)$ holds, however, since we do not know if $f''(x)$, $Af(x) \in X$ for each $x \in \mathbb{R}$, we cannot say that the equation (3.25) holds in X. A sufficient condition for $Af(x)$, $f''(x) \in X$ is $\int_{\mathbb{R}_+ \setminus N} \xi d\xi < \infty$. This can be shown by Lemma 3.1 and (3.31). Similarly sufficient conditions for the smoothness of q(x) are given by

<u>Theorem 3.2</u>. Suppose

$$\int_{\mathbb{R}_+ \setminus N} \xi^n d\xi < \infty \text{ for } n \in \{0, 1, 2, \ldots\}.$$

Then $q \in C^{2n-1}(\mathbb{R})$ and q^{2n} is essentially bounded on $\mathbb{R}$. The bound is dominated by a polynomial of $\left\{\int_{\mathbb{R}_+\setminus N} \xi^m d\xi,\ m = 0, 1, 2, \ldots, n\right\}$. This is valid for all $q \in \operatorname{supp} P$.
Now assume $\int_{\mathbb{R}_+\setminus N} \xi d\xi < \infty$. Then we can regard (3.31) as an equation in X:

$$\begin{cases} -f''(x) + q(x)f(x) = Af(x) \\ \|f(x)\| = 1 \end{cases} \tag{3.33}$$

The constrained condition comes from (3.29). (3.33) can be regarded as a constrained Hamiltonian system to the tangent bundle of $S^{\infty} = \{f \in H;\ \|f\| = 1\}$ as was done in J. Moser [3] if we introduce a Hamiltonian:

$$H = \frac{1}{2}(Af, f) + \frac{1}{2}(\|f\|^2\ \|f'\|^2 - (f, f')^2).$$

This Hamiltonian system has a one-parameter of integrals. For $\lambda \in \mathbb{C}_+$, define, for $f, g \in X$

$$Q_\lambda(f, g) = ((A - \lambda)f, g), \tag{3.34}$$

$$I_\lambda(f, g) = (1 - Q_\lambda(g, g))\, Q_\lambda(f, f) + Q_\lambda(f, g)^2.$$

Then any solution of (3.33) satisfies

$$I_\lambda(f(x), f'(x)) \equiv \frac{U(\lambda)^2}{4} \tag{3.35}$$

This can be shown directly from the equation satisfied by $g_\lambda(x, x, q)$:

$$2\cdot g(g'' - 2(q - \lambda)g) - g'^2 + 1 = 0.$$

The identity (3.30) can be obtained also from comparing the first terms of the asymptotic expansions of (3.35) as $\lambda \to -\infty$.

The complete integrability of the system (3.33) is known in finite bands case, and hence f(x) and so q(x) also are quasi-periodic functions in this case. In a special case, but sufficiently general infinitely many bands case, the almost periodicity of f(x), and q(x) will be discussed by the forthcoming paper of S. Kotani–M. Krishna.

References.

[1] S. Kotani: One-dimensional random Schrödinger operators and Herglotz functions, Proc. Taniguchi Symp. on Probablistic Methods in Mathematical Physics, Katata, to appear.

[2] ________________: Link between periodic potentials and random potentials in one-dimensional Schrödinger operators. Proc. of Conf. on Diff. Eq. in Math. Phys., Univ. of Alabama at Birmingham, March 1986, to appear.

[3] J. Moser: Integrable Hamiltonian Systems and Spectral Theory, Lezioni Fermiane, Pisa – 1981.

Department of Mathematics
Kyoto University
Kyoto, 606 Japan

RIEMANNIAN MANIFOLDS WITH STOCHASTIC INDEPENDENCE CONDITIONS ARE RICH ENOUGH

Masanori Kôzaki and Yukio Ogura

1. Introduction. In our previous papers [3] and [4], we have studied how the following two items affect the structure of a Riemannian manifold; 1) the relations between the stochastic and geometric mean values for small geodesic spheres, 2) the independence of first exit times and exit positions from small geodesic balls for a Brownian motion on the manifold (see also [5], [6] and [7] for related topics). In the course of our analysis, super-Einstein spaces and quasi-super-Einstein spaces (see Section 2 for precise definition) appeared on the scene.

The object of this article is to inspect the relations between those two manifolds in the class of direct product spaces, ensuring that they are rich enough. As a by-product, we give examples of Einstein spaces which are not super-Einsteinian other than those given in [1].

2. Statement of the results. Let (M,g) be an n-dimensional connected C^∞ Riemannian manifold with $n \geq 3$ and $B_m(\varepsilon)$ be the geodesic ball in M at center $m \in M$ with small radius $\varepsilon > 0$.

For an $m \in M$, let $(U;x^1,x^2,\dots,x^n)$ be a normal coordinate system around m, and denote by (g_{ij}) and $(R_{ijk\ell})$ the metric tensor and the curvature tensor with respect to the normal frame $(\partial/\partial x^1,\partial/\partial x^2,\dots,\partial/\partial x^n)$. The Ricci tensor and the scalar curvature are denoted by (ρ_{ij}) and τ respectively: $\rho_{ij} = R^u{}_{iuj}$, $\tau = \rho^u{}_u$. We also denote the length of a tensor $T = (T_{i_1 i_2 \dots i_p})$ by $|T|$, i.e., $|T|^2 = T_{i_1 i_2 \dots i_p} T^{i_1 i_2 \dots i_p}$. The covariant derivatives are denoted as $T_{i_1 i_2 \dots i_p; j_1 j_2 \dots j_q} = \nabla_{j_q} \cdots \nabla_{j_2} \nabla_{j_1} T_{i_1 i_2 \dots i_p}$.

We call an Einstein space *super-Einstein* if $|R|$ is constant and $\dot{R}_{ij} \equiv R_{ipqr}R_j{}^{pqr} = |R|^2 g_{ij}/n$. Similarly, we call the space (M,g) *harmonic* if, for each $m \in M$, there exist an $\varepsilon > 0$ and a function $F:(0,\varepsilon) \to \mathbf{R}$ such that the function $f(n) = F(d(m,n))$ is harmonic in $B_m(\varepsilon)\setminus\{m\}$, where d is the distance function on (M,g).

Let $X = (X(t),P_m)$ $(m \in M)$ be a Brownian motion on (M,g), and T_ε be the first exit time from the geodesic ball $B_m(\varepsilon)$. The *first mean value* $M_m(\varepsilon, f)$ for a real valued continuous function f is defined by

$$M_m(\varepsilon, f) = (\mathrm{vol}(\partial B_m(\varepsilon)))^{-1} \int_{\partial B_m(\varepsilon)} f(\omega) d\sigma(\omega),$$

where $d\sigma$ stands for the volume element on the geodesic sphere $\partial B_m(\varepsilon)$. Similarly, the *second mean value* $L_m(\varepsilon, f)$ for an f is defined by

$$L_m(\varepsilon, f) = (\mathrm{vol}(S^{n-1}(1)))^{-1} \int_{S^{n-1}(1)} (f \circ \exp_m(\varepsilon u)) du,$$

where $\exp_m$ is the exponential map at $m \in M$ and du is the usual volume element on the $(n-1)$-dimensional unit sphere $S^{n-1}(1)$.

We consider the following two conditions, which are closely related to those in [1] an [2].

$(M1)_k$ *for each* $m \in M$, *the asymptotic formula*

$$M_m(\varepsilon, f) = E_m f(X(T_\varepsilon)) + O(\varepsilon^{2k+2}) \qquad (\varepsilon \to 0)$$

holds for all functions f *of class* C^{2k+2} *near* m;

$(M2)_k$ *for each* $m \in M$, *the mean value formula*

$$M_m(\varepsilon, f) = f(m) + (E_m T_\varepsilon) \Delta f(m) + O(\varepsilon^{2k+2}) \qquad (\varepsilon \to 0)$$

holds for all bi-harmonic functions f *near* m.

In the above, k is a natural number or ∞ and, in the case of $k = \infty$, the formulas are understood to hold without remainder terms.

The conditions $(L1)_k$ and $(L2)_k$ are defined in the same way as $(M1)_k$ and $(M2)_k$ are done respectively with the first mean value $M_m(\varepsilon, f)$ replaced by the second one $L_m(\varepsilon, f)$.

We also consider the following conditions:

$(I)_k$ *for each* $m \in M$ *and* $\alpha \geq 0$, *the asymptotically independence formula*

$$E_m e^{-\alpha T_\varepsilon} f(X(T_\varepsilon)) = (E_m e^{-\alpha T_\varepsilon})(E_m f(X(T_\varepsilon))) + O(\varepsilon^{2k+2}) \qquad (\varepsilon \to 0)$$

holds for all functions f *of class* C^{2k+2} *near* m;

$(MI)_k$ *for each* $m \in M$, *the asymptotically mean independence formula*

$$E_m T_\varepsilon f(X(T_\varepsilon)) = (E_m T_\varepsilon)(E_m f(X(T_\varepsilon)) + O(\varepsilon^{2k+2}) \qquad (\varepsilon \to 0)$$

holds for all functions f *of class* C^{2k+2} *near* m.

In our previous papers [3] and [4], we obtain the following

Theorem A. 1) *Each of the conditions* $(M1)_\infty$, $(M2)_\infty$, $(L1)_\infty$ *and* $(L2)_\infty$ *is necessary and sufficient in order that* (M,g) *be a harmonic space.*

2) *Each of the conditions* $(M1)_2$, $(M2)_2$, $(L1)_2$ *and* $(L2)_2$ *is necessary and sufficient in order that* (M,g) *be an Einstein space.*

3) *Each of the conditions* $(M1)_3$, $(M2)_3$, $(L1)_3$ *and* $(L2)_3$ *is necessary and sufficient in order that* (M,g) *be a super-Einstein space.*

Theorem B. 1) *If* (M,g) *is a harmonic space, then both* $(I)_\infty$ *and* $(MI)_\infty$ *hold.*

2) *Each of the conditions* $(I)_3$ *and* $(MI)_3$ *is equivalent to that the manifold* (M,g) *is of constant scalar curvature.*

3) *Each of the conditions* $(I)_4$ *and* $(MI)_4$ *is equivalent to that the manifold* (M,g) *is of constant scalar curvature and satisfies*

(1) $$|R|^2 - |\rho|^2 = \text{constant},$$

(2) $$\dot{R}_{ij} = \frac{|R|^2-|\rho|^2}{n}g_{ij} - \rho^{pq}R_{ipjq} + 2\rho_{ip}\rho_j{}^p - \frac{3}{2}\rho_{ij;p}{}^p.$$

In order to study the relations between the manifolds which satisfy the above conditions, we shall give

Definition A manifold (M,g) is called *quasi-super-Einstein* if it satisfies the conditions in Theorem B 3).

Notice that every harmonic space is super-Einsteinian and quasi-super-Einsteinian. Also, we obtained the next assertion in [4].

Proposition A. *Let* (M,g) *be an n-dimensional Einstein space with* $n \geq 3$. *Then* (M,g) *is a quasi-super-Einstein space if and only if it is a super-Einstein space.*

We now assume that (M,g) is a direct product of an r-dimensional Riemannian manifold (M_1^r,g_1) and an s-dimensional (M_2^s,g_2) with $r,s \geq 2$; $M = M_1^r \times M_2^s$. We denote the curvature tensor, the Ricci tensor and the scalar curvature for (M_1^r,g_1) [resp. (M_2^s,g_2)] by $((R_1)_{abcd})$ [resp. $((R_2)_{\alpha\beta\gamma\delta})$], $((\rho_1)_{ab})$ [resp. $((\rho_2)_{\alpha\beta})$] and τ_1 [resp. τ_2], correspondingly. The main result in this article is the following

Theorem 1) *Let* (M_1^r,g_1) *and* (M_2^s,g_2) *be Einstein spaces. Then the direct product space* $M = M_1^r \times M_2^s$ *is Einsteinian if and only if*

(3) $$\tau_1/r = \tau_2/s.$$

2) *Let* (M_1^r, g_1) *and* (M_2^s, g_2) *be super-Einstein spaces. Then the direct product space* $M = M_1^r \times M_2^s$ *is super-Einsteinian if and only if*

(4) $$\tau_1/r = \tau_2/s \quad \text{and} \quad |R_1|^2/r = |R_2|^2/s.$$

3) *Let* (M_1^r, g_1) *and* (M_2^s, g_2) *be quasi-super-Einstein space. Then the direct product space* $M = M_1^r \times M_2^s$ *is quasi-super-Einsteinian if and only if*

(5) $$(|R_1|^2 - |\rho_1|^2)/r = (|R_2|^2 - |\rho_2|^2)/s.$$

<u>Corollary</u> *Let* (M_1^r, g_1) *and* (M_2^s, g_2) *be two spaces of constant sectional curvature.*

1) *The direct product space* $M = M_1^r \times M_2^s$ *is super-Einsteinian if and only if*

(6) $$[\ \tau_1 = \tau_2 \quad \text{and} \quad r = s\ ] \quad \text{or} \quad \tau_1 = \tau_2 = 0.$$

2) *The direct product space* $M = M_1^r \times M_2^s$ *is quasi-super-Einsteinian if and only if*

(7) $$\frac{3-r}{r^2(r-1)}\tau_1{}^2 = \frac{3-s}{s^2(s-1)}\tau_2{}^2.$$

Notice that the above Corollary 1) provides some examples of Einstein spaces which are not super-Einsteinian other than those given in [1]. Due to Proposition A, this ensures the existence of an Einstein space which is neither super-Einsteinian nor quasi-super-Einsteinian. Especially, there is an Einstein space for which T_ε and $X(T_\varepsilon)$ are not independent (see Theorem B).

Further, as a direct consequence of Corollary, we see that there are quasi-super-Einstein spaces which are not Einsteinian. Indeed, the following spaces are in that category;

$$S^p(k) \times H^p(-k),\ S^3(k) \times \mathbf{R}^p \ \text{and}\ H^3(-k) \times \mathbf{R}^p \ (p \geq 2),$$

where $S^n(k)$, $H^n(-k)$ and $\mathbf{R}^n$ denote n-dimensional spaces of constant sectional curvature $k > 0$, $-k < 0$ and 0, respectively.

<u>3. Proof of Theorem and Corollary</u>. In this section, we will prove our Theorem and its Corollary.

In the following, we let $(U_1; y^1, y^2, \ldots, y^r)$[resp. $(U_2; z^1, z^2, \ldots, z^s)$] be a normal coordinate system around $m_1 \in M_1^r$ [resp. $m_2 \in M_2^s$] and set

(8) $U = U_1 \times U_2$, $(x^1, x^2, \ldots, x^n) = (y^1, y^2, \ldots, y^r, z^1, z^2, \ldots, z^s)$.

We may adopt the normal coordinate system $(U; x^1, x^2, \ldots, x^n)$ for $m = (m_1, m_2) \in M$. It is then clear that

(9)
$$g_{ij} = \begin{cases} (g_1)_{ij}, & \text{if } 1 \le i,j \le r, \\ (g_2)_{\alpha\beta}, & \text{if } r+1 \le i = \alpha+r \le n,\ r+1 \le j = \beta+r \le n, \\ 0, & \text{otherwise.} \end{cases}$$

The similar relations for the curvature tensors and the Ricci tensors are also available.

Proof of Theorem 1). Assume that (M_1^r, g_1) and (M_2^s, g_2) are Einstein spaces, i.e.,

(10) τ_1 = constant and $(\rho_1)_{ab} = \tau_1 (g_1)_{ab}/r$,

(11) τ_2 = constant and $(\rho_2)_{\alpha\beta} = \tau_2 (g_2)_{\alpha\beta}/s$.

Let first (M, g) be an Einstein space. We then have

(12) τ = constant and $\rho_{ij} = \tau g_{ij}/n$.

Since the Ricci tensors satisfy the same relation as (9), it follows from (10)-(12) that

(13) $\tau_1/r = \tau/n = \tau_2/s$.

This proves (3).

Suppose conversely that the equality (3) holds. Then we get

$$(\tau_1+\tau_2)/(r+s) = \tau_1/r = \tau_2/s.$$

But $\tau = \tau_1+\tau_2$ and $n = r+s$. Hence the relations (10) and (11) imply that τ is also constant and

$$(\rho_1)_{ab} = \tau (g_1)_{ab}/n \quad \text{and} \quad (\rho_2)_{\alpha\beta} = \tau (g_2)_{\alpha\beta}/n.$$

Now, due to the same relation as (9) for Ricci tensors, one has the formula in (12).

Proof of Theorem 2). Assume that (M_1^r, g_1) and (M_2^s, g_2) are super-Einstein spaces. Then we have (10), (11) and

(14) $\qquad |R_1| = \text{constant} \quad \text{and} \quad (\dot{R}_1)_{ab} = |R_1|^2 (g_1)_{ab}/r.$

(15) $\qquad |R_2| = \text{constant} \quad \text{and} \quad (\dot{R}_2)_{\alpha\beta} = |R_2|^2 (g_2)_{\alpha\beta}/s.$

Suppose first that (M,g) is a super-Einstein space. We then have the relations (12), (13) and

(16) $\qquad |R| = \text{constant} \quad \text{and} \quad \dot{R}_{ij} = |R|^2 g_{ij}/n.$

Note that the tensors $((\dot{R}_1)_{ab})$, $((\dot{R}_2)_{\alpha\beta})$ and $(\dot{R}_{ij})$ also satisfy the same relation as (9). Then we obtain

$$|R_1|^2/r = |R|^2/n = |R_2|^2/s,$$

proving (4).

Suppose conversely that the equality (4) holds. Then the space (M,g) is Einsteinian by Assertion 1). Further, the relations (16) are verified in the same way as in the proof of Assertion 1). Thus (M,g) is super-Einsteinian.

<u>Proof of Theorem 3)</u>. Assume that (M_1^r, g_1) and (M_2^s, g_2) are quasi-super-Einstein spaces. Then we have

(17) $\qquad |R_1|^2 - |\rho_1|^2 = \text{constant},$

(18) $$(\dot{R}_1)_{ab} = \frac{|R_1|^2 - |\rho_1|^2}{r}(g_1)_{ab} - (\rho_1)^{cd}(R_1)_{acbd} + 2(\rho_1)_{ac}(\rho_1)_b{}^c - \frac{3}{2}(\rho_1)_{ab;c}{}^c,$$

(19) $\qquad |R_2|^2 - |\rho_2|^2 = \text{constant},$

(20) $$(\dot{R}_2)_{\alpha\beta} = \frac{|R_2|^2 - |\rho_2|^2}{s}(g_2)_{\alpha\beta} - (\rho_2)^{\gamma\delta}(R_2)_{\alpha\gamma\beta\delta} + 2(\rho_2)_{\alpha\gamma}(\rho_2)_\beta{}^\gamma - \frac{3}{2}(\rho_2)_{\alpha\beta;\gamma}{}^\gamma.$$

Further, it is not hard to see that (8) and (9) imply

(21) $$\rho^{pq}R_{ipjq} = (\rho_1)^{cd}(R_1)_{icjd}, \quad \rho_{ip}\rho_j{}^p = (\rho_1)_{ic}(\rho_1)_j{}^c, \quad \rho_{ij;p}{}^p = (\rho_1)_{ij;c}{}^c, \qquad 1 \le i,j \le r$$

$$\rho^{pq}R_{ipjq} = (\rho_2)^{\gamma\delta}(R_2)_{\alpha\gamma\beta\delta}, \quad \rho_{ip}\rho_j{}^p = (\rho_2)_{\alpha\gamma}(\rho_2)_\beta{}^\gamma, \tag{22}$$
$$\rho_{ij;p}{}^p = (\rho_2)_{\alpha\beta;\gamma}{}^\gamma, \quad r+1 \le i = \alpha+r \le n, \; r+1 \le j = \beta+r \le n.$$

Suppose now that (M,g) is a quasi-super-Einstein space, that is, the relations (1) and (2) hold. Since the tensors $((\dot{R}_1)_{ab})$, $((\dot{R}_2)_{\alpha\beta})$, and $(\dot{R}_{ij})$ satisfy the same relation as (9), we then have

$$\dot{R}_{ij} = \frac{|R|^2-|\rho|^2}{n}g_{ij} - \rho^{pq}R_{ipjq} + 2\rho_{ip}\rho_j{}^p - \frac{3}{2}\rho_{ij;p}{}^p$$
$$= \frac{|R_1|^2-|\rho_1|^2}{r}(g_1)_{ij} - (\rho_1)^{cd}(R_1)_{icjd}$$
$$+ 2(\rho_1)_{ic}(\rho_1)_j{}^c - \frac{3}{2}(\rho_1)_{ij;c}{}^c, \qquad 1 \le i,j \le r.$$

This with (21) implies

$$\frac{|R|^2-|\rho|^2}{n}g_{ij} = \frac{|R_1|^2-|\rho_1|^2}{r}(g_1)_{ij}, \qquad 1 \le i,j \le r,$$

and, with the aid of (9),

$$(|R|^2-|\rho|^2)/n = (|R_1|^2-|\rho_1|^2)/r. \tag{23}$$

Similarly, we obtain from (22) that

$$(|R|^2-|\rho|^2)/n = (|R_2|^2-|\rho_2|^2)/s. \tag{24}$$

Now (23) and (24) imply the desired (5).

Suppose conversely that the equality (5) holds. Then we get

$$\frac{(|R_1|^2 + |R_2|^2) - (|\rho_1|^2+|\rho_2|^2)}{r+s} \tag{25}$$
$$= \frac{|R_1|^2-|\rho_1|^2}{r} = \frac{|R_2|^2-|\rho_2|^2}{s}.$$

But it is clear that

$$|R|^2 = |R_1|^2 + |R_2|^2, \qquad |\rho|^2 = |\rho_1|^2 + |\rho_2|^2.$$

Combining this with (17)-(22) and (25), we obtain the relations (1) and (2). Thus (M,g) is a quasi-super-Einstein space.

Proof of Corollary. Assume that (M_1^r, g_1) and (M_2^s, g_2) are spaces of constant sectional curvature. Then we have

(26) $|\rho_1|^2 = \tau_1^2/r, \qquad |\rho_2|^2 = \tau_2^2/s.$

(27) $|R_1|^2 = 2\tau_1^2/r(r-1), \qquad |R_2|^2 = 2\tau_2^2/s(s-1).$

1) In order to prove Assertion 1), it is enough to show the equivalence of the relations (4) and (6). Assume the formulas (4) first. It then follows from (27) that

$$\tau_1/r = \tau_2/s \quad \text{and} \quad \tau_1^2/r^2(r-1) = \tau_2^2/s^2(s-1).$$

This implies, except for the case $\tau_1 = \tau_2 = 0$, the equalities $r = s$ and $\tau_1 = \tau_2$. Thus we obtain (6). The converse implication is clear.

2) Note that (26) and (27) imply

$$|R_1|^2 - |\rho_1|^2 = \tau_1^2(3-r)/r(r-1),$$

$$|R_2|^2 - |\rho_2|^2 = \tau_2^2(3-s)/s(s-1).$$

Hence the equivalence of the formulas (5) and (7) clearly follows.

References

[1] A. Gray and T. J. Willmore: Mean-value theorems for Riemannian manifolds, Proc. Roy. Soc. Edinburgh, 92A(1982), 343-364.

[2] O. Kowalski: 'The second mean-value operator on Riemannian manifolds', in Proceedings of the CSSR-GDR-Polish Conference on Differential Geometry and its Applications, Nove Mesto 1980, pp.33-45, Universita Karlova Praha, 1982.

[3] M. Kôzaki and Y. Ogura: On geometric and stochastic mean values for small geodesic spheres in Riemannian manifolds, to appear in Tsukuba J. Math. (1987).

[4] M. Kôzaki and Y. Ogura: On the independence of exit time and exit position from small geodesic balls for Brownian motions on Riemannian manifolds, preprint.

[5] M. Liao: Hitting distributions of geodesic spheres by Riemannian Brownian motion, preprint.

[6] M. Pinsky: Moyenne stochastique sur une variété riemannienne. C. R. Acad. Sci. Paris, Série I 292, 991-994(1981).

[7] M. Pinsky: Independence implies Einstein metric, preprint.

Department of Mathematics
Saga University
Saga 840, Japan

ON SOME INEQUALITIES IN THE PROBABILISTIC NUMBER THEORY

J. Kubilius

An arithmetical function $f:\mathbb{N}\to\mathbb{C}$ is called additive if $f(mn)=f(m)+f(n)$ for each pair of coprime numbers m, n. If the canonical representation of m is $m=p_1^{\alpha_1}\cdots p_s^{\alpha_s}$ then $f(m)=f(p_1^{\alpha_1})+\cdots+f(p_s^{\alpha_s})$. If $f(p^{\alpha})=f(p)$ for all prime powers p^{α} then $f(m)=f(p_1)+\cdots+f(p_s)$ and the function f is called strongly additive. One of the simplest examples of such functions is $w(m)$, the number of different prime divisors of m. The first steps of probabilistic number theory are connected with this function. In 1917 G. H. Hardy and S. Ramanujan [1] proved that for any $\varepsilon>0$

$$n^{-1}\ \mathrm{card}\{m: m\le n,\ |w(m)-\ln\ln n|\ge(\ln\ln n)^{\frac{1}{2}+\varepsilon}\}\longrightarrow 0$$

as $n\to\infty$. The proof was rather complicated. It was based on the estimation of $\mathrm{card}\{m: m\le n, w(m)=k\}$. In 1934, however, P. Turán [2] gave a very simple proof of this theorem. He noticed that the assertion of Hardy and Ramanujan is a trivial consequence of the inequality

$$\sum_{m=1}^{n}(w(m)-\ln\ln n)^2\le Cn\ln\ln n$$

C being an absolute constant. Two years later [3] he generalized this inequality. Let $f(m)$ be a real-valued strongly additive arithmetical function such that $0\le f(p)\le K$ for all primes p where K is a constant,

$$A_n(f):=\sum_{p\le n}\frac{f(p)}{p}\longrightarrow\infty$$

as $n\to\infty$. Let us denote

$$S_n(f):=\frac{1}{n}\sum_{m=1}^{n}(f(m)-A_n(f))^2. \tag{1}$$

Then $S_n(f)\le C_1(K)A_n(f)$ the quantity $C_1(K)$ depending only on K.

Turán's proof of the theorem by Hardy and Ramanujan is based on the same ideas as those of the law of large numbers in the classical probability theory. It suggests using the probabilistic interpretation of additive functions. For the sake of simplicity I shall confine myself by real-valued strongly additive functions.

Let the set of elementary events consist of natural numbers $\{1,2,\dots,n\}$ and let the probability of a random event $A=\{m_1,\dots,m_k\}$ be k/n. Let us introduce the functions $X_p(m)=f(p)$ if $p|m$ and equal to 0 otherwise. Then

$$f(m) = \sum_{p\le n} X_p(m) . \tag{2}$$

The functions X_p are in some sense weakly dependent random variables taking values $f(p)$ with probability $\frac{1}{n}\left[\frac{n}{p}\right]\simeq\frac{1}{p}$ and 0 with probability $1-\frac{1}{n}\left[\frac{n}{p}\right]\simeq 1-\frac{1}{p}$. Then $A_n(f)$ is approximately the mathematical expectation of the random variable f and the sum (1) equals approximately the second central moment.

If the variables X_p were uncorrelated (or even independent), then the second central moment of the sum (2) would be approximately equal to

$$\sum_{p\le n} \frac{f^2(p)}{p} \left(1-\frac{1}{p}\right).$$

Therefore it is natural to estimate the sum (1) in terms of this sum or in terms of

$$B_n^2(f) := \sum_{p\le n} \frac{f^2(p)}{p} .$$

In 1955 I succeeded [4] in proving that

$$S_n(f) \le c\, B_n^2(f) \tag{3}$$

where c is an absolute constant. Somewhat later I extended this inequality to arbitrary complex-valued additive functions. It turned out that this inequality was useful in the probabilistic number theory. Therefore it is of interest to improve it if it is possible.

Let us denote

$$\tau_n(f) := \frac{S_n(f)}{B_n^2(f)}, \qquad \tau_n := \sup_{f\ne 0} \tau_n(f) .$$

If the random variables X_p were uncorrelated then τ_n would be close to 1 for large n. From (3) we have that $\tau_n\le c$.

About a quarter of a century ago A. Rényi proved (unpublished) that $\tau_n \le 6$ for sufficiently large n. In 1970 P. D. T. A. Elliott [5] used the method of the large sieve to prove (3). He also obtained that $\tau_n \le 51$ for sufficiently large n. In 1973 I proved [6] that $1.47 < \tau_n < 2.08$ for large n. Some time later I improved [7] this to $\tau_n < 1.764$ for large n. The most interesting is the estimate from below. In 1976 P. Elliott gave [8] a simple proof that $\tau_n \le 2$ for large n. In 1980 I proved [9] that $1.5 + o(1) \le \tau_n \le 1.503$ for large n. I conjectured that

$$\tau_n \longrightarrow 1.5 \quad \text{as} \quad n \longrightarrow \infty . \tag{4}$$

At the end of 1980 I proved [10,11] this conjecture. Recently I obtained [12] some estimate for the rate of convergence in (4). I shall sketch the proof.

Some elementary calculations give the equality

$$\tau_n(f) = \frac{1}{n} \sum_{p \le n} \left(\left[\frac{n}{p}\right] - \left[\frac{n}{p^2}\right] \right) p x_p^2$$
$$+ \sum_{p \le n, q \le n} \left(\left[\frac{n}{pq}\right] \frac{pq}{n} - \left[\frac{n}{p}\right] \frac{p}{n} - \left[\frac{n}{q}\right] \frac{q}{n} + 1 \right) \frac{x_p x_q}{\sqrt{pq}}$$

where p and q denote prime numbers, $[y]$ is the integral part of y,

$$x_p = \frac{f(p)}{\sqrt{p} B_n(f)} , \qquad \sum_{p \le n} x_p^2 = 1 .$$

We need to find the maximal eigenvalue of the matrix M_n of the quadratic form $\tau_n(f)$. The most important term is the sum

$$- \sum_{\substack{p \le n, q \le n \\ pq > n}} \frac{x_p x_q}{\sqrt{pq}} \tag{5}$$

General methods of finding or estimating eigenvalues of matrices (like Gershgorin discs and so on) do not work.

To get some ideas I used a certain heuristic device. For a large class of additive functions the term (5) is approximately equal to

$$V(\psi) := - \frac{\displaystyle\int_0^1 \int_0^1_{x+y>1} \psi(x)\psi(y) \frac{dxdy}{xy}}{\displaystyle\int_0^1 \psi^2(x) \frac{dx}{x}}$$

where ψ's are some functions with $\psi(x)=f(p)$ for $x=\ln p/\ln n$. One can prove that if V has extremum (in some class of functions) for $\psi=\psi_0$ then

$$\int_{1-x}^{1}\psi_0(y)\frac{dy}{y} = \lambda\psi_0(x)$$

for all $x\in(0,1)$. Therefore λ should be an eigenvalue of the operator

$$\int_{1-x}^{1}\psi(y)\frac{dy}{y}\ .$$

It is easy to show that

$$\psi_r(y):=yP_{r-1}^{(1,0)}(1-2y)$$

$$=y\sum_{k=0}^{r-1}(-1)^{r-k-1}\binom{r}{k}\binom{r-1}{k}y^{r-k-1}(1-y)^k \qquad (r=1,2,\ldots)$$

are eigenfunctions of this operator. Here P's are Jacobi polynomials. The corresponding eigenvalues are $\lambda_r:=(-1)^r/r$. The functions ψ_r have the properties

$$\int_{1-x}^{1}\psi_r(y)\frac{dy}{y} = \lambda_r\psi_r(x) \qquad (r=1,2,\ldots),$$

$$\int_0^1\psi_r(y)\psi_s(y)\frac{dy}{y} = \begin{cases} 0 & \text{if } r\neq s, \\ \frac{1}{2r} & \text{if } r=s. \end{cases}$$

Now we return back to our problem. We may guess that the eigenvalue of the matrix M_n should be approximately equal to $1+\lambda_r$ ($r=1,2,\ldots$). We shall try to prove this.

Let us take strongly additive functions f_r which are defined by $f_r(p)=\psi_r(\ln p/\ln n)$. It is easy to calculate that $\tau_n(f_r)=1+\lambda_r+o(1)$ ($r=1,2,\ldots$). This gives $\tau_n\geq 1.5+o(1)$. It remains to prove the opposite inequality $\tau_n\leq 1.5+o(1)$. It is easy to show that $\tau_n\leq 1+\mu_n+o(1)$ where μ_n is the greatest eigenvalue of the matrix Q_n of the quadratic form (5).

For the estimation of μ_n from above I found two methods. The first one consists in the following. It is easy to prove that the least eigenvalue of the matrix Q_n is $-1+o(1)$. Therefore $1+o(1)+\mu_n^k\leq T_rQ_n^k$ for even k's. Hence $\mu_n\leq(T_rQ_n^k-1+o(1))^{1/k}$. We have further

$$T_r Q_n^k = \sum \frac{1}{p_1 \cdots p_k}$$

where the summation is taken over all primes $p_1, \cdots, p_k$ satisfying the inequalities $p_1 \le n, \ldots, p_k \le n$; $p_1 p_2 > n, \ldots, p_{k-1} p_k \ge n$, $p_k p_1 \ge n$. Using elementary prime number estimates we get $T_r Q_n^k = T_k + o(1)$ where

$$T_r = \int \cdots \int \frac{du_1 \cdots du_k}{u_1 \cdots u_k}$$

the domain of integration being defined by the inequalities $0 < u_1 < 1$, $\cdots, 0 < u_k < 1$; $u_1 + u_2 > 1, \cdots, u_{k-1} + u_k > 1$, $u_k + u_1 > 1$. At first the evaluation of this integral seemed to be very difficult. It occurred, however, that the properties of the functions ψ_r may be used. This gave

$$T_k = \sum_{r=1}^{\infty} \frac{(-1)^{(r-1)k}}{r^k} .$$

Hence T_k is the Riemann zeta function $\zeta(k)$ for even k. So $\mu_n \le (\zeta(k) - 1 + o(1))^{1/k}$. This gives $\mu_n \le 0.5 + o(1)$ and $\tau_n \le 1.5 + o(1)$. At last we have (4).

The second method is based on the following observation. Let Q be a $s \times s$ matrix with real elements. Let $\|U\|$ be the norm $(u_1^2 + \cdots + u_s^2)^{1/2}$ of the column-vector with elements $u_1, \ldots, u_s$. Then for each real λ and for each vector $U \ne 0$ there exists an eigenvalue ν of the matrix Q satisfying the inequality

$$|\lambda - \nu| \le \frac{\|QU - \lambda U\|}{\|U\|} .$$

We take that Q is the matrix Q_n and the vector U has the coordinates $f_r(p)/\sqrt{p}$ $(p \le n)$. It follows that for each $r = 1, 2, \ldots$ there exists an eigenvalue ν_r of the matrix Q_n such that $\nu_r = \lambda_r + o(1)$. It is easy to prove that the greatest eigenvalue of the matrix Q_n is $\nu_2 = \frac{1}{2} + o(1)$. Hence $\tau_n = 1.5 + o(1)$. This method also permits to obtain a good estimation of the rate of convergence, namely $\tau_n = 1.5 + O(1/\ln n)$.

Let us return to the estimate (3). It is an estimate from above. It is easy to show that it gives the right order for a large class of functions. However, the example of the function $\ln m$ shows that this is not always the case. We have that $S_n(\ln) \sim 1$. On the other hand, $B_n^2(\ln n) \sim \frac{1}{2} \ln^2 n$. In 1981 I. Ruzsa [13] proved that

$$S_n(f) \asymp \min_{\lambda \in R} (\lambda^2 + B_n^2(f - \lambda \ln))$$

and in 1982 A. Hildebrand [14] found an asymptotic expansion for $S_n(f)$. Hence he also obtained (4).

Some generalizations of (3) are of interest. I shall mention just one of them proved also by I. Ruzsa [15]. This is an analogue of the inequality of Burkholder. Let F be any nonnegative increasing function, $F(2x) \le c_1 F(x)$, c_1 being a constant. Then for any constant A and any real-valued additive function f

$$\sum_{m=1}^{n} F(f(m) - A) \le C_1 n M F(3|\eta_n - A|) ;$$

here C_1 is an absolute constant and M denotes the mathematical expectation;

$$\eta_n = \sum_{p \le n} x_p$$

where x_p are independent random variables,

$$P(x_p = f(p^\alpha)) = \frac{1}{p^\alpha}(1 - \frac{1}{p}) \qquad (\alpha = 0, 1, \ldots).$$

References

[1] Hardy G. H., Ramanujan S. The normal number of prime factors of n. Quart. J. Math., 48 (1917), 76-92.

[2] Turán P. On a theorem of Hardy and Ramanujan. J. London Math. Soc., 9 (1934), 274-276.

[3] Turán P. Über eine Verallgemeinerung eines Satzes von Hardy und Ramanujan. J. London Math. Soc., 11 (1936), 125-133.

[4] Kubilius J. Probabilistic methods in the theory of numbers (Russian). Uspehi Matem. Nauk, 1956, vol.11, no.2(68), 31-66.

[5] Elliott P.D.T.A. The Turán-Kubilius inequality, and limitation theorem for the large sieve. Amer. J. Math., 92 (1970), 293-300.

[6] Kubilius J. On an inequality for additive arithmetic functions. Acta Arithmetica, 27 (1975), 371-383.

[7] Kubilius J. On the law of large numbers for additive functions (Russian). Liet. Mat. Rink., 17 (1977), no.3, 113-114.

[8] Elliott P.D.T.A. The Turán-Kubilius inequality. Proc. Amer. Math. Soc., 65 (1977), 8-10.

[9] Kubilius J. On an inequality for additive arithmetical functions. Mathematisches Forschungsinstitut Oberwolfach, Tagungsberichte 48/1980, 16.

[10] Kubilius J. On the estimation of the second central moment for strongly additive arithmetical functions (Russian). Liet. Mat. Rink., 23 (1983), no.1, 122-133.

[11] Kubilius J. On the estimate of the second central moment for any additive arithmetic functions (Russian). Liet. Mat. Rink., 23 (1983), no.2, 110-117.

[12] Kubilius J. Improvement of the estimation of the second central moment for additive arithmetical functions (Russian). Liet. Mat. Rink., 25 (1985), no.3, 104-110.

[13] Ruzsa I. On the variance of additive functions. Preprint, Mathematical Institute of the Hung. Acad. of Sciences, 1981.

[14] Hildebrand A. An asymptotic formula for the variance of an additive function. Mathematische Zeitschrift, 183 (1983), 145-170.

[15] Ruzsa I. Generalized moments of additive functions. 18 (1984), 27-33.

Vilnius State University,
Vilnius

HELICES AND ISOMORPHISM PROBLEMS IN ERGODIC THEORY

Izumi Kubo, Hiroshi Murata and Haruo Totoki

§1. **Introduction.** The concept of *helix* was introduced by J. de Sam Lazaro and P. A. Meyer [3] to analyze Kolmogorov flows by using the machines of martingale theory. T. Shimano [4], [5] has used it for the classification of Kolmogorov automorphisms. We propose a new approach to isomorphism problems of automorphisms by the aid of helices. For this purpose, we will introduce an algebraic structure to a class of helices inspired by that of martingales.

Let T be a Kolmogorov automorphism on a Lebesgue space $(X, \mathcal{F}, \mu)$. Then there exists a filtration $\{\mathcal{F}_n\}_{n\in\mathbb{Z}}$ such that

$$(1) \qquad T^{-1}\mathcal{F}_n = \mathcal{F}_{n+1}, \quad \mathcal{F}_n \subset \mathcal{F}_{n+1}, \quad \bigcap_n \mathcal{F}_n = \{X, \phi\} \quad \text{and} \quad \bigvee_n \mathcal{F}_n = \mathcal{F},$$

equivalently there exists an increasing sequence of measurable partitions $\{\xi_n\}_{n\in\mathbb{Z}}$ such that $\mathcal{F}_n$ is generated by ξ_n and that

$$T^{-1}\xi_n = \xi_{n+1}, \quad \xi_n < \xi_{n+1}, \quad \bigwedge_n \xi_n = \{X\} \quad \text{and} \quad \bigvee_n \xi_n = \varepsilon,$$

where ε is the partition into individual points of X. T. Shimano called $(X, \mathcal{F}, \mu, \{\mathcal{F}_n\}, T)$ *a system*. Two systems $(X, \mathcal{F}, \mu, \{\mathcal{F}_n\}, T)$ and $(Y, \mathcal{G}, \nu, \{\mathcal{G}_n\}, S)$ are said to be *isomorphic* to each other, if there exists an isomorphism φ from X to Y such that

$$(2) \qquad \varphi T = S\varphi \quad \text{and} \quad \mathcal{F}_n = \varphi^{-1}\mathcal{G}_n \quad \text{for all } n \in \mathbb{Z}.$$

We say simply that T and S are *causally isomorphic* to each other. More weakly, if there exists a homomorphism φ from X onto Y with

$$\varphi T = S\varphi \quad \text{and} \quad \mathcal{F}_n \supset \varphi^{-1}\mathcal{G}_n,$$

we say that the system $(Y, \mathcal{G}, \nu, \{\mathcal{G}_n\}, S)$ is *a factor* of the system $(X, \mathcal{F}, \mu, \{\mathcal{F}_n\}, T)$ and simply that S is *a causal factor of* T. We are now interested in conditions for causal isomorphisms. The problem is deeply related with the Rosenblatt problem of stationary processes and that of isomorphism of exact endomorphisms (cf. the authors [1] and G. Maruyama [2]). Here, the Rosenblatt problem asks, for a given stationary process $\{\xi_n\}$, whether there exist an i.i.d. sequence $\{\eta_n\}$ and functions f, g such that

$$\xi_n = f(\cdots, \eta_{n-1}, \eta_n) \quad \text{and} \quad \eta_n = g(\cdots, \xi_{n-1}, \xi_n) \quad \text{for any } n \in \mathbb{Z}.$$

G. Maruyama discussed it in connection with Ornstein's isomorphism theorems.

§2. **Helices.** Let $\mathfrak{X} = \{X_n\}_{n\in\mathbb{Z}}$ be a sequence of integrable functions such that

(i) $X_0 = 0$ a.s.

(ii) $E[X_n|\mathcal{F}_m] = X_m$ a.s. for $n \geq m \geq 0$,

(iii) $X_n(Tx) - X_m(Tx) = X_{n+1}(x) - X_{m+1}(x)$ a.s. for $n \geq m$.

Then $\mathfrak{X} = \{X_n\}_{n\in\mathbb{Z}}$ is called a *helix*. Obviously, $\mathfrak{X}$ is a cocycle and $\mathfrak{X}^+ = \{X_n\}_{n\geq 0}$ is a martingale associated with the filtration $\{\mathcal{F}_n\}_{n\geq 0}$. In the following we will omit the term "a.s." for simplicity, if we have no confusion.

It is well known that any cocycle $\mathfrak{C} = \{C_n\}$ is generated by $C_1(x)$: $C_0(x) = 0$, $C_n(x) = \sum_{k=0}^{n-1} C_1(T^k x)$ for $n \geq 1$ and $C_n(x) = -\sum_{k=n}^{-1} C_1(T^k x)$ for $n \leq -1$. A helix $\mathfrak{X} = \{X_n\}$ is generated by the $\mathcal{F}_1$-measurable integrable function $X(x) = X_1(x)$ with $E[X|\mathcal{F}_0] = 0$. Suppose that $\mathfrak{X} = \{X_n\}$ and $\mathfrak{Y} = \{Y_n\}$ are square integrable. Denote by $\langle\mathfrak{X}, \mathfrak{Y}\rangle$ the cocycle generated by $E[X_1Y_1|\mathcal{F}_0]$ and by $\mathfrak{X}\cdot\mathfrak{Y}$ the helix generated by $X_1Y_1 - E[X_1Y_1|\mathcal{F}_0]$. Then obviously $\langle\mathfrak{X}\rangle = \langle\mathfrak{X}, \mathfrak{X}\rangle$ is an increasing process such that $\{X_n^2 - \langle\mathfrak{X}\rangle_n\}_{n\geq 0}$ is a martingale. Unfortunately, it does not coincide with $\{(\mathfrak{X}\cdot\mathfrak{X})_n\}_{n\geq 0}$;

$$(3)\qquad (\mathfrak{X}\cdot\mathfrak{X})_n = X_n^2 - \langle\mathfrak{X}\rangle_n - 2\sum_{0\leq i<j\leq n-1} X_1(T^i x)X_1(T^j x).$$

The last term is a martingale but has no cocycle property. Two helices $\mathfrak{X}$, $\mathfrak{Y}$ are said to be ***strictly orthogonal*** if $\langle\mathfrak{X}, \mathfrak{Y}\rangle = 0$. For a bounded $\mathcal{F}_0$-measurable function $f(x)$, define ***a stationary previsible sequence*** $\mathfrak{f} = \{f(T^{n-1}x)\}_{n\in\mathbb{Z}}$. Then we have a new helix $\mathfrak{f}*\mathfrak{X}$ generated by $f(x)X_1(x)$, which has the expression

$$(4)\qquad (\mathfrak{f}*\mathfrak{X})_n = \sum_{k=0}^{n-1} f(T^k x)\Big(X_{k+1}(x) - X_k(x)\Big) \quad \text{for } n \geq 1.$$

Thus we have an algebraic structure of helices;

$$\mathfrak{f}*\mathfrak{X} + \mathfrak{g}*\mathfrak{Y} \quad\text{and}\quad \mathfrak{X}\cdot\mathfrak{Y}.$$

A family $\{\mathfrak{X}^j\}$ of helices is called ***a generator*** if any square integrable $\mathfrak{X}$ is approximated in probability by helices of the form

$$\sum_j \mathfrak{f}^j * \mathfrak{X}^j.$$

T. Shimano showed that there exists a generator $\{\mathfrak{X}^j\}_{j=1}^K$ $(0 \leq K \leq \infty)$ of strictly orthogonal helices with the property

$$m_{\langle\mathfrak{X}^{j+1}\rangle} < m_{\langle\mathfrak{X}^j\rangle},$$

where $m_{\langle\mathfrak{X}\rangle}$ denotes ***the helix measure of*** $\mathfrak{X}$ defined by

$$(5)\qquad m_{\langle \mathfrak{X}\rangle}(B) \equiv \int_B X_1(x)^2 \, d\mu(x), \quad B \in \mathcal{F}_0,$$

and that such sequence of helix measures is unique up to the absolute continuity. The minimum number $M(T,\mathcal{F}_0)$ of such K is called *the multiplicity* by T. Shimano. We define a finer concept, *the multiplicity function* $M(x)$, which is similar to the multiplicity in the Hellinger-Hahn's spectral theory.

Definition. Put $M(x) \equiv 0$ on the set $\{x;\ \langle \mathfrak{X}^1\rangle_1(x) = 0\}$ and put

$$(6)\qquad M(x) \equiv \max\left\{j;\ \frac{dm_{\langle \mathfrak{X}^j\rangle}}{dm_{\langle \mathfrak{X}^1\rangle}}(x) > 0\right\} \quad \text{on } \{x;\ \langle \mathfrak{X}^1\rangle_1(x) > 0\}.$$

Then $M(x)$ is called *the multiplicity function*. If $M(x)$ is equal to a constant almost surely, we say that the multiplicity is *uniform*.

We have easily the following

Proposition 1. *Shimano's multiplicity* $M(T,\mathcal{F}_0)$ *is equal to* ess sup $M(x)$. *The multiplicity and the distribution* $\{\mu(M(x) = m)\}_{m=0}^{\infty}$ *are invariant under causal isomorphisms.*

Proposition 2. *An* $\mathcal{F}_0$*-measurable integer valued function* $M(x)$ *is the multiplicity function if and only if there exists a generator* $\{\mathfrak{X}^j\}$ *such that*

$$(7)\qquad \sum f^j * \mathfrak{X}^j = 0 \iff f^j(x) = 0 \ \textit{on}\ \{x;\ M(x) \geq j\}\ \textit{for any}\ j.$$

Proof. Let $\{\mathfrak{X}^j\}_{j=1}^K$ be a generator which defines the multiplicity function $M(x)$. Then $\langle \mathfrak{X}^j\rangle_1 > 0$ on $\{x;\ M(x) = m\}$ if and only if $j \leq m$. Since $\sum f^j * \mathfrak{X}^j = 0$ is equivalent to $\sum (f^j)^2 \langle \mathfrak{X}^j\rangle_1 = 0$, (7) holds. Conversely suppose that there exists a generator $\{\mathfrak{X}^j\}$ for which (7) holds. Applying Schmidt's method, we get strictly orthogonal generator $\{\mathfrak{Z}^k\}_{k=1}^K$ such that $m_{\langle \mathfrak{Z}^{k+1}\rangle} < m_{\langle \mathfrak{Z}^k\rangle}$ and that $\langle \mathfrak{Z}^k\rangle_1 > 0$ just on $\{x;\ M(x) \geq k\}$. □

Example 1. For a stationary mixing Markov chain with transition probability $(p_{i,j})_{i,j=1}^{K+1}$, we have $M(x) = \#\{j;\ p_{i,j} > 0\} - 1$ for any x with initial state i. In particular, if $p_{i,j} > 0$ for any i,j, then it has the uniform multiplicity K.

For a fixed square integrable generator $\{\mathfrak{Z}^k\}_{k=1}^K$, we have two kinds of quantities $\{\langle \mathfrak{Z}^k\rangle\}_{k=1}^K$ and $\{c_k^{i,j}\}_{i,j,k=1}^K$ with

$$(8)\qquad \mathfrak{Z}^i \cdot \mathfrak{Z}^j = \sum_{k=1}^{K} c_k^{i,j} * \mathfrak{Z}^k.$$

They should give more information than the multiplicity. A helix $\mathfrak{X}$ is called *closed*, if there exists a suitable stationary previsible sequence $c = \{c_j\}$ such that

$$(9) \qquad \mathfrak{X}\cdot\mathfrak{X} = c*\mathfrak{X}.$$

Proposition 3. (i) *A helix $\mathfrak{X}$ is closed if and only if $X_1(x)$ takes only two values on each element of the partition ξ_0.*

(ii) *If the partition ξ_0 has a countable generator α; that is,*

$$\xi_0 = \bigvee_{n\geq 1} T^n\alpha,$$

then there exists a generator which consists of closed helices.

Proof. (i) Suppose that $\mathfrak{X}\cdot\mathfrak{X} = a*\mathfrak{X}$. Then we have the equation

$$X_1(x)^2 - a(x)X_1(x) - E[X_1{}^2|\mathcal{F}_0] = 0,$$

which proves 'only if' part. Conversely, suppose that $X_1(x)$ takes only two values. Put $A \equiv \{x;\ X_1(x) > 0\}$. Then there exists an $\mathcal{F}_0$-measurable function f such that $X_1(x) = f(\chi_A(x) - \mu(A|\mathcal{F}_0))$. This implies 'if' part. For the proof of (ii), let $\alpha = \{A_j\}$ be a partition of $\mathbf{X}$ such that $\xi_0 = \vee_{n\geq 1}T^n\alpha$ and consider the helices $\mathfrak{X}^j$ generated by $X_1^j \equiv \chi_{A_j} - \mu(A_j|\mathcal{F}_0)$. Then $\{\mathfrak{X}^j\}$ is a generator and each $\mathfrak{X}^j$ is closed. □

§3. __Cases with the uniform multiplicity 1.__ Suppose that a system $(\mathbf{X}, \mathcal{F}, \mu, \{\mathcal{F}_n\}, T)$ has the uniform multiplicity 1 and $\mathfrak{Z} = \{Z_n\}_{n\in\mathbb{Z}}$ is a generator. Then any helix can be represented in the form $\mathfrak{f}*\mathfrak{Z}$ and hence $\mathfrak{Z}\cdot\mathfrak{Z} = c*\mathfrak{Z}$ holds; that is, $\mathfrak{Z}$ is closed. By Proposition 3, $Z_1(x)$ takes only two values, one is positive and the other is negative, on each element of ξ_0 almost surely, since the multiplicity is uniform. Put

$$A_+ \equiv \{x;\ Z_1(x) > 0\}, \qquad A_- \equiv \{x;\ Z_1(x) < 0\} \quad \text{and}$$

$$Y_1(x) \equiv \chi_{A_+}(x)\mu(A_-|\mathcal{F}_0) - \chi_{A_-}(x)\mu(A_+|\mathcal{F}_0) = \chi_{A_+}(x) - \mu(A_+|\mathcal{F}_0).$$

Then $E[Y_1|\mathcal{F}_0] = 0$ holds and the helix $\mathfrak{Y}$ generated by $Y_1(x)$ is a closed generator. Actually, we see that

$$(10) \qquad \langle\mathfrak{Y}\rangle_1 = \mu(A_+|\mathcal{F}_0)\mu(A_-|\mathcal{F}_0) \quad \text{and} \quad \mathfrak{Y}\cdot\mathfrak{Y} = \mathfrak{d}*\mathfrak{Y}$$

with the stationary previsible sequence $\mathfrak{d}$ given by $d(x) \equiv \mu(A_-|\mathcal{F}_0) - \mu(A_+|\mathcal{F}_0)$. If $d(x)$ is equal to a constant d, then $\mu(A_+|\mathcal{F}_0) = \frac{1-d}{2}$ and $\mu(A_-|\mathcal{F}_0) = \frac{1+d}{2}$ hold. Hence $\{A_+, A_-\}$ becomes a Bernoulli partition for T. Thus we have

Proposition 4. *In the above, if $\mathfrak{Y}\cdot\mathfrak{Y} = d\mathfrak{Y}$ with a constant d, then the Bernoulli shift $B(\frac{1-d}{2}, \frac{1+d}{2})$ is a causal factor of T and they have the same entropy.*

Theorem 1. *Suppose that a system* $(X, \mathcal{F}, \mu, \{\mathcal{F}_n\}, T)$ *has the uniform multiplicity* 1 *and that a given generator* $\mathfrak{Z}$ *satisfies* $\mathfrak{Z}\cdot\mathfrak{Z} = 0$. *Then the Bernoulli shift* $B(\frac{1}{2}, \frac{1}{2})$ *is a causal factor of* T *and they have the same entropy. In particular if the partition* ξ_1 *is generated by a partition with two elements, then* T *is causally isomorphic to* $B(\frac{1}{2}, \frac{1}{2})$.

Proof. The last statement can be proved in the following manner. Let $\alpha = \{A, A^c\}$ be a partition with $\xi_1 = \vee_{n\geq 0} T^n\alpha$, then the helix $\mathfrak{Y}$ generated by $Y_1(x) = \chi_A(x) - \mu(A|\mathcal{F}_0)$ satisfies $\mathfrak{Y}\cdot\mathfrak{Y} = 0$, because the generator $\mathfrak{Z}$ satisfies $\mathfrak{Z}\cdot\mathfrak{Z} = 0$. Hence $\mu(A|\mathcal{F}_0) = \frac{1}{2}$. □

Example 2. Consider a stationary mixing Markov chain with three states and the sub-σ-field $\mathcal{F}_0$ generated by past events. If the automorphism of the time shift satisfies the condition in Theorem 1, then it is causally isomorphic to $B(\frac{1}{2}, \frac{1}{2})$ by the results of Case 3.1 in page 315 of [2]. Similarly, we can see the same results for four states Markov chains.

Theorem 2. *Suppose that a system* $(X, \mathcal{F}, \mu, \{\mathcal{F}_n\}, T)$ *has the uniform multiplicity* 1 *and that a generator* $\mathfrak{Z}$ *satisfies* $\langle\mathfrak{Z}\cdot\mathfrak{Z}\rangle_1 = c\langle\mathfrak{Z}\rangle_1^2$ *with a constant* c, *then the Bernoulli shift* $B(p, 1-p)$ *with* $p = \frac{1}{2} + \frac{1}{2}\left(\frac{c}{c+4}\right)^{1/2}$ *is a causal factor of the* T *and they have the same entropy.*

Proof. Let $\mathfrak{Y}$ be the closed helix defined above. Then there exists a stationary previsible sequence $f = \{f(T^{n-1}x)\}$ such that $\mathfrak{Z} = f*\mathfrak{Y}$. Since $\langle\mathfrak{Z}\cdot\mathfrak{Z}\rangle_1 = f(x)^4\langle\mathfrak{Y}\cdot\mathfrak{Y}\rangle_1$ and $\langle\mathfrak{Z}\rangle_1 = f(x)^2\langle\mathfrak{Y}\rangle_1$ hold, $\mathfrak{Y}$ itself satisfies $\langle\mathfrak{Y}\cdot\mathfrak{Y}\rangle_1 = c\langle\mathfrak{Y}\rangle_1^2$. Put $r = \mu(A_+|\mathcal{F}_0)$. Then by (9), $(1-2r)^2 r(1-r) = cr^2(1-r)^2$ holds. Hence $\mu(A_\pm|\mathcal{F}_0) = \frac{1}{2} \pm \frac{1}{2}\left(\frac{c}{c+4}\right)^{1/2}$. Put $B_+ \equiv \{x;\ \mu(C_\alpha(x)|C_{\xi_0}(x)) = p\}$, $B_- \equiv X - B_+$ and $\beta \equiv \{B_+, B_-\}$, where $C_\alpha(x)$ is the element of the partition $\alpha = \{A_+, A_-\}$ containing x. Then $\mu(B_+|\eta_0) = p$ and $\mu(B_-|\eta_0) = 1-p$ hold with $\eta_0 \equiv \vee_{j\geq 1} T^j\beta$. This implies the assertion. □

§4. **Cases with finite multiplicity.** Suppose that a given system $(X, \mathcal{F}, \mu, \{\mathcal{F}_n\}, T)$ has a finite generator $\alpha = \{A_k\}_{k=1}^{K+1}$; that is, the σ-field $\mathcal{F}_0$ is generated by the partition $\vee_{n\geq 1} T^n\alpha$. Put $X_1^k(x) \equiv \chi_{A_k}(x) - \mu(A_k|\mathcal{F}_0)$ and let $\mathfrak{X}^k$ be the helix generated by $X_1^k(x)$. Then, by Proposition 3, $\{\mathfrak{X}^k\}_{k=1}^K$ is a generator which satisfies

(11) $\mathfrak{X}^i\cdot\mathfrak{X}^i = \hat{c}^i * \mathfrak{X}^i$ and $\mathfrak{X}^i\cdot\mathfrak{X}^j = - c^j * \mathfrak{X}^i - c^i * \mathfrak{X}^j$ for $i \neq j$,

where c^i and $\hat{c}^i$ are stationary previsible sequences given by

(12) $c^i(x) \equiv \mu(A_i | \mathcal{F}_0)$ and $\hat{c}^i \equiv 1 - 2c^i(x)$ for $1 \leq i \leq K$.

Example 3. If $c^i(x)$ and $\hat{c}^i(x)$, $1 \leq i \leq K$, are constants, then the system is isomorphic to the Bernoulli shift $B(c^1, c^2, \cdots, c^K, 1-c^1-c^2-\cdots-c^K)$.

Suppose that the multiplicity of the system is M and the multiplicity function M(x) is greater than 1 almost surely. Let $\{\mathfrak{Z}^i\}_{i=1}^M$ be a generator which satisfies (7). Suppose that (11) holds for $\{\mathfrak{Z}^i\}$ and that $c^i(x) \geq 0$, $1 \leq i \leq M$ (not necessarily given by (12)). Put $B_i \equiv \{x;\ Z_1^i(x) > 0\}$ and put

$$Y_1^i(x) \equiv \chi_{B_i}(x) - \mu(B_i | \mathcal{F}_0).$$

Then we can see that

$$\mu(B_i | \mathcal{F}_0) = \frac{c^i}{\hat{c}^i + 2c^i} \qquad \text{on } \{x;\ \mu(B_i | \mathcal{F}_0) > 0\},$$

$$\mu(B_i \cap B_j | \mathcal{F}_0) = 0 \quad \text{for } i \neq j,$$

and

$$Z_1^i(x) = (\hat{c}^i(x) + 2c^i(x))Y_1^i(x).$$

Thus we have

Theorem 3. *Suppose that a system* $(X, \mathcal{F}, \mu, \{\mathcal{F}_n\}, T)$ *has uniform multiplicity* M ($M \geq 2$) *and that* $\mathcal{F}_0$ *has a finite generator. Suppose that there exists a generator* $\{\mathfrak{Z}^i\}_{i=1}^M$ *which satisfies*

$$\mathfrak{Z}^i\cdot\mathfrak{Z}^i = \hat{c}^i * \mathfrak{Z}^i \quad \text{and} \quad \mathfrak{Z}^i\cdot\mathfrak{Z}^j = - c^j * \mathfrak{Z}^i - c^i * \mathfrak{Z}^j \quad \text{for } i \neq j,$$

with $c^i(x) \geq 0$. *If each* $\mathcal{F}_0$*-measurable function* $c^i(x)/(\hat{c}^i(x) + 2c^i(x))$ *is a constant* p_i *for any* i ($1 \leq i \leq M$), *then the Bernoulli shift* $B(p_1, p_2, \cdots, p_M, 1-p_1-\cdots-p_M)$ *is a causal factor of* T. *Moreover if* $\{B_i \equiv \{x;\ Z_1^i(x) > 0\}\}_{i=1}^M$ *generates the σ-field* $\mathcal{F}_0$, *then the system is isomorphic to the Bernoulli shift.*

§5. Remarks.

Now consider time-continuous cases. Let $\mathfrak{X} = \{X_t\}_{t\in\mathbb{R}}$ be a helix in the sense of [3]. Analogously to (3), $\mathfrak{X}\cdot\mathfrak{X}$ should be defined by

(13) $(\mathfrak{X}\cdot\mathfrak{X})_t \equiv X_t^2 - \langle\mathfrak{X}\rangle_t - 2\int_0^t X_{s-}dX_s$ for $t \geq 0$.

Actually it is a martingale for $t \geq 0$ and is extended to a cocycle for $t \in \mathbb{R}$. If $\mathfrak{X}$ has a continuous paths, then $\mathfrak{X}\cdot\mathfrak{X} = 0$. For example the flow of Brownian motion has the property, which corresponds to the Bernoulli shift $B(\frac{1}{2}, \frac{1}{2})$. For the flow of the Poisson process P_t, put $X_t = P_t - t$ for $t \geq 0$. Then we see that

$$\mathfrak{X}\cdot\mathfrak{X} = \mathfrak{X} \quad \text{and} \quad \langle\mathfrak{X}\rangle_t = t,$$

which correspond to the Bernuolli shift B(p, 1-p) with "an infinitesimally small" p.

REFERENCES

[1] I. Kubo, H. Murata & H. Totoki, On the isomorphism problem for endomorphisms of Lebesgue spaces, I, II & III, *Publ. RIMS Kyoto Univ.* **9** (1974), 285-317.

[2] G. Maruyama, Applications of Ornstein's theory to stationary processes. *Proc. 2nd Japan-USSR Symp. Prob. Theory. Lect. Notes in Math.* **330** (1973), 304-309.

[3] J. de Sam Lazaro & P.A.Meyer, Méthodes de martingales et théorie des flots, *Z. Wahrsch. Verw. Geb.* **18** (1971), 116-140.

[4] T. Shimano, An invariant of systems in the ergodic theory, *Tôhoku Math. J.* **30** (1978), 337-350.

[5] T. Shimano, The multiplicity of helices for a regularly increasing sequence of σ-fields, *Tôhoku Math. J.* **36** (1984), 141-148.

Izumi Kubo
Faculty of Integrated Arts and Sciences
Hiroshima University

Hiroshi Murata
Department of Mathematics
Naruto University of Education

Haruo Totoki
Department of Mathematics
Faculty of Sciences
Hiroshima University

A LIMIT THEOREM FOR STOCHASTIC PARTIAL DIFFERENTIAL EQUATIONS

Hiroshi Kunita

Introduction. Let us consider a family of partial differential equations with random coefficients with parameter $\varepsilon > 0$;

$$(1) \qquad \frac{\partial u^\varepsilon}{\partial t} = L_t u^\varepsilon + \sum_{i=1}^{d} f_i^\varepsilon(x,t,\omega)\frac{\partial u^\varepsilon}{\partial x_i} + f_{d+1}^\varepsilon(x,t,\omega)u^\varepsilon + f_{d+2}^\varepsilon(x,t,\omega),$$

$$u^\varepsilon(x,t)\Big|_{t=0} = \phi(x).$$

Here L_t is a second order linear differential operator of the form

$$(2) \qquad L_t u = \frac{1}{2}\sum_{i,j=1}^{d} a_{ij}(x,t)\frac{\partial^2 u}{\partial x_i \partial x_j} + \sum_{i=1}^{d} b_i(x,t)\frac{\partial u}{\partial x_i} + c(x,t)u + d(x,t),$$

whose coefficients do not depend on ω and ε, but coefficients $f_i^\varepsilon(x,t,\omega)$, $i=1,\ldots,d+2$, are stochastic processes depending on the parameter ε. The objective of this paper is to discuss the asymptotic behavior of the solutions $u^\varepsilon(x,t)$ as $\varepsilon \to 0$ when the coefficients $(f_1^\varepsilon,\ldots,f_{d+2}^\varepsilon)$ converge to a white noise or the integral $(F_i^\varepsilon(x,t) = \int_0^t f_i^\varepsilon(x,r)dr$, $i=1,\ldots,d+2)$ converges to a Brownian motion $(F_1(x,t),\ldots,F_{d+2}(x,t))$ with parameter x. We will show under certain conditions on f_i^ε, $\varepsilon > 0$, that $u^\varepsilon(x,t)$ and their derivatives $D_x^\alpha u^\varepsilon(x,t)$ converge uniformly on compact sets in the sense of the law and the limit $u(x,t)$ satisfies a stochastic partial differential equation of the form

$$(3) \qquad du = L_t u dt + \sum_{i=1}^{d} F^i(x,\circ dt)\frac{\partial u}{\partial x_i} + F_{d+1}(x,\circ dt)u + F_{d+2}(x,\circ dt).$$

A typical example of the above stochastic partial differential equation appears in the nonlinear filtering theory, describing density functions of unnormalized conditional laws of a stochastic differential system based on the observation disturbed by a white noise. It is called a Zakai equation. A basic assumption in the filtering theory is that noises governing the system and disturbing the observation are both white noises or Brownian motions. However, in the physical system noises are not exactly white but are stochastic processes with smooth sample paths. Thus equation of the form (1) could represent physical problems and equation (3) could be considered as an idealized or a limiting equation for the equation (1).

Limit theorems similar to ours are studied by Pardoux-Bouc [5] and Kushner-Huang [1]. In these works, the coefficients f_i^ε are of the form $(1/\varepsilon)h_i(x,z^\varepsilon(t))$, where $z^\varepsilon(t) = z(t/\varepsilon^2)$ and $z(t)$ is a bounded stationary process satisfying some uniform mixing condition. Hence f_i^ε, $i=1,\ldots,d+2$, are wide bandwidth noise. Our result may be regarded as a generalization and a refinement of their works in the following sense. Firstly, we treat the case where the noise processes f^ε are unbounded, including Gaussian noises. Secondly, our convergence assertion is stronger than the above mentioned works: We will show that $D_x^\alpha u^\varepsilon(x,t)$ converges to $D_x^\alpha u(x,t)$ uniformly on compact sets in (x,t) in the sense of the law. Thirdly, the characterization of the limit u by means of the stochastic partial differential equation (2) would be clearer than that of the previous works.

1. Statement of theorem. We begin with introducing some notations and definitions. Let k, d, e be positive integers and let T be a positive number. Let $C^{k,0}(R^d\times[0,T];R^e)$ or simply $C^{k,0}$ ($C^k(R^d;R^e)$ or C^k) be the totality of continuous maps $f;R^d\times[0,T]\to R^e$ ($f;R^d\to R^e$, resp.) which are k-times differentiable with respect to x and the derivatives are continuous in (x,t) (in x, resp.). The space $C^{k,0}$ is a Frechet space by semi-norms $\|\ \|_{k,N}$, $N=1,2,\dots$, defined by

$$\|f\|_{k,N} = \sum_{|\alpha|\le k} \sup_{|x|\le N, t\in[0,T]} |D_x^\alpha f(x,t)|,$$

where $D_x^\alpha = (\frac{\partial}{\partial x_1})^{\alpha_1}\cdots(\frac{\partial}{\partial x_d})^{\alpha_d}$ and $|\alpha| = \alpha_1+\dots+\alpha_d$. (The space C^k is also a Frechet space with the similar semi-norms).

Let $C_b^{k,0}$ (C_b^k) be the subspace of $C^{k,0}$ (C^k, resp.) such that f and its derivatives are all bounded functions. It is a Banach space by the norm $\|\ \|_k = \lim_{N\to\infty}\|\ \|_{k,N}$.

Let $X(x,t,\omega) \equiv (X_1(x,t,\omega),\dots X_e(x,t,\omega))$, $(x,t)\in R^d\times[0,T]$, be an R^e-valued random field defined on a probability space $(\Omega,\underline{F},P)$. For each ω, $X(\cdot,\cdot,\omega)$ can be regarded as a map from $R^d\times[0,T]$ into R^e. If $X(\cdot,\cdot,\omega)$ belongs to $C^{k,0}$ for almost all ω, $X(\cdot,t,\omega)$, $t\in[0,T]$ can be regarded as a continuous process with values in C^k. A continuous C^k-valued process $X(\cdot,t)$ is called a C^k-Brownian motion if $X(\cdot,0)$ and the increments $X(\cdot,t_{j+1}) - X(\cdot,t_j)$, $j=0,\dots,n-1$, are independent for any partition $0 = t_0 < t_1 < \dots < t_n = T$. In the following we assume that $X(\cdot,0)\equiv 0$. If the mean and the covariance are represented as

$$E[X_i(x,t)] = \int_0^t b_i(x,r)dr, \qquad Cov(X_i(x,t),X_j(y,t)) = \int_0^t a_{ij}(x,y,r)dr,$$

the pair $(b,a) = (b_i(x,t),a_{ij}(x,y,t))$ is called the characteristic of the C^k-Brownian motion $X(x,t)$.

Now let $u(x,t)$ be a continuous C^k-valued process adapted to a filtration $\underline{F}_t$, $t\in[0,T]$, of sub σ-fields of $\underline{F}$. It is called a continuous C^k-(local) martingale if $D_x^\alpha u(x,t)$, $|\alpha|\le k$, $x\in R^d$, are all continuous (local) martingales. It is called a continuous C^k-process of bounded variation if these processes are all continuous processes of bounded variations. It is called a continuous C^k-semimartingale if it is written as a sum of a continuous C^k-local martingale and a continuous C^k-process of bounded variation.

Suppose that $X(x,t)$ is a C^k-Brownian motion with the characteristic belonging to $C^{k,0}$ and $u(x,t)$ is a continuous C^k-semimartingale. Then the Stratonovich integral $\int_0^t X(x,\circ dr)u(x,r)$ is well defined for each x. Furthermore, it has a modification of a continuous C^{k-1}-semimartingale. See Kunita [3].

We shall now discuss the existence and the uniqueness of solutions of the stochastic partial differential equation (3). Concerning the coefficients of the operator L_t, we assume the following.

$(A.1)_k$ Coefficients $a_{ij}(x,t)$, $b_i(x,t)$, $c(x,t)$, $d(x,t)$ of the operator L_t are of $C_b^{k,0}$. Further, there exist functions $a_{ij}(x,y,t)$ of $C_b^{k,0}(R^{2d}\times[0,T];R^1)$ satisfying the following properties; a) $a_{ij}(x,x,t) = a_{ij}(x,t)$, b) symmetric, i.e., $a_{ij}(x,y,t) = a_{ji}(y,x,t)$, c) nonnegative definite, i.e., $\Sigma_{i,j,k,\ell}\, a_{ij}(x_k,x_\ell,t)\xi_i^k\xi_j^\ell \geq 0$ holds for any reals ξ_i^k, $i=1,\dots,d$, $k=1,\dots,n$.

Proposition (Kunita [4]). Assume that coefficients of the operator L_t satisfy $(A.1)_k$ for some $k \geq 6$ and coefficient $F(x,t) = (F_1(x,t),\dots,F_{d+2}(x,t))$ is a C^{k-1}-Brownian motion with the characteristic belonging to $C_b^{k,0}$. Then for any ϕ of C_b^{k-3}, there exists a continuous C^{k-3}-semimartingale $u(x,t)$ satisfying

$$(4)\qquad u(x,t) = \phi(x) + \int_0^t L_r u(x,r)dr + \sum_{i=1}^d \int_0^t F_i(x,\circ dr)\frac{\partial u}{\partial x_i}(x,r) + \int_0^t F_{d+1}(x,\circ dr)u(x,r) + F_{d+2}(x,t)$$

for any x, t a.s., and the growth property

$$(5)\qquad \lim_{x\to\infty} \frac{|u(x,t)|}{1+|x|^\delta} = 0 \qquad \text{a.s.}$$

for any $\delta > 0$. Furthermore, any continuous C^{k-3}-semimartingale satisfying (4) and (5) is at most unique.

We shall next consider equation (1). For the coefficient $f^\varepsilon(x,t) = (f_1^\varepsilon(x,t),\dots, f_{d+2}^\varepsilon(x,t))$, we introduce the following assumption.

$(A.2)_k$ For almost all ω, $f^\varepsilon(x,t)$ belongs to $C_b^{k,0}$.

Then, assuming $(A.1)_k$, $(A.2)_k$, $k \geq 3$ and $\phi \in C_b^k$, equation (1) has a unique solution for almost all ω by Oleinik's theorem. We denote it by $u^\varepsilon(x,t,\omega)$ or $u^\varepsilon(x,t)$. Set

$$F^\varepsilon(x,t) = \int_0^t f^\varepsilon(x,r)dr.$$

Then the pair $(F^\varepsilon,u^\varepsilon)(\omega)$ can be regarded as an element of $W_m \equiv C^{m,0}(R^d\times[0,T];R^{d+3})$ where $m \leq k-1$ for almost all ω, i.e., $(F^\varepsilon,u^\varepsilon)$ is a W_m-valued random variable. Then its law is defined on the space $(W_m,\underline{\underline{B}}(W_m))$ by setting

$$P_m^{(\varepsilon)}(A) = P\{\omega\,;(F^\varepsilon,u^\varepsilon)(\omega)\in A\},\qquad A\in \underline{\underline{B}}(W_m).$$

We shall introduce two assumptions $(A.3)_k$ and $(A.4)_k$ which ensure the tightness and the weak convergence of the family of the laws $P_m^{(\varepsilon)}$, $\varepsilon > 0$. We set

$$\bar{F}^{\varepsilon}(x,t) = E[F^{\varepsilon}(x,t)], \quad \tilde{F}^{\varepsilon}(x,t) = F^{\varepsilon}(x,t) - \bar{F}(x,t)$$

and

$$\bar{f}^{\varepsilon}(x,t) = E[f^{\varepsilon}(x,t)], \quad \tilde{f}^{\varepsilon}(x,t) = f^{\varepsilon}(x,t) - \bar{f}^{\varepsilon}(x,t).$$

$(A.3)_k$ i) There exists a positive constant K not depending on ε such that $|D_x^{\alpha}\bar{f}^{\varepsilon}(x,t)| \leq K$ holds for all x, t and $|\alpha| \leq k$.

ii) There exist positive valued processes $K_u^{\varepsilon}(\omega)$, $\varepsilon > 0$, such that

$$\sup_{\varepsilon} E[\exp \lambda \int_0^t K_u^{\varepsilon} du] < \infty, \qquad \forall\lambda > 0, \quad \forall t \in [0,T],$$

$$|E[D_x^{\alpha}\tilde{F}^{\varepsilon}(x,t) - D_x^{\alpha}\tilde{F}^{\varepsilon}(x,s)|\underline{\underline{F}}_u^{\varepsilon}]|(1 + |D_x^{\beta}\tilde{f}^{\varepsilon}(x,u)|) \leq K_u, \qquad \forall u < \forall s < \forall t,$$

are satisfied for any α, β with $|\alpha|, |\beta| \leq k$.

$(A.4)_k$ i) There exists a $C_b^{k,0}$-function $\bar{f}(x,t)$ such that

$$\bar{F}^{\varepsilon}(x,t) - \bar{F}^{\varepsilon}(x,s) \to \int_s^t \bar{f}(x,r)dr$$

uniformly on compact sets in $R^d \times [0,T]^2$.

ii) The convergence

$$E[\tilde{F}^{\varepsilon}(x,t) - \tilde{F}^{\varepsilon}(x,s)|\underline{\underline{F}}_s^{\varepsilon}] \to 0$$

holds uniformly on compact sets in L^1-sense.

iii) There exist $C_b^{k,0}$-functions $A_{ij}(x,y,t)$ symmetric and nonnegative definite such that for any α with $|\alpha| \leq 1$,

$$E[\int_s^t D_x^{\alpha}\tilde{F}_i^{\varepsilon}(x,r)d\tilde{F}_j^{\varepsilon}(y,r)|\underline{\underline{F}}_s^{\varepsilon}] \to \frac{1}{2}\int_s^t D_x^{\alpha}A_{ij}(x,y,r)dr$$

uniformly on compact sets in L^1-sense.

<u>Remark.</u> Since

$$(\tilde{F}_i^{\varepsilon}(x,t) - \tilde{F}_i^{\varepsilon}(x,s))(\tilde{F}_j^{\varepsilon}(y,t) - \tilde{F}_j^{\varepsilon}(y,s))$$

$$= \int_0^t \tilde{F}_i^{\varepsilon}(x,r)d\tilde{F}_j^{\varepsilon}(y,r) + \int_0^t \tilde{F}_j^{\varepsilon}(y,r)d\tilde{F}_i^{\varepsilon}(x,r),$$

$(A.4)_k$ iii) implies

$$E[(\tilde{F}_i^{\varepsilon}(x,t) - \tilde{F}_i^{\varepsilon}(x,s))(\tilde{F}_j^{\varepsilon}(y,t) - \tilde{F}_j^{\varepsilon}(y,s))|\underline{\underline{F}}_s^{\varepsilon}]$$

$$\to \int_s^t A_{ij}(x,y,r)dr.$$

Then for each x, $\tilde{F}^{\varepsilon}(x,t)$, $\varepsilon > 0$, will converge weakly to a martingale $\tilde{F}(x,t)$ having the joint quadratic variation $\int_0^t A_{ij}(x,y,r)dr$: The limit should be a C^{k-1}-Brownian motion with characteristic $(0,A)$.

We can now state our main result.

Theorem. Suppose $(A.1)_k$-$(A.4)_k$ for some $k \geq 6$. Let ϕ be of C_b^k. Then for any $m \leq k-6$, the family of laws $P_m^{(\varepsilon)}$ of the pairs $(F^\varepsilon, u^\varepsilon)$ converges weakly as $\varepsilon \to 0$. Further, the limit law $P_m^{(0)}$ satisfies the following properties.

i) $F(x,t)$ is a C^{k-1}-Brownian motion with characteristic $(\bar{f}, A)$.

ii) $u(x,t)$ is a continuous C^{k-1}-semimartingale satisfying the stochastic partial differential equation (4) and the growth condition (5).

Further, if F^ε, $\varepsilon > 0$, converges to F strongly in $C^{m,0}$, i.e., $\|F^\varepsilon - F\|_{m,N} \to 0$ in probability for any N, then u^ε, $\varepsilon > 0$, converges to u strongly in $C^{m,0}$.

Remark. If $A_{ij}(x,y,t)$ of $(A.4)_k$ is not symmetric, the theorem is valid with the following modification.

i') $F(x,t)$ is a C^{k-1}-Brownian motion with the characteristic $(\bar{f}, \bar{A})$ where

$$\bar{A}_{ij}(x,y,t) = \frac{1}{2}(A_{ij}(x,y,t) + A_{ji}(y,x,t)).$$

ii') Equation (4) is valid if $F_i(x,t)$ are replaced by $\bar{F}_i(x,t)$ defined by

$$\bar{F}_i(x,t) = F_i(x,t) + \frac{1}{2}\int_0^t \sum_{j=1}^d \frac{\partial}{\partial x_j}(A_{ij}(x,y,r) - A_{ji}(y,x,r))\Big|_{y=x} dr.$$

2. Associated stochastic flow. Our approach to the limit theorem is based on that of stochastic flows of diffeomorphisms. It is known that solutions of equations (1) and (3) can be represented by making use of stochastic flows of diffeomorphisms generated by suitable stochastic differential equations. At the next section it will be shown that the limit theorem for equation (1) can be reduced to the limit theorem for the associated stochastic flows. In this section we shall discuss the relationship between stochastic partial differential equations and stochastic flows following [4] and then shall discuss the weak convergence of the associated stochastic flows following [3].

Consider the partial differential equation with random coefficients (1). We assume that its coefficients satisfy $(A.1)_k$ and $(A.2)_k$, $k \geq 3$, as before. Let $(a_{ij}(x,y,t))$ be a symmetric and nonnegative definite function stated in $(A.1)_k$ and let $b(x,t)$ be the coefficient of the operator L_t. We set

$$m_i(x,t) = b_i(x,t) - \frac{1}{2}\sum_j \frac{\partial}{\partial x_j} a_{ij}(x,y,t)\Big|_{y=x}, \quad i=1,\dots,d.$$

Then on a suitable probability space $(W, \underline{B}, Q)$ we can define a C^{k-1}-Brownian motion $X(x,t) = (X_1(x,t),\dots,X_d(x,t))$ with characteristic $(m_i(x,t), a_{ij}(x,y,t))$. Let $(\Omega\times W, \underline{F}\times\underline{B}, P\times Q)$ be the product probability space. We consider a Stratonovich stochastic differential equation on the product space:

$$d\phi_t^i = -X_i(\phi_t,\circ dt) - f_i^\varepsilon(\phi_t,t)dt, \qquad i=1,\dots,d.$$

Let $\phi_{s,t}^\varepsilon(x) = (\phi_{s,t}^{\varepsilon,1}(x),\dots,\phi_{s,t}^{\varepsilon,d}(x))$, $x\in R^d$, be the solution starting at x at time s;

$$\phi_{s,t}^{\varepsilon,i}(x) = x_i - \int_s^t X_i(\phi_{s,r}^\varepsilon(x),\circ dr) - \int_s^t f_i^\varepsilon(\phi_{s,r}^\varepsilon(x),r)dr, \qquad i=1,\dots,d.$$

It is well known that $\phi_{s,t}^\varepsilon(x)$ has a modification such that it defines a stochastic flow of C^{k-2}-diffeomorphisms, i.e., it satisfies the following a)-c).

a) For almost all ω, $\phi_{s,t}^\varepsilon(x,\omega)$ is $(k-2)$-times differentiable in x and the derivatives $D_x^\alpha\phi_{s,t}^\varepsilon(x)$ are continuous in (s,t,x) for any α with $|\alpha|\le 2$.

b) For almost all ω, $\phi_{t,u}^\varepsilon(\phi_{s,t}^\varepsilon(x,\omega),\omega) = \phi_{s,u}^\varepsilon(x,\omega)$ is satisfied for all $s<t<u$ and x.

c) For almost all ω, the map $\phi_{s,t}^\varepsilon(\cdot,\omega);R^d\to R^d$ is a C^{k-2}-diffeomorphism for any $s<t$.

Set $\psi_{s,t}^\varepsilon(x) = (\phi_{s,t}^\varepsilon)^{-1}(x)$. If the initial data ϕ is of C_b^k, the unique solution of equation (1) is represented as

$$u^\varepsilon(x,t) = E_Q[\phi(\psi_{0,t}^\varepsilon(x))\xi_{0,t}^\varepsilon(x) + \eta_{0,t}^\varepsilon(x)], \tag{6}$$

where

$$\xi_{s,t}^\varepsilon(x) = \exp\int_s^t f_{d+1}^\varepsilon(\psi_{r,t}^\varepsilon(x),r)dr,$$

$$\eta_{s,t}^\varepsilon(x) = \int_s^t \xi_{r,t}^\varepsilon(x) f_{d+2}^\varepsilon(\psi_{r,t}^\varepsilon(x),r)dr.$$

Further the triple $(\psi_{s,t}^\varepsilon,\xi_{s,t}^\varepsilon,\eta_{s,t}^\varepsilon)$ satisfies the backward stochastic differential equation

$$\psi_{s,t}^{\varepsilon,i}(x) = x_i + \int_s^t X_i(\psi_{r,t}^\varepsilon(x),\circ\hat{d}r) + \int_s^t f_i^\varepsilon(\psi_{r,t}^\varepsilon(x),r)dr, \qquad i=1,\dots,d,$$

$$\xi_{s,t}^\varepsilon(x) = 1 + \int_s^t \xi_{r,t}^\varepsilon(x) f_{d+1}^\varepsilon(\psi_{r,t}^\varepsilon(x),r)dr,$$

$$\eta_{s,t}^\varepsilon(x) = 1 + \int_s^t \xi_{r,t}^\varepsilon(x) f_{d+2}^\varepsilon(\psi_{r,t}^\varepsilon(x),r)dr,$$

where $\int_s^t X(\cdot,\circ\hat{d}r)$ denotes the backward Stratonovich integral. See [4].

Because of the representation (6), the weak convergence of $u^\varepsilon(x,t)$ as $\varepsilon\to 0$ can be reduced to the weak convergence of the associated backward stochastic flows $(\psi_{s,t}^\varepsilon,\xi_{s,t}^\varepsilon,\eta_{s,t}^\varepsilon)$, $\varepsilon>0$, and the uniform L^p-boundedness of these flows. Here a family of random fields $Z_{s,t}^\varepsilon(x)$, $\varepsilon>0$, is called uniformly L^p-bounded if

$$\sup_{\varepsilon} \sup_{s,t} \sup_{|x|\leq N} E[|Z^{\varepsilon}_{s,t}(x)|^p] < \infty$$

holds for any N.

Lemma 1. $(D^{\alpha}_x \psi^{\varepsilon}_{s,t}(x), D^{\alpha}_x \xi^{\varepsilon}_{s,t}(x), D^{\alpha}_x \eta^{\varepsilon}_{s,t}(x))$, $\varepsilon > 0$, are uniformly L^p-bounded for any $p > 1$ if $|\alpha| \leq k-2$.

Proof is omitted since it is long. It will be discussed elsewhere. We next consider the weak convergence of the backward flow $(\psi^{\varepsilon}_{s,t}, \xi^{\varepsilon}_{s,t}, \eta^{\varepsilon}_{s,t})$. It is convenient to fix the time t. Then the law of $(X(s)-X(t), F^{\varepsilon}(s) - F^{\varepsilon}(t), (\psi^{\varepsilon}_{s,t}, \xi^{\varepsilon}_{s,t}, \eta^{\varepsilon}_{s,t}))$ can be defined on the space

$$\hat{W}_m = C^{m,0}(R^d \times [0,t]; R^d) \times C^{m,0}(R^d \times [0,t]; R^{d+2}) \times C^{m,0}(R^d \times [0,t]; R^{d+2})$$

where $m \leq k-2$. We denote it by $R^{(\varepsilon)}_m$. The typical element of $\hat{W}_m$ is denoted by $(\hat{X}, \hat{F}, (\hat{\psi}, \hat{\xi}, \hat{\eta}))$. Then we have the following.

Lemma 2. Assume $(A.1)_k$-$(A.4)_k$, $k \geq 3$. Then for any $m \leq k-3$, the family of measures $R^{(\varepsilon)}_m$, $\varepsilon > 0$, converges weakly as $\varepsilon \to 0$. Concerning the limit measure $R^{(0)}_m$, $(\hat{X}, \hat{F}, (\hat{\psi}, \hat{\xi}, \hat{\eta}))$ satisfies the following properties.

1) $\hat{X}(s)$ is a C^{k-1}-Brownian motion with characteristic (m,a).
2) $\hat{F}(s)$ is a C^{k-1}-Brownian motion with characteristic $(\bar{f}, A)$
3) $\hat{X}$ and $\hat{F}$ are independent.
4) $(\hat{\psi}, \hat{\xi}, \hat{\eta})$ satisfies the backward stochastic differential equation.

$$\hat{\psi}^i_s(x) = x_i + \int_s^t X_i(\hat{\psi}_r(x), \circ d\hat{r}) + \int_s^t F_i(\hat{\psi}_r(x), \circ d\hat{r}), \qquad i=1,\dots,d$$

$$\hat{\xi}_s(x) = 1 + \int_s^t \hat{\xi}_r(x) F_{d+1}(\hat{\psi}_r(x), \circ d\hat{r}),$$

$$\hat{\eta}_r(x) = 1 + \int_s^t \hat{\xi}_r(x) F_{d+2}(\hat{\psi}_r(x), \circ d\hat{r}).$$

For the proof, see [3].

3. Proof of the theorem. In order to prove the theorem, we have to show the tightness of the C^m-valued processes $(F^{\varepsilon}, u^{\varepsilon})$, $\varepsilon > 0$, where $m \leq k-5$. The tightness of F^{ε}, $\varepsilon > 0$, is shown in Kunita [2]. The tightness of u^{ε}, $\varepsilon > 0$, will follow from the following lemma by applying Kolmogorov's tightness criterion.

Lemma 3. Suppose $|\alpha| \leq k-5$. Then for any $p > 2$ and positive N, there exist positive constants K and γ such that

(7) $$E[|D^{\alpha}_x u^{\varepsilon}(x,t)|^p] \leq K,$$

(8) $$E[|D^{\alpha}_x u^{\varepsilon}(x,t) - D^{\alpha}_x u^{\varepsilon}(x,s)|^p] \leq K|t-s|^{1+\gamma}$$

hold for any $0 \leq s \leq t \leq T$, $\varepsilon > 0$ and $|x| \leq N$.

Proof. Let us observe the representation of $u^\varepsilon(x,t)$ given by (6). In view of Lemma 1,

$$\int_{|x|\leq N} E_Q[|D_x^\alpha\{\phi(\psi_{0,t}^\varepsilon(x))\xi_{0,t}^\varepsilon(x) + \eta_{0,t}^\varepsilon(x)\}|^p]dx < \infty$$

is satisfied for any α with $|\alpha| \leq k-2$ and positive N. Therefore we have

$$E_Q[\sup_{|x|\leq N} |D_x^\alpha\{\phi(\psi_{0,t}^\varepsilon(x))\xi_{0,t}^\varepsilon(x) + \eta_{0,t}^\varepsilon(x)\}|] < \infty$$

a.s. for any α with $|\alpha| \leq k-3$ and positive N by Sobolev's inequality. Hence we can interchange the order of the derivation D_x^α and the integration E_Q in the formula (6). Then we find that $u^\varepsilon(x,t)$ is a continuous C^{k-3}-process and satisfies

$$D_x^\alpha u^\varepsilon(x,t) = E_Q[D_x^\alpha\{\phi(\psi_{0,t}^\varepsilon(x))\xi_{0,t}^\varepsilon(x) + \eta_{0,t}^\varepsilon(x)\}]$$

for any x, t and $|\alpha| \leq k-3$. In view of Lemma 1, the integrands of the above are uniformly L^p-bounded with respect to $P\times Q$ for any $p>1$. Therefore $D_x^\alpha u^\varepsilon(x,t)$ is uniformly L^p-bounded with respect to P for any $p>1$. This proves (7).

We shall next prove the second inequality (8). For simplicity we shall consider the case $\alpha = 0$ only. Note that

$$\begin{aligned} u^\varepsilon(x,t) - u^\varepsilon(x,s) &= \int_s^t L_r u^\varepsilon(x,r)dr + \sum_{i=1}^d \int_s^t f_i^\varepsilon(x,r)\frac{\partial u^\varepsilon}{\partial x_i}(x,r)dr \\ &\quad + \int_s^t f_{d+1}^\varepsilon(x,r)u^\varepsilon(x,r)dr + \int_s^t f_{d+2}^\varepsilon(x,r)dr. \end{aligned}$$

We want to get the L^p-estimate of each term in the right hand side. In the following arguments, C_i, $i=1,\dots,6$, are positive constants not depending on s, t, x and ε. We have by (7)

$$E[|\int_s^t L_r^\varepsilon u^\varepsilon(x,r)dr|^p] \leq C_1|t-s|^p.$$

We will consider the third term. Set

$$A^\varepsilon(t) = \int_s^t f_{d+1}^\varepsilon(x,r)u^\varepsilon(x,r)dr.$$

Then we have

$$\begin{aligned} (9)\qquad |A^\varepsilon(t)|^p &= p\int_s^t f_{d+1}^\varepsilon(x,t_1)u^\varepsilon(x,t_1)|A^\varepsilon(t_1)|^{p-1}\operatorname{sign} A^\varepsilon(t_1)dt_1 \\ &= p\int_s^t f_{d+1}^\varepsilon(x,t_1)\Big(\int_s^{t_1}\{L_{t_2}u^\varepsilon(x,t_2) + \sum_{i=1}^d f_i^\varepsilon(x,t_2)\frac{\partial u^\varepsilon}{\partial x_i}(x,t_2) \end{aligned}$$

$$+ f^{\varepsilon}_{d+1}(x,t_2)u^{\varepsilon}(x,t_2) + f^{\varepsilon}_{d+2}(x,t_2)\}\,|A^{\varepsilon}(t_2)|^{p-1}\text{sign}\,A^{\varepsilon}(t_2)dt_2\}dt_1$$

$$+ p(p-1)\int_s^t f^{\varepsilon}_{d+1}(x,t_1)\{\int_s^{t_1} f^{\varepsilon}_{d+1}(x,t_2)u^{\varepsilon}(x,t_2)\,|A^{\varepsilon}(t_2)|^{p-2}dt_2\}\,dt_1$$

$$= I_1 + I_2.$$

The expectation of the first term is written as

$$pE[\int_s^t E[F^{\varepsilon}_{d+1}(x,t) - F^{\varepsilon}_{d+1}(x,t_2)\,|\,\underline{F}_{t_2}]L_{t_2}u^{\varepsilon}(x,t_2)|A^{\varepsilon}(t_2)|^{p-1}\text{sign}\,A^{\varepsilon}(t_2)dt_2]$$

$$+ \sum_{i=1}^{d} pE[\int_s^t E[F^{\varepsilon}_{d+1}(x,t) - F^{\varepsilon}_{d+1}(x,t_2)\,|\underline{F}_{t_2}]f^{\varepsilon}_i(x,t_2)\frac{\partial u^{\varepsilon}}{\partial x_i}(x,t_2)|A^{\varepsilon}(t_2)|^{p-1}\text{sign}\,A^{\varepsilon}(t_2)dt_2]$$

$$+ pE[\int_s^t E[F^{\varepsilon}_{d+1}(x,t) - F^{\varepsilon}_{d+1}(x,t_2)\,|\underline{F}_{t_2}]f^{\varepsilon}_{d+1}(x,t_2)u^{\varepsilon}(x,t_2)|A^{\varepsilon}(t_2)|^{p-1}\text{sign}\,A^{\varepsilon}(t_2)dt_2]$$

$$+ pE[\int_s^t E[F^{\varepsilon}_{d+1}(x,t) - F^{\varepsilon}_{d+1}(x,t_2)\,|\underline{F}_{t_2}]f^{\varepsilon}_{d+2}(x,t_2)|A^{\varepsilon}(t_2)|^{p-1}\text{sign}\,A^{\varepsilon}(t_2)dt_2].$$

By assumption $(A.3)_k$, $E[F^{\varepsilon}_{d+1}(x,t) - F^{\varepsilon}_{d+1}(x,t_2)|\underline{F}_{t_2}]f^{\varepsilon}_i(x,t_2)$, etc. are uniformly L^p-bounded for any $p>1$. By the first inequality of this lemma, $L_{t_2}u^{\varepsilon}(x,t_2)$, etc. are also uniformly L^p-bounded for any $p>1$. Therefore, by Hölder's inequality

$$|E[I_1]| \leq C_2\int_s^t E[|A^{\varepsilon}(t_2)|^p]^{\frac{p-1}{p}}dt_2 \leq C_2\{|t-s| + \int_s^t E[|A^{\varepsilon}(t_2)|^p]dt_2$$

holds for any ε. A similar estimate is valid to I_2;

$$|E[I_2]| \leq C_3\int_s^t E[|A^{\varepsilon}(t_2)|^p]^{\frac{p-2}{p}}dt_2 \leq C_3\{|t-s| + \int_s^t E[|A^{\varepsilon}(t_2)|^p]dt_2\}.$$

Consequently,

$$E[|A^{\varepsilon}(t)|^p] \leq (C_2+C_3)\{|t-s| + \int_s^t E[|A^{\varepsilon}(r)|^p]dr\}.$$

By Gronwall's inequality, we get $E[|A^{\varepsilon}(t)|^p] \leq C_4|t-s|$. Substitute this again to

$$E[|A^{\varepsilon}(t)|^p] \leq \int_s^t\{C_2E[|A^{\varepsilon}(r)|^p]^{\frac{p-1}{p}} + C_3E[|A^{\varepsilon}(r)|^p]^{\frac{p-2}{p}}\}dr.$$

Then we see that the above is bounded by $C_4|t-s|^{(2-2/p)}$.

By the similar argument, we can show that

$$E[|\int_s^t f^{\varepsilon}_i(x,r)\frac{\partial u^{\varepsilon}}{\partial x_i}(x,r)dr|^p] \leq C_5|t-s|^{2-\frac{2}{p}}, \qquad i=1,\ldots,d$$

$$E[|\int_s^t f^{\varepsilon}_{d+2}(x,r)dr|^p] \leq C_6|t-s|^{2-\frac{2}{p}}.$$

Summing up these estimations, we get the desired inequality (8) in case $|\alpha| = 0$. The proof is complete.

Proof of Theorem. We shall apply Sobolev's imbedding theorem. Let m and N be positive integers and $p>1$. We define Sobolev's seminorms $\|\ \|_{m,p,N}$, $N=1,2,\ldots$ for $f; R^d \to R^e$ by

$$\|f\|_{m,p,N} = \Big(\sum_{|\alpha|\leq m}\int_{|x|\leq N} |D^\alpha f(x)|^p dx\Big)^{\frac{1}{p}}.$$

We denote by $H^{loc}_{m,p}$ the set of all $f; R^d \to R^{d+3}$ such that $\|f\|_{m,p,N} < \infty$ for any N. Let $W^{loc}_{m,p} = C([0,T]; H^{loc}_{m,p})$ and $W_m = C([0,T]; C^m)$. Then we have $W^{loc}_{m,p} \subset W_{m-1}$ if $p>d$ by Sobolev's imbedding theorem. Furthermore, any weakly compact subset of $W^{loc}_{m,p}$ is imbedded into W_{m-1} as a strongly compact subset by Kondraseev's theorem.

Now integrate the inequalities of Lemma 3 over the ball $\{x; |x|\leq N\}$ and sum up for α with $|\alpha|\leq m$ where $m \leq k-5$. Then there exists a positive constant K' such that

$$E[\|u^\varepsilon_t\|^p_{m,p,N}] \leq K',$$

$$E[\|u^\varepsilon_t - u^\varepsilon_s\|^p_{m,p,N}] \leq K'|t-s|^{1+\gamma}$$

hold for any s, t and ε. The similar estimations are valid to F^ε. Then by Kolmogorov's tightness criterion, the laws of $(F^\varepsilon, u^\varepsilon)$, $\varepsilon>0$, defined on the space $W^{loc}_{m,p}$ are tight with respect to the weak topology of $W^{loc}_{m,p}$. Then the laws of $(F^\varepsilon, u^\varepsilon)$ defined on the space W_{m-1} are also tight with respect to the strong topology of W_{m-1} by Kondraseev's theorem.

We have thus seen that the family of laws $P^{(\varepsilon)}_m$ of $(F^\varepsilon, u^\varepsilon)$, $\varepsilon>0$, is tight if $m\leq k-6$. We wish to show that the limit law as $\varepsilon\to 0$ is unique and it satisfies the properties of the theorem. Let $P^{(\varepsilon_n)}_m$, $n=1,2,\ldots$, be a subsequence converging weakly in the space W_m. Let $(\psi^{\varepsilon_n}, \xi^{\varepsilon_n}, \eta^{\varepsilon_n})$, $n=1,2,\ldots$, be a sequence of the backward stochastic flows associated with $(F^{\varepsilon_n}, u^{\varepsilon_n})$, respectively. We fix the time t and denote by $R^{(\varepsilon_n)}_m$ the law of $(X(s)-X(t), F^{\varepsilon_n}(s)-F^{\varepsilon_n}(t), (\psi^{\varepsilon_n}_{s,t}, \xi^{\varepsilon_n}_{s,t}, \eta^{\varepsilon_n}_{s,t}))$ on the space $\hat{W}_m$. Then $R^{(\varepsilon_n)}_m$, $\varepsilon>0$, converge weakly to $R^{(0)}_m$ by Lemma 2. By Skorohod's imbedding theorem, we may assume that both $(F^{\varepsilon_n}, u^{\varepsilon_n})$ and $(\psi^{\varepsilon_n}, \xi^{\varepsilon_n}, \eta^{\varepsilon_n})$ converge almost everywhere. Let $(F(x,t), u(x,t))$ and $(\psi_{s,t}(x), \xi_{s,t}(x), \eta_{s,t}(x))$ be their limits, respectively. Then letting n tend to infinity at (6), we obtain

$$u(x,t) = E_Q[\phi(\psi_{0,t}(x))\xi_{0,t}(x) + \eta_{0,t}(x)],$$

which is the unique solution of the stochastic partial differential equation (4) satisfying the growth condition (5). See [4]. This implies that the limit of $(F^{\varepsilon_n}, u^{\varepsilon_n})$ does not depend on the choice of the sequence $\{\varepsilon_n\}$. Therefore $(F^\varepsilon, u^\varepsilon)$ converge weakly in W_m. It is now obvious that the limit measure $P_m^{(0)}$ satisfies i) and ii) of the theorem. The last assertion of the theorem is a consequence of the general property of the strong convergence. See [2].

References

[1] H. Huang and H. Kushner: Weak convergence and approximations for partial differential equations with stochastic coefficients, Stochastics 15 (1985), 209-245.

[2] H. Kunita: Convergence of stochastic flows connected with stochastic ordinary differential equations, Stochastics 17 (1986), 215-251.

[3] H. Kunita: On stochastic flows and applications, Tata institute of fundamental research, Bonbay, to appear.

[4] H. Kunita: Stochastic flows and stochastic partial differential equations, Proceedings of ICM-86, Berkeley, to appear.

[5] E. Pardoux and R. Bouc: PDE with random coefficients: Asymptotic expansion for the moments, Lecture Notes in Control and Inf. Science 42, ed. Fleming and Gorostiza, 1982, 276-289.

Department of Applied Science
Faculty of Engineering
Kyushu University 36
Fukuoka 812, Japan

SOME REMARKS ON GETZLER'S DEGREE THEOREM

Shigeo Kusuoka

Introduction. Let (μ,H,B) be an abstract Wiener space, i.e., B is a separable real Banach (or Fréchet) space, H is a separable real Hilbert space densely and continuously embedded in B, and μ is a Gaussian measure on B satisfying

$$\int_B \exp(i\cdot\langle u,z\rangle)\ \mu(dz) = \exp(\ -\frac{1}{2}\ \|u\|_H^2\ )\ ,\ u \in B^* \subset H.$$

Here we identify the dual space H^* of H with H itself.

Recently Getzler [3] introduced a notion of the degree of a map in B based on the Gaussian measure μ. Roughly speaking, he considered a nicely regular map $F:B\to H$ and defined the degree $\deg(I_B+F)$ of $I_B+F:B\to B$. He also showed that $\deg(I_B+F)$ has reasonable property if the map $F:B\to H$ satisfies some integrability conditions. The most interesting examples of maps from B into H are given through stochastic differential equations. Unfortunately it seems that the regularity conditions which Getzler imposed on the map F is too strong for such examples.

In this paper, we extend Getzler's results by weakening the regularity conditions (Section 1), and show that the solutions to some S.D.E.'s satisfy the regularity condition given in Section 1 (Section 2). We also give a remark on the Euler number (the index of a cross section) whose details will be studied in the forthcoming paper.

The reader will find that the results in Sections 2 and 3 are mere refinement of results of Getzler [3] and Bismut [1]. However, we dare to write this paper to clear what has been done.

1. A small extension of Getzler's degree theorem.

Let E be a real separable Hilbert space. Let us introduce two notions.

(1.1)Definition. We say that $F:B\to E$ is an $\mathcal{H}$-C^1 map if

(0) $F:B\to E$ is measurable,

(1) $F(z+\cdot):H\to E$ is continuous for each $z\in B$, and

(2) there is a measurable map $DF:B\to\mathcal{H}(E)$ such that for each $z\in B$, $F(z+\cdot):H\to\mathcal{H}(E)$ is continuous and

$$\|F(z+h)-F(z)-DF(z)h\|_E = o(\|h\|_H) \text{ as } \|h\|_H\to 0,$$

where $\mathcal{H}(E)$ is the Hilbert space consisting of all Hilbert-Schmidt operators from H into E with the Hilbert-Schmidt norm.

(1.2)Definition. We say that $F:B\to E$ is a compact $\mathcal{H}$-C^1 map if

(1) F is an $\mathcal{H}$-C^1 map, and

(2) for any $z\in B$ and any sequence $\{h_n\}_{n\geq1}\subset H$ with $h_n\to 0$ weakly in H as $n\to\infty$, $F(z+h_n)\to F(z)$ in E and $DF(z+h_n)\to DF(z)$ in $\mathcal{H}(E)$.

(1.3)Proposition. If $F:B\to E$ is a compact $\mathcal{H}$-C^1 map, then there is an increasing sequence $\{K_n\}_{n\geq1}$ of compact sets in B such that

(1) $K_n + \{ h\in H;\ \|h\|_H \leq 1 \} \subset K_{n+1}$,

(2) $\mu(B \setminus \bigcup_n K_n) = 0$, and

(3) $F|_{K_n}:K_n\to E$ and $DF|_{K_n}:K_n\to\mathcal{H}(E)$ are continuous for any $n\geq 1$.

Proof. Let $U_n = \{ h\in H;\ \|h\|_H \leq n \}$ and regard U_n as a topological space with the weak topology. Then U_n is a compact metric space. Note that $F(z+\cdot)|_{U_n}\in C(U_n\to E)$ and $DF(z+\cdot)|_{U_n}\in C(U_n\to\mathcal{H}(E))$. Then we can define measurable maps $\tilde{F}:B\to\prod_{n=1}^{\infty}C(U_n\to E)$ and $\widetilde{DF}:B\to\prod_{n=1}^{\infty}C(U_n\to\mathcal{H}(E))$ by

$$\tilde{F}(z) = \{ F(z+\cdot)|_{U_n} \}_{n\geq1} \quad\text{and}\quad \widetilde{DF}(z) = \{ DF(z+\cdot)|_{U_n} \}_{n\geq1} .$$

Since $C(U_n\to E)$ and $C(U_n\to\mathcal{H}(E))$ are Polish spaces, we can apply Lusin's

Theorem to $\tilde{F}$ and $\widetilde{D}F$. Therefore, there is an increasing sequence $\{K_n'\}_{n\geq 1}$ of compact sets in B such that $\tilde{F}|_{K_n'}$ and $\widetilde{D}F|_{K_n'}$ are continuous for each $n \geq 1$ and $\mu(\bigcup_{n=1}^{\infty} K_n') = 1$. Set $K_n = K_n' + U_n$. Then $\{K_n\}_{n\geq 1}$ satisfies our assertion. Q.E.D.

Let $\mathcal{P}(B^*)$ be the set of all orthogonal projections in H whose range is a finite dimensional vector subspace of B^*. Then we have the following (Corollary to Theorem 5.2 in [5]).

(1.4)<u>Proposition.</u> For any $\mathcal{H}$-C^1 map $F:B\to H$, there is a measurable map $\partial F:B\to\mathbb{R}$ such that

$$\mu(\ |({}_B\langle z,P_nF(z)\rangle_{B^*} - \operatorname{trace} P_nDF(z)\ ) - \partial F(z)| > \varepsilon\ ) \to 0,\ n\to\infty$$

for any $\varepsilon > 0$ and $\{P_n\}_{n\geq 1}\subset \mathcal{P}(B^*)$ with $P_n \uparrow I_H$.

(1.5)<u>Remark.</u> In [5], we wrote LF for ∂F.

For any $\mathcal{H}$-C^1 map $F:B\to H$, let us define $d(\cdot;F):B\to\mathbb{R}$ by

(1.6) $d(z;F) = \det_2(I_H+DF(z)) \exp(-\partial F(z) - \frac{1}{2}\|F(z)\|_H^2)$,

where $\det_2$ denotes the Carleman-Fredholm determinant.

We can show the following by using Proposition(1.3) and by almost the same argument as in the proof of Theorems 6.2 and 8.1 in [5].

(1.7)<u>Theorem.</u> For any compact $\mathcal{H}$-C^1 map $F:H\to H$, there are a measurable subset V in B and a countable family $\mathcal{K}$ of compact subsets in B such that

(1) $V + H = V$ and $\mu(V) = 1$,

(2) $\cup\,\mathcal{K} = \{\ z \in V;\ \det_2(I_H+DF(z)) \neq 0\ \}$,

(3) for each $K \in \mathcal{K}$, $I_B+F|_K:K\to B$ is one-to-one map and

$$\int_{(I_B+F)K} f(z)\ \mu(dz) = \int_K f((I_B+F)z)\ |d(z;F)|\ \mu(dz)$$

for any bounded measurable function f on B,

and

(4) the set $\{(I_B+F)z;\ z \in V,\ \det_2(I_H+DF(z)) = 0\ \}$ is measurable in B and of μ-measure zero.

Following Getzler [3], we define the degree of $I_B+F:B\to B$ as follows.

(1.8)<u>Definition.</u> For any $\mathscr{H}$-C^1 map $F:B\to H$ with $\int_B |d(z;F)|\ \mu(dz) < \infty$, we define the degree $\deg(I_B+F)$ of $I_B+F:B\to B$ by

$$\deg(I_B+F) = \int_B d(z;F)\ \mu(dz).$$

(1.9)<u>Theorem.</u> Suppose that $F:B\to H$ is a compact $\mathscr{H}$-C^1 map and $DF:B\to\mathscr{H}(H)$ is an $\mathscr{H}$-C^1 map. Suppose moreover that

$$\|F(z)\|_H\ ,\ \|DF(z)\|_{\mathscr{H}(H)}\ ,\ \|D^2F(z)\|_{\mathscr{H}^2(H)} \in \bigcap_{p\in(1,\infty)} L^p(B;d\mu),\ \text{and}$$

there are some $p \in (1,\infty)$ and $\varepsilon > 0$ such that

$$\int_B \exp(\ p\{\ \tfrac{1}{2}(\varepsilon+\ \|DF(z)\|_{\mathscr{H}(H)})^2 - \partial F(z) - \tfrac{1}{2}\ \|F(z)\|_H{}^2\})\ \mu(dz) < \infty.$$

Then we have

(1.10) $\int_B |d(z,F)|\ \mu(dz) < \infty,$

(1.11) $\int_B f((I_B+F)z)\ d(z;F)\ \mu(dz) = \deg(I_B+F)\int_B f(z)\ \mu(dz)$

for any bounded mesurable function $f:B\to\mathbb{R}$, and

(1.12) $\sum_{(I_B+F)x=z} \operatorname{sgn}(\det_2(I_H+DF(x))) = \deg(I_B+F)\quad \mu\text{-a.e.}z.$

We can prove this theorem in the same way as Getzler[3]. However, for convenience of the reader, we will give the sketch of the proof.

By easy calculation (c.f. Dunford-Schwartz [2] Chapter 11-9), we see that

(1.13) $|\det_2(I+K)| \leqq \exp(\ \tfrac{1}{2}\ \|K\|_{\mathscr{H}(H)}{}^2\ )$ and

(1.14) $\|\det_2(I+K)\ (I+K)^{-1}\|_{\text{operator}} \leqq \exp(\ \tfrac{1}{2}(\ 1 + \|K\|_{\mathscr{H}(H)}{}^2\ )\ )$

for any $K \in \mathscr{K}(H)$.

Therefore, we have (1.10) from (1.13) and the assumption. To prove (1.11), it is sufficient to show it in the case where $f(z) = f_v(z) = \exp(i\cdot {}_B\langle z,v\rangle_{B^*})$, $v \in B^*$.

By Cauchy's integral formula and (1.14), we see that

$$(1.15)\quad \| D(\ \det_2(I+DF(\cdot))\ (I_H+DF(\cdot))^{-1}v\ )(z)\|_{\mathscr{K}(H)}$$
$$\leq \varepsilon^{-1}e^{1/2}\|D^2F(z)\|_{\mathscr{K}^2(H)}\exp(\ \tfrac{1}{2}\ (\varepsilon + \|DF(z)\|_{\mathscr{K}(H)})^2\ )\ \|v\|_H\ .$$

Let $\Phi(z) = d(z;F)\cdot\exp(i\cdot {}_B\langle(I_B+F)z,v\rangle_{B^*})(I_H+DF(z))^{-1}v$. Then, by (1.15) and the results of Meyer [7], Sugita [8] and [9] (c.f. Watanabe [10]), we see that $\|D\Phi(z)\|_{\mathscr{K}(H)} \in L^p(B;d\mu)$ for some $p > 1$, and that

$$\begin{aligned} 0 &= \int_B \partial\Phi(z)\ \mu(dz) \\ &= \int_B d(z;F)\cdot(\ {}_B\langle(I_B+F)z,v\rangle_{B^*} - i\cdot\|v\|_H^2)\ f_v((I_B+F)z)\ \mu(dz). \end{aligned}$$

This implies that

$$\frac{d}{dt}\int_B f_{tv}((I_B-F)z)\cdot\exp(\ \frac{t^2}{2}\|v\|^2)\cdot d(z;F)\ \mu(dz) = 0,\ t \geq 0\ ,$$

which proves (1.11).

From Theorem(1.7), we have

$$\int_B \sum_{(I_B+F)x=z} 1\ \mu(dz) = \int_B |d(z;F)|\ \mu(dz) < \infty\ , \text{ and}$$

$$\int_B \sum_{(I_B+F)x=z} \operatorname{sgn}(\det_2(I_H+DF(x)))f(x)\ \mu(dz) = \int_B f(z)\ d(z;F)\ \mu(dz)$$

for any bounded measurable $f:B\to\mathbb{R}$. Comparing this with (1.11), we have (1.12). Q.E.D.

2. On the regularity of solutions to S.D.E.

Let $\Theta = \{ \theta = (\theta_1(\cdot),\dots,\theta_d(\cdot)) \in C([0,\infty);\mathbb{R}^d);\ \theta(0) = 0 \}$ and $\mathscr{W}$ be the Wiener measure on Θ. Also, let $V_i \in C_0^\infty(\mathbb{R}^N;\mathbb{R}^N)$, $i = 0,1,\dots,d$, and let us think of the S.D.E.:

$$(2.1)\quad X(T,x;\theta) = x + \sum_{i=1}^{d} \int_0^T V_i(X(t,x;\theta))\, d\theta_i(t) + \int_0^T V_0(X(t,x;\theta))\, dt, \quad T \geqq 0 .$$

Let $H = \{ h \in \Theta;\ h(t)$ is absolutely continuous in t and $\int_0^\infty |\dot{h}(t)|\, dt < \infty \}$.

Then $(\mathscr{W},H,\Theta)$ is an abstract Wiener space. We will show that we have a good version of the solution to (2.1) for which $X(T,x;\theta+\cdot):H\to\mathbb{R}^N$ is smooth for any $(T,x) \in [0,\infty)\times\mathbb{R}^N$.

First, remind that we can take a good version of the solution to (2.1) such that $(T,x) \in [0,\infty)\times\mathbb{R}^N \to X(T,x;\theta) \in \mathbb{R}^N$ is continuous for all $\theta \in \Theta$ and $x \to X(T,x;\theta)$ is a diffeomorphism in $\mathbb{R}^N$ for all $T \in [0,\infty)$ and $\theta \in \Theta$ (c.f. Kunita [4]). Since the support of V_i's is compact, we may assume that there is an $R > 0$ such that $X(T,x;\theta) = x$ for any $|x| \geqq R$, $T \in [0,\infty)$ and $\theta \in \Theta$. Let $J(T,x;\theta) = (\frac{\partial}{\partial x^j} X^i(T,x;\theta))_{i,j=1,\dots,N}$ and $W_i(T,x;\theta) = J(T,x;\theta)^{-1} V_i(X(T,x;\theta))$, $i = 1,\dots,d$.

The following is due to Bismut [1].

<u>(2.2)Theorem.</u> For each $h \in H$, let $Z(t,x,h;\theta)$ is the solution to the O.D.E.:

$$(2.3)\quad Z(T,x,h;\theta) = x + \sum_{i=1}^{d} \int_0^T W_i(t,Z(t,x,h;\theta));\theta)\, \dot{h}_i(t)\, dt, \quad T \geqq 0.$$

Then $\mathscr{W}[\ X(T,x;\theta+h) = Z(T,x,h;\theta)$ for all $(T,x) \in [0,\infty)\times\mathbb{R}^N\ ] = 1$ for each $h \in H$.

Now let $C_T(\theta) = \sup\{ \sum_{i=1}^{d} (|W_i(t,x;\theta)| + \sum_{j=1}^{N} |\frac{\partial}{\partial x^j} W_i(t,x;\theta)|\ ;\ (t,x) \in [0,T]\times\mathbb{R}^N \}$ ($< \infty$). Then we have the following.

(2.4) <u>Lemma.</u> For each $\theta \in \Theta$,

$$|Z(t,x,h;\theta)-Z(t,y,k;\theta)|$$
$$\leq 2(|x-y|^2 + C_T(\theta)^2\|h-k\|_H^2)^{1/2}\exp(2t\, C_T(\theta)^2(\|k\|_H\wedge\|k\|_H)^2)$$

for $t \in [0,T]$, $x,y \in \mathbb{R}^N$ and $h,k \in H$.

<u>Proof.</u> Set $Z_0(T,x,h;\theta) \equiv x$ and

$$Z_{n+1}(T,x,h;\theta) \equiv x + \sum_{i=1}^{d} \int_0^T W_i(t,Z_n(t,x,h;\theta));\theta)\, \dot{h}_i(t)\, dt \ , \ n \geq 0.$$

Then by usual argument and Schwarz's inequality, we have

$$|Z_{n+1}(t,x,h;\theta) - Z_n(t,x,h;\theta)| \leq (C_T(\theta)^2\|h\|_H^2)^n \frac{t^n}{n!} \ , \ t \in [0,T].$$

Therefore, $Z_n(T,x,h;\theta)$ converges as $n \to \infty$ and the limit is the solution to (2.3). Also, we have

$$|Z_{n+1}(t,x,h;\theta) - Z_{n+1}(t,y,k;\theta)|^2$$
$$\leq 3\{ |x-y|^2 + (\sum_{i=1}^{d}\int_0^t |W_i(\tau,Z_n(\tau,x,h;\theta),\theta)(\dot{h}_i(t)-\dot{k}_i(t))|\, d\tau)^2$$
$$+ (\sum_{i=1}^{d}\int_0^t |(W_i(\tau,Z_n(\tau,x,h;\theta),\theta)-W_i(\tau,Z_n(\tau,y,k;\theta);\theta)))\dot{k}_i(t)|\, d\tau)^2 \}$$
$$\leq 3\{ |x-y|^2 + C_T(\theta)^2\|h-k\|_H^2$$
$$+ C_T(\theta)^2\|k\|_H^2 \int_0^t |Z_n(\tau,x,h;\theta)-Z_n(\tau,y,k;\theta)|^2\, d\tau \}$$

Thus, by induction, we have

$$|Z_n(t,x,h;\theta)-Z_n(t,y,k;\theta)|^2$$
$$\leq 3(|x-y|^2 + C_T(\theta)^2\|h-k\|_H^2) \exp(3t\, C_T(\theta)^2\|k\|_H^2).$$

This proves our assertion. Q.E.D.

The following is our main result.

(2.5) <u>Theorem.</u> There are a map $\tilde{X}:[0,\infty)\times\mathbb{R}^N\times\Theta$ and a σ-compact subset Ω in Θ with $\mathscr{W}(\Omega) = 1$ such that

(1) $\tilde{X}(\cdot,x;\cdot):[0,\infty)\times\Theta\to\mathbb{R}^N$ is progressively measurable for each $x \in \mathbb{R}^N$,

(2) $\tilde{X}(t,\cdot;\theta):\mathbb{R}^N\to\mathbb{R}^N$ is smooth for each $t \in [0,\infty)$ and $\theta \in \Theta$,

(3) $\dfrac{\partial^\alpha}{\partial x^\alpha}\tilde{X}(t,x;\theta)$ is continuous in $(t,x) \in [0,\infty)\times\mathbb{R}^N$ for each $\theta \in \Theta$,

(4) $\hat{X}(t,x;\theta+h) = Z(t,x,h;\theta)$ for all $(t,x) \in [0,\infty)\times\mathbb{R}^N$, $h \in H$ and $\theta \in \Omega$, and

(5) $\hat{X}(t,x;\theta) = x$ for each $(t,x) \in [0,\infty)\times\mathbb{R}^N$ and $\theta \notin \Theta + H$.

Proof. For each $f \in L^1(\Theta;d\mathscr{W})$ and $s \in [0,\infty)$, let

$$P_s f(\theta) = \int_\Theta f(\, e^{-s}\theta + (1-e^{-2s})^{1/2}z \,)\, \mathscr{W}(dz),\ \theta \in \Theta.$$

$\{P_s\}_{s\geq 0}$ is nothing but the Ornstein-Uhlenbeck operator. As shown in [6], if $s > 0$ and $f \in L^\infty(\Theta,d\mathscr{W})$, then $P_s f(\theta+\cdot):H\to\mathbb{R}$ is continuous (actually, real analytic). Note that

(2.6) $P_s X(t,x;\cdot)(\theta+h) = P_s X(t,x;\cdot+e^s h)(\theta)$,

and so

(2.7) $P_s X(t,x;\cdot)(\theta+h) = P_s Z(t,x,e^s h;\cdot)(\theta)$, $\mathscr{W}$-a.s.θ, for each $h \in H$.

Since both sides in (2.7) are continuous in t, x and h if $s > 0$, we see that

(2.8) $\mathscr{W}[\ P_s X(t,x;\cdot)(\theta) = P_s Z(t,x,e^s h;\cdot)(\theta),\ (t,x) \in [0,\infty)\times\mathbb{R}^N,\ h \in H]$
$= 1$,

for any $s > 0$.

Let $A_{T,n} = \{\ \theta \in W;\ C_T(\theta) \leq n\ \}$, $T > 0$. Then we have

(2.9) $|P_s Z(t,x,e^s h;\cdot)(\theta) - P_s Z(t,x,e^s k;\cdot)(\theta)|$
$\leq (|x|\vee R)\cdot P_s \chi_{W\setminus A_{T,\,n}}(\theta) + 2ne^s\|h-k\|_H \exp(\ 2Tn^2e^{2s}(\|h\|_H+\|k\|_H)^2\ )$

if $t \in [0,T]$. Let V be a dense countable $\mathbb{Q}$-vector space in H. Then, we can take a sequence $s_n \downarrow 0$ and a measurable subset Ω' in Θ with $\mathscr{W}(\Omega') = 1$ such that

(2.10) $P_{s_n} X(t,x;\cdot)(\theta+h) = P_{s_n} Z(t,x,e^s h;\cdot)(\theta)$ for all $n \geq 1$,

$(t,x) \in [0,\infty)\times\mathbb{R}^N$, $h \in H$ and $\theta \in \Omega'$,

(2.11) $P_{s_n}\chi_{W\setminus A_{\ell,m}}(\theta) \to \chi_{W\setminus A_{\ell,m}}(\theta)$ as $n \to \infty$, for any $m,\ell \in \mathbb{N}$ and $\theta \in \Omega'$,

(2.12) $X(t,x;\theta+v) = Z(t,x,v;\theta)$ for all $(t,x) \in [0,\infty)\times\mathbb{R}^N$, $v \in V$ and $\theta \in \Omega'$, and

(2.13) $P_{s_n} X(\tau,\xi;\cdot)(\theta+v) \to X(\tau,\xi;\theta+v)$ as $n \to \infty$, for any $\tau \in [0,\infty)\cap\mathbb{Q}$, $\xi \in \mathbb{Q}^N$ and $v \in V$.

Take a σ-compact set Ω in Θ with $\Omega \subset \Omega'$ and $\mathbb{W}(\Omega) = 1$. By (2.9), (2.10) and (2.11), we see that

(2.14) $$\overline{\lim_{n\to\infty}} |P_{s_n} X(\tau,\xi;\cdot)(\theta+h) - P_{s_n} X(\tau,\xi;\cdot)(\theta+k)|$$

$$\leqq 2m \|h-k\|_H \exp(2\ell^2(\|h\|_H+\|k\|_H)^2)$$

for any $\tau \in [0,\ell]\cap\mathbb{Q}$, $\xi \in \mathbb{Q}^N$, $h,k \in H$ and $\theta \in \Omega\cap A_{\ell,m}$.

Since $\bigcup_{m=1}^{\infty} A_{\ell,m} = \Theta$ for each ℓ and V is dense in H, (2.13) implies that $P_{s_n} X(\tau,\xi,\cdot)(\theta+h)$ converges for any $\tau \in [0,\infty)\cap\mathbb{Q}$, $\xi \in \mathbb{Q}^N$, $h \in H$ and $\theta \in \Omega'$. Let us define $\tilde{\tilde{X}}(\tau,\xi;\cdot)$, $\tau \in [0,\infty)\cap\mathbb{Q}$ and $\xi \in \mathbb{Q}^N$, by

$$\tilde{\tilde{X}}(\tau,\xi;\theta) = \begin{cases} \lim_{n\to\infty} P_{s_n} X(\tau,\xi;\theta) & \text{if } \theta \in \Omega + H \\ \xi & \text{otherwise} \end{cases}.$$

Then by (2.12) and (2.13) we have, for any $h \in V$,

(2.15) $$\tilde{\tilde{X}}(\tau,\xi;\theta+h) = Z(\tau,\xi,h;\theta) , \quad \theta \in \Omega.$$

Because of (2.14), we see that $\tilde{\tilde{X}}(\tau,\xi;\theta+\cdot):H\to\mathbb{R}^N$ is continuous for each $\theta \in \Omega$. Therefore (2.15) is true for all $h \in H$. But $Z(t,x,h;\theta)$ is continuous in (t,x). This implies that $\tilde{X}(t,x;\theta) = \lim_{\substack{\tau\to t \\ \xi\to x}} \tilde{\tilde{X}}(\tau,\xi;\theta)$ exists for all $(t,x) \in [0,\infty)\times\mathbb{R}^N$ and $\theta \in \Theta$. These $\tilde{X}$ and Ω are our desired ones. Q.E.D.

3. A remark on the relation between degree and Euler number.

Let M be a (compact) manifold of dimension n and E be a vector bundle over M of rank n with orientation. Let E' be a vector bundle over M such that $E \oplus E'$ is a trivial bundle. Then E' is also orientable. Let $i: E \oplus E' \to M \times \mathbb{R}^N$ be an orientation preserving bundle isomorphism. For any cross section σ in E, we can define $\Phi_\sigma : E' \to \mathbb{R}^N$ by $\Phi_\sigma(u) = \pi_1 \circ i(u + \sigma(\pi_0(u)))$, $u \in E'$, where $\pi_0 : E' \to M$ and $\pi_1 : M \times \mathbb{R}^N \to \mathbb{R}^N$ are projection maps. Note that E' is a manifold of dimension N. Thus we may think of the degree $\deg(\Phi_\sigma)$ of the map $\Phi_\sigma : E' \to \mathbb{R}^N$.

Thinking of Thom class, we see that $\deg(\Phi_\sigma)$ is the Euler number of the vector bundle E in the case that $\sigma \equiv 0$ and M is compact. In general, $\deg(\Phi_\sigma)$ is regarded as the total index of the cross section σ.

Now, let $F: B \to \mathbb{R}^N$ be a compact $\mathcal{H}$-C^∞ map with $\det(DF(z)DF(z)^*) > 0$ for all $z \in B$. Then $M = \{ z \in B;\ F(z) = 0 \}$ is something like a manifold. Let $\Phi : M \times \mathbb{R}^N \to B$ be a map given by $\Phi(z,u) = z + DF(z)^* u$, $(z,u) \in M \times \mathbb{R}^N$. If one can define $\deg(\Phi)$, it is regarded as the total index of the vector field σ on M given by

$$\begin{aligned}\sigma(z) &= z - DF(z)^*(DF(z)DF(z)^*)^{-1} DF(z) z \\ &= z - DF(z)^*(DF(z)DF(z)^*)^{-1}(\partial DF(z) + \text{'trace } D^2F(z)\text{'}),\end{aligned}$$

$z \in M$.

Acknowledgement. The author is grateful to Prof. D.Elworthy and Prof. A.Cruzeiro for useful discussion at Warwick University.

References.

[1] Bismut, J.-M., Martingales, the Malliavin calculus and hypoellipticity under general Hörmander's conditions, Z. Wahr. verw. Geb. 56(1981), 469-505.

[2] Dunford, N. and J.T. Schwartz, Linear operators, Part II, Interscience, New York, 1963.

[3] Getzler, E., Degree theory for Wiener maps, to appear in J. Func. Anal.

[4] Kunita, H., Stochastic differential equations and stochastic flow of diffeomorphisms, Ecole d'Été de Prob. de Saint-Flour XII-1982, Lec. Notes in Math. 1097(1984), 144-303, Springer-Verlag, Berlin.

[5] Kusuoka, S., The nonlinear transformation of Gaussian measure on Banach space and its absolute continuity, Part I, J. Fac. Sci. Univ. Tokyo Sec. IA, 29(1982), 567-598.

[6] Kusuoka, S., Analytic functionals of Wiener process and absolute continuity, in Functional analysis in Markov process ed. by M. Fukushima, Lec. Notes in Math. 923(1982), 1-46, Springer-Verlag, Berlin.

[7] Meyer, P.A., Notes sur les processus d'Ornstein-Uhlenbeck, Séminaire de Prob. XVI, Lec, Notes in Math. 920(1982), 95-133, Springer-Verlag, Berlin.

[8] Sugita, H., Sobolev spaces of Wiener functionals and Malliavin's calculus, J. Math. Kyoto Univ. 25(1985), 31-48.

[9] Sugita, H., On a characterization of the Sobolev spaces over an abstract Wiener space, J. Math. Kyoto Univ. 25(1985), 717-725.

[10] Watanabe, S., Stochastic differential equations and Malliavin calculus, Tata Inst. of Fundamental Research Lec. on Math. and Phys. 73, 1984, Springer-Verlag, Berlin.

Department of Mathematics
Faculty of Science
University of Tokyo
Hongo, Tokyo, Japan

ON LIMIT THEOREMS FOR CONDITIONALLY INDEPENDENT RANDOM VARIABLES CONTROLLED BY A FINITE MARKOV CHAIN

Z. A. Kvatadze and T. L. Shervashidze

On the probability space $(\Omega,\mathcal{F},\mathbb{P})$ consider a stationary two-component sequence $(\xi_j,X_j)_{j\geq 1}$ where $\xi=(\xi_1,\xi_2,\dots)$ is a finite Markov chain with the states $1,\dots,s$ forming a single ergodic class (maybe with periodical subclasses). The random variables (r.v.'s) $X_1,\dots,X_n$ for every n are independent under the condition that the trajectory $\xi_{1n}=(\xi_1,\dots,\xi_n)$ is given. The conditional distribution of X_j depends only on the values of ξ_j. More precisely, if X, η are random vectors on a probability space $(\Omega,\mathcal{F},\mathbb{P})$ and P_X, $P_{X|\eta}$ denote the distribution of X and the conditional distribution of X with respect to η, then our definition means that

$$P_{(X_1,\dots,X_n)|\xi_{1n}} = P_{X_1|\xi_1}\times\cdots\times P_{X_n|\xi_n} .$$

It is convenient for our purposes to say that the r.v.'s $X_1,X_2,\dots$ are conditionally independent and controlled by a Markov chain (instead of the traditional term "r.v.'s defined on a Markov chain", see,e.g., [1]). Note also that when X_j are measurable functions of ξ_j then $X_1,X_2,\dots$ are usually called the "r.v.'s connected in a Markov chain". These latter have been intensively and efficiently studied (see, e.g. [2] with the bibliography provided and [3]).

On the basis of the results known for the sums of independent r.v.'s and r.v.'s "connected in a Markov chain" we have studied the properties of the sum $Y_n=X_1+\cdots+X_n$ proceeding from the representation

$$Y_n-EY_n = \sum_{j=1}^{n}[x_j-E(x_j|\xi_j)]+\sum_{j=1}^{n}[E(x_j|\xi_j)-EX_j] = Y_{n1}+Y_{n2}$$

partitioning the sum into two uncorrelated and asymptotically independent parts. The weak convergence of the distribution of the normalized sum $S_n=S_{n1}+S_{n2}$, $S_{ni}=Y_{ni}/\sqrt{n}$, $i=1,2$, considered in [1, 4, 5] etc. had a new interpretation in the series of papers [6-9] (see Section 1). The approach developed enabled us to cover questions of density convergence and convergence in variation [10] (the results are formulated in Section 2).

For the so-called conditional Markov chains paper [11] contains an assertion on the limit probabilities for the conditional distribution function of the sum to be in a certain zone bounded by normal distribution functions. We have obtained a similar result for S_n given in Section 3.

Finally, in Section 4 a local theorem is stated and proved in the case of integer-valued X_j.

As a simplest example of conditionally independent r.v.'s controlled by a finite Markov chain we can take

$$X_j = X_{\xi_j j}, \qquad j = 1,2,\ldots ,$$

where X_{ij}, $i=1,\ldots,s$, $j=1,2,\ldots$, is a matrix of independent r.v.'s and in each row the r.v.'s have the same distribution changing from row to row.

It should be noted here that boundary value problems for sums of r.v.'s defined on a Markov chain were investigated by H. D. Miller, A. A. Borovkov, E. L. Presman, K. A. Borovkov. The limit theorems for S_n established by Keilson and Wishart, Aleshkevichus [1] and others were proved by the method of characteristic functions and matrices. O'Brein [4] used Ibragimov's limit theorem for r.v.'s with a mixing property since if ξ has this property then X_j-s will also have it.

We finish the Introduction by specifying some notations. Let $\pi=(\pi_1,\ldots,\pi_s)$ be the vector of stationary probabilities and $Z=(z_{i,j})$, $i,j=1,\ldots,s$, be a fundamental matrix of the chain ξ.

Denote

$$\mu_i = E[X_1 | (\xi_1=i)], \quad \sigma_i^2 = E[(X_1-\mu_i)^2 | (\xi_1=i)], \quad i=1,\ldots,s,$$
$$\mu = EX_1 = \sum_{i=1}^{s} \pi_i\mu_i, \quad \sigma_0^2 = \sum_{i=1}^{s} \pi_i\sigma_i^2, \quad t = \lim_{n\to\infty} E\, S_{n2}^2 . \tag{1}$$

If t is finite then

$$t=t(\mu), \quad \mu:\{1,\ldots,s\}\longrightarrow R^1, \quad \mu(i)=\mu_i, \quad i=1,\ldots,s, \tag{2}$$

where for $f:\{1,\ldots,s\}\longrightarrow R^1$ we have the notation

$$t(f) = \sum_{i,j=1}^{s} \pi_i(2z_{ij}-\pi_i-\delta_{ij})f(i)f(j) \tag{3}$$

with δ_{ij} as Kronecker's delta. For finite σ_0^2 and t denote

$$\sigma^2 = \sigma_0^2 + t = \lim_{n\to\infty} E\, S_n^2 \ . \tag{4}$$

Let $p_X(x)$ and $p_{X|\eta}(x)$ be the densities corresponding to the distributions P_X and $P_{X|\eta}$ introduced above. The conditional distribution of X_1 given $\xi_1 = i$ is denoted by P_i, $i=1,\dots,s$. The normal distribution with the parameters $(0,\gamma)$ and its distribution function and density will be denoted by Φ_γ, $\Phi_\gamma(x)$ and $\varphi_\gamma(x)$, respectively, with $\varphi(x)=\varphi_1(x)$. The symbol $\xrightarrow{w}$ denotes the weak convergence as $n\to\infty$.

1. <u>Central limit theorem</u>. As a corollary of the central limit theorem for conditionally independent r.v.'s controlled by a general ergodic sequence [6-9] the following theorem (cf. [12]) can be proved.

<u>Theorem 1</u>. If $\sigma_0^2 < \infty$ then $t<\infty$ and

a) $P_{S_{n1}|\xi_{1n}} \xrightarrow{w} \Phi_{\sigma_0^2}$ $\mathbb{P}$-a.s. ; b) $P_{S_{n2}} \xrightarrow{w} \Phi_t$; c) $P_{S_n} \xrightarrow{w} \Phi_{\sigma^2}$,

where σ_0^2, t and σ^2 are defined by relations (1)-(4).

The following assertion, which is somewhat more general than assertion b) in Theorem 1, holds.

<u>Theorem 2</u>. If $f:\{1,\dots,s\}\longrightarrow R^1$ then for the sum $F_n = \sum_{j=1}^{n} [f(\xi_j) - Ef(\xi_j)]/\sqrt{n}$ the convergence relation $P_{F_n} \xrightarrow{w} \Phi_{t(f)}$ holds where $\lim_{n\to\infty} EF_n^2 = t(f)$ is defined by relation (4).

In the case when the chain has no periodic subclasses assertion c) of Theorem 1 was proved in [4], as we have already noted. Our theorem is a good device to prove the weak convergence when the periodic subclasses are present. In this case the $\mathbb{P}$-a.s. convergence $P_{S_{n1}|\xi_{1n}} \xrightarrow{w} \Phi_{\sigma_0^2}$ is evident and for the validity of c) it is sufficient to establish the asymptotic normality of $P_{S_{n2}}$. Therefore we must prove that $P_{F_n} \xrightarrow{w} \Phi_{t(f)}$ for such a chain, too. If ξ has d periodic subclasses then the ergodic chain $\overline{\xi} = (\xi_1, \xi_{d+1}, \xi_{2d+1}, \dots)$ has none. Now consider a two-component sequence $(\overline{\xi}_j, X_j)_{j\geq 1}$, where

$\bar{\bar{\xi}}=(\xi_{(j-1)d+1},\xi_{jd+1})$, $\quad X_j=f(\xi_{(j-1)d+1})+\cdots+f(\xi_{jd})$, $\quad j=1,2,\ldots.$

Applying a version of Theorem 1 proved in [4] we can easily obtain the convergence $P_{F_n}\xrightarrow{w}\Phi_{t(f)}$ from which it can be derived that $P_{S_n}\xrightarrow{w}\Phi_{\sigma^2}$. As proved in [4] this convergence holds for any distribution of ξ_1 distinct from the stationary one.

2. Convergence of densities and convergence in variation.

Theorem 3. If $\sigma_0^2<\infty$ and at least one of the conditional distributions $P_1,\ldots,P_s$ has a characteristic function which belongs to $L^r(R^1)$ for some positive integer r, then

a) a representation $P_{S_n}=P_{S_n}^{(1)}+P_{S_n}^{(2)}$ can be found which is , generally speaking, different from the decomposition of P_{S_n} into the absolutely continuous $(P_{S_n}^{ac})$ and singular $(P_{S_n}^{s})$ components with respect to the Lebesgue measure such that $(P_{S_n}^{s})(R^1)\le(P_{S_n}^{(2)})(R^1)\le\frac{C}{n}$, where the constant C depends on the number r and the parameters of the chain ξ;

b) the density $p_{S_n}^{(1)}(x)$ of the measure $P_{S_n}^{(1)}$ which appears when $n=r$ satisfies the following relation

$$\sup_{x\in R^1}|p_{S_n}^{(1)}(x)-\varphi_{\sigma^2}(x)| \longrightarrow 0 \qquad (n\longrightarrow\infty).$$

Theorem 4. If for some positive integer r and at least one i; $1\le i\le r$, the r-fold convolution of the conditional distribution P_i has a non-zero absolutely continuous component, then $\|P_{S_n}-\Phi_{\sigma^2}\|\to 0$ $(n\to\infty)$ (where $\|\cdot\|$ denotes the variation norm of the signed measure).

Theorems 3 and 4 are proved in [10] in the case $X_j\in R^k$. Besides, the estimates of the convergence rate in Theorems 1, 3 and 4 are given in [10]. Since the estimates are constructed on the basis of the known estimates for independent vectors and those connected in a chain, the obtained estimates imply the absence of periodic subclasses for the chain ξ in addition to a natural condition on the existence of third order conditional moments. In particular, if for some positive integer r all r-fold convolutions P_i^{r*} have bounded densities, $i=1,\ldots,s$, then $\sup|p_{S_n}(x)-\varphi_{\sigma^2}(x)|$ does not exceed $c/\sqrt{n}$. And if it appears that these convolutions have non-zero absolutely continuous components, then $\|P_{F_n}-\Phi_{\sigma^2}\|\le C((\ln n)/n)^{1/2}$. In both cases the constant C depends on the conditional distributions and chain characteristics (and also on the dimension number if $X_j\in R^k$).

3. The probability for the conditional distribution function to be in a zone. Using Theorem 1 and the properties of weak convergence (see [13]) we can obtain the following limit relations for the conditional distribution function $F_{S_n|\xi_{1n}}(x)=\mathbb{P}(S_n<x|\xi_{1n})$.

Theorem 5. If $0<\sigma_0^2<\infty$, then

a) for $t>0$, any $y>0$ and $n\to\infty$

$$\mathbb{P}(\Phi_{\sigma_0^2}(x-y) < F_{S_n|\xi_{1n}}(x) < \Phi_{\sigma_0^2}(x+y),\forall x\in R^1) \longrightarrow \Phi_t(y)-\Phi_t(-y) \ ;$$

b) for $t=0$, any $y>0$ and $n\to\infty$

$$\mathbb{P}(\Phi_{\sigma_0^2}(x-y) < F_{S_n|\xi_{1n}}(x) < \Phi_{\sigma_0^2}(x+y),\forall x\in R^1) \longrightarrow 1 \ .$$

4. Local limit theorem for integer-valued random variables. Petrov [14] considered a k-sequence $u_1,u_2,\ldots$ of integer-valued r.v.'s with finite variances; this means that among the distributions $P_{u_1},P_{u_2},\ldots$ only k distributions are different. If among these k distributions there are ℓ distributions $P_1^*,\ldots,P_\ell^*$ which are non-degenerate, they occur in the sequence infinitely and have relatively prime maximal steps $h_1,\ldots,h_\ell$ (i.e. the greatest common divisor of these numbers is equal to 1), then for the sum $V_n=u_1+\cdots+u_n$ the following relation is proved in [14]:

$$\sup_{n\in\mathbb{Z}} |\sqrt{DV_n}\ \mathbb{P}(V_n=N)-\varphi((N-EV_n)/\sqrt{DV_n})| \longrightarrow 0 \quad (n\longrightarrow\infty) \ , \tag{5}$$

where $\mathbb{Z}$ is a set of all integers.

Applying Petrov's theorem and the technique from [2] and [10] one can prove a theorem formally coinciding with a local theorem due to Gnedenko for independent identically distributed lattice r.v.'s (and there is no need to use local theorems for r.v.'s "connected in a Markov chain" [15], [2]).

Theorem 6. If the r.v.'s X_j are integer-valued, $0<\sigma_0^2<\infty$ and the maximal steps of the non-degenerate distributions $P_1^*,\ldots,P_\ell^*$

among $P_1,\dots,P_s$ are relatively prime, then for $n\to\infty$ we have

$$\sup_{n\in\mathbb{Z}} \left|\sigma\sqrt{n}\ \mathbb{P}(Y_n=N)-\varphi\left(\frac{N-nEX_1}{\sigma\sqrt{n}}\right)\right| \longrightarrow 0 .$$

<u>Proof</u>. Denote by $\nu_n(i)$ the frequency of the state i in the trajectory $\xi_{1n}=(\xi_1,\dots,\xi_n)$, $\nu_n(1)+\dots+\nu_n(s)=1$. Let $i_1,\dots,i_\ell$ be the states corresponding to the non-degenerate conditional distributions $P_1^*,\dots,P_\ell^*$. Denote $\pi_r^*=\pi_{i_r}$, $\nu_n^*(r)=\nu_n(i_r)$, $r=1,\dots,\ell$, $\nu_n^*=\nu_n^*(1)+\dots+\nu_n^*(\ell)$. Obviously, $0<\pi^*\le 1$ and $0\le\nu_n^*\le n$. Also denote

$$\underline{\sigma}^2 = \min_{1\le r\le\ell} \sigma_{i_r}^2 , \quad \bar{\sigma}^2 = \max_{1\le i\le s} \sigma_i^2 , \quad \hat{\sigma}^2 = \frac{1}{n} \sum_{i=1}^{s} \nu_n(i)\sigma_i^2 .$$

We have

$$\sigma_0^2,\ \hat{\sigma}_0^2 \in [\underline{\sigma}^2,\bar{\sigma}^2] , \qquad \mathbb{P}(\lim_{n\to\infty} \hat{\sigma}_0^2 = \sigma_0^2) = 1 \tag{6}$$

and by [16]

$$E(\nu_n(i)/n-\pi_i)^2 \le C/n , \quad i = 1,\dots,s , \tag{7}$$

where C depends on the chain parameters. By the above theorem due to Petrov (relation (5)) $\mathbb{P}$-a.s.

$$E\left\{\sup_N\left|\hat{\sigma}_0 \sqrt{n}\ \mathbb{P}(Y_n=N|\xi_{1n}) - \varphi\left(\frac{N-\sum_1^n\mu(\xi_j)}{\hat{\sigma}_0\sqrt{n}}\right)\right| \ |\xi_{1n}\right\} \longrightarrow 0 \tag{8}$$

as $n\to\infty$ (where $\mu(\xi_j)=E(X_j|\xi_j)$). This is ensured by infinitely increasing $\nu_n(i)$ for all i almost on all trajectories with $\mathbb{P}$-probability 1 and the condition that the maximal steps of the distributions P_i^*, $i=1,\dots,\ell$, are relatively prime. The conditional mathematical expectation in (8) is merely formal here since for fixed ξ_{1n} the expression under the sign sup is constant.

Now consider Rogozin's inequality from [17]. Let $\eta_1,\dots,\eta_m$ be independent r.v.'s and $p_j=\sup_x\mathbb{P}(\eta_j=x)$, $j=1,\dots,m$. Then

$$\max_x \mathbb{P}(\eta_1+\dots+\eta_m=x) \le A \left[\sum_{j=1}^{m} (1-p_j)\right]^{-1/2} ,$$

where A is an absolute constant.

If we denote $p=\max_{1\le i\le\ell}\sup_{N\in\mathbb{Z}} P_i^*(\{N\})$ ($p<1$, since P_i^*, $i=1,\dots,\ell$, are non-degenerate) and note that for fixed trajectories ξ_{1n} the sum $Y_n=X_1+\dots+X_n$ will contain ν_n^* non-degenerate summands and the remaining $n-\nu_n^*$ become constants, then, by Rogozin's inequality, we have that

$$\mathbb{P}(Y_n=N|\xi_{1n}) \le A[(1-p)\nu_n^*]^{-1/2} . \tag{9}$$

We denote the right-hand side of (8) by $G(\xi_{1n})$ and try to prove that $EG(\xi_{1n})\to 0$. Introduce the event $M=\{|\nu_n^*/n-\pi^*|<\varepsilon\}$, where ε is sufficiently small. For $\omega\in M$ we have $n/\nu_n^*<(\pi^*-\varepsilon)^{-1}$. Therefore $EG(\xi_{1n})I_M(\omega)\to 0$. By (7) we have

$$\mathbb{P}(\overline{M}) \le \sum_{i=1}^{\ell} \mathbb{P}\left\{\left|\frac{\nu_n^*(i)}{n}-\pi_i^*\right| \ge \frac{\varepsilon}{\ell}\right\} \le \frac{c\ell^3}{\varepsilon^2 n} .$$

It is clear, that $\hat{\sigma}_0=0$ as $\nu_n^*=0$ and for $\nu_n^*\ge 1$ the inequality $G(\xi_{1n})\le\sqrt{n}\bar{\sigma}+(2\pi)^{-1/2}$ holds. Hence

$$EG(\xi_{1n})I_{\overline{M}}(\omega) \longrightarrow 0 \quad \text{and} \quad EG(\xi_{1n}) \longrightarrow 0 \quad \text{as} \quad h \longrightarrow 0.$$

The assertion of Theorem 6 can be written in an equivalent form

$$\sup_{N\in\mathbb{Z}} \left|\sqrt{n}\ \mathbb{P}(Y_n=N) - \varphi_{\sigma^2}\left(\frac{N-n\mu}{\sqrt{n}}\right)\right| \longrightarrow 0 . \tag{10}$$

We have the inequalities

$$\begin{aligned}
&\sup \left|\sqrt{n}\ \mathbb{P}(Y_n=N) - \varphi_{\sigma^2}\left(\frac{N-n\mu}{\sqrt{n}}\right)\right| \\
&\le \sup \left|\sqrt{n}\ E\mathbb{P}(Y_n=N|\xi_{1n}) - E\left\{E[\varphi_{\hat{\sigma}_0^2}\left(\frac{N-\Sigma\mu(\xi_j)}{\sqrt{n}}\right)|\xi_{1n}]\right\}\right| \\
&\quad + \sup\left|E\left\{E[\varphi_{\hat{\sigma}_0^2}\left(\frac{N-\Sigma\mu(\xi_j)}{\sqrt{n}}\right)|\xi_{1n}]\right\} - E\varphi_{\sigma_0^2}\left(\frac{N-\Sigma\mu(\xi_j)}{\sqrt{n}}\right)\right| \\
&\quad + \sup\left|E\varphi_{\sigma_0^2}\left(\frac{N-\Sigma\mu(\xi_j)}{\sqrt{n}}\right) - \varphi_{\sigma_0^2+t}\left(\frac{N-n\mu}{\sqrt{n}}\right)\right| \\
&= I_1 + I_2 + I_3 .
\end{aligned}$$

The convergence $I_1\to 0$ is a consequence of the relation $EG(\xi_{1n})\to 0$. For the estimation of I_2 we can use the inequality

$$|\varphi_{b_1}(x)-\varphi_{b_2}(x)| \le (2\sqrt{2\pi}\ b^{3/2})^{-1}|b_1-b_2| ,$$

where $b=\min(b_1,b_2)>0$, which easily follows from the inversion formula. Taking (6) and (7) into account we have

$$I_2 \le E \sup_x \left|\varphi_{\hat\sigma_0^2}(x)-\varphi_{\sigma_0^2}(x)\right| \le 2\sqrt{2\pi}\underline{\sigma}^3)^{-1}\ E\ |\hat\sigma_0^2-\sigma_0^2|$$

$$\le \frac{\bar\sigma^2\underline{\sigma}^{-3}}{2\sqrt{2\pi}} \sum_{i=1}^{\ell} E\left|\frac{\nu_n^*(r)}{n}-\pi_r^*\right| \le \frac{\bar\sigma^2\underline{\sigma}^{-3}\ell\sqrt{C}}{2\sqrt{2\pi}}\ \frac{1}{\sqrt{n}}\ ,$$

where C is the constant appearing in (7).

Further,

$$I_3 \le \sup_x \left|E\ \varphi_{\hat\sigma_0^2}(x-S_{n2})-\varphi_{\sigma_0^2+t}(x)\right|$$

$$= \sup_x \left|\int \varphi_{\sigma_0^2}(x-y)(P_{S_{n2}}-\Phi_t)(dy)\right|\ .$$

If $t>0$, then by the weak convergence $P_{S_{n2}} \xrightarrow{w} \Phi_t$ which is uniform with respect to a class of functions $\{\varphi_{\sigma_0^2}(x-y), x\in R^1\}$ we obtain that the right-hand side approaches zero (see [2]). In the case when $t=0$ the sum S_{n2} goes to zero in probability and the convergence follows from the uniform continuity of $\varphi_{\sigma_0^2}(x)$. This completes the proof of Theorem 6.

Theorems 5 and 6 were announced in [18].

References

[1] Aleshkevichus G. On the central limit theorem for random variables defined on a Markov chain (in Russian). Liet. mat. rink., 1966, v.VI, No1, p.15-22.

[2] Siraždinov S. H., Formanov Sh. K. Limit theorems for sums of random vectors connected in a Markov chain (in Russian). Tashkent: FAN, 1979.

[3] Volkov I. S. On the distribution of sums of random variables given on a homogeneous Markov chain with a finite number of states (in Russian). Teor. verojatn. i primen., 1958, v.III, No4, p.413-429.

[4] O'Brein G. L. Limit theorems for sums of chain-dependent processes. J. Appl. Probab., 1974, v.11, p.582-587.

[5] Grigorescu S., Oprisan G. Limit theorems for J-X processes with a general state space. Z. Wahrscheinlichkeitstheorie verw. Gev., 1976, Bd 35, h.1, p.65-73.

[6] Bokuchava I. V. Central limit theorem for conditionally inde-

pendent sequences (in Russian). Bulletin of the Acad. of Sci. of the Georgian SSR, 1978, v.91, No2, p.305-308.

[7] Bokuchava I. V. On the central limit theorem for conditionally independent sequences (in Russian). In: Investigations in prob. theory and math. statistics. Tbilisi: Metsniereba, 1982, p.3-15.

[8] Bokuchava I. V. Limit theorems for conditionally independent sequences (in Russian). Teor. verojatn. i primen., 1984, v.XXIX No1, p.192-193.

[9] Bokuchava I. V., Chitashvili R. J., Shervashidze T. L. On a limit theorem for conditionally independent random vectors and its application in statistics (in Russian). Tbilisi: Metsniereba, 1984, p.54-70.

[10] Bokuchava I. V., Kvatadze Z. A., Shervashidze T. L. On limit theorems for random vectors controlled by a Markov chain. In: Proceedings of the IV intern. Vilnius Conf. in Probab. Theory and Math. Statist., VNU Science press. 1986 (in print).

[11] Bežaeva Z. I. Limit theorems for conditional Markov chains (in Russian). Teor. verojatn. i primen., 1971, v.XVI, No3, p.437-445.

[12] Nagayev S. V., Gizbriht N. V. On the scheme of random walk describing the phenomenon of particle transfer (in Russian). In: Limit theorems of probability theory. Novosibirsk: Nauka, Siberian Dpt., 1985, p.103-126.

[13] Billingsley P. Convergence of probability measures. New York: Wiley, 1968.

[14] Petrov V. V. Sums of independent random variables (in Russian). Moscow: Nauka, 1972.

[15] Kolmogorov A. N. Local limit theorem for classical Markov chains. Bulletin of the Acad. of Sci. of the USSR (Izvestija), Ser. Math., 1949, v.13, p.281-300.

[16] Kemeny J. G., Snell J. L. Finite Markov chains. Princeton: University press, 1959.

[17] Rogozin B. A. On a concentration function estimator (in Russian). Teor. verojatn. i primen., 1981, v.VI, No1, p.103-105.

[18] kvatadze Z. A., Shervashidze T. L. On sums of conditionally independent random variables controlled by a Markov chain (in Russian). In: XX Winter School in Probab. Theory and Math. Statist. Abstracts of Commun., Tbilisi: Metsniereba, 1986, p.19.

Razmadze Mathematical Institute
of the Academy of Sciences
of the Georgian SSR, Tbilisi

JOINT ASYMPTOTIC DISTRIBUTION OF THE MAXIMUM LIKELIHOOD ESTIMATOR AND M-ESTIMATOR

N. L. Lazrieva and T. A. Toronjadze

In the present work the concept of M-estimator is introduced in a general scheme of statistical models, and on the basis of functional limit theorems for semi-martingales under certain ergodicity conditions the consistency and joint asymptotic normality of the maximum likelihood estimators (MLE) and M-estimators are proved.

First we shall consider the examples (1° and 2°), illustrating general construction given in 3°.

$\underline{1}^\circ$. Let $X_1, X_2, \dots, X_n$, $n \geq 1$, be independent observations of the random variable X with the distribution density $f(x,\theta)$, $\theta \in R^1$ and the unknown parameter θ is to be estimated. P. Huber [1] considered the so called M-estimators $\tilde{\theta}_n = \tilde{\theta}(X_1, X_2, \dots, X_n)$, which are defined as solutions of an implicit equation

$$\sum_{i=1}^{n} \psi(X_i, \theta) = 0 , \tag{1}$$

with $E_\theta \psi(X,\theta) = 0$. In the case, when $\psi(x,\theta) = \frac{f'(x,\theta)}{f(x,\theta)}$, the estimator $\hat{\theta}_n = \hat{\theta}(X_1, X_2, \dots, X_n)$, obtained from (1), is a maximum likelihood estimator. A number of authors studied the asymptotic properties of M-estimators in various cases.

We can rewrite equation (1) in a form which is more convenient for our purposes.

To this end we shall consider a probability space with filtration $(\Omega, \underline{\underline{F}}, \underline{\underline{F}}, P_\theta, P)$, where $\Omega = R^\infty$, $\underline{\underline{F}} = \mathscr{B}(R^\infty)$, $\underline{\underline{F}} = (\underline{\underline{F}}^n)_{n \geq 1}$, $\underline{\underline{F}}^n = \mathscr{B}(R^n)$, $\frac{dP_\theta}{d\Lambda}(\omega) = \prod_{i=1}^{\infty} f(X_i, \theta)$, $\frac{d\Lambda}{dP}(\omega) = \prod_{i=1}^{\infty} \pi(X_i)$, $\Lambda \sim P$, Λ is the Lebesgue measure. Then if $P_\theta^n = P_\theta | \underline{\underline{F}}^n$, $P^n = P | \underline{\underline{F}}^n$ are restrictions of the measures P_θ, P on σ-algebra $\underline{\underline{F}}^n$, respectively, then

$$\rho_\theta^n(\omega) = \frac{dP_\theta^n}{dP^n} = \prod_{i=1}^{n} f(X_i, \theta) \pi(X_i) = \mathscr{E}_n(M_\theta) ,$$

where $M_\theta = \{M_\theta(n)\}_{n \geq 1}$ is a $(\underline{\underline{F}}, P)$-martingale with $\Delta M_\theta(n) = M_\theta(n) - M_\theta(n-1) = f(X_n, \theta)\pi(X_n) - 1$ and $\mathscr{E}(M_\theta)$ is Dolean's exponential curve of the martingale M_θ.

Now let $m, M \in \mathcal{M}_{loc}(E,P)$. Denote $L(m,M) \overset{def}{\equiv} m - \Sigma\frac{\Delta m \Delta M}{1+\Delta M} = \Sigma\frac{\Delta m}{1+\Delta M}$. By virtue of Girsanov's theorem (see, e.g., [2]) $L(m,M_\theta) \in \mathcal{M}_{loc}(\underline{F},P_\theta)$.

It can be easily seen that the likelihood equation has the following form

$$L_n(\dot{M}_\theta, M_\theta) = 0 , \tag{2}$$

where $\dot{M}_\theta = \frac{d}{d\theta}M_\theta$ and equation (1) can be written as

$$L_n(m_\theta, M_\theta) = 0 , \tag{3}$$

where $m_\theta \in \mathcal{M}_{loc}(\underline{F},P)$, $m_\theta(n) = \sum_{i=1}^{n} \psi(X_i,\theta) f(X_i,\theta)\pi(X_i)$.

$\underline{2}^\circ$. Let the unknown parameter $\theta \in R^1$ be estimated according to the observations of the diffusion type process $\xi = (\xi(t))_{t\geq 0}$ with a differential

$$d\xi(t) = a(t,\xi,\theta)dt + dW(t) .$$

What will be the form of the equations for the construction of the MLE and M-estimator in this case?

Consider a probability space with filtration $(\Omega, \underline{\underline{F}}, \underline{F}, P_\theta, P)$, where $\Omega = C_{[0,\infty)}$, $\underline{\underline{F}} = \mathcal{B}(C_{[0,\infty)})$, $\underline{F} = (\underline{F}_t)_{t\geq 0} = (\mathcal{B}(C_{[0,t)}))_{t\geq 0}$, $P_\theta = P_\theta^\xi$, $P = P^W$. Assume that $P_\theta(t) \sim P(t)$, where $P_\theta(t) = P_\theta | \underline{F}_t$, $P(t) = P | \underline{F}_t$. Then

$$\rho_\theta(t) = \frac{dP_\theta(t)}{dP(t)} = \mathcal{E}_t(M_\theta) ,$$

where $M_\theta = (M_\theta(t) = \int_0^t a(s,X,\theta)dX_s)_{t\geq 0} \in \mathcal{M}_{loc}(\underline{F},P)$, $X \in C_{[0,\infty)}$, $\mathcal{E}(M_\theta)$ is Dolean's exponential curve. Introduce Girsanov's L-transformation

$$L(m,M) = m - \langle m,M \rangle ,$$

where $m, M \in \mathcal{M}_{loc}(\underline{F},P)$. Then $L(m,M_\theta) \in \mathcal{M}_{loc}(\underline{F},P_\theta)$.

It can be easily seen, that the likelihood equation has the following form

$$L_t(\dot{M}_\theta, M_\theta) = \int_0^t \dot{a}(s,X,\theta)(dX_s - a(s,X,\theta)ds) = 0 , \tag{2'}$$

where $\dot{a}(t,X,\theta) = \frac{d}{d\theta}a(t,X,\theta)$ and the equation for the M-estimator will be

$$L_t(m_\theta, M_\theta) = \int_0^t \varphi(s,X,\theta)(dX_s - a(s,X,\theta)ds) = 0 , \qquad (3')$$

where $\varphi(t,X,\theta)$ is some function (similar to the discrete time case).

3°. The approach described in 1° and 2° enables to generalize the problem of the investigation of asymptotic properties of the MLE and M-estimator for the case of general statistical models.

In particular, consider a sequence of statistical models

$$\mathscr{E}_n = (\Omega^n, \underline{\underline{F}}^n, \underline{F}^n, P_\theta^n, P^n), \quad n \geq 1, \quad \theta \in R^1,$$

where for every $n \geq 1$, $(\Omega^n, \underline{\underline{F}}^n, \underline{F}^n)$ is a space with filtration $\underline{F}^n = (\underline{\underline{F}}_t^n)_{0 \leq t \leq T}$, satisfying the usual conditions with respect to the measure P^n for any $\theta \in R^1$, $P_\theta^n \sim P^n$, $P_\theta^n \neq P_{\theta'}^n$, when $\theta \neq \theta'$, $P_\theta^n | \underline{\underline{F}}_0^n = P^n | \underline{\underline{F}}_0^n$.

Let $P_\theta^n(t)$, $P^n(t)$ denote the restrictions of the measures P_θ^n and P^n on the σ-algebra $\underline{\underline{F}}_t^n$ and let

$$\rho_\theta^n(t) = \frac{dP_\theta^n(t)}{dP^n(t)} .$$

It is well-known (see [3]) that some P^n-local martingale $M_\theta^n = \{M_\theta^n(t)\}_{0 \leq t \leq T}$, exists such that $\rho_\theta^n(t) = \mathscr{E}_t(M_\theta^n)$, where $\mathscr{E}(M)$ is Dolean's exponential curve of the martingale M.

Let $(\Omega, \underline{\underline{F}}, \underline{F}, P)$ be a space with filtration and M, $m \in \mathscr{M}_{loc}(F,P)$ and $Q = \mathscr{E}(M) \cdot P$. Then, by virtue of Girsanov's theorem (see [2]), $L(m,M) \in \mathscr{M}_{loc}(\underline{F}, Q)$, where

$$L(m,M) \overset{def}{\equiv} m - \langle m^c, M^c \rangle - \sum \frac{\Delta m \Delta M}{1 + \Delta M} .$$

The expression $L(m,M)$ is Girsanov's L-transformation, which will be of primary importance in the forthcoming.

Suppose that the following regularity conditions are satisfied:

1) for any $n \geq 1$ and $\theta \in R^1$, $M_\theta^n \in \mathscr{M}^2(\underline{F}^n, P^n)$,

2) the family of martingales $\{M_\theta^n, \theta \in R^1\}$ is twice continuously differentiable with respect to θ in the sense of the norm $\|\cdot\|_2$ (if $M \in \mathscr{M}^2$, then $\|M\|_2 = E^{1/2}\langle M \rangle_T$),

3) for all $n \geq 1$ and $\theta \in R^1$

$$\left| \frac{\Delta \dot{M}_\theta^n(t) \Delta M_\theta^n(t)}{1 + \Delta M_\theta^n(t)} \right| \leq C_t^n ,$$

where $\sum_{s \leq t} C_s^n < \infty$.

Note that if assumptions 1)-3) are satisfied, then

$$\frac{d}{d\theta} \ln \rho_\theta^n = L(\dot{M}_\theta^n, M^n{}_\theta) .$$

Further assume that

4) Fisher's information $I_\theta^n \overset{def}{\equiv} E_\theta^n [L(\dot{M}_\theta^n, M_\theta^n)]_T$ is positive and finite for all $n\geq 1$ and $\theta \in R^1$;

5) for all $n\geq 1$ $\frac{d^2}{d\theta^2} \ln \rho_\theta^n$ exists and the equality $\frac{d^2}{d\theta^2} \ln \rho_\theta^n = L(\ddot{M}_\theta^n, M_\theta^n) - [L(\dot{M}_\theta^n, M_\theta^n)]$ holds (one can see that in this case a condition similar to 3) is sufficient for all sums entering the expression for $\frac{d^2}{d\theta^2} \ln \rho_\theta^n$). Besides, let $L(\ddot{M}_\theta^n, M_\theta^n) \in \mathcal{M}^2(\underline{F}^n, P_\theta^n)$.

Under the regularity conditions the likelihood equation has the form

$$L(\dot{M}_\theta^n, M_\theta^n) = 0 ,$$

and we shall call the M-estimator such an estimator which can be defined as a solution of the following equation

$$L(m_\theta^n, M_\theta^n) = 0 ,$$

where $m_\theta^n \in \mathcal{M}^2(\underline{F}^n, P^n)$ is some martingale.

Denote $L_{n,1}(t,\theta) = L_t(\dot{M}_\theta^n, M_\theta^n)$, $L_{n,2}(t,\theta) = L_t(m_\theta^n, M_\theta^n)$.

The following theorem holds.

Theorem. Let the following conditions be satisfied:

(a) $\lim_{n\to\infty} \varphi_n(\theta) = 0$, where $\varphi_n^{-2}(\theta) = E_\theta^n [L(\dot{M}_\theta^n, M_\theta^n)]_T$,

(b) $L(m_\theta^n, M_\theta^n) \in \mathcal{M}^2(\underline{F}, P_\theta^n)$,

(c) $P_\theta^n\text{-}\lim_{n\to\infty} \varphi_n^2(\theta) \langle L_{n,i}(\cdot,\theta), L_{n,j}(\cdot,\theta)\rangle_T = \gamma_{ij}, \gamma_{ij} > 0$,

$\Gamma = (\gamma_{ij})_{i,j=1,2}$ is a non-negatively defined matrix, ($\langle M,N\rangle$ denotes a mutual quadratic characteristic of the martingales M and N),

(d) $P_\theta^n\text{-}\lim_{n\to\infty} \varphi_n^2(\theta) \langle L(\ddot{M}_\theta^n, M_\theta^n)\rangle_T = \text{const.}$,

$P_\theta^n\text{-}\lim_{n\to\infty} \varphi_n^2(\theta) \langle L(\dot{m}_\theta^n, M_\theta^n)\rangle_T = \text{const.}$,

(e) $P_\theta^n\text{-}\lim_{n\to\infty} |x|^2 I_{\{|X|>\varepsilon\}} * \nu_T^n = 0$ for every $\varepsilon > 0$, where ν^n is a compensator of measure of jumps of a two-dimensional process $\varphi_n(\theta)(L_{n,1}(t,\theta),L_{n,2}(t,\theta))$, $0\le t\le T$,

(f) $\lim_{\eta\to 0}\overline{\lim_{n\to\infty}}\sup_{y:|y-\theta|\le\eta}\varphi_n^2(\theta)E_\theta^n\langle L_{n,i}(\cdot,\theta)-L_{n,i}(\cdot,y)\rangle_T=0$, $i=1,2$.

Then estimators $\hat\theta_n(T)$ (MLE) and $\tilde\theta_n(T)$ (M-estimator) exist such that:

I. $\lim_{n\to\infty}P_\theta^n\{L_{n,1}(T,\hat\theta_n(T))=0,\ L_{n,2}(T,\tilde\theta_n(T))=0\}=1$,

II. $P_\theta^n\text{-}\lim_{n\to\infty}(\hat\theta_n(T),\tilde\theta_n(T))=(\theta,\theta)$,

III. if $(\hat\psi_n(T),\tilde\psi_n(T))$ is another couple of estimators with properties I and II, then

$$\lim_{n\to\infty} P_\theta^n\{\hat\psi_n(T)=\hat\theta_n(T),\ \tilde\psi_n(T)=\tilde\theta_n(T)\} = 1 ,$$

and finally,

IV. $\lim_{n\to\infty}\mathscr{L}_{P_\theta^n}\text{-}\varphi_n^{-1}(\theta)(\hat\theta_n(T)-\theta,\tilde\theta_n(T)-\theta)=\mathscr{N}(0,\Sigma)$, where

$$\Sigma = T\begin{pmatrix}\gamma_{11}^{-1} & \gamma_{11}^{-1}\\ \gamma_{11}^{-1} & \gamma_{22}\gamma_{12}^{-2}\end{pmatrix},$$

(here $\mathscr{L}_{P_\theta^n}\text{-}\xi$ denotes the distribution of the random variable ξ evaluated with respect to the measure P_θ^n).

<u>Proof</u>. The proof follows the scheme proposed by Le-Breton [4], which is a generalization of the classical Dugue-Cramer's approach.

By Taylor's formula we have

$$\varphi_n^2(\theta)L_{n,i}(T,\theta_i) = \varphi_n^2(\theta)L_{n,i}(T,\theta) - (\theta_i-\theta)\gamma_{1i} + (\theta_i-\theta)\mathscr{E}_{n,i}(\theta_i), \quad i = 1,2 , \tag{4}$$

where

$$\mathscr{E}_{n,i}(\theta_i) = [\varphi_n^2(\theta)L_{n,i}^{(1)}(T,\theta)+\gamma_{1i}]+\varphi_n^2(\theta)[L_{n,i}^{(1)}(T,\theta'(\theta_i))-L_{n,i}^{(1)}(T,\theta)],$$

$$\theta'(\theta_i)=\theta_i+\alpha_i(\theta_i)(\theta_i-\theta),\quad \alpha_i\in[0,1],\quad i=1,2,\quad L_{n,i}^{(1)}(T,\theta)=\frac{d}{d\theta}L_{n,i}(T,\theta).$$

Conditions (c) and (f) imply

$$\lim_{\eta\to 0}\overline{\lim_{n\to\infty}}\sup_{\theta_i:|\theta_i-\theta|\le\eta} P_\theta^n\{|\mathscr{E}_{n,i}(\theta_i)|>\rho\} = 0, \quad \forall\rho > 0 . \tag{5}$$

Now, we can show, that using equality (4), one can construct a sequence of sets $\Omega(n,\eta)\in\underline{\underline{F}}_T^n$ such that

$$\lim_{\eta\to 0}\overline{\lim_{n\to\infty}} P_\theta^n\{\Omega(n,\eta)\} = 1 ,$$

and for any $\eta>0$, $n\geq 1$ and $\omega\in\Omega(n,\eta)$ the equations

$$L_{n,i}(T,\theta_i,\omega) = 0 , \quad i = 1,2 , \tag{6}$$

admit a unique solution $\theta^*_{n,i}(T,\omega)$ in the interval $|\theta_i-\theta|\leq\eta$, $i=1,2$.

Denote $u_i=\theta_i-\theta$. Then (4) results in

$$\varphi_n^2(\theta)L_{n,i}(T,\theta_i+u_i)u_i=u_i\varphi_n^2(\theta)L_{n,i}(T,\theta)-u_i^2\gamma_{1i}+u_i^2\mathcal{E}_{n,i}(\theta+u_i), \quad i=1,2.$$

For any $n\geq 1$ and $\eta>0$ consider the set

$$\Omega(n,\eta)=\{\omega\in\Omega^n:\varphi_n^2(\theta)|L_{n,i}(T,\theta,\omega)|<\frac{\gamma_{1i}\eta}{2}, \sup_{|u_i|\leq\eta}|\mathcal{E}_{n,i}(\theta+u_i,\omega)|<\frac{\gamma_{1i}}{2}, \ i=1,2.$$

It can be easily seen that by virtue of conditions a), c) and f)

$$\lim_{\eta\to 0}\overline{\lim_{n\to\infty}} P_\theta^n\{\Omega(n,\eta)\} = 1 ,$$

and, besides, if $|u_i|=\eta$, then

$$\varphi_n^2(\theta)L_{n,i}(T,\theta+u_i)u_i < 0 ,$$

which implies that in the interval $|u_i|\leq\eta$ a unique solution $u_{n,i}(T,\omega)$ exists for the equation

$$L_{n,i}(T,\theta+u_i,\omega) = 0 ,$$

such that $|u_{n,i}(T,\omega)|<\eta$, which leads to the existence of the solution $\theta^*_{n,i}(T,\omega)$ of (6) in the interval $|\theta_i-\theta|\leq\eta$, $i=1,2$.

Remark. It is well-known (see, e.g., [4]) that if $(\Omega,\underline{\underline{F}})$ is a measurable space and $\ell:\Omega\times R^1\to R^1$ is such that for a fixed $\omega\in\Omega$, $\ell(\omega,\cdot)$ is continuous with respect to $u\in R^1$ and for a fixed $u\in R^1$ $\ell(\cdot,u)$ is $\underline{\underline{F}}$-measurable and some $r(\omega)>0$ exists, such that in a sphere $B(h(\omega),r(\omega))=\{u:|h(\omega)-u|\leq r(\omega)\}$ the equation $\ell(\omega,u)=0$ has a unique solution $u(\omega)$, where $h(\omega)$ is $\underline{\underline{F}}$-measurable, then $u(\omega)$ is also an $\underline{\underline{F}}$-measurable mapping.

Further, assume that $\Omega_n=\bigcup_{\kappa>0}\Omega(n,\frac{1}{\kappa})$ and $(\tilde\Omega_n,\tilde{\underline{\underline{F}}}^n,\tilde{\underline{F}}^n)=(\Omega_n\cap\Omega^n, \underline{\underline{F}}^n\cap\Omega^n,\underline{F}^n\cap\Omega^n)$. It is evident, by virtue of the remark, that for any $\omega\in\Omega_n$, $\kappa(\omega)>0$ exists such that in the interval $|\theta_i-\theta|<\frac{1}{\kappa(\omega)}$ the equation $L_{n,i}(T,\theta_i,\omega)=0$ has a unique $\tilde{\underline{\underline{F}}}_T^n$-measurable solution

$\theta^*_{n,i}(T,\omega)$, $i=1,2$.

Put

$$\hat{\theta}_n(T,\omega) = \begin{cases} \theta^*_{n,i}(T,\omega), & \omega\in\tilde{\Omega}_n, \\ \theta_0, & \omega\in\tilde{\Omega}^c_n \end{cases} \qquad \tilde{\theta}_n(T,\omega) = \begin{cases} \theta^*_{n,2}(T,\omega), & \omega\in\tilde{\Omega}_n, \\ \theta_0, & \omega\in\tilde{\Omega}^c_n, \end{cases}$$

where θ_0 is some parameter value.

We shall prove that the estimators $\hat{\theta}_n(T)$ and $\tilde{\theta}_n(T)$ are consistent. In particular, we can show that for any $\rho>0$

$$\lim_{n\to\infty} P^n_\theta\{|\hat{\theta}_n(T)-\theta|\geq\rho\} = 0 .$$

Obviously, for any $\kappa>0$, such that $\frac{1}{\kappa}<\rho$, we have

$$\Omega(n,\tfrac{1}{\kappa}) \subset \{|\hat{\theta}_n(T)-\theta|\geq\rho\} .$$

Therefore

$$P^n_\theta\{|\hat{\theta}_n(T)-\theta|<\rho\} \geq P^n_\theta\{\Omega(n,\tfrac{1}{\kappa})\} ,$$

which, evidently, implies (7). Hence, property II is proved.

Further, for any $\kappa\geq1$, we have

$$\Omega(n,\tfrac{1}{\kappa}) \subset [L_{n,1}(T,\hat{\theta}_n(T))=0,\ L_{n,2}(T,\tilde{\theta}_n(T))=0] .$$

Consequently,

$$P^n_\theta\{L_{n,1}(T,\hat{\theta}_n(T))=0,\ L_{n,2}(T,\tilde{\theta}_n(T))=0\} \geq P^n_\theta(\Omega(n,\tfrac{1}{\kappa}) ,$$

which easily leads to I.

Property III is derived from the following inclusion: for any $\kappa\geq1$,

$$\{\omega:\omega\in\Omega(n,\tfrac{1}{\kappa}),\ |\hat{\psi}_n(T)-\theta|\leq\tfrac{1}{\kappa},\ |\tilde{\psi}_n(T)-\theta|\leq\tfrac{1}{\kappa},\ L_{n1}(T,\hat{\psi}_n(T),\omega)=0,$$
$$L_{n,2}(T,\tilde{\psi}_n(T),\omega)=0\} \subset \{\hat{\psi}_n(T)=\hat{\theta}_n(T),\tilde{\psi}_n(T)=\tilde{\theta}_n(T)\} .$$

Finally, from (4), we have

$$\varphi^2_n(\theta)(L_{n,1}(T,\hat{\theta}_n(T))-L_{n,1}(T,\theta)) = -\gamma_{11}(\hat{\theta}_n(T)-\theta)+\delta_{n,1}(\hat{\theta}_n(T))(\hat{\theta}_n(T)-\theta),$$
$$\varphi^2_n(\theta)(L_{n,2}(T,\tilde{\theta}_n(T))-L_{n,2}(T,\theta)) = -\gamma_{12}(\tilde{\theta}_n(T)-\theta)+\delta_{n,2}(\tilde{\theta}_n(T))(\tilde{\theta}_n(T)-\theta)$$

and, besides, by virtue of (5), (b) and I,

$$\lim_{n\to\infty} P^n_\theta\{|\delta_{n,1}(\hat{\theta}_n(T))| + |\delta_{n,2}(\tilde{\theta}_n(T))|\geq\rho\} = 0 .$$

Hence,

$$\left|\frac{1}{\gamma_{11}}\varphi_n(\theta)(L_{n,1}(T,\hat\theta_n(T))-L_{n,1}(T,\theta))-\varphi_n^{-1}(\theta)(\hat\theta_n(T)-\theta)\right|$$

$$\le \mathcal{E}_{n,1}(\hat\theta_n(T))\varphi_n^{-1}(\theta)|(\hat\theta_n(T)-\theta|\ ,$$

$$\left|\frac{1}{\gamma_{12}}\varphi_n(\theta)(L_{n,2}(T,\tilde\theta_n(T))-L_{n,2}(T,\theta))-\varphi_n^{-1}(\theta)(\tilde\theta_n(T)-\theta)\right|$$

$$\le \mathcal{E}_{n,2}(\tilde\theta_n(T))\varphi_n^{-1}(\theta)|(\tilde\theta_n(T)-\theta|\ .$$

It is known (see, e.g., [5], problem 2, Ch.I, p.46) that, if $X_n\xrightarrow{\mathcal{D}}X$ and $|X_n-Y_n|\le Z_n|Y_n|$ ($X_n\xrightarrow{\mathcal{D}}X$-convergence in distribution), where $Z_n\xrightarrow{P}0$, then $Y_n\xrightarrow{\mathcal{D}}X$.

This together with the last inequalities results in

$$\mathcal{L}_{P_\theta^n} - \lim_{n\to\infty} \varphi_n^{-1}(\theta)(\hat\theta_n(T)-\theta,\tilde\theta_n(T)-\theta)$$

$$= \mathcal{L}_{P_\theta^n} - \lim_{n\to\infty} \left(\frac{1}{\gamma_{11}}\varphi_n(\theta)L_{n,1}(T,\theta),\frac{1}{\gamma_{12}}\varphi_n(\theta)L_{n,2}(T,\theta)\right)\ .$$

By virtue of Theorem 3.2 (see [6], Ch.VIII), conditions c) and e) imply

$$\mathcal{L}_{P_\theta^n} - \lim_{n\to\infty} \left(\frac{1}{\gamma_{11}}\varphi_n(\theta)L_{n,1}(T,\theta),\frac{1}{\gamma_{12}}\varphi_n(\theta)L_{n,2}(T,\theta)\right) = \mathcal{N}(0,\Sigma)\ .$$

This completes the proof of the theorem.

References

[1] P. Huber. Robustness in statistics.-M.; "Myr", 1984 (in Russian).

[2] J. Jacod. Calcul stochastique et problemes de martingales. Lect. Note in Math., 714, Springer-Verlag, 1978.

[3] Yu. M. Kabanov, R. Sh. Liptzer, A. N. Shiryaev. Absolute continuity and singularity of locally absolutely continuous distributions. Mat. Sbornik, v.107 (449) No3 (11), 1978 (in Russian).

[4] A. Le Breton. Thesis. Grenoble, 1976.

[5] P. Billingsley. Convergence of probability measure.-M.: "Nauke", 1977 (in Russian).

[6] J. Jacod, A. N. Shiryaev. Limit theorems for stochastic processes. To appear.

Razmadze Mathematical Institute
of the Acad. of Sci.
of the Georgian SSR, Tbilisi

ON THE RESULTS OF ASYMPTOTIC ANALYSIS FOR THE RANDOM WALKS WITH TWO-SIDED BOUNDARY

V. I. Lotov

1. We study asymptotic behaviour of the values

$$EN,\ P(S_N \geq b) \tag{1}$$

as $a,b\to\infty$, where $N=N(a,b)=\min\{n: S_n\notin(-a,b)\}$, $a>0$, $b>0$, $S_n=x_1+\cdots+x_n$, $\{x_i\}_{i=1}^{\infty}$ is the sequence of i.i.d. random variables. The distribution of x_1 is always supposed to have an absolute continuous component and Cramer's condition holds: $|Ee^{\lambda x_1}|<\infty$ for $-\gamma\geq\mathrm{Re}\lambda\leq\beta$, $\gamma>0$, $\beta>0$.

The interest to this problem is accounted for a number of applications and the study of operating characteristic and average sample number of the SPRT is the first of them. One of the first approximation for (1) was suggested by Wald [1]. It is based on the neglecting of the excess of the value S_N over the boundary of the interval $(-a,b)$. The attempts to improve Wald's approximation in [2-6] are connected with various excess approximation methods and evaluating the influence of the excess. We shall refer here to Siegmund's result [5] for the exponential families, which is based on the usage of limiting value of the excess distribution instead of its exact value, the estimates of Borovkov [6] and the paper of S. Nagaev [7], where under condition $E|x_1|^3<\infty$ the estimate of the accuracy of some approximation formulae for $P(S_N\geq b)$ is obtained, which is uniform in a as $a+b\to\infty$.

The development of so-called factorization methods in the works of Borovkov (see, i.e. [8]) has showed them to be the most universal and efficient instrument in the studying of asymptotic properties of the distributions in the problems with one straight-line boundary. This technique was adapted later on to solve the problems with two straight-line boundaries [9-11]. It has allowed to account the effect of all excesses, to obtain asymptotic expansions which are complete in a number of cases, to prove the inequalities which characterized the accuracy of approximations. In section 2 the case $Ex_1=0$ is considered. The asymptotic behaviour of the distribution of S_N in this situation has been investigated in [9], therefore simple calculation leads us to the asymptotic formulae for EN which may be understood as a two-parameter complete asymptotic expansion. In the section 3 we suppose that $Ex_1<0$ and $Ee^{\beta x_1}>1$ because this situation is typical for the SPRT. Moreover we strengthen the condition on the absolute continuous component of the distribution of x_1: its "mass" must grow with the increasing of γ and β. Under this conditions the asymptotic expansion for $P(S_N\geq b)$ with the remainder term of the order $o(e^{-\beta b})+o(e^{-\gamma a-h(a+b)})$ is obtained. Here the number h satisfies the condition $Ee^{hx_1}=1$, $0<h<\beta$. We present also the asymptotic expansions for EN with the remainder term of the order $o(ae^{-\gamma a})+o((a+b)e^{-\beta b})$. The investigation of the lattice-valued random walks can be carried out by the same way without any principal changes.

2. Let $Ex_1=0$, $P(x_1\neq 0)>0$. Denote $\eta_\pm=\inf\{n:\ S_n \gtrless 0\}$,

$$\chi_\pm = S_{\eta_\pm}, \quad \tau_\pm = \pm\frac{E\chi_\pm^2}{2E\chi_\pm}, \quad \theta_\pm = \frac{E\chi_\pm^3}{3E\chi_\pm}.$$

Theorem 1. There exists $\delta>0$ such that

$$EN = (Ex_1^2)^{-1}\{ab+\tau_- b+\tau_+ a+\frac{a(\tau_+\tau_-+\theta_+-\tau_+^2)+b(\tau_+\tau_-+\theta_--\tau_-^2)+\tau_-\theta_++\tau_+\theta_-}{a+b+\tau_++\tau_-}\}$$
$$+ O(e^{-\delta a}) + O(e^{-\delta b}) \tag{2}$$

as $a\to\infty$, $b\to\infty$.

We shall obtain Wald's approximation [1] $EN\approx(Ex_1^2)^{-1}ab$ if excesses $\chi_\pm$ are neglected. Analogous formulae for the exponential families and a=0 is obtained in [5].

Proof. Denote $r_\pm(\lambda)=1-Ee^{\lambda\chi_\pm}$. The following statements are proved in [9]. For arbitrary measurable sets $A_1\subset[0,\infty)$, $A_2\subset(-\infty,0]$ holds

$$P(S_N-b\in A_1) = \frac{a+\tau_-}{a+b+\tau_++\tau_-}F_+(A_1) + \int_{A_1} e^{-\delta y}dy(O(e^{-\delta a})+O(e^{-\delta b})), \tag{3}$$

$$P(S_N+a\in A_2) = \frac{b+\tau_+}{a+b+\tau_++\tau_-}F_-(A_2) + \int_{A_2} e^{\delta y}dy(O(e^{-\delta a})+O(e^{-\delta b})). \tag{4}$$

Here $\delta>0$, the measures $F_{\pm}$ are determined by $\pm\int_0^{\pm\infty} e^{\lambda y}F_{\pm}(dy)=\frac{r_{\pm}(\lambda)}{\lambda r'_{\pm}(0)}$.

It remained to apply Wald's identity $EN\cdot Ex_1^2=ES_N^2$. Its validity in our consideration follows from the next arguments. It is known [12] that

$$r(\lambda)Q_0(\lambda) = 1-Q_1(\lambda)-Q_2(\lambda) \tag{5}$$

where $r(\lambda)=1-Ee^{\lambda x_1}$, $Q_0(\lambda)=\sum_{n=0}^{\infty} E(e^{\lambda S_n}; N>n)$, $Q_1(\lambda)=E(e^{\lambda S_N}; S_N\le -a)$, $Q_2(\lambda)=E(e^{\lambda S_N}; S_N\ge b)$. We have by l'Hospital's rule

$$EN = Q_0(0) = \lim_{\lambda\to 0} \frac{1-Q_1(\lambda)-Q_2(\lambda)}{r(\lambda)} = -\frac{Q_1''(0)+Q_2''(0)}{r''(0)} = \frac{ES_N^2}{Ex_1^2}.$$

The possibility of differentiation is provided by (3) and (4). The theorem is proved.

3. Let $Ex_1<0$, $P(x_1>0)>0$ and as before $|Ee^{\lambda x_1}|<\infty$ for $-\gamma\le\mathrm{Re}\lambda\le\beta$. We require in addition that $Ee^{\beta x_1}>1$. Then there exists unique number h, $0<h<\beta$ such that $Ee^{hx_1}=1$. Denote $\tilde{\chi}_+=-\tilde{x}_1-\cdots-\tilde{x}_{\eta_-}$ where random variables $\tilde{x}_i$ are independent and $P(\tilde{x}_i\in dy)=e^{hy}P(x_1\in dy)$, $G_1(y)=P(\chi_+<y,\eta_+<\infty)$, $G_2(y)=P(\tilde{\chi}_+<y,\eta_-<\infty)$. Suppose that

$$\int_0^{\infty} e^{\beta y}\, dG_1^0(y) < 1, \qquad \int_0^{\infty} e^{(h+\gamma)y}\, dG_2^0(y) < 1. \tag{6}$$

Here G_i^0 denotes the sum of discrete and singular components of G_i. Condition (6) obviously holds if x_1 has absolute continuous distribution.

The function $r(\lambda)=1-Ee^{\lambda x_1}$ has no zeros for $0\le\mathrm{Re}\lambda\le h$ except for $\lambda=0$ and $\lambda=h$ (see [10]). But it is possible for the function $r(\lambda)$ to have some complex zeros on the set $-\gamma\le\mathrm{Re}\lambda<0$, $h<\mathrm{Re}\lambda\le\beta$. It is clear that $r(\bar{\lambda})=0$ if the equality $r(\lambda)=0$ takes place. So, let $r(\lambda)\ne 0$ on the set $\mathrm{Re}\lambda\in[-\gamma,0]$ except for the points $\mu_0=0,-\mu_1,\ldots,-\mu_{2s}$ $(s\ge 0)$, $\mu_{2k-1}=\bar{\mu}_{2k}$, $0<\mathrm{Re}\mu_{2k}<\gamma$ $(K=1,\ldots,s)$. Also let $r(\lambda)\ne 0$ on the set $\mathrm{Re}\lambda\in[h,\beta]$ except for the points $h+\lambda_0=h, h+\lambda_1,\ldots,h+\lambda_{2r}$ $(r\ge 0)$, $0<\mathrm{Re}\lambda_{2i}<\beta-h$, $\lambda_{2i-1}=\bar{\lambda}_{2i}$ $(i=1,\ldots,r)$. Suppose for simplicity all the zeros of $r(\lambda)$ to be prime. Denote

$$r_{\pm}(\lambda) = 1-E(e^{\lambda\chi_{\pm}}; \eta_{\pm} < \infty). \tag{7}$$

It is well known that $r_+(\lambda)r_-(\lambda)=r(\lambda)$. Let for $i=0,1,\dots,2s$, $j=0,1,\dots,2r$

$$\beta_{ij} = -\frac{r_+(-\mu_i)}{r_+'(h+\lambda_j)(h+\lambda_j+\mu_i)}, \qquad \gamma_{ji} = \frac{r_-(h+\lambda_j)}{r_-'(-\mu_i)(h+\lambda_j+\mu_i)}. \tag{8}$$

Theorem 2. Let $|Ee^{\lambda x_1}|<\infty$ for $-\gamma\le \mathrm{Re}\lambda\le\beta$. Let $Ex_1<0$, $Ee^{\beta x_1}>1$ and the condition (6) holds. Then

$$P(S_N\ge b) = e^{-hb}\sum_{j=0}^{2r}\beta_{0j}e^{-\lambda_j b} - e^{-h(a+b)}\sum_{j=0}^{2r}\beta_{0j}e^{-\lambda_j(a+b)}\sum_{i=0}^{2s}\gamma_{ji}e^{-\mu_i a}$$

$$+ e^{-h(a+2b)}\sum_{j=0}^{2r}\beta_{0j}e^{-\lambda_j(a+b)}\sum_{i=0}^{2s}\gamma_{ji}e^{-\mu_i(a+b)}\sum_{k=0}^{2r}\beta_{ik}e^{-\lambda_k b} - \dots \tag{9}$$

$$+ o(e^{-\beta b}) + o(e^{-\gamma a-h(a+b)})$$

as $a\to\infty$, $b\to\infty$.

If only null-indexed terms in (9) are considered, we have

Corollary 1.

$$P(S_N\ge b) = \frac{\beta_{00}e^{-hb}(1-\gamma_{00}e^{-ha})}{1-\beta_{00}\gamma_{00}e^{-h(a+b)}} + O(e^{-(h+\beta_1)b})\,\mathrm{sgn}\,r$$

$$+ O(e^{-\gamma_1 a-h(a+b)})\,\mathrm{sgn}\,s + o(e^{-\beta b}) + o(e^{-\gamma a-h(a+b)}). \tag{10}$$

Here $\beta_1 = \min_{1\le j\le 2r}\mathrm{Re}\lambda_j$, $\gamma_1 = \min_{1\le i\le 2s}\mathrm{Re}\mu_i$ for $\tau>0$, $s>0$.

The probabilistic interpretation of the coefficients β_{00} and γ_{00} is explained by (7) and (8). Some expansion of type (10) without any estimates of accuracy of the remainder term has been obtained in [5] for exponential families.

Theorem 3. Under the conditions of Theorem 2,

$$EN = |Ex_1|^{-1}\{a+\rho_0-e^{-hb}\sum_{j=0}^{2r}((a+b+\rho_0)\beta_{0j}+\delta_j)e^{-\lambda_j b}$$

$$+ e^{-h(a+b)}\sum_{j=0}^{2r}((a+b+\rho_0)\beta_{0j}+\delta_j)e^{-\lambda_j(a+b)}\sum_{i=0}^{2s}\gamma_{ji}e^{-\mu_i a}$$

$$- e^{-h(a+2b)} \sum_{j=0}^{2r} ((a+b+\rho_0)\beta_{0j}+\delta_j) e^{-\lambda_j(a+b)} \sum_{i=0}^{2s} \gamma_{ji} e^{-\mu_i(a+b)} \sum_{k=0}^{2\tau} \beta_{ik} e^{-\lambda_k b}$$

$$+ \cdots + \sum_{j=1}^{2s}{}' \rho_j e^{-\mu_j a} - e^{-hb} \sum_{j=1}^{2s}{}' \rho_j e^{-\mu_j(a+b)} \sum_{i=0}^{2r} \beta_{ji} e^{-\lambda_i b}$$

$$+ e^{-h(a+b)} \sum_{j=1}^{2s}{}' \rho_j e^{-\mu_j(a+b)} \cdot \sum_{i=0}^{2r} \beta_{ji} e^{-\lambda_i(a+b)} \sum_{k=0}^{2s} \gamma_{ik} e^{-\mu_k a} - \cdots \}$$

$$+ o(ae^{-\gamma a}) + o((a+b)e^{-\beta b}),$$

the coefficients β_{ij}, γ_{ji} have been defined in Theorem 2,

$$\rho_0 = - \frac{r''_-(0)}{2r'_-(0)}, \qquad \rho_i = - \frac{r'_-(0)}{\mu_i r'_-(-\mu_i)}, \qquad i \geq 1,$$

$$\delta_j = - \frac{r'_+(0)}{(h+\lambda_j) r'_+(h+\lambda_j)} + \frac{\beta_{0j}}{h+\lambda_j}, \qquad j = 0,\dots,2r,$$

the sums indicated by primes disappear if $s=0$.

Neglecting the terms corresponding to the contributions of the zeros $h+\lambda_j$, $-\mu_i$ $(j\geq i,\ i\geq 1)$ we obtain

Corollary 2.

$$EN = |Ex_1|^{-1}\{(1-\beta_{00}\gamma_{00}e^{-h(a+b)})^{-1}[a+\rho_0-e^{-hb}((a+b+\rho_0)\beta_{00}+\delta_0)$$

$$+ e^{-h(a+b)}\gamma_{00}(\beta_{00}b+\delta_0)]\} + o(ae^{-\gamma a}) + o((a+b)e^{-\beta b})$$

$$+ O((a+b)e^{-(h+\beta_1)b})\operatorname{sgn} r + O(e^{-\gamma_1 a})\operatorname{sgn} s.$$

We present a short scheme of proof of Theorems 2 and 3. It follows from (5) that

$$EN = -(r'(0))^{-1}(Q_1'(0) + Q_2'(0)). \tag{11}$$

For every function g of the type $g(\lambda)=\int_{-\infty}^{\infty} e^{\lambda y} dG(y)$, $\lambda\in(0,h)$, $\operatorname{Var}G<\infty$ denote

$$(Ag)(\lambda) = r_-(\lambda)[r_-^{-1}g]^{(-\infty,-a]}(\lambda), \quad (Bg)(\lambda) = r_+(\lambda)[r_+^{-1}g]^{[b,\infty)}(\lambda).$$

We use here the notation $[g]^D(\lambda)=\int_D e^{\lambda y}dG(y)$, $D\subset R$. It is known ([12], [9],[10]) that for every integer $h\geq 0$

$$Q_1(\lambda) = \sum_{i=0}^{n} ((AB)^i(A-AB)e)(\lambda) + ((AB)^{n+1}Q_1)(\lambda), \tag{12}$$

$$Q_2(\lambda) = \sum_{i=0}^{n} ((BA)^i(B-BA)e)(\lambda) + ((BA)^{n+1}Q_2)(\lambda), \tag{13}$$

where $e(\lambda)\equiv 1$. By isolating known poles of the functions $r_\pm^{-1}(\lambda)$ we obtain the following representations

$$\begin{aligned}(Bg)(\lambda) = {} & r_+(\lambda)\sum_{j=0}^{2r}\frac{g(h+\lambda_j)e^{(\lambda-h-\lambda_j)b}}{r_+'(h+\lambda_j)(\lambda-h-\lambda_j)} \\ & + r_+(\lambda)\int_b^\infty e^{(\lambda-h)y}\,d_y\int_{-\infty}^{+0} e^{ht}\,\Delta(y-t)dG(t),\end{aligned} \tag{14}$$

$$\begin{aligned}(Ag)(\lambda) = {} & r_-(\lambda)\sum_{j=0}^{2s}\frac{g(-\mu_j)e^{-(\lambda+\mu_j)a}}{(\lambda+\mu_j)r_-'(-\mu_j)} \\ & + r_-(\lambda)\int_{-\infty}^{-a+0} e^{\lambda y}\,d_y\int_0^\infty \Delta_1(y-t)dG(t),\end{aligned} \tag{15}$$

where $\operatorname{Var}_{[x,\infty)}\Delta(y)=o(e^{-\beta x})$, $\operatorname{Var}_{(-\infty,-x]}\Delta_1(y)=o(e^{-\gamma x})$ as $x\to\infty$. Substitution of $\lambda=0$ into (13) together with (14), (15) leads us to the statement of Theorem 2. Differentiation of (14) and (15) with respect to λ together with (11)-(13) gives us the statement of Theorem 3.

References

[1] Вальд А. Последовательный анализ. М., 1960.

[2] Page E. S. An improvement to Wald's approximation for some properties of sequential tests. J. Roy. Stat. Soc., B, 1954, v.16, p.136-139.

[3] Kemp K. W. Formulae for calculating the operating characteristic and average sample number of some sequential tests. J. Roy. Stat. Soc., B, 1958, v.20, 2, p.379-386.

[4] Tallis G. M., Vagholkar M. K. Formulae to improve Wald's approximation for some properties of sequential tests. J. Roy. Stat. Soc. B, 1965, v.27, 1, p.74-81.

[5] Siegmund D. Error probabilities and average sample number of the SPRT. J. Roy. Stat. Soc., B, 1975, v.37, 5, p.394-401.

[6] Боровков А. А. Математическая статистика. Оценка параметров. Проверка гипотез. М., 1984.

[7] Нагаев С. В. Оценка скорости сходимости для вероятности поглощения. Теор. Вероятн. и ее примен., 1971, т.XVI, No.I, с.140-148.

[8] Боровков А. А. Новые предельные теоремы в граничнных задачах для сумм независимых слагаемых. Сиб. матем. ж., 1962, т.3, No.5, с.645-694.

[9] Лотов В. И. Об асимптотике распределений, связанных с выходом недискретного случайного блуждания из интервала. В кн. "Предельные теоремы теории вероятностей и смежные вопросы", Новосибирск, 1982, с.18-25.

[10] Лотов В. И. Об асимптотическом поведении характеристик последовательного критерия отношения правдоподобия. Теория вероятн. и ее примен., 1985, т.XXX, No.1, с.164-169.

[11] Лотов В. И. Об асимптотике характеристик последовательного критерия отношения правдоподобия. В кн.: Четвертая междунар. Вильнюсская конф. по теории вероятн. и мат. статистике. Тезисы докл., Вильнюс, 1985, т.II, с,115-116.

[12] Kemperman J. H. B. A Wiener-Hopf type method for a general random walk with a two-sided boundary. Ann. Math. Stat., 1963, v.34, 4, p.1168-1193.

Institute of Mathematics,
Siberian Branch of the USSR
Academy of Sciences,
Novosibirsk

GAUSSIAN LIMIT THEOREMS FOR WIENER FUNCTIONALS

G. Maruyama

We shall study central limit theorems (CLT's) and related Gaussian limit theorems for a square-integrable real stationary process $X(t)$ subordinate to the real Gaussian process

$$\xi(t) = \int_{-\infty}^{\infty} \exp[i\lambda t]d\beta(\lambda), \qquad -\infty<t<\infty,$$

with zero mean, complex spectral random measure $d\beta$, and spectral measure, which is absolutely continuous relative to Lebesgue measure,

$$d\sigma(\lambda) = f(\lambda)d\lambda, \qquad d\sigma(\lambda) = E|d\beta(\lambda)|^2.$$

$X(t)$ is represented by the Ito-Wiener expansion

$$X(t)=\sum_{k\geq 1} X_k(t), \quad X_k(t)=\int_{\mathbb{R}^k} c_k(\lambda)e_k(\lambda,t)d^k\beta, \quad e_k(\lambda,t)=\exp[i\bar{\lambda}t]$$

$$\lambda=(\lambda_1,\ldots,\lambda_k)\in\mathbb{R}^k, \quad \overline{c_k(\lambda)}=c_k(-\lambda)\in L^2(\sigma_k), \quad \bar{\lambda}=\lambda_1+\cdots+\lambda_k$$

where $c_k(\lambda)$ is symmetric, $\overline{c_k(\lambda)}=c_k(-\lambda)$, and $c_k\in L^2(\sigma_k)$, with the k-fold product measure σ_k of the spectral measure of ξ.

Put

$$S_n(t)=\sum_{k=1}^{n} X_k(t), \quad V_n(T)=V(\int_0^T S_n(t)dt), \quad V(T)=V(\int_0^T X(t)dt),$$

$$\Delta V_n(T)=V(\int_0^T [X(t)-S_n(t)]dt), \qquad \overline{X}(T)=\int_0^T X(t)dt/\sqrt{V(T)}$$

where V denotes variance. Let $\varphi_k/2$ be the spectral density of $X_k(t)$, write

$$\varphi_k(\lambda)/2(k!)=\varphi(|c_k|^2;\lambda), \qquad \delta[|c_k|^2,\alpha^2]=|c_k(\lambda)|^2-|c_k(\lambda)|^2\wedge\alpha^2,$$

and define

$$\Phi(|c_k|^2;x) = \int_0^x \varphi(|c_k|^2;\lambda)d\lambda .$$

Theorem 1. Suppose that X satisfies the following conditions:

(i) f is bounded;

(ii) let $\varphi(\lambda)/2$ be the spectral density of X, then $H(x)=\int_0^x \varphi(\lambda)d\lambda$ is regularly varying (RV) in the wide sense at zero;

(iii) $\lim_{n\to\infty} \overline{\lim_{T\to\infty}} \Delta V_n(T)/V(T)=0$;

(iv) there exists ε_0, $0<\varepsilon_0<1/2$, such that for every $k\geq 1$

$$\Phi(\delta[|c_k|^2, T^{2\varepsilon_0}]; x)=o(H(x)), \quad x=1/T, \quad \text{as} \quad T\to\infty.$$

Then $X(T) \to N(0,1)$ in distibution, as $T\to\infty$.

Theorem 2. Suppose that X satisfies the following conditions (A),(B),(C).

(A) (i),(iii),(iv) of Theorem 1.

(B) One of the conditions (B_1)-(B_3):

(B_1) $V(T)=V(\int_0^T x(s)ds)$ is RV at ∞, with $V(T)=Th_\infty(T)$, $h_\infty(T)=c(1/T)s_0(1/T)$ (a canonical expression);

(B_2) if we write $\varphi(\lambda)/2$ for the spectral density of X, $H(x)=\int_0^x \varphi(\lambda)d\lambda$ is RV at zero, $H(x)=xh_0(x)$, $h_0(x)=c(x)s_0(x)$ (a canonical expression) on some interval $(0,\delta)$ $(\delta>0)$, and $c(x)$ fulfils

$(B_2\text{-}1)$ $yc'(y)\in L(0,\delta)$, $\int_0^x |yc'(y)|dy=o(x)$, $x\to+0$,

or

$(B_2\text{-}2)$ $c(x)-c(+0)=O(x^q)$, $x\to+0, q>0$;

(B_3) $\varphi(\lambda)$ is RV at 0.

(C) $$\sum_{k\geq 1} 3^{k/2}\sqrt{k!}\,\|c_k\|_{L^2(\sigma_k)} < \infty$$

$$\sum_{k\geq 1} 3^{k/2} s_k(a)<\infty, \quad \text{for some} \quad a>0,$$

where

$$S_k^2(a) = \sup_{0<x<a} x^{-1}\int_0^x [\varphi_k(\lambda)/h_0(\lambda)]d\lambda, \quad (S_k(a)\geq 0).$$

Then

$$\overline{X}(T,t)\equiv(\int_0^{Tt} X(s)ds)/\sqrt{V(T)}, \qquad 0\leq t\leq 1,$$

converges in distribution on $C[0,1]$ to standard Brownian motion $(W(t),\ 0\leq t\leq 1)$.

Let $W_\gamma(t)$, $0\leq t<\infty$, $0<\gamma<2$, be stationary increment processes with

$$E([W_\gamma(t)-W_\gamma(s)]^2) = |t-s|^\gamma, \qquad 0<\gamma<2 .$$

Roughly speaking, the functional CLT corresponds to an almost linear growth of $V(T)$ when $T\to\infty$. As a variant of the functional CLT it is natural to consider the case when the limit process is W_γ. Under the framework of Wiener functionals, and along a similar line to the above argument it is possible to obtain requested results.

The details of the proofs of Theorems 1 and 2 will be published in a forth-coming issue of Hokkaido Math. Journal.

References

[1] D. Chambers and E. Slud, Central limit theorems for nonlinear functionals of stationary Gaussian processes, to appear.

[2] L. Giraites and Surgailis, CLT and other limit theorems for functionals of Gaussian processes, Z. Wahrscheinlichkeitstheor. Verw. Geb., 70 (1985), 191-212.

[3] I. A. Ibragimov and Yu. V. Linnik, Independent and Stationarily Connected Variables (in Russian), Moscow, Nauka, 1965.

[4] K. Itô, Multiple Wiener integrals, J. Math. Soc. Japan, 3 (1951), 157-169.

[5] K. Itô, Complex multiple Wiener integral, Jap. J. Math., 22 (1952), 63-86.

[6] G. Maruyama, Nonlinear functionals of Gaussian stationary processes and their applications, Proc. Third Japan-USSR Symp. on Prob. Theory, Lect. Notes in Math., 550, 375-378. Springer, Berliln-Heidelberg-New York, 1976.

[7] G. Maruyama, Applications of Wiener Expansions to Limit Theorems (in Japanese), Sem. on Prob., 49 (1980), 1-150.

[8] G. Maruyama, Applications of the multiplication of Ito-Wiener expansions to limit theorems, Proc. Japan Acad., 58 (1982), 388-390.

[9] G. Maruyama, Wiener functionals and probability limit theorems I: The central limit theorems, Osaka J. Math., 22 (1985), 697-732.

(Editors' note: G. Maruyama deceased on July 5, 1986. This is the lecture originally scheduled to be the first plenary talk at the symposium.)

MULTIPLICATIVE NUMBER THEORY IN PROBABILITY SPACES: AN EXAPLE

Jean Loup Mauclaire

1- In my recent work [3], I have shown that certain results in multiplicative number theory have a precise interpretation in terms of probability theory. The probability spaces induced in the above interpretation are defined as limits of projective systems of finite sets, and the key-idea is the derivation of measure.

I shall present here certain results of this kind, in relation to the Halasz's theorem [2] on the existence of the mean-value of a multiplicative arithmetical function of modulus bounded by 1.

2- For the sake of simplicity, I shall give the results and proofs only for integers, but *the whole of this section still holds if, in place of positive integers, we consider a normed semi-group* Λ *generated by a set* P *of primes* p *satisfying the following conditions* (M) *and* (C) :

There exist $L>0$ and $K\geq 0$ such that

(M):
$$\lim_{x\to+\infty}\frac{1}{x}\sum_{\substack{N(\lambda)\leq x\\ \lambda\in\Lambda}}1 = L$$

(C):
$$\sum_{\substack{p\in P\\ N(p)\leq x}} \log N(p) \leq Kx,$$

where $N(\cdot)$ denote the norm of Λ. (see [3], ch. III, § 1, 2, 3 for properties of such Λ.)

Now, we introduce some notations:

N (resp. N^*) is the set of non-negative (resp. positive) integers. P is the set of the primes.

For any prime p in P, we set

$$E_p = \{1, p, p^2, \cdots\} = \{p^\alpha\}_{\alpha\in N} ,$$

$$\mu_p(p^\alpha) = \left(1-\frac{1}{p}\right)\cdot\frac{1}{p^\alpha} , \qquad \alpha\in\mathrm{N}$$

and

$$E = \prod_p E_p .$$

An element t of E can be written by $t = (p^{V_p(t)})$, $p \in P$, where $V_p(t)$ is the p-adic valuation of t.

To any given (positive integer) n in $\mathbb{N}^*$, we associate an elementary function $I_n : E \to \{0, 1\}$ defined by :

$$I_n(t) = \begin{cases} 1 & \text{if } V_p(t) \geq V_p(n) \text{ for any } p. \\ 0 & \text{otherwise.} \end{cases}$$

We generate A, the family of the Borel sets in E, by $(I_n^{-1}(0), I_n^{-1}(1))_{n \in N^*}$. Then, a probability measure $d\mu = \bigotimes_p d\mu_p$ is defined on E and (E, A) is $d\mu$-measurable; $(I_n)_{n \in N^*}$ is dense in $L^1(E, d\mu)$, and the probability space $(E, A, d\mu)$ is separable; Moreover, $\mathbb{N}^*$ is dense in E. Till the end of this section, we assume that the following hypothesis (H) hold:

$$(H) \quad \begin{cases} f \text{ is multiplicative, i.e.: } f: \mathbb{N}^* \to \mathbb{C} \text{ and} \\ f(mn) = f(m) \times f(n) \quad \text{if } m \text{ and } n \text{ are relatively prime.} \\ |f(n)| \leq 1 \quad \text{for any } n \text{ in } \mathbb{N}^*. \\ \Pi_p \text{ is not zero, where } \Pi_p \text{ denotes the expression } \left(1-\frac{1}{p}\right)\sum_{k=0}^{+\infty} \frac{f(p^k)}{p^k}. \end{cases}$$

The method described in [3], ch. II, leads to the following result:

Theorem 1:

Suppose that f satisfies (H). Then :

1) $$\sum_p Re\left(\frac{1-f(p)}{p}\right) < +\infty$$

is equivalent to

2) the sequence $F_y(t) = \prod_{p \leq y} \frac{f(p^{V_p(t)})}{\Pi_p}$ converges for almost all t in $L^1(E, d\mu)$ and with respect to $d\mu$.

Sketch of the proof.

Let us define the Dirichlet series : $F(s) = \sum_{n=1}^{+\infty} \frac{f(n)}{n^s}$ and $\zeta(s)$ denotes the Riemann's zeta function.

We write, if $s > 1$:

$$\begin{aligned}
\zeta(s)^{-1} \times F(s) &= \zeta(s)^{-1} \sum_{n=1}^{+\infty} \frac{f(n)}{n^s} \\
&= \prod_p \left\{ \left(1-\frac{1}{p^s}\right) \times \left(1 + \frac{f(p)}{p^s} + \sum_{k\geq 2} \frac{f(p^k)}{p^{ks}}\right) \right\} \\
&= \left[\prod_p \left\{ \left(1-\frac{1}{p^s}\right) \times \left(1+\frac{f(p)}{p^s}\right) \right\} \right] \times \left[\prod_p \left(1+\frac{1}{1+\frac{f(p)}{p^s}} \times \sum_{k\geq 2} \frac{f(p^k)}{p^{ks}}\right) \right] \\
&= H_1(s) \times H_2(s)
\end{aligned}$$

(H) implies that $|H_2(s)|$ is continuous and positive if $s \geq 1$. Now, if $s>1$, we have:

$$\left| \prod_p \left\{ \left(1-\frac{1}{p^s}\right) \times \left(1+\frac{f(p)}{p^s}\right) \right\} \right|^2 = \prod_p \left\{ \left(1-\frac{1}{p^s}\right)^2 \times \left(1+\frac{2Re\, f(p)}{p^s} + \frac{|f(p)|^2}{p^{2s}}\right) \right\}.$$

Taking the logarithm, we get :

$$\left| \log\left\{ \zeta(s)^{-1} \times |F(s)| \times |H_2(s)|^{-1} \right\} - \frac{1}{2} \sum_p \frac{2Re(f(p)-1)}{p^s} \right|$$

$$= O\left(\sum \frac{1}{p^{2s}}\right) = O(1), \qquad s \geq 1.$$

Since $1-Re\, f(p) \geq 0$ for any p, 1) implies that: $\zeta(s)^{-1} \times |F(s)| \to 0$, $s \to 1^+$, and in this case, we get 2) as a special case ($A = \mathbb{N}^*$) of [3] ch. II §3 (4.1), (4.2), p. 85. If we have 2), we get 1) as a special case of [3] ch. II §3, 11 p94, for $A = \mathbb{N}^*$.

From theorem 1 we derive the following result:

Theorem 2

Suppose f satisfies (H) and $\sum_p \frac{1-Re\, f(p)}{p} < +\infty$.

We denote f_{y-} and f_{y+} the multiplicative functions defined below :

$$f_{y-}(n) = \prod_{p\leq y} f\left(p^{V_p(n)}\right)$$

and

$$f_{y+}(n) = \prod_{p>y} f\left(p^{V_p(n)}\right).$$

Then we have :

$$\lim_{y\to+\infty} \lim_{x\to+\infty} \frac{1}{x} \sum_{n\leq x} \left(\left|\frac{f_{y-}(n)}{\prod_{p\leq y} \Pi_p}\right| \times \left|1-\frac{f_{y+}(n)}{\prod_{y<p\leq x} \Pi_p}\right| \right) = 0.$$

Proof of Theorem 2 :

To prove Theorem 2, we employ the method developped in [3] ch. III § 6.

a - Let A_z be the set defined by :

$$A_z = \{n\in\mathbb{N}^* \mid p^{V_p(n)} < z \text{ for } p<z, \text{ and } V_p(n) = 1 \text{ or } 0 \text{ for } p\geq z\},$$

and denote I_{A_z} its characteristic function.

We have :

$$1 \geq M(I_{A_z}) \geq \left(1-\frac{1}{z}\right)^{w(z)} \times \prod_{p\geq z}\left(1-\frac{1}{p^2}\right) = 1+o(1), \qquad (z\to+\infty),$$

where $M(\cdot) = \lim_{x\to\infty}\frac{1}{x}\sum_{n\leq x} I_{A_z}(n)$ is the arithmetical mean-value, and

$$w(z) = \sum_{p\leq z} 1.$$

This gives that $1 - M(I_{A_z}) = o(1)$, $(z\to+\infty)$.

Hence we get :

$$\overline{\lim_{x\to+\infty}} \frac{1}{x}\sum_{n\leq x} |f(n)\times(1-I_{A_z}(n))| \leq \overline{\lim_{x\to+\infty}} \frac{1}{x}\sum_{n\leq x} |1-I_{A_z}(n)| = o(1), \quad (z\to+\infty) \quad (*)$$

Now, put $U_p(z) = \Pi_p^{-1} \times (1-\frac{1}{p}) \times \sum_{k=0}^{+\infty}\left(\frac{f(p^k)}{p^k} I_{A_z}(p^k)\right)$. Then, we have :

$$\lim_{z\to+\infty} \prod_p U_p(z) = 1,$$

which signifies together with (*), that it is sufficient to prove Theorem 2 for $f' = f \times I_{A_z}$. For making z tending to infinity, this will give the Theorem 2 for a general f.

b - Given y, $y>z$, and $x>y$, we define g by :

$$g(n) = \prod_{y<p\le x} \left(\frac{1}{\phi(p)} \times f'\ (p^{V_p(n)})\right) ,$$

$$\text{where } \phi(p) = (1-\frac{1}{p}) \times \left(1 + \frac{f'\ (p)}{p}\right).$$

The symbol $\sum_J^*$ will denote the summation $\sum_{\substack{J \\ n\in E_{y+}}}$, where J is a given set of conditions and $E_{y+} = \left\{n\in \mathbb{N}^* \mid (p\mid n) \Rightarrow p>y \text{ and } V_p(n) = 1.\right\}$

We shall evaluate the sum :

$$S_{y+,x} = \sum_{\substack{m\le x \\ m\in E_{y+}}} |1 - g(m)|^2 \log m.$$

We have

$$S_{y+,x} = \sum_{m\le x}^* |1 - g(m)|^2 \log m$$

$$= \sum_{m\le x}^* |1 - g(m)|^2 \times \sum_{p\le x}^* \log (p^{V_p(m)})$$

$$= \sum_{p\le x}^* \log p \sum_{\substack{m\le x \\ p\mid m}} |1 - g(m)|^2, \quad \text{for } V_p(m) = 1 \text{ or } 0;$$

If $p\mid m$, we have $m = pm'$, $(m', p)=1$; hence we get :

$$|1 - g(m)|^2 = |1-g(pm')|^2 = |1 - f'\ (p) \times g(m')|^2$$

$$= |(1-f(p)) + f(p)(1-g(m'))|^2$$

$$\le 2(|1-f(p)|^2 + |f(p)|^2|1-g(m')|^2)$$

$$\le 2(|1-f(p)|^2 + |1-g(m')|^2)$$

This gives :

$$S_{y+,x} \leq 2 \sum_{p\leq x}^{*}\left[(\log p)|1-f(p)|^2 \sum_{\substack{(m',p)=1\\ m'\leq x/p}}^{*} 1\right] + 2\sum_{p\leq x}^{*}\left[(\log p) \times \sum_{\substack{m'\leq x/p\\ (m',p)=1}}^{*} |1-g(m')|^2\right]$$

$$\leq 2\log x \sum_{p\leq x}^{*}\left(|1-f(p)|^2 \times \sum_{m'\leq x/p}^{*} 1\right) \qquad (I)$$

$$+\ 2\sum_{p\leq x}^{*}\left\{(\log p) \times \sum_{m'\leq x/p}^{*} |1-g(m')|^2\right\} \qquad (II)$$

(I) can be treated in the analogous way for the sum $S''_{1,y}(x)$ in [3] ch. III §6 p 120-122. Thus, we shall deal with (II). We have :

$$2\sum_{p\leq x}^{*} (\log p) \times \sum_{m'\leq x/p}^{*} |1-g(m')|^2$$

$$= 2\sum_{m'\leq x}^{*}\left(|1-g(m')|^2 \times \sum_{p\leq x/m'}^{*} \log p\right) \leq \sum_{m\in E_{y+,x-}}\left(|1-g(m)|^2 \times C\frac{x}{m}\right)$$

where $E_{y+,x-} = \{m \mid (p|m) \Rightarrow (y<p\leq x,\ V_p(m)=1)\}$ and C is an absolute constant,

$$\leq Cx \times \sum_{m\in E_{y+,x-}} \frac{|1-g(m)|^2}{m}.$$

A straightforward computation gives

$$\sum_{m\in E_{y+,x-}} \frac{|1-g(m)|^2}{m}$$

$$= \left(\prod_{y<p\leq x}(1-\tfrac{1}{p})\right)^{-1} \times \left\{\prod_{y<p\leq x}\left(1-\frac{1}{p^2}\right) + \prod_{y<p\leq x}\frac{\left(1-\frac{1}{p}\right)\left(1+\frac{|f(p)|^2}{p}\right)}{\left|\left(1-\frac{1}{p}\right)\times\left(1+\frac{f(p)}{p}\right)\right|^2} - 2\right\}$$

$$= \left(\prod_{y<p\leq x}(1-\tfrac{1}{p})\right)^{-1} \times o(1), \quad (\text{ uniformly in } x,\ y\to+\infty)$$

$$\leq \left(\prod_{p\leq y}(1-\tfrac{1}{p})\right) \times (\log x) \times o(1), \qquad (y\to+\infty),$$

by some Mertens' formula. Hence, we get :

$$S_{y+,x} \leq x\log x \times \eta_y^2 \times \prod_{p\leq y}\left(1-\frac{1}{p}\right), \qquad \text{where } \lim_{y\to+\infty}\eta_y = 0.$$

Now, we remark that :

$$\sum_{m\leq x}^{*} |1-g(m)|^2 \leq \sum_{m\leq x^{1/2}} |1-g(m)|^2 + \sum_{x^{1/2}\leq m\leq x}^{*} |1-g(m)|^2 \frac{\log m}{\log x^{1/2}}$$

$$\leq O(x^{1/2}) + 2x\,\eta_y^2 \prod_{p\leq y} (1-\tfrac{1}{p}) .$$

This implies, by Cauchy-Schwarz inequality,

$$\overline{\lim_{x\to+\infty}} \frac{1}{x} \sum_{n\leq x}^{*} |1-g(n)| \leq 2\eta_y \prod_{p\leq y} (1-\tfrac{1}{p}).$$

Now, writting $n=n_1 n_2$ where $n_1=\prod_{p\leq y} p^{V_p(n)}$, $n_2=\prod_{p>y} p^{V_p(n)}$, a straightforward computation gives the result for f'.

3- In the case of $\mathbb{N}^*$, we get the following corollary :

Corollary 1 (Halasz [2]) : If f satisfies (H) and $\sum_p \frac{Re\,(1-f(p))}{p} < +\infty$, then

$$\lim_{x\to+\infty} \left\{ \left(\frac{1}{x}\sum_{n\leq x} f(n)\right) \times \left(\prod_{p\leq x}\left\{(1-\tfrac{1}{p}) \sum_{k\geq 0} \frac{f(p^k)}{p^k}\right\}\right)^{-1} \right\} = 1.$$

An immediate computation gives :

Corollary 2 (J. Coquet [1]) : If f satisfies (H) and $\left(\sum_p \frac{1-Re\,f(p)}{p} < +\infty\right)$, then, for any m, the correlation function $\tau(m)$ defined by: $\tau(m) = \lim_{x\to+\infty} \frac{1}{x}\sum_{n\leq x} f(n)\overline{f}(n+m)$ exists, and $\{\tau(m)\}_{m\in N}$ is a limit periodic sequence, and further the support of the spectral measure associated to $\{\tau(m)\}_{m\in N}$ is in $\mathbb{Q}\cap[0,1]$.

4- Conclusion

Theorems 1, 2 and Corollary 1 still hold in more general spaces. For instance, in semi-groups satisfying conditions (M) and (C) stated at the beginning of §2. The method of proof is exactly the same as that presented here. If would be interesting to know whether these results can be obtained by applying the analytic method of Halasz ([2]).

References :

[1] J. Coquet : These de doctorat d'Etat es-sciences mathematiques (Contribution a l'etude harmonique de suites arithmetiques) 1978. Universite Paris-Sud, centre d'Orsay.

[2] G. Halasz : Uber die Mittelwerte multiplikativer Zahlen theoretischer Funktionen, Acta Math. Sci. Hung. *19* (1968) 365-403.

[3] J. L. Mauclaire. Integration et theorie des Nombres, Travaux en Cours, Hermann, Paris 1986.

Centre National de la Recherche Scientifique, Paris.

The Institute of Statistical Mathematics, 4-6-7 Minami-Azabu Minato-ku, Tokyo, Japan-106

MONTE CARLO METHODS WITH STOCHASTIC PARAMETERS

G. A. Mikhailov

1. Introduction. Various examples introducing additional randomness for constructing effective simulation algorithms can be found in the literature devoted to the Monte Carlo methods (see, e.g., [1]). This paper is concerned with the construction and the optimization of randomized algorithms for estimating probabilistic characteristics of solutions of integral equations with random parameters. In this connection, randomized models for random fields are suggested and estimates of parametric derivatives are constructed. Applications to problems of transfer theory and diffusion in stochastic media are considered.

2. Randomized estimation for the statistical moments of the solution. Consider the integral equations in L_∞:

$$\varphi_j(x,\sigma)=\int k_j(x,x';\sigma)\varphi_j(x',\sigma)dx'+h_j(x,\sigma), \quad x\in X, \quad j=1,\dots,n \qquad (2.1)$$

where σ is a random parameter (vector or function). It is supposed that the spectral radius $\rho(|K_{\sigma j}|)<1$, where $|K_{\sigma j}|$ is the integral operator with the kernel $|k_j(x,x';\sigma)|$. Let $\omega=\{x_n\}$, $n=0,1,\dots,N$, be a terminating Markov chain with the initial density function $p_0(x,\sigma)$ and the transition density function $p(x',x;\sigma)$ such that

$$q_j(x,x';\sigma)=k_j(x,x';\sigma)/p(x,x';\sigma)<+\infty , \qquad \forall x,x',\sigma,j, \qquad (2.2)$$

and $Q_{nj}(\sigma)$ are weights, determined by formulae:

$$Q_{0j}(\sigma)=f_j(x_0,\sigma)/p_0(x_0,\sigma), \quad Q_{nj}(\sigma)=Q_{(n-1)j}(\sigma)q_j(x_{n-1},x_n;\sigma).$$

Then $I_j(\sigma)=\int f_j(x,\sigma)\varphi_j(x,\sigma)dx=M\xi_j(\omega,\sigma)$,
where

$$\xi_j(\omega,\sigma)=\sum_{n=0}^{N} Q_{nj}(\sigma)h_j(x_n,\sigma).$$

The symbol M denotes the expectation, corresponding to the distribution $P(d\omega|\sigma)$, and the symbol E will denote expectation, corresponding to the distribution $P(d\sigma)$. It is possible to estimate the quantities

$$I_j=EI_j(\sigma), \quad R_{kj}=E[I_k(\sigma)I_j(\sigma)], \quad k,j=1,\dots,n, \tag{2.3}$$

by the Monte Carlo method on the basis of the evident relations:

$$I_j=M_{(\omega,\sigma)}\xi_j(\omega,\sigma), \quad R_{kj}=M_{(\omega_1,\omega_2,\sigma)}[\xi_k(\omega_1,\sigma)\xi_j(\omega_2,\sigma)],$$

where ω_1, ω_2 are conditionally independent trajectories for one fixed realization of σ, and $M_{(\omega,\sigma)}$ denotes expectation, corresponding to the distribution of (ω,σ). Thus, to estimate the quantities I_j it is sufficient to sample only one trajectory ω for fixed σ (i.e., not to solve the problem (2.1) completely), while to estimate R_{kj} it is necessary to sample at least two trajectories.

To optimize the randomization technique, it is natural to use the "splitting method" (see, e,g., [1]). In this method, the quantities I_k are estimated as follows. First, one constructs n conditionally independent trajectories (i.e., a vector $\omega=(\omega_1,\dots,\omega_n)$), σ fixed, and then the random variable

$$\xi_k^{(n)}(\omega,\sigma) = \frac{1}{n}\sum_{i=1}^{n}\xi_k(\omega_i,\sigma)$$

is used instead of $\xi_k(\omega,\sigma)$. Optimal value of n is calculated by the formula (see, e.g., [1]):

$$n = \sqrt{\frac{a_2}{a_1}\,\frac{t_1}{t_2}} \tag{2.4}$$

where $a_1=E(M\xi_k)^2-I_k^2$, $a_2=ED\xi_k$, t_1 is the average computing time for a fixed realization σ, t_2 is the average computing time for a fixed realization ω.

It is difficult to evaluate the quantities a_1, a_2, t_1, t_2 directly. However, one can obtain statistical estimates of the variances and the computing times for two values of the splitting parameter: n_1, n_2 and then solve the corresponding system of linear equation. Here, it is useful to correlate the samples of $\xi_k^{(n_1)}$ and $\xi_k^{(n_2)}$.

Let us consider now optimization of the randomized estimate of the quantity R_{kj} using the splitting technique. In this case, it is natural to use the random variable:

$$\rho_{kj}^{(n)} = \frac{\sum_{i=1}^{n}\sum_{t=i+1}^{n}[\xi_k(\omega_i,\sigma)\xi_j(\omega_t,\sigma)]}{n(n-1)/2}, \qquad M\rho_{kj}^{(n)} = R_{kj} .$$

Note, that instead of $\rho_{kj}^{(n)}$, it is possible to use another quantity which is numerically equivalent to $\rho_{kj}^{(n)}$, but more convenient for calculations. Indeed, the following relation for the covariances holds, which is easily verified:

$$K[\xi_k^{(n)}(\omega,\sigma),\xi_j^{(n)}(\omega,\sigma)] = K[I_k(\sigma),I_j(\sigma)] + \tfrac{1}{2}EK[\xi_k,\xi_j] .$$

Taking into account analogous relation for $n=1$, we obtain from this system of equations that

$$K_{jk} = K[I_k(\sigma),I_j(\sigma)] = \frac{nK[\xi_k^{(n)},\xi_j^{(n)}]-K[\xi_k^{(1)},\xi_j^{(1)}]}{n-1} . \tag{2.5}$$

It is not difficult to verify that substituting the statistical estimates of the covariances in the right-hand side of (2.5) yields an estimate $(\tilde{K}_{jk}^{(n)})$ of the quantity K_{jk} which coincides numerically with the estimate obtained on the basis of $\rho_{kj}^{(n)}$. When (2.5) is used, the average computing time is given, as in the standard splitting technique, by the formula: $T^{(n)}=T_1+nT_2$. Explicit formula for the variance $D\rho_{kj}^{(n)}$ is very cumbersome. However, it is clear that for sufficiently large n the following approximate relation holds:

$$D\,\tilde{K}_{kj}^{(n)} \approx A_1 + A_2/n .$$

Consider an optimization of the double randomization method for calculating the functionals of the form:

$$I = EI(\sigma) = EM\xi(\omega,\sigma)$$

with taking into account the cost of the approximate simulation of the field σ. Let σ_m be an approximate model for the field σ (see, Sec.4 and 5), and assume that

$$(I_m-I)^2 \lesssim cm^{-\alpha}, \qquad I_m = EI(\sigma_m), \qquad \alpha > 0.$$

Here, the symbol $\lesssim$ means that the estimate is exact enough.

Suppose also that the variance

$$d = D\xi(\omega,\sigma_m) = EM[\xi(\omega,\sigma_m)-I_m]^2$$

does not depend on m. The probabilistic error of the estimate for $I(\omega,\sigma_m)$ is defined by the quantity:

$$EM[\xi(\omega,\sigma_m)-I]^2 = d + (I_m-I)^2.$$

For $\xi_N(\omega,\sigma_m)$, the average value of N independent realizations $\xi(\omega,\sigma_m)$ we have: $D\xi_N=d/N$.

Let us measure the cost of an algorithm by average number of operations required to obtain the result within probability error ε. Suppose that the average cost of simulation of $\xi(\omega,\sigma)$ for $\sigma=\sigma_m$ is given by c_0m^β, $\beta>0$. Note that in algorithms of Sec.4 and Sec.5, β apparently equals to unity. To optimize the randomized algorithm of estimation of the quantity I, it is thus necessary to solve the following conditional minimization problem:

$$\min_{N,m} \{N\cdot m^\beta\}, \qquad \frac{d}{N} + cm^{-\alpha} = \varepsilon^2. \tag{2.6}$$

The next formula follows from (2.6):

$$N = \frac{\beta}{\alpha c} m^\alpha. \tag{2.7}$$

The main difficulty here when solving a concrete problem is to obtain sufficiently exact values of c and α. It is interesting to note that the expression (2.7) holds also if the splitting technique is used, i.e. when $\xi(\omega,\sigma_m)$ is changed with $\zeta^{(n)}(\omega,\sigma)$ provided that the average cost of one sampling of $\zeta^{(n)}(\omega,\sigma)$ for $\sigma=\sigma_m$ is also proportional to m^β.

<u>3. The estimation of parametric derivatives</u>. Suppose now that the function σ is completely determined by the finite number of scalar parameters, i.e. $\sigma=(\sigma_1,\dots,\sigma_2)$ and $\{f_j(x,\sigma)\}$ do not depend on σ. It is possible to estimate the quantities (2.3) on the basis of Taylor expansion of the solution φ with respect to σ at the point $\sigma^{(0)}=E\sigma$, and calculating corresponding derivatives $\varphi^{(n)}(x)$ by the Monte Carlo method. After the formal differentiation of equation (2.1) we get the system for the case of scalar σ:

$$\varphi^{(n)} = \sum_{i=0}^{n} c_n^i\, K^{(n-i)}\varphi^{(i)} + h^{(n)}, \qquad n = 0,1,\dots,m. \tag{3.1}$$

or in the operator form: $\Phi=K\Phi+H$. Here $K^{(n)}$ denotes the integral operator with the kernel $k^{(n)}(x,x';\sigma^{(0)})$.

<u>Theorem 1</u>. Assume

$$|k^{(n)}(x,y,\sigma)| \le k_n(x,y), \qquad \forall\sigma \in (\sigma^{(0)}-\varepsilon, \sigma^{(0)}+\varepsilon),$$

holds for some $\varepsilon>0$. Suppose also that operators K_n with the kernels $k_n(x,y)$ are bounded and $\rho(K_0)<1$. Then $\rho(K)<1$ and the functions $\varphi^{(n)}$ satisfy the system (3.1).

It is possible to use the vector Monte Carlo method for solving the system (3.1) [2]. In this method the matrix weight factor is determined by the matrix kernel $K(x,y)$ as in (2.2). The variance of the corresponding estimations are finite if the standard finiteness criterion for the variance of scalar estimations with respect to $K^{(0)}$ holds. In the case of vector σ the estimations of the partial derivatives are constructed similarly.

<u>4. Special models of non-gaussian random fields related to stationary point fluxes</u>. To solve stochastic problems by the method of statistical modeling it is necessary to construct numerically the realizations of the random processes and fields.

It is then often important to preserve the correlation function and the one-dimensional distributions, all the more there is no satisfactory information about the many-dimensional distributions, in the case of non-gaussian fields. However, it is often desired to have a set of models of random fields; on the basis of this set of models, it is possible to study a sensitivity of functionals as the many-dimensional distributions vary when converging to distributions of a random field with an appropriate (e.g., continuous) realizations. Models of this kind which are sufficiently close (in the sense of weak convergence) to the limit fields and realizable on a computer are naturally used as approximate models. The more accurate are these approximations, the larger is the expense of their realization.

In this section, we consider special random processes which are connected with stationary pointing fluxes (Palm fluxes):

$$\tau_k = \sum_{i=1}^{k} \eta_i, \qquad k = 1,2,\ldots; \qquad \tau_0 = 0, \tag{4.1}$$

where $\{\eta_i\}$ are independent non-negative random variables with distribution densities $f_i(x)$, $i=1,2,\ldots$, where

$$f_k(x) = f(x) = F'(x), \quad k = 2,3,\ldots, \quad f_1(x) = \mu^{-1}[1-F(x)] \tag{4.2}$$

We use fluxes of the form (4.1) with a given probability $p_0(t)$

$=P(k=0;t)$ on the interval $[0,t]$. It is known [3] that the Palm flux is fully defined by $p_0(t)$ so:

$$f(x) = -p_0''(x)/p_0'(0), \quad f_1(x) = -p_0'(x), \quad x \geq 0. \tag{4.3}$$

Lemma 1. Let $p_0(t)$ be a function such that $p_0(0)=1$, $p_0(+\infty)$, $|p_0'(0)|<\infty$, $p_0''(t)\geq 0$ for $t\geq 0$. If (4.2), (4.3) hold for the flux (4.1) then $P(k=0;t)=p_0(t)$.

Construct now a stationary random process $\xi(t)$, $0<t<T$ with a given one-dimensional distribution $F_\xi(x)$ using the Palm flux $\{\tau_k\}$ with $P(k=0;t)=p_0(t)$ as follows:
1. A sequence $\{\tau_k\}$ is simulated until the first exit outside $(0,T)$.
2. One takes $\xi(t)\equiv\xi_i$ in each interval (τ_{i-1},τ_i), $i=1,2,\ldots$, where $\{\xi_i\}$ are independent random variables with the distribution function $F_\xi(x)$. Denote this process by $\xi(t;p_0)$. The standardized correlation function of the process $\xi(t;p_0)$ is equal to $p_0(t)$, since

$$M[\xi(t';p_0)\xi(t'+t;p_0)] = p_0(t)M\xi^2+[1-p_0(t)](M\xi)^2.$$

Let $\mathcal{D}$ be a domain of n-dimensional Euclidean space lying in a cube $[0,T]^n$. Construct a random field $\xi(r)$, $r=(x_1,\ldots,x_n)$ as follows:
1) Sample a Palm flux $\{\tau_k^{(i)}\}$ with $p_0^{(i)}(t)=K_i(t)$ on i-th axis.
2) An independent value of $\xi(r)$ is sampled according to the distribution function for each parallelepiped of the form:

$$[\tau_{k_1-1}^{(1)}, \tau_{k_1}^{(1)}] \times \cdots \times [\tau_{k_n-1}^{(n)}, \tau_{k_n}^{(n)}].$$

Denote this field by $\xi(r,K)$ where $K=(K_1,\ldots,K_n)$. The field $\xi(r,K)$ is homogeneous and has a standardized correlation function

$$\mathcal{K}(r;K) = \prod_{i=1}^{n} K_i(|x_i|), \quad r = (x_1,\ldots,x_n). \tag{4.4}$$

Note that the set of simulated correlation functions can be extended, changing x_i in (4.4) with $\lambda_i x_i$ where $\lambda=(\lambda_1,\ldots,\lambda_n)$ is chosen at random according to the probabilistic measure $\mu(d\lambda)$ defined on $[0,+\infty]^n$ for every realization of the field. It is possible to simulate in such a manner fields with the correlation function

$$\mathcal{K}(r) = \int \exp(-\lambda_1|x_1|-\cdots-\lambda_n|x_n|)\mu(d\lambda). \tag{4.5}$$

To simulate an isotropic random field, we use a following natural technique. First one constructs a realization of the field $\xi(r,K)$. Then it is isotropically turned and oriented according to a null vector $\omega=(\omega_1,\dots,\omega_n)$. When the correlation function of such a field is evaluated, it is possible to consider the orientation as a fixed one, and r as an isotropic vector, i.e., $r=t\omega$, $t=|r|$. Therefore,

$$K(t) = \int K_1(|\omega_1|t)\cdots K_n(|\omega_n|t)d\omega.$$

Suppose now that the distribution of ξ is <u>unlimited divisible</u>, i.e. representation $\xi=\xi_1^{(m)}+\cdots+\xi_m^{(m)}$ holds for every positive integer m, where $\xi_i^{(m)}$ are independently and equally distributed according to the distribution function $F_m(x)$. Consider the process

$$\zeta_m(t) = \sum_{i=1}^{m} \xi_i^{(m)}(t;K), \tag{4.6}$$

where $\xi_i^{(m)}(t;K)$ are independent realizations of the process $\xi^{(m)}(t;K)$ distributed according to one-dimensional distribution $F_m(x)$. Obviously, the one-dimensional distribution function of $\zeta_m(t)$ is equal to $F_\xi(x)$ and the correlation function is equal to $K(t)$.

The realizations of the process $\zeta_m(t)$ are improved (as compared with $\xi(t;K)$) in the sense that they are close to continuous functions, because $\zeta_m(t)$ takes constant values in smaller domains and these values are dependent. A weak convergence of the finite-dimensional distributions corresponding to $\zeta_m(t)$ to distributions satisfying the consistency condition is obtained in [4]. Using the Chentsov-Kolmogorov criterion, the following result is obtained in [5].

<u>Theorem 2</u>. Suppose that $|K''(t)|<c<\infty$ on $[0,T]$ and $\xi\geq 0$. Then the processes $\zeta_m(t)$ are weakly (with respect to the metric D) convergent to a process which is determined by the limit finite-dimensional distributions with the correlation function $K(t)$ and one-dimensional distribution function $F_\xi(x)$.

This theorem does not include the case of gaussian distribution. However the convergence of $\zeta_m(t)$ is then easily obtained from the following known generalization of the mentioned criterion:

$$P\{|\zeta_m(t_3)-\zeta_m(t_2)|\geq\lambda, |\zeta_m(t_2)-\zeta_m(t_1)|\geq\lambda\}\leq\lambda^{-2\gamma}(t_3-t_1)^{2\alpha}, \qquad \gamma=\alpha=1.$$

It is clear enough that the considered theorem can be extended on the random fields applying the corresponding generalization of the Chentsov-Kolmogorov criterion [6]. In any case, the finite dimensional distributions of the considered field models are improved by summation of the form (4.6).

Numerical simulations of these models show that non-gaussian random fields essentially differ from the gaussian fields. Indeed, non-gaussian fields have regions of constant values such that their diameters are essentially larger than the correlation scale of the field. This fact could be interpreted as a consequence of the entropy maximality of the gaussian distribution when covariations are fixed.

5. Simulation of homogeneous gaussian fields by randomization of the spectral representation. Let $\zeta(x)$ be a real-valued random gaussian field with a correlation function $K(x)$ having spectral expansion:

$$K(x) = \int \cos(\lambda x)p(\lambda)d\lambda \ , \tag{5.1}$$

where $p(\lambda)$ is a spectral density, $\lambda\in\Lambda=R_n$. For simplicity suppose that $M\zeta(x)=0$, $D\zeta(x)=1$.

Divide the space Λ into m parts: $\Lambda_1,\dots,\Lambda_m$ and assume that random points $\lambda_1,\dots,\lambda_m$ are distributed in these domains according to probability densities

$$p_k(\lambda) = p(\lambda)[\int_{\Lambda_k} p(\lambda)d\lambda]^{-1}, \qquad \lambda \in \Lambda_k \ .$$

Then we have from (5.1)

$$K(x) = M\sum_{k=1}^{m} p_k\cos(\lambda_k x), \qquad p_k = \int_{\Lambda_k} p(x)dx.$$

Thus we obtain following method of construction of a random field with a given correlation function:

1) random values $\lambda_1,\dots,\lambda_m$ are sampled,
2) realization of the field is constructed by formula

$$\zeta_m(x) = \sum_{k=1}^{m} p_k^{1/2}[\xi_k\sin(\lambda_k x)+\eta_k\cos(\lambda_k x)] \ , \tag{5.2}$$

where (ξ_k,η_k) is a set of independent standard gaussian random values.

One-dimensional conditional distribution of the field (5.2) is normal (and standardized) provided $\lambda_1,\ldots,\lambda_m$ are fixed. Consequently absolute one-dimensional distribution of (5.2) is also normal and standardized.

Consider now the convergence of $\xi_m(x)$ to the normal field $\xi(x)$ with a given spectral density $p(\lambda)$. Let V_k be the volume of Λ_k, and let d_k be the diameter of Λ_k. We shall use uniform (with respect to the volume) partitions of the domain $\Lambda\setminus\Lambda_m$ so that

$$d_k \le cV_k^{1/n}, \quad k = 1,2,\ldots,m-1 ,$$

where $V_k\to 0$ as $m\to\infty$. Such partition can be obtained, for example, using a uniform rectangular lattice with a net size $cm^{-1/n}$.

<u>Theorem 3</u>. Let $\xi_m(x)$ be determined by formula (5.2) for $|x|<R$, $\Lambda_m=\{\lambda:|\lambda|>l_m\}$ and $d_k<cl_m m^{-1/n}$, $k=1,2,\ldots,m-1$. Then

1) $$M\int_{|x|<R}[\xi_m(x)-\xi(x)]^2dx \to 0 ;$$

if $l_m m^{-1/n}\to 0$ and $l_m\to\infty$ as $m\to\infty$.

2) $$M\int_{|x|<R}[\xi_m(x)-\xi(x)]^2dx \le cm^{-2\varepsilon/[n(2+\varepsilon)]}$$

for $l_m=cm^{2/[n(2+\varepsilon)]}$ if $\int|\lambda|^{\varepsilon}p(\lambda)d\lambda<+\infty$, for some positive ε.

Analogous statements could be formulated and proven about the convergence of the derivatives of $\xi_n(x)$ under the condition that there exist corresponding moments of the spectral density. For example,

$$M\int[\xi_m^{(k)}(x)-\xi^{(k)}(x)]^2dx \le cm^{-2\varepsilon/[n(2+\varepsilon)]} , \tag{5.3}$$

if

$$\int|\lambda|^{2+\varepsilon}p(\lambda)d\lambda < \infty , \tag{5.4}$$

where $\xi^{(k)}$ is the partial derivative of ξ with respect to $\lambda^{(k)}$, $k=1,2,\ldots,n$. Note that sometimes it is possible to prove the convergence (in probability) to zero of uniform deviation of ξ_m from ξ using imbedding theorems. In particular, for random processes (i.e., n=1) $\xi=\xi(t)$ it follows (5.3), (5.4) and by imbedding theorem ($W_2^{(1)}$ in C) that

$$M\sup_t|\xi_m(t)-\xi(t)| \le c_0m^{-\varepsilon/[n(2+\varepsilon)]}.$$

Further developments of the simulation technique presented are discussed in [7]. Particularly, [7] contains proof of the first statement of Theorem 3 without the restriction $|x|<R$. Besides, using a transformation of the weak convergence criterion due to Chentsov [6] the author [7] has obtained a simple weak convergence criterion with respect to metric C for random field models discussed. This criterion is formulated in terms of spectral moments of the random field simulated.

6. Stochastic problems of radiation transfer theory. As a first example let us consider a problem of radiation transfer in a medium represented as a random set of sphere-shaped non-homogeneities. It is assumed that the midpoints of spheres can be considered as a spatial Poisson point flux, i.e. the numbers of midpoints in non-overlapping domains are independent and distributed according to Poisson law. Intersections of spheres are allowed. Denote by σ_1 and σ_2 the total cross-sections (see [1]) of the medium in and outside the spheres, respectively. Assume that $\max(\sigma_1,\sigma_2)=\sigma_1$. Simulation of trajectories for fixed realization of such a medium can be carried out by the method of maximal cross-section (see [1]). In this method, the free path length is sampled according to formula: $l=-\ln\alpha$, where α is a random variable uniformly distributed on $(0,1)$; if the sampled collision point does not lie in a sphere then a "delta-scattering" is simulated, i.e. the particle moves in the same direction with probability $(\sigma_1-\sigma_2)/\sigma_1$. To construct a realization of the medium, it is sufficient to sample N, the number of spheres midpoints, according to Poisson distribution and then to sample all the midpoints independently and uniformly in the domain (see, e.g., [8]). To decrease the number of arithmetic operations it is possible to divide the domain into parts and sample the midpoints in a part only when the particle hits this part. Clearly, the size of a part must be larger than the radii of the spheres. The sampled portion of the point flux must be stored because the particle may repeatedly hit the corresponding part of the domain. Resampling of the portions of the flux is not in accordance with the randomization principle of Sec. 2. This result in a bias of the estimate which can be neglected only for strong scattering anisotropy.

Another general model of stochastic medium follows from the representation of the total cross-section

$$\sigma^{(m)}(r) = \sum_{i=1}^{m} \sigma_i^{(m)}(r) \tag{6.1}$$

where $\sigma_i^{(m)}$, i=1,...,m are independent realizations of a homogeneous random field related to stationary point fluxes as described in Sec. 4.

It follows from the arguments of Sec. 4 that it is useful to take a large value of m in (6.1) since the finite-dimensional distributions of the field $\sigma^{(m)}$ become absolute-continuous as $m\to\infty$. This property is natural for fields with absolute-continuous one-dimensional distributions. However in the problems of transfer theory it is possible to choose m such that the mean size of regions where $\sigma^{(m)}$ is constant is essentially less than the mean free path of the particle. It is useful to obtain such values of m in preliminary calculations. Various algorithms simulating the free path of the particle in a medium with total cross-section of the form (6.1) are considered in [9]. Perhaps the simplest but not the most effective is here an algorithm where the free path length is sampled independently for each of the term $\sigma_i^{(m)}(r)$ and then the minimum of the sampled lengths is chosen. Distribution of such a quantity coincides with the physical distribution of the free path [10].

Some results of Monte Carlo calculations of transfer problems for stochastic plane-parallel media are given in [13]. It appeared that these results are in good agreement with the asymptotics of the mean intensity of radiation propagating in such a medium (see [11]). This asymptotics shows that it is important to take into account the stochastic non-homogeneity of the actual media.

The method described in Sec. 3 was used for the solution of the stochastic transfer problems for the first order derivatives (see [12]).

There are some applications of the randomized Monte Carlo method for the solution of stochastic diffusion and elasticity problems (see [13]).

References

[1] Ermakov S. M., Mikhailov G. A. Statistical modeling. Moscow, Nauka, 1982 (in Russian).

[2] Mikhailov G. A. Investigation and increasing of the variance in weight vector Monte Carlo algorithms. Ž. Vyčisl. mat. i mat. fiz., 25 (1985), No11, p.1614-1627 (in Russian).

[3] Khinchin A. Ya. Work on mathematical theory of the queuing theory. Moscow, Fizmatgiz, 1963.

[4] Mikhailov G. A. Approximate models of random processes and fields. Ž. Vyčisl. mat. i mat. fiz., 23 (1983), No3, p.558-566 (in Russian).

[5] Chentsov N. N. The weak convergence of random processes with trajectories without jumps of the second type. Teor. verojatn. i primen., 1 (1956), No1, p.154-161 (in Russian).

[6] Chentsov N. N. Limiting theorems for some classes of random functions. In: Tellus of All-Union Symposium on probability theory and mathematical statistics. Erevan: 1960, p.280-285.

[7] Vojtishek A. V. Randomized numerical spectral model of stationary random function. Izv. vuzov, Matematika, 1985 (in Russian).

[8] Feller W. An introduction to probability theory and its application. John Wiley & Sons, Inc. New York, London, Sydney, Toronto, 1971.

[9] Trojnikov V. S. Numerical modeling of the random processes and fields on the basis of point Palm fluxes in the problems of the theory of transfer in clouds. Izv. Akad. Nauk SSSR, FAO, 20 (1984), No4, p.274-279 (in Russian).

[10] Nazaraliev M. A., Ukhinov S. A. Calculations of the brightness of complicated aerosol systems. In: Statistical modeling in mathematical physics. Novosibirsk: 1976, p.29-37 (in Russian).

[11] Mikhailov G. A. Asymptotics of the mean radiation intensity for some models of stochastic media. Izv. Akad. Nauk SSSR, FAO, 18 (1982), No12, p.1289-1295 (in Russian).

[12] Antjufeev V. S., Mikhailov G. A. Evaluation of the altitude profiles of the scattering coefficient and calculation of the correlation characteristics of the day horizon brightness by the Monte Carlo method. Izv. Akad. Nauk SSSR, FAO, 12 (1976), No5, p.485-493 (in Russian).

[13] Elepov B. S., Kronberg A. A., Mikhailov G. A., Sabelfeld K. K. Solution of boundary value problems by the Monte Carlo method. Novosibirsk, Nauka, 1980 (in Russian).

Computing Center, Siberian Division
of the USSR Academy of Sciences,
Novosibirsk 630090, USSR

SCHRÖDINGER OPERATOR WITH POTENTIAL WHICH IS THE DERIVATIVE OF A TEMPORALLY HOMOGENEOUS LÉVY PROCESS

Nariyuki Minami

§0. Introduction. Let $\{Q_\omega(t): -\infty < t < +\infty\}$ be a temporally homogeneous Lévy process, and consider the following expression - the Schrödinger operator:

$$H(Q_\omega) = -d^2/dt^2 + Q'_\omega(t) , \tag{1}$$

where $Q'_\omega(t)$ is the "derivative" of the sample function. In the typical case of the Wiener process, one is going to consider the Schrödinger operator with "white noise potential". Since sample functions of a Lévy process are not differentiable, we must give a suitable definition to the formal expression (1) in order to realize it as an operator in the Hilbert space $L^2(R) = L^2(R;dx)$. When t is restricted to a finite interval I , a formulation was given by Fukushima and Nakao [1] in terms of symmetric forms in $L^2(I)$, but their method does not work if I extends to infinity. In §1 of this note, we give a quite elementary definition to (1). Next, we will consider a class of stochastic processes $\{Q_\omega(t)\}$ with stationary ergodic increments, which includes Lévy processes as special cases, and prove in §2 that $H(Q_\omega)$ is self-adjoint with probability one. In §3, we state Kotani's support theorem for the spectrum of $H(Q_\omega)$, and using this, we will determine the exact location of the spectrum of $H(Q_\omega)$ in the case of Lévy process.

§1. The operator and its Green's function. Let $Q(t)$, $-\infty < t < +\infty$, be a real valued function which is right-continuous and has left-hand limits. Let $\underline{\underline{C}}(Q)$ be the totality of $u(t)$ which is absolutely continuous, differentiable from the right, and such that there exists a $v \in L^1_{loc}(R)$ satisfying the following equation:

$$u^+(t) - u^+(s) = Q(t)u(t) - Q(s)u(s) - \int_s^t \{Q(y)u^+(y) + v(y)\}dy , \tag{2}$$

where u^+ denotes the right-derivative of u . This v will be denoted by $H(Q)u$ for each $u \in \underline{\underline{C}}(Q)$. Set further

$$\underline{\underline{D}}(Q) = \{u \in \underline{\underline{C}}(Q) \cap L^2(R) \mid H(Q)u \in L^2(R)\} .$$

If Q is smooth, then (2) is equivalent to $-u'' + Q'u = v$, and hence our $H(Q)$ is a generalization of the usual Schrödinger operator. The integral equation:

$$\begin{aligned} &u(t) = \alpha + \int_s^t u^+(y)dy , \\ &u^+(t) = \beta + \alpha\{Q(t) - Q(s) - \lambda(t-s)\} + \int_s^t \{Q(t)-Q(y) - \lambda(t-y)\}u^+(y)dy , \end{aligned} \tag{3}$$

which can be solved uniquely, will be written in short as

$$H(Q)u = \lambda u \ , \ u(s) = \alpha \ , \ u^+(s) = \beta .$$

Let us remark that $H(Q)$ can be transformed into a usual differential operator of second order. To this end, first note that we have the following Green's formula: for any $u_1, u_2 \in \underline{\underline{C}}(Q)$,

(4) $\int_a^b \{(H(Q)u_1(t))u_2(t) - u_1(t)(H(Q)u_2(t))\}dt = [u_1,u_2](t)|_a^b$,

where $[u_1,u_2](t) = u_1(t)u_2^+(t) - u_1^+(t)u_2(t)$.

Now let $V(t)$ be the solution of $H(Q)u = iu$ with $V(0) = 1$ and $V^+(0) = 0$. Letting $u_1 = V$ and $u_2 = \overline{V}$ in (4), we see that $V(t)$ has no zero. Set

$$\tau(t) = \int_0^t |V(y)|^{-2}dy, \quad l_- = \tau(-\infty), \quad l_+ = \tau(+\infty),$$

and let $t(\tau)$, $l_- < \tau < l_+$, be the inverse function of $\tau(\cdot)$. Let further

$$m(\tau) = \int_0^\tau |V(t(\sigma))|^4 d\sigma,$$

for $\tau \in (l_-,l_+)$, and $(Tu)(\tau) = (u/V)(t(\tau))$ for $u \in L^2(R)$. This T is a unitary transformation from $L^2(R)$ onto $L^2((l_-,l_+);dm(\tau))$. Some elementary calculations show the following

Lemma 1. Let L be the differential operator

$$L = -(m'(\tau))^{-1}d^2/dt^2 + 2im(\tau)/m'(\tau)\, d/d\tau + i$$

with domain $\underline{\underline{D}}(L) = \{f \in L^2((l_-,l_+);dm(\tau)) \mid f$ and f' are absolutely continuous, and $Lf \in L^2((l_-,l_+);dm(\tau))\}$.

Then we have $TH(Q) = LT$. In other words, $H(Q)$ and L are unitarily equivalent.

From this lemma, it is clear that $\underline{\underline{D}}(Q)$ is dense in $L^2(R)$. Moreover, Weyl's classification of differential operators into limit point and limit circle types, which is valid for L, is valid also for $H(Q)$. $H(Q)$ is of limit point type at $\pm\infty$ if and only if $H(Q)$ with domain $\underline{\underline{D}}(Q)$ is self-adjoint, and in such a case, the resolvent operator $G_\lambda(Q) = (H(Q)-\lambda)^{-1}$, $\lambda \in C\backslash R$, has an integral kernel $g_\lambda(x,y;s,Q)$ which can be constructed as follows: let $\phi = \phi_\lambda(x;s,Q)$ and $\psi = \psi_\lambda(x;s,Q)$ be the solutions of (3) with $(\alpha,\beta) = (1,0)$ and $(0,1)$ respectively. Then the limits

$$h_\pm = h_\pm(\lambda;s,Q) = \mp \lim_{x \to \pm\infty} \phi(x)/\psi(x)$$

exist and $g_\lambda(x,y)$ is given by

$$g_\lambda(x,y) = -(h_+ + h_-)^{-1}f_+(x\vee y)f_-(x\wedge y),$$

where $\qquad f_\pm(x) = f_\pm(x;\lambda,s,Q) = \phi \pm h_\pm\psi$.

Now let $D(R:R)$ be the totality of real functions which are right-continuous and have left-hand limits. $D(R:R)$ will be endowed with Skorohod topology and $\underline{\underline{B}}$ will denote the associated Borel field.

For fixed $\alpha,\beta,\lambda \in C$ and $s \in R$, let $u(t;Q)$ be the solution of the equation (3).

Lemma 2. Suppose $Q_n \to Q_\infty$ in the Skorohod topology and that Q_∞ is continuous at $t = s$. Then $u^+(\cdot;Q_n) \to u^+(\cdot;Q_\infty)$ in the Skorohod topology and consequently $u(\cdot;Q_n) \to u(\cdot;Q_\infty)$ uniformly on each compact interval.

Proof. From the assumption, there exists a sequence of strictly increasing continuous functions $\{\mu_n\}$ such that $\mu_n(\pm\infty) = \pm\infty$, $\mu_n(s) = s$, and that $Q_n(\mu_n(\cdot)) \to Q_\infty$ and $\mu_n \to \mu_\infty$, uniformly on compact intervals, μ_∞ being the identity. For each Q_n , let $w_0^+(t;Q_n) = \beta + \alpha\{Q_n(t) - Q_n(s) - \lambda(t-s)\}$ and define

$$w_{j+1}^+(t;Q_n) = \int_s^t\{Q_n(t) - Q_n(y) - \lambda(t-y)\}w_j^+(y;Q_n)dy \ , \ j = 1,2,\ldots\ldots$$

inductively. Then for each finite interval I ,

$$\sup_{t\in I} |w_j^+(\mu_n(t);Q_n)| \le KM^j/j!$$

for some constants K and M which depend only on I , and we have

$$u^+(\mu_n(t);Q_n) = \sum_{j=0}^{\infty} w_j^+(\mu_n(t);Q_n) \ , \ n = 1,2,\ldots\ldots,\infty \ .$$

On the other hand, it is not difficult to show inductively that $w_j^+(\mu_n(t),Q_n) \to w_j^+(t;Q_\infty)$ uniformly on I for each j . Therefore $u^+(\mu_n(t);Q_n) \to u_n^+(t;Q\ )$ uniformly on I . □

Let S be the totality of $Q \in D(R:R)$ such that $H(Q)$ is self-adjoint or equivalently, that $H(Q)$ is of limit point type at $\pm\infty$.

Lemma 3. For any $u \in L^2(R)$ and $\lambda \in C\backslash R$, the correspondence

$$S \ni Q \mapsto (G_\lambda(Q)u,u)$$

is continuous.

Proof. Suppose $Q_n \in S$, $n = 1,2,\ldots,\infty$, and $Q_n \to Q_\infty$. Take a continuity point s of Q_∞ and use this s as that s in the construction of Green's function. Then in view of Lemma 2, it suffices to prove that $h_+(\lambda;s,Q_n) \to h_+(\lambda;s,Q_\infty)$, and for this purpose, it is sufficient to prove that $-\phi_\lambda(x;s,Q)/\psi_\lambda(x;s,Q) \to h_+(\lambda;s,Q)$ uniformly with respect to Q in the compact set $K = \{Q_n;\ n = 1,2,\ldots,\infty\}$, as $x \to \infty$. But the value of $-\phi(x)/\psi(x)$ is on a circle $C_\lambda(x;Q)$ which shrinks to the point h_+ , so that $|-\phi(x)/\psi(x) - h_+|$ is bounded by the diameter $r_\lambda^+(x;Q)$ of this circle, which is given by

$$r_\lambda^+(x;Q) = \{2|\mathrm{Im}\lambda|\int_0^x|\psi_\lambda(y)|^2dy\}^{-1} \ .$$

From Lemma 2, $K \ni Q \mapsto r_\lambda^+(x;Q)$ is continuous and $r_\lambda^+(x;Q) \downarrow 0$ monotonically as $x \to \infty$, hence uniformly on K by Dini's theorem. □

§2. Self-adjointness of $H(Q)$. In view of Green's formula (4), for each $u_1,u_2 \in \underline{\underline{D}}(Q)$, $[u_1,u_2](t)$ has a limit as $t \to +\infty$ and $t \to -\infty$ respectively. Let

$$\underline{\underline{D}}_s(Q) = \{u \in \underline{\underline{D}}(Q) \mid [u,v](t)|_{-\infty}^{+\infty} = 0 \text{ for all } v \in \underline{\underline{D}}(Q)\} \ ,$$

and let $H_s(Q)$ be the restriction of $H(Q)$ to $\underline{\underline{D}}_s(Q)$. $H_s(Q)$ is also densely defined and is symmetric. Moreover, as was done by Stone ([5], Chapter X, §3), it can be shown that $H(Q)^* = H_s(Q)$ and $H_s(Q)^* = H(Q)$. In particular, $H_s(Q)$ is a closed operator. Therefore $H_s(Q)$ is self-adjoint if and only if its defficiency indices vanish, i.e.

(5) $$d_{\pm}(Q) = \mathrm{diam}[\mathrm{Ran}(H_s(Q) \pm i)]^{\perp} = 0 ,$$

and in this case, $H(Q) = H_s(Q)$ is also self-adjoint.

Now let $(\Omega, \underline{\underline{F}}, P)$ be a probability space and $\{T_t : t \in R\}$ be an ergodic flow on Ω. Set $X = \{Q \in D(R:R) \mid Q(0) = 0\}$ and let Q_ω be an X-valued random variable. Assume that Q_ω is stationary in the sense that

(6) $$Q_{T_t\omega}(\cdot) = Q_\omega(\cdot + t) - Q_\omega(t) , \text{ for all } t \in R \text{ and } \omega \in \Omega .$$

Let further $P_\omega^{\pm}$ be the orthogonal projection onto $[\mathrm{Ran}(H_s(Q_\omega) \pm i)]^{\perp}$. Then from (6), it is clear that $P_{T_t\omega}^{\pm} = U_{-t}P_\omega^{\pm}U_t$, where U_t is the unitary operator defined by $U_t f(\cdot) = f(\cdot + t)$ for $f \in L^2(R)$. Suppose that we have shown the measurability of $\omega \mapsto (P_\omega^{\pm}u, v)$ for each $u, v \in L^2(R)$. Then from the general theory (see [2] Corollary2), we have $d_{\pm}(\omega) = 0$ with probability one.

In order to prove the weak measurability of $P_\omega^{\pm}$, we proceed as follows: the assertion is equivalent to the existence of a sequence of $L^2(R)$-valued random variables $\{\phi_n^{\pm}(\omega)\}_{n \geq 1}$ such that for each ω, $\{\phi_n^{\pm}(\omega)\}_n$ is a total sequence in $\overline{\mathrm{Ran}(H_s(Q_\omega) \pm i)}$. We will omit the detail of the construction of this sequence, but just remark that it can be done using Lemmas 1 and 2.

Thus we have shown

Theorem 1. For a stochastic process $\{Q_\omega(t)\}$ formulated as above, $H(Q_\omega)$ with domain $\underline{\underline{D}}(Q_\omega)$ is self-adjoint with probability one. In particular, it is so when $\{Q_\omega(t)\}$ is a temporally homogeneous Lévy process.

§3. Determination of the spectrum. Set $\Omega' = \{\omega \in \Omega \mid Q_\omega \in S \cap X\}$. Then Ω' is measurable and $P(\Omega') = 1$ from Theorem 1. For each $\omega \in \Omega'$, let $\{E_\omega(\lambda)\}$ be the resolution of the identity of the self-adjoint operator $H(Q_\omega)$. From Lemma 3, $\omega \mapsto E_\omega(\lambda)$ is weakly measurable on Ω' for all $\lambda \in R$. Thus we have a random self-adjoint operator on the new probability space Ω'. By the general theory ([2],[4]), there exists a closed set $\Sigma = \Sigma(P)$ such that for P-almost all $\omega \in \Omega'$, $\Sigma_\omega = \Sigma(P)$, Σ_ω being the spectrum of $H(Q_\omega)$.

Now let Supp(P) be the topological support of the image measure of P induced on $X \cap S$:

$$\mathrm{Supp}(P) = \{Q \in X \cap S \mid P(Q_\omega \in U) > 0 \text{ for every neighbourhood of } Q\} .$$

Of course, the neighbourhood is the one in the sense of Skorohod topology.

Now consider two such probability measures P_1 and P_2. Lemma 3 and the argument of Kotani ([3]) give the following support theorem.

Theorem 2. If $\mathrm{Supp}(P_1) \subset \mathrm{Supp}(P_2)$, then $\Sigma(P_1) \subset \Sigma(P_2)$.

Finally, let $\{Q_\omega(t) : -\infty < t < +\infty\}$ be a temporally homogeneous Lévy process with Lévy's canonical form:

(7) $$Q_\omega(t) = bt + B_\omega(t) + \lim_{n\to\infty} \int_{|u|>1/n} [uN_\omega((0,t]\times du) - ta(u)\nu(du)] ,$$

where b is a real constant, $\{B_\omega(t): -\infty < t < +\infty\}$ is a Wiener process with $B_\omega(t)=0$, $N_\omega(dt,du)$ is a Poisson random measure on $(-\infty,\infty)\times[R\setminus\{0\}]$ with intensity measure $dt\nu(du)$, and $a(u) = (u\wedge 1)\vee(-1)$. (More precisely, (7) is valid for $t \geq 0$, and for $t < 0$, $N_\omega((0,t]\times du)$ sould be replaced by $-N_\omega((t,o]\times du)$.)

Theorem 3. (i) If $B_\omega = 0$, $\nu((-\infty,0)) = 0$ and $\int_0^1 u\nu(du) < +\infty$, then $\Sigma(P) = [c,\infty)$, where c is given by

$$c = b - \int_0^\infty a(u)\nu(du) .$$

(ii) In all the other cases, $\Sigma(P) = (-\infty,\infty)$.

Proof of (i). The conditions imply that the sample functions of $Q_\omega(t)$ are of bounded variation on every compact interval and have positive jumps only. In this case, it is easy to see that $H(Q_\omega) - c$ is a non-negative operator. Hence $\Sigma(P) \subset [c,\infty)$.
On the other hand, the function ct belongs to $\mathrm{Supp}(P)$, hence from Theorem 2, $\Sigma(P)$ contains the spectrum of $H(ct) = -d^2/dt^2 + c$, namely $[c,\infty)$.

For the proof of (ii), we prepare the following lemma.

Lemma 4. Let $X_\omega(t)$ be a Lévy process defined by

(8) $$X_\omega(t) = \lim_{n\to\infty} \int_{|u|>1/n} [uN_\omega((0,t]\times du) - ta(u)\nu(du)] ,$$

and suppose that $0 \in \mathrm{Supp}(\nu)$. Then for any $S < T$ and $\varepsilon > 0$,

$$P(\sup_{S<t<T} |X_\omega(t)| < \varepsilon) > 0 .$$

Proof. It is sufficient to prove the assertion in the case of $\nu((-\infty,0)) = 0$. For each $\delta > 0$, define $X_\omega^\delta(t)$ by

$$X_\omega^\delta(t) = -c_\delta t + \int_\delta^\infty uN_\omega((0,t] du) ,$$

where $c_\delta = \int_\delta^\infty a(u)\nu(du)$. From the assumption, we can choose a $\eta \in (0,\varepsilon)\cap\mathrm{Supp}(\nu)$. For each $\delta \in (0,\eta)$, define $x_\delta(t) = \eta[(t+\tau)/2\tau] - c_\delta t$, $\tau = \eta/2c_\delta$. Then,

$$\sup_{-\infty<t<+\infty} |x_\delta(t)| = \eta/2 < \varepsilon/2 .$$

Now, a sample function $X_\omega^\delta(t)$ which is sufficiently near to $x_\delta(\cdot)$ clearly satisfies

$$\sup_{S<t<T} |X_\omega^\delta(t)| < \varepsilon/2 ,$$

and since $\eta \in \mathrm{Supp}(\nu)$, the probability of such sample functions is positive. On the other hand, since $\delta \in (0,\eta)$ is arbitrary, we may assume that it is sufficiently small to assure

$$P(\sup_{S<t<T} |X_\omega(t) - X_\omega^\delta(t)| < \varepsilon/2) > 0 . \qquad \square$$

Returning to the proof of (ii), we devide the argument into three cases:
(ii-1) Suppose $B_\omega \neq 0$. If $0 \in \mathrm{Supp}(\nu)$, set $X_\omega(t) - bt - B_\omega(t)$, and if $0 \notin \mathrm{Supp}(\nu)$, set

$$X_\omega(t) = \int_{-\infty}^{\infty} uN_\omega((0,t]\times du) = Q_\omega(t) - ct-B_\omega(t)$$

where $$c = b - \int_{-\infty}^{\infty} a(u)\nu(du) .$$

In any case, $X_\omega(t)$ remains in an arbitrary neighbourhood of 0 for an arbitrarily long period with positive probability. On the other hand, $Q_\omega(t) - X_\omega(t)$ is a Wiener process with constant drift, so that for any fixed $\gamma \in R$, it remains with positive probability in an arbitrary neighbourhood of the function γt for an arbitrarily long period. From these considerations, we see that the function γt belongs to Supp(P) whatever $\gamma \in R$ is. Hence $\Sigma(P) \supset [\gamma,\infty)$ for all γ.

(ii-2) Next suppose that $B_\omega = 0$, $\nu((-\infty,0)) = 0$, but $\int_0^1 u\nu(du) = +\infty$. If we rewrite (7) as

$$Q_\omega(t) = c_\delta t + \int_\delta^\infty uN_\omega((0,t]\times du) + Q_\omega^\delta(t) ,$$

with

$$c_\delta = b - \int_\delta^\infty a(u)\nu(du) ,$$

then $Q_\omega^\delta(t)$ satisfies the condition of Lemma 4 for every $\delta > 0$. Therefore, by a reasoning similar to the preceeding one, we see that the function $c_\delta t$ belongs to Supp(P) for every $\delta > 0$, and hence $\Sigma(P) \supset [c_\delta,\infty)$. But from the assumption, $c_\delta \downarrow -\infty$ as $\delta \downarrow 0$, so that we have $\Sigma(P) = (-\infty,\infty)$.

(ii-3) Finally suppose that $B_\omega = 0$ and $\nu((-\infty,0)) > 0$. Choose and fix an $\alpha \in (-\infty,0)\cap\mathrm{Supp}(\nu)$, and set $\mu = b - a(u)\nu(\{\alpha\})$ if $0 \in \mathrm{Supp}(\nu)$ and $= b - \int_{-\infty}^{\infty} a(u)\nu(du)$ if $0 \notin \mathrm{Supp}(\nu)$. Let further $q_\beta(\cdot)$ be defined by $q_\beta(t) = \mu t + \alpha[t/\beta]$ with $\beta > 0$. Then, as before, one shows that $q_\beta \in \mathrm{Supp}(P)$ for each $\beta > 0$. Note that q_β is periodic in the sense that $q_\beta(t+\beta) = q_\beta(t) + q_\beta(\beta)$, and that a periodic potential can be viewed as a process with stationary ergodic increments. Hence $\Sigma(P)\supset\cup_{\beta>0}\Sigma(q_\beta)$ by Theorem 2. On the other hand, one shows as in the theory of Hill's equation that $\Sigma(q_\beta) = \{ \lambda \in R \mid |\Delta(\lambda,\beta)| \le 2\}$. Here $\Delta(\lambda,\beta) = \phi(\beta) + \psi^+(\beta)$ and ϕ and ψ are solutions of $H(q_\beta)u = \lambda u$ with $\phi(0) = \psi^+(0) = 1$, $\phi^+(0) = \psi(0) = 0$. By an elementary calculation we obtain $\Delta(\lambda,\beta) = 2\cos(\beta\sqrt{\lambda}) + (\alpha/\sqrt{\lambda})\sin(\beta\sqrt{\lambda})$, where $\sqrt{\lambda} = i\sqrt{|\lambda|}$ if $\lambda < 0$. Now for each fixed $\lambda \in R$, $\Delta(\lambda,0) = 2$ and $(\partial/\partial\beta\Delta(\lambda,\beta))|_{\beta=0} = \alpha < 0$, hence $|\Delta(\lambda,\beta)| < 2$ for sufficiently small $\beta > 0$. This means that $(-\infty,\infty) \subset \cup\Sigma(q_\beta) \subset \Sigma(P)$. □

Acknowledgement. The author warmly thanks Professor Y. Kasahara for his valuable comments on the proofs.

References

[1] Fukushima, M. and Nakao, S.: On spectra of the Schrödinger operator with a white Gaussian potential, Z. Wahrsch. verw. Gebiete 37 (1977), 267-274.

[2] Kirsch, W. and Martinelli, F.: On the ergodic properties of the spectrum of general random operators, J. Reine Angew. Math. 334 (1982), 141-156.

[3] Kotani, S.: Support theorem for random Schrödinger operators, Comm. Math. Phys. 97 (1985), 443-452.

[4] Pastur, L. A.: Spectral properties of disordered systems in the one-body approximations, Comm. Math. Phys. 75 (1980), 179-196.

[5] Stone, M. H.: Linear transformations in Hilbert space and their applications to analysis, Amer. Math. Soc. Colloq. Publ. vol. XV, New York, (1932).

Institute of Mathematics
University of Tsukuba
Sakura-mura, Niihari-gun
Ibaraki, 305
Japan

AN EVOLUTION OPERATOR OF THE FEYNMAN-KAC TYPE

Itaru Mitoma

1. Introduction and Results

The Cauchy problem for a stochastic partial differential equation:

(1.1) $$dX(t) = dW(t) + L^*(t)X(t)dt$$

arises from a fluctuation problem for interacting diffusion particles,[1],[3],[4],[10]. Here $W(t)$ is a distribution valued Brownian motion and the operator $L^*(t)$ is the adjoint operator of a perturbed diffusion operator $L(t)$:

$$L(t) = \frac{1}{2}\alpha(t,x)^2D^2 + \beta(t,x)D + V(t,x)\cdot + J(t),$$

where $D = \frac{d}{dx}$ and $J(t)$ is a perturbation by the interaction.

Unfortunately the image of C_0^∞ under $J(t)$ is outside of C_0^∞, where C_0^∞ is the space of C^∞(infinitely differentiable)-functions with compact supports. In several cases [1],[9],[10], $(J(t)\phi)(x)$ is a polynomial of x for $\phi \in C_0^\infty$.

Therefore we need to consider the equation (1.1) on a suitable distribution space Φ^* which is the dual space of a weighted Schwartz space $\Phi = \{\phi(x) = h(x)\varphi(x);\ \varphi \in \mathcal{S}\}$. Here $\mathcal{S}$ denotes the Schwartz space of rapidly decreasing C^∞-functions, $h(x) = 1/g(x)$ and $g(x) = \int_R e^{-|y|}\rho(x-y)dy$, where $\rho(x)$ is the Friedrichs mollifier such that $\mathrm{Supp}[\rho(x)] \subset [-1,1]$. A nuclear Fréchet space Φ is metrized by the countable semi-norms:

$$\|\phi\|_n = \sup_{\substack{x\in R\\ 0\leq k\leq n}} (1+x^2)^n |D^k(g(x)\phi(x))|, \quad n=1,2,\cdots.$$

Suppose that $J(t)$ satisfies the condition: For any $T > 0$ there exists an integer $n_0 \geq 0$ such that for any integer $n \geq 0$,

$$\sup_{0 \leq t \leq T} \sup_{\substack{\|\phi\|_{n_0} \leq 1 \\ \phi \in \Phi}} \|J(t)\phi\|_n < \infty .$$

Define $A(t) = L(t) - J(t)$. Then if $A(t)$ generates the Kolmogorov evolution operator $U(t,s)$ such that

(1) for any $\phi \in \Phi$, $U(t,s)\phi$ is continuous from $\{(t,s); 0 \leq s \leq t\}$ into Φ,

(2) $U(t,t) = U(s,s) =$ identity operator,

(3) $\frac{d}{dt}U(t,s)\phi = U(t,s)A(t)\phi$ in Φ,

(4) $\frac{d}{ds}U(t,s)\phi = -A(s)U(t,s)\phi$ in Φ,

we know that $L(t)$ generates the Kolmogorov evolution operator $T(t,s)$ form Φ into itself by solving the integral equation [6]:

$$T(t,s)\phi = U(t,s)\phi + \int_s^t U(\tau,s)J(\tau)T(t,\tau)\phi d\tau .$$

By the nuclearity of the space Φ, the dual operator $T^*(t,s)$ of $T(t,s)$ is the usual evolution operator from Φ^* into itself generated by $L^*(t)$, so that the equation (1.1) has a unique solution

$$X(t) = T^*(t,0)X(0) + \int_0^t T^*(t,s)dW(s).$$

Since $T^*(t,s)$ is non-random, the above Itô integral is well defined [5].

We call the operator $U(t,s)$ the Feynman-Kac evolution operator. Inspired by [1],[9], we will consider the case where

$$\alpha(t,x) = \bar{\alpha}(t,x) + \sum_{k=1}^{p} \alpha_k(t)x^k,$$

$$\beta(t,x) = \bar{\beta}(t,x) + \sum_{k=1}^{2p+1} \beta_k(t)x^k,$$

$$V(t,x) = \bar{v}(t,x) + v(t)x.$$

Here $\bar{\alpha}(t,x)$, (resp. $\bar{\beta}(t,x)$, $\bar{v}(t,x)$), is uniformly bounded, has bounded partial derivatives of all orders with respect to x and, for any integer $n \geq 0$,

$$\lim_{t\to s}\sup_{x\in R} |D^n(\bar{\alpha}(t,x)-\bar{\alpha}(s,x))| = 0, \text{ (resp. } \bar{\beta}(t,x),\bar{v}(t,x)).$$

Further $p \geq 0$ is an integer and $\{\alpha_k(t)\}$, $\{\beta_k(t)\}$, $v(t)$ are continuous in t such that $\beta_{2p+1}(t) < 0$.

THEOREM. Suppose that $p \geq 1$ in the above case. Then $A(t)$ generates the Feynman-Kac evolution operator from Φ into itself.

If $v(t) = 0$, we remark that the conclusion of the above theorem remained valid in the case where $p = -1$ or 0 by the estimations in [8], where we use the usual convention such that $\sum_{k=1}^{p} = 0$ if $p = -1,0$ and $\sum_{k=1}^{2p+1} = 0$ if $p = -1$. But the following is an exceptional one.

COROLLARY. Suppose that $p = 0$ and that $v(t) = v$ and $b_1(t) = b < 0$ are constants such that $|v| \leq |b|$. Then the conclusion of THEOREM holds.

2. Proof of Results

We will first prove THEOREM via stochastic method inspired by Ustunel [10]. Let $\eta_{s,t}(x)$ be a stochastic flow [6] such that

$$\eta_{s,t}(x) = x + \int_s^t \alpha(r,\eta_{s,r}(x))dB(r) + \int_s^t \beta(r,\eta_{s,r}(x))dr,$$

where $B(t)$ is a 1-dimensional Brownian motion.

Since $\beta_{2p+1}(t) < 0$, we have the following integrabilities and regularities for $\eta_{s,t}(x)$ and the derivatives [9], so that $\eta_{s,t}(x)$ has no explosions.

Lemma 1. For any $\varepsilon > 0$, $T > 0$ and $M > 0$,

$$\sup_{|x|\leq M}\ \sup_{0\leq s\leq t\leq T} E[\exp(\varepsilon|\eta_{s,t}(x)|)] < \infty.$$

Lemma 2. For any integers $i \geq 1$, $j \geq 1$ and any $T > 0$,

$$\sup_{0\leq s\leq t\leq T} E[|D^i\eta_{s,t}(x)|^j] \leq C_1(T)(1+|x|)^{j(i-1)((2p-1)\vee 0)}.$$

Lemma 3. For any $T > 0$, $M > 0$, $0 \leq s \leq t \leq T$, $0 \leq s' \leq t' \leq T$ and any integers $n \geq 1$ and $m \geq 0$,

$$\sup_{|x|\leq M} E[|\eta_{s,t}(x)-\eta_{s',t'}(x)|^n] \leq C_2(n,T,M)(|t-t'|^{n/2}+|s-s'|^{n/2}),$$

$$\sup_{|x|\leq M} E[|\eta_{s,t}(x)-x|^n] \leq C_3(n,T,M)|t-s|^{n/2},$$

$$\sup_{|x|\leq M} E[|D^m\eta_{s,t}(x)-D^m\eta_{s',t'}(x)|^n] \leq C_4(n,m,T,M)(|t-t'|^{n/2}+|s-s'|^{n/2}),$$

$$\sup_{|x|\leq M} E[|D^m(\eta_{s,t}(x)-x)|^n] \leq C_5(n,m,T,M)|t-s|^{n/2}.$$

Here and in the sequel, we denote positive constants by C_i or by $C_i(\tau_1, \tau_2,\cdots)$, $i=1,2,3,\cdots$, in the case they depend on the parameters $\tau_1,\tau_2,\cdots$.

For $\phi \in \Phi$, define

$$(U(t,s)\phi)(x) = E[\phi(\eta_{s,t}(x))\exp(\int_s^t V(r,\eta_{s,r}(x))dr)].$$

Then by the Itô formula and Lemmas 1,2 and 3, we can carry out the analogy of the proof of Theorem 1 (page 73) of Gihman-Skorohod [2]. Hence we get the following pointwise equation called the forward and backward Feynman-Kac formula :

$$\frac{d}{dt}(U(t,s)\phi)(x) = (U(t,s)A(t)\phi)(x),$$

(2.1)

$$\frac{d}{ds}(U(t,s)\phi)(x) = -(A(s)U(t,s)\phi)(x).$$

Suppose that $U(t,s)\phi$ is continuous in (s,t) in Φ. Then by (2.1) and the definition of the n-th semi-norm $\|\cdot\|_n$, we have

$$\left(\int_s^t U(\tau,s)A(\tau)\phi d\tau\right)(x) = (U(t,s)\phi)(x) - \phi(x),$$

$$\left(-\int_s^t A(\tau)U(t,\tau)\phi d\tau\right)(x) = \phi(x) - (U(t,s)\phi)(x),$$

which implies the conclusion of THEOREM.

Now it is sufficient to verify the strong continuity of $U(t,s)\phi$ in (s,t) in the rest of the proof. Noticing the definition of $\|\cdot\|_n$ and Lemma 3, we will be able to complete the proof by showing that for any $T > 0$

(2.2) $$\lim_{M\to\infty} \sup_{0\leq s\leq t\leq T} I(M,s,t) = 0,$$

where $I(M,s,t) = \sup_{\substack{|x|\geq M \\ 0\leq k\leq n}} (1+x^2)^n |D^k(g(x)(U(t,s)\phi)(x))|$.

We may assume $t \neq s \in [0,T]$ and $|x| \geq M \geq 1$. Setting $\xi_{s,t}(x) = \int_s^t V(r,\eta_{s,r}(x))dr$, by the Leibniz formula we get that $I(M,s,t)$ is dominated by a finite sum of terms of the type

(2.3) $$\sup_{|x|\geq M} (1+x^2)^n |D^i g(x) E[h^{(\mu)}(\eta_{s,t}(x))\varphi^{(\nu)}(\eta_{s,t}(x))e^{\xi_{s,t}(x)}$$
$$\times (D\eta_{s,t}(x))^{n_1}(D^2\eta_{s,t}(x))^{n_2}\cdots(D^j\eta_{s,t}(x))^{n_j}(D\xi_{s,t}(x))^{m_1}(D^2\xi_{s,t}(x))^{m_2}$$
$$\cdots(D^k\xi_{s,t}(x))^{m_k}]|,$$

where $h^{(\mu)}(x) = D^\mu h(x)$, $\varphi^{(\nu)}(x) = D^\nu\varphi(x)$, $0 \leq i+j+k \leq n$, $0 \leq \mu$, $\nu \leq j$, $\sum_{\ell=1}^{j} \ell n_\ell = j$ and $\sum_{\ell=1}^{k} \ell m_\ell = k$.

Define $y_{s,t}(x) = x/(1-2p(\int_s^t \beta_{2p+1}(r)dr)x^{2p})^{1/2p}$ and $K_{s,t}(x) = \eta_{s,t}(x) - y_{s,t}(x)$. Then we have the following [9].

<u>Lemma 4</u>. For any $\varepsilon > 0$ and $T > 0$,

$$\sup_{0\leq s\leq t\leq T} E[\exp(\varepsilon|K_{s,t}(x)|)] \leq C_6(T,\varepsilon)(1+x^{2p}).$$

By Lemmas 2 and 4 and the Jensen inequality, we have

<u>Lemma 5</u>. For any integers $i \geq 1$ and $j \geq 1$,

$$\sup_{0\leq s\leq t\leq T} E[|D^i\xi_{s,t}(x)|^j] \leq C_7(T)(1+|x|)^{j(i-1)((2p-1)\vee 0)}.$$

<u>Lemma 6</u>. For any $\varepsilon > 0$ and $T > 0$,

$$\sup_{0\leq s\leq t\leq T} E[\exp(\varepsilon\int_s^t |K_{s,r}(x)|dr)] \leq C_8(T,\varepsilon)(1+x^{2p}).$$

Since

$$|D^i g(x)| \leq C_9 e^{-|x|},$$

$$|h^{(\mu)}(\eta_{s,t}(x))| \leq C_{10} e^{|\eta_{s,t}(x)|},$$

$$\sup_{0\leq r\leq T} |V(r,x)| \leq C_{11}(T)(1+|x|),$$

so by Lemmas 2, 4, 5 and 6, (2.3) is dominated by

$$\text{(2.4)}\quad C_{12}(T)\sup_{|x|\geq M}(1+x^2)^{n(p+1)+p}\exp(-|x|+|y_{s,t}(x)| + C_{11}(T)\int_s^t |y_{s,r}(x)|dr)$$
$$\times E[|\varphi^{(\nu)}(\eta_{s,t}(x))|^2]^{1/2}.$$

Define $\beta_T = \min_{0\leq r\leq T}(-2p\beta_{2p+1}(r))$ and $\lambda(s,t) = -2p\int_s^t \beta_{2p+1}(r)dr$. Then, noticing that $|x| \geq M \geq 1$, we have

$$\int_s^t |y_{s,r}(x)|dr \leq \frac{|x|}{\beta_T x^{2p}}\int_s^t -2p\beta_{2p+1}(r)x^{2p}(1+(\int_s^r -2p\beta_{2p+1}(\tau)d\tau)x^{2p})^{-1/2p}dr$$
$$\leq \frac{\lambda(s,t)|x|}{\beta_T(1+\lambda(s,t)x^{2p})^{1/2p}}.$$

Since

$$E[|\varphi^{(\nu)}(\eta_{s,t}(x))|^2]^{1/2}$$
$$= E[\frac{(1+\eta_{s,t}(x)^2)^{(4n(p+1)+8p+2)p}}{(1+\eta_{s,t}(x)^2)^{(4n(p+1)+8p+2)p}}|\varphi^{(\nu)}(\eta_{s,t}(x))|^2]^{1/2},$$

$$\le \|\phi\|_{(2n(p+1)+4p+1)p} E[(\frac{1}{1+\eta_{s,t}(x)^2})^{(4n(p+1)+8p+2)p}]^{1/2},$$

we know, setting $\gamma_{s,t}(x) = 1-(1+C_{11}(T)\lambda(s,t)/\beta_T)/(1+\lambda(s,t)x^{2p})^{1/2p}$ and $N = 2n(p+1)+4p+1$, that (2.4) is dominated by

$$(2.5)\quad C_{12}(T)\|\phi\|_{Np} \sup_{|x|\ge M} (1+x^2)^{(N-2p-1)/2} e^{-\gamma_{s,t}(x)|x|} E[(\frac{1}{1+\eta_{s,t}(x)^2})^{2Np}]^{1/2}.$$

Setting $H(s,t,x) = 1+\lambda(s,t)x^{2p}+((1+\lambda(s,t)x^{2p})^{1/2p}K_{s,t}(x)+x)^{2p}$ for the simplicity, we get

$$(2.6)\qquad E[(\frac{1}{1+\eta_{s,t}(x)^2})^{2Np}]^{1/2}$$

$$\le C_{13}(N,T)\{E[(\frac{1}{H(s,t,x)})^{2N}]^{1/2} + E[(\frac{\lambda(s,t)x^{2p}}{H(s,t,x)})^{2N}]^{1/2}\}.$$

Let $(\Omega, \mathcal{F}, P)$ be a probability space where the 1-dimensional Brownian motion $B(t)$ is defined. Setting

$$\Lambda = \{\omega\in\Omega;\ \frac{(1+\lambda(s,t)x^{2p})^{1/2p}}{|x|}|K_{s,t}(x)| < \frac{1}{2}\},$$

we have

$$E[(\frac{1}{H(s,t,x)})^{2N}] = (\int_\Lambda + \int_{\Omega\setminus\Lambda})(\frac{1}{H(s,t,x)})^{2N}dP$$

(2.7)

$$\le (\frac{1}{1+x^{2p}/4^p})^{2N} + (\frac{1}{1+\lambda(s,t)x^{2p}})^{2N}P(\Omega\setminus\Lambda).$$

By lemma 4, $\sup_{0\le s\le t\le T} E[|K_{s,t}(x)|^{4NP}] \le C_{14}(N,T)(1+x^{2p})$, and hence, by the Čebyšev inequality, we get

$$P(\Omega\setminus\Lambda) \le C_{15}(N,T)(\frac{1+\lambda(s,t)x^{2p}}{x^{2p}})^{2N}(1+x^{2p}).$$

Hence, combining this with (2.7), we have

$$(2.8)\qquad E[(\frac{1}{H(s,t,x)})^{2N}]^{1/2} \le C_{16}(N,T)\{(\frac{1}{1+x^{2p}})^N + (\frac{1}{x^{2p}})^{(2N-1)/2}\}.$$

Quite similarly we have

$$(2.9) \qquad E[(\frac{\lambda(s,t)x^{2p}}{H(s,t,x)})^{2N}]^{1/2} \le C_{17}(N,T)\lambda(s,t)^N(1+x^{2p}).$$

Define $Z = Z(s,t,x) = (1+\lambda(s,t)x^{2p})^{1/2}$ and $L_T = C_{11}(T)/\beta_T$. Then we get

$$1/\gamma_{s,t}(x) = Z/(Z-(1+L_T\lambda(s,t)))$$

$$= Z(\sum_{k=0}^{2p-1} Z^{2p-(k+1)}(1+L_T\lambda(s,t))^k)/\lambda(s,t)(x^{2p}-\sum_{k=0}^{2p-1}\binom{2p}{k}L_T(L_T\lambda(s,t))^{2p-1-k}).$$

Since $\sup_{0\le s\le t\le T}\lambda(s,t) = C_{18}(T) < \infty$, there exists a real number M_T not depending (s,t) such that

$$\frac{1}{\gamma_{s,t}(x)} \le \frac{2(1+C_{18}(T))}{\lambda(s,t)} \quad \text{if } |x| \ge M_T \text{ and } t,s \in [0,T].$$

Setting $C_{19}(T) = 2(1+C_{18}(T))$, we have for $|x| \ge M_T$,

$$(2.10) \quad (1+x^2)^{n(p+1)+p}e^{-\gamma_{s,t}(x)|x|} \le N!2^{N/2}(C_{19}(T))^N(\frac{1}{\lambda(s,t)})^N\frac{1}{(1+x^2)^{(2p+1)/}}$$

Therefore, combining (2.3),(2.4),(2.5),(2.6),(2.8),(2.9) and (2.10) together, we complete the proof of (2.2).

Next we will prove COROLLARY. Define $y_{s,t}(x) = x\exp(b(t-s))$ and define $\eta_{s,t}(x)$ and $K_{s,t}(x)$ similarly as before.

Since

$$|v|\int_s^t |y_{s,r}(x)|dr \le \frac{|v|}{|b|}\int_s^t -b|x|\exp(b(r-s))dr$$

$$\le |x|(1-\exp(b(t-s))),$$

we get

$$-|x| + |y_{s,t}(x)| + |v|\int_s^t |y_{s,r}(x)|dr \le 0.$$

Hence this, together with (2.4) and the estimations in [8], gives us that, for $0 \le k \le n$,

$$(1+x^2)^n|D^k(g(x)(U(t,s)\phi)(x))| \le C_{20}(n,T)\|\phi\|_{n+1}\frac{1}{1+x^2}, \quad t,s \in [0,T],$$

which implies (2.2). This completes the proof.

REFERENCES

[1] D.A. Dawson: Critical dynamics and fluctuations for a mean-field model of cooperative behavior. J. Statist. Phys. 31(1983), 29-85.

[2] I.I. Gihman and A.V. Skorohod: Stochastic differential equations. Berlin-Heiderberg-New York : Springer 1972.

[3] M. Hitsuda and I. Mitoma: Tightness problem and stochastic evolution equation arising from fluctuation phenomena for interacting diffusions. To appear in J. Multivariate Anal.

[4] R.A. Holley and D.W. Stroock: Generalized Ornstein-Uhlenbeck processes and infinite particle branching Brownian motions. Publ. RIMS, Kyoto Univ. 14(1978), 741-788.

[5] K. Itô: Foundations of stochastic differential equations in infinite dimensional spaces. CBMS-NSF, Regional conference series in applied Mathematics. 1984.

[6] T. Kato: Perturbation theory of linear operators. Berlin-Heiderberg-New York : Springer 1976.

[7] H. Kunita: Stochastic differential equations and stochastic flows of diffeomorphisms. Lecture Notes in Math. 1097. Berlin-Heiderberg-New York : Springer 1984.

[8] I. Mitoma: An ∞-dimensional inhomogeneous Langevin's equation. J. Funct. Anal. 61 (1985), 342-359.

[9] I. Mitoma: Generalized Ornstein-Uhlenbeck process having a characteristic operator with polynomial coefficients. (submitted).

[10] H. Tanaka and M. Hitsuda: Central limit theorem for a simple diffusion model of interacting particles. Hiroshima Math. J. 11, (1981), 415-423.

[11] A.S. Ustunel: Stochastic Feynman-Kac formula. J. D'Analyse Math. 42 (1982), 155-165.

Department of Mathematics
Hokkaido University
Sapporo 060
Japan

A THEOREM ON THE STABILITY OF NONLINEAR FILTERING SYSTEMS

Yoshio Miyahara

§1. Introduction. Suppose that a probability space (Ω,F,P) and an increasing family of σ-fields $F_t \subset F$, $t \geq 0$, are given and that F_t-adapted Wiener processes $Z(t)$ and $W(t)$, k-dimensional and d_2-dimensional respectively, are given. Let $X(t)$ and $Y(t)$ be the solution of the following stochastic differential equations

$$dX(t) = \sigma(t,X(t))dZ(t) + b(t,X(t))dt, \ X(0) = X_0, \tag{1}$$

$$dY(t) = dW(t) + h(t,X(t))dt, \ Y(0) = 0. \tag{2}$$

where $\sigma(t,x)$ is $d_1 \times k$-matrix valued function, and $b(t,x)$ and $h(t,x)$ are d_1 and d_2 vector valued functions respectively.

We study the nonlinear filtering problems for the above systems, where $X(t)$ is the signal process and $Y(t)$ is the observable process. We put the following two assumptions (A1) and (A2) on the equations (1) and (2).

Assumption (A1): (i) The initial value X_0 of the process $X(t)$ is F_0-measurable random variable, and X_0, $\{Z(t)\}$ and $\{W(t)\}$ are independent.
(ii) The functions $\sigma(t,x)$ and $b(t,x)$ are Borel measurable and satisfying

$$\|\sigma(t,x)\| + \|b(t,x)\| \leq c(1+\|x\|) \tag{3}$$

for some constant c, where $\|\sigma\|$ stands for the square root of $\Sigma|\sigma_{ij}|^2$ for $\sigma = (\sigma_{ij})$ and $\|b\|$ stands for the norm of the vector b.
(iii) The function $h(t,x)$ is of C^2-class and of polynomial order, namely it holds that for some constants c and ℓ

$$\|h(t,x)\| \leq c(1+\|x\|)^{\ell}. \tag{4}$$

Assumption (A2): The equations (1) and (2) have a unique solution.

It is well-known that, under the above assumptions, the solution $X(t)$ of the equation (1) satisfies the following inequality

$$E[\sup\{\|X(t)\|^p;\ 0 \leq t \leq T\}] \leq C(T,p)(1+E[\|X(0)\|^p]) \tag{5}$$

for any $T > 0$ and $p \geq 1$, where $C(T,p)$ is some constant depending on

T and p. Since the function h(t,x) is of polynomial order, it holds that for some constant $C_1(T,p)$

$$E[\|h(t,X(t))\|^p] \leq C_1(T,p), \quad 0 \leq t \leq T. \tag{6}$$

We here state the Bayes formula (which is known as the Kallianpur-Striebel formula) in the convenient form for our purpose. For $\underline{\underline{y}} = \{y_s,\ 0 \leq s \leq t\} \in C([0,t],\ R^{d_2})$, set

$$\Lambda_t(\omega,\underline{\underline{y}}) = \exp\{y_t \cdot h_t - \int_0^t y_s \cdot dh_s - \frac{1}{2}\int_0^t \|h_s\|^2 ds\} = \exp\{\lambda_t(\omega,\underline{\underline{y}})\}, \tag{7}$$

where $h_s = h(s,X(s,\omega))$. Then $\Lambda_t(\omega,\underline{\underline{y}})$ is the so-called unnormalized conditional density under the condition $\underline{\underline{Y}}_t \equiv \{Y_s,\ 0 \leq t\} = \underline{\underline{y}}$, namely it holds that for any $F \in C_b(C([0,t],R^{d_1}))$

$$E[F(X(\cdot,\cdot))|F^Y](\omega) = \frac{\int \Lambda_t(\omega',\underline{\underline{Y}}_t(\omega))F(X(\cdot,\omega'))dP(\omega')}{\int \Lambda_t(\omega',\underline{\underline{Y}}_t(\omega))dP(\omega')}, \quad P\text{-a.e.} \tag{8}$$

$$0 < \int \Lambda_t(\omega',\underline{\underline{Y}}_t(\omega))dP(\omega') < \infty, \quad P\text{-a.e.} \tag{9}$$

(See Kallianpur[1, p.282, Theorem 11.3.1], where we should note the fact that $Y(t)\cdot h_t - \int_0^t Y(s)\cdot dh_s = \int_0^t h_s \cdot dY(s)$.)

We next give definitions of the filtering measures. Set

$$Q(d\omega|\underline{\underline{y}}) = \Lambda_t(\omega,\underline{\underline{y}})P(d\omega), \tag{10}$$

$$\Pi(d\omega|\underline{\underline{y}}) = Q(d\omega|\underline{\underline{y}})/Q(\Omega|\underline{\underline{y}}), \quad \text{for } \underline{\underline{y}} \in C([0,t],\ R^{d_2}). \tag{11}$$

The sample paths $X(t,\omega)$, $\omega \in \Omega$, are continuous functions of $t \in [0,T]$. Therefore the mapping $X(\cdot):\Omega \to C([0,t],R^{d_1}) \equiv \Omega_X$ naturally determines measures $\tilde{Q}(\cdot|\underline{\underline{y}})$ and $\tilde{\Pi}(\cdot|\underline{\underline{y}})$ on the space Ω_X such that for any Borel subset B of Ω_X

$$\tilde{Q}(B|\underline{\underline{y}}) = Q(X^{-1}(B)|\underline{\underline{y}}) \tag{12}$$

$$\tilde{\Pi}(B|\underline{\underline{y}}) = \tilde{\Pi}(X^{-1}(B)|\underline{\underline{y}}) = \tilde{Q}(B|\underline{\underline{y}})/\tilde{Q}(\Omega_X|\underline{\underline{y}}). \tag{13}$$

The probability measure $\tilde{\Pi}(\cdot|\underline{\underline{y}})$ is the filtering measure of $X(\cdot)$ defined on the path space Ω_X, and $\tilde{Q}(\cdot|\underline{\underline{y}})$ is called the unnormalized filtering measure of $X(\cdot)$.

The filtering measure $\tilde{\Pi}(\cdot|\underline{\underline{y}})$ defined above is depending on the observed path $\underline{\underline{y}}$ and the coefficients $\sigma(t,x)$, $b(t,x)$ and $h(t,x)$ of the system (1) and (2). The purpose of this article is to investigate the continuity property of $\tilde{\Pi}(\cdot)$ with respect to $\sigma(t,x)$, $b(t,x)$, $h(t,x)$ and $\underline{\underline{y}}$ (this continuity property is called the stability of filtering systems), and to give sufficient conditions for the stability. Our main result is Theorem 1 of §3, and a good point of our result is that Theorem 1 can be applied to systems whose coefficients are not necessarily bounded. Theorem 1 is also an extension of the results of Y. Miyahara[3] in the sense that we can vary the coefficient $\sigma(t,x)$.

§2. Preliminaries and lemmas. Let $\{(X^n(t), Y^n(t)),\ n = 0,1,2,3...\}$ be a sequence of solutions of the following equations

$$dX^n(t) = \sigma^n(t,X^n(t))dZ(t) + b^n(t,X^n(t))dt,\ X^n(0)=X_0, \tag{14}$$

$$dY^n(t) = dW(t) + h^n(t,X^n(t))dt,\ Y^n(0)=0, \tag{15}$$

where we assume that the above systems satisfy the assumptions (A1) and (A2) in §1. Corresponding to the system (14) and (15), the functions $\Lambda^n(\omega|\underline{\underline{y}})$ and $\lambda^n(\omega|\underline{\underline{y}})$ (see (7)), and the measures $Q^n(\cdot|\underline{\underline{y}})$, $\Pi^n(\cdot|\underline{\underline{y}})$, $\tilde{Q}^n(\cdot|\underline{\underline{y}})$ and $\tilde{\Pi}^n(\cdot|\underline{\underline{y}})$ are defined as described in §1. Our problem can be expressed as follows. Suppose that the functions $\sigma^n(t,x)$, $b^n(t,x)$, $h^n(t,x)$ and $\underline{\underline{y}}^n$ converge to $\sigma^0(t,x)$, $b^0(t,x)$, $h^0(t,x)$ and $\underline{\underline{y}}^0$ respectively in some sense (the precise meaning shall be given later). Then, does the filtering measure $\tilde{\Pi}^n(\cdot|\underline{\underline{y}}^n)$ converge to $\tilde{\Pi}^0(\cdot|\underline{\underline{y}}^0)$?

We first investigate weak convergence of the unnormalized filtering measures $\{\tilde{Q}^n\}$. For a function $F \in C_b(C([0,t],R^{d_1}))$, it holds that

$$\begin{aligned}
&|\tilde{Q}^n(F|\underline{\underline{y}}^n) - \tilde{Q}^0(F|\underline{\underline{y}}^0)| \\
&\le \int|F(X^n(\cdot,\omega))\Lambda_t^n(\omega,\underline{\underline{y}}^n) - F(X_t^0(\cdot,\omega))\Lambda_t^0(\omega,\underline{\underline{y}}^0)|P(d\omega) \\
&\le \int|F(X^n(\cdot,\omega)) - F(X^0(\cdot,\omega))||\Lambda_t^0(\omega,\underline{\underline{y}}^0)|P(d\omega) \\
&\quad + \int|F(X^n(\cdot,\omega))||\Lambda_t^n(\omega,\underline{\underline{y}}^n) - \Lambda_t^0(\omega,\underline{\underline{y}}^0)|P(d\omega).
\end{aligned} \tag{16}$$

For the simplicity of notations, we set

$$I_1^n = \int|F(X^n(\cdot,\omega)) - F(X^0(\cdot,\omega))||\Lambda_t^0(\omega,\underline{\underline{y}}^0)|P(d\omega), \tag{17}$$

$$I_2^n = \int|F(X^n(\cdot,\omega))||\Lambda_t^n(\omega,\underline{\underline{y}}^n) - \Lambda_t^0(\omega,\underline{\underline{y}}^0)|P(d\omega) \tag{18}$$

and introduce the following assumption.

Assumption (A3): There are a continuous function $M(\underline{\underline{y}})$ defined on $C([0,t],R^{d_2})$ and a constant $p > 1$ such that $(E[|\Lambda_t^n(\cdot|\underline{\underline{y}})|^p])^{1/p} \le M(\underline{\underline{y}})$, $n = 0, 1,2,...,$ $0 \le t \le T$.

Then we obtain the following lemma.

Lemma 1. Assume that systems $\{(X^n(t),Y^n(t)),\ n=0,1,2,\ldots\}$ satisfy the assumptions (A1), (A2) and (A3). Then it holds that

(i) If $E[\sup_{0\le s\le t} \|X^n(s)-X^0(s)\|^{\gamma}] \to 0$ as $n\to\infty$ for some $\gamma>0$, then $I_1^n \to 0$.

(ii) If $\lambda_t^n(\omega|\underline{\underline{y}}^n) \to \lambda_t^0(\omega|\underline{\underline{y}}^0)$ in $L^q(\Omega,P)$ as $n\to\infty$, where q is the constant such that $\frac{1}{p}+\frac{1}{q}=1$, then $I_2^n \to 0$.

Proof. (i) By the use of Hölder's inequality, we obtain

$$I_1^n \le \|F(X^n(\cdot))-F(X^0(\cdot))\|_q \|\Lambda_t^0(\cdot\ \underline{\underline{y}}^0)\|_p,\quad \frac{1}{p}+\frac{1}{q}=1. \tag{19}$$

Therefore we have only to prove that

$$\|F(X^n(\cdot)) - F(X^0(\cdot))\|_q \to 0 \quad \text{as} \quad n\to\infty. \tag{20}$$

From the assumption of (i) of Lemma 1, it follows that $X^n(\cdot)$ converges to $X^0(\cdot)$ in probability as $C([0,t],R^{d_1})$-valued random variables. Since $F(X)$ is a bounded continuous function, the formula (20) follows directly.
(ii) From (18) it follows that

$$I_2^n \le |||F||| \int |\Lambda_t^n(\omega|\underline{\underline{y}}^n) - \Lambda_t^0(\omega|\underline{\underline{y}}^0)|P(d\omega), \tag{21}$$

where $|||F|||$ stands for the supremum norm. Using the property of exponential functions, we obtain

$$|\Lambda_t^n - \Lambda_t^0| \le (\Lambda_t^n + \Lambda_t^0)|\lambda_t^n - \lambda_t^0|. \tag{22}$$

Therefore, using Hölder's inequality, we obtain

$$\begin{aligned} I_2^n &\le |||F|||\ \|\Lambda_t^n + \Lambda_t^0\|_p \|\lambda_t^n - \lambda_t^0\|_q \\ &\le |||F|||(M(\underline{\underline{y}}^n) + M(\underline{\underline{y}}^0))\|\lambda_t^n - \lambda_t^0\|_q. \end{aligned} \tag{23}$$

By the use of the continuity of $M(\underline{\underline{y}})$ and the assumption of (ii) of Lemma 1, the conclusion $I_2^n \to 0$ follows from (23). (Q.E.D.)

Corollary 1. Suppose that all the assumptions of Lemma 1 are satisfied. Then it holds that $\tilde{Q}^n(\cdot|\underline{\underline{y}}^n) \to \tilde{Q}^0(\cdot|\underline{\underline{y}}^0)$ (weakly).

We next investigate the conditions which will certify the assumption (A3). We introduce the following assumption for a system (σ,b,h).

Assumption (A3'): There are a continuous function $M_1(y)$ and a constant $p > 1$ such that

$$p\|A_{t,s}\|^2 - 2B_{t,s} \leq M_1(\underline{\underline{y}}),\ \text{P-a.e., for } 0 \leq s \leq t \leq T,$$

where $A_{t,s}$ is the k-vector whose i-component is given by $A^i_{t,s} = \sum_j (y_s - y_t)\cdot(\frac{\partial h}{\partial x_i})_s(\sigma_{ji})_s$ and $B_{t,s} = (y_s - y_t)\cdot(Lh)_s + \frac{1}{2}\|h_s\|^2$ (L is the generator of the process X(t)).

By the same methods as we have used in Y. Miyahara[3, p.95], we can prove that if the systems $\{(\sigma^n, b^n, h^n),\ n=0,1,2,\ldots\}$ satisfy the assumption (A3') uniformly in n for a function $M_1(\underline{\underline{y}})$ and a constant $p > 1$, then the assumption (A3) is satisfied.

Now we will give a sufficient condition for the assumption (A3') to be fulfilled.

Lemma 2. Assume that the coefficients of the equations (1) and (2) satisfy the following assumptions.

Assumption (A4): For some $\ell \geq 1$, it holds that

(i) $$0 < \varliminf_{\|x\|\to\infty} [\inf_{0\leq s\leq t} \{\|h(s,x)\|/\|x\|^\ell\}] \leq \varlimsup_{\|x\|\to\infty} [\sup_{0\leq s\leq t} \{\|h(s,x)\|/\|x\|^\ell\}] < \infty$$

(ii) $$\lim_{\|x\|\to\infty} [\sup_{0\leq s\leq t} \{\|\nabla_x h(s,x)\|\ \|\sigma(s,x)\|\}/\|x\|^\ell] = 0$$

(iii) $$\lim_{\|x\|\to\infty} [\sup_{0\leq s\leq t} \{\|\nabla_x h(s,x)\|\ \|b(s,x)\| + \|\nabla_x^2 h(s,x)\|\ \|\sigma(s,x)\|^2 + \|\nabla_s h(s,x)\|\}/\|x\|^{2\ell}] = 0$$

Then the assumption (A3') is satisfied.

Proof. When $t \geq 0$ and $\underline{\underline{y}} = \{y_s,\ 0 \leq s \leq t\}$ are fixed, the following inequality (24) follows from the assumption (A4).

$$\begin{aligned} p\|A_{t,s}\|^2 &- 2B_{t,s} \\ &\leq 2p\,|||\underline{\underline{y}}|||\ \|\nabla_x h(s,x)\|\ \|\sigma(s,x)\| + 4\,|||\underline{\underline{y}}|||\ \|Lh\| - c_1\|x\|^{2\ell} \\ &\leq (2p+4)\,|||\underline{\underline{y}}|||\ \varepsilon(\|x\|)\|x\|^{2\ell} - c_1\|x\|^{2\ell} \\ &= -(c_1 - (2p+4)\,|||\underline{\underline{y}}|||\ \varepsilon(\|x\|))\|x\|^{2\ell}, \end{aligned} \tag{24}$$

where c_1 is a positive constant, $|||\underline{\underline{y}}||| = \sup\{\|y_s\|, 0 \leq s \leq t\}$ and $\varepsilon(\|x\|) \to 0$ as $\|x\| \to \infty$. The right hand side of (24) is bounded from above as a function of x, and therefore it is easy to see the existence of the function $M_1(\underline{\underline{y}})$ of Assumption (A3'). (Q.E.D.)

Corollary 2. If the assumption (A4) of Lemma 2 is satisfied for the systems $\{(\sigma^n, b^n, h^n), n=0,1,2,\ldots\}$ uniformly in n (namely, the constants ℓ, c_1 and c_2 are independent of n and the convergence of the limits in the conditions (ii) and (iii) is uniform in n as well as in s, $0 \leq s \leq t$), then the assumption (A3) is fulfilled.

Proof. By Lemma 2 the assumption (A3') is fulfilled for all systems $\{(\sigma^n, b^n, h^n), n=0,1,2,\ldots\}$ and it is easy to check that (A3') is satisfied uniformly in n. Therefore (A3) is satisfied as we have mentioned before giving Lemma 2. (Q.E.D.)

We next investigate the conditions under which the assumptions of (i) and (ii) of Lemma 1 are satisfied.

Lemma 3. Let $\{(\sigma^n(t,x), b^n(t,x)), n=0,1,2,\ldots\}$ be continuous functions and assume that they are locally Lipschitz continuous in x. If $\sigma^n(s,x) \to \sigma^0(s,x)$ and $b^n(s,x) \to b^0(s,x)$ uniformly in (s,x), $0 \leq s \leq t$, $\|x\| < K$, for any $K > 0$, then it holds that for any $q > 0$

$$E[\sup_{0 \leq s \leq t} \|X_s^n - X_s^0\|^q] \to 0 \quad \text{as } n \to \infty. \tag{25}$$

Proof. The result of Lemma 3 follows easily from Kawabata-Yamada[2, Theorem 3] and Example of Okabe-Shimizu[4, p.458]. (Q.E.D.)

For a vector valued function $f(s,x)$, $0 \leq s \leq t$, $x \in R^{d_1}$, set

$$\|f\|_K = \sup\{\|f(s,x)\|;\ 0 \leq s \leq t,\ \|x\| \leq K\}, \quad \text{for } K > 0. \tag{26}$$

Lemma 4. Let the systems $\{(\sigma^n, b^n, h^n, \underline{\underline{y}}^n), n=0,1,2,\ldots\}$ satisfy the following conditions: (i) Assumptions (A1), (A2) and (A4) are satisfied uniformly for all the above systems.
(ii) The sequence of systems $\{(\sigma^n, b^n, h^n, \underline{\underline{y}}^n), n=1,2,\ldots\}$ converges to $(\sigma^0, b^0, h^0, \underline{\underline{y}}^0)$ in the following sense;

1) $\|\sigma^n - \sigma^0\|_K \to 0$, $\|b^n - b^0\|_K \to 0$ for any $K > 0$.

2) $\|h^n - h^0\|_K \to 0$, $\|\nabla_s h^n - \nabla_s h^0\|_K \to 0$, $\|\nabla_x h^n - \nabla_x h^0\|_K \to 0$, $\|\nabla_x^2 h^n - \nabla_x^2 h^0\|_K \to 0$ for any $K > 0$.

3) $|||\underline{\underline{y}}^n - \underline{\underline{y}}^0||| \to 0$ as $n \to \infty$.

(iii) The functions $\sigma^0(s,x)$, $b^0(s,x)$, $h^0(s,x)$, $\nabla_x h^0(s,x)$ and $\nabla_x^2 h^0(s,x)$ are continuous in (s,x) and locally Lipschitz continuous in x (uniformly in s, $0 \le s \le t$).

Then it holds that for any $q > 0$

$$E[|\lambda_t^n - \lambda_t^0|^q] \to 0 \quad \text{as} \quad n \to \infty. \tag{27}$$

Proof. We first mention that by Lemma 3 it follows from the assumptions of Lemma 4 that for any $q > 0$

$$E[\sup_{0\le s\le t} \|X_s^n - X_s^0\|^q] \to 0 \quad \text{as} \quad n \to \infty. \tag{28}$$

By the definitions of λ_t^n and λ_t^0 (see (7)), we obtain

$$\|\lambda_t^n - \lambda_t^0\|_q \equiv E[|\lambda_t^n - \lambda_t^0|^q]^{1/q}$$

$$\le \|h_t^n \cdot y_t^n - h_t^0 \cdot y_t^0\|_q + \|\int_0^t y_s^n \cdot dh_s^n - \int_0^t y_s^0 \cdot dh_s^0\|_q$$

$$+ \frac{1}{2}\|\int_0^t ((y_s^n)^2 - (y_s^0)^2)ds\|_q. \tag{29}$$

Set

$$J_1^n = \|h_t^n \cdot y_t^n - h_t^0 \cdot y_t^0\|_q, \tag{30}$$

$$J_2^n = \|\int_0^t y_s^n \cdot dh_s^n - \int_0^t y_s^0 \cdot dh_s^0\|_q, \tag{31}$$

$$J_3^n = \|\int_0^t ((y_s^n)^2 - (y_s^0)^2)ds\|_q. \tag{32}$$

We will investigate J_i, $i = 1,2,3$, step by step.

Step 1. For a vector valued function $f(s,x)$, $0 \le s \le t$, $x \in R^{d_1}$, set

$$|||f|||_\beta = \sup\{\|f(s,x)\|/(1+\|x\|^\beta);\ 0 \le s \le t,\ x \in R^{d_1}\},\quad \beta > 0. \tag{33}$$

Then, from the assumptions (A1), (A4) and (ii), it follows that there exists a positive constant β such that when n tends to ∞

$$|||\sigma^n - \sigma^0|||_\beta \to 0,\ |||b^n - b^0|||_\beta \to 0,\ |||h^n - h^0|||_\beta \to 0,$$

$$|||\nabla_s h^n - \nabla_s h^0|||_\beta \to 0,\ |||\nabla_x h^n - \nabla_x h^0|||_\beta \to 0,\ |||\nabla_x^2 h^n - \nabla_x^2 h^0|||_\beta \to 0. \tag{34}$$

Step 2. From (34) we know that, for any $\varepsilon > 0$, there exists a constant N such that if $n \ge N$ then

$$\|\sigma^n(s,x) - \sigma^0(s,x)\| \le \varepsilon(1 + \|x\|^\beta). \tag{35}$$

From (35) we obtain

$$\|\sigma^n(s,\tilde{x}) - \sigma^0(s,x)\|$$
$$\leq \|\sigma^n(s,\tilde{x}) - \sigma^0(s,\tilde{x})\| + \|\sigma^0(s,\tilde{x}) - \sigma^0(s,x)\|$$
$$\leq \varepsilon(1 + \|\tilde{x}\|^\beta) + \|\sigma^0(s,\tilde{x}) - \sigma^0(s,x)\| \tag{36}$$

Therefore we get for $n \geqq N$ and any $\gamma > 0$

$$\|\sigma^n(s,X_s^n(\cdot)) - \sigma^0(s,X_s^0(\cdot))\|_\gamma \equiv E[\|\sigma^n(s,X_s^n(\cdot)) - \sigma^0(s,X_s^0(\cdot))\|^\gamma]^{1/\gamma}$$
$$\leqq \varepsilon\|1 + \|X_s^n\|^\beta\|_\gamma + \|\sigma^0(s,X_s^n) - \sigma^0(s,X_s^0)\|_\gamma. \tag{37}$$

By the Assumption (A1) and the inequality (5), the first term of the right hand side of (37) is bounded, namely

$$\sup_{n\geq 1}[\sup_{0\leqq s\leqq t}\{\|1 + \|X_s^n\|^\beta\|_\gamma\}] < \infty. \tag{38}$$

The second term of (37) is estimated as follows. Using the assumption (A1) and the inequality (5) again, we know that there is a constant K such that

$$E[\|\sigma^n(s,X_s^n) - \sigma^0(s,X_s^0)\|^\gamma;\ \|X_s^n\| > K \text{ or } \|X_s^0\| > K] < \varepsilon \text{ for } n \geqq N. \tag{39}$$

Using the assumption (iii) and the inequality (39), we obtain

$$E[\|\sigma^0(s,X_s^n) - \sigma^0(s,X_s^0)\|^\gamma]$$
$$= E[\|\sigma^0(s,X_s^n) - \sigma^0(s,X_s^0)\|^\gamma;\ \|X_s^n\| > K \text{ or } \|X_s^0\| > K]$$
$$+ E[\|\sigma^0(s,X_s^n) - \sigma^0(s,X_s^0)\|^\gamma;\ \|X_s^n\| \leq K,\ \|X_s^0\| \leq K]$$
$$\leq \varepsilon + (L(K))^\gamma E[\|X_s^n - X_s^0\|^\gamma] \quad \text{for } n \geqq N, \tag{40}$$

where $L(K)$ is the local Lipschitz constant of $\sigma(s,x)$. From the inequalities (37), (38) and (40) and the formula (28), we get the following result that for any $\gamma > 0$

$$\|\sigma^n(s,X_s^n) - \sigma^0(s,X_s^0)\|_\gamma \to 0 \text{ as } n \to \infty \text{ uniformly in S.} \tag{41}$$

<u>Step 3</u>. The discussions in Step 2 can be applied to (b^n,b^0), (h^n,h^0) $(\nabla_s h^n,\nabla_s h^0)$, $(\nabla_x h^n,\nabla_x h^0)$ and $(\nabla_x^2 h^n,\nabla_x^2 h^0)$ instead of (σ^n,σ^0), and we obtain the following results: For any $\gamma > 0$, it holds that

$$\|b^n(s,X_s^n) - b^0(s,X_s^0)\|_\gamma \to 0, \tag{42}$$

$$\|h^n(s,X_s^n) - h^0(s,X_s^0)\|_\gamma \to 0,\ \|\nabla_s h^n(s,X_s^n) - \nabla_s h^0(s,X_s^0)\|_\gamma \to 0,$$

$$\|\nabla_x h(s,X^n) - \nabla_x h^0(s,X^0)\|_\gamma \to 0,\ \|\nabla_x^2 h^n(s,X_s^n) - \nabla_x^2 h^0(s,X_s^0)\|_\gamma \to 0, \quad (43)$$

uniformly in s as $n \to \infty$.

Step 4. We are now in the position to prove that $J_i^n \to 0$ as $n \to \infty$, for $i = 1,2,3$. It is clear that

$$J_1^n \leq \|h_t^n\cdot(y_t^n - y_t^0)\|_q + \|(h_t^n - h_t^0)\cdot y_t^0\|_q$$

$$\leq |||\underline{\underline{y}}^n - \underline{\underline{y}}^0|||\ \|h_t^n\|_q + |||\underline{\underline{y}}^0|||\ \|h_t^n - h_t^0\|_q. \quad (44)$$

By the assumption (ii) 2) and (43), the right hand side of (44) tends to 0 as $n \to \infty$. Therefore we have proved that

$$J_1^n \to 0 \quad \text{as} \quad n \to \infty. \quad (45)$$

Next we will prove that

$$J_2^n \to 0 \quad \text{as} \quad n \to \infty. \quad (46)$$

From the definition of J_2^n we easily obtain

$$J_2^n \leq |||\underline{\underline{y}}^n - \underline{\underline{y}}^0|||\ \left\|\int_0^t \frac{(y_s^n - y_s^0)}{|||\underline{\underline{y}}^n - \underline{\underline{y}}^0|||}\cdot dh_s^0\right\|_q + \left\|\int_0^t y_s^n\cdot d(h_s^n - h_s^0)\right\|_q. \quad (47)$$

The first term of the right hand side of (47) converges to 0 by (ii) 3). The second term is calculated by the use of Ito-formula, and it can be easily proved that this term converges to 0 as $n \to \infty$ by (41), (42) and (43). Thus we have proved (46). Finally the fact

$$J_3^n \to 0 \quad \text{as} \quad n \to \infty \quad (48)$$

follows easily from the following inequality

$$J_3^n \leq |||\underline{\underline{y}}^n - \underline{\underline{y}}^0|||\ \left\|\int_0^t (y_s^n + y_s^0)ds\right\|_q. \quad (49)$$

Step 5. Combining all of the obtained results (29), (30), (31), (32), (45), (46) and (48), we have the results of Lemma. (Q.E.D.)

§3. Results. In this section we state our main results and give the proofs of them. Let a time interval $[0,T]$ be given and fixed, where $0 < T \leq \infty$, and in the case of $T = \infty$ the interval $[0,T]$ should be replaced by $[0,\infty)$. We introduce a set Ξ_m, $m \geq 1$, of systems (σ,b,h) as follows:

$\Xi_m = \{(\sigma,b,h);$ satisfying the following conditions (i)-(iii)$_m\}$

(i) $\sigma(t,x)$ and $b(t,x)$ are continuous functions of (t,x), and Lipschitz continuous in x (uniformly w.r.t. t).

(ii) $\lim_{\|x\|\to\infty} \{ \sup_{0\le t\le T} [\|\sigma(t,x)\|/\|x\|]\} = 0.$

(iii)$_m$ $h(t,x)$ is continuous and continuously differentiable (once w.r.t. t and twice w.r.t. x), and satisfies the following three conditions:

1) $\underline{\lim}_{\|x\|\to\infty} \{ \inf_{0\le t\le T} [\|h(t,x)\|/\|x\|^m]\} > 0,$

2) $|||h|||_m \equiv \sup_x\{ \sup_{0\le t\le T} [\|h(t,x)\|/(1+\|x\|^m)]\} < \infty,$

3) $|||\nabla_t h|||_{2m-1} < \infty,\ |||\nabla_x h|||_{m-1} < \infty,\ |||\nabla_x^2 h|||_{2m-2} < \infty.$

We next introduce a norm $|||\ |||_m$ in the space Ξ_m by the following formula

$$|||(\sigma,h,b)|||_m = |||\sigma|||_m + |||b|||_m + |||h|||_m + |||\nabla_t h|||_{2m-1} + |||\nabla_x h|||_{m-1} + |||\nabla_x^2 h|||_{2m-2}. \tag{50}$$

Now we are in the position to state our theorem.

Theorem 1. The filtering measure $\tilde{\Pi}(\cdot\,|\underline{y})$, which is given by the formula (13), is a continuous mapping of $(\sigma,b,h)\in\Xi_m$ and $\underline{y}\in C([0,t],R^{d_2})$, into the space of probability measures on $C([0,t],R^{d_1})$, where the topology of the space of probability measures is the weak topology.

Proof. Let $\{(\sigma^n,b^n,h^n)\in\Xi_m,\ n=0,1,2,\ldots\}$ be given and assume that $(\sigma^n,b^n,h^n)\to(\sigma^0,b^0,h^0)$ as $n\to\infty$ in the sense of the norm given by (50). And let $\underline{y}^n$, $n=1,2,\ldots$, be a sequence from $C([0,t],R^{d_2})$ such that $\underline{y}^n\to\underline{y}^0$ in $C([0,t],R^{d_2})$. We denote by (X^n,Y^n), $n=0,1,2,\ldots$, the processes related to the systems (σ^n,b^n,h^n) through the equations (14) and (15). In the sequel we shall prove that we can apply Lemma 1 to the above systems.

Step 1. We can easily verify that the systems $\{(\sigma^n,b^n,h^n),\ n=0,1,2,\ldots\}$ satisfy all the assumptions of Corollary 2 in §2. Therefore the assumptions (A3) is satisfied.

Step 2. It is easy to verify that $\{(\sigma^n,b^n),\ n=1,2,\ldots\}$ satisfy all the assumptions of Lemma 3. So we have obtained

$$E[\sup_{0\le s\le t} \|X_s^n - X_s^0\|^q] \to 0 \quad \text{as} \quad n \to \infty \quad \text{for any } q > 0. \tag{51}$$

Step 3. From the assumptions of Theorem 1, it follows that the systems $\{(\sigma^n, b^n, h^n, \underline{\underline{y}}^n),\ n = 0,1,2,\ldots\}$ satisfy the assumptions of Lemma 4. As a result of Lemma 4, we obtain

$$E[|\lambda_t^n - \lambda_t^0|^q] \to 0 \quad \text{as} \quad n \to \infty \quad \text{for any } q > 0. \tag{52}$$

Step 4. We prove in this step that we can apply Lemma 1 to the systems $\{(\sigma^n, b^n, h^n, \underline{\underline{y}}^n),\ n = 0,1,2,\ldots\}$. By the definition of Ξ_m it is obvious that the systems $\{(\sigma^n, b^n, h^n),\ n = 0,1,2,\ldots\}$ satisfy the assumptions (A1) and (A2). The assumption (A3) is fulfilled as proved in Step 1. The assumptions of (i) and (ii) of Lemma 1 are satisfied by the results of Step 2 and Step 3 respectively. Thus we can now apply Lemma 1 and Corollary 1 to the systems $\{(\sigma^n, b^n, h^n, \underline{\underline{y}}^n),\ n = 0,1,2,\ldots\}$, and consequently we obtain that the corresponding unnormalized filtering measures $\tilde{Q}^n(\cdot|\underline{\underline{y}}^n)$ converges weakly to the unnormalized filtering measure $\tilde{Q}^0(\cdot|\underline{\underline{y}}^0)$.

Step 5. The filtering measure $\tilde{\Pi}(\cdot|\underline{\underline{y}})$ is represented in the following form

$$\tilde{\Pi}(\cdot|\underline{\underline{y}}) = \tilde{Q}(\cdot|\underline{\underline{y}})/\tilde{Q}(\Omega_X|\underline{\underline{y}}). \tag{53}$$

Therefore the result that $\tilde{\Pi}^n(\cdot|\underline{\underline{y}}^n) \to \tilde{\Pi}^0(\cdot|\underline{\underline{y}}^0)$ as $n \to \infty$ follows from the fact that $\tilde{Q}^n \to \tilde{Q}^0$ as $n \to \infty$. We have completed the proof. (Q.E.D.)

Since $\tilde{\Pi}(\cdot|\underline{\underline{y}})$ is a version of the filtering measure on the path space $C([0,t],R^{d_1})$, the marginal distribution $\pi_t(\cdot|\underline{\underline{y}})$ of $\tilde{\Pi}(\cdot|\underline{\underline{y}})$ on $(R^{d_1},\mathcal{B})$, which is given by

$$\pi_t(B|\underline{\underline{y}}) = \tilde{\Pi}(\{\underline{\xi} \in \Omega_X;\ \xi_t \in B\}|\underline{\underline{y}}), \quad \text{for } B \in \mathcal{B}, \tag{54}$$

is a version of the filtering measure of X_t under the observation $\underline{\underline{Y}}_t = \underline{\underline{y}}$. As a corollary of Theorem 1, we obtain the following result for $\pi_t(\cdot)$.

Corollary 3. The filtering measure $\pi_t(\cdot|\underline{\underline{y}})$ is a continuous mapping of $(\sigma,b,h) \in \Xi_m$ and $\underline{\underline{y}} \in C[0,t],R^{d_2})$, into the space of probability meaures on R^{d_1}.

References

[1] G. Kallianpur, Stochastic filtering theory, Springer, 1980.

[2] S. Kawabata and T. Yamada, On some limit theorems for solutions of stochastic differential equations, Lecture Notes in Math. 920, 412-441.

[3] Y. Miyahara, A note on the stability problems of nonlinear filtering systems, OIKONOMIKA (Nagoya City Univ.) 23-1(1986), 93-100.

[4] Y. Okabe and A. Shimizu, On the pathwise uniqueness of solutions of stochastic differential equations, J. Math. Kyoto Univ. 15-2(1975), 455-466.

[5] J. Picard, Robustesse de la solution des problemes de filtrage avec bruit blanc independant, Stochastics 13(1984), 229-245.

Faculty of Economics
Nagoya City University
Mizuho-ku, Nagoya 467
Japan

LARGE DEVIATIONS FOR THE MAXIMUM LIKELIHOOD ESTIMATORS

A. A. Mogulskii

Let $a, a_1, a_2, \dots, a_n$ be i.i.d. random elements in the space $C(T)$ of real continuous functions on the closed bounded subset T of R^k. We consider random fields $A_n(t)=a_1(t)+\dots+a_n(t)$ over T and a random vector $t_n^* \in T$ such that

$$A_n(t_n^*) = \sup(A_n(t): t \in T) .$$

The vector t_n^* is the maximum likelihood estimator for the parameter t in the special case when the fields a_i have the form

$$a_i(t) = \ln f_t(x_i)$$

and are generated by a sample $(x_1, \dots, x_n)$ from a distribution P_t, $t \in T$ with the density $f_t(x)$ with respect to some measure $n(dx)$.

The vector t_n^* may not be unique in the general case. We can define the "upper distribution" for t_n^*

$$P_+(t_n^* \in U) = P(\sup_{t \in U} A_n(t) \geq \sup_{t \in T \setminus U} A_n(t))$$

and the "lower distribution"

$$P_-(t_n^* \in U) = P(\sup_{t \in U} A_n(t) > \sup_{t \in T \setminus U} A_n(t)) .$$

It is obvious that $P_+(t_n^* \in U) \geq P_-(t_n^* \in U)$ and

$$P_+(t_n^* \in U_1 \cup U_2) \leq P_+(t_n^* \in U_1) + P_+(t_n^* \in U_2) ,$$

$$P_-(t_n^* \in U_1 \cup U_2) \geq P_-(t_n^* \in U_1) + P_-(t_n^* \in U_2)$$

for $U_1 \cup U_2 = \phi$.

In this paper we study the so-called crude (logarithmic) asymptotics of the probabilities

$$P_\pm(t_n^* \in U)$$

in the case when these probabilities converge to 0. We introduce in the sequel the function $K(t)$, $t \in T$, such that for a certain class of sets $U \subseteq T$

$$\ln P_{\pm}(t_n^* \in U) \sim -n \inf_{t\in U} K(t) \tag{1}$$

The relation (1) has the same form as the asymptotic formula of large deviations for sums $S_n = x_1 + \cdots + x_n$ of i.i.d. random vectors $x_1, \ldots, x_n$ from R^k [1]:

$$\ln P_{\pm}\left(\frac{S_n}{n} \in U\right) \sim -n \inf_{t\in U} \Lambda(t) , \tag{2}$$

where

$$\Lambda(t) = -\ln \inf_{y\in R^k} M \exp\{(y, x_1) - (y, t)\}$$

is the deviations function. The formula (2) holds also for the sums $A_n = a_1 + \cdots + a_n$ of $C(T)$-valued random vectors a_i. We can write using formally the theorems of [2],

$$\ln P_{\pm}(t_n^* \in U) \sim -n \inf_{a\in G(U)_{\pm}} \Lambda(a) , \tag{3}$$

where

$$G(U)_{\pm} = \{a \in C(T): \sup_{t\in U} a(t) \gtrless \sup_{t\in T\setminus U} a(t)\} .$$

It is obvious that the formula (2) is more convenient than (3).

Let $\mathcal{P}$ denote the class of probability measures m on $(T, \mathcal{B})$, and define for $a \in C(T)$

$$\langle a, m\rangle = \int_T a(t) m(dt) .$$

Introduce the two functions $K(t)$, $K_+(t)$, $t \in T$:

$$K(t) = -\ln \inf_{u>0, m\in\mathcal{P}} M\exp(u\langle a^t, m\rangle) ,$$

$$K_+(t) = -\lim_{s\downarrow 0} \ln \inf_{u>0, m\in\mathcal{P}} M\exp(u\langle a^t, m\rangle - su) ,$$

where $a^t(v) = a(t) - a(v)$, $v \in T$. We shall need the smoothness condition

A. For any $N < \infty$

$$\limsup_{r\to 0} M\exp(Nw(r))$$

is finite, where $w(r)$ is the modulus of continuity for $a(r)$:

$$w(r) = \sup_{\substack{t_1, t_2 \in T \\ |t_1 - t_2| < r}} |a(t_1) - a(t_2)| .$$

<u>Theorem</u>. Under the condition A

1) $$\limsup_{n\to\infty} \frac{1}{n} \ln P_+(t_n^* \in U) \le -\inf_{t\in U} K(t)$$

for each closed set U in T;

2) $$\liminf_{n\to\infty} \frac{1}{n} \ln P_-(t_n^* \in V) \ge -\inf_{t\in V} K_+(t)$$

for each open set V in T;

3) the function $K(t)$ is subcontinuous on T, i.e.

$$\liminf_{t_1\to t} K(t_1) \ge K(t)$$

for each $t\in T$.

<u>Corollary</u>. Under the assumptions of the theorem, the statement (1) is correct for each measurable set U, such that

$$0 < \inf_{t\in\bar{U}} K(t) = \inf_{t\in U^\circ} K_+(t) < \infty$$

where $\bar{U}$ is the closure and U° is the interior of U.

The proof of the theorem is based on a number of lemmas.

Lemma 1. Under the condition A

$$\lim_{r\to 0} M \exp(Nw(r)) = 1$$

for each $N<\infty$.

The proof is obvious.

The statement 3 of the theorem is equivalent to

<u>Lemma 2</u>. Under the condition A

$$\limsup_{t_1\to t} \exp(-K(t_1)) \le \exp(-K(t)) .$$

Proof of Lemma 2 is based on Lemma 1 and the Hölder's inequality.

Let us denote

$$S_r(t) = \{t_1\in T: |t_1-t|<r\} .$$

<u>Lemma 3</u>. Under the condition A

$$\lim_{r\to 0} \limsup_{n\to\infty} \frac{1}{n} \ln P_+(t_n^* \in S_r(t)) \le -K(t) .$$

<u>Proof</u>. For any $m \in \mathcal{P}$, $u>0$ we can write

$$P_+(t_n^* \in S_r(t)) \le M(\exp(u\langle A_n^{t_n^*}, m\rangle);\ t_n^* \in S_r(t))$$

$$\le M(\exp(u\langle A_n^t, m\rangle + u\sum_{i=1}^{n} w_i(r))$$

$$\le (M\exp(u\langle A_1^t, m\rangle + uw_1(r)))^n ,$$

where $w_i(r)$ is the modulus of continuity of the random function $a_i(t)$. Hölder's inequality implies that for $p>0$, $q>0$, $\frac{1}{p}+\frac{1}{q}=1$,

$$\frac{1}{n} \ln P_+(t_n^* \in S_r(t)) \le \frac{1}{p} \ln M \exp(pu\langle a^t, m\rangle) + \frac{1}{q} \ln M \exp(qw(r)).$$

Lemma 1 implies the estimate

$$\limsup_{r\to 0} \limsup_{n\to\infty} \frac{1}{n} \ln P_+(t_n^* \in S_r(t)) \le \frac{1}{p} \ln M \exp(up\langle a^t, m\rangle).$$

This completes the proof.

<u>Lemma 4</u>. Under the condition A the statement 1 of the theorem is correct.

<u>Proof</u>. Let

$$S_r(U) = \{t_1 \in T: |t_1 - t| > r,\ t \in U\} .$$

It follows by Lemma 2 that for each closed set $U \subseteq T$

$$\lim_{r\to 0} \inf_{t\in S_r(U)} K(t) = \inf_{t\in U} K(t).$$

Lemma 4 is proved.

Let us pass to the part 2 of the theorem. Let $x_1, x_2, \ldots, x_n$ be i.i.d. random vectors with values in R^N,

$$S_n = x_1 + \cdots + x_n .$$

Set

$$G = \{y=(y_1,\ldots,y_N) \in R^N : y_1 \ge 0, \ldots, y_N \ge 0\} .$$

The usual technique of calculating the crude asymptotics of large deviations in R^N (cf., e.g. [1,3]) allows to prove

<u>Lemma 5</u>.

$$\liminf_{n\to\infty} \frac{1}{n} \ln P\left(\frac{S_n}{n}\in G\right) \geq -\Lambda_+ ,$$

where

$$\exp(-\Lambda_+) = \lim_{s\downarrow 0} \inf_{u\in G} M \exp((u,x_1)-(us)) .$$

<u>Lemma 6</u>. If the condition A holds then for any $N<\infty$, $s>0$ there exists $z=z(N,s)$ such that

$$P\left(\sum_{i=1}^{n} w_i(z)>ns\right) \leq \frac{1}{z} \exp(-Nn) .$$

The proof follows from the Tchebyshev's inequality.

<u>Proof of the part two of the theorem</u>. Let us consider $t\in T$, $r>0$, $S_r(t)\supset T^\circ$, where T° is the interior of T. Denote by $[T\setminus S_r(t)]_z$ a finite set of points $\{t_j\}$ in the closed bounded set $T\setminus S_r(t)$ such that $|t_i-t_j|<z$ for all i, j. Then for all $z>0$, $s>0$

$$\begin{aligned} P_-(t_n^*\in S_r(t)) &\geq P(A_n(t) > \sup_{t_1\in T\setminus S_r(t)} A_n(t_1)) \\ &\geq P(A_n(t) > \max_{t_1\in[T\setminus S_r(t)]_z} A_n(t_1)+ns,\ \sum_{i=1}^{n} w_i(z)\leq ns) \\ &\geq P(A_n(t) > \max_{t_1\in[T\setminus S_r(t)]_z} A_n(t_1)+ns) \\ &\quad - P(\sum_{i=1}^{n} w_i(z)>ns) = P_1-P_2 . \end{aligned} \tag{4}$$

Set

$$K(t,s) = -\inf_{u>0,m\in\mathcal{P}} \ln M \exp(u\langle a^t,m\rangle-us) .$$

Since $K_+(t)=\lim_{s\downarrow 0} K(t,s)$ by the definition of K_+, there exists $s_1>0$ such that for all $0<s\leq s_1$

$$K(t,s) \leq 2K_+(t) .$$

By Lemma 6 for any $s\in(0,s_1]$ there exists $z=z(s)>0$ such that

$$P_2 = o(\exp(-2nK_+(t))) \ , \tag{5}$$

if $K_+(t)>\infty$. The set $\{t_1,\dots,t_N\}=[T\setminus S_r(t)]_z$ is finite for all fixed $s\in(0,s_1]$ and $z=z(s)$. So Lemma 6 implies that for all $v>0$

$$\liminf_{n\to\infty} \frac{1}{n} \ln P_1 \geq -\Lambda_{s+v} \ ,$$

where

$$\Lambda_{s+v} = -\ln \inf_{y_1\geq 0,\dots,y_N\geq 0} M \exp\{ \sum_{i=1}^{N} y_i(a(t)-a(t_i)-(s+v))\} \ . \tag{6}$$

It is obvious that

$$\Lambda_{s+v} \leq K(t,s+v) \ .$$

Hence it follows from (6) that

$$\liminf_{n\to\infty} \frac{1}{n} \ln P_1 \geq -K(t,s+v) \ . \tag{7}$$

By (4-7) we can write, that for all $\varepsilon>0$, $v>0$

$$\liminf_{n\to\infty} \frac{1}{n} \ln P_-(t_n^*\in S_r(t)) \geq -K(t,s+v) \ .$$

We now complete the proof of the theorem by letting s, v go to 0.

References

[1] Боровков, А. А., Рогозин Б. А. О центральной предельной теореме в многомерном. Теория вероятн. и ее примен., 1965, Т.10, No1, с.61-69.

[2] Боровков, А. А., Могульский А. А. О вероятностях больших уклонений в топологических пространствах. I,II, Сиб. матем. ж., 1978, Т.19, No5, с.988-1004; 1980, Т.21, No5, с.12-26.

[3] Могульский А. А. Большие укеонения для траекторий многомерных случайных блужданий. Теория вероятн. и ее примен., 1976, Т.21, No2, с.309-323.

Institute of Mathematics,
the USSR Academy of Sciences
(Siberian Division),
Novosibirsk, USSR.

ON THE DECAY RATE OF CORRELATION FOR PIECEWISE LINEAR TRANSFORMATIONS

Makoto Mori

1. Introduction. There exist three problems when we consider one dimensional mappings. These are:

1) Is there an invariant probability measure which is absolutely continuous with respect to the Lebesgue measure? Moreover, can we get a concrete form of its density?

2) Is the dynamical system mixing?

3) Can we estimate the decay rate of the correlation?

To solve these problems, we have three ways. These are:

1) to solve the eigenvalue problems of the Perron-Frobenius operator ([2]),

2) to use the Fredholm determinant ([1],[3],[7],[8],[9],[10]), and

3) to use the renewal equation of words ([4],[5],[6]).

We will consider the above problems along the third way, that is, for an irreducible piecewise linear mapping we will characterize the density of the invariant probability measure and the decay rate of the correlation by zero points of the determinant of a matrix (we also call it the Fredholm determinant), which we will define in (3.14).

2. Definitions and Notations.

2-1. Mappings. In this paper, we will consider irreducible piecewise linear mappings F from the unit interval into itself, that is, F satisfies:

Condition 1). There exists a partition $\{I_a\}_{a\in A}$ into subintervals and positive constants $\{\lambda^a\}_{a\in A}$ such that

$$(2.1)\qquad |F'(x)| = \lambda^a \quad \text{for} \quad x\in I_a^\circ,$$

where I_a° is the inner set of I_a. We call each element $a\in A$ an alphabet.

Condition 2). (irreducibility) For any subinterval I, there exists an integer n such that $\overline{F^{(n)}(I)}\supset[0,1]$, where $F^{(n)}$ is the n-fold iterate of the mapping F and $\overline{F^{(n)}(I)}$ is the closure of the set $F^{(n)}(I)$.

Condition 3). We denote by O the set of alphabets a which satisfy $\overline{F(I_a)} \supset [0,1]$ and by C the set of other alphabets. We assume the number of the elements of the set C is finite. We denote the number of the elements by N.

2-2. Words. In the following, to simplify the notations, we denote I_a by (a). For $a \in A$, we define

$$(2.2) \qquad \operatorname{sgn} a = \begin{cases} +1 & \text{if } F'(x)>0 \text{ on } I_a^\circ, \\ -1 & \text{if } F'(x)<0 \text{ on } I_a^\circ. \end{cases}$$

We call a finite sequence of alphabets $w=a_1 \ldots a_n$ a word and define

$$(2.3) \qquad (w) = \bigcap_{i=1}^{n} F^{(i-1)}((a_i)),$$

$$(2.4) \qquad |w| = n,$$

$$(2.5) \qquad \operatorname{sgn} w = \prod_{i=1}^{n} \operatorname{sgn} a_i,$$

$$(2.6) \qquad \lambda^{-w} = \prod_{i=1}^{n} (\lambda^{a_i})^{-1}$$

and

$$(2.7) \qquad (z/\lambda)^w = z^{|w|}\lambda^{-w},$$

for $z \in \mathbb{C}$. We denote by $\langle w \rangle$ the indicator function of the set (w). We call a word w admissible if $(w) \neq \phi$. We define a point wx by $wx \in (w)$ and $F^{(|w|)}(wx)=x$. If there exists a point wx, we say that wx is admissible. We define an order on the set of alphabets: $a<b$ means $x<y$ for any $x \in (a)$ and $y \in (b)$. We also define a partial order on the set of words: $a_1 \ldots a_n < b_1 \ldots b_m$ means that there exists i such that $a_1 \ldots a_i = b_1 \ldots b_i$ and either $a_{i+1} < b_{i+1}$ and $\operatorname{sgn} a_1 \ldots a_i = +1$, or $a_{i+1} > b_{i+1}$ and $\operatorname{sgn} a_1 \ldots a_i = -1$.

By the irreducibility, we may assume

condition 3') for any $a, b \in A$

$$(2.8) \qquad \overline{(a)} = \overline{\bigcup_{w} (w)},$$

where the union is taken over all words $w=a_1 \ldots a_n$ ($a_i \in A$, $n \geq 1$) such that

(2.9) $a_1 = a$,

(2.10) $\overline{F^{(n)}((w))} \supset (b)$.

2-3. Expansion of x. For an infinite sequence of alphabets $a_1a_2\ldots$, we denote

(2.11) $a(1,n) = a_1\ldots a_n$,

(2.12) $a(n,\infty) = a_na_{n+1}\cdots$.

For $x\in[0,1]$, an expansion of x which we denote by $a^x(1,\infty)=a_1^xa_2^x\ldots$ is an infinite sequence of alphabets such that

(2.13) $F^{(i-1)}(x) \in (a_i^x)$.

For $I\in C$, + expansion of I which we denote by $I^+(1,\infty)=I_1^xI_2^x\ldots$ (- expansion of I which we denote by $I^-(1,\infty)=I_1^-I_2^-\ldots$) is the limit of the expansion of x as $x\uparrow\beta$ $(x\downarrow\alpha)$, where $\overline{I}=[\alpha,\beta]$.

3. Renewal Equation and Fredholm Determinant. For $J\in C$ and $\mathscr{C}=I^s(1,\infty)$ for some $I\in C$ and $s\in\{+,-\}$, we define

(3.1) $\alpha_1^J(\mathscr{C}) = \lambda^{-c_1}(\delta[J<c_2:\mathscr{C}c_1]+\delta[J=c_2 \text{ and } \operatorname{sgn} c_1=-1]-1/2)$

and for $m\geq 2$

(3.2)
$$\begin{aligned}\alpha_m^J(\mathscr{C}) = \lambda^{-c(1,m-1)}\{&\lambda^{-c_m}(\delta[J<c_{m+1}:\mathscr{C}c(1,m)]\\ &+\delta[J=c_{m+1} \text{ and } \operatorname{sgn} c(1,m-1)=-1]-1/2)\\ &-\Lambda(\delta[J<c_m:\mathscr{C}c(1,m-1)]+\delta[J=c_m \text{ and } \operatorname{sgn} c(1,m-1)=-1])\\ &+\sum_{a\in O}\lambda^{-a}(\delta[a<c_m:\mathscr{C}c(1,m-1)]+\delta[a=c_m]/2)\},\end{aligned}$$

where $\mathscr{C}=c(1,\infty)=c_1c_2\ldots$,

(3.3) $$\delta[\alpha] = \begin{cases}1 & \text{if } \alpha \text{ is true,}\\ 0 & \text{if } \alpha \text{ is false,}\end{cases}$$

(3.4) $$\delta[\alpha<\beta :\mathscr{C}c(1,m)] = \begin{cases}1 & \text{if } \alpha<\beta \text{ and } \operatorname{sgn}\mathscr{C}c(1,m)=+1,\\ & \text{or if } \alpha>\beta \text{ and } \operatorname{sgn}\mathscr{C}c(1,m)=-1,\\ 0 & \text{otherwise,}\end{cases}$$

(3.5) $$\operatorname{sgn}\mathscr{C}c(1,m) = \begin{cases}1 & \text{if } \mathscr{C}=I^+ \text{ for some } I \text{ and } \operatorname{sgn} c(1,m)=+1,\\ & \text{or if } \mathscr{C}=I^- \text{ and } \operatorname{sgn} c(1,m)=-1,\\ 0 & \text{otherwise,}\end{cases}$$

and

$$(3.6)\qquad \Lambda = \sum_{a\in O} \lambda^{-a}.$$

We also define $\alpha_m^x(\mathscr{C})$ $(x\in[0,1],\ \mathscr{C}=I^s(1,\infty)$ for $I\in C$ and $s\in\{+,-\})$ by changing $\delta[J<c_k:\mathscr{C}c(1,k-1)]$ and $\delta[J=c_k$ and $\operatorname{sgn} c(1,k-1)=-1]$ in (3.1) and (3.2) to $\delta[a^x(1,\infty)<c(k,\infty):\mathscr{C}c(1,k-1)]$ and $\delta[$there exists a sequence x_n such that $\lim a^{x_n}(1,\infty)=c(k,\infty)$ and $\operatorname{sgn} c(1,k-1)=-1]$, respectively. Now, we define $(2N,N)$ matrix $\phi(z)$, $(2N,2N)$ matrix $\psi(z)$, $(N,2N)$ matrix E and $2N$ dimensional vector $\chi(z;x)$ by

$$(3.7)\qquad \phi^{I^s,J}(z) = (1-\Lambda z)^{-1}\Sigma\alpha_m^J(I^s)z^m,$$

$$(3.8)\qquad \psi^{I^s,J^t}(z) = \Sigma(z/\lambda)^{I^s(m)}\delta[I_1^{s,m}=J \text{ and } st=-1],$$

$$(3.9)\qquad E^{I,J^s} = \delta[J=I]$$

and

$$(3.10)\qquad \chi^{I^s}(z;x) = (1-\Lambda z)^{-1}\Sigma\alpha_m^x(I^s)z^m,$$

where $I,J\in C$ and $s,t\in\{+,-\}$, and $I^s(m)$ and $I_1^{s,m}$ are defined by:

$$(3.11)\qquad n(m) = n \quad\text{if}\quad I_n^s \in C \quad\text{and}\quad \#\{k<n:\ I_k^s\in C\} = n,$$

$$(3.12)\qquad I^s(m) = I^s(n(m),\infty)$$

and

$$(3.13)\qquad I_1^{s,m} = I_{n(m)}^s.$$

<u>Definition 3-1</u>. Let

$$(3.14)\qquad \Phi(z) = I-E(I-\psi(z))^{-1}\phi(z),$$

where I is the identity matrix. We call its determinant the Fredholm determinant of the mapping F.

Now we will construct a renewal equation of admissible words. For $J\in C$ and $x\in[0,1]$, we define

$$(3.15)\qquad s^J(z;x) = \Sigma(z/\lambda)^w,$$

where the sum is taken over all words w such that $wx\in J$ and wx is admissible. We also define N dimensional vector

(3.16) $$s(z;x) = (s^J(z;x))_{J\in C}$$

and

(3.17) $$\bar{s}(z;x) = \sum_{J\in C} s^J(z;x),$$

that is, $\bar{s}(z;x)$ is the generating function of the sum of words for which wx are admissible. Then, for each $I\in C$, dividing words by n for which $\overline{F^{(n)}(I)\supset J}$ for some $J\in C$ for the first time, we can make the renewal equation in the following form:

Theorem 3-1. As a formal expression, we get

(3.18) $$s(z;x) = \Phi(z)^{-1}E(I-\psi(z))^{-1}\chi(z;x).$$

This equation is our main tool.

4. Decay Rate of Correlation. In this section, we will estimate the decay rate of the correlation in terms of the Fredholm determinant.

Definition 4-1. Let ξ be the infimum of the lower Lyapunov number, that is,

(4.1) $$\xi = \operatorname*{ess\,inf}_{x\in[0,1]} \varliminf_{n\to\infty} \frac{1}{n} \log |(F^{(n)})'(x)|.$$

Definition 4-2. Let η^{-1} be the minimum in modulus of e^{ξ}, zero points of $\det((z-1)^{-1}\Phi(z))$ and zero points of $\det(I-\psi(z))$.

Theorem 4-1. Assume that $\xi>0$ and $|\eta|<1$. Then the dynamical system $([0,1],\mu,F)$ is mixing, where

(4.2) $$\rho(x) = \frac{d\mu}{dx}(x) = \lim_{z\uparrow 1} (1-z)\bar{s}(z;x).$$

Proof. 1) It is not difficult to see that $\Phi(1)$ has an eigenvalue zero with the eigenvector $(\|I\|)_{I\in C}$, where $\|I\|$ is the Lebesgue measure of I. Hence, we define

(4.3) $$\Phi'(1)^{-1} = \lim_{z\uparrow 1} (1-z)\Phi(z)^{-1}$$

and

(4.4) $$\begin{aligned}\bar{\rho}(x) &= K^{-1}\ {}^t c\ \Phi'(1)^{-1}E(I-\psi(1))^{-1}\ \chi(1;x)\\ &= K^{-1} \lim_{z\uparrow 1} {}^t c\ (1-z)s(z;x),\end{aligned}$$

where

(4.5) $c = (1/\|I\|)_{I\in C}$

and K is the normalizing constant. We will show that the dynamical system $([0,1],\bar{\mu},F)$ $(\frac{d\bar{\mu}}{dx}=\bar{\rho}(x))$ is mixing in 3). Hence, it follows that the dynamical system $([0,1],\mu,F)$ is mixing.

2) We define (N,N) matrix $G^{w}(z)$ and N dimensional vector $G^{w,x}(z)$ by

(4.6) $$(G^{w}(z))_{I,J} = N^{-1} \sum_{u,v} \lambda^{-u}(z/\lambda)^{wv}\delta(uwv \text{ satisfies i)-iv)}],$$

and

(4.7) $$(G^{w,x}(z))_{I} = N^{-1} \sum_{u,v} \lambda^{-u}(z/\lambda)^{wv}\delta(uwv \text{ satisfies i)-iii),v)}],$$

where for $u=a_1\ldots a_m$ and $v=b_1\ldots b_n$

i) $a_1 = I$,

ii) if $a_{k+1} \in C$ $(1\le k\le m-1)$, then

(4.8) $$\overline{F^{(k)}((a_1\ldots a_k))} \not\supset (a_{k+1}),$$

iii) if $b_{k+1} = J$ $(1\le k\le n-1)$, then

(4.9) $$\overline{F^{(m+|w|+k)}((uwb_1\ldots b_k))} \not\supset (J),$$

iv) $\overline{F^{(m+|w|+k)}((uwv))} \supset (J)$,

v) $uwvx$ is admissible.

Then it is not difficult to see that $(G^{w}(z))_{I,J}$ and $(G^{w,x}(z))_{I}$ are analytic in $|z|<|\eta^{-1}|$.

3) It is not difficult to see that $\Phi'(1)^{-1}G^{w}(z)$ has a left eigenvector ${}^{t}c$ with its eigenvalue $\int\langle w\rangle d\mu$. Hence, for a word w and $g\in L^1$, by Theorem 3-1,

(4.10) $$\lim_{z\uparrow 1} (1-z) \sum_{n\ge|w|+1} z^n \int\langle w\rangle(x)\, g(F^{(n)}(x))d\mu$$

$$= \lim_{z\uparrow 1} (1-z) \sum_{v} (z/\lambda)^{wv} \int g(x)\, \rho(wvx)dx$$

$$= K^{-1} \int g(x) \lim_{z\uparrow 1} (1-z)\, {}^{t}c\, \Phi'(1)^{-1}\{G^{w}(x)s(z;x)+G^{w,x}(z)\}dx$$

$$= \int\langle w\rangle d\mu \int g(x)\, d\mu.$$

On the other hand, the set of words is a generator. This completes the proof.

<u>Theorem 4-2</u>. Assume that $\xi>0$ and $|\eta|<1$. Then for $f\in BV$ and $g\in L^1$, we get for any $\varepsilon>0$

$$(4.11) \qquad \lim_{n\to\infty} (|\eta|+\varepsilon)^{-n}\left\{\int f(x)g(F^{(n)}(x))d\mu - \int f\, d\mu \int g\, d\mu\right\} = 0.$$

Proof. For a function $f \in BV$, there exists a decomposition

$$(4.12) \qquad f(x) = \sum_w \alpha_w \langle w\rangle(x)$$

such that

$$(4.13) \qquad \sum_w |\alpha_w|\, \gamma^{|w|} < \infty$$

for any $0<\gamma<1$. Therefore, as in (4.10), it is not difficult to prove (4.11).

References

[1] F. Hofbauer and G. Keller: Zeta functions and transfer-operators for piecewise linear transformations, Math. Zeitschr. 180 (1982), 119-140.

[2] G. Keller: On the rate of convergence to equilibrium in one-dimensional systems, CMP. 96 (1984), 181-193.

[3] G. Keller: Markov extensions, zeta functions, and Fredholm theory for piecewise invertible dynamical systems, preprint.

[4] M. Mori: On the decay of correlation for piecewise monotonic mappings I, Tokyo J. Math. 8 (1985), 389-414.

[5] M. Mori: On the decay of correlation for piecewise monotinic mappings II, Tokyo J. Math. 9 (1986), 135-161.

[6] M. Mori: Fredholm determinant of piecewise linear transformations, preprint.

[7] Y. Oono and Y. Takahashi: Chaos, external noise and Fredholm theory, Prog. Theoret. Phys. 63 (1980), 1804-1807.

[8] Y. Takahashi: Fredholm determinant of unimodal linear maps, Sci. Papers Coll. Gen. Educ. Univ. Tokyo 31 (1981), 61-87.

[9] Y. Takahashi: Shift with orbit basis and realization of one dimensional maps, Osaka J. Math. 20 (1983), 599-629.

[10] Y. Takahashi: Gibbs variational principle and Fredholm theory for one-dimensional maps, Springer in Synergetics 24 (1984), 14-22.

Department of Mathematics
National Defense Academy
Hashirimizu, Yokosuka
Kanagawa
239 Japan

A FLUCTUATION THEOREM FOR SOLUTIONS OF CERTAIN RANDOM EVOLUTION EQUATIONS

Takehiko Morita

0. Introduction. In this paper we consider a stationary process $L(\omega,t)$ on a probability space $(\Omega, \mathcal{F}, P)$ with values in a certain class of partial differential operators on R^d. Denote by L the mean operator of $L(\omega,t)$. We assume that the following Cauchy problems are well-posed in some function space.

$$\text{(0.1)} \qquad \begin{cases} \dfrac{du(t)}{dt} = L(\omega,\dfrac{t}{\varepsilon})u(t) \\ u(0) = u_0 \end{cases}$$

and

$$\text{(0.2)} \qquad \begin{cases} \dfrac{du(t)}{dt} = Lu(t) \\ u(0) = u_0. \end{cases}$$

The aim of this paper is to prove the fluctuation of $u^{\varepsilon}(\omega,t)$ around $u^0(t)$ where $u^{\varepsilon}(\omega,t)$ and $u^0(t)$ are the solutions of (0.1) and (0.2) respectively. The random operators treated in this paper are a little more concrete than those in [4]. But we notice that an abstract version of the result of this paper can be easily obtained, see Remark 2.2.

1. The well-posed classes. Throughout the paper functions are assumed to be real valued. Let $H^p(R^d)$ $(p \in R)$, denote the Sobolev space of order p which is the completion of $C_0^{\infty}(R^d)$ by the Hilbertian norm $\|\cdot\|_p$ defined by $\|u\|_p = \|(1-\Delta)^p u\|$ where $\|\cdot\|$ denotes the usual L^2-norm of $u \in C_0^{\infty}(R^d)$ and Δ denotes the Laplacian on R^d. Let $\mathcal{S}^p(R^d)$ $(p \in R)$ denote the weighted Sobolev space which is the completion of the totality of rapidly decreasing functions $\mathcal{S}(R^d)$ by the Hilbertian norm $|||\cdot|||_p$ defined by using $|x|^2-\Delta$ instead of $1-\Delta$ where $|x|$ denotes the Euclidean norm of $x = (x_1,\ldots,x_d) \in R^d$.

Put $H^{-\infty}(R^d) = \bigcup_{p\in R} H^p(R^d)$, $H^{\infty}(R^d) = \bigcap_{p\in R} H^p(R^d)$, $\mathcal{S}^{-\infty}(R^d) = \bigcup_{p\in R} \mathcal{S}^p(R^d)$, and $\mathcal{S}^{\infty}(R^d) = \bigcap_{p\in R} \mathcal{S}^p(R^d)$. Clearly, $\mathcal{S}^{-\infty}(R^d) = \mathcal{S}'(R^d)$ and $\mathcal{S}^{\infty}(R^d) = \mathcal{S}(R^d)$. Let K be a positive constant and let $\{A_\alpha\}_\alpha$ be a family of positive numbers where α's are multi-indices. Let $\mathcal{A} = \{a(t,x) \in C(R\to C^{\infty}(R^d));\ \sup_{t,x} |\partial_x^\alpha a(t,x)| \leq A_\alpha$ for any $\alpha\}$ where $\partial_x^\alpha = \partial^{\alpha_1+\dots+\alpha_d}/\partial x_1^{\alpha_1} \dots \partial x_d^{\alpha_d}$. Consider the following classes of time dependent partial differential operators:

$$\mathcal{L}_1 = \{L(t) = \sum_{j=1}^{d} B_j(t,x)\begin{bmatrix} \partial_x^j & & 0 \\ & \ddots & \\ 0 & & \partial_x^j \end{bmatrix} + C(t,x)\ ;\ B_j(t,x) \text{ is an } r\times r\text{-}$$

symmetric matrix with entries in $\mathcal{A}$ for $j = 1, 2, \dots, d$ and $C(t,x)$ is an $r\times r$-matrix with entries in $\mathcal{A}$ }, where ∂_x^j denotes ∂_x^α with $\alpha = (0,\dots,\underset{j}{\hat{1}},\dots,0)$ and $\mathcal{L}_2 = \{L(t) = \sum_{|\alpha|\leq 2\ell} a_\alpha(t,x)\partial_x^\alpha\ ;\ a_\alpha(t,x)\in\mathcal{A}$ and $\inf_{t,x}\left((-1)^{\ell+1}\sum_{|\alpha|=2\ell} a_\alpha(t,x)\xi^\alpha\right) \geq K\left(\sum_{j=1}^{d} \xi_j^2\right)^\ell$ for any $\xi = (\xi_1,\dots,\xi_d) \in R^d\}$, where $\ell \geq 1$ is an integer and $\xi^\alpha = \xi_1^{\alpha_1}\cdots\xi_d^{\alpha_d}$. For convenience we denote by $\mathcal{L}$ both $\mathcal{L}_1$ and $\mathcal{L}_2$ and make the following convention. $H^p = \overbrace{H^p(R^d)\oplus\cdots\oplus H^p(R^d)}^{r}$, $\mathcal{S}^p = \overbrace{\mathcal{S}^p(R^d)\oplus\cdots\oplus\mathcal{S}^p(R^d)}^{r}$, $\Lambda = \overbrace{(|x|^2-\Delta)^{\frac{1}{2}}\oplus\cdots\oplus(|x|^2-\Delta)^{\frac{1}{2}}}^{r}$, and $m = 1$, if $\mathcal{L}$ denotes $\mathcal{L}_1$ and $H^p = H^p(R^d)$, $\mathcal{S}^p = \mathcal{S}^p(R^d)$, $\Lambda = (|x|^2-\Delta)^{\frac{1}{2}}$, and $m = 2\ell$ if $\mathcal{L}$ denotes $\mathcal{L}_2$. The inner products of H^p and $\mathcal{S}^p$ are denoted by $(\ ,\ )_p$ and $((\ ,\ ))_p$ respectively. In particular we often denote by $(\ ,\ )$ the inner product $(\ ,\ )_0 = ((\ ,\ ))_0$.

Remark 1.1. From the definition of $\mathcal{S}^p$, we can see that there exists a number $\delta > 0$ such that the inclusion $\mathcal{S}^{p+\delta}\subset\mathcal{S}^p$ is a Hilbert-Schmidt operator for any $p \in R$.

We summerize the properties of $\mathcal{L}$ as a proposition:

<u>Proposition 1.1.</u> For $L(\cdot) \in \mathcal{L}$, we have:

(1) For each $p \in R$, there exist a real number $q = p(q) < p$ and a positive constant C_p which depend only on the class $\mathcal{L}$ such that $L(\cdot) \in C(R \to B(H^{p+m} \to H^p))$, $L(\cdot) \in C(R \to B(\mathcal{S}^p \to \mathcal{S}^q))$, $\sup_t \|L(t)\|_{B(H^{p+m} \to H^p)} \le C_p$, and $\sup_t \|L(t)\|_{B(\mathcal{S}^p \to \mathcal{S}^q)} \le C_p$, where $B(E_1 \to E_2)$ denotes the totality of bounded linear operators from the topological vector space E_1 into E_2.

(2) For any $p \in R$, $T > 0$, and $u_0 \in H^{p+m}$, the Cauchy problem $du(t)/dt = L(t)u(t)$, $u(0) = u_0$ has a unique solution in $C([0,T] \to H^{p+m}) \cap C^1([0,T] \to H^p)$. Moreover, for each fixed t, $L(t)$ generates a strongly continuous semi-group on $\mathcal{S}^\infty$.

(3) For any p, $T > 0$, there is a positive number $C_{T,p}$ which depends only on p, T, and the class $\mathcal{L}$ such that if $v(\cdot) \in C([0,T] \to H^{p+m}) \cap C^1([0,T] \to H^p)$ satisfies the equation $dv(t)/dt = L(t)v(t) + f(t)$ for some $f(\cdot) \in C([0,T] \to H^p)$, then we have

$$\|v(t)\|_p^2 \le C_{T,p}\left(\|v(0)\|_p^2 + \int_0^t \|f(s)\|_p^2 \, ds\right) \tag{1.1}$$

for any $0 \le t \le T$.

(4) The operator $L^s(\cdot)$ belongs to $\mathcal{L}$ where $L^s(\cdot)$ is defined by $L^s(t) = L(st)$.

(5) The formal adjoint operator $L^*(\cdot)$ of $L(\cdot)$ can be extended uniquely to an element in $\mathcal{L}$.

For the proof see [3, Chapter 7] and [5, Chapter 2]. //

From the above proposition we may call $\mathcal{L}$ the <u>well-posed class.</u>

<u>2. The statement of the main theorem.</u> In this section we introduce the stationary random operators and give the main theorem.

Let $(\Omega, \mathcal{F}, P)$ be a probability space and let $\{\mathcal{F}_s^t;\ \infty \le s \le t \le \infty\}$ be a family of sub-σ-algebras of $\mathcal{F}$ with the following strong mixing property:

$$\sup_t \sup_{\xi,\eta} |E[\xi\eta] - E[\xi]E[\eta]| = \alpha(s) \downarrow 0 \quad \text{as} \quad s \uparrow \infty \tag{2.1}$$

and $\int_0^\infty s\,\alpha(s)\,ds < \infty$, where the supremum is taken over all $\mathcal{F}_{-\infty}^t$

-measurable ξ with $|\xi| \le 1$ and all $\mathcal{F}_{t+s}^{\infty}$-measurable η with $|\eta| \le 1$. Let $\{\eta(\omega,t)\ ;\ t \in R\}$ be an R^d-valued stationary process which has continuous paths and is $\mathcal{F}_t^t$-measurable for each fixed t.

Consider fixed elements $L_1 = \sum_{j=1}^{d} B_j(x) \begin{bmatrix} \partial_x^j & & 0 \\ & \ddots & \\ 0 & & \partial_x^j \end{bmatrix} + C(x) \in \mathcal{L}_1$ and $L_2 = \sum_{|\alpha| \le 2\ell} a_\alpha(x)\partial_x^\alpha \in \mathcal{L}_2$ which are independent of t, and put

$$L_1(\omega,t) = \sum_{j=1}^{d} B_j(x+\eta(\omega,t)) \begin{bmatrix} \partial_x^j & & 0 \\ & \ddots & \\ 0 & & \partial_x^j \end{bmatrix} + C(x+\eta(\omega,t))$$

and

$$L_2(\omega,t) = \sum_{|\alpha| \le 2\ell} a_\alpha(x+\eta(\omega,t))\partial_x^\alpha.$$

As before we denote by $L(\omega,t)$ both $L_1(\omega,t)$ and $L_2(\omega,t)$ according to our convention. Then $L(\omega,t)$ has the following properties:

Proposition 2.1. (1) For almost all ω, $L(\omega,\cdot)$ belongs to $\mathcal{L}$.

(2) For any $u, v \in \mathcal{S}^\infty$, $(L(\omega,t)u,v)$ is a stationary process and $\mathcal{F}_t^t$-measurable for each t.

(3) The mean operator L defined by $(Lu,v) = E[(L(\cdot,t)u,v)]$ for $u, v \in \mathcal{S}^\infty$ is uniquely extended to an element in $\mathcal{L}$. Of course, L is independent of t.

These properties can be proved in the same way as Proposition 2.3 in [4]. //

We call $L(\omega,t)$, a stationary random operator. For the stationary random operator $L(\omega,t)$, the Cauchy problems

(2.2)
$$\begin{cases} \dfrac{du(t)}{dt} = L(\omega,\frac{t}{\varepsilon})u(t) \\ u(0) = u_0 \in \mathcal{S}^\infty \end{cases}$$

and

(2.3)
$$\begin{cases} \dfrac{du(t)}{dt} = Lu(t) \\ u(0) = u_0 \in \mathcal{S}^\infty \end{cases}$$

have unique solutions $u_p^\varepsilon(t)$ and $u_p^0(t)$ in $C([0,T]\to H^{p+m}) \cap C^1([0,T]\to H^p)$ for any p respectively. Thus there occurs no confusion if we denote the solutions of (2.2) and (2.3) by $u^\varepsilon(\omega,t)$ and $u^0(t)$ respectively for any p. Now we can state the main theorem:

Theorem 2.1. For any $p \in R$ and any $T > 0$, we have

$$\sup_{0 \le t \le T} E \| u^\varepsilon(t) - u^0(t) \|_p^2 \le C\varepsilon \tag{2.4}$$

where C is a positive number independent of ε. Moreover,

the distribution of $X^\varepsilon(\omega,t) = \dfrac{u^\varepsilon(\omega,t)-u^0(t)}{\sqrt{\varepsilon}}$ on $C([0,T]\to\mathcal{S}^{-p})$

converges weakly for any $p > 0$. The limit distribution coincides with the distribution of $C([0,T]\to H^\infty)$-valued random variable $X^0(\omega,t)$ which is written as

$$X^0(\omega,t) = W^0(\omega,t) + \int_0^t LT(t-s)W^0(\omega,s)\,ds \quad \text{in } H^p \tag{2.5}$$

for any $p \in R$, where $T(t)$ denotes the semi-group generated by L and $\{W^0(\omega,t)\ ;\ 0 \le t \le T\}$ is an $\mathcal{S}^\infty$-valued continuous stochastic process with independent increments characterized by

$$E[(W^0(t),v)] = 0 \tag{2.6}$$

and

$$E[(W^0(t),v)(W^0(s),w)] = \int_0^{t\wedge s} \langle v,w\rangle(u^0(r))\,dr \tag{2.7}$$

where $$\langle v,w\rangle(u) = \int_0^\infty dr\ E[((L(t)-L)u,v)((L(0)-L)u,w) + ((L(0)-L)u,v)((L(t)-L)u,w)]$$

for any u, v and $w \in \mathcal{S}^\infty$.

Remark 2.1. In virtue of the Sobolev lemma, the limit process $X^0(\omega,t)$ takes its values in the space of smooth functions.

Remark 2.2. If one considers abstract Sobolev spaces H^p and $\mathcal{S}^p$ and an abstract class of operators $\mathcal{L}$ on $\mathcal{S}^{-\infty}$ which has the same properties as in Proposition 1.1, and if one assumes that a random operator $L(\omega,t)$ satisfies the properties in Proposition 2.1, one can obtain a more abstract version of this theorem. For example, the result in [4] can be regarded as a special case when the abstract Sobolev spaces H^p and $\mathcal{S}^p$ coincide. But we do not go into details in this paper.

3. Proof of Theorem 2.1. In this section we drop the letter ω if there occurs no confusion. Recall that $X^\varepsilon(t)$ satisfies the equation

(3.1) $$X^\varepsilon(t) = W^\varepsilon(t) + \int_0^t L(\tfrac{s}{\varepsilon})X^\varepsilon(s)\,ds \quad \text{in } H^p \text{ for any } p \in R,$$

where $W^\varepsilon(t)$ is given by the equation

(3.2) $$W^\varepsilon(t) = \frac{1}{\sqrt{\varepsilon}}\int_0^t (L(\tfrac{s}{\varepsilon})-L)u^0(s)\,ds = \frac{1}{\sqrt{\varepsilon}}\int_0^t (L(\tfrac{s}{\varepsilon})-L)T(s)u_0\,ds$$

in H^p for any $p \in R$. Let $Y^\varepsilon(t)$ be the unique solution of the equation

(3.3) $$Y^\varepsilon(t) = W^\varepsilon(t) + \int_0^t LY^\varepsilon(s)\,ds, \quad Y^\varepsilon(0) = 0 \quad \text{in } H^p$$

for any $p \in R$. Put $Z^\varepsilon(t) = X^\varepsilon(t) - Y^\varepsilon(t)$. Then

(3.4) $$Y^\varepsilon(t) = W^\varepsilon(t) + \int_0^t LT(t-s)W^\varepsilon(s)\,ds \quad \text{in } H^p$$

for any $p \in R$, and

(3.5) $$Z^\varepsilon(t) = \int_0^t (L(\tfrac{s}{\varepsilon})-L)Y^\varepsilon(s)\,ds + \int_0^t L(\tfrac{s}{\varepsilon})Z^\varepsilon(s)\,ds \quad \text{in } H^p$$

for any $p \in R$.

Let $\{h_k\}_{k=0}^\infty$ be a complete orthonormal system (CONS) of $\mathcal{S}^0 = H^0$ consisting of eigenvectors of Λ. One can show that $\{h_k^p = \Lambda^{-p}h_k\}_{k=0}^\infty$ forms a CONS of $\mathcal{S}^p$ and the linear space $[h_0, \dots, h_n]$ spanned by $\{h_k\}_{k=0}^n$ is contained in $\mathcal{S}^p$ for all p. We denote by π_p^n the orthogonal projection of $\mathcal{S}^p$ onto the orthogonal complement of $[h_0, \dots, h_n]$. From the definition, the class $\mathcal{L}$ can be regarded as a subset of a bounded closed ball $S_{-p} = L^2([0,T]\to\mathcal{H}_{-p})$ where $\mathcal{H}_{-p}$ denotes the Hilbert space of all Hilbert-Schmidt operators from $\mathcal{S}^{-p}$ into $\mathcal{S}^{-r}$ endowed with inner product $(A,B)_{HS,-p} = \sum_{k=0}^\infty ((Ae_k, Be_k))_{-r}$ where $\{e_k\}_{k=0}^\infty$ is a CONS of $\mathcal{S}^{-p}$ and the number $r\ (> p)$ is determined by p and the well-posed class $\mathcal{L}$.

In what follows, we always use the letter C to denote a positive number which is independent of $\varepsilon > 0$. We need the following lemma to prove Theorem 2.1.

Lemma 3.1. For any $p \in R$, there is a positive number $q = q(p) < p$ such that

$$(3.6) \qquad E|||\pi_q^n W^\varepsilon(t+h) - \pi_q^n W^\varepsilon(t)|||_q^4 \leq Ch^2 \Big(\sum_{k=n+1}^{\infty} ||| h_k^p |||_q^2 \Big)^2$$

and

$$(3.7) \qquad E\| X^\varepsilon(t+h) - X^\varepsilon(t) \|_p^4 \leq Ch^2 \quad \text{for any} \quad 0 \leq t,\ t+h \leq T.$$

Sketch of proof. The estimate (3.6) is proved in the same way as the proof of the estimate (4.9) of Lemma 4.3 in [4]. Next we have

$$X^\varepsilon(t)-W^\varepsilon(t) = \int_0^t L(\tfrac{s}{\varepsilon})(X^\varepsilon(s)-W^\varepsilon(s))\,ds + \int_0^t L(\tfrac{s}{\varepsilon})W^\varepsilon(s)\,ds$$

in H^p for any $p \in R$. Therefore we obtain

$\| X^\varepsilon(t)-W^\varepsilon(t) \|_p^2 \leq C\int_0^t \| W^\varepsilon(s)\|_{p+m}^2\,ds$ for any p (of course, C depends on p and T). Hence $\| X^\varepsilon(t)\|_p^2 \leq 2(\| W^\varepsilon(t)\|_p^2 + C\int_0^t \| W^\varepsilon(s)\|_{p+m}^2\,ds)$. Combining this and the equality

$X^\varepsilon(t+h)-X^\varepsilon(t) = W^\varepsilon(t+h)-W^\varepsilon(t) + \int_t^{t+h} L(\tfrac{s}{\varepsilon})X^\varepsilon(s)\,ds$, we can prove the estimate (3.7). //

Lemma 3.2. For any $v, w \in H^\infty$ and for any $0 \leq s, t \leq T$, we have $E[(W^\varepsilon(t),v)] = 0$ and

$$\lim_{\varepsilon\to 0} E[(W^\varepsilon(t),v)(W^\varepsilon(s),w)] = \int_0^{t\wedge s} \langle v,w\rangle(u^0(r))\,dr.$$

This lemma can be proved in the same manner as Lemma 3.1 in [2] or Lemma 5.1 in [4]. //

Lemma 3.3. For any $p \in R$, the distribution of $W^\varepsilon(\cdot)$ on $C([0,T]\to\mathcal{S}^p)$ converges weakly to the distribution of $W^0(\cdot)$ which appeared in the statement of Theorem 2.1.

Sketch of proof. From the estimate (3.6) in Lemma 3.1, the distributions of $W^\varepsilon(\cdot)$ on $C([0,T]\to\mathcal{S}^p)$ are tight. Moreover, in the same way as Lemma 3.3 in [2] and Step 1 of the proof of Theorem in [4], we can prove that $W^\varepsilon(\cdot)$ has asymptotically independent increments. Hence we complete the proof of Lemma 3.3 in virtue of Lemma 3.2. //

Lemma 3.4. There is a metric d on S_{-p} such that it induces the same topology as the weak topology of $L^2([0,T]\to\mathcal{H}_{-p})$ and $E[d(L^\varepsilon,L)] \to 0$ as $\varepsilon \downarrow 0$, where the operator L^ε is defined by $L^\varepsilon(t) = L(\frac{t}{\varepsilon})$.

The proof is long but it is quite similar to the proof of Step 4 in [4]. //

Now we can prove Theorem 2.1. From the equation (3.4) and Lemma 3.3, it is clear that for any $p \in R$, the distribution of $Y^\varepsilon(\cdot)$ on $C([0,T]\to H^p)$ converges weakly to the distribution of $X^0(\cdot)$ which appeared in the statement of Theorem 2.1. From the estimate (3.7) in Lemma 3.1, the distributions of $X^\varepsilon(\cdot)$ and $Z^\varepsilon(\cdot)$ are tight in $C([0,T]\to\mathcal{S}^{-p})$ for any $p > 0$, since the inclusion $H^0 \subset \mathcal{S}^{-p}$ is a compact operator. Thus it suffices to show that for some $p > 0$ the limit distribution of $Z^\varepsilon(\cdot)$ on $C([0,T]\to\mathcal{S}^{-p})$ coincides with the distribution of the random variable which vanishes almost surely. Put $S = C([0,T]\to\mathcal{S}^{-p})\times C([0,T]\to\mathcal{S}^{-p})\times S_{-p}$ then S is a complete separable metric space if we induce the metric d in Lemma 3.4 to S_{-p}. In virtue of Theorem 12.3 in [1] and Lemma 3.4, the distributions of S-valued random variables $(Z^\varepsilon, Y^\varepsilon, L^\varepsilon)$ are tight. Therefore we may assume that there is a sequence $\{\varepsilon_n\}_{n=1}^\infty$ with $\varepsilon_n \downarrow 0$ $(n \uparrow \infty)$ such that $(Z^n,Y^n,L^n) = (Z^{\varepsilon_n},Y^{\varepsilon_n},L^{\varepsilon_n})$ converges almost surely to an S-valued random variable (Z^0,X^0,L) where X^0 and L are stated before. For any $h \in \mathcal{S}^\infty$, and $0 \leq t \leq T$ we have

$$(3.8)\qquad (Z^0(t),h) = ((Z^0(t),h) - (Z^n(t),h)) + \int_0^t (L^n(s)(Z^n(s)-Z^0(s)),h)\,ds + \int_0^t ((L^n(s)-L)(Y^n(s)-X^0(s)),h)\,ds + \int_0^t ((L^n(s)-L)X^0(s),h)\,ds + \int_0^t ((L^n(s)-L)Z^0(s),h)\,ds + \int_0^t (LZ^0(s),h)\,ds$$

from the equation (3.5). It is clear that the first three terms in the right hand side of (3.8) go to 0 as n goes to ∞. Furthermore, we have

$$(3.9)\qquad \int_0^t ((L^n(s)-L)X^0(s),h)\,ds = \int_0^t (((L^n(s)-L)X^0(s),\Lambda^{2r}h))_{-r} = \int_0^T ((B^*(s)\Lambda^{2r}h,(L^n(s)^*-L^*)\Lambda^{2r}h))_{-p}$$

where $B^*(\cdot)\in L^2([0,T]\to\mathcal{H}^*_{-p})$ is defined by $B^*(s)g = cX^0(s)$ if g is written as $c\Lambda^{2r}h$ $(c\in R)$ and $0 \le s \le t$, $B^*(s) = 0$ otherwise. Here $\mathcal{H}^*_{-p}$ denotes the Hilbert space of all Hilbert-Schmidt operators from $\mathcal{S}^{-r}$ into $\mathcal{S}^{-p}$. Using the fact that $L^2([0,T]\to\mathcal{H}_{-p})$ and $L^2([0,T]\to\mathcal{H}^*_{-p})$ are isomorphic and isometric, we can show that the fourth term in the right hand side of (3.8) goes to 0 as n goes to ∞ in virtue of Lemma 3.4. The fifth term goes to 0 in the same way. Therefore we conclude that for any $h\in\mathcal{S}^\infty$ and $0 \le t \le T$, $(Z^0(t),h) = \int_0^t (LZ^0(s),h)\,ds$. Since the operators L and L^* both generate strongly continuous semi-groups on $\mathcal{S}^\infty$, it is not hard to prove that if $z(\cdot)\in C([0,T]\to\mathcal{S}^\infty)$ satisfies the equation $(z(t),h) = \int_0^t (Lz(s),h)\,ds$ and $z(0) = 0$ for any $h\in\mathcal{S}^\infty$ and $0 \le t \le T$, then $z(t) = 0$ for any $0 \le t \le T$. Hence $Z^0(\cdot)$ vanishes almost surely. This completes the proof of Theorem 2.1. //

References

[1] P. Billingsley, Convergence of probability measures, John Wiley and Sons, New York 1968.

[2] R. Z. Khas'minskii, On the stochastic process defined by differential equations with a small parameter, Theory prob. appl. 11, (1966) 211-228.

[3] H. Kumano-go, Pseudo-differential Operators, MIT Press, Cambridge, Massachusetts and London 1981.

[4] T. Morita, A fluctuation theorem associated with Cauchy problems for stationary random operators, to appear.

[5] M. Taylor, Pseudodifferential operators, Princeton University Press, Princeton 1981.

Department of Mathematics
Faculty of Science
Osaka University
Toyonaka, Osaka 560
Japan

CONVERGENCE AND UNIQUENESS THEOREMS FOR MARKOV PROCESSES ASSOCIATED WITH LÉVY OPERATORS

Akira Negoro and Masaaki Tsuchiya

Let us consider time homogeneous Lévy operators, that is, operators of the form:

$$Lf(x) = \frac{1}{2}\sum_{i,j=1}^{d} a^{ij}(x)(\partial^2/\partial x_i\partial x_j)f(x) + \sum_{i=1}^{d} b^i(x)(\partial/\partial x_i)f(x) + c(x)f(x)$$

$$+ \int_{R^d}\{f(x+y)-f(x)-(1+|y|^2)^{-1}\sum_{i=1}^{d}(\partial/\partial x_i)f(x)y_i\}\nu(x,dy),$$

where $a(x) = (a^{ij}(x))$ is an S_d-valued Borel function on R^d (S_d is the set of symmetric non-negative definite $d\times d$-matrices), $b(x)=(b^i(x))$ is an R^d-valued Borel function on R^d, $c(x)$ is a real Borel function on R^d and $\nu(x,dy)$ is a Lévy measure, i.e., it satisfies $\nu(x,\{0\}) = 0$ and $\int_{R^d} |y|^2(1+|y|^2)^{-1}\nu(x,dy) < \infty$. Then we say that L is the Lévy operator made of the data $[a, b, c; \nu]$.

Our purpose is to give convergence and uniqueness results for Markov processes associated with Lévy operators. The convergence problem for such Markov processes has been extensively studied by Skorohod [6]. Under his Convergence Condition I and Condition D, he shows the convergence of finite dimensional distributions of Markov processes. We will give another convergence condition, which is a natural extension of that for infinitely divisible distributions and is somewhat different from Skorohod's one. Furthermore, in some cases, our condition is represented in terms of data of Lévy operators. Skorohod's Condition D consists of Conditions D.I and D.II. Condition D.I means that the Markov processes have Lévy operators as pre-infinitesimal generators and that the Lévy operators map $C_b^2(R^d)$ into $C_b(R^d)$ (see §1 for the definition of these function spaces). Condition D.II is imposed only on the limit process and it means that the resolvent R_λ of the infinitesimal generator maps $C_b^2(R^d)$ into itself for sufficiently large λ. It guarantees the uniqueness of the limit distributions. He conjectures that if the data of the Lévy operator are smooth in x, then Condition D.II is fulfilled. We will give a partial positive answer to his conjecture (see Theorems 2.1 and 2.2).

1. Convergence condition.

First we introduce some notations. Let $C(R^d)$ be the space of real continuous functions on R^d. $C_b(R^d)$ and $C_0(R^d)$ stand for the subspaces of $C(R^d)$ consisting of those functions which, respectively, are bounded and vanish at infinity. Assume that these function spaces are normed by the supremum norm $\|\cdot\|_0$. The notation $\|f\|_0$ will be used also for any bounded function f. Given a positive integer n, denote by $C_b^n(R^d)$ and $C_0^n(R^d)$ the spaces of n times continuously differentiable functions on R^d whose derivatives of order up to and including n belong to $C_b(R^d)$ and $C_0(R^d)$, respectively. Define the norm $\|\cdot\|_n$ on these function spaces by $\|f\|_n = \sum_{|\alpha| \leqq n} \|f^{(\alpha)}\|_0$, where α denotes a multi-index and $f^{(\alpha)} = (\partial/\partial x)^{\alpha} f$. Let $C_b^{\infty}(R^d) = \bigcap_{n=1}^{\infty} C_b^n(R^d)$ and $C_0^{\infty}(R^d) = \bigcap_{n=1}^{\infty} C_0^n(R^d)$.

Now we introduce the conditions (L.I) and (L.II) on Lévy operators. Let L be the Lévy operator made of data $[a, b, c; \nu]$.

(L.I) (1) $a(x)$, $b(x)$ and $c(x)$ are bounded continuous.

(2) For $f \in C_b(R^d)$,

$$\int_{R^d} f(y)|y|^2(1+|y|^2)^{-1}\nu(x,dy)$$

is bounded continuous.

(3) $\|\int_{|y| \leqq \varepsilon} |y|^2\, \nu(\cdot,dy)\|_0 \to 0$ as $\varepsilon \to 0$,

$\|\int_{|y|>\ell} \nu(\cdot,dy)\|_0 \to 0$ as $\ell \to \infty$.

(L.II) (1) $a(x)$, $b(x)$ and $c(x)$ are bounded continuous.

(2) For every $x \in R^d$, $\nu(x,dy)$ is absolutely continuous with respect to a common measure $\nu(dy)$ on R^d such that $\nu(\{0\}) = 0$ and $\int_{R^d} |y|^2(1+|y|^2)^{-1}\nu(dy) < \infty$, that is, $\nu(x,dy) = d(x,y)\nu(dy)$. Furthermore, $d(x,y)$ is bounded and for each y, $d(x,y)$ is continuous in x.

It is easy to see that (L.II) implies (L.I). Under the condition (L.II), we may and do assume that $0 \leqq d \leqq 1$.

Next we introduce a convergence condition on Lévy operators. Let L_n $(n = 1, 2, \cdots)$ and L be Lévy operators with data $[a_n, b_n, c_n; \nu_n]$ $(n = 1, 2, \cdots)$ and $[a, b, c; \nu]$ respectively, and each of them satisfies (L.I). Then we say that $L_n \to L$ as $n \to \infty$ if

(1) $\lim_{\varepsilon \downarrow 0} \limsup_{n\to\infty} \|\int_{|y| \leqq \varepsilon} y_i y_j \nu_n(\cdot,dy) + a_n^{ij} - a^{ij}\|_0 = 0$

for $i, j = 1, 2, \cdots, d$;

(2) $\lim_{n\to\infty} \|b_n^i - b^i\|_0 = 0$ for $i = 1, 2, \cdots, d$;

(3) $\lim_{n\to\infty} \|c_n - c\|_0 = 0$;

(4) there is a sequence $\varepsilon_k > 0$ such that $\varepsilon_k \to 0$ and, for every $f \in C_b(R^d)$,

$$\lim_{n\to\infty} \left\| \int_{|y|>\varepsilon_k} f(y)\nu_n(\cdot,dy) - \int_{|y|>\varepsilon_k} f(y)\nu(\cdot,dy) \right\|_0 = 0.$$

In practice it would be hard to verify the condition (4). So, under (L.II), we give another condition for the condition (4). Assume that each of L_n $(n = 1, 2, \cdots)$ and L satisfies (L.II) (thus $\nu_n(x,dy) = d_n(x,y)\nu_n(dy)$, $\nu(x,dy) = d(x,y)\nu(dy)$ and $0 \leqq d_n, d \leqq 1$). Then the condition (4) is replaced by the following conditions (4-1), (4-2) and (4-3):

(4-1) $\lim_{n\to\infty} \|d_n(\cdot,y) - d(\cdot,y)\|_0 = 0$ uniformly in y on any compact set in $R^d \setminus \{0\}$;

(4-2) $d(x,y) \in C(R^d \times (R^d \setminus \{0\}))$;

(4-3) $\lim_{n\to\infty} \int_{R^d} f(y)\nu_n(dy) = \int_{R^d} f(y)\nu(dy)$

for every bounded continuous function f vanishing in some neighborhood of 0.

<u>Lemma 1.1.</u> If $L_n \to L$ as $n \to \infty$, then $\|L_n f - Lf\|_0 \to 0$ as $n \to \infty$ for every $f \in C_0^2(R^d)$.

<u>Remark.</u> Even if L_n $(n = 1, 2, \cdots)$ satisfy (L.I)(3) only, the conclusion holds provided that L satisfies (L.I) and that all the conditions of "$L_n \to L$ as $n \to \infty$" are satisfied.

Now we denote by $L^{(0)}$ the operator L with $c = 0$, that is, $L^{(0)}$ is the Lévy operator made of data $[a, b, 0; \nu]$. Then, taking $C_0^2(R^d)$ as the class of test functions, we define the martingale problem for $L^{(0)}$ on the Skorohod space $D([0,\infty), R^d)$ in the usual way (cf. Stroock [7]).

Let L_n $(n = 1, 2, \cdots)$ and L be Lévy operators which satisfy the condition (L.I) each. Then we have

<u>Theorem 1.2.</u> Let P_n be a solution to the martingale problem for $L_n^{(0)}$ and satisfy $\lim_{\ell\to\infty} \sup_n P_n[|x(0)| \geqq \ell] = 0$. Assume that $\|L_n^{(0)} f -$

$L^{(0)}f\|_0 \longrightarrow 0$ as $n \longrightarrow \infty$ for every $f \in C_0^2(R^d)$. Then $\{P_n\}$ is tight and any limit point in the weak topology is a solution to the martingale problem for $L^{(0)}$.

Remark. The remark to Lemma 1.1 is valid also to the theorem. From the theorem it follows that if the uniqueness of solutions to the martingale problem for $L^{(0)}$ holds, then P_n tends to the unique solution to the martingale problem for $L^{(0)}$ as $n \longrightarrow \infty$ in the weak topology.

Corollary 1.3. Assume that $L_n \longrightarrow L$ as $n \longrightarrow \infty$ and that the uniqueness of solutions to the martingale problems for $L_n^{(0)}$ $(n = 1, 2, \cdots)$ and $L^{(0)}$ holds, respectively. Let $\{T_n(t)\}$ and $\{T(t)\}$ be the strongly continuous non-negative semigroups on $C_0(R^d)$ associated with L_n and L, respectively. Then

$$\lim_{n\to\infty} T_n(t)f(x) = T(t)f(x)$$

for every $t \geqq 0$, $f \in C_b(R^d)$ and $x \in R^d$.

2. Uniqueness result.

In this section, we will consider the martingale problem for L with data $[a, b, c; \nu]$. So hereafter we assume that $c = 0$. Futhermore we also assume that the condition (L.II) holds for L, that is, L has the following form:

$$Lf(x) = \frac{1}{2}\sum_{i,j=1}^{d} a^{ij}(x)(\partial^2/\partial x_i\partial x_j)f(x) + \sum_{i=1}^{d} b^i(x)(\partial/\partial x_i)f(x)$$
$$+ \int_{R^d} \{f(x+y)-f(x)-(1+|y|^2)^{-1}\sum_{i=1}^{d}(\partial/\partial x_i)f(x)y_i\}d(x,y)\nu(dy).$$

Now we introduce a third condition:

(L.III) (1) $a^{ij}(\cdot),\ b^i(\cdot) \in C_b^2(R^d)$ $(i,j = 1, 2, \cdots, d)$,

$d(\cdot,y) \in C_b^2(R^d)$ for every y and

$$\int_{R^d} |y|^2(1+|y|^2)^{-1}\|d(\cdot,y)\|_2\nu(dy) < \infty.$$

(2) $$\int_{|y|\leq 1} |y|\,\|(\partial/\partial x_i)d(\cdot,y)\|_0\nu(dy) < \infty \quad (i = 1, 2, \cdots, d).$$

Here we mean that $\|d(\cdot,y)\|_n = \sum_{|\alpha|\leqq n} \|(\partial/\partial x)^\alpha d(\cdot,y)\|_0$.

Then we have the following uniqueness result which is an extension of Theorems 2 and 3.2 in Sato [4].

<u>Theorem 2.1.</u> Let L satisfy (L.II) and (L.III). Then the uniqueness of solutions to the martingale problem for L holds.

Under the same assumption with the theorem, the existence of solutions follows from Theorem (2.2) in Stroock [7]. Therefore Theorem 2.1 implies that there exists a unique strongly continuous non-negative semigroup $(T(t))$ on $C_0(R^d)$ whose infinitesimal generator is an extension of L that acts on $C_0^2(R^d)$. We say that $(T(t))$ is the semigroup associated with L. Moreover if the coefficients of L are smooth, then the semigroup has a nice regularity. Before stating it precisely, let us introduce a fourth condition:

(L.IV) (1) $a^{ij}(\cdot),\ b^i(\cdot) \in C_b^\infty(R^d)$ $(i,j = 1, 2, \cdots, d)$,

$d(\cdot,y) \in C_b^\infty(R^d)$ for every y and

$$\int_{R^d} |y|^2(1+|y|^2)^{-1}\|d(\cdot,y)\|_n \nu(dy) < \infty \quad (n = 0, 1, 2, \cdots).$$

(2) $$\int_{|y|\leqq 1} |y|\,\|(\partial/\partial x_i)d(\cdot,y)\|_0 \nu(dy) < \infty \quad (i = 1, 2, \cdots, d).$$

<u>Theorem 2.2.</u> If L satisfies (L.II) and (L.IV), then

(1) $T(t)\colon\ C_0^\infty(R^d) \to C_0^\infty(R^d)$ $(t \geqq 0)$;

(2) for any positive integer n, there exist positive constants K_n and λ_n such that

$$\|T(t)\varphi\|_n \leqq K_n e^{\lambda_n t}\|\varphi\|_n \tag{2.1}$$

for every $t \geqq 0$ and $\varphi \in C_0^\infty(R^d)$, where K_n and λ_n depend only on the supremum norm of the derivatives in x of order up to and including n of a, b and d.

To describe a key lemma for proving the theorems, we introduce a new condition on L with data $[a, b, 0; \nu]$.

(L.V) (1) a is uniformly positive definite, i.e., there exists a positive constant μ such that

$$\sum_{i,j=1}^{d} a^{ij}(x)\xi_i\xi_j \geqq \mu|\xi|^2$$

for every x, $\xi = (\xi_i) \in R^d$.

(2) $(\partial/\partial x)^\alpha d(x,y)$ is bounded for any multi-index α.

(3) The total mass of $\nu(dy)$ is finite.

Assuming (L.V) for L, we set

$$Af(x) = \frac{1}{2}\sum_{i,j=1}^{d} a^{ij}(x)(\partial^2/\partial x_i\partial x_j)f(x) + \sum_{i=1}^{d} b^i(x)(\partial/\partial x_i)f(x)$$

and

$$Bf(x) = \int_{R^d} \{f(x+y)-f(x)\}d(x,y)\nu(dy).$$

Then, by substituting $b(x)$ suitably, L is rewritten as $L = A + B$. Under the conditions (L.II), (L.IV) and (L.V), the martingale problem for L has a unique solution and moreover the associated semigroup $\{T(t)\}$ has the following property.

Lemma 2.3. Assume that L satisfies (L.II), (L.IV) and (L.V). Then, for $\varphi \in C_0^\infty(R^d)$ and $T > 0$, $u(t,x) \equiv T(t)\varphi(x)$ belongs to $C_b^\infty([0,T]\times R^d)$ and, further, the consequence of Theorem 2.2 holds.

Outline of the proof of Lemma 2.3. Let $\mathcal{L} = \partial/\partial t - L$ and $\mathcal{A} = \partial/\partial t - A$. For a non-integral positive number ℓ, following [1], we denote by $H^{\ell}(R^d)$ and $H^{\ell/2,\ell}(\bar{Q}_T)$ the Hölder spaces with exponent ℓ and exponents $\ell/2$, ℓ in the supremum norm on R^d and $\bar{Q}_T = [0,T]\times R^d$, respectively. Hereafter, γ denotes a fixed number such that $0 < \gamma < 1$. Given $f \in H^{\gamma/2,\gamma}(\bar{Q}_T)$ and $\varphi \in H^{\gamma+2}(R^d)$, we can construct a unique solution $u \in H^{(\gamma/2)+1,\gamma+2}(\bar{Q}_T)$ of the equation

$$(2.2)\qquad \begin{cases} \mathcal{L}u = f \\ u|_{t=0} = \varphi. \end{cases}$$

The construction of the solution is carried out by the use of the same idea as in §§7-8 of Chap.IV in [1] regarding B as a perturbation term. Particularly, let us consider the equation (2.2) with $f = 0$ and $\varphi \in C_0^\infty(R^d)$, that is,

$$(2.3)\qquad \begin{cases} \mathcal{L}u = 0 \\ u|_{t=0} = \varphi. \end{cases}$$

If we set $f = Bu$, then the equation (2.3) is rewritten as

$$(2.4)\qquad \begin{cases} \mathcal{A}u = f \\ u|_{t=0} = \varphi. \end{cases}$$

Since $f = Bu \in H^{(\gamma/2)+1,\gamma+2}(\bar{Q}_T)$, using Theorem 5.1 in [1], $u \in H^{(\gamma/2)+2,\gamma+4}(\bar{Q}_T)$ and so on. Therefore $u \in C_b^\infty(\bar{Q}_T)$ and this proves the first assertion by using the uniqueness of the semigroup associated with L. In the equation (2.4), we can verify that if f $(\in C_b^\infty(\bar{Q}_T))$ has the property $(\#)_0$:

$$f(t,\cdot) \in C_0(R^d) \quad (t \in [0,T]),$$

then the solution u has the property $(\#)_2$:

$$u(t,\cdot) \in C_0^2(R^d) \quad (t \in [0,T]).$$

On the other hand, checking each stage of the construction of the solution of (2.3), we see that u has the property $(\#)_2$. So, by making the operation $\partial/\partial x_k$ to the both sides of (2.4) with $f = Bu$, we have

$$\begin{cases} \mathscr{A}\tilde{u} = g \\ \tilde{u}|_{t=0} = \tilde{\varphi}, \end{cases}$$

where $\tilde{u} = (\partial/\partial x_k)u \ (\in C_b^\infty(\bar{Q}_T))$, $\tilde{\varphi} = (\partial/\partial x_k)\varphi \ (\in C_0^\infty(R^d))$ and g is a function of $C_b^\infty(\bar{Q}_T)$ with the property $(\#)_0$. Therefore $\tilde{u}$ has the property $(\#)_2$, so g has the property $(\#)_1$:

$$g(t,\cdot) \in C_0^1(R^d) \quad (t \in [0,T]).$$

Thus, repeating this argument, we see that $u(t,\cdot) = T(t)\varphi \in C_0^\infty(R^d)$. Next we show the inequality (2.1). To see this, let us consider the function

$$W_\ell = \sum_{|\alpha|=\ell} ((T(t)\varphi)^{(\alpha)})^2$$

as in the case of differential equations (cf.[8], p.75). For W_ℓ, we consider the equation

$$(\partial/\partial t)W_\ell = LW_\ell + S_\ell$$

and by using the inequality on page 76, line 9 in [8], we majorize the remainder term S_ℓ on $[0,\tau]\times R^d$ as

$$S_\ell \leqq c \sum_{k=0}^{\ell} \|W_k\|_{0,\tau}$$

with some constant c, where $\|W_k\|_{0,\tau}$ denotes the supremum norm of W_k on $[0,\tau]\times R^d$. For $p > 0$ and $q = c\sum_{k=0}^{\ell}\|W_k\|_{0,\tau}$, let

$$J_\ell(t,x) = e^{-pt}W_\ell(t,x) - qt \qquad (t \geqq 0, \quad x \in R^d).$$

Then, by using similar technique to Norman [3], we obtain the inequality

$$\sum_{\ell=0}^{n} \Big\| \sum_{|\alpha|=\ell} ((T(t)\varphi)^{(\alpha)})^2 \Big\|_0 \leqq e^{\mu_n t} \sum_{\ell=0}^{n} \Big\| \sum_{|\alpha|=\ell} (\varphi^{(\alpha)})^2 \Big\|_0$$

for some constant μ_n. This implies (2.1).

<u>Outline of the proof of Theorem 2.1.</u> Given $\varepsilon > 0$, take a mollifier ρ_ε for which the support of ρ_ε is contained in $\{|x| \leqq \varepsilon\}$. Let $a_\varepsilon^{ij} = a^{ij}*\rho_\varepsilon + \varepsilon\delta^{ij}$ (δ^{ij} is the Kronecker symbol), $b_\varepsilon^i = b^{ij}*\rho_\varepsilon$, $d_\varepsilon(\cdot,y) = d(\cdot,y)*\rho_\varepsilon$, and $\nu_\varepsilon(dy) = I_{\{|y|>\varepsilon\}}(y)\nu(dy)$. Define L_ε as the Lévy

operator made of the data $[a_\varepsilon, b_\varepsilon, 0; d_\varepsilon \nu_\varepsilon]$ and $(T_\varepsilon(t))$ denotes the semigroup associated with L_ε. Then, in the same way as in the proof of Lemma 2.3, we can show that there exist constants (independent of ε) K_2 and λ_2 such that

$$\|T_\varepsilon(t)\varphi\|_2 \leqq K_2 e^{\lambda_2 t} \|\varphi\|_2$$

for $t \geqq 0$ and $\varphi \in C_0^\infty(R^d)$. This yields that for $\lambda > \lambda_2$ the range of $\lambda - L$ acting on $C_0^2(R^d)$ is dense in $C_0(R^d)$. Therefore it follows from Theorem 1.2 in [5] that the closure of L generates a strongly continuous non-negative semigroup on $C_0(R^d)$. This shows the uniqueness of solutions to the martingale problem for L (see Proposition 1 in [9], Theorem 5.2 (p.43) in [2]).

Theorem 2.2 is proved in almost the same way as Lemma 2.3.

Acknowledgment. The authors are grateful to Professors R. Kondō, Y. Ogura and K. Sato for their valuable advice and suggestions.

References

[1] O.A. Ladyženskaja, V.A. Solonnikov and N.N. Ural'ceva: Linear and Quasilinear Equations of Parabolic Type, (English translation) Amer. Math. Soc. Providence (1968).

[2] T.M. Liggett: Interacting Particle Systems, Springer-Verlag, New York (1985).

[3] M.F. Norman: A "psychological" proof that certain Markov semigroups preserve differentiability, SIAM-AMS Proc. 13 (1981), 197-211.

[4] K. Sato: Integration of the generalized Kolmogorov-Feller backward equations, J. Fac. Sci. Univ. Tokyo Sec. I, 9 (1961), 13-27.

[5] K. Sato and T.Ueno: Multi-dimensional diffusion and Markov process on the boundary, J. Math. Kyoto Univ. 4 (1965), 529-605.

[6] A.V. Skorohod: Limit theorems for Markov processes, Theor. Prob. Appl. 3 (1958), 202-246.

[7] D.W. Stroock: Diffusion processes associated with Lévy generators, Z. Wahrsh. Verw. Geb. 32 (1975), 209-244.

[8] D.W. Stroock and S.R.S. Varadhan: Multidimensional Diffusion Processes, Springer-Verlag, New York (1979).

[9] M. Tsuchiya: Martingale problems and semigroups, Ann. Sci. Kanazawa Univ. 21 (1984), 19-22.

Akira Negoro
Faculty of Liberal Arts
Shizuoka University
Shizuoka
422 Japan

Masaaki Tsuchiya
College of Liberal Arts
Kanazawa University
Kanazawa
920 Japan

BOUNDS FOR DIFFERENCE OF TWO INTEGRALS OF A BOUNDED FUNCTION IN TERMS OF EXTENSIONS OF LÉVY METRIC

Yoshiko Nogami and James Hannan

1. Introduction.

Usual Lévy metric $d(F,G)$ is defined on the family of cumulative distribution functions (cdf's) of probability measures on the real line R (cf. e.g. Feller [2], p. 285) so that for any cdf's F and G,

$$d(F,G) = \inf\{\varepsilon \geq 0: F(x-\varepsilon) - \varepsilon \leq G(x) \leq F(x+\varepsilon) + \varepsilon, \text{ for all } x\varepsilon R\}.$$

We directly extend this definition to the family $\mathcal{F}$ of bounded nondecreasing functions on R to get a pseudo metric (cf. Dugundji [1]) L We furthermore extend this L to the family $\mathcal{M}$ of measures on R induced by the elements in $\mathcal{F}$; For μ and ν in $\mathcal{M}$,

$$\rho(\mu,\nu) = \inf_{r\varepsilon R} L(F,r+G)$$

where F and G are bounded nondecreasing functions inducing μ and ν, respectively.

As a previous work, by letting cdf's F and G be represented respective probability measures and h a bounded function, Oaten gave, in Lemmas 8 and 8' of Appendix of [5] (cf. [6] p. 1179, also), bounds for $|\int_R h\, dF - \int_R h\, dG|$ in terms of Lévy metric $d(F,G)$ and the modulus of continuity of h to be defined in Section 3.

In this paper, bounds for the similar difference of two integrals of a bounded function h to Oaten's are obtained in terms of above extensions of the Lévy metric (Theorem 1(3) and (4) and Theorem 2 in Section 3). The bounds (3) and (4) in Theorem 1 are slightly strengthened generalizations of the bounds in Lemmas 8' (corrected by replacing 3 by 4 in the bound) and 8, respectively. (Oaten's bounds are parameterized by $\lambda > L$ and $\lambda > 2L$, respectively and are improved by the $\mu(I) = \nu(I) = 1$ specialization of (3) and (4) with k taken to be the least integer greater than $(b-a)/\lambda$.) Although these theorems were used to get a rate of the risk convergence in more general compound decision problems (cf. Nogami [4]) than Oaten's [6] these bounds themselves may be interesting results to researchers in other fields.

In the next section we introduce Lemmas 1 and 2; Lemma 1 states that the infimum of the definition of ρ is attained and Lemma 2 shows that L is also defined by the supremum of the difference of two quantiles. Although these lemmas are themselves interesting results, we introduce these lemmas to prove forthcoming Theorem 1(4) and Theorem 2.

Let $\vee$ and $\wedge$ be the supremum and the infimum, respectively. We will use + (-) in the subscript position to denote the positive (negative) part, and + (-) on the line to denote right (left) limit. The symbol ∎ denotes the end of the proof. $\doteq$ denotes the defining property.

2. Extensions of Lévy Metric.

In this section we first investigate the properties of two extensions L and ρ of Lévy metric.

For F and Gε $\mathcal{F}$ L(F,G) will be defined by the infimum of $\varepsilon \geq 0$ such that $F(x-\varepsilon) - \varepsilon \leq G(x) \leq F(x+\varepsilon) + \varepsilon$ for all $x \in R$. Let $F^{\circ}(x) \doteq x+F(x)$. For every $r \in R$, let $S_r \doteq S_r(F,G)$ be the interval defined by

$$S_r \doteq \{\varepsilon \geq 0 : F^{\circ}(x-\varepsilon) \leq r + G^{\circ}(x) \leq F^{\circ}(x+\varepsilon), \text{ for all } x \in R\}.$$

Note that (i) replacement of the above inequalities by strict ones throughout would mean, at most, to subtract an end point from S_r, (ii) replacement of R by a dense subset of R would mean, at most, to add an end point to S_r. Therefore neither would affect definitions which follow. The Lévy distance L of F and G in $\mathcal{F}$ is defined by

$$(1) \qquad L(F,G) = \wedge S_0.$$

We furthermore note that (iii) for right continuous F and G, $S_r(F,G)$ is closed.

We define another distance function ρ on M as follows: for any F and G in $\mathcal{F}$,

$$\rho(F,G) = \bigwedge_{r \in R} L(F, r+G).$$

Note that ρ is invariant under translations of values of F and G.

Since functions in $\mathcal{F}$ which differ only by a constant except at discontinuity points induce the same measure, ρ is actually a metric on M:

$$\rho(\mu,\nu) = \rho(F,G) \ (\doteq \rho)$$

where μ and ν are measures in M induced by F and G, respectively. Since $\bigwedge_r (\wedge S_r) = \wedge (\bigcup_r S_r)$ for any family of subsets S_r of extended real line, we see that

(2) $\rho = \wedge (\bigcup_r S_r)$.

Lemma 1 below will be applied to prove Theorem 1(4) in Section 3.

Lemma 1. The infimum in the definition of ρ is attained.

Proof. Pick a sequence $\{\varepsilon_n\}$ of numbers which strictly decreases to ρ. Then, by (2) there exists r_n such that

$$-r_n + F^\circ(\cdot-\varepsilon_n) \le G^\circ(\cdot) \le -r_n + F^\circ(\cdot+\varepsilon_n).$$

Thus, taking $\overline{\lim}$ and $\underline{\lim}$ on the lhs and the rhs respectively, leads to

$$-\underline{\lim}\, r_n + F^\circ(\cdot-\rho-) \le G^\circ(\cdot) \le -\overline{\lim}\, r_n + F^\circ(\cdot+\rho+).$$

Therefore, for every $r \in [\underline{\lim}\, r_n, \overline{\lim}\, r_n]$, $L(F,r+G) = \rho$. ∎

For each $F \in \mathcal{F}$ and $t \in R$, let t_{F° denote the t-th quantile of F°. Note that $t \to t_{F^\circ}$ maps R onto R. Define η by

$$\eta(F,G) = \bigvee_{t \in R} |t_{F^\circ} - t_{G^\circ}|.$$

Following Lemma 2 will be furnished to prove forthcoming Theorem 1(4) and Theorem 2 in Section 3. That L is a pseudo metric will be seen from Lemma 2 below.

Lemma 2. $L = \eta$.

Proof. To show $L \le \eta$ we first have by the definition of the t-th quantiles that

$$F^\circ(t_{F^\circ}-) \le G^\circ(t_{G^\circ}+) \text{ and } G^\circ(t_{G^\circ}-) \le F^\circ(t_{F^\circ}+).$$

Hence, for every $\delta > 0$

$$F^\circ(t_{G^\circ}-\eta-\delta) \le F^\circ(t_{F^\circ}-\delta) \le G^\circ(t_{G^\circ}+\delta)$$

and

$$G^\circ(t_{G^\circ}-\delta) \le F^\circ(t_{F^\circ}+\delta) \le F^\circ(t_{G^\circ}+\eta+\delta).$$

Since the mapping $t \to t_{G^\circ}$ is onto, these inequalities show that $L(F,G) \le \eta(F,G) + 2\delta$ and thus $L \le \eta$.

On the other hand, if $L(F,G) < \varepsilon$, then $F^\circ(\cdot-\varepsilon) \le G^\circ(\cdot) \le F^\circ(\cdot+\varepsilon)$. Since by the definition of $F^\circ(\cdot+s)$, (the t-th quantile of $F^\circ(\cdot+s)$) $= t_{F^\circ} - s$ and since the t-th quantiles have the opposite ordering, above inequalities leads to $t_{F^\circ}+\varepsilon \ge t_{G^\circ} \ge t_{F^\circ} - \varepsilon$. Thus $\eta(F,G) \le \varepsilon$ and therefore $\eta \le L$. ∎

3. Bounds for Difference of Two Integrals of a Bounded Function.

The main work in this section is Theorems 1 and 2 which are extensions of Oaten's work to the family of distribution functions of measures on $(R,\mathcal{B})$. Theorem 1 is a unified generalization of Lemmas 8' and 8 of Oaten [5] and its proof evolves from those of Oaten's results. Theorem 2 gives another family of bounds for the same difference of integrals. By letting h be a measurable function, the bound in Theorem 1(4) includes $(\mu+\nu)$-modulus of continuity of h with two measures μ and ν in M. Other bounds include the modulus of continuity of h.

We shall define the modulus of continuity of h and τ-modulus of continuity of h for a measure τ, beforehand.

Definition 1. With h, a function defined on a real interval I, the modulus of continuity of h is the function given by

$$\alpha(\varepsilon) = \sup\{|h(\omega_1) - h(\omega_2)| \ : \omega_1, \omega_2 \in I, \ |\omega_1 - \omega_2| < \varepsilon\}$$

for every $\varepsilon > 0$.

Definition 2. With h measurable on a real interval I supporting a finite measure τ,

$$\tau\text{-}\sup h = \inf\{\delta \in R: \ \tau[h > \delta] = 0\},$$

$$\tau\text{-}\inf h = \sup\{\delta \in R: \ \tau[h < \delta] = 0\}$$

and, with $\tau_{r\varepsilon}$ denoting the restriction of τ to the interval $(r-\varepsilon/2, r+\varepsilon/2)$, the τ-modulus of continuity of h is the function given by

$$\tau\text{-}\alpha(\varepsilon) = \sup\{\tau_{r\varepsilon}\text{-}\sup h - \tau_{r\varepsilon}\text{-}\inf h: \ r \in I\}$$

for every $\varepsilon > 0$.

Theorem 1. Let I be a finite interval {a,b} supporting finite measures μ and ν and let h be a measurable function on I into a finite interval [c,d]. By abbreviating $\rho(\mu,\nu)$ to ρ and $L(\mu[a,\cdot], \nu[a,\cdot])$ to L, $|\int hd(\mu-\nu)|$ has the following families of upper bounds:

(3) $$\alpha((\tfrac{b-a}{k} \vee L)+)\{(k-1)L+|\mu-\nu|(I)+2(\mu(I)\wedge\nu(I))\}+((-c)\vee d)|\mu(I)-\nu(I)|,$$
$\forall$ positive integer k

(4) $$(d-c)k\rho+(\mu+\nu)\text{-}\alpha(\rho+(\tfrac{b-a}{k} \vee (2\rho))+)(\mu(I)\wedge\nu(I))+(c \vee (-d))|\mu(I)-\nu(I)|,$$
$\forall$ positive integer k

where the minimum of the bound(3) (respectively (4)), is attained for some $k < \frac{b-a}{L} + 1$ (respectively $k < \frac{b-a}{2\rho} + 1$).

To explain the bound(3) in above Theorem 1, we, for example, assume the Lipschitz condition on h; $(\alpha_h \doteq) \sup\{|h(\omega_1) - h(\omega_2)|/|\omega_1 - \omega_2| : \omega_1, \omega_2 \in I\} < +\infty$. Then we get that $\alpha(\varepsilon) \le \alpha_h \cdot \varepsilon$ and hence for an integer k_0 such that $k_0 < (b-a)/L + 1 \le k_0 + 1$, $\alpha(((b-a)/k_0) \vee L) = \alpha(L) \le \alpha_h \cdot L$. On the other hand, we get that $(k-1)L^2 \le (b-a)L$. Therefore, (3) becomes

$$\alpha_h\{(b-a) + |\mu-\nu|(I) + 2\mu(I)\}L + \sup_{x\in I}|h(x)|\cdot|\mu(I)-\nu(I)|.$$

Especially, if μ and ν are probability measures, the last term would vanish.

Proof of Theorem 1. For a given σ with $k-1 < (b-a)/\sigma < k$, let $\delta = k\sigma-(b-a)$ and let $x_j = a+j\sigma-2^{-1}\delta$ for $j = 0,1,2,\ldots,k$. Since $\sigma < (b-a)/(k-1)$, it follows that $\delta < \sigma$ and hence $(x_0+x_1)/2$ and $(x_{k-1}+x_k)/2$ both lie inside the interval I.

Proof of (3). Note $L < (b-a)/(k-1)$ and take $\sigma > ((b-a)/k) \vee L$. Let $h_j = h((x_{j-1}+x_j)/2)$ for $j = 1,2,\ldots,k$. Then, $|h(x)-h_j| \le \alpha(2^{-1}\sigma+)$ for each $x\in(x_{j-1},x_j]$, and $|h_j-h_{j+1}| \le \alpha(\sigma+)$ for each j.

Let $D_j = (\mu-\nu)(x_0,x_j]$, $j = 0,1,\ldots,k$. Then

$$\text{(5)}\qquad \begin{aligned}\int h\,d(\mu-\nu) &= \sum_{j=1}^{k}\left\{\int_{x_{j-1}+}^{x_j+}(h-h_j)\,d(\mu-\nu) + h_j(D_j - D_{j-1})\right\}\\ &\le \alpha(\tfrac{1}{2}\sigma+)(|\mu-\nu|(I)) + h_kD_k + \sum_{j=1}^{k-1}(h_j - h_{j+1})D_j.\end{aligned}$$

From $\sigma > L$,

$$\begin{aligned}(D_j)_+ &\le (\nu(x_j,x_{j+1}]+L) \wedge (\mu(x_{j-1},x_j]+L)\\ &\le \nu(x_j,x_{j+1}] \wedge \mu(x_{j-1},x_j]+L\end{aligned}$$

and, by an interchange of μ and ν,

$$(D_j)_- \le \mu(x_j,x_{j+1}] \wedge \nu(x_{j-1},x_j] + L.$$

Thus, henceforth, by abbreviating $\mu(I)$ and $\nu(I)$ to μ and ν,

$$\sum_{j=1}^{k-1}|D_j| \le 2(\mu \wedge \nu) + (k-1)L.$$

Therefore,

$$\sum_{j=1}^{k-1}(h_j-h_{j+1})D_j \le \alpha(\sigma+)\{2(\mu \wedge \nu) + (k-1)L\}.$$

Combining this with the inequality $h_k D_k \le d(\mu-\nu)_+ + (-c)(\nu-\mu)_+$, we obtain that

(6) $\quad \mathrm{lhs}(5) \le \alpha(\sigma+)\{|\mu-\nu|(I)+2(\mu\wedge\nu)+(k-1)L\}+d(\mu-\nu)_+ +(-c)(\nu-\mu)_+$.

Replacement of h by -h gives us

(7) $\quad -\mathrm{lhs}(5) \le \alpha(\sigma+)\{|\mu-\nu|(I)+2(\mu\wedge\nu)+(k-1)L\}+(-c)(\mu-\nu)_+ +d(\nu-\mu)_+$.

We obtain (3) by taking the maximum of rhs(6) and rhs(7), recognizing $(d(\mu-\nu)_+ +(-c)(\nu-\mu)_+)\vee((-c)(\mu-\nu)_+ +d(\nu-\mu)_+)=((-c)\vee d)|\mu-\nu|$ and letting σ decrease to $((b-a)/k)\vee L$.

Proof of (4). Since, by Lemma 1, $\rho = L(F,G)$ for some right-continuous bounded-nondecreasing functions F and G inducing μ and ν, it suffices to prove (4) with ρ replaced by $L \overset{\circ}{=} L(F,G)$. As in the proof of (3), note $2L < (b-a)/(k-1)$ and take $\sigma > ((b-a)/k)\vee 2L$.

By the definition of L we can find $x_0 = y_0 < y_1 < \ldots < y_k = x_k$ so that, for each j, $|x_j - y_j| \le L$ and

(8) $\quad F(y_j-) - L \le G(x_j) \le F(y_j) + L$

because

$$\bigcup_y\{[F(y-)-L,\ F(y)+L] : |y-x_j| \le L\}$$
$$= [F((x_j-L)-) - L,\ F(x_j+L) + L]$$

and the intervals $[x_j-L,\ x_j+L]$ strictly increase wrt j.

We extend the domain of h to the interval $[x_0, x_k]$ by defining $2h = c + d$ on complement of I.

For each j, let $\wedge_j \doteq x_j \wedge y_j$ and $\vee_j \doteq x_j \vee y_j$. Let $\tau \doteq \mu+\nu$, let τ_j denote the restriction of τ to the interval $[\wedge_j, \vee_{j+1}]$ and let $\underline{h}_j \doteq \tau_j\text{-inf } h$. Then, define functions h_1 and h_2 by $h_1(x_j, x_{j+1}] = h_2(y_j, y_{j+1}) = \underline{h}_j$ and $h_2(y_j) = h_2(y_j-) \vee h_2(y_j+)$. Now $h - h_1 \le \tau\text{-}\alpha(L+\sigma+)$ a.e. τ on $(x_j, x_{j+1}]$ because by definition of τ_j-inf h,

$$\forall \varepsilon > 0 \quad \tau((x_j, \vee_{j+1}]\cap\{h < \underline{h}_j+\varepsilon\})+\tau([\wedge_j, x_{j+1}]\cap\{h < \underline{h}_j+\varepsilon\}) > 0$$

so that if $\tau((x_j, x_{j+1}]\cap\{h-\underline{h}_j > \lambda\}) > 0$ then $\tau\text{-}\alpha(L+\sigma+) \ge \lambda-\varepsilon$ and $\ge\lambda$

Also $h_2 - h \le 0$ a.e. τ because $h \ge \underline{h}_{j-1}\vee\underline{h}_j \ge h_2$ a.e. τ on $[\wedge_j, \vee_j]$ and $h \ge \underline{h}_j = h_2$ a.e. τ on $(\vee_j, y_{j+1})$.

Let $r \in R$. If $h_2(y_{j-1}, y_{j+1}) \le r$, then $h_1(x_{j-1}, x_{j+1}] \le r$.

Conversely, $h_1(x_{j-1},x_{j+1}] \le r$ implies $h_2((y_{j-1},y_j)\cup(y_j,y_{j+1})) \le r$ and therefore $h_2(y_{j-1},y_{j+1}) \le r$. Hence $h_2^{-1}(-\infty,r]$ is the union of at most k/2 intervals of the form (y_i,y_j), and $h_1^{-1}(-\infty,r]$ is the union of the corresponding intervals $(x_i,x_j]$.

We note that, by two applications of (8), $\mu(y_i,y_j)\le\nu(x_i,x_j]+2L\ \forall_j$ so that

$$\mu h_2^{-1}(-\infty,r] \le \nu h_1^{-1}(-\infty,r]+kL.$$

By two usages of the Fubini representation of the integral (i.e. for a measure μ and a measurable function h, $\int h\, d\mu = \int_0^\infty \mu[h>t]dt - \int_0^\infty \mu[h<-t]dt$) of a nonnegative function in the rhs of the first equality below

$$(9)\qquad \int h_1 d\nu - \int h_2 d\mu - d(\nu(I)-\mu(I)) = \int (d-h_2)d\mu - \int (d-h_1)d\nu$$
$$= \int_0^{d-c} (\mu h_2^{-1}-\nu h_1^{-1})(-\infty,d-t]dt \le (d-c)kL.$$

Henceforth abbreviating $\tau\text{-}\alpha(\sigma+L+)$ to $\underline{\alpha}$, $\mu(I)$ to μ and $\nu(I)$ to ν, the triangle inequality and (9) bound $\int hd(\nu-\mu) - (d-c)kL$ by

$$(10)\qquad \int (h-h_1)d\nu + \int (h_2-h)d\mu + d(\nu-\mu) \le \underline{\alpha}\nu + d(\nu-\mu).$$

Applying (10) to -h with the measures interchanged gives the bound $\underline{\alpha}\mu+(-c)(\mu-\nu)$. The minimum of those bounds is the former or latter according as $\mu\ge$ or $\le\nu$ and therefore

$$(11)\qquad \int hd(\nu-\mu) \le (d-c)kL + \underline{\alpha}(\mu\wedge\nu) + c(\nu-\mu)_+ - d(\mu-\nu)_+.$$

Applying (11) to -h gives rhs(11) with c,d replaced by -d,-c and gives

$$\left|\int hd(\mu-\nu)\right| \le (d-c)kL + \underline{\alpha}(\mu\wedge\nu) + (c \vee (-d))|\mu-\nu|,$$

and (4) results on letting σ decrease to $((b-a)/k) \vee 2L$. ∎

In Theorem 2 below, the natural generalization of the inverse probability integral transformation is used to develop bounds for the same difference of integrals without recourse to partitioning.

<u>Theorem 2.</u> Let I, μ, ν and h be as in Theorem 1. Let F and G be bounded-nondecreasing functions inducing μ and ν with $\vee(a+) \le \wedge(b-)$ where $\vee$ and $\wedge$ abbreviate $F^\circ \vee G^\circ$ and $F^\circ \wedge G^\circ$. Then $|\int h\, d(\mu-\nu)|$ has the following family of bounds

$$(12)\qquad \frac{d-c}{2}\,\{|(F-G)(a+)|+|(F-G)(b-)|\}+\alpha(L(F,G)+)\{\wedge(b-)-\vee(a+)\} + \frac{d+c}{2}\,|\mu(I)-\nu(I)|.$$

Proof. Without loss of generality we can assume I is open. For, $\forall\,\varepsilon>0$ $\{a,b\}\subset(a-\varepsilon,b+\varepsilon)$ to which h is extendible with the same modulus of continuity and, if (12) holds with a,b replaced by $a-\varepsilon$, $b+\varepsilon$, then letting $\varepsilon\downarrow 0$ gives (12) with $I=\{a,b\}$.

Let $I=(a,b)$ and let f denote the map $t\to t_{F^\circ}$. Since $f^{-1}\{u\}=[F^\circ(u-), F^\circ(u+)]$ and F° is strictly increasing,

$$f^{-1}(\beta,\gamma) = \bigcup_{\beta<u<\gamma} f^{-1}\{u\} = (F^\circ(\beta+),\ F^\circ(\gamma-)).$$

Thus, Lebesgue measure and f induce the measure with F° as a corresponding bounded-nondecreasing function on the range of f. By the transformation theorem (cf. Halmos [3], p. 163),

$$(13)\qquad \int_I h\,dF^\circ = \int_{F^\circ(a+)}^{F^\circ(b-)} h(t_{F^\circ})\,dt.$$

Hereafter until the end of the proof we express a+ and b- without negative and positive signs.

Letting $\delta(H)=1$ or -1 according as $H=F$ or G, the difference of (13) for F and for G results in the following representation for $\int h\,d(\mu-\nu)$:

$$\int_{\wedge(a)}^{\vee(a)} \delta(S)h(t_{S^\circ})dt + \int_{\vee(a)}^{\wedge(b)} (h(t_{F^\circ})-h(t_{G^\circ}))dt + \int_{\wedge(b)}^{\vee(b)} \delta(T)h(t_{T^\circ})dt$$

where S and T have values in the set $\{F,G\}$ such that $S^\circ(a)=\wedge(a)$ and $T^\circ(b)=\vee(b)$. Hence, abbreviating L(F,G) to L hereafter and using Lemma 2, $\int h\,d(\mu-\nu)\le$

$$(14)\qquad d((G-F)(a))_+ + (-c)((F-G)(a))_+ + \alpha(L+)(\wedge(b)-\vee(a)) + d((F-G)(b))_+ + (-c)((G-F)(b))_+.$$

Applied to -h, (14) is altered only c,d changing to -d,-c:

$$(15)\qquad (-c)((G-F)(a)_+ + d((F-G)(a))_+ + \alpha(L+)(\wedge(b)-\vee(a)) + (-c)((F-G)(b))_+ + d((G-F)(b))_+.$$

Since $(14)+(15)=(d-c)\{|(F-G)(a)|+|(F-G)(b)|\}+2\alpha(L+)(\wedge(b)-\vee(a))$ and $(14)-(15)=(d+c)(\mu(I)-\nu(I))$, $(14)\vee(15)=(12)$. ∎

References

[1] J. Dugundji: Topology, Allyn and Bacon, INC., Boston. (1966).

[2] Williams Feller: An Introduction to Probability Theory and its Applications, Volume II (2nd ed.), Wiley, New York (1971).

[3] Paul R. Halmos: Measure Theory, Van Nostrand Reinhold, (1950).

[4] Y. Nogami: A rate of convergence for the set compound estimation in a family of certain retracted distributions, Ann. Inst. Statist. Math. 34 (1982), No. 2, 241-257.

[5] Allan Oaten: Approximation to Bayes risk in compound decision problems. RM-233, Department of Statistics and Probability, Michigan State University, (1969).

[6] Allan Oaten: Approximation to Bayes risk in compound decision problems, Ann. Statist. Math. 43 (1972), No. 4, 1164-1184.

Institute of Socio-Economic Planning
University of Tsukuba
Sakura, Niihari
Ibaraki-ken 305
Japan

Department of Statistics and Probability
Michigan State University
East Lansing
Mich. 48824
U.S.A.

ASYMPTOTIC EXPANSIONS FOR 2-SPRT

A. A. Novikov and V. P. Dragalin

1. Introduction. Let X_t be a process with independent homogeneous increments, $t\in R^+=[0,\infty)$ or $t\in Z^+=\{0,1,\cdots\}$. Suppose the distribution of X_t depends on a parameter $\theta\in\Theta\subset R$ and probability measures P_θ, $\theta\in\Theta$ generated by X_t are equivalent on σ-algebras $\mathcal{F}_t=\sigma\{X_s,\ s\le t\}$. We denote by $\frac{dp_\varphi}{dp_\theta}(t)$ the density of $P_\varphi(\cdot)$ with respect to $P_\theta(\cdot)$ on the σ-algebra $\mathcal{F}_t$.

As is well known the sequential probability ratio test (SPRT) is optimal (in proper sense) in the problem of testing of hypothesis

$$H_1 : \theta = \theta_1 \qquad \text{versus} \qquad H_2 : \theta = \theta_2 \qquad (\theta_1<\theta_2).$$

The disadvantages of the SPRT are unboundedness of the stopping time and large values (comparatively to the nonsequential Neyman-Pearson test) of $E_\theta\tau$ for $\theta\in(\theta_1,\theta_2)$ if one admits this set of parameters as "indifference" zone.

In this paper for the same problem of testing hypothesis we study asymptotic expansions of characteristics of the sequential test with stopping time

$$\tau(\varphi)=\min_{i=1,2}\tau_i(\varphi),\qquad \tau_i(\varphi)=\inf\{t:\log\frac{dp_\varphi}{dp_{\theta_i}}(t)\ge a_i\},\qquad a_i>0,\quad i=1,2.$$

The decision rule $d=d(\varphi)$ of this test is

$$d = 1 + I\left\{\log\frac{dp_\varphi}{dp_{\theta_1}}(\tau(\varphi))\ge a_1\right\},$$

where $I\{\cdot\}$ is an indicator function and $\{d=i\}=\{\text{reject } H_i\}$, $i=1,2$.

This test is called the double sequential probability ratio test (2-SPRT). Lorden [1] (for the case $t\in Z^+$) and Dragalin, Novikov [2] (for the case $t\in R^+$) proved that the 2-SPRT is an asymptotic solution of the so called modified Kiefer-Weiss problem. In addition Eisenberg [3], Huffman [4] (for the case $t\in Z^+$) and Dragalin, Novikov [2] (for the case $t\in Z^+$ and $t\in R^+$) showed that in case $P_\varphi(\cdot)$ is from the exponential family (see definition below) the same 2-SPRT is also optimal (with the proper chosen φ and a_i) for the general

Kiefer-Weiss problem. More exactly, it was shown in [2] that if $a_i = 1/\alpha_i$, α_i are given restrictions for probabilities of errors $P_{\theta_i} = \{d = i\}$, $i=1,2$ and $0 < \lim[|\log\alpha_1|/|\log\alpha_2|] < \infty$ then

$$\frac{\inf_{\theta} \sup E_\theta N}{\sup_{\theta} E_\theta \tau(\theta)} = 1 - O(|\log\alpha_1|^{-1}), \qquad \alpha_1 \to 0,$$

where the infimum is taken over all sequential tests (N,d) with given restrictions α_i for error probabilities. It is important to note that the stopping time of the 2-SPRT is bounded by some constant $t(\varphi)$ (see the proof of Theorem 1).

For practical use of this test it is necessary to have approximations of characteristics of the 2-SPRT. Some empirical results in this direction can be found in [1]. In Section 2 we present asymptotic expansions for error probabilities and expectation of stopping time of the 2-SPRT in the case of an exponential family. In Section 3 proofs of results are given. In Section 4 we discuss results of Monte-Carlo simulation.

2. Main results. We suppose that probability measures $P_\theta(t)$, belong to an exponential family, that is

$$\frac{dP_\varphi}{dP_\theta}(t) = \exp\{(\varphi-\theta)X_t - (b(\varphi)-b(\theta))t\}, \qquad \text{for all} \quad \theta,\varphi \in \Theta.$$

Here $b(\varphi)$ is a convex infinitely differentiable function on closed subintervals of Θ and such that

$$E_\varphi X_1 = b'(\varphi), \qquad \sigma^2(\varphi) = E_\varphi(X_1 - b'(\theta))^2 = b''(\varphi).$$

Denote

$$I_i = E_\varphi \log\frac{dP_\varphi}{dP_{\theta_i}}(1) = (\varphi-\theta_i)b'(\varphi) - (b(\varphi)-b(\theta_i)) \text{ (Kullback-Leibler numbers)},$$

$$\sigma^2 = \sigma^2(\varphi), \qquad \alpha_3 = E_\varphi(X_1 - b'(\varphi))^3, \qquad C_i = (\varphi-\theta_i)/I_i,$$

$$\mathfrak{X}_i(a_i) = \log\frac{dP_\varphi}{dP_{\theta_i}}(\tau_i(\varphi)) - a_i,$$

$$\gamma_i = \lim_{a_i\to\infty} E_\varphi \exp(-\mathfrak{X}_i(a_i)), \qquad \rho_i = \lim_{a_i\to\infty} E_\varphi \mathfrak{X}_i(a_i), \qquad i=1,2.$$

The existence of these limits is proved by Borovkov [5] (for the case $t\in Z^+$) and by Mogul'skii [6] (for the case $t \in R^+$). See also Woodroofe

[7], Siegmund [8]. The explicit formulae for γ_i and ρ_i can be found in [5] in terms of factorization components of the characteristic function.

Denote by $\Phi(x)$ and $f(x)$ distribution function and the density of a standard normal random variable $N(0,1)$.

Theorem 1. Let $a_1 \to \infty$ and

(1)
$$\frac{a_2}{I_2} = \frac{a_1}{I_1} + r(C_1 - C_2)\left(\frac{a_1}{I_2}\right)^{1/2},$$

where r is some constant. Then

1)
$$P_{\theta_i}\{d=i\} = \gamma_i \Phi((-1)^{i+1} r) e^{-a_i}(1+o(1)), \qquad i=1,2,$$

2)
$$E_\varphi \tau(\varphi) = \frac{a_1}{I_1} + \sigma A_1\left(\frac{a_1}{I_1}\right)^{1/2} + A_2 + o(1),$$

where $A_1 = (C_1 - C_2)(r\Phi(-r) - f(r))$,

$$A_2 = \frac{C_1 - C_2}{2}\left[C_1\sigma^2 - \frac{\alpha_3}{3\sigma^2}\right] rf(r) + \frac{\Phi(r)}{I_1}\rho_1 + \frac{\Phi(-r)}{I_2}\rho_2 .$$

The next theorem is a converse of Theorem 1.

Theorem 2. Let $a_1 \to \infty$ and

(2)
$$\frac{|\log\alpha_2|}{I_2} = \frac{|\log\alpha_1|}{I_1} + r\sigma(C_1 - C_2)\left(\frac{|\log\alpha_1|}{I_1}\right)^{1/2},$$

where r is some constant. Suppose that

(3)
$$a_i = |\log\alpha_i| + \log(\gamma_i \Phi((-1)^{i+1} r)), \qquad i=1,2.$$

Then

1')
$$P_{\theta_i}(d=i) = \alpha_i(1+o(1)), \qquad \text{as} \qquad \alpha_1 \to 0,$$

2')
$$E_\varphi \tau(\varphi) = \frac{|\log\alpha_1|}{I_1} + \sigma A_1\left(\frac{|\log\alpha_1|}{I_1}\right)^{1/2} + A_2' + o(1),$$

where $A_2' = A_2 + \frac{\Phi(r)}{I_1}\log(\gamma_1\Phi(r)) + \frac{\Phi(-r)}{I_2}\log(\gamma_2\Phi(-r))$.

Remark 1. The constant r appearing in Theorems 1 and 2 is a measure of asymmetry of the problem under consideration.

3. Proof of theorems. In notations

$$t(\varphi) = \left(\frac{a_2}{I_2} C_1 - \frac{a_1}{I_1} C_2\right)/(C_1-C_2), \qquad s(\varphi) = \left(\frac{a_2}{I_2} - t(\varphi)\right)/C_2,$$

the stopping times of 2-SPRT can be written as

$$(4) \qquad \tau_i(\varphi)=\inf\{t:(\mathrm{Sgn}C_i)(X_t-b'(\varphi)t)\geq(\mathrm{Sgn}C_i)\left(s(\varphi)+\frac{t(\varphi)-t}{C_i}\right)\}.$$

Note that $C_2<0<C_1$. Then the boundary of the continuation region for the process $\tilde{X}_t=X_t-b'(\varphi)t$ consists of two straight lines intersecting at $t=t(\varphi)$ (the maximum observation time), the value of the boundary at this point is $s(\varphi)$.

Proof of assertion 1). We consider only the case $i=1$ since the other is similar. First note that

$$(5) \quad P_{\theta_1}\{d=1\}=P_{\theta_1}\{\tau_1(\varphi)\leq\tau_2(\varphi)\}=P_{\theta_1}\{\tau_1(\varphi)\leq t(\varphi)\}-P_{\theta_1}\{\tau_2(\varphi)\leq\tau_1(\varphi)\leq t(\varphi)\}.$$

By equivalence of measures $P_{\theta_1}(\cdot)$ and $P_\varphi(\cdot)$ and by definition of the stopping time τ_1 we have

$$P_{\theta_1}\{\tau_1(\varphi)\leq\tau_2(\varphi)\} = E_\varphi \frac{dP_{\theta_1}}{dP_\varphi}(\tau_1)I\{\tau_1(\varphi)\leq t(\varphi)\}$$
$$= e^{-a_1}E_\varphi\exp\{-\mathfrak{X}_1(a_1)\}I\{\tau_1^*\leq r\} ,$$

where $\tau_1^*=(\tau_1 a_1/I_1)(|\varphi-\theta_1|\sigma(a_1/I_1^3)^{1/2})^{-1}$.

In the discrete time case, as well known, the random variables τ_1^* and $\mathfrak{X}_1(a_1)$ are asymptotically independent as $a_1\to\infty$; moreover the τ_1^* is asymptotically normal $N(0,1)$ ([8]). It is easy to obtain the analog of these results for the continuous time case. So we have

$$(6) \qquad P_{\theta_1}\{\tau_1(\theta)\leq t(\varphi)\} = e^{-a_1}\Phi(r)\gamma_1(1+o(1)), \qquad a_1\to\infty.$$

It remains only to show that the second term in (5) has a smaller order than the first one. Again by equivalence of measures $P_{\theta_1}(\cdot)$ and $P_\varphi(\cdot)$ we have

$$P_{\theta_1}\{\tau_2(\varphi)\le\tau_1(\varphi)\le t(\varphi)\} = E_\varphi \frac{dP_{\theta_1}}{dP_\varphi}(\tau_1) I\{\tau_2(\varphi)\le\tau_1(\varphi)\le t(\varphi)\}$$

$$\le e^{-a_1} P_\varphi\{\tau_2(\varphi)\le\tau_1(\varphi)\le t(\varphi)\} .$$

Choose a positive function $\psi=\psi(a_1)$ such that $\psi(a_1)\to\infty$ and $\psi(a_1)/\sqrt{a_1}\to 0$ as $a_1\to\infty$. Obviously we have

$$\begin{aligned} P_\varphi\{\tau_2(\varphi)\le\tau_1(\varphi)\le t(\varphi)\} &\le P_\varphi\{t-\psi(a_1)\le\tau_2(\varphi)\le t(\varphi)\} \\ &+ P_\varphi\{t-\tau_2(\varphi)>\psi(a_1), \tau_2(\varphi)\le\tau_1(\varphi)\le t(\varphi)\}. \end{aligned} \tag{7}$$

To estimate the first probability we note that a random variable (r.v.)

$$\tau_2^* = (\tau_2(\varphi)-a_2/I_2)\left(|\varphi-\theta_2|\sigma\left(\frac{a_2}{I_2^3}\right)^{1/2}\right)^{-1}$$

has asymptotically normal distribution (by the same reason as τ_1^* above). Since $\psi(a_1)a_1^{-1/2}\to 0$ and

$$\frac{t(\varphi)-a_2/I_2}{|\varphi-\theta_2|\sigma\sqrt{\frac{a_2}{I_2^3}}} = -r(1+o(1)) , \qquad \text{as } a_1\longrightarrow\infty ,$$

we have $P_\varphi\{t(\varphi)-\psi(a_1)\le\tau_2(\varphi)\le t(\varphi)\}=o(1)$.

The estimation of the second probability in (7) is more complicated. At the beginning let us note that

$$\{\tau_2(\varphi)\le\tau_1(\varphi)\le t(\varphi)\}$$

$$\subseteq \{\tau_2(\varphi)\le t(\varphi), \sup_{\tau_2(\varphi)\le u\le t} (\bar X_u-\bar X_{\tau_2(\varphi)})\ge(t(\varphi)-\tau_2(\varphi))\left(\frac{1}{C_1}-\frac{1}{C_2}\right)\},$$

where $\bar X_u=X_u+\frac{u}{C_1}$. This inclusion results from neglecting the excess of the process $\bar X_u$ over the boundary at time $\tau_2(\varphi)$. By homogeneity and independence of process $\bar X_u$ we have

$$P_\varphi\{t(\varphi)-\tau_2(\varphi)>\psi(a_1), \tau_2(\varphi)\le\tau_1(\varphi)\le t(\varphi)\}$$

$$\le E_\varphi[I\{t(\varphi)-\tau_2(\varphi)>\psi(a_1)\}G(t(\varphi)-\tau_2(\varphi))],$$

where

$$G(y) = P_\varphi\{ \sup_{0\le u\le y} \overline{X}_u \ge y\left(\frac{1}{C_1}-\frac{1}{C_2}\right)\}.$$

Using the mentioned result about asymptotical normality of the stopping time τ_1^* we obtain

$$(8) \qquad G(y) = \Phi(C\sqrt{y})(1+o(1)), \qquad \text{as } y \longrightarrow\infty,$$

where $C=(C_1C_2\sigma\sqrt{1-C_1/C_2})^{-1}$. Since the convergence is uniform here and $C<0$, then on the set $\{t(\varphi)-\tau_2(\varphi)>\psi(a_1)\}$ we have

$$G(t(\varphi)-\tau_2(\varphi)) = o(1), \qquad \text{as } a_1\longrightarrow\infty.$$

Hence the second probability in (7) has the same order $o(1)$ as $a_1\to\infty$. The assertion 1) of Theorem 1 is proved.

Remark 2. Huffman [4], in an analogous situation (for general Kiefer-Weiss problem for $t\in Z^+$), obtained an asymptotic result for probabilities of errors without proving asymptotic independence of the r.v.'s $\mathfrak{X}_1(a_1)$ and $I\{\alpha=1\}$. The scheme above proposed can be applied to eliminate the gap in the Huffman paper.

The proof of assertion 2) we divide into steps.

1°. We have the following representation for 2-SPRT stopping time from relation (5)

$$(9) \qquad \tau(\varphi) = t(\varphi)-\max_{i=1,2}[C_i(\widetilde{X}_{\tau(\varphi)}-s(\varphi))]+\zeta$$

where $\zeta=I\{\tau_1(\varphi)<\tau_2(\varphi)\}\dfrac{\mathfrak{X}_1(a_1)}{I_1}+I\{\tau_2(\varphi)\le\tau_1(\varphi)\}\dfrac{\mathfrak{X}_2(a_2)}{I_2}$. By conditions of the theorem we obtain the following expressions for $t(\varphi)$ and $s(\varphi)$

$$t(\varphi) = \frac{a_1}{I_1} + C_1 r\sigma\left(\frac{a_1}{I_1}\right)^{1/2}, \qquad s(\varphi) = -r\sigma\left(\frac{a_1}{I_1}\right)^{1/2}.$$

Then after some algebra we have

$$E_\varphi \max_{i=1,2}[C_i(\widetilde{X}_{\tau(\varphi)}-s(\varphi))] = \frac{C_1-C_2}{2}E_\varphi|\widetilde{X}_{\tau(\varphi)}-s(\varphi)|+\frac{C_1+C_2}{2}r\sigma\left(\frac{a_1}{I_1}\right)^{1/2}.$$

So the main difficulty is to estimate $E_\varphi|\widetilde{X}_{\tau(\varphi)}-s(\varphi)|$.

2°. Let us show that

$$\Delta = E_\varphi|\tilde{X}_{t(\varphi)}-s(\varphi)|-E_\varphi|\tilde{X}_{\tau(\varphi)}-s(\varphi)| = o(1). \tag{10}$$

Since the process $|\tilde{X}_u-s(\varphi)|$ is a submartingale and $\tau(\varphi)\le t(\varphi)$ then $\Delta\ge 0$ and hence it is necessary to obtain only the upper bound for Δ.

Let us denote $\Delta X=\tilde{X}_{t(\varphi)}-\tilde{X}_{\tau(\varphi)}$, $\eta=\tilde{X}_{\tau(\varphi)}-s(\varphi)$. Then

$$\Delta = E_\varphi|\Delta X+\eta|-E_\varphi|\eta| \le E_\varphi|\Delta X+\eta|+E_\varphi|\Delta X-\eta|-2E_\varphi|\eta|$$

since $E_\varphi|\Delta X-\eta|\ge E_\varphi|\eta|$. Note the function $\psi(x,y)=|x+y|+|x-y|-2|y|$ is even in x and y. Now it is easy to see that

$$\psi(x,y) = 2I\{|x|\ge|y|\}(|x|-|y|)$$

and so

$$\Delta \le 2E_\varphi I\{\Delta X\ge|\eta|\}(\Delta X-|\eta|)-2E_\varphi I\{\Delta X\le-|\eta|\}(\Delta X+|\eta|) = 2(\delta_1-\delta_2).$$

Integrating by parts we obtain

$$0 \le \delta_1 = E_\varphi \int_{|\eta|}^{\infty} P_\varphi\{\Delta X>y|\mathcal{F}_{\tau(\varphi)}\}dy ,$$

$$0 \le \delta_2 = E_\varphi \int_{|\eta|}^{\infty} P_\varphi\{-\Delta X>y|\mathcal{F}_{\tau(\varphi)}\}dy .$$

By Chebyshev's inequality for $\varepsilon>0$ we have

$$\delta_1 \le E_\varphi \int_{|\eta|}^{\infty} e^{-\varepsilon y}E_\varphi(e^{\varepsilon\Delta X}|\mathcal{F}_{\tau(\varphi)})dy = \frac{1}{\varepsilon} E_\varphi\exp\{(t(\varphi)-\tau(\varphi))I_\varepsilon-\varepsilon|\eta|\},$$

where $I_\varepsilon=E_\varphi\log\frac{dP_\varphi}{dP_{\varphi+\varepsilon}}(1)=\frac{\varepsilon^2}{2}b''(\varphi)+o(\varepsilon^2)$, $\varepsilon\to 0$.

Accordingly to (9)

$$t(\varphi)-\tau(\varphi) \le \max_{i=1,2} [C_i\eta] \le (C_1-C_2)|\eta|.$$

Then

$$\delta_1 \le \frac{1}{\varepsilon} E_\varphi\exp\{[C_1-C_2)I_\varepsilon-\varepsilon]|\eta|\}.$$

Note that for sufficiently small $\varepsilon>0$,

$$(C_1-C_2)I_\varepsilon < \varepsilon .$$

On the other hand $|\eta|=|\tilde{X}_{\tau(\varphi)}-s(\varphi)|\xrightarrow{P_\varphi}\infty$ as $a_1\to\infty$. So $\delta_1=o(1)$. The situation with δ_2 is similar.

3°. To prove assertion 2) now it is sufficient to find the asymptotic expansion for

$$E_\varphi|\tilde{X}_{t(\varphi)}-s(\varphi)| = \sigma\sqrt{t(\varphi)}\ E_\varphi|z_{t(\varphi)}+\beta_{t(\varphi)}|,$$

where

$$z_{t(\varphi)} = \frac{\tilde{X}_{t(\varphi)}}{\sigma\sqrt{t(\varphi)}} \xrightarrow{d(\varphi)} z \sim N(0,1),$$

$$\beta_{t(\varphi)} = r\left(\frac{a_1}{I_1 t(\varphi)}\right)^{1/2} \longrightarrow r, \qquad a_1\to\infty.$$

Using the asymptotic expansion in the CLT (see, for example, [9], (VI.3.1)) we obtain

$$E_\varphi|z_{t(\varphi)}+\beta_{t(\varphi)}| = E_\varphi|z+\beta_{t(\varphi)}| + \frac{\alpha_3}{6\sigma^3\sqrt{t(\varphi)}}\ E|z+r|(z^3-3z) + o\left(\frac{1}{a_1}\right).$$

4°. Straightforward calculations give that

$$E_\varphi|z+r| = 2f(r) - r + 2r\Phi(r),$$

$$E_\varphi|z+r|(z^3-3z) = -2rf(r),$$

$$E_\varphi|z+\beta_{t(\varphi)}| = E_\varphi|z+r| + \int_r^{\beta_{t(\varphi)}}(2\Phi(u)-1)du$$

$$= E_\varphi|z+r| - (2\Phi(r)-1)\ \frac{r^2c_1\sigma}{2}\left(\frac{I_1}{a_1}\right)^{1/2} + o\left(\frac{1}{a_1}\right).$$

5°. Now we are ready to derive assertion 2). By the representation (9) and the estimation (10) we have

$$E_\varphi\tau(\varphi) = t(\varphi) - \sigma\sqrt{t(\varphi)}\frac{C_1-C_2}{2}\ E_\varphi|z_{t(\varphi)}+\beta_{t(\varphi)}| - \frac{C_1+C_2}{2}\ r\sigma\left(\frac{a_1}{I_1}\right)^{1/2} + E_\varphi\xi + o\left(\frac{1}{a_1}\right). \tag{11}$$

By asymptotic independency of the r.v.'s $I\{d=i\}$ and $\mathfrak{X}_i(a_i)$ and asymptotic normality of the r.v.'s τ_1^* and τ_2^* we obtain

$$E_\varphi\xi = \frac{\Phi(r)}{I_1}\ \rho_1 + \frac{\Phi(-r)}{I_1}\ \rho_2 + o(1), \qquad a_1\to\infty.$$

Using the asymptotic expansions derived above we get the result of Theorem 1.

Proof of Theorem 2. By the conditions of the theorem it is easy to obtain a relation for boundaries which is similar to (1):

$$\frac{a_2}{I_2} = \frac{a_1}{I_1} + r\sigma(C_1-C_2)\left(\frac{a_1}{I_1}\right)^{1/2} + K + O(a_1^{-1/2}) , \tag{12}$$

where $K=\frac{\log[\gamma_2\Phi(-r)]}{I_2}-\frac{\log[\gamma_1\Phi(r)]}{I_1}$. A careful analysis of the proof of assertion 1) shows that in this case (7) is true also and hence 1') is also true.

Since now

$$t(\varphi) = \frac{a_1}{I_1} + r\sigma C_1\left(\frac{a_1}{I_1}\right)^{1/2} + \frac{C_1}{C_1-C_2} K , \qquad s(\varphi) = -r\sigma\left(\frac{a_1}{I_1}\right)^{1/2} - \frac{K}{C_1-C_2} ,$$

in the expansion for $E_\varphi\tau(\varphi)$ in (11) will appear the term $K\Phi(-r)$. Finally substituting the expressions for a_1 from (3) in 2) we obtain the needed expansion for $E_\varphi\tau(\varphi)$ in terms of α_1 after some simple algebra.

4. Numerical results. Anderson [10] had considered the tests with triangular boundaries for Wiener process and obtained exact formulae for error probabilities and expected sample size in the form of Mill's ratio series. In the symmetric case (i.e. $-\theta_1=\theta_2=\theta$, $\varphi=0$, $a_1=a_2=a$) we have

$$P_{-\theta}(d=1) = P_\theta(d=2) = \tfrac{1}{2}e^{-a}, \qquad E_0\tau = \frac{2a}{\theta^2} - \frac{4}{\theta}J(a)\sqrt{\frac{2a}{\theta^2}} ,$$

where $J(a)=\int_{-\infty}^{\infty}\frac{xf(x)dx}{1+\exp(\sqrt{2a}x)}$. The last expression is obtained by Siegmund in [8] (T3. 57) but there the factor $(2a/\theta^2)^{1/2}$ was missed in the second term. After expansion the integral we obtain the complete expansion for expected sample size

$$E_0\tau = \frac{2a}{\theta^2} - \frac{4}{\theta^2}\sqrt{\frac{a}{\pi}} + \frac{4}{\theta^2\sqrt{\pi}}\sum_{k=1}^{\infty}(-1)^{k-1}\frac{c_k}{a^{k-1/2}}$$

where $c_k=\frac{2^k-2^{1-k}}{2\cdot 4\cdots(2k)}\pi^{2k}B_k$ and B_k are the Bernoulli numbers. In this case the asymptotic from 2) with $I_1=\theta^2/2$ coincides with the above into the term of order $o(1)$.

For the symmetric case of normally distributed random variables

Lorden [1] proposed the empirical formula for the boundary of the 2-SPRT. In this case the Monte-Carlo results for error probabilities with the boundary given by (3) are the same as for Lorden's approximation with relative error of about 0.1%. Moreover, in non-symmetric case the boundaries (3) give the accuracy about 1% - 10% for error probabilities. Unfortunately, the asymptotic approximation for the expected stopping time, given by Theorem 2 is not enough accurate for values of α_1 and α_2, from the range 0.1÷0.001, but the first two terms of this approximation gives an accuracy about 5% - 20%. This happens since for these values of error probabilities the effect of the third term is significant and probably must be compensated by the forth term.

References

[1] Lorden G. 2-SPRT's and the modified Kiefer-Weiss problem of minimizing an expected sample size, Ann. Statist., 1976, 4, N1, 281-291.

[2] Dragalin V. P., Novikov A. A. Asymptotic solution to the Kiefer-Weiss problem for processes with independent increments, Theory Prob. Applications, 1987, 32, N1.

[3] Eisenberg B. The asymptotic solution of the Kiefer-Weiss problem, Comm. Statist.-Sequential Analysis, 1983, 1, N1, 81-88.

[4] Huffman M. D. An efficient approximate solution to the Kiefer-Weiss problem, Ann. Statist., 1983, 11, N1, 306-316.

[5] Borovkov A. A. New limit theorems in boundary problems for sums of independent terms, Selected Translation in Mathematical Statistics and Probability, 1962, 5, 315-372.

[6] Mogul'skii A. A. On the size of the first jump for a process with independent increments, Theory Prob. Applications, 1976, 21, N3, 470-481.

[7] Woodroofe M. Nonlinear renewal theory and sequential analysis. Philadelphia, SIAM, 1982.

[8] Siegmund D. Sequential Analysis. Tests and Confidence Intervals. New York-Berlin, Springer-Verlag, 1985, 272p.

[9] Petrov V. V. Sums of Independent Random Variables. New York-Berlin, Springer-Verlag, 1972.

[10] Anderson T. W. A modification of the sequential probability ratio test to reduce the sample size, Ann. Math. Statist., 1960, 31, N1, 165-197.

Steklov Mathematical Institute,
Moscow

ON DYNKIN'S STOPPING PROBLEM WITH A FINITE CONSTRAINT

Yoshio Ohtsubo

1. Introduction. Recently in [10, 11, 12] the author has formulated and investigated discrete parameter Dynkin's stopping problem with a finite constraint. Especially in [12] he has given three equivalent conditions for the closedness and has shown that the value sequence is a unique one satisfying the recursive relation and Neveu's martingale condition. In this note we represent the martingale condition equivalent to the closedness by the weaker form than [12] and we give fourth equivalent condition. We also give natural sufficient conditions for the uniqueness of the solution to the only recursive relation.

Optimal stopping games without a finite constraint was first introduced and studied by Dynkin in [3], and was generalized in [5], [9] and [4]. The continuous time version without the constraint was studied in many literature (for examples, [6], [1], [7], [8] and [13]).

2. Preliminaries. Let $\{X_n, n\in N\}$, $\{Y_n, n\in N\}$ and $\{W_n, n\in N\}$ be three random sequences defined on a probability space $(\Omega, \mathcal{F}, P)$ and adapted to a given filtration $(\mathcal{F}_n)_{n\in N}$ in $\mathcal{F}$, where $N=\{0,1,2,\dots\}$. We assume that (i) $X_n \leqq W_n \leqq Y_n$ a.s. for every $n\in N$, (ii) the r.v.'s X_n^- and Y_n^+ are integrable for every $n\in N$, and (iii) the r.v.'s $\sup_{n\in N} X_n^+$ and $\sup_{n\in N} Y_n^-$ are integrable, where $x^+ = \max(x,0)$ and $x^- = \max(0,-x)$.

For each $n\in N$, denote by Λ_n (respectively, Γ_n) the class of all $(\mathcal{F}_n)$-stopping times τ (resp. σ) with values in $N\cup\{+\infty\}$ such that $\tau \geqq n$ a.s. and $X_\tau^- I_{(\tau<\infty)}$ is integrable (resp. $\sigma \geqq n$ a.s. and $Y_\sigma^+ I_{(\sigma<\infty)}$ is integrable), where I_A is the indicator function of the set $A\in\mathcal{F}$. Also for each stopping time $\sigma\in\Gamma_n$, let $\Delta_n^1(\sigma)$ denote the set of all $\hat{\tau}\in\Lambda_n$ such that the stopping time $\hat{\tau}\wedge\sigma$ is a.s. finite, where $\xi\wedge\mu = \min(\xi,\mu)$, and for each stopping time $\tau\in\Lambda_n$, let $\Delta_n^2(\tau)$ denote the set of all $\hat{\sigma}\in\Gamma_n$ such that $\tau\wedge\hat{\sigma}$ is a.s. finite.

We define upper and lower values by, respectively,

$$\overline{v}_n = \inf_{\sigma\in\Gamma_n} \sup_{\tau\in\Delta_n^1(\sigma)} E[g(\tau,\sigma)], \qquad \underline{v}_n = \sup_{\tau\in\Lambda_n} \inf_{\sigma\in\Delta_n^2(\tau)} E[g(\tau,\sigma)],$$

where

$$g(\tau,\sigma) = X_\tau I_{(\tau<\sigma)} + Y_\sigma I_{(\sigma<\tau)} + W_\tau I_{(\tau=\sigma)}.$$

We say that the stopping game is closed if $\overline{v}_n = \underline{v}_n$ for all $n\in N$. Also we say that the integrable submartingale $\{U_n\}$ is regular if $\{U_n^+\}$ is uniformly integrable (cf. [9]).

Now we define the double essential bounds as follows:

$$\overline{V}_n=\operatorname*{ess\,inf}_{\sigma\in\Gamma_n}\ \operatorname*{ess\,sup}_{\tau\in\Delta^1_n(\sigma)} E[g(\tau,\sigma)|\mathcal{F}_n], \qquad \underline{V}_n=\operatorname*{ess\,sup}_{\tau\in\Lambda_n}\ \operatorname*{ess\,inf}_{\sigma\in\Delta^2_n(\tau)} E[g(\tau,\sigma)|\mathcal{F}_n].$$

Then let us restate the following results obtained in [10] and [12].

Proposition 1.([10]) The following assertions hold:

(i) $\overline{v}_n=E[\overline{V}_n]$, $\underline{v}_n=E[\underline{V}_n]$ and $\overline{V}_n\leqq\underline{V}_n$ a.s. for every $n\in N$.

(ii) The sequences $\{\overline{V}_n\}$ and $\{\underline{V}_n\}$ satisfy the recursive relation

$$U_n=\operatorname{med}(X_n, Y_n, E[U_{n+1}|\mathcal{F}_n]) \quad \text{a.s.} \quad n\in N, \tag{1}$$

where med(a,b,c) denotes the median of three real values a, b and c.

Proposition 2.([12]) The following four conditions are equivalent:

(a) The stopping game is closed.

(b) There exist the smallest regular supermartingale $\{\widetilde{X}_n\}$ dominating $\{X_n\}$ and the greatest regular submartingale $\{\widetilde{Y}_n\}$ dominated by $\{Y_n\}$, and both sequences $\{\overline{V}_n\}$ and $\{\underline{V}_n\}$ satisfy the inequalities

$$\widetilde{Y}_n\leqq U_n\leqq\widetilde{X}_n \quad \text{a.s.} \quad n\in N. \tag{2}$$

(c) $E[\inf_{k\geqq n} Y_k|\mathcal{F}_n]\leqq E[\sup_{k\geqq n} X_k|\mathcal{F}_n]$ a.s. for every $n\in N$.

(d) $\lim_n\inf Y_n\leqq\lim_n\sup X_n$ a.s..

As established in the following section, condition (b) really becomes the weaker form. We call the inequalities (2) Neveu's martingale condition. This is analogous to but different from a martingale condition given in Neveu [9, Prop. VI-6-9].

Proposition 3.([12]) If the stopping game is closed, then $\{\overline{V}_n=\underline{V}_n\}$ is the unique sequence satisfying the equalities (1) and the inequalities (2).

3. Neveu's martingale conditions. It is not true in general that the random sequences $\{\widetilde{X}_n\}$, $\{\widetilde{Y}_n\}$ given in Proposition 2 exist (see Example 1). However we easily see that if the random sequence $\{X_n^-\}$ is uniformly integrable then such a sequence $\{\widetilde{X}_n\}$ exists. Also even if they exist the inequalities $\widetilde{Y}_n\leqq\widetilde{X}_n$ does not necessarily hold (see Example 2).

Example 1. Let $Y_n(\omega)=n$ for every $n\in N$ and all $\omega\in\Omega$. Define random sequences $\{Z_n^k\}$, $k=1,2,\ldots.$, by $Z_n^k=Y_n$ if $n\leqq k$ and $Z_n^k=k$ if $n>k$. Then it follows that for every $k=1,2,\ldots$ $\{Z_n^k\}$ is a regular submartingale dominated by $\{Y_n\}$. If there exists a greatest regular submartingale $\{\widetilde{Y}_n\}$ dominated by $\{Y_n\}$, then $Y_n\geqq\widetilde{Y}_n\geqq\sup_k Z_n^k$, $n\in N$. Since $\sup_k Z_n^k=Y_n$, we have $\widetilde{Y}_n=Y_n$, $n\in N$.

However the integrable submartingale $\{Y_n\}$ is not regular. This is a contradiction. Thus there is no greatest regular submartingale dominated by $\{Y_n\}$.

Example 2. Let $X_0=1$, $X_n=a<0$, $n\geq 1$; $Y_0=2$, $Y_1=0$, $Y_n=b>1$, $n\geq 2$ and $Y_n=W_n$, $n\in N$, for constants a and b. Then we easily observe that $\tilde{X}_0=1$, $\tilde{X}_n=a$, $n\geq 1$, $\tilde{Y}_0=\tilde{Y}_1=0$ and $\tilde{Y}_n=b$, $n\geq 2$, so that $\tilde{X}_0>\tilde{Y}_0$ and $\tilde{X}_n<\tilde{Y}_n$, $n\geq 1$. By the way, we have $\bar{v}_0=\underline{v}_0=1$ and $\bar{v}_n=X_n<Y_n=\underline{v}_n$, $n\geq 1$.

Now we shall here give the proof of the existence of $\{\tilde{X}_n\}$ and $\{\tilde{Y}_n\}$ in the implication (a)⇒(b) of Proposition 2, which is neglected in [12].

Lemma 1. If the stopping game is closed then there exist the smallest regular supermartingale $\{\tilde{X}_n\}$ dominating $\{X_n\}$ and the greatest regular submartingale $\{\tilde{Y}_n\}$ dominated by $\{Y_n\}$ and the relations $\tilde{Y}_n\leq\tilde{X}_n$, $n\in N$, hold almost surely.

Proof. Let F (resp. G) be the family of regular supermartingales dominating $\{X_n\}$ (resp. regular submartingales dominated by $\{Y_n\}$). It is clear that F and G are nonempty. For, the sequence $\{E[\sup_{k\geq n}X_k^+|\mathcal{F}_n], n\in N\}$ belongs to F and $\{-E[\sup_{k\geq n}Y_k^-|\mathcal{F}_n], n\in N\}$ is in G.

We define random sequence $\{\tilde{X}_n\}$ by $\tilde{X}_n=\operatorname{ess\,inf}_{(U_n)\in F}U_n$. Then $\{\tilde{X}_n\}$ is $(\mathcal{F}_n)$-adapted sequence such that $\tilde{X}_n\leq E[\sup_{k\geq n}X_k^+|\mathcal{F}_n]$, $n\in N$, and we have

$$E[\tilde{X}_{n+1}|\mathcal{F}_n]\leq E[U_{n+1}|\mathcal{F}_n]\leq U_n, \quad n\in N,$$

for all $(U_n)\in F$, since $\{U_n\}$ is a supermartingale. Thus we get

$$E[\tilde{X}_{n+1}|\mathcal{F}_n]\leq \operatorname*{ess\,inf}_{(U_n)\in F} U_n=\tilde{X}_n, \quad n\in N.$$

Also since $U_n\geq X_n$ for all $(U_n)\in F$, we have $\tilde{X}_n\geq X_n$, $n\in N$. Hence $\{\tilde{X}_n\}$ is an integrable supermartingale dominating $\{X_n\}$. Similarly defining $\{\tilde{Y}_n\}$ by $\tilde{Y}_n=\operatorname{ess\,sup}_G U_n$, $n\in N$, we obtain that $\{\tilde{Y}_n\}$ is an integrable submartingale dominated by $\{Y_n\}$. Therefore in order to prove the first part of this lemma it suffices to show that these sequences $\{\tilde{X}_n\}$ and $\{\tilde{Y}_n\}$ are both regular.

Now by the definitions of $\bar{V}_n$ and $\underline{V}_n$ we have

$$\bar{V}_n\leq \operatorname*{ess\,sup}_{\tau\in\Delta_n^1(\infty)} E[g(\tau,\infty)|\mathcal{F}_n]=\operatorname*{ess\,sup}_{\tau\in\Delta_n^1(\infty)} E[X_\tau|\mathcal{F}_n]\leq E[\sup_{k\in N} X_k^+|\mathcal{F}_n],$$

and similarly $\underline{V}_n\geq -E[\sup_{k\in N}Y_k^-|\mathcal{F}_n]$. Thus since the stopping game is closed, that is, $\bar{V}_n=\underline{V}_n$, $n\in N$, we have

$$-E[\sup_{k\in N} Y_k^-|\mathcal{F}_n]\leq\bar{V}_n=\underline{V}_n\leq E[\sup_{k\in N} X_k^+|\mathcal{F}_n], \quad n\in N. \tag{3}$$

It also follows by the optional sampling theorem that for all $(U_n)\in F$

$$\overline{V}_n \leq \operatorname*{ess\,sup}_{\tau \in \Delta_n^1(\infty)} E[X_\tau | \mathcal{F}_n] \leq \operatorname*{ess\,sup}_{\tau \in \Delta_n^1(\infty)} E[U_\tau | \mathcal{F}_n] \leq U_n,$$

since $\{U_n\}$ is the regular supermartingale such that $U_n \geq X_n$, $n \in N$. Hence we have $\overline{V}_n \leq \widetilde{X}_n$, $n \in N$. By the analogous argument we get $\underline{V}_n \geq \widetilde{Y}_n$, $n \in N$, and, consequently,

$$\widetilde{Y}_n \leq \overline{V}_n = \underline{V}_n \leq \widetilde{X}_n, \quad n \in N. \tag{4}$$

These inequalities combined with (3) also yield the following boundedness :

$$\widetilde{Y}_n \leq E[\sup_{k \in N} X_k^+ | \mathcal{F}_n], \qquad \widetilde{X}_n \geq -E[\sup_{k \in N} Y_k^- | \mathcal{F}_n], \qquad n \in N.$$

Therefore by assumption (iii) $\{\widetilde{X}_n\}$ and $\{\widetilde{Y}_n\}$ are both regular, which implies the first part. From (4) the second part is also proved. □

Next we shall give the explicit representation of such sequences $\{\widetilde{X}_n\}$ and $\{\widetilde{Y}_n\}$. To this end we define the value sequences, $\{X_n^*\}$ and $\{Y_n^*\}$, of usual optimal stopping problems as follows :

$$X_n^* = \operatorname*{ess\,sup}_{\tau \in \Lambda_n} E[X_\tau | \mathcal{F}_n], \qquad Y_n^* = \operatorname*{ess\,inf}_{\sigma \in \Gamma_n} E[Y_\sigma | \mathcal{F}_n], \qquad n \in N,$$

where $X_\tau = \limsup_n X_n$ on $\{\tau = \infty\}$ and $Y_\sigma = \liminf_n Y_n$ on $\{\sigma = \infty\}$. Then it is well known (cf. [2]) that the following relations hold :

$$X_n^* = \max(X_n, E[X_{n+1}^* | \mathcal{F}_n]), \qquad Y_n^* = \min(Y_n, E[Y_{n+1}^* | \mathcal{F}_n]), \quad n \in N,$$

and hence $\{X_n^*\}$ is an integrable supermartingale dominating $\{X_n\}$ and $\{Y_n^*\}$ is an integrable submartingale dominated by $\{Y_n\}$. However the sequence $\{X_n^*\}$ and $\{Y_n^*\}$ are not always regular in the sense of the uniform integrability. For instance, in Example 1 it follows that $Y_n^* = Y_n$, $n \in N$, and $\{Y_n^*\}$ is not regular. In the lemma below we see that under the existence condition of $\{\widetilde{X}_n\}$ and $\{\widetilde{Y}_n\}$ both sequences $\{X_n^*\}$ and $\{Y_n^*\}$ are regular.

<u>Lemma 2</u>. If there exists the smallest regular supermartingale $\{\widetilde{X}_n\}$ dominating $\{X_n\}$, then $X_n^* = \widetilde{X}_n$ for all $n \in N$. If there exists the greatest regular submartingale $\{\widetilde{Y}_n\}$ dominated by $\{Y_n\}$, then $Y_n^* = \widetilde{Y}_n$ for all $n \in N$.

<u>Proof</u>. We shall show the only first statement. Since $\{\widetilde{X}_n\}$ is the regular supermartingale such that $\widetilde{X}_n \geq X_n$, $n \in N$, we first have the inequalities

$$X_n^* = \operatorname*{ess\,sup}_{\tau \in \Lambda_n} E[X_\tau | \mathcal{F}_n] \leq \operatorname*{ess\,sup}_{\tau \in \Lambda_n} E[\widetilde{X}_\tau | \mathcal{F}_n] \leq \widetilde{X}_n, \quad n \in N,$$

by the optional sampling theorem.

Next we shall show the reverse inequality $X_n^* \geq \widetilde{X}_n$. To prove this we use the same technique as in Chow et al. [2, Section 4.5]. For a constant a let

$$X_n(a) = \max(X_n, a), \qquad \gamma_n(a) = \operatorname*{ess\,sup}_{\tau \in \Lambda_n} E[X_\tau(a) | \mathcal{F}_n], \quad n \in N.$$

Then the sequence $\{\gamma_n(a)\}$ has the following properties for every a :

$$\gamma_n(a) = \max(X_n(a), E[\gamma_{n+1}(a) | \mathcal{F}_n]), \tag{5}$$

$$\limsup_n \gamma_n(a) = \limsup_n X_n(a) \quad \text{a.s..} \tag{6}$$

Since $\{\gamma_n(a)\}$ is bounded below by a, it follows by the relation (5) that $\{\gamma_n(a)\}$ is a regular supermartingale dominating $\{X_n(a)\}$ (hence $\{X_n\}$) for every a. Thus by the minimality of $\{\widetilde{X}_n\}$ we have the inequalities $\gamma_n(a) \geqq \widetilde{X}_n$, $n \in N$, for every a. However $\gamma_n(a)$ is an increasing function of a for each $n \in N$. Letting $\gamma^*_n = \lim_{a \downarrow -\infty} \gamma_n(a)$, from the above inequality we have $\gamma^*_n \geqq \widetilde{X}_n$, $n \in N$. Therefore in order to accomplish our object it suffices to obtain the equalities $X^*_n = \gamma^*_n$, $n \in N$. Now letting $a \downarrow -\infty$ in the relation (5), by the monotone convergence theorem for conditional expectations we get

$$\gamma^*_n = \max(X_n, E[\gamma^*_{n+1} | \mathcal{F}_n]) \qquad n \in N. \tag{7}$$

Also since $\gamma_n(a) \geqq \gamma^*_n \geqq X_n$ for each a and all $n \in N$ we have

$$\limsup_n \gamma_n(a) \geqq \limsup_n \gamma^*_n \geqq \limsup_n X_n \quad \text{a.s..}$$

However from (6) the r.v. $\limsup_n \gamma_n(a) = \limsup_n X_n(a) = \max(\limsup_n X_n, a)$ converges to $\limsup_n X_n$ as $a \downarrow -\infty$. Thus we have

$$\limsup_n \gamma^*_n = \limsup_n X_n \quad \text{a.s..} \tag{8}$$

Therefore by using Lemma 4.9 in [2] the relations (7) and (8) imply that the inequalites $\gamma^*_n \leqq X^*_n$ a.s. hold for all $n \in N$. Conversely, by the definition of $\gamma_n(a)$ it is easy to see that $\gamma_n(a) \geqq X^*_n$, so $\gamma^*_n \geqq X^*_n$, $n \in N$. Hence we get the equalities $\gamma^*_n = X^*_n$ and consequently $\widetilde{X}_n = X^*_n$, $n \in N$. □

Taking account of the above lemmas, we can replace the condition (b) in Proposition 2 by a simple and weaker form (b′) and we can also add other equivalent condition (e) as in the following theorem.

<u>Theorem 1</u>. The following three conditions are equivalent :

(a) The stopping game is closed.

(b′) There exist the smallest regular supermartingale $\{\widetilde{X}_n\}$ dominating $\{X_n\}$ and the greatest regular submartingale $\{\widetilde{Y}_n\}$ dominated by $\{Y_n\}$, and the inequalities $\widetilde{Y}_n \leqq \widetilde{X}_n$ almost surely hold for all $n \in N$.

(e) The inequalities $Y^*_n \leqq X^*_n$ almost surely hold for all $n \in N$.

<u>Proof</u>. From Lemmas 1 and 2 the implications (a)⇒(b′)⇒(e) are true. Since in general $X^*_n \leqq E[\sup_{k \geq n} X_k | \mathcal{F}_n]$ and $Y^*_n \geqq E[\inf_{k \geq n} Y_k | \mathcal{F}_n]$ for all $n \in N$, the implication (e)⇒(c) is also true. Thus this theorem is completely proved by Proposition 2. □

4. <u>The uniqueness relative to recursive relations</u>. In Proposition 3 we have obtained that under the closedness condition the value sequence $\{V_n\}$ is a unique solution to the recursive relation (1) and Neveu's martingale condition (2), where $\overline{V}_n=\underline{V}_n=V_n$, say. However the value sequence is not necessarily the unique one of the only relation (1).

<u>Example 3</u>. Suppose $\{U_n\}$ is the one dimentional symmetric random walk. Let $X_n=-(U_n-a)^-$ and $Y_n=(U_n+b)^+=W_n$, $n\in N$, where a and b are positive constants and let $\mathcal{F}_n=\sigma(U_k,\ k\leqq n)$, $n\in N$. Then it is easy to see that $\{X_n\}$ and $\{Y_n\}$ satisfy assumptions (i)–(iii) and, in addition, the inequalities $X_n<Y_n$ hold a.s. for all $n\in N$. Also since $\{U_n\}$ is the symmetric random walk we have $\limsup_n X_n=\liminf_n Y_n=0$ a.s.. Hence from Proposition 2 the stopping game is closed and $V_n=\overline{V}_n=\underline{V}_n=0$, $n\in N$. However $\{V_n\}$ is not a unique solution to (1). Indeed, it is not hard to show that $\{X_n\}$ is an integrable supermartingale and $\{Y_n\}$ is an integrable submartingale. Thus both $\{X_n\}$ and $\{Y_n\}$, which are different from $\{V_n\}$, satisfy the recursive relation (1).

In the argument below we shall give sufficient conditions for the uniqueness of solution to the recursive relation.

<u>Lemma 3</u>.([10, Lemmas 2.1, 2.2]) There exist the greatest sequence $\{\overline{\gamma}_n\}$ and the smallest $\{\underline{\gamma}_n\}$ satisfying the recursive relation (1), and for these sequences,

$$X_n\leqq\underline{\gamma}_n\leqq\overline{V}_n\leqq\underline{V}_n\leqq\overline{\gamma}_n\leqq Y_n \quad \text{a.s.} \quad n\in N,$$

and

$$\liminf_n \overline{\gamma}_n=\liminf_n Y_n, \qquad \limsup_n \underline{\gamma}_n=\limsup_n X_n \quad \text{a.s..}$$

For $\varepsilon\geqq0$ and $n\in N$, we define two stopping times as follows :

$$\tau_\varepsilon(n)=\inf\{k\geqq n \mid \overline{\gamma}_k\leqq X_k+\varepsilon\},$$

$$\sigma_\varepsilon(n)=\inf\{k\geqq n \mid \underline{\gamma}_k\geqq Y_k-\varepsilon\},$$

where we suppose that the infimum over empty set is equal to $+\infty$.

<u>Lemma 4</u>. Let $\varepsilon\geqq0$ and $n\in N$ be arbitrary.

(i) If the sequence $\{Y_k^+\}$ is uniformly integrable, then the stopped sequence $\{\overline{\gamma}_{\tau_\varepsilon(n)\wedge k}, k\geqq n\}$ is a regular submartingale, and consequently the inequality

$$\overline{\gamma}_n\leqq E[\overline{\gamma}_{\tau_\varepsilon(n)\wedge\sigma}\mid\mathcal{F}_n] \tag{9}$$

holds for every stopping time $\sigma\in\Delta_n^2(\tau_\varepsilon(n))$.

(ii) If the sequence $\{X_k^-\}$ is uniformly integrable, then the stopped sequence $\{\underline{\gamma}_{k\wedge\sigma_\varepsilon(n)}, k\geqq n\}$ is a regular supermartingale, and the inequality

$$\underline{\gamma}_n \geqq E[\underline{\gamma}_{\tau \wedge \sigma_\varepsilon(n)} | \mathcal{F}_n]$$

holds for every stopping time $\tau \in \Delta_n^1(\sigma_\varepsilon(n))$.

Proof. (i) Since $\{\overline{\gamma}_n\}$ satisfies the recursive relation (1) it is easy to see that $\{\overline{\gamma}_{\tau_\varepsilon(n) \wedge k}, k \geqq n\}$ is an integrable submartingale. Also if this submartingale is regular, the inequality (9) holds by the optional sampling theorem. Thus it suffices to prove the regularity of $\{\overline{\gamma}_{\tau_\varepsilon(n) \wedge k}\}$. By the definition of $\tau_\varepsilon(n)$ we have

$$\begin{aligned}\overline{\gamma}_{\tau_\varepsilon(n) \wedge k} &\leqq Y_k I_{(k<\tau_\varepsilon(n))} + (X_{\tau_\varepsilon(n)} + \varepsilon) I_{(\tau_\varepsilon(n) \leqq k)} \\ &\leqq Y_k^+ I_{(k<\tau_\varepsilon(n))} + (\sup_j X_j^+ + \varepsilon) I_{(\tau_\varepsilon(n) \leqq k)} \\ &\leqq \max(Y_k^+, \sup_j X_j^+ + \varepsilon).\end{aligned}$$

Thus for every real $a>0$,

$$E[\overline{\gamma}^+_{\tau_\varepsilon(n) \wedge k} I_{(\overline{\gamma}^+_{\tau_\varepsilon(n) \wedge k} > a)}] \leqq E[Y_k^+ I_{(Y_k^+ > a)}] + E[(\sup_j X_j^+ + \varepsilon) I_{(\sup_j X_j^+ + \varepsilon > a)}],$$

which yields the uniform integrability of $\{\overline{\gamma}_{\tau_\varepsilon(n) \wedge k}\}$.

(ii) This proof symmetrically follows by a way similar to that given in (i). □

Theorem 2. Suppose the stopping game is closed. If the sequences $\{X_n^-\}$ and $\{Y_n^+\}$ are uniformly integrable, then the value sequence $\{V_n\}$ is a unique solution to the recursive relation (1).

Proof. From Lemma 3 it follows that if $\{U_n\}$ satisfies the recursive relation (1) then $\underline{\gamma}_n \leqq U_n \leqq \overline{\gamma}_n$, $n \in N$. Since $\overline{V}_n = \underline{V}_n$, $n \in N$, it thus suffices to show that $\underline{\gamma}_n = \overline{V}_n$ and $\overline{\gamma}_n = \underline{V}_n$, $n \in N$. From Lemma 4 the inequality (9) holds as $\varepsilon = 0$. By the definition of $\tau_0(n)$ we have

$$\overline{\gamma}_n \leqq E[X_{\tau_0(n)} I_{(\tau_0(n) \leqq \sigma)} + Y_\sigma I_{(\sigma < \tau_0(n))} | \mathcal{F}_n] \leqq E[g(\tau_0(n), \sigma) | \mathcal{F}_n],$$

for every $\sigma \in \Delta_n^2(\tau_0(n))$. Hence we get

$$\overline{\gamma}_n \leqq \operatorname*{ess\,inf}_{\sigma \in \Delta_n^2(\tau_0(n))} E[g(\tau_0(n), \sigma) | \mathcal{F}_n] \leqq \underline{V}_n,$$

so that $\overline{\gamma}_n = \underline{V}_n$, since from Lemma 3 the reverse inequalities $\overline{\gamma}_n \geqq \underline{V}_n$ hold. Similarly we have the equalities $\underline{\gamma}_n = \overline{V}_n$, $n \in N$. □

The following corollary is an immediate consequence of the above theorem.

Corollary 3. ([10, Theorem 2.3]) If the stopping game is closed and the r.v.'s $\sup_N X_n^-$ and $\sup_N Y_n^+$ are integrable, then $\{V_n\}$ is a unique sequence to the recursive relation (1).

References

[1] J.-M.Bismut: Sur un probleme de Dynkin, Z.Wahrsch.Verw.Gebiete 39 (1977), 31-53.

[2] Y.S.Chow, H.Robbins and D.Siegmund: Great Expectations -The Theory of Optimal Stopping-, Houghton Mifflin, Boston (1971).

[3] E.B.Dynkin: Game variant of a problem on optimal stopping, Soviet Math. Dokl. 10 (1969), 270-274.

[4] N.V.Elbakidze: Construction of the cost and optimal policies in a game problem of stopping a Markov process, Theor.Prob.Appl. 21 (1976), 163-168.

[5] Yu.I.Kifer: Optimal stopped games, Theor.Prob.Appl. 16 (1971), 185-189.

[6] N.V.Krylov: Control of Markov processes and W-spaces, Math.USSR-Izv. 5 (1971), 233-266.

[7] J.P.Lepeltier and M.A.Maingueneau: Le jeu de Dynkin en theorie generale sans l'hypothese de Mokobodski, Stochastics 13 (1984), 25-44.

[8] H.Morimoto: Dynkin games and martingale methods, Stochastics 13 (1984), 213-228.

[9] J.Neveu: Discrete-Parameter Martingales, North-Holland, Amsterdom (1975).

[10] Y.Ohtsubo: Optimal stopping in sequential games with or without a constraint of always terminating, to appear in Math.Oper.Res. (1986).

[11] Y.Ohtsubo: The Dynkin stopping with a finite constraint and bisequential decision problems, Mem.Fac.Sci., Kochi Univ., Ser.A 7 (1986), 59-69.

[12] Y.Ohtsubo : Neveu's martingale conditions and closedness in Dynkin stopping problem with a finite constraint, Stochastic Process.Appl. 22 (1986), 333-342.

[13] Ł.Stettner: On closedness of general zero-sum stopping game, Bull.Polish Acad.Sci.Math. 32 (1984), 351-361.

Department of Mathematics
Kochi University
2-5-1 Akebono-cho
Kochi
780 Japan

ENTROPY OPERATORS AND MCMILLAN TYPE CONVERGENCE THEOREMS IN A NONCOMMUTATIVE DYNAMICAL SYSTEM

Masanori Ohya

1. INTRODUCTION : McMillan's theorem is an ergodic type convergence theorem and plays an essential role in information theory [8]. Recently, it has become important to formulate quantum information theory or quantum communication processes rigorously [9]. We introduce the entropy operator and the conditional entropy operator in a noncommutative system and study their fundamental properties. Further, we formulate McMillan's type convergence theorems and prove them in our dynamical system (cf.[3,4]). A noncommutative dynamical system (NDS for short) can be described by a von Neumann algebraic triple or, more generally, a C^*-algebraic triple denoted by $(\mathfrak{A}, \mathfrak{S}, \alpha)$. Namely, $\mathfrak{A}$ is a von Neumann algebra or C^*-algebra, $\mathfrak{S}$ is the set of all states on $\mathfrak{A}$ and α is an automorphism of $\mathfrak{A}$ describing a certain evolution of the system. A self-adjoint element A of the algebra $\mathfrak{A}$ corresponds to a random variable in usual commutative dynamical (probability) systems (CDS for short) and a state in NDS corresponds to a probability measure in CDS. Here we use a von Neumann algebraic description for simplicity. Consult the bibliography [7] for NDS and noncommutative probability theory.

2. ENTROPY OPERATORS : Let $\mathfrak{A}$ be a finite dimensional von Neumann (matrix) algebra acting on a Hilbert space $\mathcal{H}$ and $\tau \in \mathfrak{S}$ be an α-invariant faithful normal trace on $\mathfrak{A}$. Moreover, let $P(\mathfrak{M})$ be the set of all minimal finite partitions of unity I in a subalgebra of $\mathfrak{M}$; $\tilde{P} = \{P_j\}$ (i.e., $P_j \in \mathfrak{M}$, $P_i \perp P_j$ $(i \neq j)$, $\sum_j P_j = I$ and there does not exist a projection E such as $0<E<P_j$ for each j). We introduce the entropy operator and the entropy w.r.t. the subalgebra $\mathfrak{M}$ and the trace τ as follows:

$$H(\mathfrak{M}) = -\sum_k P_k \log \tau(P_k), \qquad (2.1)$$

$$S(\mathfrak{M}) = \tau(H(\mathfrak{M})). \qquad (2.2)$$

Since any two partitions in $P(\mathfrak{M})$ are unitary equivalent due to the finiteness of $\mathfrak{M}$, the entropy operator $H(\mathfrak{M})$ and the entropy $S(\mathfrak{M})$ are uniquely determined. Indeed: (1) When $\mathfrak{M}$ is a type I_n factor (i.e., the center $\mathcal{Z} \equiv \{A \in \mathfrak{M};\ AB = BA,\ \forall B \in \mathfrak{M}\}$ of $\mathfrak{M}$ is equal to the set $\mathbb{C}I$, where $\mathbb{C}$ is the set of all complex numbers), $\mathfrak{M}$ is isometrically isomorphic to $B(\mathbb{C}^n)$, the set of all nxn matrices on the n-dimensional Hibert space $\mathbb{C}^n$. Then

$$H(\mathfrak{M}) = (\log n)I$$

$$S(\mathfrak{M}) = \log n.$$

(2) When $\mathfrak{M}$ is not a factor, the center $\mathcal{Z}$ of $\mathfrak{M}$ is generated by a minimal finite partition $\{Q_j\} \in P(\mathcal{Z})$ and $\mathfrak{M}$ can be expressed as $\mathfrak{M} = \oplus_j \mathfrak{M}_j$, where $\mathfrak{M}_j = Q_j\mathfrak{M}$ is a type I_{n_j} factor (i.e., the center $\mathcal{Z}_j$ of $\mathfrak{M}_j$ is equal to $\mathbb{C}I$). Then by taking $q_j = \tau(Q_j)^{-1}$ and $\tau_j = q_j \tau \upharpoonright \mathfrak{M}_j$ (the restriction of τ to $\mathfrak{M}_j$), we have

$$H(\mathfrak{M}) = H(\mathcal{Z}) + \sum_j H_j(\mathfrak{M}_j)$$

$$S(\mathfrak{M}) = S(\mathcal{Z}) + \sum_j q_j S_j(\mathfrak{M}_j),$$

where H_j and S_j are defined by using τ_j instead of τ in (2.1) and (2.2), respectively. The above entropy $S(\mathfrak{M})$ has already been discussed in [1,2] without considering $H(\mathfrak{M})$.

For two von Neumann subalgebras $\mathfrak{M}_1$ and $\mathfrak{M}_2$, let $\mathfrak{M}_1 \vee \mathfrak{M}_2$ be the von Neumann subalgebra generated by $\mathfrak{M}_1$ and $\mathfrak{M}_2$. It is easily seen that a partition $\{P_k\} \in P(\mathfrak{M}_1)$ is not always in $P(\mathfrak{M}_1 \vee \mathfrak{M}_2)$ but there exists a partition $\{P_{kj}\}$ in $P(\mathfrak{M}_1 \vee \mathfrak{M}_2)$ with $P_k = \sum_j P_{kj}$. By this fact, we have the conditional entropy operator and the conditional entropy such as

$$H(\mathfrak{M}_1 | \mathfrak{M}_2) = - \sum_{kj} Q_{kj} \{\log \tau(Q_{kj}) - \log \tau(Q_j)\},$$

$$S(\mathfrak{M}_1 | \mathfrak{M}_2) = \tau(H(\mathfrak{M}_1 | \mathfrak{M}_2)),$$

where $\{Q_j\} \in P(\mathfrak{M}_2)$ and $\{Q_{kj}\} \in P(\mathfrak{M}_1 \vee \mathfrak{M}_2)$ with $Q_j = \sum_k Q_{kj}$. These entropies are also unique.

On the basis of the above formulations, we can prove the following two fundamental propositions [3].

Theorem 1 : _For von Neumann subalgebras_ $\mathfrak{M}$, $\mathfrak{M}_1$, $\mathfrak{M}_2$, _the following_ equalities hold : (1) $H(\mathfrak{M}) = H(\mathfrak{M}|\mathbb{C}I)$; (2) $H(\mathfrak{M}_1 \vee \mathfrak{M}_2) = H(\mathfrak{M}_1|\mathfrak{M}_2) + H(\mathfrak{M}_2)$; (3) $H(\alpha\mathfrak{M}_1|\alpha\mathfrak{M}_2) = \alpha H(\mathfrak{M}_1|\mathfrak{M}_2)$.

Sketch: (1) is immediate and (2) comes from the equality $Q_j = \sum_k Q_{kj}$. (3) is obtained from the α-invariance of τ and the fact that $\tilde{P} \in P(\mathfrak{M})$ implies $\alpha(\tilde{P}) \in P(\alpha\mathfrak{M})$. (QED)

Corollary 2 : For von Neumann subalgebras $\mathfrak{M}$, $\mathfrak{M}_1$, $\mathfrak{M}_2$, $\mathfrak{M}_3$, the following equalities and inequalities hold : (1) $S(\mathfrak{M}_1 \vee \mathfrak{M}_2) = S(\mathfrak{M}_1|\mathfrak{M}_2) + S(\mathfrak{M}_2)$; (2) $S(\mathfrak{M}_1) \le S(\mathfrak{M}_1 \vee \mathfrak{M}_2)$; (3) $S(\mathfrak{M}_1 \vee \mathfrak{M}_2) \le S(\mathfrak{M}_1) + S(\mathfrak{M}_2)$; (4) $S(\mathfrak{M}_1|\mathfrak{M}_2) \le S(\mathfrak{M}_1)$; (5) $\mathfrak{M}_2 \subset \mathfrak{M}_3$ implies $S(\mathfrak{M}_2|\mathfrak{M}_1) \le S(\mathfrak{M}_3|\mathfrak{M}_1)$; (6) $S(\alpha\mathfrak{M}) = S(\mathfrak{M})$; (7) $S(\alpha\mathfrak{M}_1|\alpha\mathfrak{M}_2) = S(\mathfrak{M}_1|\mathfrak{M}_2)$.

3. MACMILLAN'S TYPE CONVERGENCE THEOREMS : The following theorem is obtained from the previous theorem and corollary.

Theorem 3 : For a von Neumann subalgebra $\mathfrak{M}$, we have

(1) there exists an α-invariant operator $h \in \mathfrak{N}$ such that

$$\lim_{n\to\infty} H(\vee_{k=0}^{n} \alpha^k\mathfrak{M}) = h \ ;$$

(2) if α is ergodic, then there exists an integer N satisfying

$$h = S(\vee_{k=0}^{N-1} \alpha^k\mathfrak{M}).$$

Sketch: (1): Since we have

$$H(\mathfrak{M} \vee \alpha\mathfrak{M}) = H(\alpha\mathfrak{M}|\mathfrak{M}) + H(\mathfrak{M}) \ge H(\mathfrak{M}),$$
$$H(\vee_{k=0}^{n} \alpha^k\mathfrak{M}) \le H(\mathfrak{N})$$

for every n, $H(\vee_{k=0}^{n} \alpha^k\mathfrak{M})$ increasingly converges to a certain operator $h \in \mathfrak{N}$ in norm. We thus need to show the α-invariance of this operator h. For each n, we have

$$H(\vee_{k=0}^{n} \alpha^k\mathfrak{M}) = H(\vee_{k=1}^{n} \alpha^k\mathfrak{M}) + H(\mathfrak{M}|\vee_{k=1}^{n} \alpha^k\mathfrak{M}),$$

which implies

$$h = H(\vee_{k=1}^{\infty} \alpha^k\mathfrak{M}) + H(\mathfrak{M}|\vee_{k=1}^{\infty} \alpha^k\mathfrak{M})$$

Therefore, if we can prove the following facts ; (i) $\alpha(\mathfrak{M} \vee \alpha\mathfrak{M}) = \alpha\mathfrak{M} \vee \alpha^2\mathfrak{M}$, (ii) $H(\mathfrak{M}_1|\mathfrak{M}_2) = 0$ iff $S(\mathfrak{M}_1|\mathfrak{M}_2) = 0$, (iii) $S(\mathfrak{M}|\vee_{k=1}\alpha^k\mathfrak{M}) = 0$, then the above h is equal to

$$\alpha H(\bigvee_{k=0}^{\infty} \alpha^k \mathfrak{M}) + 0 = \alpha h,$$

that is, h is α-invariant.

Let us show the above facts (i),(ii) and (iii). (i) is trivial. (ii): $H(\mathfrak{M}_1|\mathfrak{M}_2) = 0$ implies $S(\mathfrak{M}_1|\mathfrak{M}_2) = 0$ from the definition. Conversely, if $S(\mathfrak{M}_1|\mathfrak{M}_2) = 0$, then the faithfulness of the trace τ implies $H(\mathfrak{M}_1|\mathfrak{M}_2) = 0$ because of the positivity of $H(\mathfrak{M}_1|\mathfrak{M}_2)$. Hence

$$H(\mathfrak{M}_1|\mathfrak{M}_2) = 0 \Longleftrightarrow S(\mathfrak{M}_1|\mathfrak{M}_2) = 0.$$

(iii): The equality $S(\mathfrak{M}|\bigvee_{k=1}^{\infty} \alpha^k \mathfrak{M}) = 0$ can be proved by the above properties and the equality (6) of Corollary 2.

(2) : Since h is α-invariant and α is ergodic, h is a multiple of identity; $h = \lambda I$. Moreover, it can be shown that, for the set $\mathfrak{M}_\alpha = \bigvee_{k=0}^{\infty}\alpha^k\mathfrak{M}$, there exists a finite integer N such that $\mathfrak{M}_\alpha = \bigvee_{k=0}^{N-1}\alpha^k\mathfrak{M}$. These facts imply

$$h = \tau(h) = S(\bigvee_{k=0}^{\infty} \alpha^k \mathfrak{M}) = S(\bigvee_{k=0}^{N-1} \alpha^k \mathfrak{M}).$$

(Q.E.D)

In the above theorem, the von Neumann subalgebra $\mathfrak{M}_\alpha = \bigvee_{k=0}^{\infty} \alpha^k \mathfrak{M}$ is a "space" constructed by moving α over $\mathfrak{M}$. This "space" can be considered as a space generated by $\mathfrak{M}$ and α. Therefore the entropy $S(\mathfrak{M}_\alpha)/N$ (denoted by $S(\mathfrak{M},\alpha)$) can be read as the averaged entropy (information) generated by $\mathfrak{M}$ and α. Furthermore it is readily seen that (1) $S(\mathfrak{M},\alpha) \le S(\mathfrak{M})$ holds and (2) $\alpha\mathfrak{M} = \mathfrak{M}$ implies $S(\mathfrak{M},\alpha) = S(\mathfrak{M})$.

Theorem 3 does not contain the averaging w.r.t. time (n), so that it is not a complete formulation of McMillan's ergodic theorem in noncommutative systems. In order to formulate and prove the McMillan ergodic type convergence theorem, we first have to set a quantum mechanical message space [4].

Since τ is faithful normal, τ might be represented by a vector x in $\mathcal{H}$ such that

$$\tau(\cdot) = \langle x, \cdot\, x \rangle,$$

where $\langle \cdot , \cdot \rangle$ is the inner product of $\mathcal{H}$. Let $\mathcal{K}$ be the infinite tensor product of $\mathcal{H}$ with respect to the above vector x in the sense of von Neumann, which is denoted by

$$\mathcal{K} = \bigotimes_{-\infty}^{\infty} (\mathcal{H}, x)$$

We define the noncommutative message space $\mathcal{A}$ as the von Neumann algebra on $\mathcal{K}$ generated by the following operators $\bar{A}_n$ $(n \in Z)$

$$\bar{A}_n(\overset{+\infty}{\underset{-\infty}{\otimes}} x_k) = \overset{+\infty}{\underset{-\infty}{\otimes}} x_k' \text{ with } x_k' = \delta_{nk}A_n x_k + (1-\delta_{nk})x_k$$

with $A_n \in \mathfrak{N}$ ($n \in Z$). This message space is called the infinite tensor product of von Neumann algebra $\mathfrak{N}$ and is denoted by

$$\mathcal{A} = \overset{\infty}{\underset{-\infty}{\otimes}} \{\mathfrak{N}, \tau\},$$

where $\mathfrak{N}$ corresponds to the alphabet space in CDS. Now, for the n-times tensor product Hilbert space $\mathcal{K}_n = \otimes^n \mathcal{K}$, every element Q in $B(\mathcal{K}_n)$ can be canonically embedded into $B(\mathcal{K}_{n+1})$ in such a way that $Q \subsetneq Q \otimes I$, so that we have the canonical embbeding j_n from $\mathfrak{N}_n = \otimes^n \mathfrak{N}$ (the n-times tensor product of von Neumann algebra $\mathfrak{N}$) into $\mathcal{A}$.

For a von Neumann subalgebra $\mathfrak{M}$ of $\mathfrak{N}$, let $\mathfrak{M}_n$ and $\mathcal{B}_n$ be

$$\mathfrak{M}_n = \overset{n}{\underset{1}{\otimes}} \mathfrak{M},$$

$$\mathcal{B}_n = j_n(\mathfrak{M}_n).$$

Then $\mathcal{B}_n$ becomes a von Neumann subalgebra (message subspace) of $\mathcal{A}$. Using a shift operator α defined as $\alpha(\otimes A_k) = \otimes A_{K+1}$, the above $\mathcal{B}_n$ is expressed by

$$\mathcal{B}_n = \vee_{k=0}^{n-1} \alpha^{-k} \mathcal{B}_1 .$$

Our "information source" is now described by $(\mathcal{K}, \mathcal{A}, \alpha)$ and an α-invariant faithful state φ on the message space $\mathcal{A}$. The state φ controls the transmission of information, so that the McMillan theorem is written in terms of φ, α and the entropy operator defined in §2. We assume that φ_n, the restriction of φ to $\mathfrak{M}_n$, is tracial. Then the entropy operator w.r.t. φ and $\mathcal{B}_n$ is given by

$$H_\varphi(\mathcal{B}_n) = -\sum_k Q_k^{(n)} \log \varphi_n(Q_k^{(n)}),$$

where $\{Q_k^{(n)}\}$ is a minimal finite partition of unity in $\mathfrak{M}_n$.

Theorem 4: <u>Under the above settings</u>, <u>we have</u> (1) <u>there exists a α-invariant operator</u> h <u>such that</u> $H_\varphi(\mathcal{B}_n)/n$ <u>converges to</u> h <u>φ-almost uniformly as</u> $n \to \infty$; (2) <u>if α is ergodic</u> (i.e., $\{A \in \mathcal{A}\ ;\ \alpha(A) = A\} = CI$), <u>then</u> $h = \varphi(h)I$.

<u>Sketch</u>: For any minimal finite partition $\{P_i; i=1,\cdots,N\}$ in $\mathfrak{M}$, the family $\{P_{i_1} \otimes P_{i_2} \otimes \cdots \otimes P_{i_n}\}$ is a minimal finite partition in $\mathfrak{M}_n$, where n indices $i_1, \cdots, i_n$ run from 1 to N. As $\mathfrak{M}_n$ is a

finite dimensional von Neumann algebra, $H_\varphi(\mathcal{B}_n)$ defined above can be expressed by

$$H_\varphi(\mathcal{B}_n) = -\sum_{i_1,\cdots,i_n=1}^{N} P_{i_1}\otimes\cdots\otimes P_{i_n}\log\varphi_n(P_{i_1}\otimes\cdots\otimes P_{i_n}).$$

Let us consider the von Neumann subalgebra $\mathcal{C}_n$ of $\mathfrak{M}_n$ generated by the family $(P_{i_1}\otimes\cdots\otimes P_{i_n})$: $\mathcal{C}_n = \{P_{i_1}\otimes\cdots\otimes P_{i_n};\ i_1,i_2,\cdots,i_n = 1,\cdots,N\}''$, and let $\mathcal{C}_1 = \{P_i\}'' = \mathcal{C}$. Then $\mathcal{C}_n$ is the n-times tensor product of $\mathcal{C}$, and $\mathcal{C}$, $\mathcal{C}_n$ are commutative von Neumann algebras, so that from Theorem 6.3 of [7], there exist compact Hausdorff spaces Ω_n, Ω and probability measures μ_n, μ such that

$$\mathcal{C}_n \simeq \mathbf{L}^\infty(\Omega_n,\mu_n),$$
$$\mathcal{C} \simeq \mathbf{L}^\infty(\Omega,\mu).$$

Moreover, $\mathcal{C}_n$ is monotonously increasing and generates the infinite tensor product $\tilde{\mathcal{C}}$ of $\mathcal{C}$. Since $\tilde{\mathcal{C}}$ is commutative, there exist a compact Hausdorff space $\tilde{\Omega}$ and a probability measure $\tilde{\mu}$ such that

$$\tilde{\mathcal{C}} \simeq \mathbf{L}^\infty(\tilde{\Omega},\tilde{\mu})$$

and

$$\varphi(P_{i_1}\otimes\cdots\otimes P_{i_n}) = \tilde{\mu}(\Delta_{i_1\cdots i_n}),$$

where $P_{i_1}\otimes\cdots\otimes P_{i_n}$ corresponds to the characteristic function $\mathbf{1}_{\Delta_{i_1\cdots i_n}}$ for some measurable set $\Delta_{i_1\cdots i_n}$ in $\prod_1^n \Omega$. Thus the commutative McMillan theorem together with Theorem 1 (based on the finite dimensionality of $\mathfrak{M}_n$) implies that our entropy operator $H_\varphi(\mathcal{B}_n)/n$ converges to some α-invariant operator h in $\mathbf{L}^1(\tilde{\Omega},\tilde{\mu})$, φ-a.u. because of φ-a.u. = $\tilde{\mu}$-a.e.

The α-invariance of h is trivial from the definition of $H_\varphi(\mathcal{B}_n)/n$.

(2) is an immediate consequence of (1).

(Q.E.D.)

For the above theorem, the Martingale convergence theorem of Umegaki [11] based on the noncommutative conditional expectation [7,10] plays an important role. We here use a pure state (projection) as a quantum mechanical signal (alphabet) as usually done, so that the proof of the noncommutative McMillan theorem is essentially traced from that of the commutative McMillan theorem, which is one of main claims of this paper. However one recently tries to use a coherent state or a mixture of pure states as an input signal. Then the situation might be different, in which more general formulation of the entropy operators and the McMillan theorem would be desirable. This will be done by dropping some of our assumptions; for instance, (1) representing each element of messsage by a projection, (2) the finite dimensionality of $\mathfrak{N}$, (3) the traciality of $\varphi_n = \varphi \restriction \mathfrak{M}_n$. Such a generalization with some applications in quantum information theory [5,9] and physical state change [6] will be discussed elsewhere.

References

[1] A.Connes and E.Størmer, Entropy for automorphisms of II_1 von Neumann algebras, Acta. Math., 134, 289-306, 1975.
[2] G.G.Emch, Positivity of the K-entropy on nonabelian K-flows, Z. Wahr. Gebiete, 29, 241-252, 1974.
[3] M.Ohya and A.Yoshikawa, Entropy operators and a convergence theorem in a noncommutative dynamical system, to appear.
[4] M.Ohya, M.Tsukada and H.Umegaki, A formulation of noncommutative McMillan convergence theorem, to appear.
[5] M.Ohya, On compound state and mutual information in quantum information theory, IEEE Information Theory, 29, 770-774, 1983.
[6] M.Ohya, State change and entropies in quantum dynamical systems, Springer Lect. Note in Math., 1136, 397-408, 1985.
[7] H.Umegaki, M.Ohya and F.Hiai, "Introduction to Operator Algebras", Kyoritsu Publishing Company, 1985.
[8] H.Umegaki and M.Ohya, "Entropies in Probabilistic Systems ", Kyoritsu Publishing Company, 1983.
[9] H.Umegaki and M.Ohya, "Quantum Mechanical Entropies", Kyoritsu Publishing Company, 1984.
[10] H.Umegaki, Conditional expectation in an operator algebra, Tohoku Math. J., 6, 177-181, 1954.
[11] H.Umegaki, Conditional expectation in an operator algebra II, Tohoku Math. J., 8, 86-100, 1956.

Department of Information Sciences
Science University of Tokyo
Noda City, Chiba 278, Japan

ON LONG TIME TAILS OF CORRELATION FUNCTIONS FOR KMO-LANGEVIN EQUATIONS

Yasunori Okabe

§1 Stokes-Boussinesq-Langevin equation

About sixteen years ago, B.Alder and T.Wainwright [1] had discovered a long time tail behavior ($\propto t^{-3/2}$) of velocity autocorrelation function for hard sphere by a computer simulation. It has become clear that this phenomenon can be explained by considering the unique stationary solution $\mathbf{X} = (X(t);t\in \mathbb{R})$ to the following Stokes-Boussinesq-Langevin equation:

$$(1.1) \qquad m^*\dot{X}(t) = -6\pi r\eta X(t) - 6\pi r^2\sqrt{\frac{\rho\eta}{\pi}}\int_{-\infty}^{t}\sqrt{t-s}^{\,-1}\dot{X}(s)ds + W(t),$$

where $m^*=m+(2/3)\pi r^3\rho$, m and r(resp.η and ρ) denote the effective mass, mass and radius of a hard sphere(resp. the viscosity and density of a viscous fluid), $\mathbf{W} = (W(t);t \in \mathbb{R})$ is a Kubo noise or a white noise. Furthermore we know that $\mathbf{X}$ has a reflection positivity ([3],[4] and [6]).

§2 KMO-Langevin equations

Conversely, we will consider a real weakly stationary process $\mathbf{X} = (X(t);t \in \mathbb{R})$ having a reflection positivity. That is,its correlation function R takes the form of

$$(2.1) \qquad R(t) = \int_0^\infty e^{-|t|\lambda}\sigma(d\lambda) \qquad (t\in \mathbb{R}),$$

where σ is a bounded Borel measure on $[0,\infty)$ satisfying

$$(2.2) \qquad \sigma(\{0\}) = 0 \quad \text{and} \quad \int_0^\infty \lambda^{-1}\sigma(d\lambda) < \infty .$$

Then we derived in [5] the second KMO-Langevin equation (2.3) describing the time evolution of $\mathbf{X}$; as random tempered distributions,

$$(2.3) \qquad \dot{\mathbf{X}} = -\beta_2\mathbf{X} - \lim_{\varepsilon\downarrow 0}\dot{\gamma}_{2,\varepsilon}*\mathbf{X} + \alpha_2\mathbf{I},$$

where

$$(2.4) \qquad \alpha_2 > 0 \quad \text{and} \quad \beta_2 > 0$$

(2.5) $\gamma_{2,\varepsilon}(t) = \chi_{(0,\infty)}(t)\int_\varepsilon^\infty e^{-t\lambda}\rho_2(d\lambda)$ with a Borel measure ρ_2 on $[0,\infty)$ satisfying

$$\rho_2(\{0\}) = 0 \quad \text{and} \quad \int_0^\infty(\lambda+1)^{-1}\rho_2(d\lambda) < \infty$$

(2.6) $\mathbf{I}$ is a weakly stationary random tempered distribution with a spectral measure $\Delta_{\mathbb{I}}$ given by

$$\Delta_{\mathbb{I}}(d\xi) = \sqrt{2/\pi}\alpha_2(\beta_2 + \int_0^\infty \xi^2(\lambda^2+\xi^2)^{-1}\rho_2(d\lambda))d\xi \ .$$

Remark 2.1. The random noise $\mathbb{I}$ is called a Kubo noise, which is related to $\mathbf{X}$ through the following relation:

(2.7) $$X(\phi) = (\sqrt{2\pi})^{-1}\int_0^\infty R(t)\, I(\phi(\cdot+t))dt \qquad (\phi\in \mathcal{S}(\mathbf{R})).$$

Furthermore, under the condition

(2.8) $$\int_0^\infty \lambda\sigma(d\lambda) < \infty ,$$

we derived in [5] the first KMO-Langevin equation (2.9) describing the time evolution of $\mathbf{X}$; as random tempered distributions,

(2.9) $$\dot{\mathbf{X}} = -\beta_1\mathbf{X} - \lim_{\varepsilon\downarrow 0} \gamma_{1,\varepsilon}*\dot{\mathbf{X}} + \alpha_1\dot{\mathbf{B}},$$

where

(2.10) $\alpha_1 > 0$ and $\beta_1 > 0$

(2.11) $\gamma_{1,\varepsilon}(t) = \chi_{(0,\infty)}(t)\int_\varepsilon^\infty e^{-t\lambda}\rho_1(d\lambda)$ with a Borel measure ρ_1 on $[0,\infty)$ satisfying

$$\rho_1(\{0\}) = 0 \quad \text{and} \quad \int_0^\infty(\lambda+1)^{-1}\rho_1(d\lambda) < \infty$$

(2.12) $\dot{\mathbf{B}}$ is a weakly stationary random tempered distribution with a spectral measure $\Delta_{\dot{\mathbf{B}}}(d\xi) = d\xi$.

Remark 2.2. The triples $(\alpha_1,\beta_1,\rho_1)$ (resp. $(\alpha_2,\beta_2,\rho_2)$) are called the first (resp. second) KMO-Langevin data associated with R.

§3 Long time tails of correlation functions

For any given quadruple $(\alpha_1,\beta_1,\rho_1,\dot{\mathbf{B}})$ (resp. $(\alpha_2,\beta_2,\rho_2,\mathbb{I})$) satisfying the conditions (2.10), (2.11), (2.12) (resp. (2.4), (2.5), (2.6)), we proved in [5] that there exists a unique weakly stationary solution $\mathbf{X}_1$ (resp. $\mathbf{X}_2$) to the first (resp. second) KMO-Langevin equation (2.9) (resp. (2.3)). We denote by R_j the correlation function of $\mathbf{X}_j$ $(j=1,2)$.

We will characterize a long time tail behavior of R_j in terms of the one of coefficient γ_j in the delayed term in the KMO-Langevin equation, where $\gamma_j=\gamma_{j,0}$ $(j=1,2)$. Suppose that we are given two positive constants $q_j\in(0,1)$ and two slowly varying functions L_j at infinity $(j=1,2)$.

Theorem 3.1. The following (i) and (ii) are equivalent;

(i) $R_2(t) \sim \sqrt{2\pi}\alpha_2\beta_2^{-2}q_2 t^{-(1+q_2)}L_2(t)$ as $t \to \infty$.

(ii) $\gamma_2(t) \sim t^{-q_2}L_2(t)$ as $t \to \infty$.

Theorem 3.2. Under the condition

(3.1) $\int_0^\infty \lambda^{-1} \rho_1(d\lambda) < \infty$,

the following (i) and (ii) are equivalent;

(i) $R_1(t) \sim \alpha_1^2 \beta_1^{-3} q_1 t^{-(1+q_1)} L_1(t)$ as $t \to \infty$

(ii) $\gamma_1(t) \sim t^{-q_1} L_1(t)$ as $t \to \infty$.

Remark 3.1. In Theorem 3.2, (ii) implies (i) without condition (3.1).

§4 Preliminaries

Suppose that we are given a positive number p and a slowly varying function L at infinity. We first recall Abelian Theorem and Tauberian Theorem.

Lemma 4.1.([7]) For any positive number T and $f \in L^1((T,\infty))$, we put $F(t) = \int_t^\infty f(s)ds$ $(t \geq T)$. If f is increasing or decreasing, then the following (i) and (ii) are equivalent;

(i) $f(t) \sim pt^{-(1+p)} L(t)$ as $t \to \infty$.

(ii) $F(t) \sim t^{-p} L(t)$ as $t \to \infty$.

Lemma 4.2.([7]) For any $g \in L^1_{loc}((0,\infty))$, we put $G(t) = \int_0^t g(s)ds$. If g is increasing or decreasing, then the following (i) and (ii) are equivalent;

(i) $g(t) \sim pt^{p-1} L(t)$ as $t \to \infty$.

(ii) $G(t) \sim t^p L(t)$ as $t \to \infty$.

Lemma 4.3.([2] and [7]) For any Borel measure m on $[0,\infty)$ such that $e^{-t\cdot} \in L^1([0,\infty)),m)$ for any $t > 0$, we put M and ω by

$$M(\lambda) = m([0,\lambda]) \text{ and } \omega(t) = \int_{[0,\infty)} e^{-t\lambda} m(d\lambda).$$

Then, the following (i) and (ii) are equivalent;

(i) $M(\lambda) \sim \Gamma(1+p)^{-1} \lambda^p L(\lambda)$ as $\lambda \to \infty$.

(ii) $\omega(t) \sim t^{-p} L(t^{-1})$ as $t \to \infty$.

Furthermore, let R be an integrable and decreasing function on $[0,\infty)$ such that for any $\eta > 0$

$$(4.1) \quad \int_0^\infty e^{-\eta t} R(t)dt = \sqrt{2\pi}\alpha \frac{1}{\beta+\eta+\eta\int_0^\infty (\lambda+\eta)^{-1} \rho(d\lambda)},$$

where $\alpha>0, \beta>0$ and ρ is a Borel measure on $[0,\infty)$ satisfying

$$\rho(\{0\}) = 0 \quad \text{and} \quad \int_0^\infty (\lambda+1)^{-1} \rho(d\lambda) < \infty.$$

Then, we will characterize a long time tail behavior of R in terms of the one of γ, where $\gamma(t)=\int_0^\infty e^{-t\lambda}\rho(d\lambda)$ $(t>0)$. The proof of the following Theorem 4.1 is due to Dr.M.Tomisaki.

Theorem 4.1. We assume that $p \in (0,1)$. Then the following (i) and (ii) are equivalent;

(i) $R(t) \sim \sqrt{2\pi}\alpha\beta^{-2}qt^{-(1+q)}L(t)$ as $t \to \infty$.

(ii) $\gamma(t) \sim t^{-q}L(t)$ as $t \to \infty$.

Proof. It follows from (4.1) that for $\eta>0$

$$(4.2)\quad \int_0^\infty e^{-\eta t}\left(\int_t^\infty R(s)ds\right)dt = \frac{\sqrt{2\pi}\alpha}{\beta}\,\frac{1+\int_0^\infty e^{-\eta t}\gamma(t)dt}{\beta+\eta+\eta\int_0^\infty(\lambda+\eta)^{-1}\rho(d\lambda)}.$$

By Lemma 4.1, (i) is equivalent to

$$\int_t^\infty R(s)ds \sim \sqrt{2\pi}\alpha\beta^{-2}t^{-q}L(t) \quad \text{as } t \to \infty$$

and which with Lemma4.2 is equivalent to

$$\int_0^t\left(\int_u^\infty R(s)ds\right)du \sim \sqrt{2\pi}\alpha\beta^{-2}(1-q)^{-1}t^{1-q}L(t) \quad \text{as } t \to \infty .$$

Therefore, it follows from Lemma4.3 that (i) is equivalent to

$$(4.3)\quad \int_0^\infty e^{-\eta t}\left(\int_t^\infty R(s)ds\right)dt \sim \sqrt{2\pi}\alpha\beta^{-2}\Gamma(2-q)(1-q)^{-1}\eta^{-(1-q)}L(\eta^{-1}) \quad \text{as } \eta\downarrow 0.$$

Since

$$\eta^{1-q}L(\eta^{-1}) \to 0 \quad \text{and} \quad \eta\int_0^\infty(\lambda+\eta)^{-1}\rho(d\lambda) \to 0 \quad \text{as } \eta\downarrow 0,$$

we see from (4.2) that (4.3) is equivalent to

$$\int_0^t\gamma(s)ds \sim (1-q)^{-1}t^{1-q}L(t) \quad \text{as } t \to \infty$$

and so which with Lemma4.2 is equivalent to (ii). (Q.E.D.)

§5 Proof of Theorems 3.1 and 3.2

By Theorem 8.5 in [6], we see that for any $\eta > 0$

$$(5.1)\quad \int_0^\infty e^{-\eta t}R_2(t)dt = \sqrt{2\pi}\alpha_2\,\frac{1}{\beta_2+\eta+\eta\int_0^\infty(\lambda+\eta)^{-1}\rho_2(d\lambda)} .$$

Therefore, immediately from Theorem 4.1, we have Theorem 3.1.

Next, we will show in Theorem 3.2 that (ii) implies (i) without condition (3.1). We denote by E_1 the canonical representation kernel of $\mathbf{X}_1$. We note from (E.4) in [6] that

$$(5.2)\quad R_1(t) = \frac{1}{2\pi}\int_0^\infty E_1(t+s)E_1(s)ds \quad \text{for any } t > 0.$$

By Theorems 2.1 and 2.2 in [6], we see that E_1 is a positive, integrable and decreasing function on $[0,\infty)$ and

(5.3) $$\int_0^\infty e^{-\eta t}E_1(t)\,dt = \sqrt{2\pi}\alpha_1 \frac{1}{\beta_1+\eta+\eta\int_0^\infty(\lambda+\eta)^{-1}\rho_1(d\lambda)} \quad \text{for any } \eta > 0.$$

Therefore, by virtue of Theorem 3.1, we find that (ii) is equivalent to

(5.4) $$E_1(t) \sim \sqrt{2\pi}\alpha_1\beta_1^{-2}q_1 t^{-(1+q_1)}L_1(t) \quad \text{as } t \to \infty.$$

Fix any $T > 0$. It follows from (5.2) that

(5.5) $$(\alpha_1\beta_1^{-2}q_1)^{-1}(t^{-(1+q_1)}L_1(t))^{-1}R_1(t) = I(t) + II(t),$$

where

$$I = \sqrt{2\pi}^{-1}\int_0^T((\sqrt{2\pi}\alpha_1\beta_1^{-2}q_1)^{-1}\frac{(t+s)^{1+q_1}}{L_1(t+s)}E_1(t+s))(\frac{t}{t+s})^{1+q_1}\frac{L_1(t+s)}{L_1(t)}E_1(s)\,ds$$

$$II = \sqrt{2\pi}^{-1}\int_T^\infty\frac{t^{1+q_1}}{L_1(t)}\frac{L_1(t+s)}{(t+s)^{1+q_1}}((\sqrt{2\pi}\alpha_1\beta_1^{-2}q_1)^{-1}\frac{(t+s)^{1+q_1}}{L_1(t+s)}E_1(t+s))E_1(s)\,ds$$

Since

$$\frac{L_1(t+s)}{L_1(t)} \text{ converges to 1 as } t \to \infty, \text{uniformly in } s \in [0,T],$$

we see from (5.4) that

(5.6) $$\lim_{t\to\infty} I(t) = \sqrt{2\pi}^{-1}\int_0^T E_1(s)\,ds.$$

Since it follows from [7] that there exists a constant $c_1 > 0$ such that

$$\left|\frac{t^{-(1+q_1)}}{L_1(t)}\sup_{y\geq t}(y^{-(1+q_1)}L_1(y))\right| \leq c_1,$$

we see from (5.4) that

(5.7) $$\overline{\lim_{t\to\infty}}\,|II(t)| \leq 2c_1\int_T^\infty E_1(s)\,ds.$$

By Lemmas 2.2 and 2.7 in [6], we have

(5.8) $$\int_0^\infty E_1(s)\,ds = \sqrt{2\pi}\alpha_1\beta_1^{-1}.$$

Therefore, by letting T in (5.6) and (5.7) to infinity, we can conclude from (5.5) and (5.8) that (i) holds.

Finally, we will show that (i) implies (ii) under condition (3.1). By (5.2) and (5.8), we have

(5.9) $$R_1(t) - \sqrt{2\pi}^{-1}\alpha_1\beta_1^{-1}E_1(t) = \frac{1}{2\pi}\int_0^\infty(E_1(t+s)-E_1(t))E_1(s)\,ds.$$

On the other hand, it follows from Theorem 2.1 in [6] that there exists a bounded Borel measure ν_1 on $[0,\infty)$ such that

(5.10) $$\nu_1(\{0\}) = 0 \quad \text{and} \quad \int_0^\infty\lambda^{-1}\nu_1(d\lambda) < \infty$$

(5.11) $E_1(t) = \int_0^\infty e^{-t\lambda}\nu_1(d\lambda)$.

By (5.3) and (5.11), we have

$$\int_0^\infty \frac{1}{\lambda+\eta}\,\nu_1(d\lambda) = \sqrt{2\pi}\alpha_1 \frac{1}{\beta_1+\eta+\eta\int_0^\infty \frac{1}{\lambda+\eta}\rho_1(d\lambda)} \quad \text{for any } \eta > 0.$$

By differentiating both hand sides at $\eta = 0$, we can see from condition (3.1) that

(5.12) $\int_0^\infty \lambda^{-2}\nu_1(d\lambda) < \infty$

and so by (5.11)

(5.13) $\int_0^\infty sE_1(s)ds < \infty$.

By substituting (5.11) into (5.9), we see that for any $t > 0$

$$t^{1+q_1}L_1(t)^{-1}(R_1(t) - \sqrt{2\pi}^{-1}\alpha_1\beta_1^{-1}E_1(t))$$

$$= (2\pi tL_1(t))^{-1}\int_0^\infty(\int_0^\infty (t\lambda)^{2+q_1}e^{-t\lambda}\frac{(e^{-s\lambda}-1)}{\lambda^{1+q_1}}\,\nu_1(d\lambda))E_1(s)ds.$$

Therefore, by (5.12) and (5.13), we see that there exists a constant c_2 such that for any $t > 0$

(5.14) $|t^{1+q_1}L_1(t)^{-1}(R_1(t)-\sqrt{2\pi}^{-1}\alpha_1\beta_1^{-1}E_1(t))| \leq c_2(tL_1(t))^{-1}$.

Since

$$tL_1(t) \to \infty \quad \text{as } t \to \infty \quad ,$$

we see from (5.14) that (i) implies (5.4) and so (ii). (Q.E.D.)

Acknowledgement. The author would like to thank Doctor M.Tomisaki for her help for the proof of Theorem 4.1.

References

[1] B.J.Alder and T.E.Wainwright : Decay of the velocity autocorrelation function, Phys. Rev. A1(1970), 18-21.

[2] W.Feller : An introduction to probability theory and its applications,vol.II,John Wiley and Sons.Inc.,New York (1966).

[3] E.H.Hauge and A.Martin-Löf : Fluctuating Hydrodynamics and Brownian Motion, J. Stat.Phys.7(1973),259-281.

[4] R.Kubo : Irreversible Processes and Stochastic Processes, RIMS, Kyoto,Oct.1979,50-93 (in Japanese).

[5] Y.Okabe : On KMO-Langevin equations for stationary Gaussian processes with T-positivity, J.Fac.Sci.Univ.Tokyo,Sect.IA 33(1986),1-56.

[6] Y.Okabe : On the theory of the Brownian motion with Alder-Wainwright effect, to appear in J.Stat.Phys.,45(1986).

[7] E.Seneta : Regularly varying functions, Lecture Notes in Math.508, Springer, Berlin-Heiderberg-New York (1976).

Department of Mathematics
Faculty of Science
Hokkaido University
Sapporo
060 Japan

ON CENTRAL LIMIT THEOREM FOR CONTINUOUS ADDITIVE FUNCTIONAL OF ZERO ENERGY

Yoichi Oshima

1. Introduction

Let X be a locally compact separable metric space and m be an everywhere dense positive probability measure on X. We assume that we are given an irreducible recurrent regular Dirichlet space $(\mathcal{E},\mathcal{F})$ on $L^2(X;dm)$. Then there corresponds an m-symmetric Markov process $M = (\Omega,\mathcal{B},X_t,P_x)$ in the sense

$$(1.1) \qquad \mathcal{E}_\alpha(R_\alpha f,v) = (f,v),$$

where R_α is the resolvent of M, $(\ ,\ )$ is the inner product in $L^2(X;dm)$ and $\mathcal{E}_\alpha(\ ,\ ) = \mathcal{E}(\ ,\ ) + \alpha(\ ,\ )$.

By Fukushima [2; Theorem 5.2.2], if we denote by $\tilde{u}$ the quasi-continuous modification of $u \in \mathcal{F}$, then there exists a martingale additive functional $M^{[u]}$ of finite energy and a continuous additive functional $N^{[u]}$ of zero energy such that

$$(1.2) \qquad \tilde{u}(X_t) - \tilde{u}(X_0) = M_t^{[u]} + N_t^{[u]}.$$

As we remark in the next section, this result can be extended to the functions u belonging to the extended Dirichlet space $(\mathcal{E},\mathcal{F}_e)$ of $(\mathcal{E},\mathcal{F})$.

The purpose of this article is to show the following central limit theorem : If $N_t = N_t^{[u]}$ for some $u \in \mathcal{F}_e$, then the distribution of the process $(\frac{1}{\sqrt{\lambda}}N_{\lambda t})_{t\geq 0}$ under P_m converges weakly to that of the Brownian motion with variance $2\mathcal{E}(u,u)$.

In [5], for an m-symmetric Markov process $M = (X_t,P_x)$ with $L^2(X;dm)$-generator L and semigroup T_t, Kipnis and Varadhan proved the central limit theorem for

$$(1.3) \qquad N_t = \int_0^t V(X_s)ds,$$

where V is a function of $L^2(X;dm)$ satisfying $(V,1) = 0$ and

$$(1.4) \qquad |(V,v)| \leq K(-Lv,v)^{1/2} \leq K\mathcal{E}(v,v)^{1/2},$$

for all $v \in \mathcal{D}(L)$.

If V satisfies these conditions, by noting that (1.4) implies

$$(1.4)' \qquad \int_0^\infty (T_tV,V)\ dt < \infty,$$

we can see that the function $u_s = \int_0^s T_t V\,dt$ is contained in F and satisfies $E(u_s,v) = (V - T_sV,v)$ for all $v \in F$. In particular, $\{u_s\}$ forms an E-Cauchy sequence as $s \to \infty$. Thus, if there exists a subsequence $\{u_{s_k}\}$ $(s_k \to \infty)$ which converges m-a.e. to some function u, then u belongs to F_e and satisfies

$$(1.5) \qquad E(u,v) = (V,v) \qquad \text{for all } v \in F_e.$$

By [2; Theorem 5.3.1], we see then that $N^{[u]}$ equals (1.3).

In general, we do not know if there exists such subsequence $\{u_{s_k}\}$ But if M satisfies the condition of Harris recurrence, in addition to our present conditions, then $(E,\hat{F}_e)$ becomes a Hilbert space, where $\hat{F}_e$ is the quotient space of F_e by the space of constant functions ([6]). Hence, under (1.4), there exists a function $u \in F_e$ satisfying (1.5) and the result of [5] can be reduced to the present one.

2. Some stochastic calculus related to (E,F_e)

As in [8], the extended Dirichlet space (E,F_e) is defined as follows : $u \in F_e$ if there exists an E-Cauchy sequence $\{u_n\}$ of functions of F which converges m-a.e. to u. In this case, $E(u,u) = \lim_{n\to\infty} E(u_n,u_n)$. In this section, we shall extend some necessary results in [2] to (E,F_e). Throughout in this section, we shall fix a function $u \in F_e$. As remarked in [6], every function of F_e has a quasi-continuous modification. Hence we shall suppose that u is quasi-continuous. Since $m(X) = 1$, $u' = (-1)\vee u \wedge 1 \in F$ and $E(u',u') \leqq E(u,u)$.

Lemma 2.1. If $0 < \varepsilon < 1$ and ν is a measure with finite energy integral, then

$$(2.1) \qquad P_\nu[\sup_{0\leqq t\leqq T} |u(X_t)| > \varepsilon] \leqq (e^T/\varepsilon)(E_1(\nu))^{1/2}\{(u',u')+E(u,u)\}^{1/2},$$

where $E_1(\nu)$ is the 1-energy integral of ν.

Proof. Since $\{\sup_{0\leqq t\leqq T}|u(X_t)| > \varepsilon\} = \{\sup_{0\leqq t\leqq T}|u'(X_t)| > \varepsilon\}$, (2.1) follows from [2; Lemma 5.1.1].

Let $\{u_n\}$ be an E-Cauchy sequence of functions of F which converges m-a.e. to u. Then $u_n' \equiv (-1)\vee u_n \wedge 1$ converges to u' m-a.e. and boundedly, and hence in $L^2(X;dm)$. Thus, by a similar argument to [2; Lemma 5.1.2], we have

Lemma 2.2. Let $\{u_n\}$ be the above approximating sequence of u, then there exists a subsequence $\{u_{n_k}\}$ satisfying

(2.2) $\quad P_x[u_{n_k}(X_t)$ converges uniformly in t on each compact interval of $[0,\infty)] = 1$,

for q.e.x.

By using Lemmas 2.1 and 2.2, we can prove the similar result to [2; Theorem 5.2.2], that is

Proposition 2.3. For any $u \in F_e$, there exists a martingale additive functional $M^{[u]}$ of finite energy and a continuous additive functional $N^{[u]}$ of zero energy such that

$$u(X_t) - u(X_0) = M_t^{[u]} + N_t^{[u]}. \tag{2.3}$$

For the argument in the next section, we need another version of Lemma 2.1. Let $\phi_p(t)$ be a non-decreasing function on R^1 such that $\phi_p(t) = 0$ for $t \leq 0$, $\phi_p(t) = 1$ for $t \geq p$ and $\frac{d}{dt}\phi_p(t) \leq \frac{c}{p}$ for all t. If $\{u_n\}$ is an E-Cauchy sequence of functions of F which converges to u m-a.e., then $\{\phi_p(u_n)\}$ converges m-a.e. to $\phi_p(u)$. Moreover, $\phi_p(u_n) \in F$ and

$$E(\phi_p(u_n)-\phi_p(u_m),\phi_p(u_n)-\phi_p(u_m)) \leq (\tfrac{c}{p})^2 E(u_n-u_m,u_n-u_m)$$

converges to zero as $m, n \to \infty$. Which shows that $\phi_p(u) \in F$.

Lemma 2.4. If $p \geq 2$, then

$$P_m[\sup_{0\leq t\leq T} |u(X_t)| \geq p] \leq e\sqrt{T}\{\tfrac{1}{T}(\phi_p(u),\phi_p(u))+(\tfrac{c}{p})^2 E(u,u)\}^{1/2}. \tag{2.4}$$

Proof. Let $E_p = \{x ; |u(x)| \geq p\}$ and $e_p^\alpha(x) = E_x[\exp(-\alpha\sigma_{E_p})]$, where σ_{E_p} is the hitting time for E_p. Then

$$\begin{aligned} P_m[\sup_{0\leq t\leq T} |u(X_t)| \geq p] &= P_m[\sigma_{E_p} \leq T] \leq e^{\alpha T} \int e_p^\alpha(x)m(dx) \\ &= e^{\alpha T} E_\alpha(R_\alpha 1, e_p^\alpha) \leq e^{\alpha T} E_\alpha(R_\alpha 1,R_\alpha 1)^{1/2} E_\alpha(e_p^\alpha,e_p^\alpha)^{1/2} \\ &= \frac{e^{\alpha T}}{\sqrt{\alpha}} E_\alpha(e_p^\alpha,e_p^\alpha)^{1/2}. \end{aligned}$$

Since $\phi_p(u(x)) = 1$ a.e. on E_p, the last term is dominated by

$$\frac{e^{\alpha T}}{\sqrt{\alpha}} E_\alpha(\phi_p(u),\phi_p(u))^{1/2} \leq \frac{e^{\alpha T}}{\sqrt{\alpha}} \{\alpha(\phi_p(u),\phi_p(u))+(\tfrac{c}{p})^2 E(u,u)\}^{1/2}.$$

Taking $\alpha = \frac{1}{T}$, we obtain the result.

3. Central limit theorem

We shall fix a function $u \in \mathcal{F}_e$ and set $N_t = N_t^{[u]}$, then our main result is the following

Theorem 3.1. The distribution of $\{\frac{1}{\sqrt{\lambda}} N_{\lambda t}\}$ under P_m converges to that of the Brownian motion with variance $2\mathcal{E}(u,u)$.

Proof. Since $N_t = u(X_t) - u(X_0) - M_t^{[u]}$, it is enough to show

$$(3.1) \qquad \lim_{\lambda\to\infty} P_m[\sup_{0\leq t\leq \lambda T} |u(X_t)-u(X_0)| > \sqrt{\lambda}\delta] = 0,$$

for all $\delta > 0$ and $T > 0$, and

$$(3.2) \qquad \lim_{\lambda\to\infty} \frac{1}{\lambda} \langle M^{[u]}\rangle_{\lambda t} = 2\mathcal{E}(u,u)t \qquad \text{a.s.} P_m,$$

for all $t > 0$ (see [4; Theorem 5.1]).

Proof of (3.1).

Since the range $R_1(L^2(X;dm))$ is $\mathcal{E}_1$-dense in $\mathcal{F}$, for all $\varepsilon > 0$, there exists a function $v = R_1 f$ $(f \in L^2(X;dm))$ such that $\mathcal{E}(u-v,u-v) < \varepsilon$. Since $\lim_{\lambda\to\infty} P_m[|u(X_0)| > \frac{\sqrt{\lambda}\delta}{3}] = 0$,

$$\begin{aligned}
&\lim_{\lambda\to\infty} P_m[\sup_{0\leq t\leq \lambda T} |u(X_t) - u(X_0)| > \sqrt{\lambda}\delta] \\
&\quad \leq \lim_{\lambda\to\infty} P_m[\sup_{0\leq t\leq \lambda T} |u(X_t) - v(X_t)| > \frac{\sqrt{\lambda}\delta}{3}] \\
&\quad + \lim_{\lambda\to\infty} P_m[\sup_{0\leq t\leq \lambda T} |v(X_t)| > \frac{\sqrt{\lambda}\delta}{3}] \\
&\quad = I + II, \qquad \text{say.}
\end{aligned}$$

Setting $p = \sqrt{\lambda}\delta/3$ in (2.4), we have

$$\begin{aligned}
I &\leq \lim_{\lambda\to\infty} e\sqrt{\lambda T}\{\frac{1}{\lambda T}(\phi_p(u-v),\phi_p(u-v)) + (\frac{c}{p})^2\mathcal{E}(u-v,u-v)\}^{1/2} \\
&= \lim_{\lambda\to\infty} e\{(\phi_p(u-v),\phi_p(u-v)) + (9c^2T/\delta^2)\mathcal{E}(u-v,u-v)\}^{1/2} \\
&= (3ce\sqrt{T}/\delta)\ \mathcal{E}(u-v,u-v)^{1/2} \\
&< 3ce\sqrt{\varepsilon T}/\delta,
\end{aligned}$$

since $\phi_p(u-v)(x)$ tends to zero a.e. and boundedly as p tends to infinity. Hence it is enough to show that $II = 0$. To this end, by noting

$$\begin{aligned}
P_m[\sup_{0\leq t\leq \lambda T} |v(X_t)| > \frac{\sqrt{\lambda}\delta}{3}] &\leq \sum_{k=1}^{[\lambda T]+1} P_m[\sup_{k-1\leq t\leq k} |v(X_t)| > \frac{\sqrt{\lambda}\delta}{3}] \\
&= ([\lambda T]+1)\, P_m[\sup_{0\leq t\leq 1} |v(X_t)| > \frac{\sqrt{\lambda}\delta}{3}]
\end{aligned}$$

$$\leq \frac{9}{\lambda\delta^2}([\lambda T]+1)E_m[\sup_{0\leq t\leq 1} |v(X_t)|^2;\ \sup_{0\leq t\leq 1} |v(X_t)|^2 > (\frac{\sqrt{\lambda}\delta}{3})^2],$$

it is enough to show that

$$E_m[\sup_{0\leq t\leq 1} |v(X_t)|^2] < \infty.$$

This follows easily from

$$v(X_t) = v(X_0) + M_t^{[v]} + \int_0^t (v-f)(X_s)ds.$$

Proof of (3.2).

Since

$$E_m[<M^{[u]}>_{t+s}] = E_m[<M^{[u]}>_t + <M^{[u]}\circ\theta_t>_s]$$

$$= E_m[<M^{[u]}>_t] + E_m[<M^{[u]}>_s]$$

and

$$\lim_{t\to 0} \frac{1}{t} E_m[<M^{[u]}>_t] = 2\mathcal{E}(u,u),$$

we have

$$E_m[<M^{[u]}>_t] = 2\mathcal{E}(u,u)t.$$

Hence (3.2) follows from [1; Theorem II.1].

References

[1] J. Azéma, M. Duflo and D. Revuz: Mesure invariante sur les classes récurrentes des processus de Markov, Z. Wahrscheinlichkeitstheorie verw. Geb. 8 (1967), 157-181.

[2] M. Fukushima: Dirichlet forms and Markov processes, Kodansha and North-Holland, Tokyo and Amsterdam (1980).

[3] M. Fukushima: On recurrence criteria in the Dirichlet space theory, to appear in Stochastic Analysis.

[4] I. Helland: Central limit theorems for martingales with discrete or continuous time, Scand. J. Stat. 9 (1982), 79-94.

[5] C. Kipnis and S. R. S. Varadhan: Central limit theorem for additive functionals of reversible Markov processes and applications to simple exclusions, Comm. Math. Phys. 104 (1986), 1-19.

[6] Y. Oshima: Potential of recurrent symmetric Markov processes and its associated Dirichlet spaces, Functional Analysis in Markov Processes, Lecture Notes in Math. 923 (1982), Springer.

[7] R. Rebolledo: La méthode des martingales appliquée a l'étude de la convergence en loi de processus, Bull. Soc. Math. France, Mémoire 62 (1979).

[8] M. Silverstein: Symmetric Markov processes, Lecture Notes in Math. 426 (1974), Springer.

Department of Mathematics
Faculty of Engineering
Kumamoto University
Kumamoto
860 Japan

ERGODIC PROPERTIES OF PRODUCT TYPE ODOMETERS

Motosige Osikawa

1. Introduction. For each positive integer k, let n_k be a positive integer, $\Omega_k = \{0,1,\cdots,n_k-1\}$, $\mathcal{B}_k$ the σ-albebra consisting of all subsets of Ω_k, P_k a probability measure defined on $\mathcal{B}_k$ and G_k the permutation group of Ω_k. We assume that $P_k(\{j\}) > 0$ for any j in Ω_k and $k \in N$. Let $(\Omega,\mathcal{B},P)$ be the infinite direct product measure space of $(\Omega_k,\mathcal{B}_k,P_k)$, $k \in N$. We may consider that each G_k acts on Ω. Each element g of the union of G_k, $k \in N$ is a non-singular transformation, that is, $P(gA) = 0$ if and only if $P(A) = 0$. By the well-known 0-1 law it is easy to see that the group G generated by the union of G_k, $k \in N$, is ergodic, that is, every G-invariant measurable function is constant a.e. For a point $\omega = (\omega_k)$ in $\Omega(i)$, $i \in N$, $T\omega = (\omega_k')$ is defined by

$$\omega_k' = \begin{cases} 0 & (k=1,2,\cdots,i-1) \\ \omega_i+1 & (k=i) \\ \omega_k & (k=i+1,i+2,\cdots), \end{cases}$$

where $\Omega(i) = \{\omega \in \Omega\colon \omega_k = n_k-1,\ k=1,2,\cdots,i-1,\ \omega_i \neq n_i-1\}$. We consider only a case that an infinite product measure P does not have an atomic point. Then T is an ergodic non-singular transformation of $(\Omega,\mathcal{B},P)$ and called a product type odometer. For positive numbers(positive integers) $a(i,j)$ we define a function $\xi(\omega)$ on Ω by $\xi(\omega) = a(i,j)$ for ω in $\Omega(i,j)$, where $\Omega(i,j) = \{\omega \in \Omega\colon \omega_k = n_k-1,\ k=1,2,\cdots,i-1,\ \omega_i = j\}$, $j=1,2,\cdots,n_i-2$, $i \in N$. A flow(a transformation) built under function $\xi(\omega)$ based on a product type odometer T is called an AC-flow(an AC-transformation).

For an ergodic non-singular flow $\{T_s,\ s \in R\}$ (an ergodic non-singular transformation T) of a measure space $(\Omega,\mathcal{B},P)$ L^∞-point spectrum $Sp(\{T_s\})$ $(Sp(T))$ is defined as the set of all real numbers t such that there exists a measurable function $f(\omega)$ with $|f(\omega)| = 1$ and $f(T_s\omega) = e^{its}f(\omega)$ for a.e.ω and $s \in R$ ($f(T\omega) = e^{it}f(\omega)$ for a.e.ω). L^∞-point spectrum $Sp(\{T_s\})$ $(Sp(T))$ is a subgroup of real numbers and may be uncountable if $\{T_s\}$ (T) is not finite measure preserving. In fact, there exists an AC-flow(an AC-transformation) whose L^∞-point spectrum is uncountable([9]). Y.Ito, T.Kamae and I.Shiokawa [5] showed that for any $0 \leq \alpha \leq 1$ there exists an AC-transformation whose L^∞-point spectrum has Hausdorff dimension α.

For a countable group G of non-singular transformations of a Lebesgue space $(\Omega,\mathcal{B},P)$ and a point ω $\{ g\omega: g \in G \}$ is called a G-orbit of ω and denoted by $Orb_G(\omega)$. Countable groups G and G' of non-singular transformations of Lebesgue spaces $(\Omega,\mathcal{B},P)$ and $(\Omega',\mathcal{B}',P')$ respectively are said to be mutually orbit equivalent if there exists a non-singular one-to-one mapping ϕ from $(\Omega,\mathcal{B}, P)$ onto $(\Omega',\mathcal{B}',P')$ such that $\phi Orb_G(\omega) = Orb_{G'}(\phi\omega)$ for a.e.ω in Ω. For example, a product type odometer group G and a group generated by a product type odometer T of the same infinite direct product space is mutually orbit equivalent.

Let $(R,\mathcal{B}_R,m)$ be the Lebesgue space of real line, G a countable group of non-singular transformations of a Lebesgue space $(\Omega,\mathcal{B},P)$ and $\tilde{G}$ the group consisting of all transformations $\tilde{g}$ of the direct product measure space $(\Omega\times R,\mathcal{B}\times\mathcal{B}_R,P\times m)$ defined by

$$\tilde{g}(\omega,u) = (g\omega,u - \log\frac{dP g}{dP}(\omega)), \quad (\omega,u) \in \Omega\times R, \quad \text{for } g \text{ in } G.$$

Then there exists a measure space $(X,\mathfrak{F},\mu)$ unique up to isomorphism and a mapping π from $\Omega\times R$ onto X satisfying the following conditions:

(1) $\pi^{-1}A$ is in $\mathcal{B}\times\mathcal{B}_R$ if and only if A is in $\mathfrak{F}$,

(2) $P\times m(\pi^{-1}A) = 0$ if and only if $\mu(A) = 0$, $A \in \mathfrak{F}$,

(3) π is $\tilde{G}$-invariant, that is, $\pi\tilde{g}(\omega,u) = \pi(\omega,u)$ for a.e.(ω,u) and $g \in G$,

(4) for a $\tilde{G}$-invariant measurable function $f(\omega,u)$ there exists a measurable function $\overline{f}(x)$ on X such that $f(\omega,u) = \overline{f}(\pi(\omega,u))$ for a.e.(ω,u).

$(X,\mathfrak{F},\mu)$ is nothing but the space of all $\tilde{G}$-ergodic components equipped with a quotient measure. The mapping π is called a $\tilde{G}$-factor map. Let $\{T_s\}$ be a flow on $\Omega\times R$ defined by $T_s(\omega,u) = (\omega,u+s)$ for $(\omega,u) \in \Omega\times R$ and $s \in R$. The π-image of the flow $\{T_s\}$ is a flow on $(X,\mathfrak{F},\mu)$ and called an associated flow of G. Associated flows of mutually orbit equivalent groups of non-singular transformations are mutually isomorphic. Any ergodic flow can be an associated flow of an ergodic group of non-singular transformations ([5],[6]). An ergodic group G is said to be of type II, III_λ $(0 < \lambda < 1)$, III_1 and III_0 if its associated flow is isomorphic to the transitive flow on the real line, a periodic flow on an interval $[0,|\log\lambda|)$, the trivial flow on a one point set and an ergodic aperiodic conservative flow, respectively.An ergodic group G is of type II if and only if there is a σ-finite G-invariant measure equivalent to P. Ergodic groups of type II are divided into ones of type II_1 and of type II_∞ according as a G-invariant measure is finite or infinite. Ergodic group G is of type III_1 if and only if $\tilde{G}$ is ergodic. This type classification corresponds to a type classification of von Neumann algebras ([2]). An ergodic group G generates a von Neumann algebra by a so-called group measure space construction. Type of an ergodic group G coincides with type of a von Neumann algebra generated by G. A product type odometer generates an infinite tensor product of finite factors of type I (ITPFI) factor, which was studied by H.Araki and J.Woods [1]. A necessary and sufficient condition for a flow to be an associated flow of a product type odometer group has been given by A.Connes and J.Woods [3]. We are interested in concrete computations of associated flows of product type odometer groups. The following lemma is useful to compute associated flows of ergodic groups of non-singular transformations.

Lemma 1. Let G be an ergodic countable group of non-singular transformations of a Lebesgue space $(\Omega,\mathcal{B},P)$, T an ergodic non-singular transformation of a measure space $(X,\mathfrak{F},\mu)$, $\xi(x)$ a positive measurable function on X, H a subgroup of the group consisting of all non-singular transformations h such that $h\omega \in Orb_G(\omega)$ and $\frac{dPh}{dP}(\omega) = 1$ for a.e.ω, and θ an H-factor map from Ω onto X satisfying the following conditions:

(1) $\{T^n\theta(\omega) : n \in Z\} = \{ \theta(g\omega) : g \in G\}$, a.e.$\omega$,

(2) $\{\xi(n,\theta(\omega)) : n \in Z\} = \{\log\frac{dPg}{dP}(\omega) : g \in G\}$, a.e.$\omega$,

(3) $T^n\theta(\omega) = \theta(g\omega)$ if $\log\frac{dPg}{dP}(\omega) = \xi(n,\theta(\omega))$,

where $\xi(n,x)$ is a cocycle function defined by

$$\xi(n,x) = \begin{cases} \sum_{i=0}^{n-1} \xi(T^i x) & (n=1,2,\cdots) \\ 0 & (n=0) \\ -\sum_{i=1}^{-n} \xi(T^{-i}x) & (n=-1,-2,\cdots). \end{cases}$$

Proof. Put $\psi(x,u) = (T^n x, u - \xi(n,x))$ if $\xi(n,x) \leq u < \xi(n+1,x)$, and $\pi(\omega,u) = \psi(\theta(\omega),u)$ for (ω,u) in $\Omega\times R$. Then π is a $\tilde{G}$-factor map from $\Omega\times R$ onto a set $\{(x,u) : x \in X, 0 \leq u < \xi(x)\}$.

Using Lemma 1 we can show that any AC-flow is an associated flow of a product type odometer group ([9]). If n_k, $k \in N$, are bounded positive integers a product type odometer group G is said to have bounded states. T.Giordano and G.Skandalis [4] showed that a product type odometer group with bounded states is orbit equivalent to a product type odometer group with two points state space ($n_k=2$). In the next section we restrict ourselves to determination of types of product type odometer groups with two points state space.

2. Product type odometer groups with two points state space. We compute an associated flow of a product type odometer group with

two points state space.

Example. Let $\Omega_k = \{0,1\}$, $P_k(\{0\}) = \frac{1}{1+\lambda}$ and $P_k(\{1\}) = \frac{\lambda}{1+\lambda}$, $k \in N$, for $0 < \lambda \leq 1$. Then a product type odometer group G of the infinite direct product measure space $(\Omega,\mathcal{B},P) = \prod_{k=1}^{\infty} (\Omega_k,\mathcal{B}_k,P_k)$ is of type III_λ in case $0 < \lambda < 1$, and of type III_1 in case $\lambda = 1$.

Proof. In case $0 < \lambda < 1$, let H be the group consisting of all transformations h with $h\omega \in Orb_G(\omega)$ and $\frac{dPh}{dP}(\omega) = 1$ for a.e.ω in Ω. Let A be an H-invariant measurable set and $\varepsilon > 0$. Then there exist a positive integer n and a $\bigvee_{k=1}^{n}\mathcal{B}_k$-measurable set A' with $P(A\Delta A') < \varepsilon$. A transformation h defined by

$$h(\omega_1,\omega_2,\cdots,\omega_n,\omega_{n+1},\cdots) = (\omega_{n+1},\omega_{n+2},\cdots,\omega_{2n},\omega_1,\omega_2,\cdots,\omega_n,\omega_{2n+1},\cdots)$$

is in H and hA' is $\bigvee_{k=n+1}^{\infty}\mathcal{B}_k$-measurable. Since $P(A\Delta hA') < \varepsilon$ and ε is any positive number, A is in $\bigvee_{k=n+1}\mathcal{B}_k$. By 0-1 law P(A) is 0 or 1, and hence H is ergodic. A mapping θ from Ω onto a one point set is an H-factor map. Let T be the trivial transformation on a one point set and ξ a function on the one point set which takes value $|\log\lambda|$. Then conditions of Lemma 1 are satisfied and the associated flow of G is a periodic flow on an interval $[0,|\log\lambda|)$. Hence G is of type III_λ. In case $\lambda = 1$, P is a finite G-invariant measure and G is of type II_1.

The following lemma due to Araki-Woods [1] is fundamental to determination of types of product type odometer groups with two points state space. We give an ergodic theoretical proof of the lemma.

Lemma 2. Let $\Omega_k = \{0,1\}$, $P_k(\{0\}) = \frac{1}{1+\lambda e^{\varepsilon_k}}$, $P_k(\{1\}) = \frac{\lambda e^{\varepsilon_k}}{1+\lambda e^{\varepsilon_k}}$, $k \in N$, and G be a product type odometer group of the infinite direct product measure space $(\Omega,\mathcal{B},P) = \prod_{k=1}^{\infty} (\Omega_k,\mathcal{B}_k,P_k)$.

(1) If $\sum_{k=1}^{\infty} \varepsilon_k^2 < \infty$, G is of type III_λ in case $0 < \lambda < 1$ and of type II_1 in case $\lambda = 1$.

(2) If $\lim_{k\to\infty} \varepsilon_k = 0$ and $\sum_{k=1}^{\infty}\varepsilon_k^2 = \infty$, G is of type III_1.

Proof. (1) Under the condition the infinite product measure P is equivalent to one of the above example by Kakutani's criterion. Hence we have the conclusion.

(2) We choose infinite subcoordinates such that the condition of (1) is satisfied, and let G_1 be a product type odometer group determined by these subcoordinates and G_2 a product type odometer group determined by the remaining coordinates. From (1) G_1 is orbit equivalent to

the product type odometer group of the above example, which we denote G_λ. Since G_λ is orbit equivalent to the product group $G_\lambda \times G_\lambda$, G_1 is orbit equivalent to $G_1 \times G_\lambda$. Hence $G = G_2 \times G_1$ is orbit equivalent to $G_2 \times G_1 \times G_\lambda = G \times G_\lambda$. Therefore it is enough to show that $\widetilde{G \times G_\lambda}$ is ergodic. We consider that $G \times G_\lambda$ acts on $(\prod_{k=1}^{\infty}\{0,1\}, \mathcal{B}, P)$ and put $\varepsilon_k = 0$ for a coordinate k on which G_λ acts. For subsets $\tilde{A}_i \in \mathcal{B} \times \mathcal{B}_R$, $i=1,2$, with $P \times m(\tilde{A}_i) > 0$, let $\tilde{A}_i(\omega) = \{u \in R : (\omega,u) \in \tilde{A}_i\}$, $i=1,2$, and $A_i = \{\omega \in \Omega : m(\tilde{A}_i(\omega)) > 0\}$, $i=1,2$. Then $P(A_i) > 0$, $i=1,2$, and by the ergodicity of $G \times G_\lambda$ there exists a transformation g in $G \times G_\lambda$ such that $P(gA_1 \cap A_2) > 0$. There exist a subset A of $gA_1 \cap A_2$, real numbers a_1, a_2 and a positive number ε such that $P(A) > 0$, $m((\tilde{g}\tilde{A}_1)(\omega) \cap [a_1-\varepsilon, a_1+\varepsilon])) > \frac{3}{4}\varepsilon$ and $m(\tilde{A}_2(\omega) \cap [a_2-\varepsilon, a_2+\varepsilon])) > \frac{3}{4}\varepsilon$ for a.e. ω in A. We may assume $a_1 - a_2 > 0$. Put $p = a_1 - a_2$, $c = (1/\sqrt{2\pi})\exp(-(\frac{p}{2}+1)^2/2)$ and $d = \min\{ce^{-p^2-1}, ce^{-\varepsilon^2-\varepsilon-1}\}$, and let K be a positive integer such that $K > \max\{\frac{p+2}{\varepsilon}, \frac{1}{p}\}$. There exist a positive integer N and a subset γ of $\prod_{k=1}^{N}\{0,1\}$ such that

$$P(A \cap \gamma) > (1 - (1 - e^{2\varepsilon}/2)d)P(\gamma),$$

where we denote $\prod_{k=1}^{N}\{0,1\}$ the family of all cylinder sets $[w_1, w_2, \cdots, w_N]$ determined by coordinates from 1 to N. Put

$$Y_k(\omega) = \begin{cases} 0 & \text{if } \omega_k = 0 \\ \varepsilon_k & \text{if } \omega_k = 1, \end{cases}$$

then $Y_k(\omega)$, $k \in N$ are independent random variables with expectation $E(Y_k) = \varepsilon_k e^{\varepsilon_k}/(1+\lambda e^{\varepsilon_k})$ and variance $V(Y_k) = \varepsilon_k^2 \lambda e^{\varepsilon_k}/(1+\lambda e^{\varepsilon_k})^2$.

Since $\displaystyle\int_{-\frac{p}{2}+\frac{j}{2}}^{-\frac{p}{2}+\frac{j+1}{K}} \frac{1}{\sqrt{2\pi}} e^{-\frac{x^2}{2}} dx > \frac{c}{K}$ and $\displaystyle\int_{\frac{p}{2}+\frac{j}{K}}^{\frac{p}{2}+\frac{j+1}{K}} \frac{1}{\sqrt{2\pi}} e^{-\frac{x^2}{2}} dx > \frac{c}{K}$

for $j=1,2,\cdots,K-1$, by central limit theorem there exist positive integers m and n $(m > n > N)$ satisfying the following conditions:

$$P(\alpha_j) > \frac{c}{K} \qquad (j=0,1,\cdots,K-1),$$

$$P(\beta_j) > \frac{c}{K} \qquad (j=0,1,\cdots,K-1),$$

$$\left|\sqrt{\sum_{k=n}^{m} V(Y_k)} - 1\right| < \frac{1}{K}$$

and $P_m([w_n, w_{n+1}, \cdots, w_m]) < \frac{c}{K}(e^{-p-\frac{p+2}{K}} - e^{-p-1})$ for $[w_n, w_{n+1}, \cdots, w_m]$ in $\prod_{k=n}^{m}\{0,1\}$, where

$$\alpha_j = \{[w_n, w_{n+1}, \cdots, w_m] \in \prod_{k=n}^{m}\{0,1\} : -\frac{p}{2}+\frac{j}{K} < \sum_{k=n}^{m}(Y_k(w_k) - E(Y_k))/\sqrt{\sum_{k=n}^{m} V(Y_k)} < -\frac{p}{2}+\frac{j+1}{K}\},$$

and

$$\beta_j = \{[w_n, w_{n+1}, \cdots, w_m] \in \prod_{k=n}^{m}\{0,1\} : \frac{p}{2}+\frac{j}{K} < \sum_{k=n}^{m}(Y_k(w_k) - E(Y_k))/\sqrt{\sum_{k=n}^{m} V(Y_k)} < \frac{p}{2}+\frac{j+1}{K}\}.$$

Since

$$\log\frac{P([w'_n,w'_{n+1},\cdots,w'_m])}{P([w_n,w_{n+1},\cdots,w_m])} = \sum_{k=n}^{m}(Y_k(w'_k) - Y_k(w_k)) + \sum_{k=n}^{m}(w'_k - w_k)$$

for $[w_n,w_{n+1},\cdots,w_m]$ and $[w'_n,w'_{n+1},\cdots,w'_m]$ in $\prod_{k=n}^{m}\{0,1\}$, we have

$$p - \frac{p+1}{K} < \log\frac{P([w'_n,w'_{n+1},\cdots,w'_m])}{P([w_n,w_{n+1},\cdots,w_m])} - \sum_{k=n}^{m}(w'_k - w_k)\log\lambda < p + \frac{p+2}{K}$$

for $[w_n,w_{n+1},\cdots,w_m] \in \alpha_j$ and $[w'_n,w'_{n+1},\cdots,w'_m] \in \beta_j$, $j=0,1,\cdots,K-1$. From a condition of size of cylinder sets in $\prod_{k=n}^{m}\{0,1\}$ there is a subset α'_j of α_j such that $\frac{c}{K}\exp(-p-1) < P(\alpha'_j) < \frac{c}{K}\exp(-p-\frac{p+2}{K})$, $j=0,1,2,\cdots,K-1$. Since $P(\alpha'_j)\exp(p+\frac{p+2}{K}) < \frac{c}{K} < P(\beta_j)$ there exists a one-to-one map ϕ from α'_j into β_j, $j=0,1,\cdots,K-1$. Put $\alpha = \bigcup_{j=0}^{K-1}\alpha'_j$ and let h_1 be a transformation in $G\times G_\lambda$ such that for ω in $[w_n,w_{n+1},\cdots,w_m] \in \alpha$ with $\phi[w_n,w_{n+1},\cdots,w_m] = [w'_n,w'_{n+1},\cdots,w'_m]$

$$(h_1\omega)_k = \begin{cases} w'_k & (k=n,n+1,\cdots,m) \\ \omega_k & (k=1,2,\cdots,n-1,m+1,m+2,\cdots). \end{cases}$$

Let h_2 be a transformation in G_λ such that $\log(dPh_2/dP)(\omega) = -\sum_{k=n}^{m}(w'_k - w_k)\log\lambda$. Then $h = h_1h_2$ is a transformation in $G\times G_\lambda$ with $p - \varepsilon < \log\frac{dPh}{dP}(\omega) < p + \varepsilon$ for ω in α. Furthermore we have $P(\alpha) \geq ce^{-p-1} \geq d$ and $P(h\alpha) > ce^{-p-1}e^{p-\varepsilon} \geq d$. Then we have $P(A\cap\gamma\cap\alpha) > \frac{e^{2\varepsilon}}{2}P(\gamma)P(\alpha)$. Because, if it is not true $P(A^c\cap\gamma) > P(A^c\cap\gamma\cap\alpha) > (1-\frac{e^{2\varepsilon}}{2})P(\gamma)P(\alpha) > (1-\frac{e^{2\varepsilon}}{2})dP(\gamma)$, which contradicts to the choice of γ. By the same way we have $P(A\cap\gamma\cap h\alpha) > e^{2\varepsilon}P(\gamma)P(h\alpha) > \frac{1}{2}P(\gamma)P(h\alpha)$. Then we have $P(hA\cap\gamma\cap h\alpha) > e^{p-\varepsilon}P(A\cap\gamma\cap\alpha) > e^{p-\varepsilon}(e^{2\varepsilon}/2)P(\gamma)P(\alpha) > \frac{1}{2}P(\gamma)P(h\alpha)$. Two inequalities above imply $P(A\cap hA) > 0$. Let ω be an element in $h^{-1}A\cap A$ and put $\delta(\omega) = \log\frac{dPh}{dP}(\omega) - p$. Since $(\tilde{h}\tilde{g}\tilde{A}_1)(h\omega) = (\tilde{g}\tilde{A}_1)(\omega) - \log\frac{dPh}{dP}(\omega)$ we have $m((\tilde{h}\tilde{g}\tilde{A}_1)(h\omega)\cap[a_2+\delta(\omega)-\varepsilon,a_2+\delta(\omega)+\varepsilon]) > \frac{3}{4}\varepsilon$. Since $h\omega$ is in A we have $m((\tilde{A}_2)(h\omega)\cap[a_2-\varepsilon,a_2+\varepsilon]) > \frac{3}{4}\varepsilon$. Two inequalities above and $|\delta(\omega)| < \varepsilon$ imply that $m((\tilde{h}\tilde{g}\tilde{A}_1)(h\omega)\cap\tilde{A}_2(h\omega)) > 0$. Hence we have $P\times m(\tilde{h}\tilde{g}\tilde{A}_1\cap\tilde{A}_2) > 0$. This shows that $\widetilde{G\times G_\lambda}$ is ergodic.

Theorem. Let $\Omega_k = \{0,1\}$, $P_k(\{0\}) = 1/(1+\lambda_k)$, $P_k(\{1\}) = \lambda_k/(1+\lambda_k)$, $(0 < \lambda_k \leq 1)$, $k \in N$, and G be a product type odometer group of the infinite direct product measure space $\prod_{k=1}^{\infty}(\Omega_k,\mathcal{B}_k,P_k)$.

(1) If there exist a subsequence k_n of positive integers and a number λ $(0 < \lambda \leq 1)$ such that $\lim_{n\to\infty}\lambda_{k_n} = \lambda$ and $\sum_{k=1}^{\infty}|\log\lambda_{k_n} - \log\lambda|^2 = \infty$, then G is of type III_1.

(2) If the condition of (1) are not satisfied and if λ_k, $k \in N$, are bounded, then there exist a finite partition of infinite subsets $N^{(1)},N^{(2)},\cdots,N^{(q)}$ of positive integers N and numbers $\lambda^{(1)},\lambda^{(2)},\cdots,\lambda^{(q)}$ $(0 < \lambda^{(j)} \leq 1,\ j=1,2,\cdots,q)$ such that $\sum_{k\in N^{(j)}}^{\infty}|\log\lambda_k - \log\lambda^{(j)}|^2 < \infty$, $j=1,2,\cdots,q$. In this case let Λ be the group generated by $\log\lambda^{(1)}$,

$\log\lambda^{(2)},\cdots,\log\lambda^{(q)}$. Then type of G is the following :
G is of type III_1 if Λ is dense in R.
G is of type III_ζ if Λ is generated by $\log\zeta$ $(0 < \zeta < 1)$.
G is of type II_1 if $\Lambda = \{0\}$.

The theorem follows from Lemma 2 and the following lemma:

Lemma 3. ([1],[7]). Let G_1 and G_2 be ergodic countable groups of non-singular transformations.
(1) If G_1 is of type II_1, $G_1 \times G_2$ is of the same type as G_2.
(2) If G_1 is of type III_1, so is $G_1 \times G_2$.
(3) Let G_1 and G_2 be of type III_λ $(0 < \lambda < 1)$ and III_η $(0 < \eta < 1)$ respectively and Λ be the group generated by $\log\lambda$ and $\log\eta$. $G_1 \times G_2$ is of type III_1 and III_ζ if Λ is dense in R and is generated by $\log\zeta$ $(0 < \zeta < 1)$, respectively.

If there exists a subsequence k_n of positive integers such that $\log\lambda_{k_n}$ converges to $\log\lambda$ as $n \to \infty$ for a number λ $(0 < \lambda < 1)$ a product type odometer group G with two point state space is orbit equivalent to the product group of a group of type III_λ and a product type odometer group, and hence is of type III_η where $\eta = \lambda^{(1/p)}$ for an integer $p > 0$ or of type III_1. A product type odometer G with two points state space may be of type III_0 only if the sequence $\log\lambda_k$ does not have a convergent subsequence except a subsequence $\log\lambda_{k_n}$ such that $\sum_{n=1}^{\infty}|\log\lambda_{k_n}|^2 < \infty$. If $\lim_{k\to\infty}\lambda_k = -\infty$, a product type odometer group G with two points state space can be of type III_0. In fact, for any countable group Γ of rational numbers and a positive number α, a pure point spectrum flow with $\alpha\Gamma$ as its point spectrum is an associated flow of a product type odometer group with two point state space ([8]).

References

[1] H. Araki and J. Woods: A classification of factors, Publ. RIMS, Kyoto Univ. 3 (1967), 51-130.

[2] A. Connes: Une classification des facteurs de type III, Ann. Sci. Ec. Norm. Sup. 6 (1973), 133-252.

[3] A. Connes and J. Woods: Approximately transitive flows and ITPFI factors, Ergod. Th. and Dynam. Syst. 5 (1985), 203-236.

[4] T. Giordano and G. Scandalis: On infinite tensor products of factors of type I, Ergod. Th. and Dyndm. Syst. 5 (1985), 565-586.

[5] Y. Ito, T. Kamae and I. Shiokawa: Point spectrum and Hausdorff dimension, Number Th. and Combinatorics (1985), 209-227.

[6] W. Krieger: On ergodic flows and the isomorphism of factors, Math. Ann. 223 (1976), 19-70.

[7] T. Hamachi, Y. Oka and M. Osikawa: Flows associated with ergodic non-singular transformation groups, Publ. RIMS, Kyoto Univ. 11 (1975), 31-50.

[8] T. Hamachi and M. Osikawa: Computation of the associated flows of ITPFI factors of type III, to appear in Proceedings of U.S.-Japan Symposium on Geometric Methods in Operator Algebras.

[9] M. Osikawa: Point spectra of non-singular flows, Publ. RIMS, Kyoto Univ. 13 (1977), 167-172.

[10] M. Osikawa: Flows associated with product type odometers, Ergod. Th. and Dynam. Syst. 3 (1983), 601-612.

Department of Mathematics
College of General Education
Kyushu University
Ropponmatsu Fukuoka 810
Japan

MEASURING PROCESSES AND REPEATABILITY HYPOTHESIS

Masanao Ozawa

1. Introduction. The problem of extending the von Neumann-Lüders collapse postulate [4,5] to observables with continuous spectrum is one of the major problems of the quantum theory of measurement. Recently, Srinivas [11] posed a set of postulates which gave an answer to this problem. However, it does not seem to be a complete solution. The following two problems remain.

(1) The Srinivas collapse postulate is not consistent with the σ-additivity of probability distributions and it requires ad hoc treatment of calculus of probability and expectation. How can we improve his set of postulates in order to retain the consistency with the σ-additivity of probability.

(2) His collapse postulate depends on a particular choice of an invariant mean. What is the physical significance of employing different invariant means? Can we characterize the various different ways of measuring the same observable [11;p.149]?

The purpose of this paper is to resolve the second question by constructing different measuring processes of the same observable satisfying the Srinivas collapse postulate corresponding to the given invariant means. In our construction, the pointer position of the apparatus is the position observable and the given invariant mean corresponds to the momentum distribution at the initial state of the apparatus. Thus the choice of the invariant mean characterizes the state preparation of the apparatus.

For the general theory of quantum measurements of continuous observables, we shall refer to Davies [1], Holevo [3] and Ozawa [6-10]. The entire discussion including the solution of the first question above will be published elsewhere.

2. Formulation of the problem. In this paper, we shall deal with quantum systems with finite degrees of freedom. In the conventional formulation, the states of a system are represented by density

operators on a separable Hilbert space H and the observables are represented by self-adjoint operators on H. In this formulation, however, as shown in [7;Theorem 6.6], we cannot construct measuring processes satisfying the repeatability hypothesis, which follows from the Srinivas postulates; hence some generalization of the framework of quantum mechanics is necessary. We adopt the formulation that the states of a system are represented by norm one positive linear functionals on the algebra $L(H)$ of bounded operators on H; states corresponding to density operators will be called <u>normal states</u>. For any state σ and compatible observables X, Y we shall denote by $Pr[X\in dx, Y\in dy \| \sigma]$ the joint distribution of the outcomes of the simultaneous measurement of X and Y. Our basic assumption is that $Pr[X\in dx, Y\in dy \| \sigma]$ is a σ-additive probability distribution on $\bar{R}^2$ uniquely determined by the relation

$$\int_{\bar{R}^2} f(x,y)\, Pr[X\in dx, Y\in dy \| \sigma] = \langle f(X,Y), \sigma \rangle, \tag{2.1}$$

for all $f \in C(\bar{R}^2)$, where $\bar{R} = R\cup\{\infty\}\cup\{-\infty\}$ and $C(\bar{R}^2)$ stands for the space of continuous functions on $\bar{R}^2$. If σ is a normal state, Eq.(2.1) is reduced to the usual statistical formula. Apart from classical probability theory, we can consider another type of joint distributions in quantum mechanics. Let $\langle X,Y\rangle$ be an ordered pair of any observables. We shall denote by $Pr[X\in dx; Y\in dy \| \sigma]$ the joint distribution of the outcomes of the successive measurement of X and Y, performed in this order, in the initial state σ. Let η be a fixed invariant mean on the space $CB(R)$ of continuous bounded functions on R. Let X be an observable. Denote by $E^X{}_\eta$ the norm one projection from $L(H)$ onto $\{X(B);\, B \in B(R)\}'$ such that

$$Tr[E^X{}_\eta[A]\rho] = \eta_u\, Tr[e^{iuX}Ae^{-iuX}\rho], \tag{2.2}$$

for all normal state ρ and $A \in L(H)$, where $B(R)$ stands for the Borel σ-field of R and $'$ stands for the operation making the commutant in $L(H)$. Then by a slight modification, the Srinivas collapse postulate asserts the following relation for the successive measurement of X and any bounded observable Y,

$$\int_R y\, Pr[X\in B; Y\in dy \| \rho] = Tr[X(B)E^X{}_\eta[Y]\rho], \tag{2.3}$$

for all normal state ρ and $B \in B(R)$. Obviously, this relation implies the following generalized Born statistical formula [11]: If X and Y are compatible then

$$\Pr[X \in B; Y \in C \| \rho] = \mathrm{Tr}[X(B)Y(C)\rho], \tag{2.4}$$

for all normal state ρ and $B, C \in B(R)$. Our purpose is to construct a measuring process of X which satisfies the Srinivas collapse postulate Eq.(2.3).

Throughout this paper, we shall fix an invariant mean η which is, by a technical reason, a topological invariant mean on CB(R) (cf.[2;p.24]).

3. Dirac state. In this section, we shall consider a quantum system with a single degree of freedom. Denote by Q the position observable and by P the momentum observable. A state δ on $L(L^2(R))$ is called an η-Dirac state if it satisfies the following conditions (D1)-(D2):

(D1) For each $f \in CB(R)$, $\langle f(Q), \delta \rangle = f(0)$.

(D2) For each $f \in CB(R)$, $\langle f(P), \delta \rangle = \eta(f)$.

Lemma 3.1. For any $f \in CB(R)$, $E^Q{}_\eta(f(P)) = \eta(f)1$.

Proof. Let ξ be a unit vector in $L^2(R)$. Letting $g(p) = |\xi(-p)|^2$, g is a density function on R. Then for any $f \in CB(R)$, we have

$$\begin{aligned}\langle \xi | E^Q{}_\eta(f(P)) | \xi \rangle &= \eta_u \langle \xi | e^{iuQ} f(P) e^{-iuQ} | \xi \rangle \\ &= \eta_u \langle \xi | f(P + u1) | \xi \rangle \\ &= \eta_u \int_R f(p + u) |\xi(p)|^2 \, dp \\ &= \eta_u (f * g)(u) = \eta(f),\end{aligned}$$

where $f * g$ stands for the convolution of f and g. It follows that $E^Q{}_\eta(f(P)) = \eta(f)1$. QED

Theorem 3.2. For every topological invariant mean η, there exists an η-Dirac state.

Proof. Let ϕ be a state on $\{Q(B);\ B \in B(R)\}'$ such that $\langle f(Q),\phi\rangle = f(0)$ for all $f \in CB(R)$ and δ a state on $L(L^2(R))$ such that $\langle A,\delta\rangle = \langle E^Q{}_\eta(A),\phi\rangle$ for all $A \in L(L^2(R))$. Then by Lemma 3.1, δ is obviously an η-Dirac state. QED

4. Canonical measuring processes. Let X be an observable of a quantum system I described by a Hilbert space H. We consider the following measuring process of X by an apparatus system II. The apparatus system II is a system with a single degree of freedom described by the Hilbert space $K = L^2(R)$. Thus the composite system I+II is described by the Hilbert space $H\otimes K$, which will be identified with the Hilbert space $L^2(R;H)$ of all norm square integrable H-valued functions on R by the Schrödinger representation of K. The pointer position of the apparatus system is the position observable Q. The interaction between the measured system I and the apparatus system II is given by the following Hamiltonian:

$$H_{int} = \lambda(X\otimes P), \tag{4.1}$$

where P is the momentum of the apparatus. The strength λ of the interaction is assumed to be sufficiently large that other terms in the Hamiltonian can be ignored. Hence the Schrödinger equation will be $(h = 2\pi)$

$$\frac{\partial}{\partial t}\Psi_t(q) = -\lambda\{X\otimes\frac{\partial}{\partial q}\}\Psi_t(q), \tag{4.2}$$

in the q-representation, where $\Psi_t \in H\otimes K$. The measurement is carried out by the interaction during a finite time interval from $t = 0$ to $t = 1/\lambda$. The outcome of this measurement is obtained by the measurement of Q at time $t = 1/\lambda$. The statistics of this measurement depends on the initially prepared state σ of the apparatus. According to [7;Theorem 6.6], if σ is a normal state then this measurement cannot satisfy Eq.(2.3). Now we assume that the initial state of the apparatus is an η-Dirac state δ and we shall call this measuring process as a canonical measuring process of X

with preparation δ.

In order to obtain the solution of Eq.(4.2), assume the initial condition

$$\Psi_0 = \psi\otimes\alpha, \tag{4.3}$$

where $\psi \in H$ and $\alpha \in K$. The solution of the Schrödinger equation is given by

$$\Psi_t = e^{-it\lambda(X\otimes P)}\psi\otimes\alpha, \tag{4.4}$$

and hence for any $\phi \in H$ and $\beta \in K$, we have

$$\begin{aligned}
&\int_R \langle\phi\otimes\beta(q)|\Psi_t(q)\rangle\, dq \\
&= \int_{R^2} e^{-it\lambda xp} \langle\phi\otimes\beta|X(dx)\otimes P(dp)|\psi\otimes\alpha\rangle \\
&= \int_R \langle\beta|e^{-it\lambda xP}|\alpha\rangle\langle\phi|X(dx)|\psi\rangle \\
&= \int_R \{\int_R \beta(q)^*\alpha(q - t\lambda x)dq\} \langle\phi|X(dx)|\psi\rangle \\
&= \int_{R^2} \alpha(q - t\lambda x) \langle\beta(q)\phi|X(dx)|\psi\rangle\, dq \\
&= \int_R \langle\phi\otimes\beta(q)|\alpha(q1 - t\lambda X)\psi\rangle\, dq.
\end{aligned}$$

It follows that

$$\Psi_t(q) = \alpha(q1 - t\lambda X)\psi. \tag{4.5}$$

For $t = 1/\lambda$, we have

$$\Psi_{1/\lambda}(q) = \alpha(q1 - X)\psi. \tag{4.6}$$

<u>Theorem 4.1.</u> For any $f \in L^\infty(R)$, we have

$$U_t^*(1\otimes f(Q))U_t = f(t\lambda(X\otimes 1) + 1\otimes Q), \tag{4.7}$$

where $U_t = e^{-it\lambda(X\otimes P)}$.

Proof. By Eq.(4.5), for any $\psi \in H$ and $\alpha \in K$, we have

$$\langle \psi\otimes\alpha | U_t^*(1\otimes f(Q))U_t | \psi\otimes\alpha \rangle$$

$$= \int_R f(q) \langle \psi | \alpha(q1 - t\lambda X)^* \alpha(q1 - t\lambda X) | \psi \rangle \, dq$$

$$= \int_{R^2} f(q) | \alpha(q - t\lambda x) |^2 \, dq \, \langle \psi | X(dx) | \psi \rangle$$

$$= \int_{R^2} f(q + t\lambda x)) | \alpha(q) |^2 \, dq \, \langle \psi | X(dx) | \psi \rangle$$

$$= \langle \psi\otimes\alpha | f(t\lambda(X\otimes 1) + 1\otimes Q) | \psi\otimes\alpha \rangle.$$

Thus the assertion holds. QED

5. Statistics of measurement. Suppose that the state of the measured system I at $t = 0$ is a normal state ρ. We shall denote by $\rho\otimes\delta$ the state at $t = 0$ of the composite system I+II, which is defined by the relation $\langle T, \rho\otimes\delta \rangle = \langle E_\rho(T), \delta \rangle$ for all $T \in L(H)\otimes L(K)$, where E_ρ: $L(H)\otimes L(K) \to L(K)$ is a normal completely positive map such that $E_\rho(A_1\otimes A_2) = Tr[A_1\rho]A_2$ for all $A_1 \in L(H)$, $A_2 \in L(K)$. Thus letting $U = e^{-i(X\otimes P)}$, the state at $t = 1/\lambda$ of the composite system I+II is $U(\rho\otimes\delta)U^*$.

Let Y be a bounded observable of the system I. By the argument similar with [7;§.3], the joint distribution $Pr[X\in dx; Y\in dy \| \rho]$ of the outcomes of the successive measurement of X and Y coincides with the joint distribution $Pr[Q\in dq, Y\in dy \| U(\rho\otimes\delta)U^*]$ of the simultaneous measurement of the pointer position Q and Y at time $t = 1/\lambda$, i.e.,

$$Pr[X\in dx; Y\in dy \| \rho] = Pr[Q\in dx, Y\in dy \| U(\rho\otimes\delta)U^*]. \tag{5.1}$$

The rest of this section will be devoted to proving Eq.(2.3) for this measuring process. Denote by E_δ the completely positive map $E_\delta: L(H)\otimes L(K) \to L(H)$ defined by $Tr[E_\delta[T]\rho] = \langle T, \rho\otimes\delta \rangle$ for all normal state ρ and $T \in L(H)\otimes L(K)$. From Eq.(2.1), for any $f, g \in C(R^-)$, we have

$$\int_{R^{-2}} f(x)g(y) \, Pr[Q\in dx, Y\in dy \| U(\rho\otimes\delta)U^*]$$

$$= \langle g(Y)\otimes f(Q), U(\rho\otimes\delta)U^*\rangle$$

$$= \mathrm{Tr}[E_\delta[U^*(g(Y)\otimes f(Q)U]\rho], \tag{5.2}$$

for all normal state ρ.

Lemma 5.1. Let X be an observable of the system I. Then for any $f \in CB(R^2)$,

$$E_\delta[f(X\otimes 1, 1\otimes Q)] = f(X,0).$$

Proof. Let $\psi \in H$. For any $\alpha \in K$, we have

$$\langle\alpha| E_{|\psi\rangle\langle\psi|}[f(X\otimes 1, 1\otimes Q)]|\alpha\rangle$$

$$= \langle\psi\otimes\alpha| f(X\otimes 1, 1\otimes Q)|\psi\otimes\alpha\rangle$$

$$= \int_R\int_R f(x,q)\ \langle\psi|X(dx)|\psi\rangle\langle\alpha|Q(dq)|\alpha\rangle$$

$$= \langle\alpha|F(Q)|\alpha\rangle,$$

where $F(q) = \int_R f(x,q)\ \langle\psi|X(dx)|\psi\rangle$. Thus $E_{|\psi\rangle\langle\psi|}[f(X\otimes 1, 1\otimes Q)] = F(Q)$. It is easy to see that $F \in CB(R)$ and hence $\langle F(Q), \delta\rangle = F(0)$ by (D1). We see that

$$\langle F(Q), \delta\rangle = \langle E_{|\psi\rangle\langle\psi|}[f(X\otimes 1, 1\otimes Q)], \delta\rangle,$$

$$= \langle\psi| E_\delta[f(X\otimes 1, 1\otimes Q)]|\psi\rangle,$$

and

$$F(0) = \langle\psi| f(X,0)|\psi\rangle.$$

It follows that $E_\delta[f(X\otimes 1, 1\otimes Q)] = f(X,0)$. QED

Theorem 5.2. For any $f \in CB(R)$, we have

$$E_\delta[U^*(1\otimes f(Q))U] = f(X).$$

Proof. From Theorem 4.1, $U^*(1\otimes f(Q))U = f(X\otimes 1 + 1\otimes Q)$ and hence the assertion follows from applying Lemma 5.1 to $g \in CB(R^2)$ such that $g(x,y) = f(x + y)$. QED

Theorem 5.3. For any $Y \in L(H)$, we have

$$E_\delta[U^*(Y\otimes 1)U] = E^X{}_\eta(Y).$$

Proof. Let $\psi \in H$. For any $\alpha \in K$, we have

$$\langle\alpha| E_{|\psi\rangle\langle\psi|}[U^*(Y\otimes 1)U]|\alpha\rangle$$

$$= \langle\psi\otimes\alpha|U^*(Y\otimes 1)U|\psi\otimes\alpha\rangle$$

$$= \int_R \langle\alpha(p)\psi|e^{ipX}Ye^{-ipX}|\alpha(p)\psi\rangle\, dp$$

$$= \int_R \langle\psi|e^{ipX}Ye^{-ipX}|\psi\rangle\langle\alpha|P(dp)|\alpha\rangle$$

$$= \langle\alpha|F(P)|\alpha\rangle,$$

where $F(p) = \langle\psi|e^{ipX}Ye^{-ipX}|\psi\rangle$. Consequently, $E_{|\psi\rangle\langle\psi|}[U^*(Y\otimes 1)U] = F(P)$. Since $F \in CB(R)$, we have from (D2), $\langle F(P),\delta\rangle = \eta(F)$. We see that

$$\langle F(P),\delta\rangle = \langle E_{|\psi\rangle\langle\psi|}[U^*(Y\otimes 1)U],\delta\rangle$$

$$= \langle\psi|E_\delta[U^*(Y\otimes 1)U]|\psi\rangle,$$

and

$$\eta(F) = \eta_p \langle\psi|e^{ipX}Ye^{-ipX}|\psi\rangle$$

$$= \langle\psi|E^X{}_\eta(Y)|\psi\rangle.$$

Thus, $E_\delta[U^*(Y\otimes 1)U] = E^X{}_\eta(Y)$. QED

Theorem 5.4. Let $Y \in L(H)$ and $f \in CB(R)$. Then we have

$$E_\delta[U^*(Y\otimes f(Q))U] = f(X)E^X{}_\eta[Y].$$

Proof. By the Stinespring theorem [12], there is a Hilbert space W, an isometry $V\colon H\otimes K \to W$ and a *-representation $\pi\colon L(H)\otimes L(K) \to L(W)$ such that $E_\delta[U^*AU] = V^*\pi(A)V$ for all $A \in L(H)\otimes L(K)$.

By Theorem 5.2, $V^*\pi(1\otimes f(Q))V = f(X)$. Thus by easy computations,

$$(\pi(1\otimes f(Q))V - Vf(X))^* (\pi(1\otimes f(Q))V - Vf(X)) = 0.$$

It follows that $\pi(1\otimes f(Q))V = Vf(X)$, and hence from Theorem 5.3, we have

$$E_\delta[U^*(Y\otimes f(Q))U] = V^*\pi(Y\otimes f(Q))V = V^*\pi(Y\otimes 1)Vf(X)$$

$$= E^X{}_\eta[Y]f(X) = f(X)E^X{}_\eta[Y].$$

QED

Now we can prove that the canonical measuring process of X with preparation δ, where δ is an η-Dirac state, satisfies the Srinivas collapse postulate for the given invariant mean η.

Theorem 5.5. For any bounded observable $Y \in L(H)$ and $B \in B(R)$, we have

$$\int_R y \Pr[X\in B; Y\in dy \| \rho] = \mathrm{Tr}[X(B)E^X{}_\eta[Y]\rho],$$

for all normal state ρ.

Proof. Denote by $C_0(R)$ the space of continuous functions on R vanishing at infinity. Let Y be a bounded observable and ρ a normal state. From Eqs.(5.1) and (5.2) and from Theorem 5.4, for any $f, g \in C_0(R)$ we have

$$\int_{R^2} f(x)g(y) \Pr[X\in dx; Y\in dy \| \rho] = \mathrm{Tr}[f(X)E^X{}_\eta[g(Y)]\rho].$$

By the bounded convergence theorem and the normality of the state ρ, the set of all Borel functions f satisfying the above equality is closed under bounded pointwise convergence and contains $C_0(R)$. Thus the equality holds for all bounded Borel functions f. Since Y is bounded, there is a function $h \in C_0(R)$ such that $h(y) = y$ on the spectrum of Y. Letting $f = \chi_B$ and $g = h$, we have $f(X) = X(B)$ and $f(Y) = Y$ so that we obtain the desired equality. QED

References

[1] E.B. Davies: Quantum Theory of Open Systems, Academic Press, London (1976).

[2] F.P. Greenleaf: Invariant Means on Topological Groups, van Nostrand, New York (1969).

[3] A.S. Holevo: Probabilistic and Statistical Aspects of Quantum Theory, North-Holland, Amsterdam (1982).

[4] G. Lüders: Über die Zustandsänderung durch den Messprozess, Ann. Physik 8 (1951), 322-328.

[5] J. von Neumann: Mathematical Foundations of Quantum Mechanics, Princeton U.P., Princeton (1955).

[6] M. Ozawa: Conditional expectation and repeated measurements of continuous quantum observables, Lecture Notes in Math. 1021 (1983), 518-525.

[7] M. Ozawa: Quantum measuring processes of continuous observables, J. Math. Phys. 25 (1984), 79-87.

[8] M. Ozawa: Conditional probability and a posteriori states in quantum mechanics, Publ. RIMS, Kyoto Univ. 21 (1985), 279-295.

[9] M. Ozawa: Concepts of conditional expectations in quantum theory, J. Math. Phys. 26 (1985), 1948-1955.

[10] M. Ozawa: On information gain by quantum measurements of continuous observables, J. Math. Phys. 27 (1986), 759-763.

[11] M.D. Srinivas: Collapse postulate for observables with continuous spectra, Commun. Math. Phys. 71 (1980), 131-158.

[12] W.F. Stinespring: Positive functions on C^*-algebras, Proc. Amer. Math. Soc. 6 (1955), 211-216.

Department of Mathematics
College of General Education
Nagoya University
Chikusa-ku, Nagoya 464
Japan

ESTIMATES OF THE RATE OF CONVERGENCE IN THE CENTRAL LIMIT THEOREM IN BANACH SPACES

V. J. Paulauskas

Let B be separable Banach space with norm $\|\cdot\|$, B^* be conjugate space, $\xi, \xi_1, \dots, \xi_n, \dots$ - sequence of i.i.d. random elements (r.e.) with values in B. We assume that $E\xi=0$, $E\|\xi\|^2<\infty$, and by F we denote the distribution of r.e. ξ and by T-covariance operator of ξ, which is defined by means of formula $\langle Tf,g\rangle=E\langle f,\xi\rangle\langle g,\xi\rangle$, $f,g\in B^*$. If $A\subset B$, then $\bar{A}$ stands for closure of A and $\mathring{A}$ - for interior points of A, $\partial A=\bar{A}\setminus\mathring{A}$, $A_\varepsilon=\{x: \|x-y\|\le\varepsilon, y\in A\}$, $A_{-\varepsilon}=B\setminus((B\setminus A)_\varepsilon)$. Let $S_n=n^{-1/2}\sum_{i=1}^{n}\xi_i$, $F_n(A)=P\{S_n\in A\}$. We say that r.e. ξ satisfies the central limit theorem in the space B ($\xi\in CLT(B)$) if there exists Gaussian measure $\mathcal{N}$ with mean zero and covariance operator T such, that $F_n\Rightarrow\mathcal{N}$, where $\Rightarrow$ denotes weak convergence. It is well-known that relation $F_n\Rightarrow\mathcal{N}$ is equivalent, among others, to the following two statements

$$|F_n(A)-\mathcal{N}(A)| \to 0 \quad \text{for every set } A \text{ for which } \mathcal{N}(\partial A) = 0;$$

$$\rho(F_n,\mathcal{N}) \to 0,$$

where ρ-any metric on the space of all probability measures on B, metrizing the weak convergence. Therefore it is natural to investigate at what rate the quantities $\rho(F_n,\mathcal{N})$ and $\Delta_n(\mathcal{A})=\sup_{A\in\mathcal{A}}\Delta_n(A)$ tends to zero. Here $\Delta_n(A)=|F_n(A)-\mathcal{N}(A)|$ and $\mathcal{A}$-some class of $\mathcal{N}$-uniform sets; we recall that the class $\mathcal{A}$ is called $\mathcal{N}$-uniform if $\Delta_n(\mathcal{A})$ tends to zero for any sequence of F_n, which weakly converges to $\mathcal{N}$.

The aim of this paper is to present a survey of results, concerning the estimates of $\Delta_n(\mathcal{A})$. The first result in this direction belongs to N. P. Kandelaki and N. N. Vakhanija [1] [2], where the estimate of the quantity $\Delta_n(\mathcal{V}_a(H))$ was given. Here $\mathcal{V}_a(H)$ denotes the class of balls of Hilbert space H with a fixed center $a\in H$. This first estimate had logarithmic order with respect to n. For summands in H with special structure (namely, considering the rate of convergence in ω^2-criterion) V. V. Sazonov [3] obtained the estimate $\Delta_n(\mathscr{W}_0(H))=O(n^{-1/6+\varepsilon})$, $\varepsilon>0$. Essential progress was made by J. Kuelbs and T. Kurtz in [4], where in Hilbert space under natural moment condition the estimate with power order was obtained. Their result in the case of i.i.d. can be formulated as follows.

Theorem 1. [4]. If $E\|\xi\|^3<\infty$ then $\Delta_n(\mathscr{W}_0(H))=O(n^{-1/8})$ and if $E\|\xi\|^{7/2}<\infty$ then $\Delta_n(\mathscr{W}_0(H))=O(n^{-1/6+\varepsilon})$, $\varepsilon>0$.

Here and in the sequel as a moment condition we consider the existence of the third absolute moment or pseudomoment. This is done for the simplicity, since usually the generalization to the case of moments of order $2+\delta$, $0<\delta<1$, does not cause great difficulties.

Now we formulate the general result which shows what condition on space B, set A and Gaussian r.e. η ensure the rate of convergence of order $O(n^{-1/8})$. For this aim we need some additional definitions. Let $A_1\subset A_2\neq\phi$ and let $\mathscr{F}(A_1,A_2)$ stand for the class of functions $f:B\to R$ such, that for all $x\in B$ $0\le f(x)\le 1$, $f(x)=1$ for $x\in A_1$ and $f(x)=0$ for $x\in B\setminus A_2$. The class of functions $f:B\to R$, having three continuous derivatives we denote by $Q^3(B,R)$. Let η be Gaussian r.e. with distribution $\mathscr{N}$.

Theorem 2. Let Banach space B, the Borel set $A\subset B$ and Gaussian r.e. η be such, that the following two conditions are fulfilled:

(i) there exist constant $C(T)>0$ such that for any $\varepsilon>0$

$$P\{\eta\in(\partial A)_\varepsilon\} \le C(T)\cdot\varepsilon \; ; \tag{1}$$

(ii) for any $\varepsilon>0$ there exist functions $g_i\in Q^3(B,R)$ $i=1,2$, $g_1\in\mathscr{F}(A,A_\varepsilon)$, $g_2\in\mathscr{F}(A_{-\varepsilon},A)$ such, that the derivatives (in the sense of Fréchet) satisfy

$$\|g_j^{(i)}(x)\| \le \bar{C}\,\varepsilon^{-i}, \qquad i=1,2,3, \quad j=1,2. \tag{2}$$

Then for any r.e. ξ, belonging to the domain of normal attraction of r.e. η and for any $n\ge 1$

$$\Delta_n(A) \le 2\,\bar{C}^{1/4}\,C^{3/4}(T)\,\nu_3^{1/4}\,n^{-1/8}, \tag{3}$$

where

$$\nu_3 = \int_B \|x\|^3\,|F-\mathscr{N}|\,(dx) .$$

In the case of set $A=\{x\in B: \|x-a\|\le r\}$ this result is contained in [5], [8] and almost in such generality as above - in [7].

In order to get better estimate of the quantity $\Delta_n(A)$ it is insufficient to consider single set A, and one must consider a class of sets with some properties. More over $C(s\cdot T)$ from (1) as a function of s must satisfy some additional conditions. In [5] for balls with fixed center in Banach spaces with sufficiently smooth norm there was obtained the estimate of order $n^{-1/6}$.

Theorem 3. [5]. Let the function $\varphi: B\setminus\{0\}\to R$, $\varphi(x)=\|x\|$ be three times differentiable and satisfy the estimates

$$(4) \qquad \|\varphi^{(i)}(x)\| \le C_1\|x\|^{1-i}, \quad i = 1,2,3, \quad x \ne 0 ;$$

Let Gaussian measure $\mathcal{N}$ for any $\varepsilon>0$ and $a\in B$ satisfy

$$\mathcal{N}(x:\ r\le\|x-a\|\le r+\varepsilon) \le C(T)(1+\|a\|^2)\cdot\varepsilon$$

and $C(\alpha\cdot T)\le\alpha^{-\varepsilon}C(T)$. Then there exist constant $\overline{C}(T)$, depending on C_1 and C_2, too, such that

$$\Delta_n(\mathcal{W}_a(B)) \le \overline{C}(T)(1+\|a\|^2)\nu_3^{1/3-\varepsilon_n}\, n^{-1/6}, \quad \varepsilon_n \sim n^{-2} .$$

In [7] this result was generalized for class of sets of the form $A=\{x:\ f(x)\le r\}$ with sufficiently smooth function f.

One can notice, that part of conditions in Theorems 2 and 3 are expressed in terms of Fréchet differentiability, and in infinite dimensional spaces this restricts the application of presented estimates. For example it is known [9], that in the case of $B=C[0,1]$ there does not exist the bounded set A, satisfying condition (ii) of Theorem 2. Therefore it is natural to change strong differentiability by weaker notion - differentiability in the directions from some subspace of B. The idea fits very well with the fact, that in the spaces with "rough" norm, such as $C[0,1]$ and c_0, r.e., satisfying the CLT in these spaces, as a matter of fact, are concentrated on smaller subspaces, for example, in $\mathrm{Lip}_\alpha[0,1]$ in the case of the space $C[0,1]$ and $c_{0,\lambda}$ in the case of c_0. Here $c_{0,\lambda}=\{x\in c_0:\ \lim x_i\lambda_i^{-1}=0\}$, $\|x\|_\lambda=\sup_i|x_i\lambda_i^{-1}|$. This idea was realized by V. Bentkus and A. Račkauskas in series of papers [10]. [11]. In order to formulate their main result we need some more notations. Let $E\subset B$ be another Banach space with norm $|\cdot|$. We assume that identical inclusion map is linear and continuous. If $f:B\to R$ then the derivative in direction h is defined as follows

$$d_h f(x) = \lim_{t\to 0} t^{-1} (f(x+th) - f(x)) .$$

The iterated derivatives $d_{h_1}\dots d_{h_k} f(x)$ are defined by means of induction (if $h_1=h_2=\dots=h_k$, then $d_h^k f(x)=d_h\dots d_h f(x)$). Function $f:B\to R$ is called s times differentiable in directions from subspace E if for all $x\in B$, $h_1,\dots,h_s\in E$ there exists derivative $d_{h_1}\dots d_{h_s} f(x)$, which for fixed x is continuous symmetric s-linear form of variables $h_1,\dots,h_s$. The class of functions $f:B\to R$, which are s-times continuously differentiable in directions from E, we denote by $Q_E^s(B,R)$.

We consider the class of sets $\mathscr{W}=\{W_t, t\in R\}$ indexed by one-dimensional parameter t and we always shall assume that $W_t\subseteq W_{t'}$ if $t<t'$. We say that class $\mathscr{W}$ is absorbing if for all $x\in B$, $t\in R$, $\delta>0$ following relations hold:

$$W_{t-\|x\|} \subset W_t + x \subset W_{t+\|x\|}$$

$$\overline{W}_t \subset W_{t+\delta} , \quad \mathring{W} \supset W_{t-\delta} .$$

For example, for any function $f:B\to R$, satisfying Lipschitz condition $|f(x)-f(y)|\le\|x-y\|$, the class of sets $\mathscr{W}_f=\{W_t(f), t\in R\}$ is absorbing, where $W_t(f)=\{x\in B;\ f(x)\le t\}$. Particularly, any class of balls $\mathscr{W}_a(B)$ with fixed center is absorbing.

<u>Theorem 4</u>. [10]. Let the pair of spaces $E\subset B$, class $\mathscr{W}$ and Gaussian r.e. η satisfy the following conditions:

(i) class $\mathscr{W}$ is absorbing and for any $t\in R$ and $\varepsilon>0$ there exists function $f_{t,\varepsilon}\in\mathscr{F}(W_t, W_{t+\varepsilon})\cap Q_E^3(B,R)$ such that

$$\|d_h^i f_{t,\varepsilon}(x)\| \le C\, \varepsilon^{-i}, \quad i = 1,2,3, \quad h\in E, \quad |h| = 1 ;$$

(ii) for any $\varepsilon>0$ and $0<s<1$

$$\sup_t P\{\sqrt{s}\,\eta \in W_{t+\varepsilon} \setminus W_{t-\varepsilon}\} \le C(T)\, s^{-1/2}\, \varepsilon .$$

Then for any r.e. ξ belonging to the domain of normal attraction of η and any $n\ge 1$

$$\Delta_n(\mathscr{W}) \le \tilde{C}(T)\, \nu_{3,E}^{1/3}\, n^{-1/6},$$

where

$$\nu_{3,E} = \int_B |x|_E^3 \, |F-\mathcal{N}| \, (dx) .$$

As a corollary of this general result we shall bring the estimate in particular case $B=C[0,1]$, $E=\mathrm{Lip}_\alpha[0.1]$. Let η be Gaussian r.e. concentrated in $\mathrm{Lip}_\alpha[0,1]$; $0<\alpha\le 1$. By $D_{3,\alpha}(\eta)$ we denote class of all $C[0,1]$-valued r.e. ξ, satisfying conditions: $E\xi=0$, $E\xi(s)\xi(t)=E\eta(s)\eta(t)$ for all $s,t\in[0,1]$,

$$\nu_{3,\alpha} = \int_{C[0,1]} |x|^3_{\mathrm{Lip}_\alpha[0,1]} \, |F - \mathcal{N}| \, (dx) < \infty .$$

Here F and $\mathcal{N}$ stand for the distribution of r.e. ξ and η, respectively.

<u>Theorem 5</u>. Let $P(\eta\in\mathrm{Lip}_\alpha[0,1])=1$, $\frac{1}{2}<\alpha\le 1$, $f:C[0,1]\to R$ satisfy Lipschitz condition. If random variable $f(\eta)$ has bounded density, then there exists constant $C(\alpha,T)$ such, that for any $\xi\in D_{3,\alpha}(\eta)$ and all $n\ge 1$

$$\Delta_n(\mathcal{W}_f) \le C(\alpha,\ T)\ \nu_{3,\alpha}^{1/3}\ n^{-1/6} . \tag{5}$$

It should be noted, that both conditions of the Theorem 4 are essential in the sense, that if we even strengthen moment condition, we could not omit any of these two conditions.

Corresponding examples are constructed in [12], [13], [14]. Moreover, it turned out, that there exists the pair of spaces $E\subset B$ such, that the exponent 1/6 in (5) generally can not be increased. Now we shall formulate two results concerning the estimates from below.

<u>Theorem 6</u>. [12]. Let $B=c_0$, $E=\ell_2$ and $f:c_0\to R$, $f(x)=\sup_i x_i$. For the pair $\ell_2\subset c_0$ the condition (i) of Theorem 4 is fulfilled, nevertheless for any monotone sequence $b_n>0$, $\lim b_n=0$ there exist ℓ_2-valued Gaussian r.e. η and r.e. ξ such that $P(|\xi|=1)=1$ and

$$\Delta_n(\mathcal{W}_f) \to 0 \quad \text{as} \quad n\to\infty ,$$

$$\limsup b_n^{-1} \cdot \Delta_n(\mathcal{W}_f) = \infty .$$

<u>Theorem 7</u>. [13]. Let $\theta_n>0$ be sequence, tending to zero. There exist Hilbert space E, Banach space B and E - valued r.e. ξ

and η such that

(i) for the pair $E \subset B$ and class of balls $\mathfrak{M}_0(B)$ conditions (i) and (ii) of Theorem 4 are fulfilled,

(ii) $P\{|\xi| < C\} = 1$,

(iii) for infinitely many indices n

$$\Delta_n(\mathfrak{M}_0(B)) \geq \theta_n \, n^{-1/6}.$$

As a space B one can take ℓ_1 and as E - the space $\ell_{2,\lambda}$ with specific $\lambda = (\lambda_1, \ldots, \lambda_n, \ldots)$. Here

$$\ell_{2,\lambda} = \{x : |x|_{2,\lambda} = (\sum_{i=1}^{\infty} x_i^2 \, \lambda_i^{-1})^{1/2} < \infty\} .$$

Theorem 7 shows that generally the result of Theorem 4 can not be improved. It remains the open question about the optimality of the exponent 1/6 in Theorem 3 - at present we don't know if it is possible to construct a space B, satisfying (4) and for which one could estimate from below $\Delta_n(\mathfrak{M}_0(B))$ by $n^{-1/6}$.

In the case when the set A has sufficiently smooth boundary one can improve the estimate given in Theorem 4. We shall not touch in detail this question and restrict ourselves with some results in the case of the class $\mathfrak{M}_a(H)$. The estimate

$$\Delta_n(\mathfrak{M}_0(\ell_2)) = O(n^{-1/2}) \tag{6}$$

was obtained by S. V. Nagajev and V. I. Čebotarev [15] but with strong condition of independence of coordinates of ℓ_2-valued r.e. ξ Later this condition was weakened by Ju. V. Borovskich and A. Račkauskas [16]. The essential step was made by F. Götze [17] who obtained (6) assuming $E\|\xi\|^6 < \infty$. V. V. Jurinskii [18] reduced the moment condition to natural condition $E\|\xi\|^3 < \infty$, and this result can be formulated as follows.

<u>Theorem 8</u>. Let ξ be H-valued r.e. with $E\xi = 0$, $E\|\xi\|^2 = 1$, $\beta_3 = E\|\xi\|^3 < \infty$. Then for all $n \geq 1$

$$\Delta_n(\mathfrak{M}_a(H)) \leq C(T)(1 + \|a\|^3) \, \beta_3 \, n^{-1/2} .$$

For further investigations in the case of sets with smooth boundary we refer to papers of F. Götze [19], V. V. Sazonov, B. A. Zalesskii [20], T. R. Vinogradova [21].

References

[1] Kandelaki N. P. On limit theorem in Hilbert space. Trans. Comput. Center Acad. Sci. Georgian SSR 1, 1965, 46-55 (in Russian)

[2] Vakhanija N. N., Kandelaki N. P. On the estimation of the speed of convergence in the central limit theorem in Hilbert space. ibid. 10, 1 (1965), 150-160 (in Russian).

[3] Sazonov V. V. On w^2-criterion. Sankhya, Ser.A, 30, 1968, No 2, 205-209.

[4] Kuelbs J., Kurtz T. Berry-Esseen estimates in Hilbert space and an application to the law of the iterated logarithm. Ann. Probab., 1974, v.2, No 3, 387-407.

[5] Paulauskas V. J. On the rate of convergence in the central limit theorem in some Banach spaces. Teor. verojat. i primen., 1976, v.21, No 4, 775-791.

[6] Paulauskas V. J. On convergence of some functionals of sums of independent random variables in a real Banach space. Liet. matem. rink., 1976, v.16, No 3, 103-121 (in Russian).

[7] Paulauskas V. J. Limit theorems for sums of independent random elements in Banach spaces. Doct. Dissertation, Vilnius, 1978.

[8] Zolotarev V. M. Approximation of distributions of sum of independent random variables with values in infinite-dimensional spaces. Teor. verojat. i primen., 1976, v.21, No 4, 741-757.

[9] Bonic R., Frampton J. Smooth functions on Banach manifolds. J. of Math. and Mech., 1966, v.15, 877-893.

[10] Bentkus V. J., Račkauskas A. J. Closeness of sums of independent random variables in Banach spaces. I, II. Liet. matem. rink., 1982, 22, No 3, 12-28, No 4, 8-20 (in Russian).

[11] Bentkus V. J., Račkauskas A. J. Estimates of distances between sums of independent random elements in Banach spaces. Teor. verojat. i primen., 1984, 29, 1, 49-64 (in Russian).

[12] Bentkus V. J. Lower estimates of the Gaussian approximation in Banach spaces. Liet. matem. rink., 1984, 24, 1, 12-18.

[13] Bentkus V. J. Lower bounds for the convergence rate in the central limit theorem in Banach spaces. Liet. matem. rink., 1985, 25, 4, 10-21.

[14] Rhee W. S., Talagrand M. Bad rates of convergence for the central limit theorem. Ann. Probab., 1984, v.12, No 3, 843-850.

[15] Nagaev S. V., Čebotarev V. I. On estimates of the speed of convergence in the central limit theorem for random vectors with

values in ℓ_2. Mathematical analysis and related topics, Nauka, Novosibirsk, 1978, 153-182.

[16] Borovskih Yu. V., Račkauskas A. Asymptotic of distributions in Banach spaces. Liet. matem. rink., 1979, 19, 1, 39-54.

[17] Götze F. Asymptotic expansion for bivariate von Mises functionals. Z. Wahrscheinlich. verw. Geb., 1979, 50, 333-355.

[18] Jurinskii V. V. On the accuracy of Gaussian approximation for the probability of hitting a ball. Teor. verojat. i primen., 1982, 27, 2, 270-278.

[19] Götze F. On the rate of convergence in the central limit theorem in Banach spaces. Preprints in Statistics, University of Cologne, 1981, June, No 68, 34.

[20] Sazonov V. V., Zalesskii B. A. On the central limit theorem in Hilbert space. Tech. Report No 35, University of Carolina, 1983, 32 p.

[21] Vinogradova T. R. On the accuracy of normal approximation on sets defined by a smooth function. I, II. Teor. verojat. i primen., 1985, 30, No 3, 554-557, No 2, 219-229.

V. Kapsukas university,
Vilnius, USSR

SIMPLE METHOD OF OBTAINING ESTIMATES IN THE INVARIANCE PRINCIPLE

A. I. Sakhanenko

The well-known method of Lindeberg [1, Ch.VIII] is the simplest way to prove the central limit theorem. It also yields the simplest way to obtain estimates in this theorem in terminology of the Ljapunov relation L_a of the order $2<a\leq 3$. But thus obtained estimate $cL_a^{1/(a+1)}$ is rather rough.

The Lindeberg's method, due to its simplicity and naturalness, is one of the basic means to study the rate of convergence in the central limit theorem for martingales [2,3] and to obtain estimates for the Prokhorov distance in the infinite-dimensional central limit theorem [4,5].

In this paper we show that this method being slightly modified is applicable for obtaining estimates in the invariance principle. In particular, it turns out to be the simplest known way to prove the unimprovable estimate of Borovkov [6]

$$d(S,W) \leq cL_a^{1/(a+1)}, \qquad 2<a\leq 3, \tag{1}$$

where $d(S,W)$ is the Prokhorov distance between the distributions in $C[0,1]$ of the Wiener process $W(t)$ and a random broken line $S(t)$ constructed in a standard way.

The proposed way of obtaining estimates in the invariance principle in the general form is described in Section 2. For comparison, in Section 1 we describe what the Lindeberg method implies, and in Section 3 we give a number of additional results which were obtained earlier when realizing this method. The immediate deduction of estimates in the generalizations of the classical invariance principle of Donsker - Prokhorov is presented in Section 4, and similar results concerning the Strassen invariance principle are collected in Section 5.

Note that the method of Section 2 allows one to avoid the use of the complicated techniques concerning the method of a common probability space which often requires [6-8] that sufficiently exact estimates should be obtained in the central limit theorem. In particular, the proof of Corollaries 1 and 2 from Section 4 containing inequality (1) only uses simple enough Theorems 1 and 2 and Lemma 5. All the results obtained in the paper are immediately transferred onto the infinite-dimensional invariance principle for martingales (see Corollary 3). However, without making it more complicated the method under consideration does not allow us to deduce estimate (1) for $a>3$. At the same time, in the one-dimensional case [9-11] the Skorohod method proves (1) for $a<5$, and for independent variables the strengthening [8] of the Komlos - Major - Tusnady method allowed us to obtain all the statements of Corollaries 1 and 2 for any $a>2$ with the constant c depending on a only.

1. Preliminary remarks. Let us first give the necessary information about the Lindeberg idea being used to obtain estimates in various generalizations of the central limit theorem. Let $X_1,\dots, X_n$, $Y_1,\dots,Y_n$ and V be random variables with values in some linear normed space $\mathfrak{X}$ with norm $|\cdot|$ and with σ-algebra of Borel sets $\mathcal{A}$. Put

$$S_k = X_1 + \cdots + X_k, \qquad Z_k = Y_1 + \cdots + Y_k, \tag{2}$$

and A^r for $r>0$ is said to denote the r-neighbourhood of the set $A\subset\mathfrak{X}$. We have to estimate the closeness of the distribution of the sum S_n to another distribution which may be represented as the distribution of a sum Z_n of specially chosen random elements $Y_1,\dots, Y_n$. The simplest way to obtain estimates in this problem is based on the following obvious statement.

Lemma 1. If a number r and a random variable V satisfy the condition

$$P(|V|<r) = 1, \qquad r > 0, \tag{3}$$

then for any set $A\in\mathcal{A}$ the inequality

$$\begin{aligned} &P(S_n\in A)-P(Z_n\in A^{2r}) \le P(S_n+V\in A^r)-P(Z_n+V\in A^r) \\ &= \textstyle\sum_{k=1}^n [P(U_k+X_k+V\in A^r)-P(U_k+Y_k+V\in A^r)], \end{aligned} \tag{4}$$

is valid, where $U_k=X_1+\cdots+X_{k-1}+Y_{k+1}+\cdots+Y_n$.

Let us now consider the problem of estimating the right-hand part of (4). Put

$$q_k(r,A,S_{k-1}) = \sup_{x\in\mathfrak{X}} |P(x+X_k+V\in A^r\,|\,S_{k-1})-P(x+Y_k+V\in A^r)|. \tag{5}$$

Lemma 2. If random variables $V, Y_1, \ldots, Y_n$ are independent and if V is independent of the sequence $\{X_k\}$, then

$$|P(U_k+X_k+V\in A^r)-P(U_k+Y_k+V\in A^r)| \le Eq_k(r,A,S_{k-1}).$$

This simple lemma is proved in analogy with the below Theorem 2.

Let us now touch upon a similar but a more difficult problem when we are to estimate the closeness of the distribution in $\mathfrak{X}^n$ of the whole sequence $S_1, \ldots, S_n$ by distribution in $\mathfrak{X}^n$ of the sequence $Z_1, \ldots, Z_n$.

2. The basic idea. Let T be some set on a real axis in which we fix points $0=t_0<t_1<\cdots<t_n<\infty$. By the sequence $S_1, \ldots, S_n$ defined in (2) let us construct a random process $S=S(t)$ putting $S(t_0)=S_0=0$ and

$$S(t) = S_{k-1} + X_k e_k(t) \qquad \text{for} \qquad t_{k-1} \le t \le t_k, \tag{6}$$

$k=1,\ldots,n$ where $e_k(t)$ are some monotone non-random functions satisfying the condition

$$e_k(t)=0 \quad \text{for} \quad t\le t_{k-1} \quad \text{and} \quad e_k(t)=1 \quad \text{for} \quad t\ge t_k. \tag{7}$$

Remark 1. If $T=\{t_0,t_1,\ldots,t_n\}$ then S is a trajectory of the random walk $(0,S_1,\ldots,S_n)$. If $T=[0,t_n]$ and $e_k(t)=(t-t_{k-1})/(t_k-t_{k-1})$ for $t\in[t_{k-1},t_k]$, then S may be called a random broken line. But if $T\supset[0,t_n]$ and $e_k(t)=0$ for $t<t_k$, then S is the analogy of the random stepwise function continuous from the right.

Due to (6) and (7) the process S may be represented in the form

$$S = X_1e_1 + \cdots + X_ne_n . \tag{8}$$

We have to estimate how close the distribution of the process S is to the distribution of the process

$$Z = Y_1e_1 + \cdots + Y_ne_n \tag{8'}$$

which is similar in structure.

Denote by $\mathfrak{X}^T$ the space of all functions $f=f(t)$ defined on T

and taking values in $\mathfrak{X}$ with the norm $\|f\|=\sup_{t\in T}|f(t)|<\infty$ and with the σ-algebra of Borel sets $\mathcal{B}$. In this case the trajectories of the processes S and Z and of the processes

$$U^k = X_1e_1+\cdots+X_{k-1}e_{k-1}+Y_{k+1}e_{k+1}+\cdots+Y_ne_n, \qquad k=1,\ldots,n \tag{9}$$

117ong to the space $\mathfrak{X}^T$.

For $r>0$ denote by B^r an r-neighbourhood of the set $B\subset\mathfrak{X}^T$, i.e. $B^r=\{x\in\mathfrak{X}^T:\exists y\in B,\|x-y\|<r\}$. Let us introduce a sequence $B^{r,k}$ of enlargements of the set B where $B^{r,n+1}=B$, $e_{n+1}(t)\equiv 0$ and

$$B^{r,k} = \{x=y+v(e_k-e_{k+1}):y\in B^{r,k+1},\ v\in(-r,r)\} \tag{10}$$

for $k=1,\ldots,n$. In this case obviously $B\subset B^{r,n}\subset\cdots\subset B^{r,1}\subset B^r$. We put

$$Q^k(B) = P(U^k+X_ke_k+Ve_k\in B^{r,k})-P(U^k+Y_ke_k+Ve_k\in B^{r,k}); \tag{11}$$

$$D_0(B,r) = P(S\in B)-P(Z\in B^r), \qquad B\in\mathcal{B}. \tag{12}$$

The following generalization of Lemma 1 is the basis for the proposed method of obtaining estimates in functional limit theorems.

Theorem 1. If a number r and a random variable V satisfy condition (3) then for any set $B\in\mathcal{B}$ the inequality

$$D_0(B,2r) \le \sum_{k=1}^n Q^k(B) \tag{13}$$

is valid.

Proof. First of all note upon the inclusion of events

$$\{U^k+Y_ke_k+Ve_k\in B^{r,k}\}\subset\{U^{k-1}+X_{k-1}e_{k-1}+Ve_{k-1}\in B^{r,k-1}\}, \tag{14}$$

$k=2,\ldots,n$. This fact follows from (3) and (10) since

$$U^{k-1}+X_{k-1}e_{k-1}+Ve_{k-1} = (U^k+Y_ke_k+Ve_k)+V(e_{k-1}-e_k)$$

due to (9). Similarly

$$\{S\in B\}\subset\{S+Ve_n\in B^{r,n}\} = \{U^n+X_ne_n+Ve_n\in B^{r,n}\}, \tag{15}$$

$$\{U^1+Y_1e_1\in B^{r,1}\} = \{Z+Ve_1\in B^{r,1}\}\subset\{Z\in B^{2r}\}. \tag{16}$$

From (12), (15) and (16) we obtain

$$D_0(B,2r) \le P(S+Ve_n \in B^{r,n})-P(Z+Ve_1 \in B^{r,1}) = \sum_{k=1}^{n} [Q^k(B)+Q_0^k(B)] \qquad (17)$$

where $Q^k(B)$ is defined in (11) and

$$Q_0^k(B)=P(U^k+Y_k e_k+Ve_k \in B^r)-P(U^{k-1}+X_{k-1}e_{k-1}+Ve_{k-1} \in B^{r,k-1}) \qquad (18)$$

for $k=2,\dots,n$ and $Q_0^1(B)=0$. From (17) and (18) there follows (13) since $Q_0^k(B) \le 0$ for all k due to (14).

Now consider the problem of estimating the value

$$D_0(r) = \sup_{B \in \mathscr{B}} D_0(B,r). \qquad (19)$$

Denote by P_k and E_k, respectively, the conditional probability and conditional expectation with respect to the σ-algebra generated by random variables $X_1,\dots,X_k$. Put

$$q_k(r) = \sup_{A \in \mathscr{A}} |P_{k-1}(X_k+V \in A)-P(Y_k+V \in A)|. \qquad (20)$$

<u>Theorem 2</u>. If random variables $V,Y_1,\dots,Y_n$ are independent and if V is independent of the sequence $\{X_k\}$ then

$$D_0(2r) \le \sum_{k=1}^{n} Eq_k(r) \qquad (21)$$

provided condition (3) holds.

<u>Proof</u>. Without loss of generality let us assume that the variables $V,Y_1,\dots,Y_n$ and the sequence $\{X_k\}$ are independent. Let us introduce random sets

$$A(B,U^k) = \{x:\ U^k+xe_k \in B^{r,k}\}.$$

In this case from the definition (11) we obviously have

$$\begin{aligned} Q^k(B) &= P(X_k+V \in A(B,U^k))-P(Y_k+V \in A(B,U^k)) \\ &\le E \sup_{A \in \mathscr{A}} |P(X_k+V \in A|U^k)-P(Y_k+V \in A|U^k)|. \end{aligned} \qquad (22)$$

To obtain (21) from (22) and (20) it suffices to remark that

$$P(X_k+V \in A|U^k) = P_{k-1}(X_k+V \in A),$$

$$P(Y_k+V \in A|U^k) = P(Y_k+V \in A)$$

with probability 1 due to the assumption about joint independence of the variables $V,Y_1,\dots,Y_n$ and sequence $\{X_k\}$.

__Remark 2__. Let $W=W(t)$ be some process whose trajectories with probability 1 belong to the space $\mathfrak{X}^T$. Suppose that we have to estimate the closeness of distributions of the processes S and W, i.e. to obtain estimates for the value

$$D(r) = \sup_{B\in\mathscr{B}}[P(S\in B)-P(W\in B^r)] . \tag{23}$$

In this case we define the process Z assuming $Y_k=W(t_k)-W(t_{k-1})$, $k=1,\dots,n$ in formula (8'). In such a construction obviously

$$\|Z-W\| \le \max_{1\le k\le n} \sup_{t\in T_k} |W(t)-W(t_{k-1})|, \qquad T_k=T\cap[t_{k-1},t_k].$$

To estimate the value $D(r)$ it is convenient to use the inequality

$$D(r+r_0) \le D_0(r)+P(\|Z-W\|>r_0) \qquad \forall r,r_0>0$$

which easily follows from definitions (12), (19) and (23).

Note that in the notations introduced in (19) and (23) the values

$$d(S,Z) = \inf\{r: D_0(r)\le r\}, \quad d(S,W) = \inf\{r: D(r)\le r\} \tag{24}$$

coincide with the Prokhorov distances between distributions of the correspondidng processes.

__3. Auxiliary estimates__. Let X, Y and V be independent random variables with values in a space $\mathfrak{X}$. The problem of estimating values q_k defined in (6) and (20) is reduced to the problem of obtaining uniform in A estimates for the difference

$$q(A) = P(X+V\in A)-P(Y+V\in A).$$

For any non-random $u\in[0,1]$ and $x\in\mathfrak{X}$ put

$$f(u,x,A) = P(ux+V\in A)$$

and let all the derivatives below be taken by u. The following simple statement was widely used in [4,5].

__Lemma 3__. Let there exist a random variable V such that condition (3) holds and for all $x\in\mathfrak{X}$ and $u\in[0,1]$

$$\sup_{A\in\mathscr{A}} |f^{(j)}(u,x,A)| \le C_j!h^j(x/r)$$

for $j=m-1$ and $j=m$ where $h(x)$ is some non-negative function, $h(x/r)=h(x)/r$. Then

$$|q(A)| \le EH_m(X/r)+EH_m(Y/r) + \sum_{j=1}^{m-1}|Ef^{(j)}(0,X,A)-Ef^{(j)}(0,Y,A)|/j!$$

where $H_m(x)=2C\min(h^m(x),h^{m-1}(x))$.

To prove this statement it suffices to note that $q(A)=f(1)-f(0)$ for $f(u)=P(uX+V\in A)-P(uY+V\in A)$ and to use the Taylor formula.

Lemma 4. Let a space $\mathfrak{X}$ consist of all bounded sequences $x=(x^{(1)},x^{(2)},\ldots)$ and

$$|x| = \sup_i |x^{(i)}|, \quad h(x) = \left(\sum_{i=1}^{\infty}(x^{(i)})^2\right)^{1/2} . \tag{25}$$

In this case all the conditions of Lemma 3 are satisfied.

This statement was proved in [5] (see also Theorem 25 in [4]). In this case as V one chooses a random variable whose coordinates $v^{(1)},v^{(2)},\ldots$ are independent, identically distributed and have infinitely differentiable density.

Now let us consider the simplest case, $\mathfrak{X}=R$. Take, as V, a random variable which is the sum of $(m+1)$ independent random variables uniformly distributed on the interval $[-r/(m+1),r/(m+1)]$. Denote by $p(x)$ the density of distribution of V and put

$$f(r,x) = -\int_A p'(v-ux)dv, \qquad G(x) = P(X<x)-P(Y<x).$$

In this case

$$|f^{(j-1)}(r,x)| = |x|^{j-1}\left|\int_A p^{(j)}(v-ux)dv\right|$$

$$\le |x|^{j-1}\int_{-\infty}^{\infty}|p^{(j)}(v-ux)|dv = C_{j,m}|x|^{j-1}/r^j .$$

for $j=1,\ldots,m$. Integrating by parts we obtain

$$q(A) = \int_A [Ep(v-X)-Ep(v-Y)]dv = \int_{-\infty}^{\infty} f(1,x)G(x)dx .$$

If we decompose now the function $f(u,x)$ into the Taylor series by u, we can easily see that there holds

Lemma 5. If $\mathfrak{X}=R$ and $EX^j=EY^j$ for $j=1,\ldots,m-1$, then for all $r>0$ there exists a random variable V satisfying (3) and such that

$$\sup_{A\in\mathcal{A}}|q(A)| \le \int_{-\infty}^{\infty} h_{m-1}\left(\frac{x}{r}\right)|G(x)|\frac{dx}{r} \le Eh_m\left(\frac{X}{r}\right)+Eh_m\left(\frac{Y}{r}\right)$$

where $h_m(x)=c_m\min\{|x|^m, |x|^{m-1}\}$.

4. Estimates in the invariance principle of Donsker-Prokhorov. In this section we assume that $T=[0,t_n]$ and we investigate the estimates for generalizations $D_0(r)$ and $D(r)$ of the Prokhorov distances.

Theorem 3. Let $\mathfrak{X}=R$ and let random variables $Y_1,\dots,Y_n$ be independent. Suppose that the distribution of the sequence $\{X_k\}$ satisfies the following condition

$$E_{k-1}X_k^j = EY_k^j \qquad \text{a.s.} \qquad \text{for} \quad j=1,\dots,m-1.$$

In this case

$$D_0(2r) \le \Sigma_{k=1}^n L_{k,m}(r) \le \Sigma_{k=1}^n [Eh_m(X_k/r)+Eh_m(Y_k/r)],$$

where $L_{k,m}(r)=E\int_{-\infty}^{\infty} r^{-1}h_{m-1}(x/r)|G_k(x)|dx$ and $G_k(x)=P_{k-1}(X_k<x)-P(Y_k<x)$.

Thus, Theorem 3 yields the estimate in the invariance principle in terms of pseudomoments $L_{k,m}(r)$ taking into account the closeness of distributions of the variables X_k and Y_k. Let us now bring this estimate to a more classical form. From Theorem 3 for $m=3$ and from Remark 1 for normally distributed variables $\{Y_k\}$ there directly follows

Corollary 1. Let random variables $X_1,\dots,X_n$ form a martingale difference sequence with non-random conditional variances, i.e.

$$E_{k-1}X_k = 0, \qquad E_{k-1}X_k^2 = EX_k^2 \qquad \text{a.s.} \tag{26}$$

Suppose that the process $S=S(t)$ defined in Section 2 is constructed by the points

$$t_k = \Sigma_{i=1}^k EX_i^2 , \qquad k = 1,\dots,n$$

and $W=W(t)$ is a standard Wiener process. Then

$$D(r) \le c\ \Sigma_{k=1}^n\ E\min\{|X_k|^3/r^3,\ X_k^2/r^2\} , \tag{27}$$

where $c<\infty$ is an absolute constant.

Remind that the distance $D(r)$ between the distributions of the processes S and W is defined in (23). Let us now give the estimate for the Prokhorov distance $d(S,W)$ between these processes. From (24) and (27) there obviously follows

Corollary 2. If the conditions of Corollary 1 hold, then for $2 \le a \le 3$

$$D(r) \le c \sum_{k=1}^{n} E\,|X_k|^a / r^a \qquad \forall r>0$$
$$d(S,W) \le c\Big(\sum_{k=1}^{n} E|X_k|^a\Big)^{1/(a+1)} . \tag{28}$$

For independent random variables $X_1,\dots,X_n$ and $t_n=1$ inequality (28) coincides with Borovkov's estimate (1). If the second condition in (26) holds and $a \le 3$, then inequality (28) may also be more exact than the similar result in [11] where the case $a<5$ was considered under somewhat different assumptions. However, our proof is simpler and admits the following infinite-dimensional generalizations.

Theorem 4. Let a space $\mathfrak{X}$ and a number m satisfy the conditions of Lemma 3. Suppose that random variables $Y_1,\dots,Y_n$ are independent and let the distribution of the sequence $\{X_k\}$ be such that

$$E_{k-1} f^{(j)}(0,X_k,A) = Ef^{(j)}(0,Y_k,A) \qquad \text{a.s.}$$

for all $A \in \mathcal{A}$, $j=1,\dots,m-1$ and $k=1,\dots,n$. In this case for all $r>0$ and $m-1 \le a \le m$

$$D_0(2r) \le \sum_{k=1}^{n} [EH_m(X_k/r)+EH_m(Y_k/r)] \le c(a)\sum_{k=1}^{n} [Eh^a(X_k)+Eh^a(Y_k)]/r^a .$$

Remind that the functions h and H_m are defined in Lemma 3. From Theorem 4 and Lemma 4 there follows

Corollary 3. Let a space $\mathfrak{X}$ consist of all bounded sequences so that $X_k=(X_k^{(1)},X_k^{(2)},\dots)$ and $Y_k=(Y_k^{(1)},Y_k^{(2)},\dots)$. Suppose that random variables $Y_1,\dots,Y_n$ are independent and let the distribution of the sequence $\{X_k\}$ satisfy the conditions

$$E_{k-1}X_k^{(i)} = EY_k^{(i)}, \qquad E_{k-1}X_k^{(i)}X_k^{(j)} = EY_k^{(i)}Y_k^{(j)} \qquad \text{a.s.}$$

for all i, j and k. Then the statement of Theorem 4 is valid for $m=3$ and for the functions $|x|$ and $h(x)$ defined in (25).

Thus, Corollary 3 extends estimate (1) to the case of infinite-dimensional martingales. Note that in Corollary 3, unlike [12], no assumptions are made upon the structure of covariance matrices of the vectors $\{Y_k\}$ and upon the dependence character of coordinates of the vectors $\{X_k\}$.

Theorems 3 and 4 directly follow from Theorem 2 and from the estimates for q_k given below.

Lemma 6. If the conditions of Theorem 3 hold, then

$$Eq_k(r) \le L_{k,m}(r) \le Eh_m(X_k/r)+Eh_m(Y_k/r) \quad . \tag{29}$$

To prove this it suffices to put $Y=Y_k$ in Lemma 5 and to take as a distribution of the variable X the conditional distribution P_{k-1} of the variable X_k. Similarly, Lemma 4 yields

Lemma 7 . If the conditions of Theorem 4 hold, then

$$Eq_k(r) \le EH_m(X_k/r)+EH_m(Y_k/r) \quad . \tag{30}$$

Remark 3. Due to the inequality

$$P(S_n \in A) - P(Z_n \in A^r) \le D_0(r) \qquad \forall A \in \mathscr{A} \qquad \forall r>0$$

from Theorem 4 one can also obtain estimates in the infinite-dimensional central limit theorems for martingales. In particular, Corollary 3 extends to martingales Theorem 25 from [4]. This estimates can be strengthened if instead of Theorem 2 one uses in proofs the simpler Lemma 2.

5. Estimates in the Strassen invariance principle. Now let us suppose that only distributions of each of the sequences $\{X_k\}$ and $\{Y_k\}$ are given, and let the joint distribution of these sequences be not chosen yet. From the well-known Strassen's theorem in analogy with [7] we obtain

Lemma 8. Suppose that for any number $\varepsilon>0$ and for all k there exist random variables $X_{k,\varepsilon}$ and $Y_{k,\varepsilon}$ taking values in some complete and separable subspaces of the space $\mathfrak{X}$, such that

$$P(|X_k-X_{k,\varepsilon}|>\varepsilon) \le \varepsilon, \qquad P(|Y_k-Y_{k,\varepsilon}|>\varepsilon) \le \varepsilon.$$

In this case for all $\varepsilon>0$ there exists a joint distribution of sequences $\{X_k\}$ and $\{Y_k\}$ such that

$$P(\|S-Z\|>r+2n\varepsilon) \le D_0(r)+2n\varepsilon \qquad \forall r,\varepsilon>0.$$

In particular, if the values $Y_1,\ldots,Y_n$ are independent, then due to Theorem 2

$$P\left(\max_{k\le n}|S_k-Z_k|>2r+2n\varepsilon\right) \le \sum_{k=1}^n Eq_k(r)+2n\varepsilon . \tag{31}$$

Below we suppose the infinite sequences $\{X_k\}$ and $\{Y_k\}$ to be given. Let us consider the problem if there exists their joint distribution such that

$$|S_n-Z_n| = O(b_n) \qquad \text{a.s.} \qquad \text{for} \qquad n\longrightarrow\infty \tag{32}$$

or

$$|S_n-Z_n| = o(b_n) \qquad \text{a.s.} \qquad \text{for} \qquad n\longrightarrow\infty \tag{33}$$

where $\{b_n\}$ is some sequence of numbers, $0<b_n\uparrow\infty$.

Theorem 5. Let a sequence $\{Y_k\}$ consist of independent random variables. If for some $r>0$

$$\sum_{k=1}^{\infty} E\delta_k(rb_k) < \infty \tag{34}$$

then there exists a joint distribution of the sequences $\{X_k\}$ and $\{Y_k\}$ such that relation (32) is valid. But if condition (34) holds Lor all $r>0$, then for some joint distribution (33) is valid too.

Proof. Introduce the notations

$$N(m) = \max\{k:b_k\le 2^m\}, \qquad I(m) = (N(m-1),N(m)]$$
$$r_m = r2^{m-1} < r\min\{b_k:k\in I(m)\}, \qquad \varepsilon_m = r2^{-m}/N(m), \tag{35}$$

$$d_m = \max_{k\in I(m)} |(S_k-S_{N(m-1)})-(Z_k-Z_{N(m-1)})|, \tag{36}$$

where $m=1,2,\ldots$, $N(0)=0$. If the joint distribution of the variables X_k and Y_k for $k\le N(m-1)$ is defined, then for every fixed value of the variables $\{X_k,Y_k,\ k\le N(m-1)\}$ we can, due to statement (31) of Lemma 8, choose a conditional joint distribution of variables $\{X_k, Y_k,\ k\in I(m)\}$ so that

$$P_{N(m-1)}(d_m>2r_m+2r2^{-m}) \le \sum_{k\in I(m)} E_{N(m-1)}q_k(r_m)+2r2^{-m} . \tag{37}$$

Repeating this argument for $m=1,2,\ldots$ we define the joint distribution of the sequences $\{X_k\}$ and $\{Y_k\}$. In this case, from (35) and (37) we have

$$\sum_{m=1}^{\infty} P(d_m>r2^{m+1}) \le \sum_{k=1}^{\infty} Eq_k(rb_k)+2r . \tag{38}$$

From (34), (36), (38) and from traditional arguments based on the lemma of Borel and Cantelli there follows (32). But if the series (34) converges for all $r>0$, then $\sum_{k=1}^{\infty} Eq_k(r_kb_k)<\infty$ for some $r_k\downarrow 0$. Replacing in (32) b_k for $r_kb_k=o(b_k)$ we obtain (33).

<u>Corollary 4</u>. Let the conditions of Theorem 3 hold for all n and let $\sum_{k=1}^{\infty} L_{k,m}(b_k)<\infty$. Then relation (33) is valid.

Thus we obtained the sufficient condition in the Strassen invariance principle in terms of pseudomoments. To prove this statement it suffices to substitute estimate (29) in (34) and to see that

$$Eq_k(rb_k) \le L_{k,m}(r,b_k) \le \max\{r^{-m},1\}L_{k,m}(b_k).$$

Similarly, (30) and (34) yield the following statement for infinite-dimensional martingales.

<u>Corollary 5</u>. Let the conditions of Corollary 3 hold for all n and let

$$\sum_{k=1}^{\infty} [EH_3(X_k/b_k)+EH_3(Y_k/b_k)] < \infty .$$

Then (33) is valid for some joint distribution of $\{X_k\}$ and $\{Y_k\}$.

Remind that here $H_3(x)=2C\min\{h^3(x),h^2(x)\}$ and the function $h(x)$ is defined in (25). Let us consider a more special case.

Corollary 6. Let the assumptions of Corollary 5 be valid and let there exist a random variable $X \geq 0$ such that

$$EX^a < \infty , \qquad 2 < a < 3 , \tag{39}$$

$$P(h(X_k)>x)+P(h(Y_k>x)) \leq \mathcal{K}P(X\geq x) \tag{40}$$

where $\mathcal{K}<\infty$. Then (33) is valid for $b_n=n^{1/a}$.

To obtain this statement from Corollary 5 it suffices to note that

$$\sum_{k=1}^{\infty}[EH_3(X_k/k^{1/a})+EH_3(Y_k/k^{1/a})] \leq 2C\mathcal{K}\sum_{k=1}^{\infty}EH(X/k^{1/a})$$

$$\leq 2C\mathcal{K}E\int_0^{\infty}H(X/u^{1/a})du = 2C\mathcal{K}a(a-2)^{-1}(3-a)^{-1}EX^a < \infty$$

provided assumptions (39) and (40) hold where $H(x)=\min\{x^3,x^2\}$.

Note that condition (39) and (40) hold if each of the sequences $\{X_k\}$ and $\{Y_k\}$ consists of identically distributed variables and

$$Eh^a(X_1)+Eh^a(Y_1) < \infty \qquad \text{for} \qquad 2<a<3 .$$

References

[1] Feller W. An introduction to probability theory and its applications. V.2. New York: Wiley, 1971.

[2] Lévy P. Théorie de l'addition des variables aléatoires. Paris: Gauthier-Villars, 1937.

[3] Bolthausen E. Exact convergence rates in some martingale central limit theorems. Ann. Probability, 1982, v.10, N3, 672-688.

[4] Bentkus V., Rac̆kauskas A. Estimates of distances between sums of independent random elements in Banach spaces. Teor. Verojatn. i Primenen., 1984, v.29, N1, 49-64 (in Russian).

[5] Bentkus V. Asymptotic analysis of sums of independent random elements of Banach space. Doctoral dissertation: Vilnius, 1985 (in Russian).

[6] Borovkov A. A. On the rate of convergence for the invariance principle. Theory Probab. Appl., 1973, v.18, N2, 207-225.

[7] Berkes I., Philipp W. Approximation theorems for independent and weakly dependent random vectors. Ann. Probability, 1979, v.7, N1, 29-54.

[8] Sakhanenko A. I. Estimates in invariance principle. Proc. Inst. Math. Novosibirsk, 1985, v.5, 37-44 (in Russian).

[9] Hall P., Heyde C. C. Martingale limit theory and its application. New York: Academic Press, 1980.

[10] Utev S. A. A remark on the rate of convergence in the invariance principle. Sibirsk. Mat. Ẑ., 1981, v.22, N5, 206-208 (in Russian).

[11] Haeusler E. An exact rate of convergence in the functional central limit theorem for special martingale difference arrays. Z. Wahrscheinlichkeitstheorie verw. Gebiete, 1984, v.65, N4, 523-534.

[12] Borovkov K. A. On the invariance principle in Hilbert space. Teor. Verojatn. i Primenen., 1983, v.28, N3, 603 (in Russian).

Institute of Mathematics,
Novosibirsk, 630090, USSR

MUTUALLY REPELLING PARTICLES OF m TYPES

Yasumasa Saisho

In this paper we construct a model for the random motion of mutually repelling n particles of m types in R^d by using the multi-dimensional Skorohod equation: the number of particles of type k is n_k ($\sum_{k=1}^m n_k = n$), all particles are undergoing Brownian motion and when the distance between two particles of different type attains a given value ρ (> 0), they repel each other instantly.

The problem can be formulated as follows. Let W be the set of continuous functions on R^d and Λ_k be the set of indexes of the particles of type k ($1 \leqq k \leqq m$). We denote the type of the i-th particle by $\tau(i)$. Given $w_1, w_2, \cdots, w_n \in W$ with $|w_i(0) - w_j(0)| \geqq \rho$, $\tau(i) \neq \tau(j)$, solve the equation

$$(1) \quad \xi_i(t) = w_i(t) + \sum_{\substack{\ell=1 \\ (\neq \tau(i))}}^{m} \sum_{j \in \Lambda_\ell} \int_0^t (\xi_i(s) - \xi_j(s))\, d\phi_{ij}(s), \quad 1 \leqq i \leqq n,$$

under the following conditions (2)-(3).

$$(2) \quad \xi_i, \xi_j \in W, \ |\xi_i(t) - \xi_j(t)| \geqq \rho, \quad \tau(i) \neq \tau(j), \quad t \geqq 0.$$

(3) ϕ_{ij}'s are continuous non-decreasing functions with $\phi_{ij}(0) = 0$, $\phi_{ij}(t) = \phi_{ji}(t)$, $t \geqq 0$ and

$$\phi_{ij}(t) = \int_0^t \mathbb{1}(|\xi_i(s) - \xi_j(s)| = \rho)\, d\phi_{ij}(s), \quad \tau(i) \neq \tau(j).$$

In [6] we constructed the motion of mutually reflecting n hard balls in R^d by solving the Skorohod equation for the domain

$$D = \{x = (x_1, x_2, \cdots, x_n) : |x_i - x_j| > \rho, \ 1 \leqq i \neq j \leqq n\}$$

in R^{nd}. The purpose of this paper is to show that the equation (1) can be solved uniquely following the idea of [6]. More precisely, we make use of the results of [5] on Skorohod equations for general domains. We show that the equation (1) is equivalent to the Skorohod equation for the domain

$$(4) \quad D = \{x = (x_1, x_2, \cdots, x_n) : |x_i - x_j| > \rho, \quad \tau(i) \neq \tau(j)\}.$$

Then we see the domain D satisfies Conditions (A) and (B) (see § 1) which assure the existence of the unique solution of the Skorohod equation. We note that if each type consists of only one particle, our problem reduces to the problem of [6].

In § 1, we state briefly the theory of the (multi-dimensional) Skorohod equation following [5]. In § 2, we show that the domain D

given by (4) satisfies the conditions so that our Skorohod equation can be solved. The main theorem is given in § 3.

§ 1. Skorohod equation In this section we state the known results on Skorohod equation given in [5] so that we can make use of the results in § 3. The multi-dimensional Skorohod equation for a convex domain was discussed by Tanaka [7] and then by Lions and Sznitman [3] when a domain D satisfies Conditions (A) and (B) (stated later) together with the additional condition that D is admissible, which means roughly that D can be approximated by smooth domains. Recently, Frankowska [1] and Saisho [5] extended independently the result of Lions and Sznitman by removing the additional condition.

Denote by $B(z,r)$ the open ball in R^N with center z and radius r. We define $\mathcal{N}_{x,r}$ by

$$\mathcal{N}_{x,r} = \{\mathbf{n} \in R^N : |\mathbf{n}| = 1,\ B(x - r\mathbf{n}, r) \cap D = \phi\},$$

and let $\mathcal{N}_x$ be the union of $\mathcal{N}_{x,r}$ as r runs over all positive numbers. In general it can happen that $\mathcal{N}_x = \phi$. In what follows $\langle \cdot , \cdot \rangle$ denotes the usual inner product in R^N. We introduce two conditions on the domain D.

Condition (A). There exists a constant $r_0 > 0$ such that $\mathcal{N}_x = \mathcal{N}_{x,r_0} \neq \phi$ for any $x \in \partial D$.

Condition (B). There exist constants $\delta > 0$ and β $(1 \leqq \beta < \infty)$ with the following property: for any $x \in \partial D$ there exists a unit vector ℓ_x such that

$$\langle \ell_x, \mathbf{n} \rangle \geqq 1/\beta \quad \text{for any} \quad \mathbf{n} \in \bigcup_{y \in B(x,\delta) \cap \partial D} \mathcal{N}_y .$$

We note that the following two statements for a unit vector $\mathbf{n}$ are equivalent (see [3: Remark 1.2]).

(i) $\mathbf{n} \in \mathcal{N}_{x,r}$.

(ii) $\langle y - x, \mathbf{n} \rangle + \frac{1}{2r}|y - x|^2 \geqq 0$ for any $y \in \bar{D}$.

Denote by $W(R^N)$ (resp. $W(\bar{D})$) the space of continuous paths in R^N (resp. $\bar{D}$). Skorohod equation for D with reflecting boundary is written in the form

$$(1.1) \qquad \xi(t) = w(t) + \int_0^t \mathbf{n}(s)\, d\phi(s),$$

where $w \in W(R^N)$ is given and satisfies $w(0) \in \bar{D}$; a solution (ξ, ϕ)

of (1.1) should be found under the following conditions.

(1.2) $\xi \in W(\bar{D})$.

(1.3) ϕ is a continuous non-decreasing function such that $\phi(0) = 0$ and

$$\phi(t) = \int_0^t \mathbb{1}(\xi(s) \in \partial D)\, d\phi(s).$$

(1.4) $\mathbf{n}(s) \in \mathcal{N}_{\xi(s)}$ if $\xi(s) \in \partial D$.

The following theorem is the result of Saisho [5].

Theorem 1.1. If D satisfies Conditions (A) and (B), then, for any $w \in W(R^N)$ with $w(0) \in \bar{D}$, there exists a unique solution of the equation (1.1). In particular, if D is a convex domain, then, for any $w,w' \in W(R^N)$ with $w(0), w'(0) \in \bar{D}$, we have

(1.5) $$|\xi(t) - \xi'(t)| \leq \| w - w' \|_t + |w(0) - w'(0)|,$$

where ξ, ξ' are the solution of (1.1) for w,w', respectively, and

$$\| u \|_t = \text{the total variation of } u \in W(R^N) \text{ on } [0,t]$$
$$= \sup \sum_{k=1}^{n} |u(t_k) - u(t_{k-1})|,$$

the supremum being taken over all partitions $0 = t_0 < t_1 < \cdots < t_n = t$.

§ 2. D satisfies Conditions (A) and (B) In this section, we show that the domain D in R^{nd} given by (4) satisfies Conditions (A) and (B) assuming that $\rho > 0$.

We use the following notation:

$$\Lambda = \{1,2,\cdots,n\} = \sum_{k=1}^{m} \Lambda_k, \quad (n = \sum_{i=1}^{m} n_i),$$

$$\Lambda_k = \{s_k + 1, s_k + 2, \cdots, s_k + n_k\},$$

$$s_k = \sum_{i=1}^{k-1} n_i \quad (k \geq 2), \quad s_1 = 0,$$

$$x(I) = \{x_i : i \in I\},$$

$$\Gamma_k = \Gamma_k(x) = \{i \in \Lambda_k : |x_i - x_j| \geq 2\rho, \ \forall j \in \bigcup_{\substack{1 \leq \ell \leq m \\ \ell \neq k}} \Lambda_\ell\}$$

where $I\ (\neq \emptyset) \subset \Lambda$, $x = (x_1, x_2, \cdots, x_n) \in R^{nd}$ and $1 \leq k \leq m$.

Definition 2.1. Suppose $I, I'\ (\neq \emptyset) \subset \Lambda$. (i) $x(I)$ and $x(I')$ are said to be separated if

$$|x_i - x_j| \geq 2\rho, \quad \forall i \in I \cap \Lambda_k, \ \forall j \in I' \cap \Lambda_\ell, \ 1 \leq k \neq \ell \leq m.$$

(ii) $x(I)$ is called a <u>cluster</u> if there exist $i,j \in I$ such that $\tau(i) \neq \tau(j)$ and, for $\forall i,j \in I$ with $\tau(i) \neq \tau(j)$, there exist $i_0(=i), i_1, \cdots, i_{h-1}, i_h(=j)$ in I such that $\tau(i_q) \neq \tau(i_{q-1})$ and

$$|x_{i_q} - x_{i_{q-1}}| < 2\rho, \quad 1 \leqq q \leqq h.$$

<u>Remark 2.1.</u> (i) Notice that some of $i_1, i_2, \cdots, i_h$ in the above (ii) may be the same.
(ii) For any $x \in R^{nd}$, we can write

$$\{x_1, x_2, \cdot \cdot \cdot, x_n\} = \bigcup_{k=1}^{p} x(I_k) \cup \bigcup_{k=1}^{m} x(\Gamma_k) \tag{2.1}$$

where $x(I_k)$'s are mutually separated clusters.

<u>Remark 2.2.</u> If $x(I)$ is a cluster, it is easy to see

(i) $$|x_i - x_j| < 2\rho(\#I - 1), \quad \forall i,j \in I,$$

(ii) $$|x_i - x_I| = (\#I)^{-1} |\sum_{j \in I} (x_i - x_j)|$$
$$< 2\rho(\#I - 1)^2 (\#I)^{-1} < 2\rho(\#I - 1), \quad \forall i \in I,$$

where $\#I$ is the number of elements in I and $x_I = (\#I)^{-1} \sum_{j \in I} x_j$.

For $x \in \partial D$, we denote

$$L_x = \{(i,j) : |x_i - x_j| = \rho, \ i < j, \ \tau(i) \neq \tau(j)\}$$

and, in what follows, we fix $x \in \partial D$, so I_k, $1 \leqq k \leqq p$, Γ_k, $1 \leqq k \leqq m$ are also fixed. Let $\delta > 0$ be a constant determined later and $y \in B(x,\delta) \cap \partial D$. Next define $u = (u_1, u_2, \cdot \cdot \cdot, u_n)$ by

$$u_i = \begin{cases} x_i & \text{if } i \in \Gamma_k \text{ for some } 1 \leqq k \leqq m, \\ 2x_i - x_{I_k} & \text{if } i \in I_k \text{ for some } 1 \leqq k \leqq p \end{cases}$$

and $\ell_x = (u - x)/|u - x|$.

The purpose of this section is to show the following proposition.

<u>Proposition 2.1.</u> (i) D satisfies Condition (A) with $r_0 = \rho/\sqrt{2}\beta$ and for any $x \in \partial D$ we have

$$\mathcal{N}_x = \{\mathbf{n} = \sum_{(i,j) \in L_x} c_{ij} \mathbf{n}_{ij}(x) : c_{ij} \geqq 0, \ |\mathbf{n}| = 1\},$$

where

$$\mathbf{n}_{ij}(x) = (0, \cdot \cdot \cdot, 0, \underset{(i\text{-th})}{\frac{x_i - x_j}{\sqrt{2}\rho}}, 0, \cdot \cdot \cdot, 0, \underset{(j\text{-th})}{\frac{x_j - x_i}{\sqrt{2}\rho}}, 0, \cdot \cdot \cdot, 0)$$

and $\beta = 4\sqrt{2}(n-1)^{3/2}$.
(ii) D satisfies Condition (B) with any $\delta \in (0, \rho/2\sqrt{2})$, β and $\ell_x = (u-x)/|u-x|$.

We begin with the following lemma.

Lemma 2.1.

$$(2.2)\qquad |u - x| < 2\rho(n-1)^{3/2}.$$

Proof. If $i \in \Gamma_k$ for some $1 \leqq k \leqq m$, it immediately follows from the definition of u that $|u_i - x_i| = 0$. On the other hand, if $i \in I_k$ for some $1 \leqq k \leqq p$, Remark 2.2 yields

$$(2.3)\qquad |u_i - x_i| = |x_i - x_{I_k}| < 2\rho(\# I_k - 1)^2(\# I_k)^{-1}.$$

Therefore, we have

$$|u-x|^2 = \sum_{k=1}^{p} \sum_{i \in I_k} |u_i - x_i|^2$$
$$< 4\rho^2 \sum_{k=1}^{p} (\# I_k - 1)^4 (\# I_k)^{-1} \leqq 4\rho^2(n-1)^3,$$

completing the proof.

Lemma 2.2. Suppose $\delta < \rho/\sqrt{2}$. Then if $i \in \Lambda_k$, $j \in \Lambda_\ell$ $(k \neq \ell)$ satisfy $|x_i - x_j| \geqq 2\rho$, we have $(i,j) \notin L_y$ for $\forall y \in B(x,\delta) \cap \partial D$.

Proof. For each i,j satisfying the assumption we have

$$|y_i - y_j| = |y_i - x_i + x_i - x_j + x_j - y_j|$$
$$\geqq |x_i - x_j| - |y_i - x_i| - |y_j - x_j|$$
$$\geqq 2\rho - \sqrt{2}\delta > \rho,$$

completing the proof.

The following lemma is immediate from the previous lemma and the definition of I_k.

Lemma 2.3. Suppose $\delta < \rho/\sqrt{2}$. Then $(i,j) \in L_y$ for some $y \in B(x,\delta) \cap \partial D$ implies $i,j \in I_k$ for some $1 \leqq k \leqq p$.

Lemma 2.4. If $\delta < \rho/2\sqrt{2}$, then for $y \in B(x,\delta) \cap \partial D$, we have

$$(2.4)\qquad \langle \ell_x, \mathbf{n}_{ij}(y) \rangle \geqq 1/\beta, \quad (i,j) \in L_y.$$

Proof. Since $i,j \in I_k$ for some $1 \leqq k \leqq p$ by Lemma 2.3, we have

$$\langle \ell_x, \mathbf{n}_{ij}(y)\rangle = \frac{1}{\sqrt{2}\rho|u - x|}\langle x_i - x_j, y_i - y_j\rangle,$$

$$\begin{aligned}\langle x_i - x_j, y_i - y_j\rangle &= \langle x_i - x_j, x_i - x_j\rangle \\ &\quad + \langle x_i - x_j, y_i - x_i\rangle + \langle x_i - x_j, x_j - y_j\rangle \\ &\geqq |x_i - x_j|^2 - |x_i - x_j|(|y_i - x_i| + |x_j - y_j|) \\ &\geqq |x_i - x_j|^2 - |x_i - x_j|\sqrt{2}\delta \\ &= (|x_i - x_j| - \delta/\sqrt{2})^2 - \delta^2/2.\end{aligned}$$

Since

$$|x_i - x_j| - \delta/\sqrt{2} > \rho - \rho/4 = 3\rho/4 \quad (> 0),$$

we have

$$\langle x_i - x_j, y_i - y_j\rangle > \rho^2/2.$$

Thus, by Lemma 2.1 we have

$$\langle \ell_x, \mathbf{n}_{ij}(y)\rangle > \frac{1}{\sqrt{2}\rho|u - x|}\rho^2/2 > 1/\beta.$$

The proof of Lemma 2.4 is finished.

The following lemma is found in [6:Lemma 3.1] and the proof is omitted.

Lemma 2.5 ([6]). For any $(i,j) \in L_x$, we have

$$B(x - \frac{1}{\sqrt{2}}\rho\mathbf{n}_{ij}(x), \frac{1}{\sqrt{2}}\rho) \cap D = \phi.$$

Lemma 2.6. Setting

$$\mathcal{N}'_x = \{\mathbf{n} = \sum_{(i,j)\in L_x} c_{ij}\mathbf{n}_{ij}(x) : c_{ij} \geqq 0, \ |\mathbf{n}| = 1\}, \quad x \in \partial D,$$

we have $\mathcal{N}'_x \subset \mathcal{N}_{x,r_0}$.

Proof. By Remark 1.1, it is enough to show that

$$\langle y - x, \mathbf{n}\rangle + \frac{1}{2r_0}|y - x|^2 \geqq 0, \quad \forall y \in \bar{D}, \quad \forall \mathbf{n} \in \mathcal{N}'_x.$$

By Lemma 2.4, we have

$$\langle \ell_x, \mathbf{n}_{ij}(x)\rangle \geqq 1/\beta \quad \text{for any} \quad (i,j) \in L_x.$$

Since $\mathbf{n} \in \mathcal{N}'_x$ can be written as

$$\mathbf{n} = \sum_{(i,j)\in L_x} c_{ij}\mathbf{n}_{ij}(x), \quad c_{ij} \geqq 0,$$

we have

$$(2.5)\qquad 1 \geq \langle \ell_x, \mathbf{n} \rangle = \sum_{(i,j)\in L_x} c_{ij} \langle \ell_x, \mathbf{n}_{ij}(x) \rangle \geq \sum_{(i,j)\in L_x} c_{ij}/\beta,$$

$$1/2r_0 = \beta/\sqrt{2}\rho \geq \sum_{(i,j)\in L_x} c_{ij}/\sqrt{2}\rho,$$

and hence by Lemma 2.5,

$$\langle y - x, \mathbf{n} \rangle + \frac{1}{2r_0}|y - x|^2$$

$$\geq \sum_{(i,j)\in L_x} c_{ij}\langle y - x, \mathbf{n}_{ij}(x)\rangle + \frac{1}{\sqrt{2}\rho}\sum_{(i,j)\in L_x} c_{ij}|y - x|^2$$

$$= \sum_{(i,j)\in L_x} c_{ij}\{\langle y - x, \mathbf{n}_{ij}(x)\rangle + \frac{1}{\sqrt{2}\rho}|y - x|^2\} \geq 0.$$

The proof is finished.

The following is Lemma 3.3 of [6] and the proof is omitted.

Lemma 2.7. For any ε $(0 < \varepsilon < 1)$ and $x \in \partial D$, there exists a positive constant δ' such that

$$\{\bigcap_{(i,j)\in L_x} C_{ij}(x,\varepsilon)\} \cap B(x,\delta') \subset D \cup \{x\},$$

where

$$C_{ij}(x,\varepsilon) = \{y \in R^{nd}: \langle y - x, n_{ij}(x)\rangle \geq \varepsilon|y - x|\},\ (i,j)\in L_x .$$

Proof of Proposition 2.1. (i) Employing the same argument as in the proof of Proposition 3.1 of [6], we immediately have $\mathcal{N}_x \subset \mathcal{N}'_x$. Therefore, combining this with Lemma 2.6, we get the assertion.

(ii) By (i), any $\mathbf{n} \in \mathcal{N}_y$, $y \in B(x,\delta) \cap \partial D$ can be written in the form

$$\mathbf{n} = \sum_{(i,j)\in L_y} c_{ij}\mathbf{n}_{ij}(y), \quad c_{ij} \geq 0.$$

Thus, repeating a similar calculation to (2.5), we have

$$\langle \ell_x, \mathbf{n} \rangle \geq \sum_{(i,j)\in L_y} c_{ij}/\beta \geq 1/\beta.$$

The proof of Proposition 2.1 is finished.

§ 3. Motion of mutually repelling particles of m types In § 2 we showed that the domain D satisfies Conditions (A) and (B). Therefore Theorem 1.1 guarantees the existence of the unique solution of the Skorohod equation for D:

$$(3.1)\qquad \xi(t) = w(t) + \int_0^t \mathbf{n}(s)\,d\phi(s),$$

where $w = (w_1, w_2, \cdots, w_n)$, $w_i \in W(R^d)$, $1 \leq i \leq n$, and $|w_i(0) - w_j(0)|$

$\geqq \rho,\ \ \tau(i) \neq \tau(j)$. The component-wise expression of (3.1) is

$$\xi_i(t) = w_i(t) + \int_0^t \mathbf{n}_i(s)\,d\phi(s), \quad 1 \leqq i \leqq n. \tag{3.2}$$

In this section we prove the following theorem by showing that (3.2) is equivalent to the equation (1).

Theorem 3.1. For any $w = (w_1, w_2, \cdots, w_n) \in W(R^{nd})$ with $|w_i(0) - w_j(0)| \geqq \rho,\ \ \tau(i) \neq \tau(j)$, there exists a unique solution of (1).

Proof. Set $\mathbf{n}(s) = (n_1(s), n_2(s), \cdots, n_n(s))$. Then Proposition 2.1 yields that $\mathbf{n}(s)$ can be written in the form

$$\mathbf{n}(s) = \sum_{1\leqq k<\ell\leqq m} \sum_{\substack{i\in\Lambda_k \\ j\in\Lambda_\ell}} c_{ij}(s)\mathbf{n}_{ij}(s), \quad c_{ij}(s) \geqq 0$$

with $c_{ij}(s) = 0$ for $(i,j) \notin L_{\xi(s)}$, $\xi(s) \in \partial D$, so if we define $c_{ij}(s) = c_{ji}(s)$ for $j < i$, we have

$$n_i(s) = \frac{1}{\sqrt{2}\rho} \sum_{\substack{\ell=1 \\ (\neq \tau(i))}}^{m} \sum_{j\in\Lambda_\ell} c_{ij}(s)(\xi_i(s) - \xi_j(s)), \quad 1 \leqq i \leqq n,$$

Therefore if we set

$$\phi_{ij}(t) = \frac{1}{\sqrt{2}\rho} \int_0^t c_{ij}(s)\,d\phi(s), \quad 1 \leqq i,j \leqq n,\ \tau(i) \neq \tau(j),$$

(3.2) yields (1). Converse is also easy (cf. [6: § 4]). The proof of Theorem 3.1 is finished.

Remark 3.1. In case $\rho = 0$, our method does not work. But if $d = 1$, $m = 2$ and $\rho = 0$ we can formulate the following problem. Given $w_1, w_2, \cdots, w_n \in W(R)$ with $\max_{i\in\Lambda_1} w_i(0) \leqq \min_{j\in\Lambda_2} w_j(0)$, solve the equation

$$\begin{cases} \xi_i(t) = w_i(t) + \sum_{j=n_1+1}^{n} \int_0^t (\xi_i(s) - \xi_j(s))\,d\phi_{ij}(s), & 1 \leqq i \leqq n_1, \\ \xi_j(t) = w_j(t) - \sum_{i=1}^{n_1} \int_0^t (\xi_i(s) - \xi_j(s))\,d\phi_{ji}(s), & n_1+1 \leqq j \leqq n, \end{cases} \tag{3.3}$$

under the following conditions (3.4) - (3.5).

$$\xi_i, \xi_j \in W(R), \quad \max_{i\in\Lambda_1} \xi_i(t) \leqq \min_{j\in\Lambda_2} \xi_j(t), \quad t \geqq 0. \tag{3.4}$$

(3.5) ϕ_{ij}'s are continuous non-decreasing functions with $\phi_{ij}(0) = 0$, $\phi_{ij}(t) = \phi_{ji}(t)$, $t \geqq 0$ and

$$\phi_{ij}(t) = \int_0^t \mathbb{1}\{\max_{i\in\Lambda_1} \xi_i(s) = \min_{j\in\Lambda_2} \xi_j(s)\}\,d\phi_{ij}(s).$$

Using the same argument as in the proof of Theorem 3.1, we can show that the equation (3.3) is equivalent to the Skorohod equation for the domain D given by

(3.6) $$D = \{x = (x_1, x_2, \cdots, x_n) : \max_{i\in\Lambda_1} x_i < \min_{j\in\Lambda_2} x_j\}.$$

It is easy to see that D is a convex domain and so, Condition (A) holds trivially with

$$\mathcal{N}_x = \{\mathbf{n} = \sum_{(i,j)\in L_x} c_{ij}\mathbf{n}_{ij} : c_{ij} \geq 0, \ |\mathbf{n}| = 1\}, \quad x \in \partial D,$$

where

$$L_x = \{(i,j) : x_i = x_j, \ i \in \Lambda_1, \ j \in \Lambda_2\},$$

$$\mathbf{n}_{ij} = (0,\cdots,0,\underset{(i\text{-th})}{-1/\sqrt{2}},0,\cdots,0,\underset{(j\text{-th})}{1/\sqrt{2}},0,\cdots,0).$$

We can also prove that D satisfies Condition (B) with $\beta = \sqrt{n/2}$,

$$\ell = (\underbrace{\frac{-1}{\sqrt{n}},\cdots,\frac{-1}{\sqrt{n}}}_{n_1},\underbrace{\frac{1}{\sqrt{n}},\cdots,\frac{1}{\sqrt{n}}}_{n_2}) \quad \text{(independent of } x \in \partial D)$$

and any positive δ. Thus, we have the same assertion as Proposition 2.1 and we get the following theorem, which is also discussed in [4]. The last assertion (3.7) follows from (1.5).

<u>Theorem 3.2.</u> For any $w = (w_1, w_2, \cdots, w_n) \in W(R^n)$ with $\max_{i\in\Lambda_1} w_i(0) \leq \min_{j\in\Lambda_2} w_j(0)$, there exists a unique solution $\xi = (\xi_1, \xi_2, \cdots, \xi_n)$ of (3.3). Moreover, if ξ, ξ' are solutions of (3.3) for $w, w' \in W(R^n)$ respectively, we have

(3.7) $$\sum_{i=1}^{n} |\xi_i(t) - \xi_i'(t)| \leq \sum_{i=1}^{n} \{\|w_i - w_i'\|_t + |w_i(0) - w_i'(0)|\}.$$

Finally we construct the stochastic version of the motion of mutually repelling particles. Let $B_i(t)$, $1 \leq i \leq n$, be independent d-dimensional $\mathcal{F}_t$-Brownian motions with $B_i(0) = 0$ defined on a probability space $(\Omega, \mathcal{F}, P)$ with a right-continuous filtration $\{\mathcal{F}_t\}$. We assume that each $\mathcal{F}_t$ contains all P-null sets. Given

$$\sigma: R^d \to R^d \otimes R^d, \quad b: R^d \to R^d,$$

we consider the following Skorohod SDE:

(3.8) $$dX_i(t) = \sigma(X_i(t))dB_i(t) + b(X_i(t))dt + \sum_{\substack{\ell=1 \\ (\neq \tau(i))}}^{m} \sum_{j\in\Lambda_\ell} (X_i(t) - X_j(t))d\Phi_{ij}(t), \quad 1 \leq i \leq n,$$

where the initial values are assumed to be $\mathcal{F}_0$-measurable random variables with $|X_i(0) - X_j(0)| \geq \rho$ if $\tau(i) \neq \tau(j)$. The solution $X_i(t)$, $1 \leq i \leq n$ should be solved under the following conditions (3.9)-(3.10).

(3.9) $X_i(t)$'s are $\mathcal{F}_t$-adapted continuous processes with

$$|X_i(t) - X_j(t)| \geqq \rho, \quad \tau(i) \neq \tau(j), \quad t \geqq 0.$$

(3.10) $\Phi_{ij}(t)$'s are $\mathcal{F}_t$-adapted continuous non-decreasing processes with $\Phi_{ij}(0) = 0$, $\Phi_{ij}(t) = \Phi_{ji}(t)$, $t \geqq 0$ and

$$\Phi_{ij}(t) = \int_0^t \mathbb{1}(|X_i(s) - X_j(s)| = \rho)\, d\Phi_{ij}(s), \quad \tau(i) \neq \tau(j).$$

Then using Theorem 5.1 of [5], we can show the following theorem. The meaning of a strong solution is the same as in Definition IV-1.6 of [2].

Theorem 3.3. Suppose that σ and b are bounded and Lipschitz continuous. Then for any initial values $X_i(0)$'s such that $|X_i(0) - X_j(0)| \geqq \rho$, $i \in \Lambda_k$, $j \in \Lambda_\ell$, $k \neq \ell$ there exists a unique strong solution of the SDE (3.8).

Acknowledgement The author would like to express his gratitude to Professor H.Tanaka for his helpful suggestions and valuable comments.

References

[1] H.Frankowska, A viability approach to the Skorohod problem, Stochastics, 14 (1985), 227-244.

[2] N.Ikeda and S.Watanabe, Stochastic Differential Equations and Diffusion Processes, North Holland-Kodansha, Amsterdam-Tokyo, (1981).

[3] P.L.Lions and A.S.Sznitman, Stochastic differential equations with reflecting boundary conditions, Comm. Pure Appl. Math., 37 (1984), 511-537.

[4] M.Nagasawa and H.Tanaka, Diffusion with interactions and collisions between coloured particles and the propagation of chaos, to appear in Probab. Th. Rel. Fields.

[5] Y.Saisho, Stochastic differential equations for multi-dimensional domain with reflecting boundary, to appear in Probab. Th. Rel. Fields.

[6] Y.Saisho and H.Tanaka, Stochastic differential equations for mutually reflecting Brownian balls, Osaka J. Math. 23 (1986), 725-740.

[7] H.Tanaka, Stochastic differntial equations with reflecting boundary condition in convex regions, Hiroshima Math. J. 9 (1979), 163-177.

Department of Mathematics
Faculty of Science and Technology
Keio University,
Hiyoshi, Kohoku-ku,
Yokohama, 223 Japan

SOME CLASSES GENERATED BY EXPONENTIAL DISTRIBUTIONS

Ken-iti Sato

1. Introduction and main results. We define some classes of distributions generated by exponential distributions on $\mathbb{R}_+ = [0, +\infty)$ and on $\mathbb{R}_+ = (-\infty, 0]$ and study their properties. The distribution of a sojourn time (with weight not necessarily positive) of a birth-and-death process up to a first passage time belongs to the classes that we consider.

For $a > 0$, let μ_a denote the exponential distribution on $\mathbb{R}_+$ with mean $1/a$, that is, μ_a is supported on $\mathbb{R}_+$ and $\mu_a(dx) = ae^{-ax}dx$ on $\mathbb{R}_+$. For $a < 0$, the exponential distribution μ_a on $\mathbb{R}_-$ with mean $1/a$ is defined as a distribution supported on $\mathbb{R}_-$ such that $\mu_a(dx) = -ae^{-ax}dx$ on $\mathbb{R}_-$. Let us introduce classes ME^k_{++}, ME^k_{+0}, ME^k_{+}, CE^k_{+} and ME^k_{--}, ME^k_{-0}, ME^k_{-}, CE^k_{-} for any positive integer k. The class ME^k_{++} [resp. ME^k_{--}] is the set of mixtures of k distinct exponential distributions on $\mathbb{R}_+$ [resp. $\mathbb{R}_-$]. That is, $\mu \in ME^k_{++}$ [resp. ME^k_{--}] iff $\mu = \Sigma^k_{i=1} p_i \mu_{a_i}$ where $p_i > 0$ for $1 \leq i \leq k$, $\Sigma^k_{i=1} p_i = 1$, and $a_1, \ldots, a_k$ are distinct positive [resp. negative] numbers. The $a_1, \ldots, a_k$ are uniquely determined by μ; we call them the spectra of μ. The class ME^k_{+0} [resp. ME^k_{-0}] is defined as the set of $\mu = p_0 \delta_0 + \Sigma^k_{i=1} p_i \mu_{a_i}$ where δ_0 is the delta distribution at 0, $p_i > 0$ for $0 \leq i \leq k$, $\Sigma^k_{i=0} p_i = 1$, and $a_1, \ldots, a_k$ are distinct positive [resp. negative] numbers. Let us call $a_1, \ldots, a_k, +\infty$ [resp. $a_1, \ldots, a_k, -\infty$] the spectra of μ. Define

$$ME^k_{+} = ME^k_{++} \cup ME^k_{+0}, \quad ME^k_{-} = ME^k_{--} \cup ME^k_{-0}.$$

The class CE^k_{+} [resp. CE^k_{-}] is the set of convolutions of k (not necessarily distinct) exponential distributions on $\mathbb{R}_+$ [resp. $\mathbb{R}_-$]. A distribution μ in CE^k_{+} or CE^k_{-} is said to have spectra $a_1, \ldots, a_k$ if $\mu = \mu_{a_1} * \cdots * \mu_{a_k}$. The spectra are uniquely determined by μ. We

---

This research was partly done while the author was visiting Institute for Mathematics and its Applications, University of Minnesota, in November 1985.

adopt the convention that ME_{+0}^0, ME_+^0, ME_{-0}^0, ME_-^0, CE_+^0, and CE_-^0 denote the set consisting of a single element δ_0. For any two class $A^{(i)}$, $i=1,2$, denote by $A^{(1)}*A^{(2)}$ the class of μ such that $\mu=\mu^{(1)}*\mu^{(2)}$, $\mu^{(1)}\in A^{(1)}$, and $\mu^{(2)}\in A^{(2)}$; we call $\mu^{(1)}*\mu^{(2)}$ an expression of μ in $A^{(1)}*A^{(2)}$. Finally we define

$$E(k, \ell, r, s) = ME_+^k * CE_+^\ell * ME_-^r * CE_-^s$$

for nonnegative integers k, ℓ, r, s.

Remarkable properties of distributions in the class $E(k, \ell, r, s)$ are infinite divisibility and unimodality. As is well-known, exponential distributions on $\mathbb{R}_+$ and $\mathbb{R}_-$ are infinitely divisible and the class of infinitely divisible distributions is closed under convolution. It is not closed under mixture, but mixtures of exponential distributions on $\mathbb{R}_+$ are infinitely divisible, which is proved by C. Goldie [1] (see also F.W. Steutel [9]). Hence, distributions in $E(k,\ell,r,s)$ are infinitely divisible. It is not hard to see that they are unimodal. We consider their strict unimodality and bounds of their modes.

Theorem 1. Let μ be in CE_+^k for some $k\geq 1$ and let $a_1\leq a_2\leq \cdots \leq a_k$ be the spectra of μ. Then μ is strictly unimodal and has a real-analytic density on $(0,+\infty)$. The mode M of μ satisfies

$$\max_{1\leq i\leq k} \frac{i-1}{a_i} \leq M \leq \frac{k-1}{a_1} \tag{1.1}$$

Note that, in case $a_1=a_2=\cdots=a_k$ (gamma distribution), the bound (1.1) gives the exact mode $(k-1)/a_1$. Let us denote by $\mu|_I$ the restriction of μ to an interval I. It is obvious that, if $\mu\in ME_+^k$ for some $k\geq 0$, then μ is strictly unimodal with mode 0 and $\mu|_{(0,+\infty)}$ has a real-analytic density.

Theorem 2. Let $\mu=\mu^{(1)}*\mu^{(2)}$ where $\mu^{(1)}\in ME_+^k$ and $\mu^{(2)}\in CE_+^\ell$ for some $k\geq 1$ and $\ell\geq 1$. Then μ is strictly unimodal and $\mu|_{(0,+\infty)}$ has a real-analytic density. Let $a_1,\ldots,a_k$ or $a_1,\ldots,a_k,+\infty$ be the spectra of $\mu^{(1)}$ in increasing order, and let $b_1,\ldots,b_\ell$ be the spectra of $\mu^{(2)}$ in nondecreasing order. Let M be the mode of μ. If $\mu^{(1)}\in ME_{++}^k$, then

$$\max_{1\leq i\leq \ell+1} \frac{i-1}{b_i'} \leq M \leq \frac{\ell}{\min\{a_1,b_1\}}, \tag{1.2}$$

where $b_1'\leq b_2'\leq\cdots\leq b_{\ell+1}'$ is the rearrangement of $a_k, b_1,\cdots,b_\ell$ in the nondecreasing order. If $\mu^{(1)}\in ME_{+0}^k$, then

$$(1.3)\qquad \max_{1\le i\le \ell} \frac{i-1}{b_i} \le M \le \frac{\ell}{\min\{a_1, b_1\}} .$$

Theorem 3. Let $\mu \in E(k, \ell, r, s)$ with some nonnegative integers k, ℓ, r, s, that is, let

$$\mu = \mu^{(1)} * \mu^{(2)} * \mu^{(3)} * \mu^{(4)},$$

$$\mu^{(1)} \in ME_+^k, \quad \mu^{(2)} \in CE_+^\ell, \quad \mu^{(3)} \in ME_-^r, \quad \mu^{(4)} \in CE_-^s .$$

Suppose that $k + \ell \geq 1$ and $r + s \geq 1$. Let a be the minimum of the spectra of $\mu^{(1)}$ and $\mu^{(2)}$, and let b be the maximum of the spectra of $\mu^{(3)}$ and $\mu^{(4)}$. Then μ is strictly unimodal with mode M satisfying

$$(1.4)\qquad \frac{s}{b} \le M \le \frac{\ell}{a} .$$

Moreover, $\mu|_{(0,+\infty)}$ and $\mu|_{(-\infty,0)}$ have real-analytic densities.

In the statement above, we have used the words unimodal and strictly unimodal. A distribution μ on the real line is called unimodal if there exists a point M (called a mode) such that its distribution function is convex on $(-\infty, M)$ and concave on $(M, +\infty)$. Thus μ is unimodal if and only if there is a point M such that $\mu|_{(-\infty,M)}$ has a nondecreasing density and $\mu|_{(M,+\infty)}$ has a nonincreasing density. A distribution μ is called strictly unimodal if it is unimodal with a mode M and the density is increasing on (a, M) and decreasing on (M, b), where a and b are the left and the right extremities, respectively, of the support of μ. Here we are using the words "increase" and "decrease" in the strict sense.

We remark that, in some cases, our result in [7] can give better bounds of modes in terms of absolute moments or central absolute moments. For example, in place of the bound of μ in Theorem 2, bounds depending on the mixing weight in $\mu^{(1)}$ can be given.

Let $X(t)$, $t \geq 0$, be a birth-and-death process on the nonnegative integers with reflection at 0. Let $m < n$ and let σ^{mn} be the first passage time to n given that $X(0) = m$. Given a real-valued function $f(i)$, consider a random variable T defined by

$$(1.5)\qquad T = \int_0^{\sigma^{mn}} f(X(t))dt.$$

We call T the sojourn time up to σ^{mn} with weight function f. Let μ be the distribution of T. Infinite divisibility of μ is proved by Kent [4]. We can show that μ belongs to the class $E(k, \ell, r, s)$. How the numbers k, ℓ, r, s are related to the function f will be formulated in Theorem 4 in Section 4. It follows that μ is strictly unimodal and that $\mu|_{(-\infty,0)}$ and $\mu|_{(0,+\infty)}$ have real-analytic densities. We note that Rösler [6] and Keilson [3] show that σ^{mn} has a unimodal distribution, and Yamazato [10] determines the class of distributions of σ^{mn}.

It is known that expression of a distribution in $ME_+^k * CE_+^\ell$ is not always unique. Also, a distribution may belong to two classes $ME_+^k * CE_+^\ell$ and $ME_+^{k'} * CE_+^{\ell'}$. For example, suppose that μ has Laplace transform

$$\frac{a_1 a_2}{b} \frac{s+b}{(s+a_1)(s+a_2)} \qquad (0 < a_1 < a_2,\ a_1 < b).$$

Then $\mu = (p_0\delta_0 + p_1\mu_{a_1}) * \mu_{a_2} \in ME_{+0}^1 * CE_+^1$ for some $p_0, p_1 > 0$, $p_0+p_1 = 1$. However, if $b < a_2$ then $\mu \in ME_{++}^2$; if $b = a_2$, then $\mu = \mu_{a_1}$, a single exponential distribution; if $b > a_2$, then μ has another expression in $ME_{+0}^1 * CE_+^1$, namely, $\mu = (p_0'\delta_0 + p_1'\mu_{a_2}) * \mu_{a_1}$ for some $p_0', p_1' > 0$, $p_0' + p_1' = 1$. Such redundancy in expressions of distributions of σ^{mn} is thoroughly studied by Yamazato [10].

2. Lemmas. We prepare four lemmas.

Lemma 1. *An exponential distribution μ_a on $\mathbb{R}_+$ or $\mathbb{R}_-$ is strongly unimodal. Namely, for any unimodal distribution ν, the convolution $\nu * \mu_a$ is unimodal.*

Proof. Since the density of μ_a is log-concave, this is a consequence of Ibragimov's theorem [2].

Lemma 2. *Let μ be unimodal with a mode M and let $\mu = p\nu + (1-p)\delta_M$, where ν is absolutely continuous, δ_M is the delta distribution at M, and $0 \leqq p \leqq 1$. If the density $f(x)$ of ν is real-analytic except at one point, then μ is strictly unimodal.*

Proof. We may assume $p > 0$. Let $f(x)$ be real-analytic on $(-\infty, a)$ and $(a, +\infty)$. Suppose that $a < M$. Since $f(x)$ is non-decreasing on $(-\infty, M)$, it is 0 or increasing on $(-\infty, a)$, and constant or increasing on (a, M). It cannot be constant on (a, M) because of the real-analyticity on $(a, +\infty)$. Since $f(x)$ is non-increasing on $(M, +\infty)$, it must be decreasing there. Hence μ is strictly unimodal. The cases $a = M$ and $a > M$ are similar.

The following lemma is a repetition of Lemma 6.1 of Sato-Yamazato [8].

Lemma 3. Let μ be unimodal with a mode M and let λ be a distribution supported on $\mathbb{R}_+$. Then, on $(-\infty, M)$, $\mu*\lambda$ has a nondecreasing density. Thus, if $\mu*\lambda$ is unimodal, then it has a mode bigger than or equal to M.

Proof. Let $\mu = p\nu + (1-p)\delta_M$ as in Lemma 2. The density $f(x)$ of ν is nondecreasing on $(-\infty, M)$. We have

$$\mu * \lambda = p\nu * \lambda + (1-p)\delta_M * \lambda$$

and $\nu*\lambda$ is absolutely continuous with density

$$\int_{\mathbb{R}_+} f(x-y)\lambda(dy),$$

which is nondecreasing on $(-\infty, M)$. Since $\delta_M*\lambda$ is supported on $[M, +\infty)$, the assertion follows.

Lemma 4. Let $\mu^{(1)}$ and $\mu^{(2)}$ be unimodal with modes M_1 and M_2, respectively. Suppose that $\mu^{(1)}$ is supported on $\mathbb{R}_+$ and $\mu^{(2)}$ is supported on $\mathbb{R}_-$. Then, $\mu^{(1)}*\mu^{(2)}$ has a nondecreasing density on $(-\infty, M_1)$ and a nonincreasing density on $(M_2, +\infty)$. If $\mu^{(1)}*\mu^{(2)}$ is unimodal, then it has a mode in $[M_1, M_2]$.

This is a direct comsequence of Lemma 3.

3. Proof of Theorems.

Proof of Theorem 1. We use some elementary facts in the calculus of finite differences (see Milne-Thomson [5] p. 6-7). Given a function $g(x)$, let

$$g_1(x_1,x_2) = (g(x_1)-g(x_2))/(x_1-x_2) = g_1(x_2,x_1),$$

$$g_2(x_1,x_2,x_3) = (g_1(x_1,x_2)-g_1(x_2,x_3))/(x_1-x_3),$$

$$g_{n-1}(x_1,x_2,\ldots,x_n) = (g_{n-2}(x_1,x_2,\ldots,x_{n-1})-g_{n-2}(x_2,x_3,\ldots,x_n))/(x_1-x_n)$$

for any distinct points $x_1, x_2,\ldots,x_n$.

(i) The follwoing identity holds.

$$g_{n-1}(x_1,x_2,\ldots,x_n) = \frac{g(x_1)}{(x_1-x_2)(x_1-x_3)\ldots(x_1-x_n)} + \frac{g(x_2)}{(x_2-x_1)(x_2-x_3)\ldots(x_2-x_n)} + \ldots + \frac{g(x_n)}{(x_n-x_1)(x_n-x_2)\ldots(x_n-x_{n-1})}.$$

(ii) Let a and b be the minimum and the maximum, respectively, of $x_1, x_2,\ldots,x_n$. If g is n-1 times differentiable, then

$$g_{n-1}(x_1,x_2,\ldots,x_n) = \frac{g^{(n-1)}(\xi)}{(n-1)!}$$

for some $\xi \in (a, b)$.

First, the distribution μ in the theorem is unimodal by Lemma 1. Since the theorem is obvious for $k=1$, let $k \geq 2$. Suppose that $a_1 < a_2 < \ldots < a_k$. Then μ has Laplace transform $\varphi(s) = \int e^{-sx}\mu(dx)$ for $s > -a_1$ and we have

$$\varphi(s) = a_1 \cdots a_k/((s+a_1)\cdots(s+a_k)). \tag{3.1}$$

Hence

$$\varphi(s) = b_1/(s+a_1) + \cdots + b_k/(s+a_k),$$

where

$$b_1 = (-1)^{k-1}a_1 \cdots a_k/((a_1-a_2)(a_1-a_3)\cdots(a_1-a_k)),$$

$$b_2 = (-1)^{k-1}a_1 \cdots a_k/((a_2-a_1)(a_2-a_3)\cdots(a_2-a_k)),$$

$$\ldots\ldots$$

$$b_k = (-1)^{k-1}a_1 \cdots a_k/((a_k-a_1)(a_k-a_2)\cdots(a_k-a_{k-1})).$$

Therefore we have, for the density $f(x)$ of μ on $(0, +\infty)$,

$$f(x) = b_1e^{-a_1x} + b_2e^{-a_2x} + \cdots + b_ke^{-a_kx}.$$

Thus $f(x)$ is real-analytic and μ is strictly unimodal by Lemma 2. Let M be the mode of μ. Fix j and let $a_1/a_j = \alpha_1, \ldots, a_k/a_j = \alpha_k$, and let $g(x) = xe^{-(k-1)x}$. Then we have

$$f'\left(\frac{k-1}{a_j}\right) = (-1)^k a_j^2\alpha_1 \cdots \alpha_k\left(\frac{g(\alpha_1)}{(\alpha_1-\alpha_2)(\alpha_1-\alpha_3)\ldots(\alpha_1-\alpha_k)}\right.$$
$$\left. + \frac{g(\alpha_2)}{(\alpha_2-\alpha_1)(\alpha_2-\alpha_3)\ldots(\alpha_2-\alpha_k)} + \cdots + \frac{g(\alpha_k)}{(\alpha_k-\alpha_1)(\alpha_k-\alpha_2)\ldots(\alpha_k-\alpha_{k-1})}\right).$$

By the properties (i) and (ii) we have

$$f'\left(\frac{k-1}{a_j}\right) = (-1)^k a_j^2\alpha_1 \cdots \alpha_k \frac{g^{(k-1)}(\xi)}{(k-1)!}$$

for some ξ satisfying $\alpha_1<\xi<\alpha_k$. Since

$$g^{(k-1)}(x) = (-1)^k(k-1)^{k-1}e^{-(k-1)x}(1-x),$$

the value $(-1)^k g^{(k-1)}(\xi)$ is positive for $\xi < 1$ and negative for $\xi > 1$. Thus

$$f'\left(\frac{k-1}{a_k}\right) > 0 \quad \text{and} \quad f'\left(\frac{k-1}{a_1}\right) < 0.$$

Therefore

$$\frac{k-1}{a_k} < M < \frac{k-1}{a_1}.$$

Next we drop the assumption that $a_1,\ldots,a_k$ are distinct. We still have (3.1) and hence, by the partial fraction expansion, $\varphi(s)$ is a linear combination of the Laplace transforms of gamma distributions with integer exponents. Thus the density $f(x)$ of μ is a linear combination of gamma densities, and is real-analytic on $(0,+\infty)$. Therefore μ is strictly unimodal by Lemma 2. Since μ is approximated by distributions in CE_+^k with nonoverlapping spectra, we have

$$\frac{k-1}{a_k} \leq M \leq \frac{k-1}{a_1}$$

for the mode M of μ. For each $i \geq 2$ the mode M_i of the distribution $\mu_{a_1} * \mu_{a_2} * \cdots * \mu_{a_i}$ satisfies $(i-1)/a_i \leq M_i$. It follows from Lemma 3 that $M_i \leq M$. Thus we have (1.1), completing the proof.

Proof of Theorem 2. Since $\mu^{(1)}$ is unimodal with mode 0, μ is unimodal by Lemma 1. We have $\mu^{(1)} = p_0\delta_0 + \Sigma_{j=1}^k p_j\mu_{a_j}$ where $p_0 \geq 0$, $p_j > 0$ $(1 \leq j \leq k)$, and $\Sigma_{j=0}^k p_j = 1$. Hence

$$\mu = p_0\mu^{(2)} + \sum_{j=1}^{k} p_j\mu_{a_j} * \mu^{(2)}, \tag{3.2}$$

that is, μ is a finite mixture of distributions in CE_+^ℓ or $CE_+^{\ell+1}$. Hence, by Theorem 1, μ has a real-analytic density on $(0,+\infty)$. Thus strict unimodality follows from Lemma 2. Let $M^{(2)}$ be the mode of $\mu^{(2)}$, and M_j be the mode of $\mu_{a_j} * \mu^{(2)}$. Theorem 1 says that

$$\max_{1\leq i\leq \ell} \frac{i-1}{b_i} \leq M^{(2)} \leq \frac{\ell}{b_1}, \tag{3.3}$$

$$\max_{1\leq i\leq \ell+1} \frac{i-1}{b_i(j)} \leq M_j \leq \frac{\ell}{b_1(j)}, \tag{3.4}$$

where $b_1(j) \leq b_2(j) \leq \cdots \leq b_{\ell+1}(j)$ is the rearrangement of $a_j, b_1, \ldots, b_\ell$ in the nondecreasing order. Noting that $b_i(j) \leq b_i(k)$ for $1 \leq i \leq \ell+1$, $b_i(j) \leq b_i$ for $1 \leq i \leq \ell$, and $b_1(1) \leq b_1(j)$, we get

$$\max_{1\leq i\leq \ell} \frac{i-1}{b_i} \leq \max_{1\leq i\leq \ell+1} \frac{i-1}{b_i(k)} \leq M_j \leq \frac{\ell}{b_1(1)}. \tag{3.5}$$

If $\mu^{(1)} \in ME_{++}^k$, then $p_0 = 0$ and (1.2) is derived from (3.2) and (3.5). If $\mu^{(1)} \in ME_{+0}^k$, then $p_0 > 0$ and (1.3) is obtained from (3.2), (3.3), and (3.5).

Proof of Theorem 3. By Lemma 4, the convolution $\mu^{(1)} * \mu^{(3)}$ is unimodal with a mode 0. Hence, by Lemma 1, μ is unimodal. As we show in the proof of Theorems 1 and 2, the density of $\mu^{(1)} * \mu^{(2)}$ on $(0,+\infty)$ is a linear combination of gamma densities on $\mathbb{R}_+$ with integer components, and simlarly for the density of $\mu^{(3)} * \mu^{(4)}$ on $(-\infty, 0)$. Let $\nu^{(1)}$ [resp. $\nu^{(2)}$] be a gamma density on $\mathbb{R}_+$ [resp. $\mathbb{R}_-$] with exponent n [resp. m] and location parameter $a > 0$ [resp. $b < 0$]. Then $\nu^{(1)} * \nu^{(2)}$ has a density

$$g(x) = \frac{a^n(-b)^m}{(n-1)!(m-1)!}\int_{-\infty}^{\min\{x,0\}} (x-y)^{n-1}e^{-a(x-y)}(-y)^{m-1}e^{-by}dy.$$

We see that $g(x)$ is, on $(0,+\infty)$, a linear combination of $x^i e^{-ax}$, $0 \le i \le n-1$, and that it is, on $(-\infty, 0)$, a linear combination of $(-x)^i e^{-bx}$, $0 \le i \le m-1$. It follows that the distribution μ has a real-analytic density on $(0,+\infty)$ and $(-\infty, 0)$. Thus μ is strictly unimodal by Lemma 2. The bound (1.4) of the mode M follows from Theorem 2 combined with Lemma 4. The proof is complete.

4. Sojourn times for birth-and-death processes. Let us consider, for a birth-and-death process $X(t)$, $t \ge 0$, starting at m with reflection at 0, a sojourn time T with a weight function f up to a first passage time σ^{mn}. Definition is given by (1.5). We assume that $m < n$. Let N_k^+ [resp. N_k^-] be the number of i such that $0 \le i \le k-1$ and $f(i) > 0$ [resp. $f(i) < 0$]. Thus $k - N_k^+ - N_k^-$ is the number of i such that $0 \le i \le k-1$ and $f(i) = 0$. Let us use the following notations for some subclasses of $E(k, \ell, r, s)$:

$$E((k)_{++}, \ell, (r)_{-0}, s) = ME_{++}^k * CE_+^\ell * ME_{-0}^r * CE_-^s,$$

$$E((k)_{+0}, \ell, (r)_{--}, s) = ME_{+0}^k * CE_+^\ell * ME_{--}^r * CE_-^s,$$

$$E((k)_{+0}, \ell, (r)_{-0}, s) = ME_{+0}^k * CE_+^\ell * ME_{-0}^r * CE_-^s.$$

Theorem 4. Let $k = N_{m+1}^+$, $\ell = N_n^+ - N_{m+1}^+$, $r = N_{m+1}^-$, and $s = N_n^- - N_{m+1}^-$. Then the distribution μ of T belongs to the class $E(k, \ell, r, s)$. More precisely,

$\mu \in E((k)_{++}, \ell, (r)_{-0}, s)$ if $f(m) > 0$;

$\mu \in E((k)_{+0}, \ell, (r)_{--}, s)$ if $f(m) < 0$;

$\mu \in E((k)_{+0}, \ell, (r)_{-0}, s)$ if $f(m) = 0$.

Corollary. The distribution μ is strictly unimodal. The restrictions $\mu|_{(-\infty,0)}$ and $\mu|_{(0,+\infty)}$ have real-analytic densities. If $n = m+1$, then the mode of μ is 0.

This is a consequence of Theorem 4 combined with Theorem 3.

For the proof of Theorem 4, we seek the representation of the Laplace transform (in a neighborhood of the origin) of μ and its meromorphic extension. It is the quotient of two polynomials, which are generalization of the polynomials employed by Ledermann, Reuter, Karlin, and McGregor. To analyze the zeros of the polynomials, we adopt the method of Yamazato [10] and use another system (adjoint in a sense) of polynomials. All zeros of the polynomials are real and simple, although not always positive. They have an interlacing property in some sense. Details will be published elsewhere.

In some cases there is redundancy in the expression of μ in Theorem 4. Namely, the numbers k, ℓ, r, s of the class $E(k,\ell,r,s)$ to which μ belongs may possibly be reduced. In order to make use in the next theorem, let us define parameters of a distribution of our classes. If μ is in CE_+^k or CE_-^k and has spectra $a_1,\dots,a_k$, then we say that μ has parameters $a_1,\dots,a_k$. Suppose that μ belongs to ME_+^k and has spectra $a_1,\dots,a_k$ or $a_1,\dots,a_k,+\infty$ in increasing order. If $\mu \in ME_{++}^k$, then its Laplace transform is of the form

$$\frac{(s+b_1)(s+b_2)\cdots(s+b_{k-1})}{(s+a_1)(s+a_2)\cdots\cdots(s+a_k)} \quad \text{for} \quad s > -a_1,$$

where $a_1 < b_1 < a_2 < b_2 < \dots < a_{k-1} < b_{k-1} < a_k$; let us call these $a_1, b_1, a_2, \dots, b_{k-1}, a_k$ parameters of μ. If $\mu \in ME_{+0}^k$, then it has Laplace transform

$$\frac{(s+b_1)(s+b_2)\dots(s+b_k)}{(s+a_1)(s+a_2)\dots(s+a_k)} \quad \text{for} \quad s > -a_1,$$

where $a_1 < b_1 < a_2 < b_2 < \dots < a_{k-1} < b_{k-1} < a_k < b_k$, and we call these $a_1, b_1, a_2, \dots, b_{k-1}, a_k, b_k$ parameters of μ. Parameters of μ in ME_-^k are defined similarly.

Theorem 5. Let $\ell = N_n^+ - N_{m+1}^+$ and $s = N_n^- - N_{m+1}^-$. There uniquely exist nonnegative integers k and r that have the following properties (i) and (ii):

(i) $\mu \in E((k)_{++}, \ell, (r)_{-0}, s)$ in case $f(m) > 0$;

$\mu \in E((k)_{+0}, \ell, (r)_{--}, s)$ in case $f(m) < 0$;

$\mu \in E((k)_{+0}, \ell, (r)_{-0}, s)$ in case $f(m) = 0$.

(ii) If $k'+\ell' < k+\ell$ or $r'+s' < r+s$, then μ does not belong to $E(k', \ell', r', s')$.

Moreover, these k and r have the following properties:

(iii) If $\mu = \mu^{(1)} * \mu^{(2)} * \mu^{(3)} * \mu^{(4)}$ is an expression of μ in the class described in (i), then the set of all parameters of $\mu^{(1)}, \mu^{(2)}, \mu^{(3)}$, and $\mu^{(4)}$ has no overlapping.

(iv) $$\max\{0, 2N_{m+1}^+ - N_n^+\} \le k \le N_{m+1}^+,$$
$$\max\{0, 2N_{m+1}^- - N_n^-\} \le r \le N_{m+1}^-.$$

References

[1] C. Goldie: A class of infinitely divisible random variables, Proc. Cambridge Phil. Soc. 63 (1967), 1141-1143.

[2] I.A. Ibragimov: On the composition of unimodal distributions, Theor. Probab. Appl. 2 (1957), 117-119.

[3] J. Keilson: On the unimodality of passage time densities in birth-death processes, Statistica Neerlandica 35 (1981), 49-55.

[4] J.T. Kent: The appearance of a multivariate exponential distribution in sojourn times for birth-death and diffusion processes, in Probability, Statistics and Analysis, ed. by J.F.C. Kingman et al. (London Math. Soc. Lecture Notes Series, No. 79, 1983), 161-179.

[5] L.M. Milne-Thomson: The calculus of finite differences, Second ed., Chelsea, 1981.

[6] U. Rösler: Unimodality of passage times for one-dimensional strong Markov processes, Ann. Probab. 8 (1980), 853-859.

[7] K. Sato: Modes and moments of unimodal distributions, to appear in Ann. Statist. Math.

[8] K. Sato and M. Yamazato: On distribution functions of class L, Zeit.Wahrsch. Verw. Geb. 43 (1978), 273-308.

[9] F.W. Steutel: Note on the infinite divisibility of exponential mixtures, Ann. Math. Statist. 38 (1967), 1303-1305.

[10] M. Yamazato: Characterization of the class of upward first passage time distributions of nonnegative birth and death processes and related results, to appear.

Department of Mathematics
College of General Education
Nagoya University
Nagoya, 464 Japan

REMARKS ON THE CANONICAL REPRESENTATION OF STATIONARY LINEAR SYMMETRIC α-STABLE PROCESSES $(0<\alpha<1)$

Yumiko Sato

1. Introduction. Urbanik [6] studies linear prediction problem for strictly stationary processes with finite first moments by introducing a class called "admitting a prediction". He proves that in completely non-deterministic case the process has a canonical representation, using the technique of Banach space. P.Lévy treated the essentially same (but not necessarily stationary) class "à corrélation linéaire" [2] and also Hida-Ikeda [1]. So, we use the terminology "linear" in this paper. When the first absolute moment is infinite,the author presents in [4] a sufficient condition for a linear process $\{X_t;\ -\infty<t<\infty\}$ to have a canonical representation, assuming that the process is a symmetric α-stable process $(1/2<\alpha<1)$. Let $[X_t]$ be the closed linear span of $\{X_t\}$ by convergence in probability. It is a Fréchet space. Difficulty exists in taking innovation out of $[X_t]$. By technical reason the results in [4] are proved only when $\alpha>1/2$.

The aim of this paper is to show that, under some strong condition, existence of canonical representation is proved in the same way also for $0<\alpha\leq 1/2$. Our condition is so strong that the class we treat here turns out to be identical with the class of Ornstein-Uhlenbeck type processes

$$X_t = \int_{-\infty}^{t} e^{-c(t-u)}\, M(du),$$

where $c>0$ and $M(du)$ is a symmetric α-stable homogeneous random measure, $0<\alpha<1$. However, we believe that our discussion on Riemann type integrability for functions taking values in a Fréchet space will be useful in future.

2. Definitions and results. Let $\{X_t\}$ be a strictly stationary process. We use brackets to denote closed linear span with respect to convergence in probability.

Definition 2.1. We call $\{X_t\}$ linear if there exists a continuous linear operator A_0 from $[X_t]$ onto $[X_t;\ t\leq 0]$ such that (i) $A_0x = x$ whenever $x\in[X_t;\ t\leq 0]$, (ii) if $x\in[X_t]$ and, for every $y\in[X_t;\ t\leq 0]$, x and y are independent, then $A_0x=0$, (iii) if $x\in[X_t]$ and $y\in[X_t;\ t\leq 0]$, then $x-A_0x$ and y are independent.

This definition is due to Urbanik. If $A_0x = x$ for any $x\in[X_t]$, then

$\{X_t\}$ is called deterministic. Let A_t be defined by $A_t = T_t A_0 T_{-t}$, where T_t is the shift operator of the strictly stationary process. If $\lim_{t\to-\infty} A_t x = 0$ for any $x \in [X_t]$, then $\{X_t\}$ is called completely non-deterministic. Urbanik proves that a linear process $\{X_t\}$ is decomposed into two independent parts, $X_t = X_t^1 + X_t^2$, where each of $\{X_t^1\}$ and $\{X_t^2\}$ is a linear strictly stationary process and $\{X_t^1\}$ is deterministic, while $\{X_t^2\}$ is completely non-deterministic. His definition and argument are in L^1-setting, but are easily transferred to the setting of convergence in probability. Let $\{x(t)\}$ be a symmetric α-stable process. By this we mean that, for any positive integer n and real numbers a_i and t_i, $i=1,2,\ldots,n$, the linear combination $\sum_{i=1}^n a_i x(t_i)$ has a symmetric α-stable distribution, where α is a constant number, $0<\alpha\leq 2$. Any element x in $[x(t)]$ has a characteristic function of the form

$$(2.1) \qquad \varphi_x(u) = E\, e^{iux} = \exp(-a_x |u|^\alpha)$$

where a_x is a non-negative number depending on x. We study the case $0<\alpha<1$ (x(t) has infinite first absolute moment if and only if $0<\alpha\leq 1$). When the characteristic function of x has the form (2.1) we define

$$(2.2) \qquad \|x\| = a_x .$$

Then, $\|\ \|$ is a quasi-norm in $[x(t)]$ and

$$(2.3) \qquad \|\lambda x\| = |\lambda|^\alpha \|x\| ,$$

$$(2.4) \qquad \|x + x'\| = \|x\| + \|x'\| \quad \text{if } x \text{ and } x' \text{ are independent.}$$

With this quasi-norm $[x(t)]$ becomes a Fréchet space and convergence in probability coincides with convergence in this quasi-norm [5].

From now on assume that $\{X_t\}$ is a non-trivial symmetric α-stable $(0<\alpha<1)$ completely non-deterministic linear process. Here non-trivial means that X_t is not identically zero. We will study construction of canonical representation for this process.

<u>Definition 2.2.</u> We say that $\{X_t\}$ has a canonical representation if $\{X_t\}$ is represented in the form

$$(2.5) \qquad X_t = \int_{-\infty}^t f(t-u) M(du) ,$$

where M(du) is a $\{T_t\}$-homogeneous random measure taking values in $[X_t]$, satisfying

$$(2.6) \qquad [X_t;\ t\leq t_0] = [M(I);\ I\in \mathcal{R}_0,\ I\subset(-\infty, t_0]]$$

for any t_0, and f(u) is a measurable function such that $\int_0^\infty |f(u)|^\alpha du < \infty$. By $\mathcal{R}_0$ we denote the class of all bounded intervals of the form (c,d]

When we construct canonical representation, we have to take innovation M(du) out of $[X_t]$. For this purpose two integrals $\int_I T_t y\, dt$ and $\int_I e^{-t} X_t\, dt$ play an important role, where I is a bounded interval and y is an element of $[X_t]$. First we need to define integral in our Fréchet space. [3].

Let X be the Fréchet space $[X_t]$ defined above and let $\xi(t)$ be a function defined on a bounded closed interval $I = [a,b]$ and taking values in X.

Definition 2.3. Let γ, δ_0, K be positive numbers. We say that $\xi(t)$ satisfies Condition $C_\gamma(\delta_0, K)$ if

$$(2.7) \qquad \|\xi(t)-\xi(s)\| \leq K|t-s|^\gamma$$

for any $t,s \in I$ such that $|t-s| \leq \delta_0$.

Let $\{I_i;\ 1 \leq i \leq n\}$ be a partition of I such that $a = a_0 < a_1 < \dots < a_n = b$, $I_i = [a_{i-1}, a_i]$ and let $\{t_i;\ 1 \leq i \leq n\}$ be such that $t_i \in I_i$. The pair of $\{I_i\}$ and $\{t_i\}$ is denoted by $S = (\{I_i\}, \{t_i\})$ and length of I_i is denoted by $|I_i|$.

Definition 2.4. We say that $\xi(t)$ is Riemann type integrable over I if there is an element $\mathcal{J}$ in X with the following property: For each $\varepsilon > 0$, there is $\delta > 0$ such that

$$(2.8) \qquad \|\sum_{i=1}^n |I_i|\, \xi(t_i) - \mathcal{J}\| < \varepsilon$$

whenever $S = (\{I_i\},\{t_i\})$ satisfies $\max_{1\leq i\leq n} |I_i| < \delta$. We call $\mathcal{J}$ the Riemann type integral of $\xi(t)$ over I and write $\mathcal{J} = \int_I \xi(t)dt$.

Theorem 2.1. Suppose that g(t) is a continuous real function on I and y(t) is a function on I taking values in X and satisfying Condition $C_\gamma(\delta_0, K)$ for some δ_0, K and γ such that $1 \geq \gamma > 1-\alpha$. Suppose, further, that y(t) is of bounded variation in the sense of the quasi-norm, that is, there exists C such that, for any $t_1 < t_2 < \dots < t_n$, $t_i \in I$,

$$(2.9) \qquad \sum_{i=2}^n \|y(t_i) - y(t_{i-1})\| \leq C.$$

Let $\xi(t) = g(t)y(t)$. Then $\xi(t)$ is Riemann type integrable over I and we have

$$(2.10) \quad \|\int_I g(t)y(t)dt\| \leq (\int_I |g(t)|^\alpha dt)\ \{C + \|y(a)\|\}.$$

Given $\varepsilon_0 > 0$, choose an integer $M > 2|I|/\delta_0$ such that $|g(t)-g(s)| < \varepsilon_0$ whenever $|t-s| < |I|/M$. Then

$$(2.11) \quad \|\int_I g(t)y(t)dt\| \leq |I|^\alpha\{M^{1-\alpha}\sup\|g(t)y(t)\| + \varepsilon_0^\alpha\,(C+\|y(a)\|)\} + C_1\,|I|^{\alpha+\gamma}\,K\,\sup|g(t)|^\alpha M^{1-(\alpha+\gamma)}.$$

Here C_1 is a constant depending only on α and γ.

Theorem 2.2. Suppose that the process X(t) has a form $X(t)=g(t)Y(t)$, where g(t) is a real continuous function and Y(t) is a process with independent increments, that is, $Y(t_1)$, $Y(t_2)-Y(t_1)$, ..., $Y(t_n)-Y(t_{n-1})$ are independent for any $t_1 < t_2 < \dots < t_n$. Suppose that, for every bounded closed interval $[c, d]$, Y(t) satisfies Condition $C_\gamma(\delta_0, K)$ on $[c, d]$

with some δ_0, K, and γ satisfying $1 \geq \gamma > 1-\alpha$. Then X_t has a canonical representation.

If X_t is an Ornstein-Uhlenbeck type process, $X_t = \int_{-\infty}^{t} e^{-c(t-u)} M(du)$, then $X_t = g(t)Y(t)$ with $g(t) = e^{-ct}$ and $Y(t) = \int_{-\infty}^{t} e^{cu} M(du)$. These g(t) and Y(t) satisfy the conditions in Theorem 2.2. The following theorem shows that our conditions characterize Ornstein-Uhlenbeck type processes.

Theorem 2.3. If X_t satisfies the assumption in Theorem 2.2, then it is an Ornstein-Uhlenbeck type process.

3. Connection with the previous results. In [3] the author gives the following sufficient condition for Riemann type integrability. The value of $A_{\alpha\gamma}$ in [3] should be changed to the value given below.

Theorem 3.1. If $\xi(t)$ satisfies Condition $C_\gamma(\delta_0, K)$ for some δ_0, K and γ such that $1 \geq \gamma > 1-\alpha$, then $\xi(t)$ is Riemann type integrable over I and, for any integer $M > 2|I|/\delta_0$, we have

$$(3.1) \qquad \left\| \int_I \xi(t)dt \right\| \leq M^{1-\alpha}|I|^\alpha \sup_{t\in I} \|\xi(t)\| + M^{-\rho}|I|^{\alpha+\gamma} K\, A_{\alpha\gamma},$$

where $\rho = \alpha+\gamma-1 > 0$ and $A_{\alpha\gamma} = 2^{1+\gamma-2\alpha} 2^\rho/(2^\rho - 1) + 2^\gamma$.

Note that, even if y(t) is Riemann type integrable over I and g(t) is a real continuous function, we do not know whether the multiple g(t)y(t) is Riemann type integrable over I. A sufficient condition is as follows (see [4] Lemma 4.4):

If y(t) satisfies Condition $C_\gamma(\delta_0, K)$ for some δ_0, K and $1 \geq \gamma > 1-\alpha$ and g(t) satisfies

$$(3.2) \qquad |g(t)-g(s)|^\alpha \leq \text{const}\ |t-s|^\gamma \quad \text{whenever } |t-s| \leq \delta_0,$$

then g(t)y(t) satisfies Condition $C_\gamma(\delta_0, K')$ for some $K' > 0$, and hence is Riemann type integrable.

In case $g(t) = e^{-t}$, we have an inequality

$$(3.3) \qquad |e^{-t} - e^{-s}|^\alpha \leq e^{-t_1\alpha} |t-s|^\alpha \qquad \text{for } t,s \geq t_1.$$

Thus, if $\alpha \geq \gamma > 1-\alpha$, then $e^{-t}y(t)$ satisfies Condition $C_\gamma(\delta_0, K')$ for some $K' > 0$. The restriction $\alpha > 1/2$ in the following Theorem 3.2 comes from this circumstance. In our Theorem 2.1 we do not assume (3.2) on g(t), but we make the new restriction (2.9) on y(t). The case $g(t) = e^{-t}$ is now included for all $0 < \alpha < 1$.

For $\{X_t\}$ satisfying Condition $C_\gamma(\delta_0, K)$, let L be the class of elements x in $[X_t;\ t \leq 0]$ such that x is the limit in probability of a sequence $\{x_n\}$ where each x_n is a finite linear combination of $\{X_t;\ t \leq 0\}$

and $T_t x_n$ satisfies Condition $C_\gamma(\delta_0, K_n)$ with $\sup_n K_n < \infty$. Lemma 4.7 and Theorems 5.1 and 5.2 of [4] need a supplementary condition that $A_0 X_t \in L$ for every $t>0$. Whether this condition is indispensable or not is left for further study. Thus Theorem 5.2 of [4] should be as follows.

Theorem 3.2. If $\alpha > 1/2$ and if the process $\{X_t\}$ satisfies Condition $C_\gamma(\delta_0, K)$ for some δ_0, K, and $1 \geq \gamma > 1-\alpha$, and, further, $A_0 X_t \in L$ for every $t>0$, then $\{X_t\}$ has a canonical representation.

4. Proof of Theorems.

Proof of Theorem 2.1. Let $\mathcal{I}_S = \sum_{i=1}^n |I_i| g(t_i) y(t_i)$ for $S = (\{I_i\}, \{t_i\})$. We prove that, when $\max|I_i|$ tends to zero, $\mathcal{I}_S$ converges in X.

Step 1. First we prove boundedness of $\|\mathcal{I}_S\|$. Since

$$\begin{aligned}\mathcal{I}_S = &\{|I_1|g(t_1) + |I_2|g(t_2) + \dots + |I_n|g(t_n)\} y(t_1)\\ &+\{|I_2|g(t_2) + \dots + |I_n|g(t_n)\}\{y(t_2)-y(t_1)\} + \dots\\ &+ |I_n|g(t_n)\{y(t_n)-y(t_{n-1})\},\end{aligned}$$

we have

$$\begin{aligned}\|\mathcal{I}_S\| \leq &\left\||I_1|g(t_1) + \dots + |I_n|g(t_n)\right|^\alpha \|y(t_1)\|\\ &+\left\||I_2|g(t_2) + \dots + |I_n|g(t_n)\right|^\alpha \|y(t_2)-y(t_1)\| + \dots\\ &+\left\||I_n|g(t_n)\right|^\alpha \|y(t_n)-y(t_{n-1})\|\\ \leq &\{|I_1||g(t_1)| + \dots + |I_n||g(t_n)|\}^\alpha\{\|y(t_1)\| + \|y(t_2)-y(t_1)\| + \dots\\ &+ \|y(t_n)-y(t_{n-1})\|\}\\ \leq &\{|I_1||g(t_1)| + \dots + |I_n||g(t_n)|\}^\alpha\{C + \|y(a)\|\}.\end{aligned}$$

Thus, for given $\varepsilon > 0$, we have

$$(4.1)\qquad \|\mathcal{I}_S\| \leq \left(\int_I |g(t)|\,dt + \varepsilon\right)^\alpha\{C + \|y(a)\|\}$$

if $\max|I_i|$ is small enough.

Step 2. Let $S = (\{I_i; 1\leq i\leq n\}, \{t_i\})$ and $S' = (\{J_j; 1\leq j\leq m\}, \{s_j\})$. We claim that $\|\mathcal{I}_S - \mathcal{I}_{S'}\|$ tends to zero if $\max|I_i|$ and $\max|J_j|$ tend to zero. Without loss of generality, we give a proof when $\{J_j\}$ is a refinement of $\{I_i\}$.

We have

$$\mathcal{I}_S = \sum_{i=1}^n |I_i| g(t_i) y(t_i) \quad\text{and}\quad \mathcal{I}_{S'} = \sum_{j=1}^m |J_j| g(s_j) y(s_j).$$

Each I_i is divided into k_i subintervals belonging to $\{J_j\}$ as follows:

$I_i = J_1^i \cup J_2^i \cup \dots \cup J_{k_i}^i$, $k_1+k_2+\dots+k_n = m$, $s_\ell^i \in J_\ell^i$, $\{J_\ell^i\} = \{J_j\}$ and $\{s_\ell^i\} = \{s_j\}$.

Then

$$\mathcal{I}_{S'} - \mathcal{I}_S = \sum_{j=1}^m |J_j|\{g(s_j)y(s_j) - g(s_j')y(s_j')\}$$

where $s_j' = t_i$ if $J_j \subset I_i$. Hence

(4.2) $$\mathcal{J}_{S'}-\mathcal{J}_S = \sum_{j=1}^m |J_j|\{(g(s_j) - g(s_j'))y(s_j)\} + \sum_{j=1}^m |J_j| g(s_j')\{y(s_j) - y(s_j')\}.$$

We write the first term of the right hand side in (4.2) as $\mathcal{J}_1(S, S')$ and the second term as $\mathcal{J}_2(S, S')$.

Step 3. Set $g(s_j) - g(s_j') = f_j$. Then

$$\mathcal{J}_1(S, S') = \{|J_1| f_1 + |J_2| f_2 + \dots + |J_m| f_m\} y(s_1) + \{|J_2| f_2 + \dots + |J_m| f_m\} \cdot \{ y(s_2) - y(s_1)\} + \dots + |J_m| f_m \{y(s_m) - y(s_{m-1})\},$$ and hence

$$\|\mathcal{J}_1(S, S')\| \le \{|J_1||f_1| + \dots + |J_m||f_m|\}^\alpha \{C + \|y(a)\|\}$$

Suppose that $\varepsilon_0 > 0$ is given. If we take $\max|I_i|$ small enough, then $|f_i| < \varepsilon_0$ because of continuity of $g(t)$, and we have

(4.3) $$\|\mathcal{J}_1(S, S')\| \le \varepsilon_0^\alpha |I|^\alpha (C + \|y(a)\|).$$

Step 4. From the definition of $\mathcal{J}_2(S, S')$,

$$\mathcal{J}_2(S, S') = \sum_{i=1}^n g(t_i)\Big(\sum_{j=1}^{k_i} |J_j^i| y(s_j^i) - |I_i| y(t_i)\Big).$$

Since $y(t)$ satisfies Condition $C_\gamma(\delta_0, K)$, we can make the same calculation as in [3]. It will be done in Lemma 4.1. Take $|I_i|$ so small that $\max|I_i| < \delta_0/2$ and apply Lemma 4.1. Consider $y(t)$, I_i, and J_j^i, $1 \le j \le k_i$, as $z(t)$, I, and I_i, $1 \le i \le n$, in Lemma 4.1, respectively. Let $M = 1$ in Lemma 4.1. Then $\mathcal{J}_S^0$ in Lemma 4.1 turns out to be $|J_1^i| y(s_1^i) + (|I_i| - |J_1^i|) y(s_2^i)$. Thus, if we choose N large enough, then the estimate (4.7) shows that

(4.4) $$\Big\|\{ |J_1^i| y(s_1^i) + (|I_i| - |J_1^i|) y(s_2^i)\} - \sum_{j=1}^{k_i} |J_j^i| y(s_j^i)\Big\| \le 2^{1+\gamma-2\alpha} K |I_i|^{\alpha+\gamma} 2^\rho/(2^\rho - 1).$$

Since

$$\begin{aligned}&\big\| |I_i| y(t_i) - \{ |J_1^i| y(s_1^i) + (|I_i| - |J_1^i|) y(s_2^i)\} \big\| \\ &= \big\| |J_1^i| \{y(t_i) - y(s_1^i)\} + (|I_i| - |J_1^i|)\{y(t_i) - y(s_2^i)\} \big\| \\ &\le (|J_1^i|^\alpha + (|I_i| - |J_1^i|)^\alpha) K |I_i|^\gamma \le 2^{1-\alpha} K |I_i|^{\alpha+\gamma},\end{aligned}$$

we have, from (4.4), that

(4.5) $$\begin{aligned}\|\mathcal{J}_2(S, S')\| &\le C_1 \sum_{i=1}^n |g(t_i)|^\alpha K |I_i|^{\alpha+\gamma} \\ &\le C_1 \sup_{t\in I} |g(t)|^\alpha K \sum_{i=1}^n |I_i|^{\alpha+\gamma},\end{aligned}$$

where $C_1 = 2^{1+\gamma-2\alpha} 2^\rho/(2^\rho - 1) + 2^{1-\alpha}$. Since $\alpha + \gamma > 1$, we see that, when $\max|I_i|$ tends to zero, $\sum_{i=1}^n |I_i|^{\alpha+\gamma}$ tends to zero. Hence the assertion at the beginning of Step 2 is now proved.

Step 5. Now we proved that $\mathcal{J}_S$ converges in X when $\max|I_i|$ tends to zero. The estimate (2.10) is a consequence of (4.1). Let us prove(2.11). Take any integer M so big that $|I|/M<\delta_0/2$ and $|g(t)-g(s)|<\varepsilon_0$ for $|t-s|\leq|I|/M$. Let $S_0 = (\{I_i\}, \{t_i\})$ where $|I_i| = |I|/M$ and $S = (\{J_j\}, \{s_j\})$ where $\max|J_j|<|I|/M$ and $\{J_j\}$ is a refinement of $\{I_i\}$. Since $\|\mathcal{J}_S\|\leq\|\mathcal{J}_{S_0}\| + \|\mathcal{J}_S - \mathcal{J}_{S_0}\|$, and

(4.6) $$\|\mathcal{J}_{S_0}\| = \|\textstyle\sum_1^M(|I|/M)g(t_i)y(t_i)\|\leq M^{1-\alpha}|I|^\alpha\sup_{t\in I}\|g(t)y(t)\| ,$$

together with the inequalities (4.3) and (4.5) of Step 3 and Step 4, we have (2.11).

Lemma 4.1. Let z(t) be a function on $I=[a,b]$ taking values in X and satisfying Condition $C_\gamma(\delta_0, K)$ for some γ, δ_0, K. Suppose that we are given a pair $S = (\{I_i\},\{t_i\})$, $1\leq i\leq n$, and an integer M such that $\max_{1\leq i\leq n}|I_i|<|I|/M<\delta_0/2$. Make the following partitions.

i) For each nonnegative integer p, let $J_k^p = [c_{k-1}^p, c_k^p]$ where $c_k^p = a + k|I|/(2^pM)$, that is, $\{J_k^p\}$ is the partition of I into 2^pM subintervals of equal length.

ii) Let $\{I_j^p\}$, $1\leq j\leq p'$, be the superposition of $\{I_i\}$ and $\{J_k^p\}$. Numbering of $I_1^p,\ldots, I_{p'}^p$ is from left to right.

iii) Each J_k^p is the union of some intervals from $\{I_j^p\}$. Denote $J_k^p = I_{k'}^p\cup I_{k'+1}^p\cup\ldots\cup I_{k'+k''}^p$. In case $k''\geq 1$, let $F_{k1}^p = I_{k'}^p$ and $F_{k2}^p = I_{k'+1}^p\cup\ldots\cup I_{k'+k''}^p$. In case $k''= 0$, let $F_{k1}^p = J_k^p$ and $F_{k2}^p = \phi$.

Let $\mathcal{J}_S^p = \sum_{k=1}^{2^pM}\{|F_{k1}^p|z(s_{k1}^p) + |F_{k2}^p|z(s_{k2}^p)\}$, where s_{k1}^p and s_{k2}^p are chosen from $\{t_i\}$ as follows: $s_{k1}^p = t_i$ if $F_{k1}^p\subset I_i$, $s_{k2}^p = t_i$ if $I_{k'+1}^p\subset I_i$. Then we have, for any N,

(4.7) $$\|\mathcal{J}_S^0 - \mathcal{J}_S^N\|\leq 2^{1+\gamma-2\alpha}K|I|^{\alpha+\gamma}M^{-\rho}2^\rho/(2^\rho-1)$$

where $\rho=\alpha+\gamma-1$. If N is large enough, then $\mathcal{J}_S^N = \sum_{i=1}^n|I_i|z(t_i)$.

Proof. We calculate $\mathcal{J}_S^p - \mathcal{J}_S^{p+1} = \sum_{k=1}^{2^pM} I(k,p)$, where

$$I(k, p) = |F_{k1}^p|z(s_{k1}^p)+|F_{k2}^p|z(s_{k2}^p)-\{|F_{\ell1}^{p+1}|z(s_{\ell1}^{p+1})+|F_{\ell2}^{p+1}|z(s_{\ell2}^{p+1})$$
$$+|F_{\ell+1,1}^{p+1}|z(s_{\ell+1,1}^{p+1}) + |F_{\ell+1,2}^{p+1}|z(s_{\ell+1,2}^{p+1})\},\ \ell=2k-1.$$

Let $I_i= [\tau_{i-1},\tau_i]$ and call the points τ_i, $0\leq i\leq n$, (τ)-points. For $J_k^p=[c_{k-1}^p, c_k^p]$ let τ_k^p be the smallest (τ)-point satisfying $c_{k-1}^p\leq\tau<c_k^p$, if such a (τ)-point exists. One of the following three situations

occurs: (1) No (τ)-point exists in (c_{k-1}^p, c_k^p); (2) τ_k^p is in $(c_{k-1}^p, c_\ell^{p+1}]$; (3) τ_k^p is in (c_ℓ^{p+1}, c_k^p). In Situation (1), F_{k2}^p, $F_{\ell 2}^{p+1}$, and $F_{\ell+1,2}^{p+1}$ are empty and hence $I(k,p) = 0$. In Situation (2), $F_{k1}^p = F_{\ell 1}^{p+1}$ and we have

$$I(k,p) = |F_{\ell+1,1}^{p+1}|(z(s_{k2}^p)-z(s_{\ell+1,1}^{p+1}))+|F_{\ell+1,2}^{p+1}|(z(s_{k2}^p)-z(s_{\ell+1,2}^{p+1})).$$

In Situation (3), we have $I(k,p) = 0$, because $F_{k1}^p = J_\ell^{p+1} \cup F_{\ell+1,1}^{p+1}$, $F_{\ell 2}^{p+1} = \phi$, and $F_{k2}^p = F_{\ell+1,2}^{p+1}$. Let us estimate $|s_{k2}^p - s_{\ell+1,1}^{p+1}|$ and $|s_{k2}^p - s_{\ell+1,2}^{p+1}|$ in Situation (2). If there exists a (τ)-point in $(c_\ell^{p+1}, c_k^p]$, then $|s_{k2}^p - s_{\ell+1,1}^{p+1}| \leq |J_k^p|$. If there are two ($\tau$)-points in $(c_\ell^{p+1}, c_k^p]$, then, further, $|s_{k2}^p - s_{\ell+1,2}^{p+1}| \leq |J_k^p|$. If there is only one ($\tau$)-point in $(c_\ell^{p+1}, c_k^p]$ and it is not c_k^p, then we look at the smallest (τ)-point which is bigger than c_k^p. Suppose that it is in $J_{k+\nu+1}^p$. Then we have $|s_{k2}^p - s_{\ell+1,2}^{p+1}| \leq (\nu+2)|J_k^p|$ and hence

$$|s_{k2}^p - s_{\ell+1,2}^{p+1}|^\gamma \leq 2^\gamma(1+\nu/2)^\gamma |J_k^p|^\gamma < 2^\gamma(1+\nu)|J_k^p|^\gamma.$$

Situation (1) occurs for ν intermediate intervals $J_{k+1}^{p+1}, \ldots, J_{k+\nu}^{p+1}$. If there is no ($\tau$)-point in (c_ℓ^{p+1}, c_k^p) and c_k^p is a (τ)-point, then $F_{\ell+1,2}^{p+1} = \phi$. If there exists no ($\tau$)-point in $(c_\ell^{p+1}, c_k^p]$, then $F_{\ell+1,2}^{p+1} = \phi$ and we have a similar estimate $|s_{k2}^p - s_{\ell+1,1}^{p+1}| \leq (\nu+2)|J_k^p|$. With all consideration above, we have

$$\|\mathcal{J}_S^p - \mathcal{J}_S^{p+1}\| \leq \sum{}'\{|F_{\ell+1,1}^{p+1}|^\alpha K |s_{k2}^p - s_{\ell+1,1}^{p+1}|^\gamma + |F_{\ell+1,2}^{p+1}|^\alpha K |s_{k2}^p - s_{\ell+1,2}^p|^\gamma\}$$
$$\leq \sum{}' K\, 2^\gamma(1+\nu)|J_k^p|^\gamma\{|F_{\ell+1,1}^{p+1}|^\alpha + |F_{\ell+1,2}^{p+1}|^\alpha\},$$

where $\sum'$ is taken over all k for which Situation (2) occurs. Recalling the meaning of ν, we get

$$\|\mathcal{J}_S^p - \mathcal{J}_S^{p+1}\| \leq \sum_{k=1}^{2^p M} K\, 2^\gamma |J_k^p|^\gamma \{|F_{\ell+1,1}^{p+1}|^\alpha + |F_{\ell+1,2}^{p+1}|^\alpha\}.$$

Since $|J_k^p| = |I|/(2^p M)$ and $|F_{\ell+1,1}^{p+1}|^\alpha + |F_{\ell+1,2}^{p+1}|^\alpha \leq 2^{1-\alpha}|J_{\ell+1}^{p+1}|^\alpha = 2|I|^\alpha/(2^{p+2}M)^\alpha$, we have

$$\|\mathcal{J}_S^p - \mathcal{J}_S^{p+1}\| \leq 2^{1+\gamma-2\alpha} K\, M^{-\rho}|I|^{\alpha+\gamma}\, 2^{-\rho p}.$$

Therefore we get (4.7). As the last sentence in the Lemma is obvious, proof of Lemma 4.1 is complete.

Proof of Theorem 2.2. Since the process $Y(t)$ has independent increments, it is of bounded variation on any bounded interval in the sense of the quasi-norm. Using Theorem 2.1 we can show that $M(a,b] = A(a,b]\int_{a-1}^b T_t y_0 dt$ for a suitable $y_0 \in [X_t]$ is well defined and prove Theorem 2.2 in a way similar to the proof of Theorem 3.2. Note that both $X_t = g(t)Y(t)$ and

$e^{-t}X_t$ are now Riemann type integrable for any $0<\alpha<1$, and that the condition $A_0 X_t \in L$ for $t>0$ is satisfied since $A_0 X_t = (g(t)/g(0))X_0$.

Proof of Theorem 2.3. Since $X(t)$ is non-trivial strictly stationary, the function $g(t)$ does not vanish for any t. By Theorem 2.2, $X(t)$ is represented in a canonical form,

$$(4.8)\qquad X(t) = g(t)Y(t) = \int_{-\infty}^{t} f(t-u)M(du),$$

where $M(du)$ is a $\{T_t\}$-homogeneous symmetric α-stable random measure. We have

$$(4.9)\qquad Y(t) = \int_{-\infty}^{t} \frac{f(t-u)}{g(t)} M(du) .$$

Let $s' < t' \leq s < t$. We write

$$Y(t)-Y(s) = \int_{-\infty}^{s}\Big(\frac{f(t-u)}{g(t)} - \frac{f(s-u)}{g(s)}\Big)M(du) + \int_{s}^{t} \frac{f(t-u)}{g(t)} M(du),$$

$$Y(t')-Y(s') = \int_{-\infty}^{s'}\Big(\frac{f(t'-u)}{g(t')} - \frac{f(s'-u)}{g(s')}\Big)M(du) + \int_{s'}^{t'} \frac{f(t'-u)}{g(t')} M(du).$$

Independence of $Y(t)-Y(s)$ and $Y(t')-Y(s')$ is equivalent to the condition that

$$(4.10)\qquad \Big(\frac{f(t-u)}{g(t)} - \frac{f(s-u)}{g(s)}\Big)\Big(\frac{f(t'-u)}{g(t')} - \frac{f(s'-u)}{g(s')}\Big) = 0$$

for a.e.$u \in (-\infty, s']$ and

$$(4.11)\qquad \frac{f(t'-u)}{g(t')}\Big(\frac{f(t-u)}{g(t)} - \frac{f(s-u)}{g(s)}\Big) = 0$$

for a.e.$u \in (s', t']$ (see [5]). Since s' is any number less than t', (4.11) holds for a.e.$u \in (-\infty, t']$. Hence, if we fix $s<t$, then, for a.e.$u \in (-\infty, s]$, (4.11) holds for a.e.$t' \in [u,s]$.

Let $\xi_0 = \sup\{\xi\,;\, f(u) = 0 \text{ a.e. on } (0,\xi)\}$. If $\xi_0 > 0$, then $X(t) = \int_{-\infty}^{t-\xi_0} f(t-u)M(du)$ and $X(t) \in [M(I);\, I \in \mathcal{R}_0,\, I \in (-\infty, t-\xi_0]]$ $=[X(t);\, t \leq t-\xi_0]$, which is contradiction. Therefore $\xi_0 = 0$. It follows that, for a.e.$u<s$, there exists $t' \in [u,s]$ such that $f(t'-u) \neq 0$ and (4.11) holds. So we have, for any fixed $s<t$ and for a.e. $u<s$,

$$(4.12)\qquad \frac{f(t-u)}{g(t)} = \frac{f(s-u)}{g(s)} .$$

Hence, for any fixed t and for a.e.$u<t$, we have (4.12) for a.e.$s \in (u,t)$. By continuity of $g(s)$, this shows that $f(v)$ has a continuous version $\widetilde{f}(v)$ for $v>0$. It follows from (4.12) that

$$(4.13)\qquad \frac{\widetilde{f}(t-u)}{g(t)} = \frac{\widetilde{f}(s-u)}{g(s)}$$

for any t,s,u, satisfying $u<s<t$. Here letting s tend to u, we have

finite $\tilde{f}(0+) = \frac{\tilde{f}(t-u)}{g(t)} g(u)$. In case $t > 0$, we take $u = 0$ and we have

$$\tilde{f}(t) = \frac{\tilde{f}(0+)}{g(0)} g(t).$$

Then for all $u < s < t$, $\frac{g(t-u)}{g(t)} = \frac{g(s-u)}{g(s)}$. Again letting s tend to u, we get $g(t) = \frac{g(u)}{g(0)} g(t-u)$. Thus we have $g(t) = c_1 e^{-c_2 t}$ for some constants c_1, c_2. Hence $\tilde{f}(t) = c_3 e^{-c_2 t}$, $t > 0$, for some constant $c_3 \neq 0$. Since $f(u)$ is L^α-function, we know $c_2 > 0$. This shows that X_t is of Ornstein-Uhlenbeck type.

References

[1] T.Hida and N.Ikeda, Note on linear processes, J. Math. Kyoto Univ., 1(1961), 75-86.

[2] P.Lévy, Fonctions aléatoires à corrélation linéare, Illinois J. Math., 1(1957), 217-258.

[3] Y.Sato, On Riemann type integral of functions with values in a certain Fréchet space, Proc. Japan Acad. Ser. A, 61(1985),249-251.

[4] Y.Sato, A linear prediction problem for symmetricα-stable processes with $1/2<\alpha<1$, J. Math. Soc. Japan, Vol. 39, No. 1(1987),33-49.

[5] M.Schilder, Some structure theorems for the symmetric stable laws, Ann. Math. Statist.,41(1970), 412-421.

[6] K.Urbanik, Some prediction problems for strictly stationary processes, Proc. Fifth Berkeley Symp. on Math. Statist. and Probab. Vol.II, Part 1, Univ. California Press,1967, 235-258.

Department of General Education
Aichi Institute of Technology
Yakusa-cho, Toyota 470-03
Japan

ASYMPTOTICS OF THE MEAN OF A FUNCTIONAL OF A RANDOM WALK

S. H. Sirazdinov and M. U. Gafurov

Let $X_1,\dots,X_n$ be independent identically distributed random variables (r.v.), $S_0=0$, $S_n=X_1+\dots+X_n$, $n\geq 1$.

With any nonnegative, monotonically increasing function $H(x)$ and any nonnegative function $\varphi(x)$ we shall associate the functional

$$\nu(\varepsilon) = \sum_{n=1}^{\infty} \varphi(n) I(S_n \geq \varepsilon H(n)) \tag{1'}$$

and its two-sided analogue

$$\bar{\nu}(\varepsilon) = \sum_{n=1}^{\infty} \varphi(n) I(|S_n| \geq \varepsilon H(n)) \tag{1''}$$

where $\varepsilon>0$ is a parameter and $I(\cdot)$ is the indicator function.

There are many papers devoted to the investigation of the means of $\nu(\varepsilon)$ and $\bar{\nu}(\varepsilon)$ or in other words of the series

$$\sum_{n=1}^{\infty} \varphi(n) P(S_n \geq \varepsilon H(n)) \quad \text{and} \quad \sum_{n=1}^{\infty} \varphi(n) P(|S_n| \geq \varepsilon H(n))$$

respectively, in the case of identically distributed r.v.s X_i. From [1] and [2] it follows that for a wide class of functions φ and H,

$$E\nu(\varepsilon)<\infty \quad \forall\varepsilon>0 \Leftrightarrow E[H^{-1}(X_1^+)]^2 \varphi(H^{-1}(X_1^+)) < \infty .$$

Here $H^{-1}(x)$ is the inverse function of $H(x)$, $X^+=\max(0,X)$.

For the last years the problem of the finding asymptotic of $E\nu(\varepsilon)$ as well as of $E\bar{\nu}(\varepsilon)$ when $\varepsilon\downarrow 0$ attracts the attention of many authors. The meaning of this problem becomes clear if, for example, the case when $\varphi(x)\equiv 1$, $H(x)=x$ is considered.

It is obvious, that in this case $\nu(\varepsilon)$ $(\bar{\nu}(\varepsilon))$ is the number of exits of a random walk S_n from the one-sided (two-sided) boundary εn $(\pm\varepsilon n)$. It is not difficult to understand that if $\varepsilon\downarrow 0$, then $E\nu(\varepsilon)\uparrow (E\bar{\nu}(\varepsilon)\uparrow)$ and so the asymptotic finding problem $E\nu(\varepsilon)$ and $E\bar{\nu}(\varepsilon)$ when $\varepsilon\downarrow 0$ is of interest.

Let us note that this problem in the case of $\varphi(x)\equiv 1$, $H(x)=x$ under rather strict conditions on the distribution function (d.f.) of X_1 was considered by C. Wu [3], also by J. Slivka, N. S. Severo [4] in the case of a normally distributed summands with zero mean and unit variance.

For arbitrary sequence of r.v.'s with finite second moments the asymptotics of $E\bar{\nu}(\varepsilon)$ was studied by C. C. Heyde [5] and A. V. Nagaev [6] (in multidimensional case) C. C. Heyde proved that if

$$EX_1 = 0 \quad \text{and} \quad EX_1^2 = \sigma^2 < \infty \ , \text{ then } \quad \varepsilon^2 E\bar{\nu}(\varepsilon) \longrightarrow \sigma^2 \qquad (1.1)$$

as $\varepsilon\downarrow 0$. Further results corresponding to the case

$$\varphi(x) = x^{\ell-2} \qquad \ell \geq 1 \ , \qquad H(x) = x^{\alpha} \qquad \alpha > 1/2$$

have been obtained by the authors of this paper in [7] (see [9] as well). One-sided generalization of the results obtained by the above mentioned authors with some additional remarks you can find in paper [2].

Thus, the results of the papers [1]-[7] give complete asymptotic analysis of the values $E\nu(\varepsilon)$ and $E\bar{\nu}(\varepsilon)$ when $\varepsilon\downarrow 0$ in case of power functions.

Extension of the results described above to more general situations, in particular, to nonidentically distributed r.v., discrete argument random fields is of interest.

In this connection V. V. Petrov's work [8] should be mentioned where the analogue of the relation (1.1) for nonidentically distributed r.v. has been obtained.

Let $X_1,\dots,X_n$ be independent r.v.

$$EX_k = 0, \qquad EX_k^2 = \sigma_k^2, \qquad B_n = \sum_{k=1}^{n} \sigma_k^2.$$

Let us consider the functional $\bar{\nu}(\varepsilon)$, where $\varphi(x)\equiv 1$,

$$H(n) = \sqrt{B_n}\, n^{\alpha} \qquad \text{and} \qquad \alpha \in (0,1/2] \ .$$

Let G be a set of functions $g(x)$ defined for all real x and satisfying the following conditions:

a) $g(x)\geq 0$, $g(x)=g(-x)$, $g(x)$ is not decreasing for $x>0$ and $g(x)\neq 0$ when $x\neq 0$; b) the function $x/g(x)$ is not decreasing for $x>0$.

V. V. Petrov proved that if $EX_k=0$, $EX_k^2 g(X_k)<\infty$, $k=1,2,\dots$ for some function $g\in G$

$$\limsup_{n\to\infty} \frac{1}{B_n} \sum_{j=1}^{n} EX_j^2 g(X_j) < \infty$$

and

$$\sum_{n=1}^{\infty} \frac{1}{n^{2\alpha} g(B_n^{1/2})} < \infty \qquad \text{for some} \qquad \alpha \in (0,1/2]$$

then

$$\lim_{\varepsilon \downarrow 0} \varepsilon^{1/\alpha} E\bar{\nu}(\varepsilon) = \pi^{-1/2}\, 2^{1/2\alpha}\, \Gamma(1/2+1/2\alpha) . \qquad (1.2)$$

Let us note, that the above formulated result is a generalization and strengthening of a theorem by Chen [9], in which the relationship (1.2) was obtained under the conditions

$$EX_k = 0 , \qquad EX_k^2 = 1 \qquad \forall k \geq 1 ,$$

$$\limsup_{n \to \infty} \frac{1}{n} \sum_{j=1}^{n} EX_j^2 g(X_j) < \infty$$

$$\sum_{n=1}^{\infty} \frac{\log n}{n^{2\alpha} g(\sqrt{n})} < \infty$$

for some positive constant $\alpha \leq 1/2$ and some function $g \in G$.

It is clear that the relation (1.1) does not follow from V. V. Petrov's result. In this connection V. V. Petrov asked if it is possible to get a generalization of (1.2) so that it should contain the relation (1.1) as an immediate consequence. A similar question arises in connection with the recently published paper by Z. A. Logodowsky and Z. Rychlik [10], where V. V. Petrov's and R. Chen's theorems were extended for the discrete argument fields.

A significant point in the method of proof of the relation (1.2) is the application of the estimate of the rate of convergence in the central limit theorem for nonidentically distributed independent r.v., which gives us the possibility to study the behaviour of $E\bar{\nu}(\varepsilon)$ as $\varepsilon \to 0$.

In the present paper the investigation of the above mentioned authors are continued. In particular, the answer to V. V. Petrov's question is given.

We have obtained one-sided generalization of the relation (1.2). The method of proof is different from that of [8]-[10] and is based on S. V. Nagaev's and Sakoyan's probability inequality [11].

Results.

Let $X_1,\ldots,X_n$ be independent r.v. with $EX_k=0$,

$$EX_k^2 = \sigma_k^2, \quad k = 1,2,\dots, \quad B_n = \sum_{j=1}^{n} \sigma_j^2, \quad Z_n = B_n^{-1/2} \sum_{j=1}^{n} X_j.$$

In (1'), let $\varphi(n) \equiv 1$, $H(n) = \sqrt{B_n}\, n^{\alpha}$, $\alpha \in (0,1/2]$ so that $\nu(\varepsilon) = \sum_{n=1}^{\infty} I(Z_n \geq \varepsilon n^{\alpha})$. Denote $C_\alpha = \pi^{-1/2} 2^{1/2\alpha - 1} \Gamma(1/2 + 1/2\alpha)$, in particular $C_{1/2} = 1/2$.

<u>Theorem 2.1</u>. Let the following conditions be satisfied:

1) $\liminf_{n \to \infty} \dfrac{B_n}{n^{2\alpha}} > 0$; (2.1)

2) There exist constants $a>0$, $b>0$ and r.v. X, $EX=0$, $EX^2=\sigma^2<\infty$ such that for sufficiently large n and x,

$$\sum_{j=1}^{n} P\,(X_j \geq x) \leq an\, P\,(X \geq bx). \tag{2.2}$$

Then

$$\lim_{\varepsilon \downarrow 0} \varepsilon^{1/\alpha}\, E\nu(\varepsilon) = C_\alpha .$$

<u>Proof</u>. We restrict ourselves to the case $\alpha = 1/2$ as the proof in the case $\alpha \in (0,1/2)$ is analogous.

First of all let us notice that under conditions (2.1) and (2.2) the central limit theorem holds:

$$\lim_{n \to \infty} \sup_x |P(Z_n < x) - \Phi(x)| = 0 \tag{2.3}$$

where $\Phi(x)$ is the distribution function of normal r.v. with parameters (0,1).

It is obvious that

$$P(Z_n \geq \varepsilon\sqrt{n}) = P(Z_n \geq \varepsilon\sqrt{n}) - \Phi(-\varepsilon\sqrt{n}) + \Phi(-\varepsilon\sqrt{n})$$

$$\leq \Phi(-\varepsilon\sqrt{n}) + |P(Z_n \geq \varepsilon\sqrt{n}) - \Phi(-\varepsilon\sqrt{n})|.$$

Due to the Euler-Macloren's summation formula, it is not difficult to obtain that (see [7])

$$\lim_{\varepsilon \downarrow 0} \varepsilon^2 \sum_{n=1}^{\infty} \Phi(-\varepsilon\sqrt{n}) = 1/2 . \tag{2.4}$$

Hence, for the proof of Theorem 2.1 it is sufficient to show that when $\varepsilon \downarrow 0$

$$\varepsilon^2 \sum_{n=1}^{\infty} |P(Z_n \geq \varepsilon\sqrt{n}) - \Phi(-\varepsilon\sqrt{n})| \longrightarrow 0. \tag{2.5}$$

Let $\delta>0$ be an arbitrary number, which we choose later on. We have

$$\begin{aligned} &\sum_{n=1}^{\infty} |P(Z_n \geq \varepsilon\sqrt{n}) - \Phi(-\varepsilon\sqrt{n})| \\ &= \sum_{n\varepsilon^2 \leq \delta} + \sum_{\delta < n\varepsilon^2 \leq 1/\delta} + \sum_{n\varepsilon^2 > 1/\delta} = \Sigma_1^{(\varepsilon)} + \Sigma_2^{(\varepsilon)} + \Sigma_3^{(\varepsilon)}. \end{aligned} \tag{2.6}$$

It is obvious that

$$\varepsilon^2 \Sigma_1^{(\varepsilon)} \leq \delta . \tag{2.7}$$

Then, using the same methods which were applied in [7], by (2.3) we have

$$\lim_{\varepsilon \downarrow 0} \varepsilon^2 \Sigma_2^{(\varepsilon)} = 0 . \tag{2.8}$$

For the estimation of $\Sigma_3^{(\varepsilon)}$ we shall use a result of S. V. Nagaev and S. K. Sakoyan which we formulate below. For any $\Delta \geq 2$, let us define the class $G(\Delta)$ of all functions $f(u)$ on $[0,\infty)$ with a continuous nonincreasing derivative, satisfying the following conditions

1) $f'(u) \longrightarrow 0$ as $u \longrightarrow \infty$;

2) $f'(u) \geq \frac{\Delta}{u}$.

Let

$$B_{fn} = \sum_{j=1}^{n} \int_0^{\infty} e^{f(x)}\, dP(X_j < x)$$

Let γ, γ_1, γ_2, γ_3 and β be positive constants such that

$$\sum_{j=1}^{3} \gamma_j = 1 , \qquad \beta = 1 - \frac{\gamma_1}{2} + \frac{\gamma_2}{\Delta} , \qquad \gamma < 1 .$$

<u>Theorem</u> (S. V. Nagaev, S. K. Sakoyan). For any function $f \in G(\Delta)$ and $n \geq 1$ the inequality holds:

$$\begin{aligned} P(X_1+\cdots+X_n \geq x) \leq e^{\gamma_3/\gamma} &\left[\exp\left(-\frac{\gamma_1 \beta x^2}{2B_n}\right) + \exp\left\{-\frac{\beta x}{S^{-1}\left(\frac{\gamma_2 x}{eB_{fn}}\right)}\right\} \right. \\ &\left. + \left(\frac{\gamma}{\gamma_3}\right)^{\beta/\gamma} B_{fn}^{\beta/\gamma} \exp\left(-\frac{\beta}{\gamma} f(\gamma x)\right) \right] + \sum_{j=1}^{n} P(X_j \geq \gamma x) \end{aligned} \tag{2.9}$$

where $S^{-1}(u)$ is the inverse function to $S(u)=e^{-f(u)}f'(u)u^2$.

In our case $f(x)=2\log x$, $B_{fn}=\sum_{j=1}^{n}\int_0^\infty x^2 dP(X_j<x)$. We have

$$\Sigma_3^{(\varepsilon)} \le \sum_{n\varepsilon^2>1/\delta} P(Z_n \ge \varepsilon\sqrt{n}) + \sum_{n\varepsilon^2>1/\delta} \Phi(-\varepsilon\sqrt{n}) = \Sigma_{31}^{(\varepsilon)} + \Sigma_{32}^{(\varepsilon)} . \tag{2.10}$$

Let's denote by C a positive constant which may be not always the same. By (2.9) we have

$$\begin{aligned}\Sigma_{31}^{(\varepsilon)} &\le C \sum_{n\varepsilon^2>1/\delta} \exp\{-Cn\varepsilon^2\} + C \sum_{n\varepsilon^2>1/\delta} \exp\left(-\frac{C\sqrt{n}B_n\varepsilon}{S^{-1}\left[C\frac{\sqrt{n}B_n\varepsilon}{B_{fn}}\right]}\right) \\ &+ C \sum_{n\varepsilon^2>1/\delta} B_{fn}^{\beta/\gamma}\exp(-\frac{\beta}{\gamma}\log(nB_n\varepsilon)) + \sum_{n\varepsilon^2>1/\delta} \sum_{j=1}^{n} P(X_j>\gamma\sqrt{n}B_n) \\ &= A_1(\varepsilon) + A_2(\varepsilon) + A_3(\varepsilon) + A_4(\varepsilon) .\end{aligned} \tag{2.11}$$

It is clear that

$$\lim_{\varepsilon\downarrow 0} \varepsilon^2 A_1(\varepsilon) = 0 . \tag{2.12}$$

By the definition of function $S(x)$, we have

$$S(x) = 2/x$$

and hence

$$S^{-1}(x) = 2/x .$$

The considerations above imply that

$$\varepsilon^2 A_2(\varepsilon) \le C\varepsilon^2 \sum_{n\varepsilon^2>1/\delta} \exp(-Cn\varepsilon^2) = o(1) . \tag{2.13}$$

As γ is arbitrary, we choose $\gamma<\beta$. As $\delta>0$ is arbitrary, so

$$\begin{aligned}\varepsilon^2 A_3(\varepsilon) &\le C\varepsilon^2 \sum_{n\varepsilon^2>1/\delta} n^{-\beta/\gamma} = C\varepsilon^2 \sum_{n\varepsilon^2>1/\delta} \varepsilon^{2\beta/\gamma}(\varepsilon^2 n)^{-\beta/\gamma} \\ &= C\varepsilon^{2\beta/\gamma} \sum_{n\varepsilon^2>1/\delta} \frac{\varepsilon^2}{(n\varepsilon^2)^{\beta/\gamma}} \le C\varepsilon^{2\beta/\gamma}\int_{1/\gamma}^{\infty} x^{-\beta/\gamma}dx = o(1)\end{aligned} \tag{2.14}$$

as $\varepsilon \to 0$.

To estimate the last summand in (2.11), we use the conditions (2.1) and (2.2). Then

$$\varepsilon^2 A_4(\varepsilon) = o(1) . \qquad (2.15)$$

From (2.11)-(2.15), as $\varepsilon \downarrow 0$, we have

$$\varepsilon^2 \Sigma_{31}^{(\varepsilon)} = o(1) , \qquad (2.16)$$

and hence

$$\varepsilon^2 \Sigma_{32}^{(\varepsilon)} = o(1) . \qquad (2.17)$$

So, from (2.16), (2.17) we have

$$\varepsilon^2 \Sigma_{3}^{(\varepsilon)} = o(1) . \qquad (2.18)$$

The relations (2.7), (2.8) and (2.18) imply (2.5). Theorem 2.1 is proved.

Let us make some remarks about Theorem 2.1.

1. Theorem 2.1 can be strengthened if the existence of moments greater than 2 of the summands is supposed. Assume that

$$EX_k = 0, \qquad EX_k^2 = \sigma_k^2 < \infty, \qquad E(X_k^+)^\ell < \infty, \qquad \ell \geq 2$$

and the conditions (2.1), (2.2) with $\alpha=1/2$ are fulfilled. Then

$$\lim_{\varepsilon \downarrow 0} \varepsilon^{2(\ell-1)} E \sum_{n=1}^{\infty} n^{\ell-2} I(Z_n \geq \varepsilon\sqrt{n}) = 2^{\ell-2}\Gamma(\ell-\tfrac{1}{2})/\sqrt{\pi}(\ell-1) .$$

2. This method of the proof works in the multivariate case as well, i.e. when r.v. take values in the Euclidian space R^d. This method gives the possibility to extend Theorem 2.1 to the discrete argument random fields (i.e. to sequences of r.v. with multivarious indices).

3. The case when $\varphi(n)=\frac{1}{n}$, $n\geq 1$, in the definition (1') of $\nu(\varepsilon)$, i.e.

$$\nu(\varepsilon) = \sum_{n=1}^{\infty} \frac{1}{n} I(Z_n \geq \varepsilon\sqrt{n})$$

is of interest.

The average $\nu(\varepsilon)$ is an analogue of the well-known F. Spitzer's series for nonidentically distributed summands and in this case under the conditions of the Theorem 2.1 it is possible to prove that

$$\lim_{\varepsilon \downarrow 0} \frac{E\nu(\varepsilon)}{\log 1/\varepsilon} = 1 .$$

References

[1] Gafurov M. U., Slastnikov A. D. To the distribution of the last exit time, the excess and the number of exit of a random walk a curved boundary. Sov. Math. Dok. 1981, v.23, N2, pp.285-288.

[2] Gafurov M. U. Estimates of probability characteristics for the exit of random walk over curved boundary. Doctorate disser., Tashkent, 1981, 281p.

[3] Wu. C. F. A note on the convergence rate of the strong law of large numbrs. Bull. Inst. Math. Acad. Sinica, 1973, 1, 121-124.

[4] Slivka J., Severo N. C. On the strong law of large numbers. Proc. Amer. Math. Soc. 1970, 24, 4, 729-734.

[5] Heyde C. C. A supplement to the strong law of large numbers. J. Appl. Prob., 1975, 12, 173-175.

[6] Nagaev A. V. A note on large number law in R^d, $d\geq1$. Summary of the III Soviet-Japanese symposium of the theory of probability and mathematical statistics. Tacshkent, 1975.

[7] Gafurov M. U., Sirazdinov S. H. Some generalizations of Erdos-Katz results related to strong laws of large numbers and their applications. Kybernetica, 1979, v.15, N4, pp.272-292.

[8] Petrov V. V. One limited theorem for independent nonidentically distributed random variables. Study on the probability distributions theory, IV, Leningrad, "Nauka", 1979, pp.188-192.

[9] Chen R. A remark on the strong law of large nlumbers. Proc. Amer. Math. Soc. 1976, 61, 1, pp.112-116.

[10] Logodowski Z. A., Rychlik Z. Some remarks on the strong law for random fields. Bull. Acad. Polon. Sci. Ser. Math., 1984.

[11] Nagaev S. V., Sakoyan S. K. On the estimate for large deviation probabilities. Proceeding of the Limited Theories of Mathematical Statistics., Tashkent, "Fan", 1976, pp.132-140.

Tashkent, USSR

LONG TIME ASYMPTOTICS OF THE RATIO OF MEASURES OF SMALL TUBES AND A LARGE DEVIATION RESULT

Y. Takahashi

1. Introduction.

In [8] a heuristic argument is presented, using which one can actually reproduce the level three large deviation result of Donsker and Varadhan [2] for Markov processes from the long time asymptotics of the quantities

(1) $$\frac{1}{T} \log \frac{\mu(B_T(x,\delta))}{m(B_T(x,\delta))} \qquad \text{as} \quad T \to \infty$$

where m is the law of an original Markov process on the path space X, μ is the law of another Markov process and

(2) $$B_T(x,\delta) = \{y \in X;\ d(F^t y, F^t x) \leqq \delta \quad \text{for} \quad 0 \leqq t \leqq T\} .$$

Here the dynamical system F^t is taken as the time shift and d is the pseudo-metric on X induced from a metric on the state space.

In the present paper we shall restrict ourselves to the study of compact dynamical systems and clarify the meaning of quantities (1).

First we need some known results.

Theorem 1. (Local entropy theorem of Katok[3], Brin-Katok[1]) Let (X,d) be a compact metric space and $(F^t)_{t \geqq 0}$ a continuous or discrete semigroup of continuous maps of X to itself. Take an (F^t)-invariant probability measure μ and define the (δ,T)-tube $B_T(x,\delta)$ by (2). Then the following statements are true:

(i) $$\lim_{\delta \to 0} \limsup_{T\to\infty} -\frac{1}{T} \log \mu(B_T(x,\delta))$$
$$= \lim_{\delta\to 0} \liminf_{T\to\infty} -\frac{1}{T} \log \mu(B_T(x,\delta))$$
$$=: h_\mu(x) .$$

(ii) $$\int h_\mu(x)\mu(dx) = h_\mu(X,(F^t)) = h_\mu(X,F^1)$$

the metrical entropy of $(X,(F^t),\mu)$.

(iii) If, in addition, μ is ergodic, then,

$$h_\mu(x) = h_\mu(X,(F^t)) \quad \mu\text{-a.e.}x.$$

We shall give a proof of Theorem 1 in the next section because it is brief and new and its idea is also necessary for our results stated below. In the proof we use an interesting estimate, namely, the following uniform estimate related to the Shannon-McMillan theorem:

Proposition. Let α be a finite measurable partition of X with N atoms. Then for every positive ε

$$\limsup_{n\to\infty} \frac{1}{n} \log \mu(\alpha_n^\varepsilon(x)) \leqq -I_\mu(x,\alpha) + J(\varepsilon,N) \quad \mu\text{-a.e.}x \tag{3}$$

where

$$I_\mu(x,\alpha) = \lim_{n\to\infty} -\frac{1}{n} \log \mu(\alpha_n(x)) , \tag{4}$$

$$J(\varepsilon,N) = \max\{0, -\varepsilon \log \varepsilon - (1-\varepsilon)\log(1-\varepsilon) + \varepsilon \log(N-1)\} , \tag{5}$$

$\alpha_n(x)$ denotes the atom of the refinement of partitions α, $F^{-1}\alpha$, ..., $F^{-n+1}\alpha$ to which x belongs and $\alpha_n^\varepsilon(x)$ is the ε-neighborhood of $\alpha_n(x)$ (w.r.t. the Hamming distance) whose definition will be given in the next section by (35).

Remark 1. As a consequence of our proof one obtains the equality

$$h_\mu(x) = I_\mu^\star(x) := \sup\{I_\alpha(x);\ \alpha \text{ finite},\ \mu(\partial\alpha) = 0\} \quad \mu\text{-a.e.}x \tag{6}$$

which is obscure in the original proofs in [3] and [1] where $\partial\alpha$ is the union of boundaries of atoms of α, i.e.,

$$\partial\alpha = \bigcup\{\partial c;\ c\in\alpha\} \tag{7}$$

Remark 2. The quantity $J(\varepsilon,N)$ appearing in (3) is the growth rate of the numbers $L_n(\varepsilon,N)$ of the ε-neighborhoods w.r.t. the Hamming distance as n goes to infinity. By definition (cf. Section 2 (34))

$$\begin{aligned} L_n(\varepsilon,N) &= \sum_{k\leqq N\varepsilon} \binom{n}{k}(N-1)^k \\ &= N^k \sum_{k\leqq N\varepsilon} \binom{n}{k}\left(\frac{1}{N}\right)^{n-k}\left(1-\frac{1}{N}\right)^k \end{aligned} \tag{8}$$

and it is immediate to obtain (5) from a large deviation estimate for $(\frac{1}{N}, 1-\frac{1}{N})$-coin tossing since the right hand side of (5) is equal to

$$-(1-\varepsilon)\log\frac{1-\varepsilon}{1/N} - \varepsilon \log\frac{\varepsilon}{1-1/N} + \log N .$$

Next let us state the formulation of the large deviation in [6] which can be applied to quite general situation.

Let X be a compact Hausdorff space and (F^t) a semigroup of continuous maps of X to itself. As usual let us denote by $\underline{\underline{C}}(X)$ and $\underline{\underline{M}}(X)$ the space of all continuous functions on a compact space X endowed with the uniform topology and the space of all probability Borel measures endowed with the weak topology, respectively.

Take an $m \in \underline{\underline{M}}(X)$ and fix it from now on and put

$$Q(G) = \limsup_{T\to\infty} \frac{1}{T} \log m\{x \in X;\ \xi_T \in G\} \quad (G \subset \underline{\underline{M}}(X)) \tag{9}$$

and

$$q(\mu) = \inf\{Q(G);\ G \text{ a neighborhood of } \mu\} \tag{10}$$

where ξ_T denotes the empirical distribution

$$\xi_T = \frac{1}{T}\int_0^T \delta_{F^t x}\, dt \quad \text{or} \quad \frac{1}{T}\sum_{t=0}^{T-1} \delta_{F^t x} . \tag{11}$$

Similarly let us define $\underline{q}(\mu)$ through $\underline{Q}(G)$ by taking 'lim inf' in place of 'lim sup' in (9).

Note that $q(\mu) = \underline{q}(\mu) = -\infty$ unless μ is (F^t)-invariant.

One of the merits of our formulation lies in the following direct consequence of the definitions:

<u>Theorem 2</u>. ([6], Lemma 1 p.442)

(i) q and $\underline{q}$ are upper semicontinuous.

(ii) For a compact subset C of $\underline{\underline{M}}(X)$,

$$Q(C) \leqq \max\{q(\mu);\ \mu \in C\} . \tag{12}$$

(iii) For an open subset G of $\underline{\underline{M}}(X)$,

$$\underline{Q}(G) \geqq \sup\{\underline{q}(\mu);\ \mu \in G\} . \tag{13}$$

<u>Corollary</u>. The large deviation functional exists iff the functionals q and $\underline{q}$ coincide.

<u>Remark</u>. One can find a counter-example by simple dynamical systems on the plane for the existence problem of large deviation functionals in Section 4 of [6].

The following is immediate from the results of [6].

<u>Theorem 3</u>. (Variational principle of Gibbs type)
Let $X, (F^t)$ and m be as above. For $\Phi \in \underline{\underline{C}}(\underline{\underline{M}}(X))$ put

$$P(\Phi) = \limsup_{T\to\infty} \frac{1}{T} \log \int_X m(dx) \exp T\Phi(\xi_T) . \tag{14}$$

Then

(15) $$P(\Phi) = \max\{q(\mu) + \Phi(\mu);\ \mu \in \underline{\underline{M}}(X)\}$$

and

(16) $$-\infty \leqq \underline{q}(\mu) \leqq q(\mu) \leqq \inf\{P(\Phi_U) - \Phi_U(\mu);\ U \in \underline{\underline{C}}(X)\} \leqq 0 .$$

where

(17) $$\Phi_U(\mu) = \int_X U d\mu \qquad (U \in \underline{\underline{C}}(X)) .$$

Now we need the following:

Theorem 4. (Multiplicative ergodic theorem of Oseledec [4]). Let X be a compact Riemannian manifold and (F^t) a semigroup of diffeomorphisms of X. Denote by dF^t the map on the tangent bundle TX of X induced by F^t.

(i) There exists the limit

(18) $$\lim_{t\to\infty} \frac{1}{t} \log \|dF^t u\| =: \chi(x,u) , \quad u \in T_x X \qquad \text{u.a.e.} \quad x \in X$$

where u.a.e. stands for 'universally almost every' that means 'almost every' for every (F^t)-invariant probability Borel measure.

(ii) For u.a.e. $x \in X$ there exist a filtration of the tangent space $T_x X$ by vector subspaces

(19) $$L_0(x) = \{0\} \subsetneqq L_1(x) \subsetneqq \ \cdots\cdots \ \subsetneqq L_{s(x)}(x) = T_x M$$

and real numbers $\chi_i(x)$, $i = 1 , \ldots, s(x)$ such that

(20) $$\chi(x,u) = \chi_i(x) \quad \text{whenever} \quad u \in L_i(x) \cap L_{i-1}(x)^c \quad (1 \leqq i \leqq s(x))$$

(iii) If μ is an ergodic (F^t)-invariant measure, then $\chi_i(x)$'s and $s(x)$ are constant for μ-a.e.x.

It is not so difficult to see the following:

Corollary. Let m be the Riemannian volume measure on the compact Riemannian manifold X . Then

$$\lim_{\delta\to 0} \limsup_{T\to\infty} \frac{1}{T} \log m(B_T(x,\delta))$$

(21) $$= \lim_{\delta\to 0} \liminf_{T\to\infty} \frac{1}{T} \log m(B_T(x,\delta))$$

$$= -\chi^+(x) \qquad \text{u.a.e.}$$

where

(22) $$\chi^+(x) = \sum_i m_i(x) \max\{\chi_i(x), 0\}$$

with

(23) $$m_i(x) = \dim L_i(x) - \dim L_{i-1}(x) \qquad (1 \leqq i \leqq s(x)) .$$

Finally let us state our main result:

Theorem 5. Let X be a compact Riemannian manifold, and $(F^t)_{t \geqq 0}$ a semigroup of diffeomorphisms of X. Take the Riemannian volume measure m on X as the reference measure m in the definition of the functionals q and $\underline{q}$.

(i) Then for any (F^t)-invariant probability Borel measure μ

$$\lim_{\delta\to 0} \limsup_{T\to\infty} \frac{1}{T} \log \frac{\mu(B_T(x,\delta))}{m(B_T(x,\delta))}$$

(24) $$= \lim_{\delta\to 0} \liminf_{T\to\infty} \frac{1}{T} \log \frac{\mu(B_T(x,\delta))}{m(B_T(x,\delta))}$$

$$= \chi^+(x) - I^*_\mu(x) =: -f(x,\mu;m) \qquad \mu\text{-a.e.x.}$$

(ii) If, in addition, μ is ergodic, then

(25) $$f(x,\mu;m) = h_\mu(X,(F^t)) - \chi^+_\mu \qquad \mu\text{-a.e.x}$$

and

(26) $$\underline{q}(\mu) \geqq h_\mu(X,(F^t)) - \chi^+_\mu$$

where $\chi^+_\mu = \int \chi^+(x)\mu(dx)$ is the μ-a.e. constant value of $\chi^+(x)$.

Corollary (Pesin's inequality). For every (F^t)-invariant probability Borel measure μ on X,

(27) $$h_\mu(X,(F^t)) \leqq \chi^+_\mu .$$

2. Proof of Theorem 1.

We shall prove the equality (i) of Theorem 1 as the following two inequalities for μ-a.e.x.

(28) $$\bar{h}_\mu(x) := \lim_{\delta\to 0} \limsup_{T\to\infty} -\frac{1}{T} \log \mu(B_T(x,\delta)) \leqq I^*_\mu(x)$$

(29) $$\underline{h}_\mu(x) := \lim_{\delta\to 0} \liminf_{T\to\infty} -\frac{1}{T} \log \mu(B_T(x,\delta)) \geqq I^*_\mu(x) .$$

Here we shall give the proofs only for the discrete time case since in continuous time case one can use the approximation of $B_T(x,\delta)$ by the $B_n(x,\delta)$'s for discrete time case under the assumption that $F^t x$ is jointly continous in t and x.

The compactness assumption on X is necessary only for the proof of (28) which is the easier part of the proofs (cf.[1]). In fact, one can then find a finite Borel partition α for any given $\delta > 0$ such that

$$\text{mesh}(\alpha):=\max\{\text{diam}(c);\ c\in\alpha\}<\delta \tag{30}$$

$$\mu(\partial\alpha)=0\ . \tag{31}$$

Then,

$$B_n(x,\delta)\supset\alpha_n(x) \qquad (x\in X) \tag{32}$$

and so

$$-\frac{1}{n}\log\mu(B_n(x,\delta))\leqq-\frac{1}{n}\log\mu(\alpha_n(x))\ . \tag{33}$$

Hence (28) follows immediately by (31).

Let us proceed to the proof of (29). First of all we shall give the difinition of the set $\alpha_n^\varepsilon(x)$. Given a partition α put

$$D_H^n(x,y):=\frac{1}{n}\sum_{i=0}^{n-1}1(\alpha(F^i x)\neq\alpha(F^i y)) \qquad \text{(Hamming distance)} \tag{34}$$

$$\alpha_n^\varepsilon(x):=\{y\in X;\ D_H^n(x,y)\leqq\varepsilon\}\ . \tag{35}$$

Note that $\alpha_n^\varepsilon(x)$ is α_n-measurable and that if $\#\alpha=N$ then,

$$\#\{c\in\alpha_n;\ c\subset\alpha_n^\varepsilon(x)\}=L_n(\varepsilon,N):=\sum_{k\leqq N\varepsilon}\binom{n}{k}(N-1)^k \tag{36}$$

independently of $x\in X$. Hence, as is mentioned in Remark 2,

$$\begin{aligned} J(\varepsilon,N)&=\lim_{n\to\infty}\frac{1}{n}\log L_n(\varepsilon,N)\\ &=\max\{0,-\varepsilon\log\varepsilon-(1-\varepsilon)\log(1-\varepsilon)+\varepsilon\log(N-1)\}\ . \end{aligned} \tag{37}$$

Next we prove Proposition. Let $\#\alpha=N$ and $s>0$. Then by Chebyshev's inequality and (36)

$$\begin{aligned} P_n(\varepsilon,s)&:=\mu\{x\in X;\ \mu(\alpha_n^\varepsilon(x))/\mu(\alpha_n(x))\geqq 2^{ns}\}\\ &\leqq 2^{-ns}\int\mu(dx)\,[\mu(\alpha_n^\varepsilon(x))/\mu(\alpha_n(x))]\,1(\mu(\alpha_n(x))>0)\\ &=2^{-ns}\sum_{c\in\alpha_n,\mu(c)>0}\mu(c)[\mu(c^\varepsilon)/\mu(c)] \end{aligned}$$

$$\leq 2^{-ns} \sum_{c \in \alpha_n} \mu(c^\varepsilon) = 2^{-ns} L_n(\varepsilon, N) .$$

Hence it follows from (37) that

$$\sum_n p_n(\varepsilon, s) < \infty \quad \text{if} \quad s > J(\varepsilon, N) .$$

By the Borel-Cantelli lemma

$$\limsup_{n \to \infty} \frac{1}{n} \log [\mu(\alpha_n^\varepsilon(x))/\mu(\alpha_n(x))] \leq J(\varepsilon, N).$$

Lemma. For $\delta > 0$, denote

(38) $$\delta\text{-}\partial\alpha := \bigcup\{\delta\text{-}\partial c;\ c \in \alpha\}$$

where δ-• denotes the δ-neighborhood. Assume that a point $x \in X$ satisfies the inequality

(39) $$\frac{1}{n} \sum_{i=0}^{n-1} 1_{\delta\text{-}\partial\alpha}(F^i x) < \varepsilon .$$

Then,

(40) $$B_n(x, \delta) \subset \alpha_n^\varepsilon(x) .$$

Proof. Take $y \in B_n(x, \delta)$. Then,

$$1(\alpha(F^i y) \neq \alpha(F^i x)) \leq 1(\mathrm{dist}(F^i x, \partial\alpha) < \delta) = 1_{\delta\text{-}\partial\alpha}(F^i x) .$$

Hence,

$$D_H^n(y, x) \leq \frac{1}{n} \sum_{i=0}^{n-1} 1_{\delta\text{-}\partial\alpha}(F^i x) < \varepsilon .$$

Now we are going to complete the proof of the lower estimate (29).

Let μ be an (F^t)-invariant probability Borel measure and, following [3], take any finite measurable partition α such that

(41) $$\mu(\partial\alpha) = 0 .$$

Then for any given $\varepsilon > 0$ and $\eta > 0$ one can find $\delta > 0$ such that (39) holds for any sufficiently large n on a set of measure greater than $1-\eta$ because of Birkhoff's ergodic theorem, since (41) implies

$$\lim_{\delta \to 0} \mu(\delta\text{-}\partial\alpha) = \mu(\partial\alpha) = 0 .$$

On this set one obtains the estimate

$$\limsup_{n\to\infty} -\frac{1}{n}\log \mu(B_n(x,\delta))$$

$$\geqq \limsup_{n\to\infty} -\frac{1}{n}\log \mu(\alpha_n^{\varepsilon}(x)) \geqq I_\mu(x,\alpha) - J(\varepsilon,N)$$

by Lemma and Proposition. Since δ can be chosen arbitrarily small, the inequality

$$\underline{h}_\mu(x) \geqq I_\mu(x,\alpha) - J(\varepsilon,N)$$

holds on a set of measure greater than $1-\eta$ for any $\varepsilon,\eta > 0$. Consequently, one obtains the desired inequality

$$\underline{h}_\mu(x) \geqq I_\mu^*(x) \qquad \mu\text{-a.e.}x.$$

The rest of the proof of Theorem 1 follows from a standard argument using the Fatou theorem and an inequality

$$h_\mu(x) \geqq h_\mu(Fx)$$

which comes from the inclusion $B_n(x,\delta) \subset B_{n-1}(Fx,\delta)$.

3. Proof of Theorem 5.

The assertions (i), (ii) and (25) of (iii) follow from Theorem 1 and Corollary to Theorem 4. The rest, the assertion (26), is a lower large deviation estimate. We shall use the Chebyshev's inequality technique and Lemma in the previous section again.

Let μ be an ergodic invariant probability Borel measure and take an arbitrary neighborhood G of μ in the space $\underline{\underline{M}}(X)$. Then one can find a positive number ε , a finite measurable partition α and a family of real numbers $a(c)$, $b(c)$, $c \in \alpha$ with the following properties:

(41') $$m(\partial\alpha) = \mu(\partial\alpha) = 0$$

(42) $$G \supset G(s) := \{\nu\in \underline{\underline{M}}(x);\ a(c)-s<\nu(c)<b(c)+s\} \quad \text{if } 0 \leqq s \leqq \varepsilon .$$

In fact, if $\mu(\partial\alpha) = 0$, then the sets of the form $G(s)$ are neighborhoods of μ by the definition of the weak topology on $\underline{\underline{M}}(X)$. The condition $m(\partial\alpha) = 0$ will make it possible to apply Lemma.

Put

(43) $$A_n(s) := \{x\in X;\ \xi_n(x) \in G(s)\} , \quad \xi_n(x) = \frac{1}{n}\sum_{i=0}^{n-1} \delta_{F^i x} .$$

Note that $A_n(s)$, $n \geqq 0$, $0 \leqq s \leqq \varepsilon$, are α_n-measurable. Furthermore

(44) $$\alpha_n^\varepsilon(x) \subset A_n(\varepsilon) \quad \text{if} \quad x \in A_n(0) .$$

In fact, let $x \in A_n(0)$ and take $y \in \alpha_n^\varepsilon(x)$. Then, for $c \in \alpha$

$$\begin{aligned} \xi_n(y)(c) &= \frac{1}{n} \sum_{i=0}^{n-1} 1_c(F^i y) \\ &\leqq \frac{1}{n} \sum_{i=0}^{n-1} 1_c(F^i x) + \frac{1}{n} \sum_{i=0}^{n-1} 1(\alpha(F^i y) = \alpha(F^i x)) \quad < b(c) + \varepsilon \end{aligned}$$

and, similarly, $\xi_n(y)(c) > a(c) - \varepsilon$. Hence, $\xi_n(y) \in G(\varepsilon)$ and so $y \in A_n(\varepsilon)$.

Now put

$$\#\alpha = N , \quad A_n^* = \{x \in A_n(0);\ \mu(\alpha_n(x)) > 0\} .$$

Then, by (44)

(45) $$\begin{aligned} &m\{x \in X;\ \xi_n(x) \in G\} \\ &\geqq m(A_n(\varepsilon)) \geqq \sum_{c \in \alpha_n} 1(c \subset A_n(0)) L_n(\varepsilon,N)^{-1}\, m(c^\varepsilon) \\ &\geqq L_n(\varepsilon,N)^{-1} \int_{A_n^*} \mu(dx) m(\alpha_n^\varepsilon(\varepsilon))/\mu(\alpha_n(x)) . \end{aligned}$$

Finally take any $\eta > 0$ and let $\delta > 0$ be such that (39) holds except on a set of measure less than η . Let us denote by E_n the subset of A_n^* consisting of points x which satisfy (39) and the inequalities

(46) $$\frac{1}{n} \log m(B_n(x,\delta)) \geqq -\chi_\mu^+ - \eta$$

(47) $$-\frac{1}{n} \log \mu(B_n(x,\delta)) \geqq h - \eta ,$$

where h denotes the metrical entropy $h_\mu(X,F^1,\alpha)$ of F . Then it follows from (45) and (40) that

(48) $$\begin{aligned} &m\{x \in X;\ \xi_n(x) \in G\} \\ &\geqq L_n(\varepsilon,N)^{-1} \int_{E_n} \mu(dx) m(\alpha_n^\varepsilon(x))/\mu(\alpha_n(x)) \\ &\geqq L_n(\varepsilon,N)^{-1} \exp n(h-\chi_\mu^+-2\eta)\mu(E_n) . \end{aligned}$$

Now applying the Shannon-McMillan theorem and the Oseledec theorem

(Corolary to Theorem 4) one finds that

$$\liminf_{n\to\infty} \mu(E_n) > 0 \quad (\text{in fact, } \geqq 1-\eta) .$$

Therefore for $0 < \eta < 1$ it follows from (48) that

$$\underline{Q}(G) \geqq -J(\varepsilon,N) + h - \chi_\mu^+ - 2\eta .$$

Since ε and η are arbitrary, one obtains

$$\underline{Q}(G) \geqq h - \chi_\mu^+ , \quad h = h_\mu(X,F^1,\alpha)$$

for an arbitrary finite measurable partition α satisfying (41'). Consequently,

$$\underline{Q}(G) \geqq h_\mu(X,F^1) - \chi_\mu^+$$

and the proof is completed.

References

[1] M.Brin and A.Katok, On local entropy, Geometric Dynamics, Springer Lecture Notes in Math. 1007(1983), 30-38.

[2] M.D.Donsker and S.R.S.Varadhan, Asymptotic evaluation of certain Markov process expectations for large time I, IV, Comm. Pure Appl. Math. 28(1975), 1-45, 36(1983), 183-212.

[3] A.Katok, Lyapunov exponents, entropy and periodic points for diffeomorphisms, I.H.E.S. Publ. Math. 51(1980), 137-174.

[4] V.I.Oseledec, Multiplicative ergodic theorem. Lyapunov numbers for dynamical systems, Trudy Moskov. Obšč. 19(1968), 179-210.

[5] J.Pesin, Characteristic exponents and smooth ergodic theory, Uspehi Mat. Nauk 32(1977), 55-112.

[6] Y.Takahashi, Entropy functional (free energy) for dynamical systems and their random perturbations, Proc. Taniguchi Symp. on Stochastic Analysis, Katata and Kyoto 1982, Kinokuniya/North Holland (1984) , 437-467.

[7] Y.Takahashi, Observable chaos and variational formalism for one dimensional maps, Proc. 4th Japan-USSR Symp. on Probability Theory and Mathematical Statistics, Springer Lecture Notes in Math. 1021 (1983), 676-686.

[8] Y.Takahashi, Two aspects of large deviation theory for large time, to appear in Proc. Taniguchi Symp. on Probabilistic Methods in Mathematical Physics, Katata and Kyoto 1985.

Dept. Pure and Applied Sciences
College of Arts and Sciences
University of Tokyo
Komaba, Meguro, Tokyo 153
Japan

ON CORNISH-FISHER TYPE EXPANSION OF LIKELIHOOD RATIO STATISTIC IN ONE PARAMETER EXPONENTIAL FAMILY

Kei Takeuchi and Akimichi Takemura

1. Introduction and main result.

In our previous article (Takemura and Takeuchi[9]) we studied properties of Cornish-Fisher expansion based on normal distribution. Under general condition on asymptotic cumulants, we proved that the term of order $n^{-k/2}$ in the Cornish-Fisher expansion is a polynomial of degree k+1. Here we study Cornish-Fisher type expansion of a random variable converging in distribution to chi-square distribution. Hill and Davis[6] discuss a recursive algorithm for obtaining Cornish-Fisher type expansion based on chi-square distribution from the corresponding Edgeworth expansion of the distribution function. General algebraic structure of Cornish-Fisher type expansion based on chi-square distribution is considerably more complicated than the normal case. However the actual expression of Cornish-Fisher expansion is often remarkably simple because of cancellation of higher degree terms. In this article we restrict our attention to the aymptotic null distribution of log likelihood ratio statistic in continuous one parameter exponential family. We show that the Cornish-Fisher type expansion of twice the log likelihood ratio has a very special form. The same result seems to hold more generally for likelihood ratio statistic. Likelihood ratio statistic in more general family of distributions will be discussed in our subsequent papers.

Let the density of continuous random variable X be of one parameter exponential family type:

$$f(x,\theta) = h(x) \exp\{\theta t(x)-c(\theta)\} . \tag{1.1}$$

Let λ be the likelihood ratio statistic for testing null hypothesis

$$H: \theta = \theta_0 , \tag{1.2}$$

where θ_0 is an inner point of the natural parameter space Θ. Under the null hypothesis $2\log\lambda$ can be expanded into Cornish-Fisher type expansion

$$2 \log \lambda = Y \left(1 + \frac{1}{n} B_1(Y) + \frac{1}{n^2} B_2(Y) + \dots\right) , \tag{1.3}$$

where Y is distributed according to chi-square distribution with one degree of freedom, n is the sample size, and B_1, $B_2, \dots$, are polynomials in Y. Our main result is the following theorem:

Theorem 1.

(1.4) $\deg B_i = i - 1$

for all i.

Proof of this theorem is given in Section 2.

Before going into proof we discuss two points concerning this theorem.

First, note that the term of order 1/n in (1.3) is some constant c since $\deg B_1=0$. The null distribution of $2\log\lambda$ agrees with the distribution of $(1+c/n)Y$ except for terms of order n^{-2} or smaller. This particular feature of the log likelihood ratio statistic has been found more generally: when twice the log likelihood ratio is adjusted by a scale factor of the form $(1+c/n)$ with a suitable constant c, its distribution agrees with chi-square distribution up to order 1/n. The factor $(1+c/n)$ has been called Bartlett correction factor. See Box[3], Lawley[7], Hayakawa[5], Barndorff-Nielsen and Cox[1], McCullagh[8] for relevant results. Although our discussion here is restricted to the very regular situation of one parameter exponential family, our result covers terms of <u>all</u> orders of n, thus generalizing the earlier results on the term of order 1/n.

Second, we briefly discuss the validity of Cornish-Fisher expansion (1.3). Our theorem is of algebraic nature and will be proved by manipulating formal infinite series. However, the validity of the expansion can be established based on the validity of the corresponding Edgeworth expansion of $2\log\lambda$ given in Section 3 of Chandra and Ghosh[4]. In the course of our proof in Section 2 we point out how each step can be justified, although we omit the details.

2. Proofs and other results.

Let $X_1,\ldots,X_n$ be n independent continuous random variables having the density (1.1) and let

(2.1) $$T = T_n = \sum_{i=1}^{n} t(X_i)/n$$

be the sufficient statistic. Then $L=L_n=2\log\lambda$ is expressed as

(2.2) $$L = 2\log\{\Pi f(x_i,\hat\theta)/\Pi f(x_i,\theta_0)\}$$
$$= 2\,[n(\hat\theta-\theta_0)T - n(c(\hat\theta)-c(\theta_0))]\ ,$$

where $\hat\theta$ is the maximum likelihood estimate defined by

(2.3) $$T = c'(\hat\theta).$$

The j-th cumulant $\kappa_j(T_n)$ of T_n is given as

$$\kappa_j(T_n) = c^{(j)}(\theta_0) \;/\; n^{j-1} \;.$$

For simplicity of notation let the variance of T_n be normalized as

$$c^{(2)}(\theta_0) = 1,$$

which can be assumed without loss of generality by reparametrization if necessary. Write

$$\gamma_j = c^{(j)}(\theta_0).$$

The first step of our proof is concerned with expressing L as infinite series in the sufficient statistic T. Note that $\hat{\theta}=\hat{\theta}(T)$ is a function of T by (2.3). Therefore L is a function of T as well:

Lemma 1.

(2.4)
$$\frac{\partial L}{\partial T} = 2n\,(\,\hat{\theta} - \theta_0)\;.$$

Proof.

$$\frac{\partial L}{\partial T} = 2n(\hat{\theta}-\theta_0) + 2n\,\frac{\partial \hat{\theta}}{\partial T}(T-c'(\hat{\theta}))$$
$$= 2n(\hat{\theta}-\theta_0).$$

QED

Define

$$z = \sqrt{n}(T-\gamma_1), \quad v = \sqrt{n}(\hat{\theta}-\theta_0)\;.$$

Rewriting (2.4) we have

(2.5)
$$\frac{\partial L}{\partial z} = 2v\;.$$

Now the sufficient statistic T can be easily expressed by infinite series in $\hat{\theta}$ as

(2.6)
$$T - \gamma_1 = c'(\hat{\theta}) - c'(\theta_0)$$
$$= (\hat{\theta}-\theta_0) + \frac{\gamma_3}{2}(\hat{\theta}-\theta_0)^2 + \frac{\gamma_4}{3!}(\hat{\theta}-\theta_0)^3 + \ldots$$

or

(2.7)
$$z = v + \frac{\gamma_3}{2n^{1/2}}v^2 + \frac{\gamma_4}{3!n}v^3 + \ldots$$
$$= \sum_{r=1}^{\infty} \frac{\gamma_{r+1}}{n^{(r-1)/2}r!}\,v^r.$$

Expressing v in terms of z we have

$$(2.8)\qquad v = z - \frac{\gamma_3}{2n^{1/2}}z^2 + \frac{1}{6n}(3\gamma_3^{\,2}-\gamma_4)z^3 - \frac{1}{24n^{3/2}}(15\gamma_3^{\,3}-10\gamma_3\gamma_4+\gamma_5)z^4+\ldots$$

Actually an explicit expression of all terms of (2.8) can be given by Lagrange inversion formula:

$$(2.9)\qquad v = z + \sum_{\ell=1}^{\infty} \frac{(-1)^\ell}{\ell!}\left(\frac{d}{dz}\right)^{\ell-1}\left(\frac{\gamma_3}{2n^{1/2}}z^2 + \frac{\gamma_4}{6n}z^3 + \ldots\right)^\ell$$

$$= z + \sum_{\ell=1}^{\infty} \frac{(-1)^\ell}{\ell!}\left(\frac{d}{dz}\right)^{\ell-1}\left(\sum_{r=2}^{\infty}\frac{\gamma_{r+1}}{n^{(r-1)/2}r!}z^r\right)^\ell .$$

Expanding (2.9) we obtain the following lemma.

Lemma 2. Let $\alpha_i=\gamma_i/[(i-1)!n^{(i-2)/2}]$, $i=3,4,\ldots$. Then

$$(2.10)\qquad v = z + \sum_{\ell=1}^{\infty}\frac{(-1)^\ell}{\ell!}\sum_{i_1,\ldots,i_\ell\ge 3}\Big[\alpha_{i_1}\cdots\alpha_{i_\ell} \times \prod_{t=0}^{\ell-2}(i_1+\ldots+i_\ell-\ell-t)\, z^{i_1+\ldots+i_\ell-2\ell+1}\Big] .$$

Now let $\beta_i=\alpha_i/i$ and

$$(2.11)\qquad c(i_1,\ldots,i_\ell) = (-1)^{\ell-1} i_1\cdots i_\ell \prod_{t=0}^{\ell-3}(i_1+\ldots+i_\ell-\ell-t).$$

By Lemma 1 term by term integration of (2.10) yields the following expression of L in terms of z:

Lemma 3.

$$(2.12)\qquad L = z^2 - 2\sum_{\ell=1}^{\infty}\frac{1}{\ell!}\sum_{i_1,\ldots,i_\ell} c(i_1,\ldots,i_\ell)\beta_{i_1}\cdots\beta_{i_\ell} z^{i_1+\ldots+i_\ell-2(\ell-1)} .$$

At this point we consider justification of above steps. First, since $t(X)$ has moment generating function, we can find c_n such that $c_n^{\,m}/\sqrt{n}\to 0$ for any m and

$$P(|T-\gamma_1| > c_n/\sqrt{n}\,) = o(n^{-k}) \qquad \text{for any } k.$$

Now because the relation between $\hat\theta$ and T is monotone and smooth by (2.3) we also have

$$P(|\hat\theta-\theta_0| > c_n/\sqrt{n}\,) = o(n^{-k}) \qquad \text{for any } k.$$

Therefore ignoring a subset of the sample space of probability $o(n^{-k})$ for any k, we have $|z|<c_n$ and $|v|<c_n$. Since (2.7) and (2.10) are power series expressions of analytic functions c' and c'^{-1}, term by term integration is justified in deriving Lemma 3.

The second step of our proof is concerned with obtaining the asymptotic expansion of the characteristic function of L. Let $f_n(z)$ denote the density function of z. Then

$$(2.13)\qquad E\{\exp(itL)\} = E\{\exp(itL(z))\} = \int \exp\{it(z^2-\gamma_3 z^3/3\sqrt{n} + \dots)\}\, f_n(z)\, dz\ .$$

Since z is a normalized sum of i.i.d. random variables, $f_n(z)$ can be expanded into Edgeworth series. Actually we consider the Edgeworth expansion of $\log f_n(z)$. From Theorem 2.3 of Takemura and Takeuchi[9] we have the following lemma.

Lemma 4.

$$(2.14)\qquad \log f_n(z) = \log \phi(z) + \sum_{\ell=1}^{\infty} \frac{1}{\ell!} \sum_{i_1,\dots,i_\ell} c(i_1,\dots,i_\ell) \times \beta_{i_1}\cdots\beta_{i_\ell} P_{i_1\dots i_\ell}(z)\ ,$$

where $\phi(z)$ is the standard normal density and

$$(2.15)\qquad P_{i_1\dots i_\ell}(z) = c(i_1,\dots,i_\ell)\, z^{i_1+\dots+i_\ell-2(\ell-1)} + \text{lower degree terms}$$

is a polynomial in z of degree $i_1+\dots+i_\ell-2(\ell-1)$ with the leading coefficient $c(i_1,\dots,i_\ell)$ given by (2.11).

Therefore

$$(2.16)\qquad E\{\exp(itL)\} = \int \frac{1}{\sqrt{2\pi}} \exp[-\frac{1-2it}{2} z^2 + \sum_{\ell=1}^{\infty} \frac{1}{\ell!}\beta_{i_1}\cdots\beta_{i_\ell} c(i_1,\dots,i_\ell) \times \{(1-2it)z^{i_1+\dots+i_\ell-2(\ell-1)} + Q_{i_1\dots i_\ell}(z)\}]\ ,$$

where $Q_{i_1\dots i_\ell}(z)$ is a polynomial in z of degree $i_1+\dots+i_\ell-2\ell$. This follows from the fact that $P_{i_1\dots i_\ell}(z)$ contains either even degree terms only or odd degree terms only. Now

$$(2.17)\qquad \beta_{i_1}\cdots\beta_{i_\ell} = O(n^{(i_1+\dots+i_\ell-2\ell)/2})\ .$$

Therefore (2.16) can be written as

$$(2.18)\qquad E\{\exp(itL)\} = \int \frac{1}{\sqrt{2\pi}} \exp\{-\frac{1}{2s} z^2 + \sum_{k=1}^{\infty} n^{-k/2}(b_k z^{k+2}/s + R_k(z))\}\, dz ,$$

where $s=1/(1-2it)$, b_k is some constant, and $R_k(z)$ is a polynomial of degree k in z.

Now to evaluate the integral we expand

$$(2.19)\qquad \exp\{\sum_{k=1}^{\infty} n^{-k/2}[b_k z^{k+2}/s + R_k(z)]\} = 1 + \sum_{k=1}^{\infty} n^{-k/2} W_k(z) .$$

Then we see that $W_k(z)$ is a polynomial in z, whose general term is of the following form:

$$(2.20)\qquad (z^2/s)^i z^j \qquad \text{with} \quad j \le k .$$

Now

$$(2.21)\qquad \int \frac{1}{\sqrt{2\pi}} z^{\ell} \exp\{\frac{1}{2s} z^2\}\, dz$$

$$= 0 \qquad\qquad \text{if } \ell \text{ is odd}$$

$$= s^{(\ell+1)/2} 1\cdot 3\cdot 5\cdots(\ell-1) \quad \text{if } \ell \text{ is even.}$$

Noting that W_k contains either odd degree terms only or even degree terms only, term by term integration of (2.18) then leads to the following asymptotic expansion of the characteristic function of L.

Lemma 5.

$$(2.22)\qquad \phi(t) = E\{\exp(itL)\} = s^{1/2}(1 + \frac{1}{n} h_1(s) + \frac{1}{n^2} h_2(s) + \dots) ,$$

where $s=1/(1-2it)$ and $h_i(s)$ is a polynomial of degree i in s.

Again we take a brief look at validity of the above steps. Although (2.14) is an infinite series in z, it is an asymptotic expansion and actually should be considered to be finite series terminating at the term of order $n^{-M/2}$. M is fixed but arbitrary. From the validity of the Edgeworth expansion in the variational sense (Batthacharya and Ghosh[2]) we can replace f_n by its Edgeworth expansion in evaluating (2.13). Then we evaluate the integral in the range $|z|<c_n$. In this range we can expand the exponential (2.19) and integrate term by term. Integration of each term in the range $|z|<c_n$ then can be replaced by integration over the whole real line. In each of these steps, the error in the approximation is $o(n^{-M/2})$. Hence (2.22) is valid as an asymptotic expansion.

It might be of some interest to write down some h_i's. h_1, h_2, h_3 are given as follows. Note that $h_i(s)$ has a factor $(s-1)$ because $\phi(0)=1$ implies $h_i(1)=0$.

$$(2.23)\qquad h_1(s)/(s-1) = \frac{5}{24}\gamma_3^{\,2} - \frac{1}{8}\gamma_4 ,$$

$$h_2(s)/(s-1) = \frac{1465s-385}{1152}\gamma_3^{\,4} - \frac{341s-105}{192}\gamma_3^{\,2}\gamma_4 + \frac{18s-7}{48}\gamma_3\gamma_5$$
$$+ \frac{99s-35}{384}\gamma_4^{\,2} - \frac{2s-1}{48}\gamma_6\ ,$$

$$h_3(s)/(s-1) = \frac{1234205s^2-585970s+85085}{82944}\gamma_3^{\,6} - \frac{296065s^2-153930s+25025}{9216}\gamma_3^{\,4}\gamma_4$$
$$+ \frac{8790s^2-5285s+1001}{1152}\gamma_3^{\,3}\gamma_5 + \frac{145935s^2-82390s+15015}{9216}\gamma_3^{\,2}\gamma_4^{\,2}$$
$$- \frac{1440s^2-1025s+231}{1152}\gamma_3^{\,2}\gamma_6 - \frac{1682s^2-1085s+231}{384}\gamma_3\gamma_4\gamma_5$$
$$+ \frac{13s^2-11s+3}{96}\gamma_3\gamma_7 - \frac{3105s^2-1890s+385}{3072}\gamma_4^{\,3}$$
$$+ \frac{110s^2-83s+21}{384}\gamma_4\gamma_6 + \frac{116s^2-84s+21}{640}\gamma_5^{\,2}$$
$$- \frac{3s^2-3s+1}{384}\gamma_8\ .$$

The final step of our proof is to investigate the properties of polynomials B in the Cornish-Fisher type expansion by equating the characteristic function. Let $L=L_n$ be expanded into Cornish-Fisher type expansion

$$L_n = Y + \frac{1}{n}\tilde{B}_1(Y) + \frac{1}{n^2}\tilde{B}_2(Y) + \dots\ . \qquad (2.24)$$

Note that in (1.3) there is no constant term. At this point we do not know this yet and we allow constant term in $\tilde{B}_i$. In the course of our proof we show that in fact $\tilde{B}_i$ does not have constant term and hence can be written as $YB_i(Y)$. Now using the Cornish-Fisher type expansion the characteristic function of L_n can be expanded as

$$E\{\exp(itL_n)\} = E[\exp\{it(Y + \frac{1}{n}\tilde{B}_1(Y) + \frac{1}{n^2}\tilde{B}_2(Y) + \dots)\}] \qquad (2.25)$$
$$= E[\exp(itY)\{1 + \frac{it}{n}\tilde{B}_1(Y) + \frac{1}{n^2}(it\,\tilde{B}_2(Y) + \frac{(it)^2}{2}\tilde{B}_1^{\,2}(Y)) + \dots\}]\ .$$

We evaluate this and equate it with (2.22) Then we obtain the following lemma, which completes our proof of Theorem 1.

Lemma 6. For all i, $\tilde{B}_i(Y)$ is a polynomial in Y of degree i without a constant term.

Proof. Note that
$$E(e^{itY}Y^k) = \frac{\Gamma(1/2+k)}{\Gamma(1/2)}\,2^k\,s^{k+1/2}\ .$$
Equating the term of order 1/n we have

$$\frac{s-1}{2s}E(\exp(itY)\tilde{B}_1(Y)) = h_1(s)s^{1/2},$$

where $it=(s-1)/2s$. Now $h_1(s)$ is a polynomial of degree 1 in s. This implies that $\tilde{B}_1(Y)$ is a polynomial of degree 1 in Y without a constant term.

Now we argue by induction. For induction assume that each $\tilde{B}_j(Y)$, $j\leq k$, is a polynomial of degree j without constant term. Then by considering the term of order $n^{-(k+1)}$ it is easy to see that

$$\frac{s-1}{2s} E(\exp(itY)\tilde{B}_{k+1}(Y)) = (h_{k+1}(s) - \tilde{h}_{k+1}(s))s^{1/2} ,$$

where $\tilde{h}_{k+1}(s)$ is again a polynomial in s of degree k+1 arising from expectation of products of $\tilde{B}_j(Y)$, $j\leq k$. This implies that $\tilde{B}_{k+1}(Y)$ is a polynomial of degree k+1 without a constant term. QED

From the viewpoint of validity it remains to show that the asymptotic expansion of the characteristic function (2.22) agrees with the expansion of the characteristic function of a random variable defined by the Cornish-Fisher type expansion. Let a random variable $\tilde{L}_{n,M}$ be defined by

$$\tilde{L}_{n,M} = Y\,(1 + \frac{1}{n} B_1(Y) + \dots + \frac{1}{n^M} B_M(Y)). \tag{2.26}$$

Let g_n denote the density function of L_n and $\tilde{g}_{n,M}$ denote the density function of $\tilde{L}_{n,M}$. Then from the validity of the Edgeworth expansion (Section 3 of Chandra and Ghosh[5]) in the variational sense we can justify Cornish-Fisher expansion in the variational norm as

$$\int |g_n(x) - \tilde{g}_{n,M}(x)|\, dx = o(n^{-M}) . \tag{2.27}$$

(2.27) guarantees that the characteristic function of $\tilde{L}_{n,M}$ agrees with that of L_n up to order n^{-M}. Proving (2.27) from the corresponding validity of the Edgeworth expansion is essentially the same as jutification of Cornish-Fisher expansion based on normal distribution. For the normal case we gave a full proof in Section 5 of Takemura and Takeuchi[9]. Since the proof is similar we omit it for the present case.

Finally we give explicit expression of some B_i's. B_1, B_2, B_3 are as follows.

$$B_1(Y) = \frac{5}{12}\gamma_3^2 - \frac{1}{4}\gamma_4 , \tag{2.28}$$

$$B_2(Y) = \frac{695Y-465}{864}\gamma_3^4 - \frac{163Y-135}{144}\gamma_3^2\gamma_4 + \frac{6Y-7}{24}\gamma_3\gamma_5 + \frac{15Y-13}{96}\gamma_4^2 - \frac{2Y-3}{72}\gamma_6 ,$$

$$B_3(Y) = \frac{56404Y^2-116005Y+51945}{31104}\gamma_3^6 - \frac{13622Y^2-31575Y+15765}{3456}\gamma_3^4\gamma_4 + \frac{834Y^2-2365Y+1344}{864}\gamma_3^3\gamma_5 + \frac{6684Y^2-16975Y+9585}{3456}\gamma_3^2\gamma_4^2 - \frac{139Y^2-480Y+324}{864}\gamma_3^2\gamma_6 - \frac{796Y^2-2435Y+1575}{1440}\gamma_3\gamma_4\gamma_5$$

$$+ \frac{13Y^2 - 55Y + 45}{720}\gamma_3\gamma_7 - \frac{44Y^2 - 111Y + 73}{384}\gamma_4^{\,3}$$

$$+ \frac{10Y^2 - 35Y + 27}{288}\gamma_4\gamma_6 + \frac{116Y^2 - 420Y + 315}{4800}\gamma_5^{\,2}$$

$$- \frac{Y^2 - 5Y + 5}{960}\gamma_8 \quad .$$

As an example let $\tau X_1,\ldots,\tau X_n$ be independent random variables having chi-square distribution with one degree of freedom and consider the likelihood ratio test for testing the null hypothesis $H:\tau=\tau_0$. The standardized cumulants γ_k's are give as

$$\gamma_k = 2^{k/2-1}(k-1)! \ . \tag{2.29}$$

Substituting this into (2.28) we obtain the following Cornish-Fisher type expansion (terms up to n^{-5} are shown):

$$L = Y\ [\ 1 + \frac{1}{3n} - \frac{1}{n^2}(\ \frac{Y}{54} - \frac{1}{18}\ ) - \frac{1}{n^3}(\ \frac{8Y^2}{6075} + \frac{61Y}{2430} + \frac{31}{810}\ ) \tag{2.30}$$

$$+ \frac{1}{n^4}(\ \frac{131Y^3}{510300} + \frac{53Y^2}{72900} - \frac{229Y}{29160} - \frac{139}{9720}\ )$$

$$+ \frac{1}{n^5}(\ \frac{344Y^4}{24111675} + \frac{21803Y^3}{32148900} + \frac{21229Y^2}{4592700} + \frac{10529Y}{612360} + \frac{9871}{204120}\ ) + o(n^{-5})$$

These as well as (2.23) were obtained using REDUCE language.

REFERENCES

[1] O.E. Barndorff-Nielsen and D.R. Cox: Bartlett adjustments to the likelihood ratio statistic and the distribution of the maximum likelihood estimator, J. R. Statist. Soc. B 46 (1984), 483-495.

[2] R.N. Bhattacharya and J.K. Ghosh: On the validity of the formal Edgeworth expansion, Ann. Statist 6 (1978), 431-451.

[3] G.E.P. Box: A general distribution theory for a class of likelihood criteria, Biometrika 36 (1949), 317-346.

[4] T.K. Chandra and J.K. Ghosh: Valid asymptotic expansions for the likelihood ratio statistic and other perturbed chi-square variables, Sankhya A 41 (1979), 22-47.

[5] T. Hayakawa: The likelihood ratio criterion and the asymptotic expansion of its distribution, Ann. Inst. Statist. Math. 29 (1977), 359-378.

[6] G.W. Hill and A.W. Davis: Generalized asymptotic expansion of Cornish-Fisher type, Ann. Math. Statist. 39 (1968), 1264-1273.

[7] D.N. Lawley: A general method for approximating to the distribution of the likelihood ratio criteria, Biometrika 43 (1956), 295-303.

[8] P. McCullagh: Tensor notation and cumulants of polynomials, Biometrika 71 (1984), 461-476.

[9] A. Takemura and K. Takeuchi: Some results on univariate and multivariate Cornish-Fisher expansion: algebraic properties and validity, Discussion paper 85-F-18, Faculty of Economics, Univ. of Tokyo (1986).

Faculty of Economics
University of Tokyo
Bunkyo, Tokyo
113 Japan

STOCHASTIC PROCESS FOR AN INFINITE HARD CORE PARTICLE SYSTEM IN R^d

Hideki Tanemura

§ 0. Introduction

In this paper, we consider a system of infinitely many hard balls with the same diameter r moving discontinuously in R^d. The system is described as follows: A ball at x[1] waits an exponential holding time with mean one which is independent of the motion of the other balls and then jumps to the position y where y is distributed according to $p_x(dy) = p(|x - y|)dy$[2] independently of the holding time and the motion of the other balls; however, if there are some other balls in $B_r(y)$[3] at the jump time then the jump is suppressed. Thus if $x_i(t)$ denotes the center of the i-th ball then $|x_i(t) - x_j(t)| \geq r$, $t \geq 0$, $i \neq j$ (hard core condition). We can construct the Markov process $\xi_t = \{x_i(t), i = 1,2,\cdots\}$ describing our model.

Let μ be a grand canonical Gibbs state associated with hard core potential Φ (see the definition of § 1). Then, μ is a stationary measure for ξ_t. The main result of this paper is to show ergodicity of the stationary Markov process ξ_t in the case that density of balls is sufficiently small. In § 1, we construct the Markov process ξ_t by using Liggett's theorem [5] and show that grand canonical Gibbs states are reversible measure for the process. In § 2, we show ergodicity of the process. In § 3, we shall study a tagged particle of our process. Kipnis-Varadhan [3] proved the central limit theorem of a tagged particle of simple exclusion process. Using the same argument as [3], we shall discuss the central limit theorem of a tagged particle of our process but the non-degeneracy of the covariance matrix is an open problem.

§ 1. Construction of a Markov process

To give a precise description of the model we introduce the space of configurations

$$\mathfrak{X} = \{ \xi = \{x_i\} : |x_i - x_j| \geq r, \quad i \neq j \},$$

where $x_i \in \xi$ denotes the position of a hard ball and $r > 0$ denotes the diameter of a hard ball. Let $\mathfrak{M}$ be the set of all countable locally finite subsets of R^d. We regard $\xi \in \mathfrak{M}$ as a non-negative integer valued Radon measure on R^d: $\xi(\cdot) = \Sigma\, \delta_{x_i}(\cdot)$, where $\delta_*(\cdot)$ denotes

1) A ball is said to be at x if the center is x.
2) $p(\cdot)$ is a given non negative function on $[0,\infty)$ such that $\int_{R^d} p(|x|)dx = 1$.
3) $B_r(y)$ is the r-neighborhood of y.

the Dirac measure corresponding to the point $*$. $\mathfrak{X}$ is a compact subset of $\mathfrak{M}$ with the vague topology.

For any $\xi \in \mathfrak{M}$ and $y \in R^d$ we denote $\xi \cup \{y\}$ by $\xi\cdot y$. Also we denote $\xi\setminus\{z\}$ by $\xi\setminus z$ for $z \in \xi$. ξ_K stands for the restricted configuration of ξ on K.

Let $C(\mathfrak{X})$ be the space of all real valued continuous functions on $\mathfrak{X}$ with supremum norm $\|\ \|_\infty$. We denote by $C_0(\mathfrak{X})$ the set of functions of $C(\mathfrak{X})$ each of which depends only on the configurations in some compact set K;

$$C_0(\mathfrak{X}) = \{ f \in C(\mathfrak{X}) : f(\xi) = f(\xi_K) \text{ for some compact set } K \}$$

It is easily seen that $C_0(\mathfrak{X})$ is dense in $C(\mathfrak{X})$. We define σ-fields

$$\mathcal{B}(\mathfrak{X}) = \sigma(N_A : A \in \mathcal{B}(R^d)),$$

$$\mathcal{B}_K(\mathfrak{X}) = \sigma(N_A : A \in \mathcal{B}(R^d),\ A \subset K),$$

where $N_A(\xi)$ is the number of particles of ξ in A. The σ-field $\mathcal{B}(\mathfrak{X})$ coincides with the topological Borel field on $\mathfrak{X}$.

Before defining a linear operator on $C(\mathfrak{X})$ which generates a Markov process, we prepare some terminologies. For $\underline{x} = \{x_1, x_2, \cdots, x_n\}$ and $\xi \in \mathfrak{M}$ we write

$$\xi^{x,y} = \begin{cases} (\xi\setminus x)\cdot y, & \text{if } x \in \xi,\ y \notin \xi, \\ \xi, & \text{otherwise}, \end{cases}$$

$$\chi(\underline{x}|\xi) = \begin{cases} 1, & \text{if } \xi \cup \underline{x} \in \mathfrak{X},\ \xi \cap \underline{x} = \phi, \\ 0, & \text{otherwise}. \end{cases}$$

Introducing the hard core potential Φ,

$$\Phi(y,z) = \begin{cases} 0, & \text{if } |y-z| \geqq r, \\ +\infty, & \text{otherwise}, \end{cases}$$

the function $\chi(\underline{x}|\xi)$ is rewritten in the form,

$$\chi(\underline{x}|\xi) = \exp\{ -\frac{1}{2} \sum_{\substack{y,z\in\underline{x} \\ y\neq z}} \Phi(y,z) - \sum_{y\in\underline{x},\, z\in\xi} \Phi(y,z) \}.$$

Now, we shall define a linear operator on $C_0(\mathfrak{X})$ by

$$Lf(\xi) = \sum_{x\in\xi} \int_{R^d} \{f(\xi^{x,y}) - f(\xi)\}\chi(y|\xi\setminus x)p(|x-y|)dy,$$

where $p(\cdot)$ is a non-negative function on $[0,\infty)$ such that

(1.1) $$\int_{R^d} dx\, p(|x|) = 1,$$

(1.2) $$\int_{R^d} dx\, |x|^2 p(|x|) < \infty,$$

(1.3) $p(\cdot) > 0$ on $[0,2h)$ for some $h > 0$.

Since L is dissipative and $C_0(\mathfrak{X})$ is dense in $C(\mathfrak{X})$, L has a minimal closed extension $\bar{L}$. With a slight modification of Liggett's theorem [5], $(\bar{L}, \mathfrak{D}(\bar{L}))$ generates a unique strongly continuous Markov semigroup T_t on $C(\mathfrak{X})$.

We denote by (ξ_t, P_μ) the Markov process generated by $\bar{L}$ with initial distribution μ. By strong continuity of T_t, we can take a version of ξ_t which is right continuous with left limits.

For any compact subset $K \subset R^d$, we denote by $\mathfrak{M}(K)$ and $\mathfrak{M}(K,n)$ the set of all finite subsets of K and the set of all subsets of K having n points ($\mathfrak{M}(K,0) = \{\phi\}$), respectively.

The set $\mathfrak{M}(K,n)$ can be written in the form,

(1.4) $$\mathfrak{M}(K,n) = (K^n)'/S_n,$$

where $(K^n)' = \{(x_1,\cdots,x_n) \in K^n : x_i \neq x_j \text{ for all } i \neq j\}$ and S_n is the symmetric group of degree n. By means of the factorization (1.4) we introduce a measure $\lambda_{K,z}$ into $\mathfrak{M}(K,n)$ so that $\lambda_{K,z}(\phi) = 1$ and

$$\lambda_{K,z}(\Lambda) = \frac{z^n}{n!} \int_{\tilde{\Lambda}} dx_1 dx_2 \cdots dx_n, \qquad n = 1, 2, \cdots,$$

where $z \geqq 0$ and $\tilde{\Lambda}$ is a preimage of Λ by factorization (1.4). We also introduce a measure $\lambda_{K,z}$ into the space $\mathfrak{M}(K) = \bigcup_{n=0}^{\infty} \mathfrak{M}(K,n)$ in a natural way. The integral of a measurable function f on $\mathfrak{M}(K)$ with respect to this measure is denoted by $\int f(\underline{x})\, d^z\underline{x}$.

Now, we are going to define a Gibbs state. We will see that this Gibbs state is a stationary measure of our process ξ_t.

Definition 1.1 ([2]). A probability measure μ on $\mathfrak{X}$ is called a grand canonical Gibbs state with activity $z \geqq 0$, if for any compact set K, the restriction of μ on $\mathfrak{B}_K(\mathfrak{X})$ is absolutely continuous with respect to $d^z\underline{x}$ with density function $\sigma_K(\underline{x}) = \int_{\eta(K)=0} \mu(d\xi)\, \chi(\underline{x}|\eta)$.

Denote by $\mathcal{G}(z)$ the convex set of all grand canonical Gibbs states with activity $z \geqq 0$. This set $\mathcal{G}(z)$ is convex and compact in the topology of weak convergence, so that the element of $\mathcal{G}(z)$ is represented by the extremal points of $\mathcal{G}(z)$. We denote by $\mathrm{ex}\mathcal{G}(z)$ the set of all extremal points of $\mathcal{G}(z)$.

Remark 1.1 ([7]). Let $\mu \in \mathcal{G}(z)$. Then,

$$\rho(z) = \lim_{K \uparrow R^d} \frac{1}{|K|} \int_{\mathfrak{X}} \xi(K)\, \mu(d\xi)$$

exists and it is called the particle density of μ. And the following equation holds. For any $\varepsilon > 0$,

$$\mu\left(\left|\frac{\xi(K)}{|K|} - \rho(z)\right| \geq \varepsilon \right) \to 0 \qquad \text{as} \quad K \uparrow R^d.$$

Remark 1.2 ([4]). If z is sufficiently small (i.e. $\rho(z)$ is sufficiently small), then $\#\mathcal{G}(z) = 1$.

From the definition of a Gibbs state, the following lemma is easily obtained.

Lemma 1.2. If μ is a grand canonical Gibbs state, then μ is a reversible measure for ξ_t, i.e.

$$\langle T_t f, g \rangle_\mu = \langle f, T_t g \rangle_\mu \qquad \text{for any} \quad f, g \in C(\mathfrak{X}),\ t \geq 0,$$

where $\langle \cdot , \cdot \rangle_\mu$ is an L^2 inner product with respect to μ.

§ 2. Ergodicity of (ξ_t, P_μ)

In this section, we shall give the proof of the following theorem.

Theorem 2.1. Let $z > 0$ be sufficiently small. Then, for any $\mu \in \mathrm{ex}\mathcal{G}(z)$, the reversible Markov process (ξ_t, P_μ) is ergodic.

Let $\hat{T}_t$ be the strongly continuous semigroup on $L^2(\mathfrak{X}, \mu)$ associated with ξ_t and $\hat{L}$ be the generator for $\hat{T}_t$. To prove the ergodicity of the process (ξ_t, P_μ) it is sufficient to prove the following condition,

(C.1) If $\hat{T}_t f = f$ for any $t \geq 0$, then f is constant.

From a relation between canonical Gibbs states and grand canonical Gibbs states, we have that if $\mu \in \mathrm{ex}\mathcal{G}(z)$ for some $z \in [0, \infty)$, then $\mu(\Lambda) = 0$ or 1 for any $\Lambda \in \mathcal{E}_\infty(\mathfrak{X}) = \bigcap \sigma(N_K, \mathcal{B}_{K^c}(\mathfrak{X}))$, where the intersection runs over all compact sets K (see Georgii [1]). Using this property, we have the following lemma.

Lemma 2.1. Let $f \in L^2(\mathfrak{X}, \mu)$ and $\mu \in \mathrm{ex}\mathcal{G}(z)$. If there exists $\varepsilon > 0$ such that for almost all $\xi \in \mathfrak{X}$ and all natural numbers m and n such that $n < (\rho(z) + \varepsilon)|K_m|$

$$\int_{\Lambda(\xi, K_m, n)} d^1\underline{x} \int_{\Lambda(\xi, K_m, n)} d^1\underline{y}\ |f(\underline{x}\cdot\xi_{K_m^c}) - f(\underline{y}\cdot\xi_{K_m^c})| = 0, \tag{2.1}$$

then f is constant, where $\Lambda(\xi, K_m, n)$ is the set of all interior points of $\{ \underline{x} \in \mathfrak{M}(K_m, n) : \chi(\underline{x}|\xi_{K_m^c}) = 1 \}$ and $K_m = \{ x \in R^d : |x| \leq \sqrt{d}2^m r \}$.

Take any function f satisfying the assumption of (C.1). From Lemma 2.1 and (C.1), if we prove that

(C.1)' f satisfies (2.1) for almost all $\xi \in \mathfrak{X}$ and all natural numbers m and n such that $n < c|K_m|$ for some constant c,

then (ξ_t, P_μ) is ergodic for $\mu \in ex\mathcal{G}(z)$ with $\rho(z) < c$.

To check the condition (C.1)' we have to study the topological property of $\Lambda(\xi, K_m, n)$. Now, we shall introduce the notion about a generalization of connectivity for the configuration space.

Definition 2.1. i) $\xi \in \mathfrak{X}$ and $\eta \in \mathfrak{X}$ are said to be in h-communication $(\xi \leftarrow h \rightarrow \eta)$, if there exist $x \in \xi$ and $y \in \eta$ such that $|x-y| \leq h$ and $\xi^{x,y} = \eta$.

ii) $\Gamma \subset \mathfrak{X}$ and $\Lambda \subset \mathfrak{X}$ are said to be in h-communication $(\Gamma \leftarrow h \rightarrow \Lambda)$, if there exist $\xi \in \Gamma$ and $\eta \in \Lambda$ such that $\xi \leftarrow h \rightarrow \eta$.

iii) $\{\Lambda(j)\}_{j \in J}$ are said to be in h-communication, if for any $j', j'' \in J$, there exists a sequence $\{j_1, j_2, \cdots, j_q\}$ such that

$$\Lambda(j') \leftarrow h \rightarrow \Lambda(j_1) \leftarrow h \rightarrow \Lambda(j_2) \leftarrow h \rightarrow \cdots \leftarrow h \rightarrow \Lambda(j_q) \leftarrow h \rightarrow \Lambda(j'').$$

Let $\{\Lambda_j(\xi, m, n)\}_{j \in J}$ be the set of all connected components of $\Lambda(\xi, K_m, n)$. Then, our key lemma is the following (See [8] for the proof of this lemma).

Lemma 2.2. There exists a positive constant $c(r,h)$ such that for any natural numbers m, n such that $n < c(r,h)|K_m|$ and any $\xi \in \mathfrak{X}$, $\{\Lambda_j(\xi, m, n)\}_{j \in J}$ are in h-communication.

To complete the proof of Theorem 2.1 we shall check the condition (C.1)' for $c = c(r,h)$.

From the definition of L and Lemma 1.2, for $g \in C_0(\mathfrak{X})$ we have

$$-2\langle Lg, g\rangle_\mu$$

$$= \int_{\mathfrak{X}} \mu(d\xi) \sum_{x \in \xi} \int_{R^d} \{g(\xi^{x,y}) - g(\xi)\}^2 \chi(y|\xi \backslash x) p(|x-y|) dy.$$

Since f is $\hat{T}_t$-invariant for any $t \geq 0$, we see that $\hat{L}f = 0$. Since $\hat{L}$ is a minimal closed extention of L, we have

$$\int_{\mathfrak{X}} \mu(d\xi) \sum_{x \in \xi} \int_{R^d} \{f(\xi^{x,y}) - f(\xi)\}^2 \chi(y|\xi \backslash x) p(|x-y|) dy = 0.$$

From the definition of Gibbs state and (1.3), we have

$$(2.2)\qquad \int_{\Lambda(\xi,K_m,\underline{n})} d^1\underline{x} \sum_{x\in\underline{x}} \int_{B_{2h}(x)} dy\, |f(\underline{x}^{x,y}\cdot\xi_{K_m^c}) - f(\underline{x}\cdot\xi_{K_m^c})|\, 1_{\Lambda(\xi,K\ ,n)}(\underline{x}^{x,y})$$
$$= 0$$

for all natural numbers m and n, and almost all ξ. From (2.2) we obtain

$$(2.3)\qquad \int_{\Lambda(\xi,K_m,n)} d^1\underline{x} \int_{\Lambda(\xi,K_m,n)} d^1\underline{y}\, |f(\underline{x}\cdot\xi_{K_m^c}) - f(\underline{y}\cdot\xi_{K_m^c})|\, H(\underline{x},\underline{y}) = 0,$$

where H is non-negative function on $\mathfrak{M}(K_m,n) \times \mathfrak{M}(K_m,n)$ defined by

$$H(\underline{x},\underline{y}) = \sum \prod_{i=1}^{n} 1_{\Lambda(\xi,K_m,n)}(x_1\cdots x_{i-1}\cdot y_i\cdots y_n)\, 1_{B_{2h}(x_i)}(y_i)\,.$$

The above sum runs over all ordered 2n-tuples $(x_1,\cdots,x_n,y_1,\cdots,y_n)$ such that $\{x_1,\cdots,x_n\} = \underline{x}$ and $\{y_1,\cdots,y_n\} = \underline{y}$.

Since $\Lambda(\xi,K_m,n)$ is open, for any $\underline{x}^1$, $\underline{x}^2 \in \Lambda(\xi,K_m,n)$, we can choose $\varepsilon(\underline{x}^1,\underline{x}^2) \in (0,h)$ such that $I(\underline{x}^1,\varepsilon(\underline{x}^1,\underline{x}^2)) \cup I(\underline{x}^2,\varepsilon(\underline{x}^1,\underline{x}^2)) \subset \Lambda(\xi,K_m,n)$, where

$$I(\underline{x},\varepsilon) = \{\{y_1,\cdots,y_n\} \in \mathfrak{M}(K_m,n) : |y_i - x_i| < \varepsilon\}.$$

We write $I(\underline{x}^1,\underline{x}^2)$ for $I(\underline{x}^1,\varepsilon(\underline{x}^1,\underline{x}^2)) \cup I(\underline{x}^2,\varepsilon(\underline{x}^1,\underline{x}^2))$. If $\underline{x}^1 \leftarrow h \rightarrow \underline{x}^2$, H is positive on $I(\underline{x}^1,\underline{x}^2) \times I(\underline{x}^1,\underline{x}^2)$. From (2.2), we have

$$(2.4)\qquad \int_{I(\underline{x}^1,\underline{x}^2)} d^1\underline{x} \int_{I(\underline{x}^1,\underline{x}^2)} d^1\underline{y}\, |f(\underline{x}\cdot\xi_{K_m^c}) - f(\underline{y}\cdot\xi_{K_m^c})| = 0.$$

If $\Gamma_1,\ \Gamma_2 \subset \Lambda(\xi,K_m,n)$ such that $|\Gamma_1 \cap \Gamma_2| > 0$ and

$$\int_{\Gamma_i} d^1\underline{x} \int_{\Gamma_i} d^1\underline{z}\, |f(\underline{x}\cdot\xi_{K_m^c}) - f(\underline{z}\cdot\xi_{K_m^c})| = 0 \quad \text{for } i = 1,2,$$

then

$$|\Gamma_1 \cap \Gamma_2| \int_{\Gamma_1} d^1\underline{x} \int_{\Gamma_2} d^1\underline{y}\, |f(\underline{x}\cdot\xi_{K_m^c}) - f(\underline{y}\cdot\xi_{K_m^c})|$$
$$\leq |\Gamma_2| \int_{\Gamma_1} d^1\underline{x} \int_{\Gamma_1} d^1\underline{z}\, |f(\underline{x}\cdot\xi_{K_m^c}) - f(\underline{z}\cdot\xi_{K_m^c})|$$
$$+ |\Gamma_1| \int_{\Gamma_2} d^1\underline{z} \int_{\Gamma_2} d^1\underline{y}\, |f(\underline{z}\cdot\xi_{K_m^c}) - f(\underline{y}\cdot\xi_{K_m^c})|$$
$$= 0,$$

and so

$$\int_{\Gamma_1\cup\Gamma_2} d^1\underline{x} \int_{\Gamma_1\cup\Gamma_2} d^1\underline{z}\, |f(\underline{x}\cdot\xi_{K_m^c}) - f(\underline{z}\cdot\xi_{K_m^c})| = 0.$$

Repeating this procedure, from (2.4) we have that if $\Lambda_j(\xi,m,n) \leftarrow h \rightarrow \Lambda_{j'}(\xi,m,n)$, then

$$\int_{\Lambda_j(\xi,m,n)} d^1\underline{x} \int_{\Lambda_{j'}(\xi,m,n)} d^1\underline{y}\, |f(\underline{x}\cdot\xi_{K_m^c}) - f(\underline{y}\cdot\xi_{K_m^c})| = 0.$$

Therefore, using Lemma 2.2 we obtain (C.1)' for $c = c(r,h)$, and this completes the proof of Theorem 2.1.

§ 3. Asymptotics for a tagged particle

In this section, we study the behavior of one of the balls in our process. We call this ball the tagged particle. In order to follow the motion of the tagged particle it is convenient to regard the process ξ_t as a Markov process (x_t, η_t) on the locally compact space $R^d \times \mathfrak{X}_0$, where

$$\mathfrak{X}_0 = \{ \eta \in \mathfrak{X} : \eta \cap B_r(0) = \phi \}.$$

x_t is the position of the tagged particle and η_t is the entire configuration seen from the tagged particle. We can see that η_t is a Markov process whose generator $\bar{\mathfrak{L}}$ is the closure of the operator given by

$$\mathfrak{L}f(\eta) = \int_{R^d} \{f(\tau_{-u}\eta) - f(\eta)\}\chi(u|\eta)p(|u|)du$$
$$+ \sum_{z\in\eta} \int_{R^d\setminus B_r(0)} \{f(\eta^{z,y}) - f(\eta)\}\chi(y|\eta\setminus z)p(|z-y|)dy,$$

where

$$\tau_u\eta = \{x_i + u\}, \qquad \text{if } \eta = \{x_i\}.$$

We denote by S_t the semigroup for $\bar{\mathfrak{L}}$ and (η_t, P^0_ν) the Markov process generated by $\bar{\mathfrak{L}}$ with initial distribution ν.

From Remark 1.2, for sufficiently small $z > 0$ $\#\mathcal{G}(z) = 1$ and so $\mu \in \mathcal{G}(z)$ is shift invariant. For any shift invariant $\mu \in \mathrm{ex}\mathcal{G}(z)$ we define

$$\mu_0(d\eta) = \frac{1}{\mu(\mathfrak{X}_0)} \chi(0|\eta)\mu(d\eta).$$

Using the same argument as Lemma 1.2 and Theorem 2.1, we have the following lemma.

Lemma 3.1. Let $z > 0$ be sufficiently small. Then, for any $\mu \in \mathrm{ex}\mathcal{G}(z)$, $(\eta_t, P^0_{\mu_0})$ is an ergodic reversible Markov process.

The process x_t is driven by the process η_t. We introduce measurable sets

$$\Delta = \{(\eta,\zeta) \in \mathfrak{X}_0 \times \mathfrak{X}_0 : \eta = \zeta\},$$

$$\Gamma_A = \{(\eta,\zeta) \in \mathfrak{X}_0 \times \mathfrak{X}_0 \setminus \Delta : \zeta = \tau_{-u}\eta \text{ for some } u \in A\},$$

for $A \in \mathcal{B}(R^d)$ and define

$$\mathfrak{F}_t = \sigma(\eta_s : s \in (-\infty, t]).$$

Then,

$$N((t_1,t_2] \times A) = \sum_{s\in(t_1,t_2]} 1_{\Gamma_A}(\eta_{s-}, \eta_s) \qquad \text{for } 0 < t_1 < t_2,$$

is an $\mathfrak{F}_t$-adapted σ-finite random measure and we have

$$x_t = x_0 + \int_{(0,t]} \int_{R^d} u\, N(dsdu).$$

Using the same argument as Theorem 2.4 of [3], we have the following result.

Theorem 3.1. For sufficiently small $z > 0$ if $\mu \in ex\mathcal{G}(z)$ then

$$\lambda x_{t/\lambda^2} \to D\cdot B_t \qquad \text{as } \lambda \to 0$$

in the sense of distribution in the Skorohod space, where B_t is d-dimensional Brownian motion and D is $d\times d$ matrix such that

$$(D^2)_{ij} = \int_{R^d} du \int_{\mathfrak{X}_0} d\mu_0\ u_i u_j \chi(u|\cdot)p(|u|) - 2\int_{[0,\infty)} dt\ \langle S_t F_i, F_j\rangle_{\mu_0}.$$

$$F(\eta) = \int_{R^d} du\ u\ p(|u|)\chi(u|\eta).$$

Unfortunately, we haven't proved the non-degeneracy of D.

Acknowledgement. The author would like to express his thanks to Professor H.Tanaka for helping him with valuable suggestion and constant encouragement. The author also expresses his thanks to Dr.K.Kuroda and Dr.Y.Tamura for their encouragements.

References

[1] H. O. Georgii: Canonical and Grand Canonical Gibbs States for continuum systems, Commun. Math. Phys. 48, (1976), 31-51.

[2] H. O. Georgii: Canonical Gibbs Measures, Springer Lecture Notes in Mathematics, Vol. 760, (1979).

[3] C. Kipnis and S. R. S. Varadhan: Central limit theorem for additive functional of reversible Markov processes and applications to simple exclusions, Commun. Math. Phys. 104, (1986), 1-9.

[4] D. Klein: Dobrushin uniqueness technique and the decay of correlations in continuum statistical mechanics, Commun. Math. Phys. 86, (1982), 227-246.

[5] T. M. Liggett: Existence theorems for infinite particle systems, Trans. Amer. Math. Soc. 165, (1972), 471-481.

[6] T. M. Liggett: Interacting Particle Systems, Springer-Verlag, Berlin, (1985).

[7] R. A. Minlos: Lectures on statistical physics, Russian Math. Surveys 23, no. 1, (1968), 137-196.

[8] H. Tanemura: Ergodicity for an infinite particle system in R^d of jump type with hard core interaction, to appear.

Department of Mathematics
Keio University
Hiyoshi, Kohoku-ku
Yokohama
223 Japan

POWER ORDER DECAY OF ELEMENTARY SOLUTIONS OF GENERALIZED DIFFUSION EQUATIONS

Matsuyo Tomisaki

1. Introduction. Let $\mathfrak{G} = \frac{d}{dm}\frac{d}{dx}$ be a generalized diffusion operator on an interval S and $p(t,x,y)$ the elementary solution of the generalized diffusion equation

$$(1.1) \qquad \partial u(t,x)/\partial t = \mathfrak{G}u(t,x), \qquad t > 0,\ x \in S,$$

in the sense of McKean [8]. The aim of this article is to give a criterion, in terms of m, for the convergence of the integral $P_\gamma \equiv \int^{+\infty} t^\gamma p(t,x,y)dt$, which yields $p(t,x,y) = o(t^{-\gamma-1})$ as $t \to \infty$.

In the case where $\mathfrak{G}$ is *recurrent*, that is, $P_0 = \infty$, the author has already obtained a criterion in order that $P_\gamma < \infty$ with $-1 < \gamma < 0$ in [10]. On the other hand, $p(t,x,y)$ decays exponentially if and only if m satisfies a condition for the positivity of the principal eigenvalue of $\mathfrak{G}$, due to Kotani [6], whence $P_\gamma < \infty$ for every $\gamma \in \mathbf{R}$ (see Remark 2.3 below and also [5]). Thus our study is concentrated on the *transient* case except the case where m satisfies Kotani's condition. Through the spectral expansion of $p(t,x,y)$, its asymptotic behavior for large t is closely connected to the behavior near the origin of spectral function of $\mathfrak{G}$. In order to get our criterion, we will show some criteria, in terms of m, for the behavior near the origin of the spectral function. They are similar to I. S. Kac's ones [2], [3] for its behavior at infinity.

In §2 we will state our main results. The precise definition of $p(t,x,y)$ will be given in §3. §4 will be devoted to asymptotic theorems for Krein's correspondence. The proof of our main results will be given in §5.

2. Statement of results. Let $S = (\ell_1, \ell_2)$ be an open interval with $-\infty \le \ell_1 < 0 < \ell_2 \le +\infty$ and $m(x)$ be a real valued nontrivial right continuous nondecreasing function on it with $m(0) = 0$. We denote the induced measure by $dm(x)$. For a function u on S, we set

$$u(\ell_i) = \lim_{x \to \ell_i, x \in S} u(x), \qquad i = 1, 2,$$

$$u^+(x) = \lim_{\varepsilon \downarrow 0}(u(x+\varepsilon) - u(x))/\varepsilon,$$

if there exist the limits . Let $D(\mathfrak{G})$ be the space of all functions $u \in L^2(S,m)$ which have continuous versions u (we use the same symbol) satisfying the following conditions. a) There are two complex constants A, B and a function $\mathfrak{G}u \in L^2(S,m)$ such that

$$u(x) = A + Bx + \int_{0+}^{x+} (x-y)\mathfrak{G}u(y)dm(y), \qquad x \in S,$$

where the integral $\int_{a+}^{b+}$ is read as $\int_{(a,b]}$ or $-\int_{(b,a]}$ according as $a \le b$ or $a > b$. b) For each $i = 1, 2$, if $\ell_i + m(\ell_i)$ is finite, then $u(\ell_i) = 0$. We then define the generalized diffusion operator $\mathfrak{G}$ from $D(\mathfrak{G})$ into $L^2(S,m)$ by $D(\mathfrak{G}) \ni u \longmapsto \mathfrak{G}u \in L^2(S,m)$. With the aid of S. Watanabe's argument [11], the above setting includes all cases of sticky elastic boundary conditions for regular boundaries. Indeed, if ℓ_1 is the regular boundary with the sticky elastic boundary condition $\theta_1 u(\ell_1) - \theta_2 u^+(\ell_1) + \theta_3 \mathfrak{G}u(\ell_1) = 0$, where $\theta_1 + \theta_2 + \theta_3 = 1$, $\theta_i \ge 0$, $i = 1, 3$ and $\theta_2 > 0$, then we reset $S = (\ell_1 - \theta_2/\theta_1, \ell_2)$ and extend $m(x)$ by setting $m(x) = m(\ell_1) - \theta_3/\theta_2$ for $\ell_1 - \theta_2/\theta_1 < x \le \ell_1$. It is the same with the regular boundary ℓ_2 with a sticky elastic boundary condition. Here and hereafter, $1/\infty = 0$, $\pm a/0 = \pm\infty$, $\infty \pm a = \infty$ and $-\infty \pm a = \infty$ for $a > 0$. Now we can define the elementary solution $p(t,x,y)$ of the diffusion equation (1.1) following McKean [8], whose precise definition will be given in §3. In order to state our result we need the following conditions. Let $i = 1, 2$, $\gamma \ge -1$.

$\mathrm{P}^{(\gamma)}$: $\quad \displaystyle\int^{\infty} t^{\gamma} p(t,x,y)dt < \infty, \qquad x, y \in S.$

$\mathrm{A}_i^{(\gamma)}$: $\quad |\ell_i| < \infty$ and $\displaystyle\left|\int_0^{\ell_i} |\ell_i - x|^{-\gamma} \left|\int_0^x dy \int_0^y m(z)dz\right|^{\gamma} dx\right| < \infty.$

$\mathrm{B}_i^{(\gamma)}$: $\quad |\ell_i| < \infty$ and $\displaystyle\left|\int_0^{\ell_i} \left|\int_0^x m(y)dy\right|^{\gamma} dx\right| < \infty.$

$\mathrm{C}_i^{(\gamma)}$: $\quad |\ell_i| = \infty$ and $\displaystyle\left|\int_0^{m(\ell_i)} |m(\ell_i) - x|^{-\gamma} \left|\int_0^x dy \int_0^y m^{-1}(z)dz\right|^{\gamma} dx\right| < \infty.$

$\mathrm{D}_i^{(\gamma)}$: $\quad |\ell_i| = \infty$ and $\displaystyle\left|\int_0^{m(\ell_i)} \left|\int_0^x m^{-1}(y)dy\right|^{\gamma} dx\right| < \infty.$

$\mathrm{E}_i^{(\gamma)}$: $\quad |\ell_i| = |m(\ell_i)| = \infty$ and $\displaystyle\left|\int_0^{m(\ell_i)} \left|\int_0^x m^{-1}(y)dy\right|^{\gamma} dx\right| < \infty.$

Here $m^{-1}(x)$ stands for the inverse function of $m(x)$, i.e. $m^{-1}(x) = \sup\{y\colon m(y) \le x\}$, $m(\ell_1) < x < m(\ell_2)$. Note that $\mathrm{C}_i^{(\gamma)}$ or $\mathrm{D}_i^{(\gamma)}$ with $\gamma \ge 0$ yield $|m(\ell_i)| < \infty$.

The main result of this paper is the following

Theorem 1. $P^{(0)}$ *holds if and only if* $\ell_1 > -\infty$ *or* $\ell_2 < \infty$. *For* $0 < \gamma < 1$, $P^{(\gamma)}$ *holds if and only if one pair of the conditions* $(A_1^{(\gamma)}, A_2^{(\gamma)})$, $(A_1^{(\gamma)}, C_2^{(0)})$, $(C_1^{(0)}, A_2^{(\gamma)})$, $(A_1^{(\gamma)}, E_2^{(\gamma-1)})$, $(E_1^{(\gamma-1)}, A_2^{(\gamma)})$ *holds. For* $n \in \mathbf{N}$, $P^{(n)}$ *follows if and only if one of* $(B_1^{(n)}, B_2^{(n)})$, $(B_1^{(n)}, D_2^{(n-1)})$, $(D_1^{(n-1)}, B_2^{(n)})$ *is valid.*

Remark 2.1. We also consider the following conditions.

$B_i^{(\infty)}$: $\quad |\ell_i| < \infty$ and $\left|\int_0^{\ell_i}\left|\int_0^x m(y)dy\right|^n dx\right| < \infty$, $n \in \mathbf{N}$.

$B_i^{(S)}$: $\quad |\ell_i| < \infty$ and $\sup_{\ell_i \wedge 0 < x < \ell_i \vee 0} |(\ell_i - x)m(x)| < \infty$.

$B_i^{(EX)}$: $\quad |\ell_i| < \infty$ and $\left|\int_0^{\ell_i} m(y)dy\right| < \infty$.

$B_i^{(AB)}$: $\quad |\ell_i| < \infty$ and $|m(\ell_i)| < \infty$.

$D_i^{(\infty)}$: $\quad |\ell_i| = \infty$ and $\left|\int_0^{m(\ell_i)}\left|\int_0^x m^{-1}(y)dy\right|^n dx\right| < \infty$, $n \in \mathbf{N}$.

$D_i^{(S)}$: $\quad |\ell_i| = \infty$ and $\sup_{\ell_i \wedge 0 < x < \ell_i \vee 0} |x(m(\ell_i) - m(x))| < \infty$.

$D_i^{(EN)}$: $\quad |\ell_i| = \infty$ and $\left|\int_0^{\ell_i} y dm(y)\right| < \infty$.

Here $a \wedge b = \min\{a,b\}$, $a \vee b = \max\{a,b\}$. $B_i^{(AB)}$ implies that ℓ_i is the regular boundary with the absorbing boundary condition. $B_i^{(EX)}$ [$D_i^{(EN)}$] means that ℓ_i is an exit [resp. entrance] point (cf. [1]). There are the following implications among the conditions. Let p, $q \in \mathbf{N}$, $p < q$ and $0 \leq \alpha < \beta < 1$. Then $B_i^{(AB)} \Rightarrow B_i^{(EX)} \Rightarrow B_i^{(S)} \Rightarrow B_i^{(\infty)} \Rightarrow B_i^{(q)} \Rightarrow B_i^{(p)} \Rightarrow A_i^{(\beta)} \Rightarrow A_i^{(\alpha)}$. $D_i^{(EN)} \Rightarrow D_i^{(S)} \Rightarrow D_i^{(\infty)} \Rightarrow D_i^{(q)} \Rightarrow D_i^{(p)} \Rightarrow C_i^{(\beta)} \Rightarrow C_i^{(\alpha)}$. $E_i^{(\beta-1)} \Rightarrow E_i^{(\alpha-1)}$. No converse implication is true. Further, for $0 < \gamma < 1$, $B_i^{(\gamma)} \Rightarrow A_i^{(\gamma)}$, $D_i^{(\gamma)} \Rightarrow C_i^{(\gamma)}$.

Remark 2.2 ([9],[10]). Assume $\ell_1 = -\infty$ and $\ell_2 = \infty$. Then $P^{(0)}$ does not occur. Further, $\lim_{t\to\infty} p(t,x,y) > 0$, x, $y \in S$ if and only if $C_1^{(0)}$ and $C_2^{(0)}$ are satisfied, whence $P^{(\gamma)}$ does not hold for any $\gamma \geq -1$. Given $-1 < \gamma < 0$, in order that $P^{(\gamma)}$ holds it is necessary and sufficient that one of the conditions (a) and (b) is satisfied. (a) $C_i^{(0)}$, $|\ell_j| = |m(\ell_j)| = \infty$ and $\left|\int_0^{\ell_j}\left|\int_0^x m(y)dy\right|^\gamma dx\right| < \infty$, $i = 1, 2$, $j = 3 - i$.

(b) $|\ell_i| = |m(\ell_i)| = \infty$, $i = 1, 2$ and $\int_0^\infty \{\int_0^x (m(y) - m(-y))dy\}^\gamma dx < \infty$.

Remark 2.3 (Kotani[6], see also [5]). $p(t,x,y)$ decays exponentially as $t \to \infty$, x, $y \in S$, that is, $\lim_{t\to\infty} e^{ct}p(t,x,y) = 0$, x, $y \in S$ for some $c > 0$, if and only if one pair of the conditions $(B_1^{(S)}, B_2^{(S)})$, $(B_1^{(S)}, D_2^{(S)})$, $(D_1^{(S)}, B_2^{(S)})$ holds. Since $B_i^{(\infty)}$ $[D_i^{(\infty)}]$ does not imply $B_i^{(S)}$ [resp. $D_i^{(S)}$] as in Remark 2.1, we find an example such that $\lim_{t\to\infty} e^{ct}p(t,x,y) = \infty$ and $\lim_{t\to\infty} t^n p(t,x,y) = 0$, x, $y \in S$ for $c > 0$, $n \in \mathbf{N}$ (see §5).

Though it is expected that, for each $\gamma \geq 0$, $P^{(\gamma)}$ occurs if and only if one of $(B_1^{(\gamma)}, B_2^{(\gamma)})$, $(B_1^{(\gamma)}, D_2^{(\gamma-1)})$, $(D_1^{(\gamma-1)}, B_2^{(\gamma)})$ holds, we can only prove it for $\gamma \in \mathbf{N}$ and the following special case.

Theorem 2. *Let* $1 < \gamma < 2$. *If one of* $(B_1^{(AB)}, B_2^{(AB)})$, $(B_1^{(AB)}, C_2^{(\gamma-1)})$, $(C_1^{(\gamma-1)}, B_2^{(AB)})$ *holds, then* $P^{(\gamma)}$ *follows.*

When $m(x)$ varies regularly near $\ell_1 > -\infty$ or $\ell_2 < \infty$, we can show that so does $p(t,x,y)$ as $t \to \infty$. It will be written in another paper.

3. Elementary solution. We define the elementary solution $p(t,x,y)$ of the generalized diffusion equation (1.1) following [1], [8], [12]. Let S and $m(x)$ be those in §2. For each $i = 1, 2$, let $\varphi_i(x,\alpha)$, $x \in S$, $\alpha \in \mathbf{C}$, be the solution of the integral equation

$$\varphi_i(x,\alpha) = 2 - i + (i - 1)x + \alpha\int_{0+}^{x+} (x - y)\varphi_i(y,\alpha)dm(y). \tag{3.1}$$

Then there exist the limits

$$h_i(s) = (-1)^i \lim_{x\to\ell_i, x\in S} \varphi_2(x,s)/\varphi_1(x,s), \quad i = 1, 2, \ s > 0. \tag{3.2}$$

We set

$$\begin{aligned} &1/h(s) = 1/h_1(s) + 1/h_2(s), \ h_{11}(s) = h(s), \\ &h_{22}(s) = -(h_1(s) + h_2(s))^{-1}, \ h_{12}(s) = h_{21}(s) = - h(s)/h_2(s). \end{aligned} \tag{3.3}$$

$h_{ij}(s)$, i, $j = 1, 2$, can be analytically continued to $\mathbf{C}\setminus(-\infty,0]$. The spectral measures σ_{ij}, i, $j = 1, 2$, are defined by

$$\sigma_{ij}([\lambda_1,\lambda_2]) = \lim_{\varepsilon\downarrow 0} \frac{1}{\pi}\int_{\lambda_1}^{\lambda_2} \mathcal{I}m\, h_{ij}(-\lambda-\sqrt{-1}\varepsilon)d\lambda, \tag{3.4}$$

for all continuity points $\lambda_1 < \lambda_2$. The matrix valued measure $[\sigma_{ij}]_{i,j=1,2}$ is symmetric nonnegative definite. Now the elementary solution of the generalized diffusion equation is given by

(3.5) $$p(t,x,y) = \sum_{i,j=1,2} \int_{0-}^{\infty} e^{-\lambda t}\varphi_i(x,-\lambda)\varphi_j(y,-\lambda)\sigma_{ij}(d\lambda), \quad t > 0,\ x,\ y \in S.$$

In particular, if $B_1^{(AB)}$ is satisfied, then we have the following representation. Let

$$\psi_1(x,\alpha) = \varphi_2^+(\ell_1,\alpha)\varphi_1(x,\alpha) - \varphi_1^+(\ell_1,\alpha)\varphi_2(x,\alpha),$$
$$\psi_2(x,\alpha) = -\varphi_2(\ell_1,\alpha)\varphi_1(x,\alpha) + \varphi_1(\ell_1,\alpha)\varphi_2(x,\alpha), \quad x \in S,\ \alpha > 0.$$

Further we set

(3.6) $$h_0(s) = \lim_{x\uparrow\ell_2} \psi_2(x,s)/\psi_1(x,s), \quad s > 0,$$

(3.7) $$\sigma_0([\lambda_1,\lambda_2]) = -\lim_{\varepsilon\downarrow 0} \frac{1}{\pi}\int_{\lambda_1}^{\lambda_2} \mathscr{Im}\{1/h_0(-\lambda-\sqrt{-1}\varepsilon)\}d\lambda,$$

for all continuity points $\lambda_1 < \lambda_2$. Note that $\sigma_0(\{0\}) = 0$. Then (3.5) is reduced to

(3.8) $$p(t,x,y) = \int_{0+}^{\infty} e^{-\lambda t}\psi_2(x,-\lambda)\psi_2(y,-\lambda)\sigma_0(d\lambda), \quad t > 0,\ x,\ y \in S.$$

By means of the expressions (3.5) and (3.8), we get immediately the following proposition. So the proof will be omitted. Let

$\sigma_i^{(\gamma)}$: $\int_{0-} \lambda^{-\gamma-1}\sigma_{ii}(d\lambda) < \infty,$

$\sigma_0^{(\gamma)}$: $\int_{0+} \lambda^{-\gamma-1}\sigma_0(d\lambda) < \infty.$

<u>Proposition 3.1</u>. *For* $\gamma > -1$, $P^{(\gamma)}$ *and the pair of conditions* $(\sigma_1^{(\gamma)},\sigma_2^{(\gamma)})$ *are equivalent each other. In the case of* $B_1^{(AB)}$, *these are also equivalent with* $\sigma_0^{(\gamma)}$.

4. <u>Krein's correspondence</u>. In this section we give some asymptotic theorem for Krein's correspondence. The arguments of Krein's correspondence are due to [4] and [6]. Let $m(x)$ be a nontrivial right continuous nondecreasing function on $[0,\infty]$ such that $m(\infty) = \infty$. We always set $m(0-) = 0$ and denote the totality of such m by $\mathcal{M}$. For $m \in \mathcal{M}$ we can consider the solutions $\varphi_i(x,\alpha)$, $i = 1, 2$, $0 \le x < \ell$, $\alpha \in \mathbf{C}$, of the integral equation (3.1), where $\ell = \sup\{x: m(x) < \infty\}$. Set

(4.1) $$k(s) = \lim_{x\uparrow\ell} \varphi_2(x,s)/\varphi_1(x,s) = \int_0^{\ell} \varphi_1(x,s)^{-2}dx, \quad s > 0.$$

k is called the *characteristic function* of m and the correspondence $m \in \mathcal{M} \longrightarrow k$ is called *Krein's correspondence*. Let $\mathcal{K}$ be the class of functions k on $(0,\infty)$ such that

$$(4.2)\qquad k(s) = c + \int_{0-}^{\infty} \frac{\sigma(d\lambda)}{s+\lambda}, \qquad s > 0,$$

for some $c \geq 0$ and some nonnegative Borel measure σ on $[0,\infty)$ satisfying $\int_{[0,\infty)}(1+\lambda)^{-1}\sigma(d\lambda) < \infty$.

Theorem (*M.G.Krein* [4]). *Krein's correspondence* $m \in \mathcal{M} \longrightarrow k \in \mathcal{K}$ *is one to one and onto.*

From now on we denote Krein's correspondence by $m \in \mathcal{M} \leftrightarrow k \in \mathcal{K}$. It is easy to see that

$$c = \inf\{x > 0\colon m(x) > 0\}, \quad \ell = \lim_{s\downarrow 0} k(s) = c + \int_{0-}^{\infty} \lambda^{-1}\sigma(d\lambda).$$

Further, for $0 < \gamma < 1$,

$$(4.3)\qquad \int_0 s^{-\gamma}k(s)ds < \infty \iff \int_{0-} \lambda^{-\gamma}\sigma(d\lambda) < \infty.$$

By means of [6], we also get that

$$(4.4)\qquad k(s) \asymp U(1/s), \qquad s > 0,$$

where $U(s)$ is the inverse function of $s \longmapsto \int_0^s m(y)dy$, and the symbol $a(s) \asymp b(s)$, $s \in I$, means that $C_1 b(s) \leq a(s) \leq C_2 b(s)$, $s \in I$, for some positive constants C_i, $i = 1, 2$, I being an interval. If $\ell < \infty$, then (4.4) is reduced to that $k(s) \asymp \ell$ as $s \downarrow 0$. We note the following result in the case that $\ell < \infty$. Let $V(s)$ be the inverse function of $s \longmapsto s\int_0^{\ell-1/s} dx \int_0^x m(y)dy$.

Proposotion 4.1. *Let* $m \in \mathcal{M} \leftrightarrow k \in \mathcal{K}$ *and assume* $\ell < \infty$. *Then*

$$(4.5)\qquad \ell - k(s) \asymp 1/V(1/s), \quad s > 0.$$

Proof. First we note that $\varphi_1(x,s)$ is nondecreasing in $x \geq 0$. Since $\varphi_1(x,s) \geq 1$, we have by (4.1)

$$(4.6)\qquad \int_0^{\ell}\{1 - \varphi_1(x,s)^{-1}\}dx \leq \ell - k(s) \leq 2\int_0^{\ell}\{1 - \varphi_1(x,s)^{-1}\}dx.$$

In view of (3.1),

$$s\int_0^x m(y)dy \leq \varphi_1(x,s) - 1 = s\int_0^x dy\int_{0+}^{y+}\varphi_1(z,s)dm(z) \leq s\varphi_1(x,s)\int_0^x m(y)dy.$$

Therefore

$$(4.7)\qquad \{\varphi_1(\ell-1/V(1/s),s)V(1/s)\}^{-1}$$

$$= \varphi_1(\ell-1/V(1/s),s)^{-1}\ s\int_0^{\ell-1/V(1/s)}dx\int_0^x m(y)dy$$

$$\le \int_0^{\ell-1/V(1/s)}\{1-\varphi_1(x,s)^{-1}\}dx$$

$$\le s\int_0^{\ell-1/V(1/s)}dx\int_0^x m(y)dy = 1/V(1/s).$$

On the other hand,

$$(4.8)\qquad \{1-\varphi_1(\ell-1/V(1/s),s)^{-1}\}/V(1/s)$$

$$\le \int_{\ell-1/V(1/s)}^{\ell}\{1-\varphi_1(x,s)^{-1}\}dx \le 1/V(1/s).$$

(4.6) with (4.7) and (4.8) leads us to

$$1/V(1/s) \le \ell - k(s) \le 4/V(1/s),\quad s > 0.$$

This is the desired result. q.e.d.

Proposition 4.2. *Let* $m \in \mathcal{M} \leftrightarrow k \in \mathcal{K}$. *Then, for* $n \in \mathbf{N}$, *the following* (i) - (iii) *are mutually equivalent.*

(i) $\int_0^\ell\left(\int_0^x m(y)dy\right)^n dx < \infty.$

(ii) $k^{(n)}(0) = \lim_{s\downarrow 0} d^n k(s)/ds^n$ *is finite.*

(iii) $\int_{0-}^\infty \lambda^{-n-1}\sigma(d\lambda) < \infty.$

Proof. First of all we note that each condition implies $\ell < \infty$. Let $[X(t)\colon t \ge 0,\ P_x\colon x \in \mathbf{R}]$ be a one-dimensional Brownian motion and $t(t,\xi)$ the local time at ξ. Put $\tau_x = \int_{[0,\ell]} t(\sigma_x,\xi)dm(\xi)$, where $\sigma_x = \inf\{t > 0\colon X(t) = x\}$. Then by means of [1; §4.6 and §5.3],

$$(4.9)\qquad \varphi_1(x,s)^{-1} = E_0\left[\exp(-s\tau_x)\right],\qquad 0 \le x < \ell,\quad s > 0.$$

It follows from (3.1) that

$$E_0\left[\tau_x^n\right] = (-1)^n n!\sum_{1\le p\le n}(-1)^p\sum_{i_1+\cdots+i_p=n,\,i.\ge 1}\Pi_{1\le j\le p}M_{i_j}(x),$$

for $0 < x < \ell$, $n \in \mathbf{N}$, where

$$M_1(x) = \int_0^x m(y)dy,$$

$$M_n(x) = \int_0^x dy \int_{0+}^{y+} M_{n-1}(z)dm(z), \quad n \geq 2.$$

Since $E_0\left[\tau_x\right]^n \leq E_0\left[\tau_x^n\right]$ and $M_n(x) \leq M_1(x)^n$, the condition (i) is equivalent with

$$(4.10) \qquad \int_0^{\ell} E_0\left[\tau_x^j\right] E_0\left[\tau_x^{n-j}\right] dx < \infty, \quad 0 \leq j \leq n.$$

Noting (4.1) and (4.9), we find

$$(-1)^n k^{(n)}(s) = \int_0^{\ell} \sum_{0 \leq j \leq n} \binom{n}{j} E_0\left[\tau_x^j \exp(-s\tau_x)\right] E_0\left[\tau_x^{n-j} \exp(-s\tau_x)\right] dx.$$

The integrand of the right hand side is nonnegative and nonincreasing in s. Therefore the condition (ii) is equivalent with (4.10). Thus we obtain (i) <=> (ii). (ii) <=> (iii) is obvious from (4.2). q.e.d.

We next note the dual string. For $m \in \mathcal{M}$, we set $m^{-1}(x) = \sup\{y: m(y) \leq x\}$, $0 \leq x < m(\ell)$. If $m \in \mathcal{M} \leftrightarrow k \in \mathcal{K}$, then by virtue of [7]

$$(4.11) \qquad m^{-1} \in \mathcal{M} \leftrightarrow 1/sk(s) \in \mathcal{K}.$$

m^{-1} is called the *dual string* of m. Let $U^*(s)$ and $V^*(s)$ be the inverse functions of $s \longmapsto \int_0^s m^{-1}(y)dy$ and $s \longmapsto s\int_0^{m(\ell)-1/s} dx \int_0^x m^{-1}(y)dy$, respectively. It follows from the above observations that

$$(4.12) \qquad 1/sk(s) \asymp U^*(1/s), \quad s > 0 \qquad \text{if } m(\ell) = \infty;$$

$$(4.13) \qquad m(\ell) - 1/sk(s) \asymp 1/V^*(1/s), \quad s > 0 \qquad \text{if } m(\ell) < \infty;$$

$$(4.14) \qquad \int_0^{m(\ell)} \left(\int_0^x m^{-1}(y)dy\right)^n dx < \infty \iff \lim_{s \downarrow 0} \frac{\partial^n}{\partial s^n}(1/sk(s)) \in \mathbf{R}, \quad n \in \mathbf{N}.$$

5. <u>Proof of theorems</u>. By the change of the role of ℓ_1 and ℓ_2, it is enough to prove Theorems 1 and 2 in the case that $\ell_1 > -\infty$. So we assume $\ell_1 > -\infty$ in this section.

Let us recall (3.2), (3.3) and (3.4). We set

$$m_1(x) = -m(-x), \qquad 0 \leq x < -\ell_1,$$

$$m_2(x) = m(x), \qquad 0 \leq x < \ell_2.$$

Taking the right continuous modification of m_1 and putting $m_i(x) = \infty$ for $|\ell_i| \leq x \leq \infty$, $i = 1, 2$, we find that $m_i \in \mathcal{M} \leftrightarrow h_i \in \mathcal{K}$ is Krein's correspondence, $i = 1, 2$. For each $i = 1, 2$, let U_i^* and V_i be the inverse functions of $s \longmapsto \int_0^s m_i^{-1}(y)dy$ and $s \longmapsto s\int_0^{|\ell_i|-1/s} dx\int_0^x m_i(y)dy$, respectively. We also set $h_i^*(s) = 1/sh_i(s)$. Then, by means of (4.5), (4.12), (4.14) and Proposition 4.2,

(5.1) $|\ell_i| - h_i(s) \asymp 1/V_i(1/s), \quad s > 0 \qquad$ if $|\ell_i| < \infty$;

(5.2) $h_i^*(s) \asymp U_i^*(1/s), \quad s > 0 \qquad$ if $|m(\ell_i)| = \infty$;

(5.3) $\int_0^{|\ell_i|}\left(\int_0^x m_i(y)dy\right)^n dx < \infty \iff |h_i^{(n)}(0)| < \infty, \qquad n \in \mathbf{N}$;

(5.4) $\int_0^{|m(\ell_i)|}\left(\int_0^x m_i^{-1}(y)dy\right)^n dx < \infty \iff |h_i^{*(n)}(0)| < \infty, \qquad n \in \mathbf{N}$.

By virtue of [9; Lemma 1], we now have

(5.5) $$h(s) = \xi + \int_{0-}^{\infty} \frac{\sigma_{11}(d\lambda)}{s + \lambda}, \quad s > 0,$$

where $a = \sup\{x < 0: m(x) < 0\}$, $b = \inf\{x > 0: m(x) > 0\}$, and $\xi = 0$ if $a = b = 0$, or $= 1/(1/b - 1/a)$ otherwise. Since $h(0) = 1/(1/\ell_2 - 1/\ell_1) < \infty$, we see that

(5.6) $$H_1(s) \equiv (h(0) - h(s))/s = \int_{0-}^{\infty} \frac{\sigma_{11}(d\lambda)}{\lambda(s + \lambda)}, \quad s > 0.$$

In view of dual string, we also have

(5.7) $$H_2(s) \equiv 1/s(h_1(s) + h_2(s)) - 1/(\ell_2 - \ell_1)s$$
$$= \int_{0+}^{\infty} \frac{\sigma_{22}(d\lambda)}{\lambda(s + \lambda)}, \qquad s > 0.$$

We note that $\sigma_{11}(\{0\}) = \sigma_{22}(\{0\}) = 0$. By using (5.6), (5.7), (4.3) and Proposition 4.2, we get the following (5.8) and (5.9) immediately. Let $i = 1, 2$, $0 < \gamma < 1$ and $n \in \mathbf{N}$. Then

(5.8) $\sigma_i^{(\gamma)} \iff \int_0 s^{-\gamma}H_i(s)ds < \infty$,

(5.9) $\sigma_i^{(n)} \iff |H_i^{(n-1)}(0)| < \infty$.

We will show the following basic lemma.

<u>Lemma</u>. $\sigma_i^{(0)}$, $i = 1, 2$ *always hold. For* $0 < \gamma < 1$ *and* $n \in \mathbf{N}$,

$(\sigma_1^{(\gamma)}, \sigma_2^{(\gamma)}) \iff (A_1^{(\gamma)}, A_2^{(\gamma)})$ *or* $(A_1^{(\gamma)}, C_2^{(0)})$ *or* $(A_1^{(\gamma)}, E_2^{(\gamma-1)})$,

$$(\sigma_1^{(n)},\sigma_2^{(n)}) \iff (\mathrm{B}_1^{(n)},\mathrm{B}_2^{(n)}) \quad or \quad (\mathrm{B}_1^{(n)},\mathrm{D}_2^{(n-1)}).$$

Proof. The first assertion is obvious by (5.6) and (5.7). We will prove the second one.

Assume $\ell_2 < \infty$. Then

$$(5.10) \qquad H_1(s) = \{(1/h_1(s) + 1/\ell_1)/s + (1/h_2(s) - 1/\ell_2)/s\} \times \{(-1/\ell_1 + 1/\ell_2)(1/h_1(s) + 1/h_2(s))\}^{-1}.$$

By means of (5.7) and (5.10),

$$H_i(s) \asymp (-\ell_1 - h_1(s) + \ell_2 - h_2(s))/s, \quad 0 < s < s_1, \ i = 1, 2,$$

for some $s_1 > 0$. Noting (5.1), we have, for each $i = 1, 2$,

$$(5.11) \qquad \int_0 s^{-\gamma}H_i(s)ds < \infty \iff \int_0 (s^{\gamma+1}V_j(1/s))^{-1}ds < \infty, \ j = 1, 2 \iff (\mathrm{A}_1^{(\gamma)},\mathrm{A}_2^{(\gamma)}).$$

Also by (5.7), (5.10) and (5.3), and by the fact $h(0) + h_1(0) + h_2(0) < \infty$, we see that for each $i = 1, 2$

$$(5.12) \qquad |H_i^{(n-1)}(0)| < \infty \iff |h_j^{(n)}(0)| < \infty, \ j = 1, 2 \iff (\mathrm{B}_1^{(n)},\mathrm{B}_2^{(n)}).$$

Let $\ell_2 = m(\ell_2) = \infty$. Then

$$(5.13) \quad H_1(s) = \{(-\ell_1 - h_1(s))/s - \ell_1h_1(s)/sh_2(s)\}h_2(s)/(h_1(s) + h_2(s)).$$

This and (5.7) lead us to

$$\left.\begin{aligned} &H_1(s) \asymp (-\ell_1 - h_1(s))/s + 1/sh_2(s), \\ &H_2(s) \asymp 1/sh_2(s), \end{aligned}\right\} \quad 0 < s < s_2,$$

for some $s_2 > 0$. Hence, by (5.1) and (5.2),

$$(5.14) \qquad \int_0 s^{-\gamma}H_1(s)ds < \infty \iff \int_0 (s^{\gamma+1}V_1(1/s))^{-1}ds < \infty \ \text{and} \ \int_0 s^{-\gamma}U_2^*(1/s)ds < \infty \iff (\mathrm{A}_1^{(\gamma)},\mathrm{E}_2^{(\gamma-1)}),$$

$$(5.15) \qquad \int_0 s^{-\gamma}H_2(s)ds < \infty \iff \int_0 s^{-\gamma}U_2^*(1/s)ds < \infty \iff \mathrm{E}_2^{(\gamma-1)}.$$

(5.7) and (5.13) also imply that

(5.16) $H_1(0) = H_2(0) = \infty$.

Suppose that $\ell_2 = \infty$ and $m(\ell_2) < \infty$. Namely $C_2^{(0)}$ holds. In this case (5.13) is also valid and

$$H_1(s) \asymp (-\ell_1 - h_1(s))/s, \qquad 0 < s < s_3,$$

for some $s_3 > 0$. $H_2(0) < \infty$ as in (ii). Consequently by (5.1)

(5.17) $$\int_0 s^{-\gamma}H_1(s)ds < \infty \iff \int_0 (s^{\gamma+1}V_1(1/s))^{-1}ds < \infty \iff (A_1^{(\gamma)}, C_2^{(0)}).$$

Further, by virtue of (5.13), (5.3) and (5.4),

(5.18) $$|H_1^{(n-1)}(0)| < \infty$$

$$\iff |h_1^{(n)}(0)| < \infty \text{ and } |h_2^{*(n-1)}(0)| < \infty \iff (B_1^{(n)}, D_2^{(n-1)}).$$

Since $H_2(0) < \infty$, it follows from (5.3), (5.4) and (5.7) that for $n \geq 2$

(5.19) $$|H_2^{(n-1)}(0)| < \infty \iff \lim_{s\downarrow 0}|\partial^{n-1}(sh_i(s))/\partial s^{n-1}| < \infty, \; i = 1, 2$$

$$\iff |h_1^{(n-2)}(0)| < \infty \text{ and } |h_2^{*(n-1)}(0)| < \infty \iff (B_1^{(n-2)}, D_2^{(n-1)}).$$

The second assertion of the lemma follows from (5.8), (5.9), (5.11), (5.12), (5.14) - (5.19). q.e.d.

Now Theorem 1 is obtained by the Lemma and Proposition 3.1.

Finally we give

Proof of Theorem 2. We may assume that $\ell_1 = 0$, $B_1^{(AB)}$ and $C_2^{(0)}$ without loss of generality. If $B_2^{(AB)}$ holds, then the assertion of the theorem is obvious by Remark 2.3. We use the representation (3.8). Note that

(5.20) $$1/sh_o(s) = \int_{0+}^{\infty} \frac{\sigma_o(d\lambda)}{\lambda(s+\lambda)}, \qquad s > 0.$$

Since $\lim_{s\downarrow 0}(1/sh_o(s)) = m(\ell_2) < \infty$, we have

$$H_o(s) \equiv (m(\ell_2) - 1/sh_o(s))/s = \int_{0+}^{\infty} \frac{\sigma_o(d\lambda)}{\lambda^2(s+\lambda)}.$$

In view of Proposition 4.1 and (4.11),

$$m(\ell_2) - 1/sh_o(s) \asymp 1/V_o(1/s), \quad s > 0,$$

where $V_o(s)$ is the inverse function of $s \longmapsto s\int_0^{m(\ell_2)-1/s}\int_0^x m^{-1}(y)dy$.

Therefore by virtue of (4.3), for $0 < \delta < 1$,

$$\int_{0+} \lambda^{-\delta-2}\sigma_0(d\lambda) < \infty \iff \int_0 s^{-\delta}H_0(s)ds < \infty$$

$$\iff \int_0 (s^{\delta+1}V_0(1/s))^{-1}ds < \infty \iff C_2^{(\delta)}.$$

Thus we get Theorem 2. q.e.d.

Example. Assume that, for each $i = 1, 2, m$ satisfies one of the following conditions (i) - (iii).

(i) $|\ell_i| < \infty$ and $\lim_{x\to\ell_i, x\in S}|m(x)|/|\ell_i - x|^{-\varepsilon_i}L_i(1/|\ell_i - x|) = 1$.

(ii) $|\ell_i| = \infty$ and $\lim_{x\to\ell_i, x\in S}|m(x)|/(1 - |x|^{-\varepsilon_i}/L_i(|x|)) = 1$.

(iii) $|\ell_i| = \infty$ and $\lim_{x\to\ell_i, x\in S}|m(x)|/|x|^{\varepsilon_i}L_i(|x|) = 1$.

Here $\varepsilon_i > 0$ and L_i is a slowly varying function at ∞. Let K_i be a slowly varying function such that

$$\lim_{x\to\infty}K_i(x)^{\varepsilon_i}L_i(x^{1/\varepsilon_i}K_i(x)) = 1.$$

Let $0 < \gamma < 1$ and $n \in \mathbf{N}$. If (i) is satisfied, then

$A_i^{(\gamma)} \iff \varepsilon_i < (\gamma+1)/\gamma$, or $\varepsilon_i = (\gamma+1)/\gamma$ and $\int^\infty x^{-1}L_i(x)^\gamma dx < \infty$,

$B_i^{(n)} \iff \varepsilon_i < (n+1)/n$, or $\varepsilon_i = (n+1)/n$ and $\int^\infty x^{-1}L_i(x)^n dx < \infty$,

$B_i^{(\infty)} \iff \varepsilon_i \le 1$,

$B_i^{(S)} \iff \varepsilon_i < 1$, or $\varepsilon_i = 1$ and $L_i(x)$ is bounded.

If (ii) is satisfied, then

$C_i^{(\gamma)} \iff \varepsilon_i > \gamma/(\gamma+1)$, or $\varepsilon_i = \gamma/(\gamma+1)$ and $\int^\infty x^{-1}K_i(x)^\gamma dx < \infty$,

$D_i^{(n)} \iff \varepsilon_i > n/(n+1)$, or $\varepsilon_i = n/(n+1)$ and $\int^\infty x^{-1}K_i(x)^n dx < \infty$,

$D_i^{(\infty)} \iff \varepsilon_i \ge 1$,

$D_i^{(S)} \iff \varepsilon_i > 1$, or $\varepsilon_i = 1$ and $L_i(x)$ is bounded.

If (iii) is satisfied, then

$E_i^{(\gamma-1)} \iff \varepsilon_i < (1-\gamma)/\gamma$, or $\varepsilon_i = (1-\gamma)/\gamma$ and $\int^\infty x^{-1}K_i(x)^{\gamma-1}dx < \infty$.

References

[1] K. Itô and H. P. McKean, Jr.: Diffusion Processes and their Sample Paths, Springer, Berlin-Heidelberg-New York (1965).

[2] I. S. Kac: On the behaviour of spectral functions of second order differential systems, Dokl. Acad. Nauk SSSR 106 (1956), 183-186.

[3] I. S. Kac: On the growth of spectral functions of differential systems of second order, Izv. Akad. Nauk SSSR 23 (1959), 257-274.

[4] I. S. Kac and M. G. Krein: On the spectral functions of the string, Amer. Math. Transl. Ser. 2, 103 (1974), 19-102.

[5] J. Kaneko, Y. Ogura and M. Tomisaki: On the principal eigenvalue for one-dimensional generalized diffusion operators (preprint).

[6] S. Kotani and S. Watanabe: Krein's spectral theory of strings and generalized diffusion processes, Functional Analysis in Markov Processes (M. Fukushima, ed.), Lecture Notes in Math. 923, 235-259, Springer, Berlin-Heidelberg-New York (1982).

[7] M. G. Krein: On some cases of the effective determination of the densities of a non-homogeneous string from its spectral function, Dokl. Acad. Nauk SSSR 93 (1953), 617-620.

[8] H. P. McKean, Jr.: Elementary solutions for certain parabolic differential equations, Trans. Amer. Math. Soc. 82 (1956), 519-548.

[9] N. Minami, Y. Ogura and M. Tomisaki: Asymptotic behavior of elementary solutions of one-dimensional generalized diffusion equations, Ann. Prob. 13 (1985), 698-715.

[10] M. Tomisaki: On the asymptotic behaviors of transition probability densities of one-dimensional diffusion processes, Publ. RIMS, Kyoto Univ. 12 (1977), 819-834.

[11] S. Watanabe: On time inversion of one-dimensional diffusion processes, Z. Wahrsch. verw. Geb. 31 (1975), 115-124.

[12] K. Yoshida: Lectures on Differential and Integral Equations, Interscience, New York (1960).

Department of Mathematics
Saga University
Saga 840, Japan

LORD'S PARADOX ON MEAN ABSOLUTE DEVIATION

N. N. Vakhania

§1. Introduction.

If a fixed numerical parameter θ is measured under the additive influence of a random noise ξ, then we get $\hat{\theta}=\theta+\xi$ as a result of the measurement. Different criteria can be taken to evaluate the accuracy of the measurement. Among those the criterion of minimality of mean value (mathematical expectation) of some function Φ of the absolute deviation $|\hat{\theta}-\theta|$ is most natural to use. So we get $E\Phi(|\xi|)$ as a measure of discrepancy. The functions $\Phi(t)=t$ and $\Phi(t)=t^2$ are usually in use, and they give the concepts of mean absolute deviation (MAD) and mean squared error, correspondingly. In [1] F. Lord noticed that MAD criterion has an undesirable property: it has not necessarily to be reduced if one of two mutually independent symmetric sources of noise is eliminated. This is a consequence of the possibility of the equality $E|\xi+\eta|=E|\xi|$ for mutually independent symmetric random variables ξ and η. As an example of this situation F. Lord indicated mutually independent Bernoulli random variables ε_1 and ε_2, $P\{\varepsilon_i=+1\}=P\{\varepsilon_i=-1\}=\frac{1}{2}$, $i=1,2$. For mean squared error criterion this phenomenon can not occur because we have strict inequality $E(\xi+\eta)^2>E\xi^2$ for any pair of mutually independent symmetric random variables ξ and η provided η is non-degenerated.

Soon after that in a note [2] containing the expression "Lord's paradox" in the title, D. Hildebrand considered the phenomenon for the case of arbitrary mutually independent symmetric random variables ξ and η. For this case it is easy to see that we have

$$E|\xi+\eta| \geq E|\xi| \tag{1}$$

and D. Hildebrand showed that Lord's paradox takes place (i.e. we have equality in (1)) if $|\eta|\leq|\xi|$ with probability 1.

In the present paper we continue to study questions connected with Lord's paradox. At first we obtain the precise relation between $E|\xi+\eta|$ and $E|\xi|$ for the case of arbitrary mutually independent random variables ξ and η (Theorem 1). This relation contains Hildebrand's result and sheds light on the phenomenon giving the full description of the conditions on ξ and η under which Lord's paradox takes place. Then we show that Lord's paradox can occur not only for MAD criterion (Theorem 2). Furthermore, having in mind the vector version of Lord's paradox we give more general (but less precise) result concerning the relation between $E\Phi(\|\xi+\eta\|)$ and $E\Phi(\|\xi\|)$, where ξ and η are mutually independent random vectors in a finite or infinite dimensional normed space X, and $\Phi:[0,\infty)\to[0,\infty)$ is an increasing convex function. This function appears here in a natural way if we want to evaluate the measurement accuracy by means of the magnitude of $E\Phi(\|\hat{\theta}-\theta\|)$ thus generalizing the MAD criterion. In the one-dimensional case $(X=R^1)$ the character of the criterion depends on Φ. In the multidimensional case, as it should be expected, the geometry of the space X is also important. It is shown that the vector version of Lord's paradox can take place for mean squared error criterion as well, i.e. for the case $\Phi(t)=t^2$. This paradox will not occur, if (and only if) the function Φ and the norm in X are strictly convex (Theorem 3).

The main results of this paper were announced in [3].

§2. One dimensional case.

Let $\operatorname{sgn} a$ denote the sign of a, if $a\neq 0$ and $\operatorname{sgn}0=0$.

Theorem 1. For any pair of mutually independent random variables ξ and η having mathematical expectations the following relation holds

$$(2)\qquad E|\xi+\eta| = E|\xi| + (P\{\xi>0\}-P\{\xi<0\})E\eta + \int_A (|\eta|-|\xi|)(1-\operatorname{sgn}\xi\eta)dP ,$$

where A denotes the event $\{|\eta|>|\xi|\}$.

Proof. The idea of the proof is quite simple and consists in presentation of the sure event (the set of all elementary events) as a union of A and its complement A^c. Having in mind that $|a+b|=(a+b)\operatorname{sgn}a$ if $|a|\geq|b|$ we get the equality

$$E|\xi+\eta| = \int_A (\xi+\eta)\operatorname{sgn}\eta\, dP + \int_{A^c}(\xi+\eta)\operatorname{sgn}\xi\, dP$$

$$= E|\xi| + \int_A (|\eta|-|\xi|)dP + \int_{A^c}\eta\operatorname{sgn}\xi\, dP + \int_A \xi\operatorname{sgn}\eta\, dP$$

and it remains to note that

$$\int_{A^c} \eta \,\mathrm{sgn}\xi \, dP = E(\eta \,\mathrm{sgn}\xi) - \int_A |\eta| \,\mathrm{sgn}\xi\eta \, dP$$

and to use the mutual independence of ξ and η.

Now we want to make two simple remarks.

Remark 1. The equality (2) shows immediately that if ξ and η are symmetric and $P\{|\eta|\le|\xi|\}=1$, then $E|\xi+\eta|=E|\xi|$ (the Hildebrand's result). Moreover this equality shows that Lord's paradox for MAD criterion occurs in a wider case, namely, if the following two conditions are satisfied: (a) either median of ξ or mean of η is zero (symmetry of ξ and η is not necessary), (b) if ξ and η have different signs, then $|\eta|\le|\xi|$, $\xi=0$ implies $\eta=0$ (for the argument in which ξ and η have same signs the restriction $|\eta|\le|\xi|$ is not necessary). The conditions (a) and (b) which are sufficient for equality in (1) are not necessary (counter-example: $\xi=\varepsilon_1+1$, $\eta=\varepsilon_2-1$). However if one of those conditions is fulfilled the other one follows of course from the equality $E|\xi+\eta|=E|\xi|$.

Remark 2. In virtue of independence of ξ and η the Hildebrand's condition $P\{|\eta|\le|\xi|\}=1$ is equivalent to the following one: there exists a positive constant c such that

$$P\{|\xi|\ge c\} = P\{|\eta|\le c\} = 1 .$$

Indeed let $C>0$ be a number satisfying the condition $P\{|\xi|\le C\}>0$. We have $P\{|\xi|\le C, |\eta|>C\}=0$ because of $P\{|\eta|\le|\xi|\}=1$ and so we get $P\{|\eta|>C\}=0$. Now it is straightfforward to see that the greatest lower bound of such numbers C can be taken as c in the above equality. The converse is trivial.

A function $\Phi:[0,\infty)\to[0,\infty)$ is called convex, if $\Phi(\alpha t_1+\beta t_2)\le \alpha\Phi(t_1)+\beta\Phi(t_2)$ for all pairs of arguments t_1, t_2 and all positive numbers α, β such that $\alpha+\beta=1$. A convex function Φ is called strictly convex, if in this relation equality holds only for $t_1=t_2$. A typical example of convex function is $\Phi(t)=t^p$, $p\ge1$. For $p>1$ this function is strictly convex.

The following theorem asserts that Lord's paradox can take place not only for MAD criterion.

Theorem 2. For any convex but not strictly convex function Φ

there exists a pair of mutually independent non-degenerate symmetric random variables ξ and η such that

$$E\Phi(|\xi+\eta|) = E\Phi(|\xi|).$$

Proof. Since Φ is convex but not strictly convex we have $\Phi(\alpha t_1+\beta t_2)=\alpha\Phi(t_1)+\beta\Phi(t_2)$ for some t_1, t_2 and α, β such that $t_1>t_2$ and $\alpha>0$, $\beta>0$, $\alpha+\beta=1$. This equality means that at the point $\alpha t_1+\beta t_2=t$ the value of Φ coincides with the value of the linear (more precise - affine) function passing through $\Phi(t_1)$ and $\Phi(t_2)$. So the convex function Φ coincides with a linear function at three points $t_1<t<t_2$ and hence everywhere in the segment $[t_1,t_2]$. Therefore there exist points $u_2>u_1\geq 0$ such that

$$\Phi\left(\frac{u_1+u_2}{2}\right) = \frac{1}{2}\Phi(u_1) + \frac{1}{2}\Phi(u_2)$$

and we can construct ξ and η with the needed properties as follows

$$\xi = \frac{u_1+u_2}{2}\,\varepsilon_1, \qquad \eta = \frac{u_1-u_2}{2}\,\varepsilon_2$$

where ε_1 and ε_2 denote, as above, mutually independent Bernoulli random variables $(P(\varepsilon_i=+1)=P(\varepsilon_i=-1)=\frac{1}{2},\ i=1,2)$.

§ 3. Multidimensional case.

A normed space X (finite or infinite-dimensional) is called strictly convex, if the conditions $\|x\|=\|y\|=1$ and $\|x+y\|=2$ imply $x=y$, $x,y\in X$, or, equivalently, $\|x+y\|=\|x\|+\|y\|$ implies $x=\alpha y$, $\alpha>0$. Geometrically that means that every point on the surface of the unit ball of X is an extreme point of the ball, i.e. it is not an interior point of any line segment entirely belonging to the ball, or, equivalently, it is not a middle point of such a line segment.

We remind also that if ζ is a random vector with values in a normed space X, then its mean (or mathematical expectation) $E\zeta$ is defined as weak integral (Pettis integral) of ζ (details can be found in [4]). In finite-dimensional case $\zeta=(\zeta_1,\dots,\zeta_n)$, $E\zeta$ is a vector $(E\zeta_1,\dots,E\zeta_n)$.

The proof of the theorem below is based on th following lemma which is intuitively quite evident. For the case of compact (convex) sets in locally convex spaces this lemma has been proved by G. Bauer (see [5]).

<u>Lemma</u>. Let X be a normed space, and let ξ be a random vector concentrated on a convex closed separable set $B\subset X$. If $E\xi$ exists and is an extreme point of the set B, then $P\{\xi=E\xi\}=1$.

<u>Proof</u>. Supposing opposite we will assume that $P\{\xi\in B\setminus E\xi\}>0$ and arrive at a contradiction. Let (x_n) be a countable set dense in B. Consider the system of all closed balls with centers at x_n, $n=1,2,\ldots$ and rational radii such that neither of these balls contains $E\xi$ (we assume that the dense set (x_n) does not contain $E\xi$). It is quite easy to see that the union of these balls contains the set $B\setminus E\xi$ and so one of those balls, say V, is such that $P\{\xi\in V\}=\alpha>0$. The number α can not be one as otherwise we would have $E\xi\in V$ because the mean (Pettis integral) of a random vector belongs to the closed convex hull of the range on any event of full probability (see, for example, [4], p.99).

Now we have

$$E\xi = \int \xi dP = \alpha\int_{\{\xi\in V\}} \xi dP' + (1-\alpha)\int_{\{\xi\bar{\in} V\}} \xi dP'' ,$$

where

$$P' = \frac{1}{\alpha} P, \qquad P'' = \frac{1}{1-\alpha} P \tag{3}$$

and using the notations

$$m_1 = \int_{\{\xi\in V\}} \xi dP', \qquad m_2 = \int_{\{\xi\bar{\in} V\}} dP'' \tag{4}$$

we get

$$E\xi = \alpha m_1 + (1-\alpha)m_2, \qquad 0<\alpha<1, \tag{5}$$

Because of the normalization (3) P' and P'' are usual probability measures on the sets $\{\xi\in V\}$ and $\{\xi\bar{\in} V\}$ correspondingly, with induced classes of events; according to (4), m_1 and m_2 are their means. Therefore by the same reasons as just mentioned $m_1, m_2\in B$ since B is itself closed and convex. By the very same reason we have $m_1\in V$. But $E\xi\bar{\in} V$, so $m_1\neq E\xi$ and hence $m_1\neq m_2$. Therefore the relation (5) contradicts with the given assumption that $E\xi$ is an extreme point for B.

<u>Remark</u>. The proof given here as well as the proof in [4] utilizes the following property of the set B: (o) for every covering

of B by the system of open sets of zero measure the union of the all open sets has again zero measure. Therefore the lemma remains valid for the case when X is an arbitrary locally convex space if instead of being separable B has some restriction which guarantees the property (o) (for example the Lindelof condition for B or τ-smoothness of the probability distribution in question).

<u>Theorem 3</u>. Let X be a separable normed space, $\Phi:[0,\infty)\to[0,\infty)$ be an increasing convex function, ξ and η be mutually independent random vectors in X such that $E\eta=0$. Then we have

(6) (a) $E\Phi(\|\xi+\eta\|) \geq E\Phi(\|\xi\|)$;

(b) the inequality (6) is strict for any nonzero η and any independent of it ξ if and only if both the space X and the function Φ are strictly convex.

<u>Proof</u>. For any $x\in X$ we have $x=E(x+\eta)$ as $E\eta=0$, and therefore

(7) $\|x\| \leq E\|x+\eta\|$

Noting first that Φ is increasing and using then Jensen's inequality for convex functions ($\Phi E \leq E\Phi$) we get

(8) $\Phi(\|x\|) \leq E\Phi(\|x+\eta\|)$.

If we integrate this inequality by the probability measure μ_ξ which is the distribution of the random vector ξ we come directly to the statement (a) (see also [4], p.222). Indeed after the integration in (8) we get $E\Phi(\|\xi\|)$ on the left-hand side and $E\Phi(\|\xi+\eta\|)$ on the right-hand side because we have

$$E\Phi(\|x+\eta\|) = \int\Phi(\|x+y\|)d\mu_\eta(y).$$

where μ_η is the distribution of η, and it is enough to use Fubini's theorem and mutual independence of ξ and η.

Now we prove (b). If we assume the equality in (6), then according to (8) we would have equality in (8) a.e. by μ_ξ. Therefore in Jensen's inequality

$$\Phi(E\|x+\eta\|) \leq E\Phi(\|x+\eta\|)$$

the equality actually would hold (use (7)). But this is possible only for degenerate random variables, as Φ is strictly convex. So we have $P\{\|x+\eta\|=C_x\}=1$, where C_x is a number depending only on x. This implies $E\Phi(\|x+y\|)=\Phi(C_x)$, and hence $\Phi(C_x)=\Phi(\|x\|)$ (use equality in (8)). This gives $C_x=\|x\|$ since strictly convex increasing function is strictly increasing. Therefore the random vector $\zeta=x+\eta$ is concentrated on the closed ball of radius $\|x\|$ and $E\zeta=x$ is a boundary point of this ball. But in a strictly convex normed space every boundary point of a ball is an extreme point of it and so the lemma above gives that $P\{\zeta=E\zeta\}=P\{\eta=0\}=1$ in contradiction with non-degeneracy of η.

Conversely, if the inequality (6) is strict, then the function Φ should be strictly convex, this is an immediate consequence of Theorem 2. The space X also should be strictly convex. Indeed, if this is not the case, then there exist two different points $x,y\in X$ such that $\|x\|=\|y\|=1$ and $\|x+y\|=2$. Let

$$\xi = \frac{x+y}{2}\varepsilon_1, \qquad \eta = \frac{x-y}{2}\varepsilon_2$$

where, as above, ε_1 and ε_2 are mutually independent Bernoulli random variables $(P\{\varepsilon_i=+1\}=P\{\varepsilon_i=-1\}=\frac{1}{2},\ i=1,2)$. Then ξ and η are mutually independent, η is non-degenerate, and still we would have $E\Phi(\|\xi+\eta\|)=E\Phi(\|\xi\|)$ in contradiction with the assumption.

<u>Corollary</u>. If we introduce the norm in the n-dimensional vector space R^n as $\|x\|=\max(|x_1|,|x_2|,\dots,|x_n|)$, or as $\|x\|=\sum_{i=1}^{n}|x_i|$ then Lord's paradox can occur if mean squared error is taken as a measure of discrepancy.

<u>Remark</u>. The both statements of Theorem 3 remain valid, if we require that the random variables $\|\xi+\eta\|$ and $\|\xi-\eta\|$ are identically distributed instead of ξ and η are mutually independent and $E\eta=0$. The proof can be done analogously proceeding from the obvious inequality $2\|\xi\|\le\|\xi+\eta\|+\|\xi-\eta\|$.

<u>References</u>

[1] Lord F. M. MAD Query. The American Statistician, 1983, v.37, no.1, 343-344.

[2] Hildebrand D. K. Lord's MAD paradox and Jensen's Inequality. The American Statistician, 1984, v.38, no.4, 296-297.

[3] Vakhania N. N. Towards the Lord's paradox on mean absolute

deviation. Dokl. Akad. Nauk SSSR, 1986, v.287, no.2, 265-268.
[4] Vakhania N. N., Tarieladze V. I., Chobanian S. A. Probability Distributions in Banach Spaces. Moscow: Nauka, 1985, 368p.
[5] Phelps R. R. Lectures on Choquet's Theorem. Princeton: D. Van Nostrand, 1966, 110p.

Institute of Computer Mathematics,
Academy of Sciences of Georgian SSR,
Tbilisi, USSR

APPROXIMATION OF STATIONARY PROCESSES AND THE CENTRAL LIMIT PROBLEM

Dalibor Volný

In dealing with the central limit problem for stationary (we shall always mean strictly stationary) sequences of random variables it is often useful to approximate the process by some special ones. First, we shall approximate an α-mixing process by sequences of independent and identically distributed random variables. In the second case we shall choose a sequence of martingale differences as the approximating process.

1. Approximation of α-mixing processes by i.i.d. Let (X_n) be a stationary sequence of r.v. with values from $\mathbb{R}^d$, $1 \leq d < \infty$. $\mathcal{F}_m^n$ denotes the σ-algebra generated by $X_m, X_{m+1}, \ldots, X_n$, $m \leq n$.

Let $\alpha(n) = \sup\{|\mu(A \cap B) - \mu(A) \cdot \mu(B)| : A \in \mathcal{F}_0^m,\ B \in \mathcal{F}_{m+n}^{\infty},\ m \geq 0\}$, $n \geq 0$.

If $\alpha(n) \to 0$, we say that (X_n) is α-mixing.

Let $1 \leq \nu < \infty$; $\| X_n \|$ denotes $(E|X_n|^{\nu})^{1/\nu}$ where $|.|$ is the norm in $\mathbb{R}^d$. We shall assume that $EX_n = 0$ and $\| X_n \| < \infty$. S_n means $\sum_{j=1}^n X_j$ and s_n^{ν} means $E|S_n|^{\nu}$.

Theorem 1. Let (X_n) be a stationary α-mixing process, $s_n \to \infty$ and $|S_n|^{\nu}/s_n^{\nu}$ be uniformly integrable (where $1 \leq \nu < \infty$ is fixed). Then there exist random variables $Z_{n,1}, \ldots, Z_{n,k(n)}$ where $k(n) \to \infty$ such that

i) for each $n = 1, 2, \ldots$, $Z_{n,1}, \ldots, Z_{n,k(n)}$ are i.i.d.,

ii) $|Z_{n,j}|^{\nu}$, $1 \leq j \leq k(n)$, $n = 1, 2, \ldots$ are uniformly integrable and,

iii) there exist $p(n)$ such that $\frac{p(n) \cdot k(n)}{n} \to 1$ and

$$\frac{1}{s_n} \cdot \| S_n - s_{p(n)} \cdot \sum_{j=1}^{k(n)} Z_{n,j} \| \to 0.$$

Proof. For an arbitrary positive integer k and $\eta > 0$ given we shall prove that for any n sufficiently large $Z_{n,1}, \ldots, Z_{n,k}$ exist such that (i) is satisfied and $\frac{1}{s_n} \cdot \| S_n - s_{p(n)} \cdot \sum_{j=1}^{k} Z_{n,j} \| < \eta$ where $\frac{n}{p(n)} \to k$. The construction will guarantee (ii) as well.

Let us put $n = k\cdot[p(n)+q(n)]+r(n)$, $0\leq r(n)\leq k$ (sometimes we shall write p,q,r): $S_{n,p,j} = \sum_{i=(j-1)\cdot(p+q)+1}^{j\cdot(p+q)-q} X_i$, $j=1,\dots,k$; $p(n)$ and $q(n)$ will be chosen so that $\frac{1}{s_n}\cdot\| S_n - \sum_{j=1}^{k} S_{n,p,j}\| \to 0$ for $n\to\infty$ (where $p>>q$).

Let us denote $S'_{n,p,j} = \frac{1}{s_p}\cdot S_{n,p,j}$. $|S'_{n,p,j}|^{\nu}$ are uniformly integrable hence for any $\delta_1>0$ there exists a $K(\delta_1)$ such that for $S''_{n,p,j} = S'_{n,p,j}\cdot\chi\{|S'_{n,p,j}|<K\}$ we have $\| S''_{n,p,j}-S'_{n,p,j}\| < \delta_1$. For any $\delta_2>0$ there exists a finite partition $\underset{\sim}{P}$ of $\{x\in\mathbb{R}^d : |x|<K\}$ such that for any $A\in\underset{\sim}{P}$, $\operatorname{diam}(A)<\delta_2$. Let $\psi: \underset{\sim}{P}\to \mathbb{R}^d$ assign to each $A\in\underset{\sim}{P}$ a $\psi(A)\in A$ and let $U(S''_{n,p,j}) = \sum_{A\in P}\psi(A)\cdot\chi\{S''_{n,p,j}\in A\}$. We shall denote $S^{\wedge}_{n,p,j} = U(S''_{n,p,j})$. Given $\delta_3>0$ we can thus find $S^{\wedge}_{n,p,j}$ such that $\|S'_{n,p,j}-S^{\wedge}_{n,p,j}\| < \delta_3$ for all n.

We shall borrow a notion from ergodic theory now. Let $\varepsilon>0$, $\underset{\sim}{P} = \{P^i:i=1,\dots,s\}$ and $\underset{\sim}{Q} = \{Q^j:j=1,\dots,t\}$ be finite measurable partitions. We say that P and Q are <u>ε-independent</u>, if there exists a partition $\{N_1,N_2\}$ of $\{1,\dots,t\}$ such that $\sum_{i=1}^{s}|\mu(P^i|Q^j)-\mu(P^i)| < \varepsilon$ for each $j\in N_1$ and $\mu(\bigcup_{j\in N_2}Q^j)<\varepsilon$. A sequence $\underset{\sim}{P}_1,\dots,\underset{\sim}{P}_n$ of finite partitions is ε-independent if $\underset{\sim}{P}_j$ is ε-independent of $\bigvee_{i=1}^{j-1}\underset{\sim}{P}_i$ for each $j\geq 2$. Let $\underset{\sim}{P}_j = \{P_j^1,\dots,P_j^M\}$ be the partition generated by $S^{\wedge}_{n,p,j}$, $j=1,\dots,k$. We shall prove that for q large enough, the sequence $\underset{\sim}{P}_1,\dots,\underset{\sim}{P}_k$ is ε-independent: For $2\leq m\leq k$ and $Q\in \bigvee_{i=1}^{m-1}\underset{\sim}{P}_i$, it is $\mu(Q)\cdot|\mu(P_m^j|Q)-\mu(P_m^j)|\leq\alpha(q)$ for each $1\leq j\leq M$. Let N be the set of $Q\in\bigvee_{i=1}^{m-1}\underset{\sim}{P}_i$ such that $\mu(Q)<\varepsilon\cdot M^{1-k}$; it holds $\mu(\bigcup N)<\varepsilon$. Choosing q so large that $\alpha(q)\cdot M^k < \varepsilon^2$ we have $\sum_{j=1}^{M}\mu|(P_m^j|Q)-\mu(P_m^j)| < \varepsilon$ for each $Q\in \bigvee_{i=1}^{m-1} P_i\setminus N$. Hence, $\underset{\sim}{P}_1,\dots,\underset{\sim}{P}_k$ are ε-independent. Following [10] there exists a sequence $\bar{\underset{\sim}{P}}_1,\dots,\bar{\underset{\sim}{P}}_k$ of independent partitions with the same distributions as $\underset{\sim}{P}_1,\dots,\underset{\sim}{P}_k$ such that $\sum_{i=1}^{M}\mu(P_j^i\Delta\bar{P}_j^i) < 3\cdot\varepsilon$, $j=1,\dots,k$. Choosing $\varepsilon>0$ sufficiently small we can find independent r.v. $Z_{n,j}$ equally distributed as $S^{\wedge}_{n,p,j}$ such that $E|Z_{n,j}-S^{\wedge}_{n,p,j}|^{\nu}$ is small.

Properties (i) and (ii) follow from the construction. We shall verify (iii). By the construction, we can make $\|\frac{S_n}{s_p} - \sum_{j=1}^{k}S^{\wedge}_{n,p,j}\|$ small, hence $\|\frac{S_n}{s_p} - \sum_{j=1}^{k} Z_{n,j}\|$ small.

Thus, we have $|\frac{s_n}{s_p} - \| \sum_{j=1}^{k} Z_{n,j} \| |$ small. As $Z_{n,j}$ are uniformly integrable, $s_n/s_{p(n)}$ cannot be unbounded (for k fixed). Thus, $\frac{1}{s_n} \cdot \| S_n - s_p \cdot \sum_{j=1}^{k} Z_{n,j} \|$ can be done arbitrarilly small.

Another approximation theorem for mixing processes was given by I.Berkes and W.Philipp in [2].

Theorem 2. (Corollary.) Let (X_n) be a stationary α-mixing sequence of real square integrable random variables. If $s_n \to \infty$, then S_n/s_n converge in distribution to $N(0,1)$ if and only if S_n^2/s_n^2 are uniformly integrable.

Proof. The sufficiency follows from Theorem 1 and from the Lindeberg theorem (see [3], Theorem 7.2). By the definition $ES_n^2/s_n^2 = 1$ for all n, hence the necessity follows from [3], Theorem 5.4 (see [8]).

A different proof of Theorem 2 was given by T.Mori and K.Yoshihara in [8].

2. Approximating martingales. In the sequel $(\Omega, \mathcal{A}, T, \mu)$ will denote a dynamical system; $(\Omega, \mathcal{A}, \mu)$ is a probability space with σ-algebra $\mathcal{A}$ and measure μ, T is a 1-1 bimeasurable mapping of Ω onto itself and $\mu(T^{-1}A) = \mu(A)$ for each $A \in \mathcal{A}$. $\mathcal{J}$ denotes the σ-algebra of sets $A \in \mathcal{A}$ such that $A = TA$. If it holds $\mu(A) = 0$ or $\mu(A) = 1$ for each $A \in \mathcal{J}$ we say that μ is ergodic.

For any measurable function f on Ω, the process $(f \circ T^i)$ is stationary and for any stationary process (X_i) a dynamical system and a process $(f \circ T^i)$ with the same distribution as (X_i) can be found.

The central limit problem case begins with the martingale difference sequences. By $s_n(f)$ we shall denote the sum $\frac{1}{\sqrt{n}} \cdot \sum_{j=1}^{n} f \circ T^j$.

Theorem 3. Let $(f \circ T^i)$ be a square integrable martingale difference sequence (m.d.s.) and $E(f^2 | \mathcal{J}) = \eta^2$. Then the measures $\mu s_n^{-1}(f)$ weakly converge to a probability measure with characteristic function $E \exp(-\frac{1}{2}\eta^2 t^2)$.

The theorem can be easily derived from Theorem 3.2 in [6]. For μ ergodic, η^2 is constant and we obtain the Billingsley-Ibragimov central limit theorem (see [6], Chapter 5). Other limit theorems can be obtained by approximation of stationary processes by martingale difference sequences. Before formulating the results let us give some definitions.

A σ-algebra $\mathcal{M}\subset\mathcal{A}$ is invariant if $\mathcal{M}\subset T^{-1}\mathcal{M}$. By $\mathcal{M}_i$ we shall denote $T^{-i}\mathcal{M}$, $i\in\mathbb{Z}$. The family of projection operators onto $L^2(\mathcal{M}_{i+1})\ominus L^2(\mathcal{M}_i)$ will be denoted by (P_i) and called the family of difference projection operators (d.p.o.) generated by $\mathcal{M}$. U denotes the operator sending $f\in L^2(\mathcal{A})$ to $f\circ T$.

Lemma 1. $$U\circ P_i = P_{i+1}\circ U.$$

Proof. For any $f\in L^2(\mathcal{A})$, $P_i f = E(f|\mathcal{M}_{i+1})-E(f|\mathcal{M}_i)$ and $UE(f|\mathcal{M}_j) = E(Uf|\mathcal{M}_{j+1})$.

As a corollary we obtain that for $f=P_j f$, $(f\circ T^i)$ is a martingale difference sequence. On the other hand, if $(f\circ T^i)$ is a m.d.s. and P_j are difference projection operators generated by invariant σ-algebra $\sigma\{f\circ T^i: i<0\}$, we have $f=P_0 f$.

The set of $f\in L^2(\mathcal{A})$ such that $f = \sum_{i=-n}^{n} P_i f$ for some family of d.p.o. (P_i) and nonnegative n will be denoted by K; the set of $f=P_0 f$ will be denoted by K_0 (i.e. $f\in K_0$ iff $(f\circ T^i)$ is a m.d.s. in $L^2(\mathcal{A})$).

For $f\in L^2(\mathcal{A})$, let $\rho(f) = \limsup_{n\to\infty}\|s_n(f)\|$ ($\|.\|$ is an $L^2(\mathcal{A})$ norm here). We can easily see that ρ is a pseudonorm with values in $[0,1]$. Let G be the set of functions $f\in L^2(\mathcal{A})$ such that $f_j \xrightarrow[\rho]{} f$ for some sequence (f_j) of functions from K. We shall call G the set of finitely approximable functions.

Theorem 4. Let $f\in L^2(\mathcal{A})$ be a finitely approximable function. Then there exists a function $\eta^2\in L^1(\mathcal{J})$ such that $E(s_n^2(f)|\mathcal{J}) \to \eta^2$ in $L^1(\mathcal{J})$ and, the measures $\mu s_n^{-1}(f)$ weakly converge to a probability measure with characteristic function $E\exp(-\frac{1}{2}\eta^2 t^2)$.

Proof. From Lemma 1 we can derive that for any $g\in K$ there exists an $h\in K_0$ such that $\rho(g-h)=0$. Hence, $f_j \xrightarrow[\rho]{} f$ for some sequence (f_j) from K_0. By Theorem 3, $\mu s_n^{-1}(f_j)$ converge to some probability measure ν_j for each j. It follows that there exists a probability measure ν such that $\nu_j\to\nu$ weakly and $\mu s_n^{-1}(f)\to\nu$ weakly. For any f_j, there exists an invariant σ-algebra $\mathcal{M}$ such that $f_j = E(f_j|T^{-1}\mathcal{M})-E(f_j|\mathcal{M})$. According to Theorem 6 we can assume $\mathcal{J}\subset\mathcal{M}$. Thus, $E(s_n^2(f_j)|\mathcal{J}) = E(f_j^2|\mathcal{J})$ for each $n=1,2,\ldots$. Put $\eta_j^2 = E(f_j^2|\mathcal{J})$. There exists $\eta^2\in L^1(\mathcal{J})$ such that $\eta_j^2\to\eta^2$ and

$E(s_n^2(f)|\mathcal{J}) \to \eta^2$ in $L^1(\mathcal{J})$. From this we obtain that $E\exp(-\frac{1}{2}\eta^2 t^2)$ is the characteristic function of ν.

The details of the proof are given e.g. in [12]. Moreover, the convergence is stable in the sense of [1].

An ergodic version of Theorem 4 was found by M.I.Gordin ([5]); his results were communicated at the First USSR-Japan Symposium in Khabarovsk. A nonergodic version of Gordin's theorem was given by G.K.Eagleson ([4]). The computation of η^2 was left as an open problem there. The formulation of the result was affected by a claim that $\eta^2>0$ a.s., which does not hold.

Let $f\in L^2(\mathcal{A})$. If $f = \sum_{i\in\mathbb{Z}} P_i f$ for some family (P_i) of d.p.o., we shall say that f is <u>difference decomposable</u>. We shall prove the following statement:

<u>Theorem 5</u>. If f is finitely approximable, then there exist functions $f', f'' \in L^2(\mathcal{A})$ such that $f = f'+f''$, f' is difference decomposable and $\rho(f'') = 0$.

First, we shall introduce some notions and give two theorems. We shall say that a σ-algebra $\mathcal{Q}\subset\mathcal{A}$ is <u>stiff</u> if $\mathcal{Q}$ is invariant and for any invariant σ-algebra $\mathcal{C}\subset\mathcal{Q}$, $T^{-1}\mathcal{C} = \mathcal{C} \bmod \mu$. For example, $\mathcal{J}$ is stiff. In any dynamical system the maximal stiff σ-algebra (with respect to set inclusion) exists. We shall call it Pinsker σ-algebra and denote it by $\mathcal{P}$. Thus, $\mathcal{C}$ is a stiff σ-algebra iff it is an invariant sub-σ-algebra of $\mathcal{P}$. The usual definition of Pinsker σ-algebra and the proof of existence are done by means of entropy (see e.g. [9]); the equivalence of definitions is easy to prove (see [12]).

<u>Theorem 6</u>. The set of all difference decomposable functions from $L^2(\mathcal{A})$ is equal to $L^2(\mathcal{A})\ominus L^2(\mathcal{P})$.

<u>Theorem 7</u>. Let $\mathcal{M}$ be an invariant σ-algebra and $\mathcal{Q}$ be a stiff σ-algebra. Let (P_i) be the family of d.p.o. generated by $\mathcal{M}$ and $(\bar{P}_i)$ be family of d.p.o. generated by $\mathcal{M}\vee\mathcal{Q} = \sigma(\mathcal{M}\cup\mathcal{Q})$.
If $f\in L^2(\mathcal{A})$ is $\mathcal{M}_\infty$-measurable, then $E(f|\mathcal{M}_{-\infty}) = E(f|\mathcal{Q}\vee\mathcal{M}_{-\infty})$ and $P_i f = \bar{P}_i f$ for all $i\in\mathbb{Z}$.

<u>Proof</u>. Using entropic ergodic theory it can be proved that for any $i\in\mathbb{Z}$, $\mathcal{M}_i$ and $\mathcal{Q}$ are conditionally independent w.r. to $\mathcal{M}_{i-1}$ (see [12]).

For any $g \in L^2(\mathcal{M}_i)$ we thus have $E(g|\mathcal{M}_{i-1}) = E(g|\mathcal{Q} \vee \mathcal{M}_{i-1})$. Hence, $\bar{P}_i P_i f = P_i f$ for each i. From this, Theorem 7 can be easily derived.

Proof of Theorem 6. Let $f \in L^2(\mathcal{A})$, $\mathcal{A}_f = \sigma\{f \circ T^i : i \in \mathbb{Z}\}$ and μ_f be the restriction of μ to $\mathcal{A}_f$. $(\Omega, \mathcal{A}_f, T, \mu_f)$ is then a dynamical system with Pinsker σ-algebra $\mathcal{P}_f$. By Rohlin-Sinai theorem (see [9]) there exists an invariant σ-algebra $\mathcal{M}$ such that $\mathcal{M}_\infty = \mathcal{A}_f$ and $\mathcal{M}_{-\infty} = \mathcal{P}_f$. Thus, any function $f \in L^2(\mathcal{A})$ can be expressed in the form $f = \sum_{i \in \mathbb{Z}} P_i f + E(f|\mathcal{P}_f)$. From Theorem 7 it follows that for any difference decomposable function $g \in L^2(\mathcal{A})$ it holds $E(g|\mathcal{P}) = 0$. Hence $E(f|\mathcal{P}_f) = E(f|\mathcal{P})$.

Proof of Theorem 5. Let f_j be functions from K (the definition of K was given before Theorem 4) and $\rho(f-f_j) \to 0$. Let $g_j = f-f_j$, $g_j' = E(g_j|\mathcal{P})$, $g_j'' = g_j - g_j'$. It holds $Es_n^2(g_j) = Es_n^2(g_j') + Es_n^2(g_j'')$, hence $\rho(g_j') \leq \rho(g_j)$. From this we obtain $\rho(g_j') \to 0$ and from Theorem 6 we get $g_j' = E(f|\mathcal{P})$.

Any function $f \in L^2(\mathcal{A})$ such that $Ef=0$ and $(f \circ T^i)$ is an α-mixing process, is difference decomposable. This function, however, need not be finitely approximable (if, for example, $\rho(f)=\infty$).

It is natural to raise the question which difference decomposable functions are not finitely approximable. From the following theorem and from Theorem 4 it follows that there exist a lot of them.

Theorem 8. Let the Hilbert space of difference decomposable functions be nonzero. Then it contains a dense set of functions f such that the measures $\mu s_n^{-1}(f)$ weakly converge to a probability measure and it contains a dense set of functions f such that the sequence $\mu s_n^{-1}(f)$ has at least two different weak limit points.

Proof. The set K (defined before Theorem 4) is dense in the Hilbert space of difference decomposable functions and by definition, each function from K is finitely approximable.

To find a dense set of difference decomposable functions for which the measures $\mu s_n^{-1}(f)$ fail to converge is more difficult. The ergodic case is treated in [11], the general case in [12]. The construction is based on the fact that for any $f = P_i f + P_j f$, $i,j \in \mathbb{Z}$, $s_n(f)$ is close to $s_n(P_i f + P_i U^{i-j} f)$ (which follows from Lemma 1).

Very little is known about the central limit problem for stationary processes from $L^2(\mathcal{P})$ (processes of zero entropy) yet. However, if

$g \in L^2(\mathcal{P})$ and the measures $\mu s_{k(n)}^{-1}(g)$ converge to a probability measure ($k(n)\to\infty$ and μ is ergodic), we can deduce the limit behavior of $\mu s_{k(n)}^{-1}(f+g)$ for any finitely approximable f.

Theorem 9. Let μ be ergodic, $g \in L^2(\mathcal{P})$ and f a finitely approximable function, $\rho^2(f)=\sigma^2$. If $\mu s_{k(n)}^{-1}(g)$ weakly converge to a probability measure ν and $k(n)\to\infty$, then $\mu s_{k(n)}^{-1}(f+g)$ weakly converge to the convolution $\nu * \nu_{0,\sigma^2}$ where ν_{0,σ^2} is the normal distribution with mean 0 and variance σ^2.

Proof. First, let us suppose that $f \in K_0$ (the definition of K_0 is given before Theorem 4) and $\sigma^2=1$. Let us denote $X_{n,j} = \frac{1}{\sqrt{n}}\cdot f \circ T^j$, $Y_{n,j} = \frac{1}{\sqrt{n}}\cdot g \circ T^j$, $I_n'(t) = \exp(it\cdot\sum_{j=1}^{k(n)} X_{n,j})$, $I_n''(t) = \exp(it\cdot\sum_{j=1}^{k(n)} Y_{n,j})$. For a real number x, $|x|<1$, it is $\exp(ix) = (1+ix)\cdot\exp(-\frac{1}{2}x^2+r(x))$ where $|r(x)|\leq|x|^3$.
We have $I_n'(t) = T_n(t)\cdot U_n(t)$ where

$T_n(t) = \Pi_{j=1}^{k(n)}(1+it\cdot X_{n,j})$ and

$U_n(t) = \exp(-\frac{1}{2}\cdot\sum_{j=1}^{k(n)} X_{n,j}^2 + \sum_{j=1}^{k(n)} r(t\cdot X_{n,j}))$;

$I_n'(t) = T_n(t)\cdot(U_n(t)-\exp(-\frac{1}{2}t^2)) + T_n(t)\cdot\exp(-\frac{1}{2}t^2)$.

It can be easily derived that $\max_{1\leq j\leq k(n)}|X_{n,j}| \to 0$ in probability and $\|\max_{1\leq j\leq k(n)}|X_{n,j}|\,\|$ is bounded (see [12]); from the Birkhoff ergodic theorem we get $\sum_{j=1}^{k(n)} X_{n,j}^2 \to 1$ a.s. Following McLeish's arguments we can thus derive that $E(T_n(t)\cdot(U_n(t) - \exp(-\frac{1}{2}t^2))) \to 0$. From Theorem 7 we have $E(T_n(t)|\mathcal{P}) = 1$, hence $E(I_n'(t)|\mathcal{P}) \to \exp(-\frac{1}{2}t^2)$. $I_n''(t)$ is $\mathcal{P}$-measurable so $E(I_n'(t)\cdot I_n''(t)) = E(E(I_n'(t)|\mathcal{P})\cdot I_n''(t)) \to \phi(t)\cdot\exp(-\frac{1}{2}t^2)$ where $\phi(t) = \lim_{n\to\infty} E\exp(I_n''(t))$ is the characteristic function of ν.

If f is finitely approximable then there exists a sequence (f_j) of functions from K_0 such that $f_j \underset{\rho}{\to} f$ (we made use of this fact in the proof of Theorem 4). For each j, $\mu s_{k(n)}^{-1}(g+f_j)$ weakly converge to $\nu * \nu_{0,\sigma_j^2}$ where $\sigma_j^2 = E(f_j^2)$. It holds $\rho((g+f_j) - (g+f)) \to 0$. Hence, $\sigma_j^2 \to \sigma^2$ where $\sigma^2 = \lim_{n\to\infty} E\, s_n^2(f)$, $\nu * \nu_{0,\sigma_j^2} \to \nu * \nu_{0,\sigma^2}$ weakly and $\mu s_{k(n)}^{-1}(f+g)$ weakly converge to $\nu * \nu_{0,\sigma^2}$.

Acknowledgement. The author thanks to many Japanese mathematicians who prepared excellent conditions for his research during the stay in Japan.

References

[1] D. J. Aldous and G. K. Eagleson: On mixing and stability of limit theorems, Ann. Probab. 6 (1978), 325-331.

[2] I. Berkes and W. Philipp: Approximation theorems for independent and weakly dependent random vectors, Ann. Probab. 7 (1979), 29-54.

[3] P. Billingsley: Convergence of Probability Measures, Wiley, New York (1968).

[4] G. K. Eagleson: On Gordin's central limit theorem for stationary processes, J. Appl. Probab. 12 (1975), 176-179.

[5] M. I. Gordin: The central limit theorem for stationary processes, Soviet Math. Doklady 10 (1969), 1174-1176.

[6] P. Hall and C. C. Heyde: Martingale Limit Theory and Its Application, Academic Press, New York (1980).

[7] D. L. McLeish: Dependent central limit theorems and invariance principles, Ann. Probab. 2 (1974), 620-628.

[8] T. Mori and K. Yoshihara: A note on the central limit theorem for stationary strong mixing sequences, to appear in Yokohama Math. J.

[9] W. Parry: Topics in Ergodic Theory, Cambridge Univ. Press, Cambridge (1980).

[10] P. Shields: The Theory of Bernoulli Shifts, The Univ. Chicago Press, Chicago (1973).

[11] D. Volný: A negative answer to the central limit problem for strictly stationary processes, Proceedings 3rd Prague Symp. Asymptot. Statistics, Elsevier Sci. Publ., Amsterdam (1984).

[12] D. Volný: The central limit problem for strictly stationary sequences (in Czech): Ph.D. thesis, Math. Institute Charles Univ. Prague, Prague (1984).

Department of Mathematics
Osaka Kyoiku University
Tennoji, Osaka
543 Japan

Mathematical Institute of
Charles University
Sokolovska 83
186 00 Praha 8
Czechoslovakia

GENERALIZED WIENER FUNCTIONALS AND THEIR APPLICATIONS

Shinzo Watanabe

1. Introduction.

In this note, a (d-dimensional) Wiener space is denoted by (W,P) where W is the space of continuous paths $w \in \mathbf{C}([0,\infty) \to R^d)$ such that $w(0)=0$ and P is the standard d-dimensional Wiener measure on W. As usual, we denote by E(F) the integral $\int_W F(w)P(dw)$ for an integrable Wiener functional F.

Since the Wiener space was introduced by N. Wiener in 1923, the theory of Wiener functional integration has played an important role in many problems in mathematics and mathematical physics. For an important example, M. Kac (e.g. [6],[7]) discussed several applications of Wiener functional integration to problems related to Schrödinger equations through the Feynman-Kac formula. The case discussed by Kac was restricted to Schrödinger operators on Euclidean space but if we want to consider a similar problem on a curved space, we have to make use of another important calculus on the Wiener space, namely, Itô's stochastic calculus: If we consider an initial value problem

(1.1) $$\partial u/\partial t = \frac{1}{2} \Delta u - V u \quad , \qquad u|_{t=0} = f$$

on a Riemannian manifold M where Δ is the Laplacian on M and V (potential) is a given function on M, then the solution u(t,x) is represented by a Wiener functional integration as

(1.2) $$u(t,x) = E\{M(t,w)\, f(X_{\mathbf{e}}(t,x,w))\} \quad .$$

Here $X_{\mathbf{e}}(t,x,w)$ is a Brownian motion on M starting at $x \in M$ which is obtained as the trace (=projection) on M of the <u>stochastic moving frame</u> starting at a frame (i.e. an ONB) $\mathbf{e}$ of $T_x(M)$. This stochastic moving frame can be constructed, as we know ([3]), by using Itô's calculus on the Wiener space, that is, by solving a stochastic differential equation defined by the system of canonical horizontal vector fields on the orthonormal frame bundle over M. M(t,w) is the Feynman-Kac functional given by $M(t,w) = \exp\{-\int_0^t V(X_{\mathbf{e}}(s,x,w))ds\}$. Note that $X_{\mathbf{e}}(t,x,w)$ depends on a choice of the initial frame $\mathbf{e}$ but the integral (1.2) is independent of such a choice (for details, we refer to Chapter V, Section 4 of [3]).

In many problems related to the Schrödinger operator $\frac{1}{2}\Delta - V$, it is necessary to treat the fundamental solution $p(t,x,y)$ of the above heat equation rather than the solution $u(t,x)$ of an initial value problem. An obvious heuristic expression for $p(t,x,y)$ would be

$$(1.3) \qquad p(t,x,y) = E\{M(t,w)\, \delta_y(X_{\mathbf{e}}(t,x,w))\}$$

where δ_y is the Dirac δ-function at $y \in M$ but $\delta_y(X_{\mathbf{e}}(t,x,w))$ is no longer a Wiener functional in an ordinary sense. As we discussed in, e.g. [11],[4],[10],[5], this heuristic expression (1.3) can be given a correct mathematical sense by introducing a family of Sobolev spaces $\mathbf{D}_p^s$, $s \in R$, $1 < p < \infty$, over the Wiener space which is a refinement of the family of usual L_p-spaces in the sense $\mathbf{D}_p^0 = L_p$: s denotes the order of differentiation so that, roughly, $\mathbf{D}_p^s$ consists of Wiener functionals whose derivatives up to s-th order belong to L_p. We have $\mathbf{D}_p^s \subset \mathbf{D}_{p'}^{s'}$ if $s' \leq s$ and $p' \leq p$ and, if $s < 0$, $\mathbf{D}_p^s$ contains elements which are no longer Wiener functionals in the sense of measurable functions on the Wiener space. Thus there is a clear analogy with the Schwartz distribution theory and it is natural to call such an element a generalized Wiener functional. Since the Malliavin covariance of M-valued Wiener functional $X_{\mathbf{e}}(t,x,w)$, for fixed $t > 0$, $x \in M$ and a frame $\mathbf{e}$, is non-degenerate in the sense of Malliavin, the pull-back $\delta_y(X_{\mathbf{e}}(t,x,w))$ of the Dirac δ-function δ_y by the M-valued Wiener map $w \longmapsto X_{\mathbf{e}}(t,x,w)$ can be defined as an element in $\bigcap_{1<p<\infty} \mathbf{D}_p^{-s}$ for some $s > 0$. On the other hand, under a reasonable conditions on the potential V, $M(t,w) \in \widetilde{\mathbf{D}}^{\infty} = \bigcap_{s>0} \bigcup_{1<p<\infty} \mathbf{D}_p^s$ and the expectation in (1.3) is now understood as the coupling of the generalized Wiener fuctional $\delta_y(X_{\mathbf{e}}(t,x,w))$ and the test Wiener functional $M(t,w)$. By the continuity or differentiability, described in terms of Sobolev spaces, of $\delta_y(X_{\mathbf{e}}(t,x,w))$ in x and y, we can deduce the regularity of $p(t,x,y)$ in x and y. Also, we can obtain the asymptotic expansion as $t \downarrow 0$ of $p(t,x,y)$ through the asymptotic expansion of the functional in the sense of Sobolev spaces, cf. [11].

In this note, we will illustrate an example of application of the representation (1.3) for $p(t,x,y)$ by a generalized Wiener functional integration in proving the following asymptotic property: Let M be compact with dimension d and the potential V be given by

$$(1.4) \qquad V(x) = \lambda^2 V_1(x) + \lambda V_2(x), \qquad \lambda > 0$$

where $V_i(x)$, $i = 1,2$, are smooth, $V_1(x) \geq 0$ everywhere on M and $V_1(x) = 0$ only at two different points a and b in M where the Hessians $\nabla^2 V_1$ are non-degenerate. Let $p^\lambda(t,x,y)$ be the fundamental solution to heat equation

(1.5) $\partial u/\partial t = \frac{1}{2}\Delta u - \lambda^2 V_1 u - \lambda V_2 u.$

Then, for every $T > 0$, the following holds:

(1.6) $\lim_{\lambda\uparrow\infty} \int_M p^\lambda(T/\lambda, x, x)dx$ (dx: Riemannian volume)

$$= \exp\{-TV_2(a)\} \Pi_{k=1}^{d}\{\exp(T\omega_k(a)/2) - \exp(-T\omega_k(a)/2)\}^{-1}$$

$$+ \exp\{-TV_2(b)\} \Pi_{k=1}^{d}\{\exp(T\omega_k(b)/2) - \exp(-T\omega_k(b)/2)\}^{-1}$$

where $\omega_k(a)^2$ $(\omega_k(a) > 0)$ and $\omega_k(b)^2$ $(\omega_k(b) > 0)$, $k = 1,2,\ldots,d$, are eigen-values of the Hessians $\nabla^2 V_1$ at a and b, respectively.

2. A proof of (1.6).

We set $\lambda = \varepsilon^{-2}$, $\varepsilon > 0$. Let $O(M)$ be the orthonormal frame bundle over M and $L_1,\ldots,L_d$ be the system of the canonical horizontal vector fields on $O(M)$ (cf.[3], Chapter V, 4 for these notions). Let $r = (x,\mathbf{e}) \in O(M)$ and $r^\varepsilon(t)$ be the solution to the stochastic differential equation (SDE)

(2.1) $dr(t) = \varepsilon \Sigma_{i=1}^{d} L_i(r(t))\circ dw^i(t)$, $r(0) = r$

where $w = (w^i(t)) \in W$. Let $X_{\mathbf{e}}^\varepsilon(t,x,w)$ be the projection of $r^\varepsilon(t) \in O(M)$ onto M. Then, noting the self-similarity of Wiener measure, $\{X_{\mathbf{e}}^\varepsilon(t,x,w)\} \overset{\mathcal{L}}{\sim} \{X_{\mathbf{e}}^1(\varepsilon^2 t,x,w)\}$ and hence

(2.2) $$p^{\varepsilon^{-2}}(\varepsilon^2 T,x,x) = E\Big[\exp\{-\varepsilon^{-2}\int_0^T V_1(X_{\mathbf{e}}^\varepsilon(t,x,w))dt - \int_0^T V_2(X_{\mathbf{e}}^\varepsilon(t,x,w)) dt\}\, \delta_x(X_{\mathbf{e}}^\varepsilon(T,x,w))\Big]$$

in the sense of a generalized Wiener functional expectation explained above. For disjoint neighborhoods $U(a)$ and $U(b)$ of a and b, respectively, we set

(2.3) $$\int_M p^{\varepsilon^{-2}}(\varepsilon^2 T,x,x)dx = \int_{M\setminus[U(a)\cup U(b)]}'' + \int_{U(a)\cup U(b)}'' = I_1 + I_2 .$$

First we note that a constant $K_1 > 0$ exists such that

(2.4) $I_1 = O(\exp(-K_1\varepsilon^{-2}))$

We may assume that $V_1(x) \geq c > 0$ for every $x \in M\setminus(U(a)\cup U(b))$.

By a standard estimate, $K_2 > 0$ exists such that

(2.5) $P\left[\frac{1}{T}\int_0^T (V_1(X_\varepsilon^\varepsilon(t,x,w)) - V_1(x))^2 dt > (c/2)^2\right] \leq \exp(-K_2\varepsilon^{-2})$.

Also, by an integration by parts on the Wiener space and by a standard estimate of the Malliavin covariance (cf. [8], [11]), we see that the right-hand side (RHS) of (2.2) is equal to

(2.6) $E\left[\exp\{-\varepsilon^{-2}\int_0^T V_1(X_\varepsilon^\varepsilon(t,x,w))dt\} F(\varepsilon,w)\right]$

where $F(\varepsilon,w) \in L_2$ with an estimate

(2.7) $E(F(\varepsilon,w)^2) \leq K_3\, \varepsilon^{-n}$

for some $K_3 > 0$ and $n > 0$. Noting that $\int_0^T V_1(X_\varepsilon^\varepsilon(t,x,w))dt \geq (Tc)/2$ if $\frac{1}{T}\int_0^T (V_1(X_\varepsilon^\varepsilon(t,x,w)) - V_1(x))^2 dt \leq (c/2)^2$ and $x \in M\setminus(U(a)\cup U(b))$, it is easy to conclude (2.4) from (2.5),(2.6) and (2.7).

We have

$$I_2 = \int_{U(a)} p^{\varepsilon^{-2}}(\varepsilon^2 T,x,x)dx + \int_{U(b)} p^{\varepsilon^{-2}}(\varepsilon^2 T,x,x)dx$$

and we may consider each of these terms separately. We may assume that $U(a) \Subset U_1(a) \Subset U_2(a)$ and $U_2(a)$ is a coordinate neighborhood of M. We extend the components $(g_{ij}(x))$ of the metric tensor in this coordinate system restricted to $U_1(a)$, $V_1(x)|_{U_1(a)}$ and $V_2(x)|_{U_1(a)}$ to R^d so that $(g_{ij}(x))$ is uniformly positive definite and, outside $U_2(a)$, $g_{ij}(x) = \delta_{ij}$, $V_1(x) = V_2(x) = 0$. It is a standard result (which can also be proved by a probabilistic method) that if $\tilde{p}^{\varepsilon^{-2}}(t,x,y)$ is the fundamental solution to the heat equation on R^d corresponding to these extended $(g_{ij}(x))$, $V_1(x)$ and $V_2(x)$, then

$$\sup_{x \in U(a)} \left|\tilde{p}^{\varepsilon^{-2}}(\varepsilon^2 T,x,x) - p^{\varepsilon^{-2}}(\varepsilon^2 T,x,x)\right| = O(e^{-K_4\varepsilon^{-2}}).$$

In this way, the problem is reduced to the case of R^d: It is sufficient to show that $\lim_{\varepsilon \downarrow 0} \int_{U(a)} \tilde{p}^{\varepsilon^{-2}}(\varepsilon^2 T,x,x)dx$ is given by the first term in the RHS of (1.6). In the following, we may and therefore actually assume that our Euclidean coordinate on R^d is a normal coordinate for $(g_{ij}(x))$ around a so that $a = 0$, $g_{ij}(a) = \delta_{ij}$ and $\Gamma_{jk}^i(a) = 0$: Also, we may assume that $U(a) = \{|x| < \delta\}$, $\delta > 0$.

<u>Lemma 1</u>. If $\xi \in R^d$, $|\xi| < \delta/\varepsilon$,

(2.8) $\varepsilon^d \tilde{p}^{\varepsilon^{-2}}(\varepsilon^2 T, \varepsilon\xi, \varepsilon\xi) \leq K_5 e^{-K_6\varepsilon^{-2}} + K_7 e^{-K_8|\xi|^2}$

where $K_5, \ldots, K_8$ are positive constants independent of $\varepsilon \in (0,1]$ and ξ.

Proof. Let $X^\varepsilon(t,x,w)$ be the solution to SDE

(2.9) $$dX_t^i = \varepsilon \Sigma_{j=1}^d \sigma^{ij}(X_t)dw^j(t) - \varepsilon^2/2 \; \Sigma_{k,j=1}^d g^{kj}(X_t)\Gamma_{kj}^i(X_t)dt$$

$$X_0 = x$$

where $(\sigma^{ij}(x))$ is the square root of $(g^{ij}(x)) = (g_{ij}(x))^{-1}$. Then

$$J := \varepsilon^d \tilde{p}^{\varepsilon^{-2}}(\varepsilon^2 T,x,x) = E\Big[\exp\{-\varepsilon^{-2}\int_0^T V_1(X^\varepsilon(t,x,w))dt$$
$$- \int_0^T V_2(X^\varepsilon(t,x,w))dt\} \; \delta_0((X^\varepsilon(T,x,w)-x)/\varepsilon)\Big]$$

and the Wiener functional $(X^\varepsilon(T,x,w)-x)/\varepsilon$ is easily seen to be non-degenerate in the sense of Malliavin uniformly in ε and $x \in R^d$. Hence, by an integration by parts on the Wiener space,

$$J = E\Big[\exp\{-\varepsilon^{-2}\int_0^T V_1(X^\varepsilon(t,x,w)dt\} \; G(\varepsilon,x,w)\Big]$$

with an estimate $E(G(\varepsilon,x,w)^2) = O(\varepsilon^{-k})$ for some $k > 0$ uniformly in x and also, if $\eta > 0$, $E(G(\varepsilon,x,w)^2; \sup_{0\le t\le T} |X^\varepsilon(t,x,w)| \le \eta)$ $\le |P(|x|/\varepsilon)|$ where P is some polynomial. Then

$$J = E\Big[\exp\{-\varepsilon^{-2}\int_0^T V_1(X^\varepsilon(t,x,w))dt\} \; G(\varepsilon,x,w); \; \sup_{0\le t\le T}|X^\varepsilon(t,x,w)$$
$$- x| \le \eta\Big] + E\Big[\quad // \quad ; \; \sup_{0\le t\le T}|X^\varepsilon(t,x,w) - x| > \eta\Big] \quad (\eta > 0)$$
$$= J_1 + J_2 .$$

By a standard estimate,

$$|J_2| \le K_5 e^{-K_6 \varepsilon^{-2}} .$$

We may choose $U(a) = \{|x| < \delta\}$ and $\eta > 0$ so that, if $U_\eta(a) = \{|x| < \delta + \eta\}$,

$$\inf_{x \in U_\eta(a)} \langle \nabla^2 V_1(x)\theta, \theta\rangle \geqq c_1 |\theta|^2 , \quad \theta \in R^d$$

where c_1 is a positive constant. Note that

$$V_1(y) = V_1(y) - V_1(0) = \tfrac{1}{2}\langle \nabla^2 V_1(z)y, y\rangle \ge \tfrac{c_1}{2} |y|^2$$

if $y \in U_\eta(a)$.

Hence, if $x \in U(a)$,

$$|J_1| \le E\Big[\exp(-c_1\varepsilon^{-2}\int_0^T |X^\varepsilon(t,x,w)|^2 dt)\Big]^{1/2} E(G^2; \sup_{0\le t\le T}|X^\varepsilon(t,x,w)| \le \delta + \eta)^{\frac{1}{2}}.$$

Setting $x = \varepsilon\xi$ and applying the following lemma, it is easy to conclude that

$$|J_1| \leqq K_7 e^{-K_8|\xi|^2} .$$

Lemma 2. If $x \in R$ and $\{B(t), \mathcal{F}_t\}$ is a one-dimensional $\{\mathcal{F}_t\}$-Brownian motion such that $B(0) = 0$ and if ϕ_t is an $\{\mathcal{F}_t\}$- adapted strictly increasing process such that $d\phi_t \le c\, dt$ and $\phi_T^{-1} \le c$ for some constant $c > 0$, then a constant $K > 0$ exists such that

$$E(\exp\{-\int_0^T (x+B(\phi_s^{-1}))^2 ds\}) \le e^{-K|x|^2} .$$

Proof is easily reduced to the known case of $\phi_t \equiv t$.

Now,

$$H := \int_{|x|<\delta} \tilde{p}^{\varepsilon^{-2}}(\varepsilon^2 T, x, x)dx = \varepsilon^d \int_{|\xi|<\delta/\varepsilon} \tilde{p}^{\varepsilon^{-2}}(\varepsilon^2 T, \varepsilon\xi, \varepsilon\xi)d\xi$$

$$= \varepsilon^d \int_{|\xi|<\delta/\varepsilon,\ |\xi|\le N} + \varepsilon^d \int_{|\xi|<\delta/\varepsilon,\ |\xi|>N}$$

$$= H_1 + H_2.$$

By Lemma 1, it is easy to deduce that

$$H_2 \le A_1 e^{-A_2 N^2} + A_3 e^{-A_4 \varepsilon^{-2}} \tag{2.10}$$

where $A_1, \ldots, A_4$ are positive constants independent of N and ε. We know (cf.[11]) that

$$X^\varepsilon(t,\varepsilon\xi,w) = \varepsilon\xi + \varepsilon w(t) + O(\varepsilon^2) \quad \text{in } D^\infty \text{ as } \varepsilon \downarrow 0$$

and this estimate is uniform in ξ, $|\xi| \le N$, if $N > 0$ is fixed. Hence

$$\delta_{\varepsilon\xi}(X^\varepsilon(T,\varepsilon\xi,w)) = \varepsilon^{-d}[\delta_0(w(T)) + O(\varepsilon)] \quad \text{in } \tilde{D}^{-\infty} \text{ as } \varepsilon \downarrow 0$$

uniformly in $|\xi| \le N$. Also,

$$\exp\{-\varepsilon^{-2}\int_0^T V_1(X^\varepsilon(t,\varepsilon\xi,w))dt - \int_0^T V_2(X^\varepsilon(t,\varepsilon\xi,w))dt\}$$

$$= \exp\{-\int_0^T \langle \frac{\nabla^2 V_1(0)}{2}(\xi+w(s)), \xi+w(s)\rangle ds - TV_2(0)\} + O(\varepsilon)$$

in D^∞ as $\varepsilon \downarrow 0$ uniformly in $|\xi| \leq N$.

Hence,

$$\lim_{\varepsilon\downarrow 0} H_1 = \int_{|\xi|\le N} E\Big[\exp\{-\int_0^T \langle \frac{\nabla^2 V_1(0)}{2}(\xi+w(s)),\xi+w(s)\rangle ds - TV_2(0)\} \times \delta_0(w(T))\Big] d\xi$$

and, combining this with (2.10), we can now conclude that

$$\lim_{\varepsilon \downarrow 0} H = \int_{R^d} E\Big[\exp\{-\int_0^T \Big\langle \frac{\nabla^2 V_1(0)}{2}(\xi+w(s)),\ \xi+w(s)\Big\rangle ds - TV_2(0)\}\ \delta_0(w(T))\Big]\ d\xi$$

The RHS can be computed by the following lemma, completing the proof of (1.6).

Lemma 3. If $\beta > 0$ and $w \in W$ is one-dimensional,

$$(2.11)\qquad \int_{-\infty}^{\infty} E(\exp\{-\beta^2/2\int_0^T (x+w(s))^2 ds\}\ \delta_0(w(T)))dx = \{\exp(\beta T/2) - \exp(-\beta T/2)\}^{-1}\ .$$

Proof is easily provided by the following formula (cf.[3]):

$$(2.12)\qquad E(\exp\{-\alpha\int_0^1 (x+w(s))^2 ds\}\delta_0(w(1))) = \frac{\exp\{-(2\alpha)^{1/2}\coth\{(2\alpha)^{1/2}\}(1-\operatorname{sech}\{(2\alpha)^{1/2}\})x^2\}}{(2\pi\sinh\{(2\alpha)^{1/2}\}(2\alpha)^{-1/2})^{1/2}},\quad \alpha > 0.$$

3. Concluding remarks.

(i) The asymptotic formula (1.6) is known in the problem of double wells, cf. Simon [9], from which we can obtain semi-classical asymptotics for eigenvalues of the Schrödinger operator

$$-\frac{1}{2}\Delta + \lambda^2 V_1 + \lambda V_1 \quad \text{as} \quad \lambda \uparrow \infty.$$

(ii) We know that a heat equation on differential forms can also be solved by a Wiener functional integration using the stochastic moving frame $r(t)$ as a basic tool, cf [3]. Hence the fundamental solution $p(t,x,y)$, which is a linear mapping from the fibre $\wedge T_y^*(M)$ on y of the exterior product of the cotangent bundle to the fibre $\wedge T_x^*(M)$ on x, is obtained as above by a generalized Wiener functional integration. If the same technique as we discussed above is applied to the fundamental solution of a heat equation on the Witten complex, we can prove the strong Morse inequalities as was first found by Bismut [2].

For similar applications of the Malliavin calculus to index theorems and fixed point formulas, we would refer to Bismut [1] and also, Ikeda-Watanabe [5].

References.

[1] J.-M. Bismut: The Atiyah-Singer theorems: a probabilistic approach, I,the index theorem, II, the Lefschetz fixed point formulas, J. Funct. Anal.57(1984), 56-99 and 329-348.

[2] J.-M. Bismut: The Witten complex and the degenerate Morse inequalities, J.Differential Geometry, 23(1986), 207-240

[3] N. Ikeda and S. Watanabe: Stochastic differential equations and diffusion processes, North-Holland/Kodansha, 1981

[4] N. Ikeda and S. Watanabe: An introduction to Malliavin calculus, Taniguchi Symp. SA, Katata 1982, Kinokuniya/North-Holland, 1984, 1-52

[5] N. Ikeda and S. Watanabe: Malliavin calculus of Wiener functionals and its applications, to appear in Proc.Stochastic Analysis Year, University of Warwick.

[6] M. Kac: On some connections between probability theory and differential and integral equations, Proc. Second Berkeley Symp., Univ. California Press, 1951, 189-215.

[7] M. Kac: Integration in function spaces and some of its application Fermi Lecture, Pisa, 1980.

[8] S. Kusuoka and D. W. Stroock: Applications of the Malliavin calculus, Part II, J. Fac. Sci. Univ. Tokyo, Sect. IA Math. 32 (1985), 1-76.

[9] B. Simon: Semiclassical analysis of low lying eigenvalues I. Non-degenerate minima: Asymptotic expansions, Ann.Inst. Henri-Poincare, Section A, 38(1983), 295-307.

[10] S. Watanabe: Lectures on stochastic differential equations and Malliavin calculus, Tata Institute of Fundamental Research/ Springer, 1984.

[11] S.Watanabe: Analysis of Wiener functionals (Malliavin calculus) and its applications to heat kernels, to appear in Ann. Probab.

Department of Mathematics
Kyoto University
Kyoto
606, Japan

A HEAVY TRAFFIC LIMIT THEOREM FOR G/M/∞ QUEUEING NETWORKS

Keigo Yamada

1. Introduction Let us consider a queueing network which consists of several service stations. Each station consists of several service channels which perform their services independently each other according to FCFS (First Come, First Served) service discipline with a common service time distribution. A customer, after the completion of service, may leave from the network or go to other channels according to fixed probabilities. Our concern here is to obtain an approximation for the process of the numbers of customers at service stations when at each station, the number of service channels is very large and the intensity of the customer arrival from outside of the network is also very large while the service time distribution at each station is fixed.

When the network consists of only a single station and there is no feedback, there is a work by Borovkov [1, Chap.2,2.2, Theorem 1] which can be stated roughly in the following way. He considers a sequence of such queueing systems in which we let $X_n(t)$ be the number of customers at time t in the n-th system and $A_n(t)$ be the number of arrivals of customers until time t. Suppose that $X_n(0)=0$ and there exists a nondecreasing deterministic function m(t) such that the processes $\tilde{A}_n(t)$, $n\geq 1$, defined by $\tilde{A}_n(t)=(A_n(t)-nm(t))/\alpha_n$, converge weakly to a continuous process $\xi(t)$ as $n\to\infty$, where α_n is a suitable normalization constant. Then putting $Q(t)=\int_0^t(1-F(t-u))dm(u)$ where $F(\cdot)$ is the distribution of service times, if we let the number of channels go to infinity in an appropriate way, suitably normalized processes $Y_n(t)$, $n\geq 1$, of the form $Y_n(t)=(X_n(t)-nQ(t))/\beta_n$ converge weakly to process Y(t) which can be represented as

$$(1)\qquad Y(t)=C\int_0^t(1-F(t-u))d\xi(u)+\theta(t),\quad C \text{ a constant},$$

where $\theta(t)$ is a Gaussian process independent of $\xi(t)$.

In this paper we try to obtain an analogous result for queueing networks mentioned above. The essential restriction on our model is that service times are exponentially distributed. (See, however, Section 5 where the network in which distributions of service times are phase-type is treated.) By making such a restriction, the network is almost Markovian and, by a device due to Bremaud [2, V], we are able to express the numbers of customers at each station as a solution of a stochastic equation, and this enables us to use martingale theoretical approach to our problem.

The same problem as ours was considered in Whitt [3]. Assuming that all but one external arrival process is Poisson, with the non-Poisson arrival process being a renewal process, embedded Markov chains are obtained by looking at the system at renewal epochs. Whitt then proposes to apply some general limit theorems to obtain diffusion limits This approach seems, however, not appropriate for revealing the detailed structure of limit processes as given in (1).

We denote by $D([0, T], R^1)$ the space of right continuous functions $f:[0, T]\to R^1$ having limits on the left with the Skorohod J_1 topology. We denote by $Y_n \overset{P}{\to} Y$ the convergence of the corresponding random variables Y_n and Y in probability and by $Y_n(t)\Rightarrow Y(t)$ the weak convergence of distributions of processes $Y_n(t)$ to the distribution of the process $Y(t)$. For any element z in $D([0, \infty), R^1)$ we will define $\Delta z(s)$ by $\Delta z(s) = z(s)-z(s-)$. The k-tuple product of $D([0, T], R^1)$ with the product topology is denoted by $D^k([0, T], R^1)$.

2. Basic model Let us consider a sequence of queuing networks introduced in Section 1. For the n-th network, we introduce the following data:

- K : the number of service stations,
- $A_n^k(t)$: the number of customers arriving at service station k until time t from outside the network,
- S_n^k : the number of service channels at service station k, possibly infinite,
- p_{ij} : the probability with which a customer leaving station i goes to station j, not depending on index n,
- $D_k^\ell(t)$, $\ell=1,2,\ldots$: a sequence of independent Poisson processes with intensity parameter μ_k,
- $X_n^k(t)$: the number of customers at station k at time t.

We shall assume the following condition (A1):

(A1) For each $n\geq 1$, $X_n^k(0)$, $\{D_k^\ell(t), \ell\geq 1\}$, $A_n(t)=(A_n^1(t),\ldots,A_n^K(t))$, $k=1,\ldots,K$ are defined on a probability space $(\Omega, \mathcal{F}, P)$ and are mutually independent. $\{D_k^\ell(t), \ell\geq 1\}$, $A_n^k(t)$, $k=1,\ldots,K$, have no common discontinuities with probability one. Furthermore, for $1\leq k\leq K$, $E(X_n^k(0))^2<\infty$ and

$E(A_n^k(t))^2<\infty$ for each t, and $\int_0^T E(A_n^k(t)^2 dt<\infty$ for each $T>0$.

Let, for $t\geq 0$, $\mathcal{F}_n(t)$ be the sub-σ-field of $\mathcal{F}$ generated by $X_n^k(0)$, $A_n^k(s)$, $\{D_k^\ell(s), \ell\geq 1\}$, $s\leq t$, $k=1,\ldots,K$. Following Bremaud [2, V], let us define the processes $X_n^k(t)$, $1\leq k\leq K$, as the unique solution of the following equation:

$$X_n^k(t)=X_n^k(0)+A_n^k(t)+\sum_{i=1}^{K} p_{ik}B_n^i(t)-B_n^k(t), \tag{2}$$

$$B_n^k(t)=\sum_{\ell=1}^{S_n^k}\int_0^t I(X_n^k(s-)\geq\ell)dD_k^\ell(s), \quad 1\leq k\leq K.$$

Note that the process $D_k^\ell(t)$ can be written as

$$D_k^\ell(t)=\mu_k t+m_k^\ell(t), \quad 1\leq k\leq K, \quad \ell\geq 1$$

where $m_k^\ell(t)$ is a square integrable $F_n(t)$-martingale with the quadratic variational process $\langle m_k^\ell\rangle(t)=\mu_k t$.
Then equation (2) can be written as

$$X_n^k(t)=X_n^k(0)+A_n^k(t)+\sum_{i=1}^{K} q_{ik}\int_0^t \mu_i . S_n^i\wedge X_n^i(s)ds \tag{3}$$

$$+\sum_{i=1}^{K} q_{ik}\sum_{\ell=1}^{S_n^i}\int_0^t I(X_n^i(s-)\geq\ell)dm_i^\ell(s), \quad 1\leq k\leq K$$

where $q_{ik}=p_{ik}$ $(i\neq k)$, $=p_{kk}-1$ $(i=k)$.

A fundamental fact which we will use is the following.

<u>Lemma 1</u> Let

$$M_n^k(t)=\sum_{\ell=1}^{S_n^k}\int_0^t I(X_n^k(s-)\geq\ell)dm_k^\ell(s), \quad 1\leq k\leq K.$$

Then $M_n^k(t)$, $1\leq k\leq K$, are square integrable $F_n(t)$-martingales with

$$\langle M_n^i, M_n^j\rangle(t)=\delta_{ij}\mu_i\int_0^t S_n^i\wedge X_n^i(s)ds$$

where $\delta_{ij}=1$ $(i=j)$, $=0$ $(i\neq j)$.

<u>Proof</u> First note that even if $S_n^k=\infty$, $M_n^k(t)$ is well defined as

$$M_n^k(t)=\lim_{L\to\infty}\sum_{\ell=1}^{L}\int_0^t I(X_n^k(s-)\geq\ell)dm_k^\ell(s)$$

$$=\lim_{L\to\infty}\Big[\sum_{\ell=1}^{L}\int_0^t I(X_n^k(s-)\geq\ell)dD_k^\ell(s)-\sum_{\ell=1}^{L}\int_0^t I(X_n^k(s-)\geq\ell)\mu_k ds\Big]$$

$$=\sum_{\ell=1}^{\infty}\int_0^t I(X_n^k(s-)\geq\ell)dD_k^\ell(s)-\int_0^t \mu_k X_n^k(s)ds.$$

In view of (A1), the assertion of the lemma is trivial when S_n^i and S_n^j are both finite. Hence let us suppose that $S_n^i=S_n^j=\infty$. To see that $M_n^i(t)$ is a square integrable $F_n(t)$-martingale, it suffices to show that $\sup_L E|\tilde{M}_i^L(t)|^4<\infty$ for each t, where

$$\tilde{M}_i^L(t)=\sum_{\ell=1}^{L}\int_0^t I(X_n^i(s-)\geq\ell)dm_i^\ell(s).$$

Since $|\Delta\tilde{M}_i^L(t)|\leq 1$ for any t, we have the following inequality:

$$E(\sup_{0\leq t\leq T}|\tilde{M}_i^L(t)|^4)\leq\gamma_1 E<\tilde{M}_i^L>(T)+\gamma_2 E(<\tilde{M}_i^L>(T))^2$$

(Jacod[4, Lemma 2.23]), where constants γ_1, γ_2 do not depend on $\tilde{M}_i^L(t)$. But

$$<\tilde{M}_i^L>(t)=\int_0^t\mu_i\cdot L\wedge X_n^i(s)ds\leq\int_0^t\mu_i X_n^i(s)ds.$$

Then since

$$E(<\tilde{M}_i^L>(t))^2\leq\mu_i^2t^2\int_0^t E(X_n^i(s))^2ds\leq Ct^2\mu_i^2\sum_{k=1}^{K}\int_0^t E(|X_n^k(0)|^2+|A_n^k(s)|^2)ds$$

(note that $X_n^i(s)\leq\sum_{k=1}^{K}(X_n^k(0)+A_n^k(s))$), C a constant, we have $\sup_L E|\tilde{M}_i^L(t)|^4<\infty$. To see the second assertion, note that since $\tilde{M}_i^L(t)$ and $\tilde{M}_j^L(t)$ have no common discontinuities with probability one,

$$<\tilde{M}_i^L,\ \tilde{M}_j^L>(t)=\delta_{ij}\int_0^t\mu_i\cdot L\wedge X_n^i(s)ds.$$

Hence

$$\tilde{M}_i^L(t)\tilde{M}_j^L(t)-\delta_{ij}\int_0^t\mu_i\cdot L\wedge X_n^i(s)ds$$

is an $F_n(t)$-martingale. Then again noting $\sup_L E|\tilde{M}_k^L(t)|^4<\infty$, $1\leq k\leq K$ and $E\int_0^t(X_n^i(s))^2ds<\infty$, and letting L to infinity,

$$\tilde{M}_n^i(t)\tilde{M}_n^j(t)-\delta_{ij}\int_0^t\mu_i X_n^i(s)ds$$

is an $F_n(t)$-martingale, which implies the desired result.

3. Basic results In addition to condition (A1), we further assume

the following conditions (A2-4):

(A2) For each k ($1\leq k\leq K$), there exists a sequence of constants $\{\alpha_n^k, n\geq 1\}$ and a deterministic function $a_k(t)$ such that $\alpha_n^k\to\infty$ ($n\to\infty$) and $\tilde{A}_n(t)\Rightarrow\xi(t)$ in $D^K([0, T], R^1)$, T arbitrary. Here $\tilde{A}_n(t)=(\tilde{A}_n^1(t),\dots,\tilde{A}_n^K(t))$ is defined as

$$\tilde{A}_n^k(t)=\frac{A_n^k(t)-na_k(t)}{\alpha_n^k},$$

and $\xi(t)=(\xi_1(t),\dots,\xi_K(t))$ is a continuous process. Further there exists a measurable function b(t) such that for all n,

$$\sum_{k=1}^{K}\frac{1}{(\alpha_n^k)^2}(EX_n^k(0)+EA_n^k(t))\leq b(t) \tag{4}$$

$$\int_0^t b(s)ds<\infty \quad \text{for any } t>0.$$

(A3) As $n\to\infty$, $\sqrt{n}/\alpha_n^k \to \alpha_k$ ($1\leq k\leq K$) and

$$\left(\frac{X_n^1(0)}{(\alpha_1\ \alpha_n^1)^2},\dots,\frac{X_n^K(0)}{(\alpha_K\ \alpha_n^K)^2}\right)\Rightarrow(Q_1(0),\dots,Q_K(0))\equiv Q(0),$$

where Q(0) is a constant vector.

(A4) Define deterministic processes $\{Q_k(t), 1\leq k\leq K\}$ as the unique solution of

$$Q_k(t)=Q_k(0)+a_k(t)+\sum_{i=1}^{K} q_{ik}\mu_i\int_0^t Q_i(s)ds, \quad 1\leq k\leq K.$$

Then for each k ($1\leq k\leq K$), $(S_n^k-nQ_k(s))/\alpha_n^k\geq 0$ for all n and s and, as $n\to\infty$,

$$\frac{S_n^k-nQ_k(s)}{\alpha_n^k}\to\infty \text{ for each s.}$$

Let us define a sequence of the processes $\{Y_n^k(t), n\geq 1\}$, $1\leq k\leq K$, by

$$Y_n^k(t)=\frac{X_n^k(t)-nQ_k(t)}{\alpha_n^k}, \quad 1\leq k\leq K.$$

Our concern is the limiting behavior of the K-dimensional processes $Y_n(t)=(Y_n^1(t),\dots,Y_n^K(t))$, $n\geq 1$, as n tends to infinity.

To state our result, let us assume

(A5) $Y_n(0)\Rightarrow Y(0)$.

Further let $B(t)=(B_1(t),\dots,B_k(t))$ be a standard K-dimensional Brownian motion which is independent of Y(0) and $\xi(t)$. Then we have the following

<u>Theorem 1</u> Assume Conditions (A1-5). We also assume that the equation

$$(5)\qquad Y^k(t)=Y^k(0)+\xi_k(t)+\sum_{i=1}^{K} q_{ik}\mu_i(\alpha_i/\alpha_k)\int_0^t Y^i(s)ds + \sum_{i=1}^{K} q_{ik}(\alpha_i^2/\alpha_k)\int_0^t \sqrt{\mu_i Q_i(s)}dB_i(s),\quad 1\le k\le K$$

has a unique (in the law sense) solution $Y(t)=(Y^1(t),\dots,Y^K(t))$. Then $Y_n(t)\Rightarrow Y(t)$ in $D^K([0,T],R^1)$, T arbitrary.

Remark 1 Condition (4) of (A2) is usually satisfied in applications. As a typical case, let $A_n^k(t)=A_k(nt)$, $1\le k\le K$, where $A_k(t)$ is a renewal process with mean interarrival time $1/\lambda_k$. Then it is well known that if σ_k^2 is the variance of interarrival time.

$$\frac{A_n^k(t)-n\lambda_k t}{\alpha_n^k}\Rightarrow \xi_k(t),\quad \alpha_n^k=\sigma_k(1/\lambda_k)^{-\frac{3}{2}}\sqrt{n}$$

in $D([0,T],R^1)$, where $\xi_k(t)$ is a standard Wiener process. In this case we know that

$$(EA_k(nt)+1)\mu_M\le(E\bar{A}_k(nt)+1)\mu_M\le nt+M,$$

where $\bar{A}_k(t)$ is a renewal process generated from $A_k(t)$ by truncating interarrival time by a constant M and μ_M is the mean inter arrival time of $\bar{A}_k(t)$ (Karlin and Taylor [5, p.189]). Then

$$\frac{EA_k(nt)}{n}\le\frac{t}{\mu_M}+\frac{M}{n\mu_M},$$

and hence, as long as $\sup_n EX_n^k(0)/(\alpha_n^k)^2<\infty$, condition (4) of (A2) is satisfied.

Remark 2 As for condition (A4), in order that $(S_n^k-nQ_k(s))/\alpha_n^k\ge 0$ for all n and s, it is necessary that $Q_k(t)$ is bounded if $S_n^k<\infty$. It is not hard to see that a sufficient condition for $Q_k(t)$ to be bounded is that 1) real parts of all eigen values of the matrix A, where $A_j^i=q_{ij}\mu_i$, are negative and 2) $\int_0^t e^{-cs}|da(s)|<\infty$, where $|da(s)|=|da_1(s)|+\cdots+|da_K(s)|$ and $|da_k(s)|$ denotes the measure induced by the total variation of $a_k(\cdot)$ and

$$\mathrm{Re}\lambda_i<-c<0\quad (i=1,\dots,K)$$

$(\lambda_1,\dots,\lambda_K$ are eigenvalues of A). Now suppose $Q_k(t)$, $1\le k\le K$, are bounded and $a>Q_k(t)$. Then as long as $S_n^k/(na)\ge 1$

$$\frac{S_n^k-nQ_k(s)}{\alpha_n^k}\to\infty\quad \text{for each } s$$

as $n\to\infty$ by condition (A3).

Remark 3 Suppose that $\xi(t)=(\xi_1(t),\dots,\xi_K(t))$ is a semimartingale. Then (5) has a unique solution (Ikeda and Watanabe [6, III, Theorem

2.1]), and using Ito's formula we can easily show that the solution Y(t) is given by

$$Y(t)=e^{Bt}Y(0)+e^{Bt}\int_0^t e^{-Bs}d\xi(s)+e^{Bt}\int_0^t e^{-Bs}d\eta(s), \tag{6}$$

where B is a K-dimensional matrix with $B_j^i=q_{ji}(\alpha_j/\alpha_i)$ and $\eta(t)=(\eta_1(t),\ldots,\eta_K(t))$ with

$$\eta_k(t)=\sum_{i=1}^K q_{ik}(\alpha_i^2/\alpha_k)\int_0^t\sqrt{\mu_i Q_i(s)}dB_i(s), \quad 1\le k\le K.$$

The expression (6) shows that the limit process Y(t) can be expressed as a sum of two independent components one of which (i,e., $e^{Bt}\int_0^t e^{-Bs}d\eta(s)$) is a K-dimensional Gaussian process, which is an analogue of Borovkov's result [1].

4. Proof of Theorem 1. The proof consists of two steps. In the first step we show the tightness of the family of processes $\{Y_n(t), n\ge 1\}$. In the second step we must identify any weak limit of $\{Y_n(t), n\ge 1\}$ as the unique solution of (5). For the first step, we need a lemma.

Lemma 2 For each $t\ge 0$ and $1\le k\le K$,

$$\sup_n P(\sup_{0\le s\le t}|Y_n^k(s)|\ge N)\to 0$$

as $N\to\infty$.

proof Putting $Q_n^i(t)=nQ_i(t)$, from (3) we have

$$Y_n^k(t)=Y_n^k(0)+\tilde{A}_n^k(t)+\sum_{i=1}^K\frac{q_{ik}}{\alpha_n^k}M_n^i(t) \tag{7}$$

$$+\sum_{i=1}^K q_{ik}\int_0^t\mu_i\frac{1}{\alpha_n^k}[S_n^i\wedge(\alpha_n^iY_n^i(s)+Q_n^i(s))-Q_n^i(s)]ds$$

Then noting the trivial inequality

$$|(S_n^i-Q_n^i(s))\wedge(\alpha_n^iY_n^i(s))|\le\alpha_n^i|Y_n^i(s)|$$

(since $S_n^i-Q_n^i(s)\ge 0$), we have

$$|Y_n^k(t)|\le|Y_n^k(0)|+|\tilde{A}_n^k(t)|+\sum_{i=1}^K|q_{ik}|\int_0^t\mu_i\frac{\alpha_n^i}{\alpha_n^k}|Y_n^i(s)|ds+\sum_{i=1}^K\frac{|q_{ik}|}{\alpha_n^k}|M_n^i(t)|.$$

Let

$$Z_n(t)=\sum_{k=1}^K|Y_n^k(t)|.$$

Then from the above inequality we have

$$Z_n(t)\leq\alpha_n(t)+C\int_0^t Z_n(s)ds,\quad C>0$$

where C is a constant (hereafter C is used for various constants) and

$$\alpha_n(t)=\sup_{0\leq s\leq t}\sum_{k=1}^{K}(|Y_n^k(0)|+|\tilde{A}_n^k(s)|)+C\sup_{0\leq s\leq t}\sum_{k=1}^{K}\frac{|M_n^k(s)|}{\alpha_n^k}.$$

Hence by Gronwall's inequality,

$$Z_n(t)\leq\alpha_n(t)+C\int_0^t e^{C(t-s)}\alpha_n(s)ds$$

and since $\alpha_n(t)$ is increasing,

$$Z_n(t)\leq C(t)\alpha_n(t),\quad (C(t)=1+Ce^{Ct}t).$$

Then

$$P(\sup_{0\leq s\leq t} Z_n(s)\geq N)\leq P(\alpha_n(t)\geq\frac{N}{C(t)})$$

$$\leq P(\sum_{k=1}^{K}|Y_n^k(0)|\geq\frac{N}{3C(t)})+P(\sup_{0\leq s\leq t}\sum_{k=1}^{K}|\tilde{A}_n^k(s)|\geq\frac{N}{3C(t)})$$

$$+\sum_{i=1}^{K}P(\sup_{0\leq s\leq t}\frac{|M_n^i(s)|}{\alpha_n^i}\geq\frac{N}{3C(t)KC})$$

By Doob-Kolmogorov's inequality

$$P(\sup_{0\leq s\leq t}\frac{|M_n^i(s)|}{\alpha_n^i}\geq\frac{N}{C})\leq\frac{C^2}{N^2}E\frac{(M_n^i(t))^2}{(\alpha_n^i)^2}$$

$$=\frac{C^2}{N^2}\frac{1}{(\alpha_n^i)^2}E\int_0^t\mu_i\cdot S_n^i\Lambda X_n^i(s)ds$$

$$\leq\frac{C^2}{N^2}\mu_i\int_0^t\frac{1}{(\alpha_n^i)^2}(E\sum_{k=1}^{K}(X_n^k(0)+A_n^k(s)))ds$$

$$\leq\frac{C^2}{N^2}\mu_i\int_0^t b(s)ds.$$

From these, noting that

$$\sup_n P(\sum_{k=1}^{K}|Y_n^k(0)|\geq\frac{N}{3C(t)})\to 0$$

$$\sup_n P(\sup_{0\leq s\leq t}\sum_{k=1}^{K}|\tilde{A}_n^k(s)|\geq\frac{N}{3C(t)})\to 0$$

as $N\to\infty$ since $Y_n(0)\Rightarrow Y(0)$ and $\tilde{A}_n(t)\Rightarrow\xi(t)$ in $D^K([0, T], R^1)$, T arbitrary, we reach the conclusion.

Step 1 In this step we shall prove that the family of processes $\{Y_n(t),$

$n\geq 1\}$ is tight in $D^K([0, T], R^1)$ and any weak limit is a continuous process. For this it suffices to show this fact for each component of $Y_n(t)$. Let

$$I_N(x)=\begin{cases}1 & |x|\leq N\\ 0 & |x|>N, \qquad x\in R^K,\end{cases}$$

and let us define the process $\tilde{Y}_n(t)=(\tilde{Y}_n^1(t),\ldots,\tilde{Y}_n^K(t))$ as the unique solution of the following equation:

$$\tilde{Y}_n^k(t)=\tilde{Y}_n^k(0)+\tilde{A}_n^k(t)$$
$$+\sum_{i=1}^K q_{ik}\int_0^t \mu_i\frac{1}{\alpha_n^k}[S_n^i\wedge(\alpha_n^i\tilde{Y}_n^i(s)+Q_n^i(s))-Q_n^i(s)]I_N(\tilde{Y}_n(s))ds$$
$$+\sum_{i=1}^K \frac{q_{ik}}{\alpha_n^k}\sum_{\ell=1}^{S_n^i}\int_0^t I(\alpha_n^i\tilde{Y}_n^i(s-)+Q_n^i(s-)\geq\ell)I_N(\tilde{Y}_n(s-))dm_i^\ell(s), \quad 1\leq k\leq K.$$

Note that the solution of the above equation is unique since equation (7) has a unique solution. Let

$$\tau_n=\inf\{t;\ |Y_n(t)|>N\}.$$

Then $\tilde{Y}_n(t)=Y_n(t)$, $t<\tau_n$. To see the tightness of $\{Y_n^k(t),\ n\geq 1\}$ in $D([0, T], R^1)$, it suffices to show the following (Gihman and Skorohod [7, p.186]):

1) $\sup_n P(\sup_{0\leq t\leq T}|Y_n^k(t)|>N)\to 0$ as $N\to\infty$,

2) $\{\tilde{Y}_n^k(t),\ n\geq 1\}$ is tight in $D([0, T], R^1)$.

1) is direct from Lemma 2. To see 2), let us write

$$\tilde{Y}_n^k(t)=Y_n^k(0)+\tilde{A}_n^k(t)+Z_n^1(t)+Z_n^2(t)$$

where $Z_n^i(t)$, $i=1,2$, are obviously defined. Then

$$|Z_n^1(t)-Z_n^1(u)|\leq\sum_{i=1}^K(|q_{ik}|\mu_i\alpha_n^i/\alpha_n^k)\int_u^t|\tilde{Y}_n^i(s)\wedge\frac{S_n^i-Q_n^i(s)}{\alpha_n^i}|I_N(\tilde{Y}_n(s))ds$$
$$\leq C\sum_{i=1}^K\int_u^t|\tilde{Y}_n^i(s)|I_N(\tilde{Y}_n(s))ds$$
$$\leq C(N)(t-u),$$

where constants C and $C(N)$ are not dependent on n.
Thus $\{Z_n^1(t),\ n\geq 1\}$ is C-tight in $D([0, T], R^1)$ i.e., $\{Z_n^1(t),\ n\geq 1\}$ is tight in $D([0, T], R^1)$ and any weak limit of $\{Z_n^1(t),\ n\geq 1\}$ is a continuous process. Next we will show that

$$E[(Z_n^2(t)-Z_n^2(u))^2|F_n(u)]\leq C(t-u),\quad u\leq t$$

where C is not dependent on n. Indeed

$$E[(Z_n^2(t)-Z_n^2(u))^2|F_n(u)]$$

$$=\sum_{i=1}^{K}(q_{ik}\frac{\alpha_n^i}{\alpha_n^k})^2E[\int_u^t\frac{1}{(\alpha_n^i)^2}\cdot S_n^i\Lambda(\alpha_n^i\tilde{Y}_n^i(s)+Q_n^i(s))\cdot I_N(\tilde{Y}_n(s))\mu_i ds|F_n(u)]$$

$$\leq C\sum_{i=1}^{K}E[\int_u^t(\frac{1}{\alpha_n^i}|\tilde{Y}_n^i(s)|+\frac{Q_n^i(s)}{(\alpha_n^i)^2})I_N(\tilde{Y}_n(s))\mu_i ds|F_n(u)]$$

$$\leq C(t-u).$$

From this we can easily verify a tightness condition in Varadhan [8, p. 51] for $\{Z_n^2(t),\ n\geq 1\}$. Now since $\{Y_n^k(0)+\tilde{A}_n^k(t),\ n\geq 1\}$ and $\{Z_n^1(t),\ n\geq 1\}$ are C-tight, $\{\tilde{Y}_n^k(t),\ n\geq 1\}$ is tight in $D([0,\ T],\ R^1)$ (Jacod [4, p.310, Corollary 2.18]), and this completes 2). To see that any weak limit of $\{Y_n^k(t),\ n\geq 1\}$ is a continuous process, it suffices to note that

$$\sup_{0\leq s\leq T}|\Delta Y_n^k(s)|\leq\sup_{0\leq s\leq T}|\Delta\tilde{A}_n^k(s)|+\sup_{0\leq s\leq T}\max_{1\leq i\leq K}|(q_{ik}/\alpha_n^k)\Delta M_n^i(s)|$$

$$\leq\sup_{0\leq s\leq T}|\Delta\tilde{A}_n^k(s)|+\max_{1\leq i\leq K}|q_{ik}/\alpha_n^k|,$$

the weak limit of $\{\tilde{A}_n(t),\ n\geq 1\}$ is continuous, and the mapping $x\to\sup_{0\leq s\leq T}|\Delta x(s)|$, $x\in D([0,\ T],\ R^1)$, is continuous in $D([0,\ T],\ R^1)$.

Step 2 In this step we shall prove that any weak limit Y(t) of $\{Y_n(t),\ n\geq 1\}$ has the same distribution as the unique solution of (5). To this end, let

$$\psi_n(t)=(\tilde{A}_n^1(t),\ldots,\tilde{A}_n^K(t),\frac{M_n^1(t)}{\alpha_n^1},\ldots,\frac{M_n^K(t)}{\alpha_n^K})$$

$$\psi(t)=(\xi_1(t),\ldots,\xi_K(t),\alpha_1\int_0^t\sqrt{\mu_1Q_1(s)}dB_1(s),\ldots,\alpha_K\int_0^t\sqrt{\mu_KQ_K(s)}dB_K(s)),$$

and we first show that $\psi_n(t)\Rightarrow\psi(t)$ in $D^{2K}([0,\ T],\ R^1)$. Let us define the processes $\tilde{\psi}_n(t)$, $n\geq 1$, by

$$\tilde{\psi}_n(t)=(\tilde{A}_n^1(t),\ldots,\tilde{A}_n^K(t),\ \hat{M}_n^1(t),\ldots,\hat{M}_n^K(t)),$$

where

$$\hat{M}_n^k(t)=\frac{1}{\alpha_n^k}\sum_{\ell=1}^{S_n^k}\int_0^t I(Q_n^k(s-)\geq\ell)dm_k^\ell(s).$$

To see that $\psi_n\Rightarrow\psi$, it suffices to show the following (Billingsley [9, I, Theorem 4.1]):

1) $\tilde{\psi}_n(t) \Rightarrow \psi(t)$ in $D^{2K}([0, T], R^1)$,

2) $\sup_{0\leq t\leq T} |\psi_n(t)-\tilde{\psi}_n(t)| \overset{P}{\to} 0.$

Since $(\tilde{A}_n^1(t),\ldots,\tilde{A}_n^K(t))$, $\hat{M}_n^1(t),\ldots,\hat{M}_n^K(t)$ are independent, 1) follows if we can show that

$$\hat{M}_n^k(t) \Rightarrow \alpha_k\int_0^t \sqrt{\mu_k Q_k(s)}dB_k(s) \equiv M_k(t).$$

We have

$$<\hat{M}_n^k>(t)=\frac{1}{(\alpha_n^k)^2}\int_0^t S_n^k \Lambda Q_n^k(s)\cdot\mu_k ds$$

$$\to(\alpha_k)^2\int_0^t Q_k(s)\mu_k ds=<M_k>(t)$$

and

$$\sup_{0\leq s\leq T} |\Delta M_n^k(s)|<\frac{1}{\alpha_n^k} \to 0 \quad (n \to \infty),$$

and these imply the above desired convergence (Liptser and Shiryaev [10]). To see 2), it suffices to show that

$$\sup_{0\leq t\leq T} |\frac{1}{\alpha_n^k} M_n^k(t)-\hat{M}_n^k(t)| \overset{P}{\to} 0, \quad 1\leq k\leq K$$

But for this it is sufficent to show that

$$<(1/\alpha_n^k)M_n^k-\hat{M}_n^k>(T) \overset{P}{\to} 0$$

(Lenglart [11]). We have

$$<(1/\alpha_n^k)M_n^k-\hat{M}_n^k>(T)$$

$$=\frac{1}{(\alpha_n^k)^2} \sum_{\ell=1}^{S_n^k}\int_0^T\{I(\alpha_n^k Y_n^k(s)+Q_n^k(s)\geq\ell)-2I(\alpha_n^k Y_n^k(s)+Q_n^k(s)\geq\ell)I(Q_n^k(s)\geq\ell)$$

$$+I(Q_n^k(s)\geq\ell)\}\mu_k ds$$

$$=\int_0^T\{\frac{S_n^k}{(\alpha_n^k)^2}\Lambda(\frac{Y_n^k(s)}{\alpha_n^k} + \frac{Q_n^k(s)}{(\alpha_n^k)^2})-2\cdot\frac{S_n^k}{(\alpha_n^k)^2}\Lambda(\frac{Y_n^k(s)}{\alpha_n^k} + \frac{Q_n^k(s)}{(\alpha_n^k)^2})\Lambda\frac{Q_n^k(s)}{(\alpha_n^k)^2}$$

$$+\frac{S_n^k}{(\alpha_n^k)^2}\Lambda\frac{Q_n^k(s)}{(\alpha_n^k)^2}\}\mu_k ds \overset{P}{\to} 0.$$

In the above we have used the fact that

$$\frac{1}{\alpha_n^k} \sup_{0\le s\le T} |Y_n^k(s)| \overset{P}{\to} 0 \quad (n \to \infty),$$

which follows from the tightness of $\{Y_n^k(t), n\ge 1\}$. This completes 2). Now since weak limits $Y(t)$ and $\psi(t)$ are continuous, we may assume that $\sup_{0\le s\le T} |Y_n(s)-Y(s)|\to 0$ and $\sup_{0\le s\le T} |\psi_n(s)-\psi(s)|\to 0$ with probability one. Then

$$\int_0^t \frac{1}{\alpha_n^i}[S_n^i \Lambda(\alpha_n^i Y_n^i(s)+Q_n^i(s))-Q_n^i(s)]ds$$

$$=\int_0^t \frac{S_n^i-Q_n^i(s)}{\alpha_n^i} \Lambda Y_n^i(s)ds \to \int_0^t Y_i(s)ds$$

with probability one. Here note that

$$\left|\frac{S_n^i-Q_n^i(s)}{\alpha_n^i} \Lambda Y_n^i(s)\right| \le |Y_n^i(s)| \le \sup_{0\le s\le T} |Y_n^i(s)-Y_i(s)|+|Y_i(s)|$$

and hence the bounded convergence theorem applies in the above convergence. Thus by letting n tend to infinity in (7), we see that $Y(t)$ satisfies

$$Y_k(t)=Y_k(0)+\xi_k(t)+\sum_{i=1}^K q_{ik}\mu_i(\alpha_i/\alpha_k)\int_0^t Y_i(s)ds$$

$$+\sum_{i=1}^K q_{ik}(\alpha_i/\alpha_k)M_i(t), \quad 1\le k\le K$$

and this completes Step 2 owing to the uniqueness of the solution of (5).

5. An extension In the previous sections we assumed that service times are exponentially distributed. However our result can be extended to the case where service time distributions are phase-type by extending our queuing networks to bigger ones. (This idea is due to Whitt [3]). A (service time) distribution $H(t)$ is said to be phase-type, if it can be written in the form:

$$1-H(t)=\sum_{i=1}^{M} \prod_{k=1}^{k=i-1} (1-p_k)e^{-\beta t}\frac{(\beta t)^{i-1}}{(i-1)!}, \quad t\ge 0, \tag{8}$$

where $M,\beta>0$ and $0\le p_k<1$ $(1\le k\le M-1)$. It is significant that any service-time distribution can be approximated (in the weak convergence sense) by a sequence of phase-type distributions. The phasetype distribution $H(t)$ given in (8) is identical with the distribution of the staying time of a customer until leaving a network in which there exist M stations in a series. Each customer, after completing the service at phase k, leaves the network with probability p_k and moves on to phase k+1 with $1-p_k$ and receives the service without deley (i.e., there exist infinite number of service channels) according to an exponential distribution with parameter β.

Fig. 1.

Now consider a sequence of networks as in Section 2, but this time we assume that at station k, the service time distribution $H_k(t)$ is assumed to be phase-type, i.e.,

$$(9)\qquad 1-H_k(t)=\sum_{i=1}^{M_k}\prod_{j=1}^{j=i-1}(1-p_j^k)e^{-\beta_k t}\frac{(\beta_k t)^{i-1}}{(i-1)!}, \quad t\geq 0$$

We extend the n-th network by replacing the k-th ($1\leq k\leq K$) service station by the subnetwork representing the phase-type distribution $H_k(t)$ which consists of M_k stations (Fig. 2.)

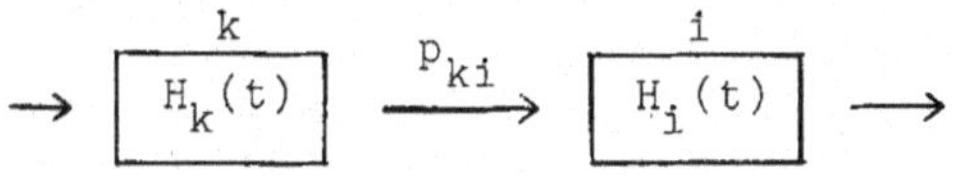

original n-th network

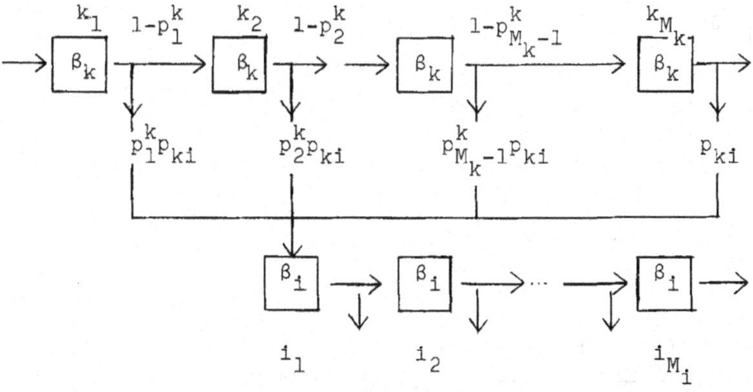

extended n-th netowrk

Fig. 2.

In the extended network, the i-th station of the subnetwork of the original k-th station will be denoted by ki ($1 \le i \le M_k$). Then the transition probability for the extended network is given by

$$\tilde{p}_{ki,k\ell} = \begin{cases} 1-p_i^k & \ell=i+1 \\ p_i^k p_{kk} & \ell=1 \\ 0 & \ell\ne 1,\ i+1 \end{cases} \qquad \tilde{p}_{ki,j\ell} = \begin{cases} p_i^k p_{kj} & \ell=1 \\ 0 & \ell\ne 1 \quad (k\ne j) \end{cases}$$

We let

$$\tilde{A}_n^{ki}(t) = \begin{cases} A_n^k(t) & i=1 \\ 0 & 1<i\le M_k, \end{cases} \qquad \tilde{a}_{ki}(t) = \begin{cases} a_k(t) & i=1 \\ 0 & 1<i\le M_k \end{cases}$$

and $\tilde{X}_n^{ki}(t)$ be the number of customers at the station ki in the n-th extended network with the initial condition $\tilde{X}_n^{ki}(0)=X_n^k(0)$, if $i=1$, $=0$ if $1<i\le M_k$. Then assuming that $S_n^k=\infty$ for all n and k, we have

$$X_n^k(t) = \sum_{i=1}^{M_k} \tilde{X}_n^{ki}(t).$$

(Note that in the extended network we assume the number of service channels at each station is infinite.) We also define $\tilde{Q}_{ki}(t)$, $1\le k\le K$, $1\le i\le M_k$, as the unique solution of

$$\tilde{Q}_{ki}(t)=\tilde{Q}_{ki}(0)+\tilde{a}_{ki}(t)+\sum_{j=1}^{K}\sum_{\ell=1}^{M_j}\tilde{q}_{j\ell,ki}\beta_j\int_0^t\tilde{Q}_{j\ell}(s)ds$$

where

$$\tilde{q}_{j\ell,ki} = \begin{cases} \tilde{p}_{j\ell,ki} & j\ell\ne ki \\ \tilde{p}_{j\ell,ki}-1 & j\ell=ki \end{cases}$$

$$\tilde{Q}_{ki}(0) = \begin{cases} Q_k(0) & i=1 \\ 0 & 1<i\le M_k. \end{cases}$$

We also define

$$(10) \qquad \tilde{Y}_n^{ki}(t) = \frac{\tilde{X}_n^{ki}(t)-n\tilde{Q}_{ki}(t)}{\alpha_n^k}, \qquad 1\le k\le K,\quad 1\le i\le M_k,$$

$$Q_k(t)=\sum_{i=1}^{M_k}\tilde{Q}_{ki}(t), \qquad Y_n^k(t)=\frac{X_n^k(t)-nQ_k(t)}{\alpha_n^k},$$

$$\tilde{\xi}_{ki}(t)=\begin{cases}\xi_k(t) & i=1\\ 0 & 1<i\le M_k\end{cases}$$

Let $\tilde{B}(t)=(\tilde{B}_{ki}(t),\ 1\le k\le K,\ 1\le i\le M_k)$ be a $M_1+\cdots+M_K$-dimensional standard Brownian motion independent of $\tilde{\xi}_{ki}(t)$ and $\tilde{Y}_{ki}(0)$, $1\le k\le K$, $1\le i\le M_k$, and let $\tilde{Y}_{ki}(t)$, $1\le k\le K$, $1\le i\le M_k$, be the unique solution of

$$(11)\qquad \tilde{Y}_{ki}(t)=\tilde{Y}_{ki}(0)+\tilde{\xi}_{ki}(t)+\sum_{j=1}^{K}\sum_{\ell=1}^{M_j}\tilde{q}_{j\ell,ki}\beta_j(\alpha_j/\alpha_k)\int_0^t\tilde{Y}_{j\ell}(s)ds$$
$$+\sum_{j=1}^{K}\sum_{\ell=1}^{M_j}\tilde{q}_{j\ell,ki}(\alpha_j^2/\alpha_k)\int_0^t(\beta_j\tilde{Q}_{j\ell}(s))^{\frac{1}{2}}dB_{j\ell}(s),\ 1\le k\le K,\ 1\le i\le M_k.$$

We know from Theorem 1 that $\tilde{Y}_n(t)\Rightarrow\tilde{Y}(t)$ in $D^{\tilde{K}}([0,\ T],\ R^1)$, T arbitrary, where $\tilde{K}=M_1+\cdots+M_K$. Since $Y_n^k(t)=\sum_{i=1}^{M_k}\tilde{Y}_n^{ki}(t)$, we have $(Y_n^1(t),\ldots,Y_n^K(t))\Rightarrow$ $(Y_1(t),\ldots,Y_K(t))$ in $D^K([0,\ T],\ R^1)$, T arbitrary, where

$$(12)\qquad Y_k(t)=\sum_{i=1}^{M_k}\tilde{Y}_{ki}(t).$$

Summarizing the above, we have

Theorem 2 In Theorem 1 let us assume that the service time distribution $H_k(t)$ at station k is given by (9) and $S_n^k=\infty$ $(1\le k\le K)$. We also assume that equation (11) has the unique (in the law sense) solution. Then $(Y_n^1(t),\ldots,Y_n^K(t))\Rightarrow(Y_1(t),\ldots,Y_K(t))$ in $D^K([0,\ T],\ R^1)$, T arbitrary, where $Y_n^k(t)$ and $Y_k(t)$ were defined in (10) and (12).

Remark From our discussion it will be apparent that Remark 3 still applies to Theorem 2. That is, if we suppose $\xi(t)=(\xi_1(t),\ldots,\xi_K(t))$ is a semimartingale, the limit process $Y(t)$ can be expressed as a sum of two independent components one of which is a K-dimensional Gaussian process which can be expressed as a Wiener integral. For the case where service time distributions are general and not necessarily phase-type, since these distributions can be approximated by phase-type distributions, we conjecture that the limit process $Y(t)$ still can be expressed as a sum of two independent components one of which is a Gaussian process. As is remarked in Borovkov [1, 2.2 Remark 1], this Gaussian part will not be expressed as a Wiener integral any more, but will be limit of a sequence of Wiener integrals.

References

[1] A.A. Borovkov, Asymptotic Method in Queuing Theory, John Wiley and Sons, 1984

[2] P. Brémaud, Point Processes and Queues, Martingale Dynamics, Springer-Verlag, 1981

[3] W. Whitt, On the heavy traffic limit theorem for GI/G/∞ Queues, Adv. Appl. Probab. 14 (1982), 171-190

[4] J. Jacod, Théorémes limite pour les processus, in École d'Été de Probabilités de Saint-Flour XIII-1983, Lecture Notes in Math., Vol. 1117, Springer-Verlag, 1985

[5] S. Karlin and H.M. Taylor, A First Course in Stochastic Processes, Academic Press, 1976

[6] N. Ikeda and S. Watanabe, Stochastic Differential Equations and Diffusions, North Holland/Kodansha, 1981

[7] I.I. Gihman and A.V. Skorohod, The Theory of Stochastic Processes III, Springer-Verlag, Berlin 1979

[8] S.R.S. Varadhan, Stochastic Processes, Lecture Notes, Courant Institute of Math. Sci. New York University, 1968

[9] P. Billingsley, Convergence of Probability Measures, Wiley, 1968

[10] R.S. Liptser and A.N. Shiryaev, A functional central limit theorem for semimartingales, Theory of Probab. and its Appl., 25 (1980) 667-689

[11] E. Lenglart, Relation de domination entre deux processus, Ann. Inst. H. Poincaré (B) XIII, (1977), 171-179

Institute of Information
Sciences and Electronics
University of Tsukuba
Ibaraki 305, Japan

AN UPPER BOUND TO THE CAPACITY OF DISCRETE TIME GAUSSIAN CHANNEL WITH FEEDBACK

Kenjiro Yanagi

1. Introduction. The following model for the N-discrete time Gaussian channel with feedback is considered:

$$Y_n = U_n + Z_n, \quad n = 1,2,\ldots,N, \tag{1}$$

where $Z = \{Z_n;\ n = 1,\ldots,N\}$ is a non-degenerate, zero mean Gaussian process representing the noise and $U = \{U_n;\ n = 1,\ldots,N\}$ and $Y = \{Y_n;\ n = 1,\ldots,N\}$ are stochastic processes representing input signals and output signals, respectively. The channel is with noiseless feedback, so U_n is a function of a message X to be transmitted and the output signals $Y_1,\ldots,Y_{n-1}$. We assume that a constraint, given in terms of the covariance matrix, is imposed on the input signals. Rigorously speaking, we assume the following constraints:

(A.1) A message X to be transmitted is a random variable, taking values in an arbitrary measurable space and independent of Z. However, we may regard messages $X = \{X_n;\ n = 1,\ldots,N\}$ as stochastic processes.

(A.2) U_n is $\mathfrak{F}(X) \vee \mathfrak{F}_{n-1}(Y)$-measurable, where $\mathfrak{F}(X)$ and $\mathfrak{F}_{n-1}(Y)$ are the σ-fields generated by X and $\{Y_k;\ k = 1,\ldots,n-1\}$, respectively, and $\mathfrak{F} \vee \mathcal{G}$ denotes the σ-field generated by σ-fields $\mathfrak{F}$ and $\mathcal{G}$.

(A.3) $$\sum_{n=1}^{N} E[U_n^2] \leqq P.$$

Denote by Ω the class of all pairs (X,U) of a message X and an input U which satisfy the conditions (A.1)-(A.3), and denote by $I_N(X,Y)$ the mutual information quantity between a message X and the output $Y = \{Y_n;\ n = 1,\ldots,N\}$. Then the capacity C of the channel is defined as

$$C = \sup\{I_N(X,Y);\ (X,U) \in \Omega\}.$$

We introduce a Gaussian subclass Ω^0 of the class Ω by

$$\Omega^0 = \{(X,U) \in \Omega;\ (X,U) \text{ satisfies (B.1) and (B.2)}\},$$

where the conditions (B.1) and (B.2) are as follows:

(B.1) $X = \{X_n;\ n = 1,\ldots,N\}$ is a Gaussian process.

(B.2) U_n is a linear combination of $X_1,\ldots,X_n$ and $Y_1,\ldots,Y_{n-1}$.

By Ihara [5], the capacity C of the channel (1) under the conditions (A.1)-(A.3) is achieved in the Gaussian system:

$$C = \sup\{I_N(X,Y);\ (X,U) \in \Omega^0\}.$$

By considering the optimal coding, we can assume the following:

$$Y = X - T\cdot Y + Z,$$

where $Z = \{Z_n;\ n = 1,\ldots,N\}$ is a non-degenerate Gaussian process representing the noise, $X = \{X_n;\ n = 1,\ldots,N\}$ and $Y = \{Y_n;\ n = 1,\ldots,N\}$ are Gaussian processes representing messages and output signals, respectively, and $T = \{t_{ij};\ i,j = 1,\ldots,N,\ t_{ij} = 0 \text{ for } i \leq j\}$ is a Volterra type matrix. A constraint is imposed in the following:

$$E[\|X - T\cdot Y\|^2] \leq P, \tag{2}$$

where $\|\cdot\|$ is N-dimensional norm.
By (2) and the independence of X and Z, we obtain

$$\mathrm{Tr}[(I+T)^{-1}R_X(I+T^*)^{-1} + (I+T)^{-1}TR_ZT^*(I+T^*)^{-1}] \leq P, \tag{3}$$

where R_X, R_Z are the covariance matrices of X, Z, respectively, and Tr denotes the trace.
The mutual information quantity between a Gaussian message X and the corresponding Gaussian output Y is given as follows:

$$I_N(X,Y) = \frac{1}{2}\,\mathrm{Tr}[\log(I + R_Z^{-1}R_X)].$$

We remark that R_Z is non-degenerate covariance matrix of Z. For any Volterra type matrix T, there exists a Volterra type matrix S such that $(I+T)^{-1} = I+S$. Then we can also represent (3) in the following:

$$\mathrm{Tr}[(I+S)R_X(I+S^*) + SR_ZS^*] \leq P.$$

In general, it is not easy to calculate the capacity C concretely. However, for a constraint of a special type, some formulas for the capacity were obtained (cf. Butman [1,2], Ihara [5,6], Tierman and Schalkwijk [8], Wolfowitz [9]).
In this paper, we give upper and lower bounds to the capacity of N-discrete time Gaussian channel with feedback by operator methods.

2. Main theorem. We denote $C_0(P)$ and $C_f(P)$ the capacities of Gaussian channels without and with feedback, respectively, when the noise process is $Z = \{Z_n;\ n = 1,\ldots,N\}$.
The capacity of N-discrete time Gaussian channel without feedback was

obtained in the following (see [3]).

Proposition 2.1.

$$C_0(P) = \frac{1}{2}\sum_{i=1}^{K}\log\frac{P+r_1+\cdots+r_K}{K\cdot r_i},$$

where $0 < r_1 \leqq r_2 \leqq \cdots \leqq r_N$ are eigenvalues of R_Z and $K(K \leqq N)$ is the largest integer satisfying $P+r_1+\cdots+r_K \geqq K\cdot r_K$.

It is known that the capacity of the continuous time white Gaussian channel is not increased by feedback ([7]).

The capacity of 2-discrete time Gaussian channel with feedback was obtained by the author. We set the 2-dimensional matrices R_X, R_Z and T in the following:

$$R_X = \begin{pmatrix} a & c \\ c & b \end{pmatrix}, \quad \text{where } a,\ b \geqq 0, \quad ab \geqq c^2,$$

$$R_Z = \begin{pmatrix} k & m \\ m & \ell \end{pmatrix}, \quad \text{where } k,\ \ell > 0, \quad k\ell > m^2,$$

$$T = \begin{pmatrix} 0 & 0 \\ t & 0 \end{pmatrix}.$$

Proposition 2.2 (Yanagi [10]). We assume that $m \neq 0$. Then

$$\exp(2C_f(P))$$

$$= \max_{0 \leqq a \leqq P}\left\{1 + \frac{Pk}{k\ell-m^2} + \frac{(P+\ell-k)a}{k\ell-m^2} + \frac{a^2}{k\ell-m^2} + \frac{2|m|\sqrt{a(P-a)(a+k)}}{(k\ell-m^2)\sqrt{k}}\right\}.$$

The maximum is attained when

$$b = (P-a)(a+k)/k,$$

$$c = -\sqrt{ab} \text{ (in the case of } m > 0),$$

$$c = \sqrt{ab} \text{ (in the case of } m < 0),$$

$$t = c/(a+k).$$

Now we obtain an upper bound to the capacity of N-discrete time Gaussian channel with feedback and not necessarily white noise.

Theorem 2.3.

$$C_0(P) \leqq C_f(P) \leqq C_0(P_1),$$

where $P_1 = \sup\{\|(I+S)^{-1}\|^2(P - Tr[SR_Z S^*]);$ S are Volterra type matrices$\}$ and $P \lneqq P_1 < \infty$.

3. Proof and remarks.

Proof of Theorem 2.3. We set $R = (I+S)R_X(I+S^*) + SR_Z S^*$. Then $R_X = (I+S)^{-1}(R - SR_Z S^*)(I+S^*)^{-1}$. Hence

$$\begin{aligned} Tr[R_X] &= Tr[(I+S)^{-1}(R - SR_Z S^*)(I+S^*)^{-1}] \\ &= Tr[(I+S^*)^{-1}(I+S)^{-1}(R - SR_Z S^*)] \\ &\leq \|(I+S^*)^{-1}(I+S)^{-1}\| \, Tr[R - SR_Z S^*] \\ &= \|(I+S)^{-1}\|^2(Tr[R] - Tr[SR_Z S^*]). \end{aligned}$$

Since $Tr[R] \leq P$, we obtain

$$Tr[R_X] \leq \|(I+S)^{-1}\|^2(P - Tr[SR_Z S^*]).$$

We denote $P_1 = \sup\{\|(I+S)^{-1}\|^2(P - Tr[SR_Z S^*]);$ S are Volterra type matrices$\}$. Then $C_f(P) \leq C_0(P_1)$.
We estimate P_1.

$$\begin{aligned} P_1 &= \sup_S \|I - S + S^2 - \cdots + (-1)^{N-1}S^{N-1}\|^2 (P - Tr[SR_Z S^*]) \\ &\leq \sup_S (1 + \|S\| + \|S\|^2 + \cdots + \|S\|^{N-1})^2 (P - Tr[SR_Z S^*]) \\ &\leq \sup_{0 \leq \|S\| \leq \sqrt{P/r_1}} (1 + \|S\| + \|S\|^2 + \cdots + \|S\|^{N-1})^2 (P - r_1\|S\|^2) \\ &< \infty, \end{aligned}$$

where r_1 is the smallest eigenvalue of R_Z. It is not difficult to show that $P \lneqq P_1$. Thus we obtain

$$C_0(P) \leq C_f(P) \leq C_0(P_1).$$

q.e.d.

Remark 3.1. When $R_Z = \begin{pmatrix} k & m \\ m & \ell \end{pmatrix}$, it is easy to show that $P_1 = P + P^2/(4k)$.

Remark 3.2. $C_0(P_1)$ is the optimal upper bound of $C_f(P)$. We give an example of $C_f(P) = C_0(P_1)$. Let

$$R_Z = \begin{pmatrix} 1+\alpha & 1 \\ 1 & 1 \end{pmatrix},$$

where $0 < \alpha \leqq \alpha_0$ and α_0 ($\fallingdotseq 0.4450517$) is the smaller positive root of the equation

$$x^3 - x^2 - 2x + 1 = 0.$$

If $\alpha = \alpha_0$ and $P = 2\sqrt{\alpha(1+\alpha)}$, then

$$P_1 = P + \alpha,$$

$$C_0(P) = \frac{1}{2} \log\{1 + 4\sqrt{\alpha(1+\alpha)}/(2+\alpha-\sqrt{\alpha^2+4})\},$$

$$C_f(P) = C_0(P_1) = -\frac{3}{2} \log \alpha.$$

If $0 < \alpha < \alpha_0$ and $P = \alpha(1+\alpha)(\sqrt{\alpha^2+4} + \alpha)$, then

$$P_1 = P\{1+(\sqrt{\alpha^2+4} + \alpha)/4\},$$

$$C_0(P) = \frac{1}{2} \log\{1 + 2\alpha(1+\alpha)(\sqrt{\alpha^2+4} + \alpha)/(2+\alpha-\sqrt{\alpha^2+4})\},$$

$$C_f(P) = C_0(P_1) = \frac{1}{2} \log\{1 + (1+\alpha)(2+\alpha)(\sqrt{\alpha^2+4} + \alpha)/2 + (1+\alpha)(2+\alpha+\alpha^2)(\sqrt{\alpha^2+4} + \alpha)^2/4\}.$$

4. Additional result. We consider the capacity of non-Gaussian channel with linear feedback. The channel is represented by

$$Y = X - T\cdot Y + Z,$$

where $X = \{X_n;\ n = 1,\dots,N\}$, $Y = \{Y_n;\ n = 1,\dots,N\}$ and $Z = \{Z_n;\ n = 1, \dots,N\}$ are stochastic processes, X and Y denote a channel input and the channel output, respectively, Z is the noise independent of channel inputs, and $T = \{t_{ij};\ i,j = 1,\dots,N,\ t_{ij} = 0 \text{ for } i \leqq j\}$ denotes a Volterra type matrix. We assume the following constraint:

$$E[\|X - T\cdot Y\|^2] \leqq P. \tag{4}$$

Let Z_0 be the Gaussian noise with the same covariance matrix as that of Z. We denote by $C_f^*(P)$ the capacity of channel with noise Z with linear feedback under constraint (4) and denote by $C_0(P)$ the capacity of channel with noise Z_0 without feedback under constraint (4) where $T = 0$. We will give bounds of the capacity $C_f^*(P)$ in terms of the capacity $C_0(P)$ and $C_0(P_1)$ and the relative entropy $H_{Z_0}(Z)$ of Z with respect to Z_0.

Theorem 4.1. $C_0(P) \leqq C_f^*(P) \leqq C_0(P_1) + H_{Z_0}(Z)$.

The proof is given by our main theorem, Theorem 2.3, and the result of Ihara [4].

Acknowledgement. The author wishes to thank Professor H. Umegaki of Science University of Tokyo and Professor C. R. Baker of University of North Carolina at Chapel Hill for many detailed and helpful comments in this paper. And he also thanks Professor S. Ihara of Kochi University for many useful comments in preparing this paper.

References

[1] S. Butman: A general formulation of linear feedback communication systems with solutions, IEEE Trans. Information Theory IT-15 (1969) 392-400.

[2] S. Butman: Linear feedback rate bounds for regressive channels, IEEE Trans. Information Theory IT-22 (1976) 363-366.

[3] R. G. Gallager: Information theory and reliable communication, John Wiley & Sons (1968).

[4] S. Ihara: On the capacity of channels with additive non-Gaussian noise, Information and Control 37 (1978) 34-39.

[5] S. Ihara: On the capacity of the discrete time Gaussian channel with feedback, Trans. Eighth Prague Conference, Czecho. Academy Sci., Vol. C(1979) 175-186.

[6] S. Ihara: Stochastic processes and entropy, Iwanami-Shoten (1984) in Japanese.

[7] T. T. Kadota, M. Zakai and J. Ziv: Mutual information of the white Gaussian channel with and without feedback, IEEE Trans. Information Theory IT-17 (1971) 368-371.

[8] J. C. Tierman and J. P. M. Schalkwijk: An upper bound to the capacity of band-limited Gaussian autoregressive channel with noiseless feedback, IEEE Trans. Information Theory IT-20 (1974) 311-315.

[9] J. Wolfwitz: Signaling over a Gaussian channel with feedback and autoregressive noise, J. Appl. Prob. 12 (1975) 713-723.

[10] K. Yanagi: An example for Gaussian channel whose capacity is increased by feedback, (1985) preprint.

Department of Mathematics
Faculty of Science
Yamaguchi University
Yoshida 1677-1
Yamaguchi 753
Japan

ON THE VALUE FOR OLA-OPTIMAL STOPPING PROBLEM BY POTENTIAL THEORETIC METHOD

Masami Yasuda

Summary. The stopping problem on Markov process with OLA(One-stage Look Ahead) policy is considered. Its associated optimal value could be expressed explicitly by a potential for a charge of the positive part of the difference between the immediate reward and the one-period-after reward. As application to the best choice problem, the optimal value of three problems: the classical secretary problem, the problem with a refusal probability and the one with random number of objects are calculated.

1. Introduction. The optimal stopping problem is a special case of Markov decision processes. The decision maker can either select to stop, in which case he receives reward and the process terminates, or to pay cost and continue observing the state. If the decision is made to continue, then he proceeds to the next state according to the given transition probability. The objective is to choose the policy which maximizes the expected value. A policy for the decision process means to take the adaptation of a stopping time of the process.

Let x_n, $n=0,1,2,..$ be a Markov chain over a state space S in R^1. We assume that S is countable, but this is inessential for our discussion. In the last section 3 the cases of the unit interval are considered. The optimal stopping problem is to find a stopping time τ which maximizes the expectation of payoff $v(i;\tau)$ starting at i. Let us denote the optimal value by

$$(1.1)\qquad v(i) = \sup_{\tau<\infty} v(i;\tau) = \sup_{\tau<\infty} E[r(x_\tau) - \sum_{n=0}^{\tau-1} c(x_n) \mid x_0=i], \qquad i\in S$$

where r(i) means an immediate reward and c(i) a paying cost. The admissible class of the policies is the set of all finite stopping times. The detailed analyses are discussed by many authors such as Chow/Robbins/Siegmund[1], Shiryaev[11] and so on.

Consider the set of states for which stopping immediately is at least as good as stopping after exactly one more period. Denote this set by

(1.2) $\quad B = \{i \in S \; ; \; r(i) \geq Pr(i) - c(i)\}$

where $Pr(i) = \sum_{j \in S} P(i,j)r(j)$ and $P(i,j)$, $i,j \in S$, is a stationary transition probability. The policy, defined by the hitting time of this set B, is called a "One-stage Look Ahead"(abridged by OLA) policy by Ross[10], since it compares stopping immediately with stopping after one period. Under certain conditions, He shows that the OLA policy is optimal. The OLA policy is useful for many problems and also extended to the continuous parameter process by Prabhu[8].

Our aim of this note is to obtain, by applying the potential operator, the explicit optimal value associated with the OLA policy for stopping problems. With this result, we will give, in the section 3, the explicit solution of the various versions for the best choice problem in the asymptotic form. The first is the classical problem and the second is the case with the refusal probability. Also the solution for the case of random number of objects is calculated. The last case was reduced to a functional optimality equation by an ad hoc method in Yasuda[13].

The motivation for this approach arose in connection with the results of Darling[2] and that of Hordijk[5]. The former gave the upper bound of the optimal value by a potential operator and the latter gave a sufficient condition to find an optimal stopping time.

2. Markov potential and the optimal value. For a transition probability $P=P(i,j)$, $i,j \in S$, a function $f=f(i)$, $i \in S$, is called a charge if $\lim [\{I+P+...+P^k\}f](i)$ for each $i \in S$, exists and is finite-valued. Function $g=g(i)$, $i \in S$ is a potential if there exists a charge f such that $g(i) = \lim [\{I+P+...+P^k\}f](i)$, $i \in S$. We shall use the notation $g(i) = Nf(i)$, $i \in S$.

The relation between Markov Potential Theory and Dynamic Programming or Markov Decision Process is discussed by Hordijk[5]. Since, as in the section 1, the optimal stopping problem is a special case of Markov Decision Process, the optimality equation of the stopping problem is reduced as follows.

(2.1) $\quad v(i) = \max\{r(i), \; Pv(i) - c(i)\}, \quad i \in S.$

Fundamental in such an investigation of Dynamic Programming is the uniqueness of the equation (2.1) and the determination of the optimal policy and the optimal value.

In this note we consider the optimal stopping problem, for which the OLA policy is optimal. To give a sufficient condition, we prepare the next two propositions. The subset B of S in (1.2) is closed if

(2.2) $\quad P(i,j) = \emptyset \quad$ for $i \in B$, $j \in \bar{B}$

where $\bar{B}$ denotes the complement of the set B. The process is stable if the sequence defined by $v_\emptyset(i)=r(i)$,

$$v_n(i) = \max\{r(i),\ Pv_{n-1}(i) - c(i)\} \text{ for } n \geq 1,$$

converges uniformly and $\lim_{n \to \infty} v_n(i) = v(i)$ for $i \in S$.

Proposition 2.1 (Ross[1∅]). If the process is stable and the set B is closed, then the OLA policy is optimal.

Although this situation occurs in many applications and is useful to determine the stopping region, the proposition does not state the optimal value. The stability is somewhat less satisfactory to check in the application because the optimal value is unknown. So we impose assumption on a potential of the chain. Our aim is to calculate the optimal expected value under the closedness and the following equalization assumption in stead of the stability assumption, and express it explicitly by using a potential. We call the problem where the OLA policy is optimal as the OLA-optimal stopping problem.

Proposition 2.2 (Hordijk[5]). Suppose that

(2.3) $\quad v(i) = Pv(i) - c(i) \quad$ when $i \notin \Gamma = \{i \in S\ ;\ v(i) = r(i)\}$.

If the value function is a potential, then the hitting time of set Γ becomes optimal.

Proof. This is a special case in Theorem 4.1 of Hordijk[5]. Q.E.D.

Let P_A denote the restriction to a subset A of the transition probability P, i.e.,

(2.4) $\quad P_A(i,j) = P(i,j)1_A(j) \quad$ if $i,j \in S$,

where 1_A denotes the indicator of set A.

When the stability is dropped, as Ross shows, a stopping problem does not imply the optimality of the OLA policy. So we must impose a condition so as to preserve the OLA principle. The condition of equalizing for the reward function due to a potential notion is considered.

Assumption 2.1. We assume that

(2.5) $\quad \lim_{k \to \infty}[(P_{\bar{B}})^k r](i) = \emptyset \quad$ for $i \in \bar{B}$

where $\bar{B}$ denotes the complement of the set B defined by (1.2), and that the potential for $P_B r - c$ is finite-valued with respect to $P_{\bar{B}}$, that is,

(2.6) $[N_{\bar{B}}(P_B r-c)](i) < \infty$ for $i \in \bar{B}$.

The assumption (2.5) is equivalent to

$$\lim_{k \to \infty} E[r(x_k) 1_{\{\tau > k\}}; x_n \notin B, n=1,..,k-1] = \emptyset$$

where τ is the hitting time of B. The property $\lim_{k \to \infty} (P_{\bar{B}})^k v = \emptyset$ for the optimal value is called equalizing in Optimal Gambling(Dubins /Savage[3] or Hordijk[5]). One might say that here the actually received in the time period up to N and the promised earnings equalize as N tends to infinity. If the optimal value satisfies this property, the assumption (2.5) holds since $v(i) \geqq r(i)$, $i \in S$ by (2.1).

<u>Theorem 2.1.</u> Under the assumptions (2.5) and (2.6), the OLA policy τ_B is optimal, that is, the optimal policy is the hitting time of B. The optimal value $v(i) = v(i;\tau_B)$ is given by

$$(2.7) \qquad v(i) = r(i) + N(Pr-r-c)^+(i) = \begin{cases} r(i) & \text{on } B \\ N_{\bar{B}}(P_B r-c)(i) & \text{on } \bar{B} \end{cases}$$

where + is the positive part of the function, and N and $N_{\bar{B}}$ are potentials with respect to P and $P_{\bar{B}}$ respectively.

Proof. The proof for the optimality of the OLA policy is similar to Ross[10] and Hordijk[5]. In generally if $v(i)$, $i \in S$ is a solution of the optimality equation(2.1), then $v(i) \geqq r(i)$ and $v(i) \geqq Pv(i) - c(i)$ for $i \in S$. So

$$v(i) \geqq P^n r(i) - \sum_{k=\emptyset}^{n-1} P^k c(i) \text{ for each } n.$$

It yields that

$$v(i) \geqq v(i;\tau) \text{ for any policy } \tau < \infty.$$

Therefore it is sufficient to assert the followings to show the optimality. One is that, for the OLA policy τ_B, its value equals the right hand side of (2.7), that is, $v(i;\tau_B) = r(i)$ on B and $N_{\bar{B}}(P_B r-c)(i)$ on $\bar{B}$, and the second is that it satisfies the optimality equation (2.1).

Because of the definition of the OLA policy, $v(i) = r(i)$ on B and $v(i) = Pv(i)-c(i) = P_{\bar{B}}v(i) + P_B r(i) - c(i)$ on $\bar{B}$. Hence we get the first assertion immediately.

Nextly we show that it satisfies the optimality equation (2.1). From $Pf(i) = P_B f(i)$, $i \in B$ for any $f=f(i)$, we have

$$Pv(i) - c(i) = P_B v(i) - c(i) = P_B r(i) - c(i) = Pr(i) - c(i)$$

on B. Hence

$$\max\{r(i),\ Pv(i) - c(i)\}$$
$$= \max\{r(i),\ Pr(i) - c(i)\}$$
$$= r(i) \quad \text{for } i\varepsilon B.$$

On the other hand, by substitution of (2.7),

$$Pv(i) - c(i) = P_{\bar{B}}v(i) + P_B v(i) - c(i)$$
$$= [P_{\bar{B}}N_{\bar{B}}(P_B r-c)](i) + P_B r(i) - c(i)$$
$$= [N_{\bar{B}}(P_B r-c)](i), \quad i\varepsilon S.$$

On $i\varepsilon\bar{B}$, that

$$r(i) < Pr(i) - c(i) = P_{\bar{B}}r(i) + P_B r(i) - c(i)$$

implies $r(i) - P_{\bar{B}}r(i) < P_B r(i) - c(i)$. And, by the assumption (2.6), $r(i) \leqq [N_{\bar{B}}(P_B r - c)](i)$ for $i\varepsilon\bar{B}$. Combining the above assertions,

$$\max\{r(i),\ Pv(i) - c(i)\}$$
$$= \max\{r(i),\ [N_{\bar{B}}(P_B r-c)](i)\}$$
$$= [N_{\bar{B}}(P_B r-c)](i) \quad \text{for } i\varepsilon\bar{B}.$$

Thus the value $v(i)=v(i;\tau_B)$ satisfies the optimality equation.

It now remains to calculate the potential $N(Pr-r-c)^+(i)$, $i\varepsilon S$. From the definition of the set B in (1.2), we have $(Pr-r-c)^+(i)=\emptyset$ on B. So the support of charge is the complement of B and hence

$$N(Pr-r-c)^+(i)=\emptyset \quad \text{on B.}$$

On other hand, since $(Pr-r-c)^+(i) = (Pr-r-c)(i) = (P_{\bar{B}}r+P_B r-r-c)(i)$ for $i\varepsilon\bar{B}$, we have that

$$P(Pr-r-c)^+(i) = P_{\bar{B}}(Pr-r-c)(i)$$
$$= ((P_{\bar{B}})^2 r+P_{\bar{B}}P_B r-P_{\bar{B}}r-P_{\bar{B}}c)(i), \quad i\varepsilon\bar{B}.$$

Repeating this procedure to take the expectation up to k times and adding these,

$$\{I+P+P^2+\ldots+P^k\}(Pr-r-c)^+(i)$$
$$= (P_{\bar{B}})^k r(i) + [\{(P_{\bar{B}})^{k-1}+(P_{\bar{B}})^{k-2}+..+I_{\bar{B}}\}(P_B r-c)](i) - r(i), \quad i\varepsilon\bar{B}.$$

Hence

$$N(Pr-r-c)^+(i) = [N_{\bar{B}}(P_B r-c)](i) - r(i), \ i\varepsilon\bar{B}$$

follows immediately from the assumptions (2.5) and (2.6). Q.E.D.

We remark that the upper bound on the optimal value in Theorem 3.6 of Darling[2] equals exactly the optimal value in this case. That is, the bound holds with equality when the OLA policy is optimal and it is equalizing. This explicit solution and the proposition 2.1, 2.2 determine the optimal value and policy completely in the OLA-optimal stopping problem.

3. The Best Choice Problem. In this section we apply the previous method to the typical stopping problem known as the best

choice problem. By taking a scale limit, the asymptotic form of the problem is considered in order to get an analytical explicit solution. In the asymptotic form the state space of the problem is not countable but the unit interval. However immediately the previous method can be applied to the case. Some of results are well known and the optimal values are already obtained by some method, for example, by using differential equation. Here we intend to illustrate it for the application which optimal value can be determined by the unified previous method. Each of these are cases whose problem is the OLA-optimal stopping one.

3.1. The Classical Secretary Problem. The secretary problem, variously called dowry problem or Googol, is an optimal stopping problem based on relative ranks for objects arriving in a random fashion; the objective is to find the stopping rule that maximizes the probability of stopping at the best object of the sequence. The optimality equation for the optimal value $v(i)$, $i \in S$, the maximal probability of win, on the state space $S=\{1,2,..,n\}$ is as follows.

$$v(i) = \max\{i/n,\ i \sum_{j=i+1}^{n} v(j)/((j-1)j)\}, \qquad i \in \{1,2,...,n-1\}, \qquad v(n)=1$$

where there are no costs. This formulation of the problem as Markov Process is given by Dynkin and Yushkevitch[4]. Also refer to Chow/Robbins/Siegmund[1] or Shiryaev[11]. To consider the problem in the asymptotic form, take the scale limit by $i/n \to x$ as $i,n \to \infty$. It becomes that

$$(3.1) \qquad v(x) = \max\{\ x\ ,\ x \int_x^1 v(y)y^{-2}dy\}, \qquad x \in S=[0,1].$$

The solution of (3.1) is well known as

$$(3.2) \qquad v(x) = \begin{cases} e^{-1} & \text{on } [0,\ e^{-1}], \\ x & \text{on } [e^{-1},\ 1]. \end{cases}$$

Now we extend the problem by changing the reward function $r(x)=x$, which means probability of choosing the best object in the classical secretary problem, to a general reward $r(x)$. To ensure the OLA policy is still optimal, we assume, for a function $h(x)$ defined by

$$(3.3) \qquad h(x) = r(x) - x \int_x^1 r(y)y^{-2}dy \qquad \text{on } [0,1],$$

that

i) each term of the function $h(x)$ is finite-valued on $[0,1]$, and

ii) it changes its sign from - to + only once as x increases 0 to 1 and so the equation $h(x)=0$ has a unique solution α.

The optimality equation (3.1) with the reward function $r(x)$ is

$$(3.4) \qquad v(x) = \max\{r(x),\ x\int_x^1 v(y)y^{-2}dy\}, \qquad x\in[0,1].$$

provided that the underlying Markov chain is unchanged. The solution of this equation (3.4) is given by

$$(3.5) \qquad v(x) = \begin{cases} r(\alpha) & \text{on } [0,\alpha], \\ r(x) & \text{on } [\alpha,1] \end{cases}$$

where α is defined by (3.3ii).

In fact, one can show this (3.5) by applying the previous result. Straightforward calculation yields that the set B becomes $[\alpha,1]$ and it is closed with respect to the transition probability:

$$P(x,dy) = \begin{cases} xy^{-2}dy & \text{for } 0<x<y<1, \\ 0 & \text{otherwise.} \end{cases}$$

We have $P_B r(x) = r(\alpha)(x/\alpha)$ and $(P_{\bar{B}})^n P_B r(x) = r(\alpha)(x/\alpha)\log^n(\alpha/x)/n!$ for $x\in[0,\alpha]$, $n=0,1,2,\ldots$ Hence

$$N_{\bar{B}}P_B r(x) = r(\alpha)(x/\alpha)\{1+\log(\alpha/x)+2^{-1}\log^2(\alpha/x)+\ldots +(n!)^{-1}\log^n(\alpha/x)+\ldots\}$$

$$= r(\alpha)$$

for $x\in[0,\alpha]$ and thus the assumption (2.6) is satisfied. Also we can check the condition of (2.5):

$$(P_{\bar{B}})^n r(x) = x\int_x^{\alpha}\{r(y)y^{-2}\log^{n-1}(y/x)/(n-1)!\}dy$$

tends to zero as $n \longrightarrow \infty$ for $x\in[0,\alpha]$.

3.2. <u>A Problem with Refusal Probability.</u> A variant of the best choice problem is a case with a refusal probability discussed by Smith[12]. We can also formulate the problem as optimal stopping on a Markov chain. The asymptotic form of the transition probability is obtained immediately and we have

$$P(x,dy) = \begin{cases} py^{-1}(x/y)^p dy & 0<x<y<1, \\ 0 & \text{otherwise} \end{cases}$$

where p is a given parameter $0 < p \leqq 1$, which quantity $1-p$ means a probability of the refusal. When there is no refusal, i.e., $p=1$, it reduces to the classical secretary problem discussed in the section 3.1. Similarly as before, the optimality equation for the stopping problem with refusal probability p is

$$v(x) = \max\{r(x), \int_x^1 py^{-1}(x/y)^p v(y)dy\}, \qquad x \varepsilon [0,1]. \tag{3.6}$$

Define a function h(x) by

$$h(x) = r(x) - \int_x^1 \{py^{-1}(x/y)^p r(y)\}dy, \qquad x \varepsilon [0,1]. \tag{3.7}$$

Under the same assumptions as (3.3i) and (3.3ii), we obtain the optimal value with refusal probability as follows.

$$v(x) = \begin{cases} r(\alpha) & \text{on } [0,\alpha], \\ r(x) & \text{on } [\alpha,1] \end{cases} \tag{3.8}$$

where α is a unique solution of $h(x)=0$ in (3.7).

Another method to solve the best choice problem is given by Mucci[7], which method reduces the value to the solution of a differential equation.

Let

$$V(x) = \int_x^1 \{py^{-1}(x/y)^p r(y)\}dy, \qquad x \varepsilon [0,1]$$

which means a conditional optimal value. This satisfies

$$\begin{cases} dV(x)/dx = -px^{-1}(r(x)-V(x))^+, & x \varepsilon [0,1] \\ V(1)=0. \end{cases} \tag{3.9}$$

The optimal value at the beginning $v^* = V(0)$ equals $r(\alpha)$.

3.3. <u>A Problem with Random Number of Objects.</u> The discussion of the problem for variant on the random number of objects is given by Presman/Sonin[9]. The random environment in the problem means that there is a distribution $\Phi(x)$ over $x \varepsilon [0,1]$ which denotes the random number of objects. If we adapt the approach by the differential equation(3.9), the following functional equation is obtained by Yasuda[13]. For $x \varepsilon [0,1]$,

$$\begin{cases} dV(x) = V(x)[1-\Phi(x)]^{-1}d\,\Phi(x) - x^{-1}(R(x)-V(x))^+dx, \\ V(1) = 0 \end{cases} \tag{3.10}$$

where we set

$$(3.11)\qquad R(x) = x(1-\Phi(x))^{-1} \int_x^1 y^{-1} d\,\Phi(y).$$

When the distribution is absolutely continuous, (3.10) reduces to a differential equation such as (3.9). Let us define

$$g(x) = \int_x^1 y^{-1} d\,\Phi(y) \quad \text{and} \quad h(x) = g(x) - \int_x^1 y^{-1} g(y)\,dy \quad \text{for } x\varepsilon[0,1].$$

We assume conditions on the distribution $\Phi(x)$ so that these functions $R(x)$, $g(x)$ and $h(x)$ are well defined. The following result is obtained already by Yasuda[13].

If $h(x)$ changes its sign only once from - to + as x varies from 0 to 1, and if $\Phi(x)$ is continuous for $0 < x < 1$, then the optimal value at the beginning $v^* = V(0)$ is given by

$$(3.12)\qquad v^* = (1-\Phi(\alpha))V(\alpha) = \alpha g(\alpha)$$

where α is a unique solution of $h(x)=0$ for $x\varepsilon[0,1]$.

This can be also obtained by applying the previous method. Since the optimality equation with random number of objects in the asymptotic form is, for $x\varepsilon[0,1]$,

$$(3.13)\qquad v(x) = \max\{R(x),\ x(1-\Phi(x))^{-1} \int_x^1 (1-\Phi(y))y^{-2} v(y)\,dy\},$$

we have

$$(3.14)\qquad v(x) = \begin{cases} R(\alpha) & \text{on } [0,\alpha], \\ R(x) & \text{on } [\alpha,1]. \end{cases}$$

In fact, it is equivalent to the classical secretary problem

$$w(x) = \max\{r(x),\ x \int_x^1 y^{-2} w(y)\,dy\}$$

with the function

$$(3.15)\qquad w(x) = (1-\Phi(x))v(x)$$

and with the reward function

$$(3.16)\qquad r(x) = x \int_x^1 y^{-1} d\Phi(y).$$

Hence the solution (3.14) is immediately obtained by the result of (3.5). This method is simpler than the ad hoc treatment of the functional equation (3.10).

References

[1] Chow,Y.S., Robbins,H. & Siegmund,D. : Great Expectations: The Theory of Optimal Stopping, Houghton Mifflin, Boston (1971).

[2] Darling, D.A. : Contribution to the optimal stopping problem, Z. Wahr. verw. Gebiete 70 (1985), 525-533.

[3] Dubins,L.E. and Savage,L.J. : How to Gamble if You Must: Inequalities for Stochastic Processes, McGrow-Hill, New York (1965).

[4] Dynkin,E.B. & Yushkevitch,A.A. : Theorems and Problems on Markov Processes, Translation: Plenum Press, New York (1969).

[5] Hordijk, A. : Dynamic Programming and Markov Potential Theory, Mathematisch Centrum, Amsterdam (1974).

[6] Kemeny,J.G., Snell,J.L. & Knapp,A,W. : Denumerable Markov Chains, 2nd eds., Springer-Verlag, Berlin (1976).

[7] Mucci, A.G. : Differential equations and optimal choice problem, Ann. Statist. 1 (1973), 104-113.

[8] Prabhu, N.U. : Stochastic control of queueing systems, Nav. Res. Logist. Q. 21 (1974), 411-418.

[9] Presman,E.L. & Sonin,I.M. : The best choice problem for a random number of objects, T. Prob. Appl. 17 (1972), 657-668.

[10] Ross, S.M. : Introduction to Stochastic Dynamic Programming, Academic Press, New York (1983).

[11] Shiryaev, A.N. : Statistical Sequential Analysis, Translation: American Mathematical Society, Providence (1973).

[12] Smith, M.H. : A secretary problem with uncertain employment, J. Appl. Prob. 12 (1975), 620-624.

[13] Yasuda, M. : Asymptotic results for the best-choice problem with a random number of objects, J. Appl. Prob. 21 (1984), 521-536.

College of General Education
Chiba University
Chiba
260 Japan

FIXED POINT THEOREM FOR MEASURABLE FIELD OF OPERATORS WITH AN APPLICATION TO RANDOM DIFFERENTIAL EQUATION

T. A. Ždanok

The existence of solution of Cauchy problem for differential equation with random parameters is usually proved with the help of the fixed point theorem for the corresponding random operator (see for example [1],[9]). Such an approach leads to rather restrictive assumptions on the right-hand side of the equation and the initial condition. In this paper we show that these assumptions may be avoided when we use instead of a "classical" random operator the more general notion of a measurable field of operators investigated by the author in [10].

Let $(\Omega,\mathcal{F})$ be a measurable space. An argument $\omega\in\Omega$ of a function we shall write everywhere as an index. Let $(X_\omega)_{\omega\in\Omega}$ be a family of metric spaces with the metric d_ω. A function $x:\Omega\to \bigcup_{\omega\in\Omega} X_\omega$ such that $x_\omega \in X_\omega$ for all $\omega\in\Omega$ is called a section of this family. Let $\mathfrak{X}$ be a set of sections and $\mathfrak{X}_0$ be its countable subset such that the following axioms are satisfied:

(1) for all $x,y\in\mathfrak{X}$ the function $\omega\to d_\omega(x_\omega,y_\omega)$ is measurable;

(2) for all $\omega\in\Omega$ the set $\{x_\omega:\ x\in\mathfrak{X}_0\}$ is dense in X_ω;

(3) if $x_n\in\mathfrak{X}$ $(n=1,2,\dots)$ and $\lim\limits_{n\to\infty} x_{n\omega}=x_\omega\in X_\omega$ for all $\omega\in\Omega$; then $x\in\mathfrak{X}$;

(4) if $y,z\in\mathfrak{X}$, $E\in\mathcal{F}$ and $x_\omega=y_\omega$ for $\omega\in E$, $x_\omega=z_\omega$ for $\omega\in\Omega\setminus E$, then $x\in\mathfrak{X}$.

The triple $((X_\omega)_{\omega\in\Omega},\mathfrak{X},\mathfrak{X}_0)$ is called <u>a separable measurable field of metric spaces</u>. The notion of a measurable field of spaces was studied in the case of Hilbert spaces by Dixmier [4] but in the case of metric spaces by Delode, Arino, Penot [3] and by Evstigneev, Kuznecov [12].

The family $(T_\omega)_{\omega\in\Omega}$ of operators $T_\omega:X_\omega\to X_\omega$ will be called <u>a measurable field of operators</u> if the function $\omega\to T_\omega(x_\omega)$ belongs to $\mathfrak{X}$ for every $x\in\mathfrak{X}$.

The family $(M_\omega)_{\omega\in\Omega}$ of sets $M_\omega\subset X_\omega$ will be called a <u>measurable family of sets</u> if the function $\omega\to d_\omega(x_\omega,M_\omega)$ is measurable for any $x\in\mathfrak{X}$. The following criterion of measurability of the family of sets is due to Evstigneev and Kuznecov [12].

Proposition 1. The family $(M_\omega)_{\omega\in\Omega}$ of closed sets in a separable measurable field of complete metric spaces $((X_\omega)_{\omega\in\Omega},\mathfrak{X},\mathfrak{X}_0)$ is measurable if and only if there exists a sequence (x_i) of elements of $\mathfrak{X}$ such that the set $\{x_{i\omega}: i=1,2,\ldots\}$ is dense in M_ω for all $\omega\in\Omega$.

Now we give sufficient conditions for existence of a fixed-point for a measurable field of operators. The theorem below is the generalization of a similar theorem for random operator due to Sehgal and Waters [11].

Theorem 1. Let X_ω be a closed subset of a Banach space Y_ω normed by $\|\cdot\|_\omega$. Let $((X_\omega)_{\omega\in\Omega},\mathfrak{X},\mathfrak{X}_0)$ be a separable measurable field of metric spaces (with metric induced by the norm) such that if $y,z\in\mathfrak{X}$, $\alpha,\beta\in R$ and $x_\omega=\alpha y_\omega+\beta z_\omega\in X_\omega$ for all $\omega\in\Omega$, then $x\in\mathfrak{X}$. Let $(T_\omega)_{\omega\in\Omega}$ be a measurable field of continuous operators $T_\omega:X_\omega\to X_\omega$. Let the following conditions be satisfied:

a) the set $M_\omega=\{x\in X_\omega: T_\omega(x)=x\}$ of fixed points of T_ω is non-void and compact for all $\omega\in\Omega$;

b) if $y_n\in\mathfrak{X}$ $(n=1,2,\ldots)$ and $\lim_{n\to\infty}(Y_\omega(y_{n\omega})-y_{n\omega})=0$ for all $\omega\in\Omega$, then for any $\omega\in\Omega$ the set $\{y_{n\omega}: n=1,2,\ldots\}$ is relatively compact.

Then there exists a sequence (x_i) of elements of $\mathfrak{X}$ such that the set $\{x_{i\omega}: i=1,2,\ldots\}$ is dense in M_ω for all $\omega\in\Omega$.

Proof. It is enough to establish the measurability of the family $(M_\omega)_{\omega\in\Omega}$ and then to apply Proposition 1. Let $x\in\mathfrak{X}$, $c>0$ and B_ω be the closed ball of radius c with the centre x_ω in X_ω. As M_ω is compact the condition $d_\omega(x_\omega,M_\omega)\le c$ is equivalent to the condition $B_\omega\cap M_\omega\ne\phi$. Since the family $(B_\omega)_{\omega\in\Omega}$ is obviously measurable, we can find a sequence (y_i) of elements of $\mathfrak{X}$ such that the set $\{y_{i\omega}: i=1,2,\ldots\}$ is dense in B_ω for all $\omega\in\Omega$ by Proposition 1. No we prove the equality

$$\{\omega: B_\omega\cap M_\omega\ne\phi\} = \bigcap_{m=1}^{\infty}\bigcup_{i=1}^{\infty}\{\omega: \|T_\omega(y_{i\omega})-y_{i\omega}\|_\omega<\frac{1}{n}\},$$

and then utilize the measurability of the field of operators $(T_\omega)_{\omega\in\Omega}$. It is sufficient to show that the right-hand part of this equality is contained in the left-hand part because the converse inclusion is obvious. Let ω belong to the right-hand part. It means that there exists such a sequence $(y_{i_n\omega})$ that $\lim_{n\to\infty}(T_\omega(y_{i_n\omega})-y_{i_n\omega})=0$. By condition (b) we find a convergent subsequence $(y_{i_{n_k}\omega})$ of this sequence. Let $\lim_{k\to\infty} y_{i_{n_k}\omega}=y_\omega$. Then $T_\omega(y_\omega)=y_\omega$, hence $y_\omega\in B_\omega\cap M_\omega$. The theorem is proved.

As the corollaries, we have the fixed-point theorem for a measurable field of condensing operators and the fixed-point theorem of Schauder's type. An operator $T:X\to X$, X being a metric space, is called condensing if T is continuous and for each bounded subset B of X with $\gamma(B)>0$, $\gamma(T(B))<\gamma(B)$, where the number $\gamma(B)=\inf\{\varepsilon>0$: B can be covered by a finite number of subsets of X with diameter $\leq \varepsilon\}$ is called the measure of noncompactness of the set B.

<u>Corollary 1</u>. Let $((X_\omega)_{\omega\in\Omega},\mathfrak{X},\mathfrak{X}_0)$ be such as in theorem 1 and X_ω be convex for all $\omega\in\Omega$. If $(T_\omega)_{\omega\in\Omega}$ is a measurable field of condensing operators $T_\omega:X_\omega\to X_\omega$ with bounded range, then the assertion of Theorem 1 is true.

<u>Proof</u>. By Furi and Vignoli [7],[8], the set of fixed points of condensing operator T_ω is compact, and the condition $\lim_{n\to\infty}(T_\omega(y_n)-y_n)=0$ implies that the set $\{y_n: n=1,2,\dots\}$ is relatively compact. Therefore, all the assumptions of Theorem 1 are fulfilled.

<u>Corollary 2</u>. Let $((X_\omega)_{\omega\in\Omega},\mathfrak{X},\mathfrak{X}_0)$ be such as in Theorem 1, and X_ω be convex for all $\omega\in\Omega$. If $(T_\omega)_{\omega\in\Omega}$ is a measurable field of continuous operators $T_\omega:X_\omega\to X_\omega$ with relatively compact range, then the assertion of theorem 1 is true.

<u>Proof</u>. It is enough to notice that the operator T_ω is condensing. So all the assumptions of Corollary 1 are fulfilled.

Corollaries 1 and 2 are generalizations of fixed-point theorems for random operators due to Itoh [9] and Bharucha-Reid [2] respectively. Using Corollary 2, we prove the existence theorem of Peano type for random differential equation.

For what follows we need some additional notions.

Let X be a metric space and 2^X be the family of all nonempty subsets of X. A function $\Gamma:\Omega\to 2^X$ will be called: a) p-measurable if $\{\omega: x\in\Gamma_\omega\}\in\mathcal{F}$ for any $x\in X$; b) separable if there exists such a countable set $Z \subset X$ that for any $\omega \in \Omega$ the set $\Gamma_\omega \cap Z$ is dense in Γ_ω. The graph of Γ is the set $Gr(\Gamma)=\{(\omega,x)\in\Omega\times X: x\in\Gamma_\omega\}$. A (Borel) measurable function $x:\Omega\to X$ is called a measurable selector of Γ if $x_\omega \in \Gamma_\omega$ for all $\omega\in\Omega$. Let Y be a metric space and $\Gamma:\Omega\to 2^X$ be a p-measurable function. A function $f:Gr(\Gamma)\to Y$ is called a random function if for any $x\in X$ the function $\omega\to f_\omega(x)$ defined on the set $\{\omega: x\in\Gamma_\omega\}$ and taking values from Y is measurable. The random function f will be called continuous, bounded, etc., if for every $\omega\in\Omega$ the function $f:\Gamma_\omega\to Y$ has the corresponding properties.

<u>Proposition 2</u> ([6], lemma 10). Let X and Y be metric spaces, $\Gamma:\Omega\to 2^X$ be a separable p-measurable function, $f:Gr(\Gamma)\to Y$ be a random continuous function. Then for any measurable selector x of Γ the function $\omega\to f_\omega(x_\omega)$ is measurable.

Notice that by assumptions of lemma 10 in [6], $X=Y$ is separable Banach space but Γ takes the closed values and satisfies some stronger property of measurability. Nevertheless the proof of lemma 10 can be transferred to the more general situation formulated above.

<u>Proposition 3</u> ([10], proposition 1.3). Let X be a metric space, $\Gamma:\Omega\to 2^X$ be a separable p-measurable function, $f:Gr(\Gamma)\to R$ be a random continuous function such that $\alpha_\omega=\sup_{x\in\Gamma_\omega} f_\omega(x)<\infty$. Then the function $\omega\to\alpha_\omega$ is measurable.

Now we present an existence theorem for random differential equation as follows.

<u>Theorem 2</u>. Let $y_0:\Omega\to R^n$ and $r:\Omega\to]0,+\infty[$ be measurable functions; $B_\omega=\{y\in R^n: \|y-y_{0\omega}\|\le r_\omega\}$, $\omega\in\Omega$; J_0 be the segment of a real line and t_0 be an interior point of J_0. Let $f: \{(\omega,t,y)\in\Omega\times J_0\times R^n: y\in B_\omega\}\to R^n$ be a random continuous function. Let $\alpha_\omega = \sup_{(t,y)\in J_0\times B_\omega} \|f_\omega(t,y)\|$; $c_\omega=r_\omega/\alpha_\omega$; $J_\omega=J_0\cap[t_0-c_\omega; t_0+c_\omega]$; $Gr(J)=\{(\omega,t): t\in J_\omega\}$. Then the Cauchy problem

$$\frac{d}{dt} x_\omega(t) = f_\omega(t,x_\omega(t)), \qquad t\in J_\omega ;$$
$$x_\omega(t_0) = y_{0\omega} \tag{1}$$

has a solution $x:Gr(J)\to R^n$ which is a random continuous function.

Proof. We first notice that the function $\omega\to\alpha_\omega$ is measurable by Proposition 3 since the multifunction $\omega\to J_0\times B_\omega$ is obviously p-measurable and separable. Hence the function $\omega\to c_\omega$ is measurable and the multifunction $\omega\to J_\omega$ is p-measurable.

Denote by Y_ω the Banach space of all continuous functions defined on J_ω and taking values from R^n with the supremum norm $\|\cdot\|_\omega$. Let $\mathcal{Y}$ be the set of all sections y of the family $(Y_\omega)_{\omega\in\Omega}$ such that for any fixed $t\in J_0$ the function $\omega\to y_\omega(t)$ defined on the set $\{\omega: t\in J_\omega\}$ is measurable. Let W be the family of all polynomials w with coefficients being n-dimensional vectors with rational coordinates. Let $\omega\to w_\omega$ be the function whose value of any point $\omega\in\Omega$ is the restriction of polynomial w on J_ω, and let $\mathcal{Y}_0$ be the family of all such functions. It is easily seen that $\mathcal{Y}_0\subset\mathcal{Y}$. In order to show that $((Y_\omega)_{\omega\in\Omega},\mathcal{Y},\mathcal{Y}_0)$ is a separable measurable field of metric spaces it is enough to verify the axiom 1 because the other axioms are obvious. If $x,y\in\mathcal{Y}$ then the function $\omega\to d_\omega(x,y)=\|x_\omega-y_\omega\|_\omega = \sup_{t\in J_\omega}\|x_\omega(t)-y_\omega(t)\|$ is measurable by Proposition 3. Since the multifunction $\omega\to J_\omega$ is separable and p-measurable.

Now consider the set X_ω of all elements $x\in Y_\omega$ satisfying the inequality $\|x-y_{0\omega}\|_\omega\le r_\omega$ (here $y_{0\omega}$ is the constant function taking the value $y_{0\omega}$ of any point from J_ω). Evidently, X_ω is closed and convex. Let $\mathfrak{X}=\{x\in\mathcal{Y}: x_\omega\in X_\omega$ for all $\omega\in\Omega\}$; $\mathfrak{X}_0=\{x\in\mathcal{Y}_0: x_\omega\in X_\omega$ for all $\omega\in\Omega\}$. Then obviously $((X_\omega)_{\omega\in\Omega},\mathfrak{X},\mathfrak{X}_0)$ is a separable measurable field of metric spaces satisfying all the conditions of Corollary 2.

For every $\omega\in\Omega$ define an operator $T_\omega: X_\omega\to X_\omega$ by

$$T_\omega(x)(t) = y_{0\omega} + \int_{t_0}^{t} f_\omega(s,x(x))ds\ , \qquad x\in X_\omega\ .$$

The proof of continuity of T_ω and relative compactness of $T_\omega(X_\omega)$ can be given in the same way as in deterministic case (see for ex. [5]). We show that $(T_\omega)_{\omega\in\Omega}$ is a measurable field of operators. By proposition 2.14 of [10] it is sufficient to prove that for any $x\in\mathfrak{X}_0$ the function $\omega\to T_\omega(x_\omega)$ belongs to $\mathfrak{X}$. The value of the function $\omega\to x_\omega$ of the point $\omega\in\Omega$ is the restriction of some polynomial $w\in W$ on J_ω. For any fixed $t\in J_0$ the function $\omega\to x_\omega(t)$ of Ω into R^n is measurable since the inverse image of a set $U\subset R^n$ is either empty (if $w(t)\in U$) or $D_t=\{\omega: t\in J_\omega\}$ (if $w(t)\in U$). Hence the function $\omega\to (t,x_\omega(t))$ of Ω into $J_0\times R^n$ is measurable. Since the multifunction $\omega\to J_0\times B_\omega$ is separable and p-measurable, the function $\omega\to f_\omega(t,x_\omega(t))$ of Ω into R^n is measurable by Proposition 2. Consider the Banach space $C(J_0)$ of all continuous functions defined on J_0 and taking values from R^n. By proposition 4.1 of [9], the function $\omega\to f_\omega(\cdot,x_\omega(\cdot))$ of Ω into $C(J_0)$ is measurable. Since the integral operator is continuous, the function $\omega\to\int_{t_0}^{(\cdot)} f_\omega(s,x_\omega(s))ds$ of Ω into $C(J_0)$ is measurable. Then by proposition 4.1 of [9] for any fixed $t\in J_0$ the function $\omega\to\int_{t_0}^{t} f_\omega(s,x_\omega(s))ds$ of Ω into R^n is measurable. Hence the measurability of the function $\omega\to T_\omega(x_\omega(t))$ defined on D_t and taking values from R^n.

Thus we have verified all the assumptions of Corollary 2. By this corollary, there exists such $x\in\mathfrak{X}$ that $T_\omega(x_\omega)=x_\omega$ for all $\omega\in\Omega$. It is clear that for any $\omega\in\Omega$ the function $x_\omega(\cdot)$ is the solution of the Cauchy problem (1). Since $x\in\mathfrak{X}$, for every $t\in J_0$ the function $\omega\to x_\omega(t)$ of D_t into R^n is measurable. The theorem is proved.

References

[1] Bharucha-Reid A. T. Random integral equations. Academic Press, New York, 1972.

[2] Bharucha-Reid A. T. Fixed point theorems in probabilistic analysis. Bull. Amer. Math. Soc., 1976, v.82, N5, p.641-657.

[3] Delode C., Arino O., Penot J. P. Champs mesurables et multi-sections. Ann. Inst. H. Poincaré, 1976, v.12, N1, p.11-42.

[4] Dixmier J. Les algèbres d'opérateurs dans l'espace Hilbertien (algèbres de Von Neuman). Cahier Scientifique, XXV. Paris, Gauthier-Villars, 1969.

[5] Edwards R. E. Functional analysis. Theory and applications. New York - London, 1965.

[6] Engl Heinz W. Some random fixed point theorems for strict contractions and nonexpansive mappings. Nonlinear Analysis, Theory, Meth. and Appl. 1978, v.2, N5, p.619-626.

[7] Furi M., Vignoli A. Fixed points for densifying mappings. Atti Accad. Naz. Lincei Rend. Cl. Sci. Fis. Mat. Natur., 1969, v.47, p.465-467.

[8] Furi M., Vignoli A. On α-nonexpansive mappings and fixed points. Atti Accad. Naz. Lincei Rend. Cl. Sci. Fis. Mat. Natur., 1970, v.48, p.195-198.

[9] Itoh Shigeru. Random fixed-point theorems with an application to random differential equations in Banach spaces. J. Math. Anal.

and Appl., 1979, v.67, N2, p.261-273.

[10] Jdanok T. Opérateurs et fonctionnelles aleatoires dans les champs mesurables. Seminaire d'analyse convexe, Montpellier 1983, N2, p.1-36.

[11] Sehgal V. M., Waters Charlie. Some random fixed point theorems. Contemp. Math., 1983, v.21, p.215-218.

[12] Евстигнеев И. В., Кузнецов С. Е. "Косые произведения" измеримых пространств. В кн. "Избранные вопросы теории вероятностей и математической экономики". М.: ЦЭМИ АН СССР, 1977, с.28-37.

Latvian State University
226098 Riga USSR

RECORDS OF MEETINGS

Speakers in special morning sessions

A.A. Borovkov, S. Watanabe, J. Kubilius, H. Kunita, A.A. Novikov, M. Akahira, N.N. Vakhania, H. Tanaka, A.N. Shiryayev, M. Osikawa, S.H. Sirazdinov, S. Kotani, G.A. Mikhailov, T. Kamae, S. Kusuoka, V. Statulevicius

List of sections

Limit theorems, Martingales and Markov processes, Information and control, Stationary processes and random fields, Statistics, Infinite dimensional analysis, Probabilistic methods in mathematical physics, Probabailistic number theory, Ergodic theory

ORGANIZING COMMITTEE

Japanese side

G. Maruyama (chairman), S. Watanabe (vice-chairman), M. Fukushima (secretary), N. Kôno (executive office), T. Hida, M. Hitsuda, N. Ikeda, K. Itô, H. Kunita, H. Morimoto, M. Nisio, T. Onoyama, K. Sato, T. Sirao, Y. Takahashi, K. Takeuchi, H. Totoki, H. Watanabe

Soviet side

Yu.V. Prohorov (chairman), A.N. Shiryayev (vice-chairman), A.A. Novikov (secretary)

Local editorial committee of the Proceedings

S. Watanabe, M. Fukushima, K. Sato